Radar
Weather
History
c. Collies
Imperial
Coll,
25/09/99

Ch 8

Can I
tell Peter
Tony
or
Martin Lord.

NOWCASTING
Nowcast   p. 325   Anderson
Nowcasting   p. 348   Zipser (1983)

HYDROLOGICAL APPLICATIONS
OF WEATHER RADAR

HYDROLOGICAL APPLICATIONS OF WEATHER RADAR

Editors

I. D. CLUCKIE B.Sc., M.Sc., Ph.D., C.Eng., F.I.C.E., F.I.W.E.M., F.R.Met.Soc.
Professor of Water Resources
Department of Civil Engineering, University of Salford

C. G. COLLIER B.Sc., A.R.C.S., F.R.Met.Soc.
Assistant Divisional Director
and Head of Nowcasting and Satellite Imagery Research
The Meteorological Office, Bracknell
and Visiting Professor, University of Salford

ELLIS HORWOOD
NEW YORK LONDON TORONTO SYDNEY TOKYO SINGAPORE

First published in 1991 by
ELLIS HORWOOD LIMITED
Market Cross House, Cooper Street,
Chichester, West Sussex, PO19 1EB, England

A division of
Simon & Schuster International Group
A Paramount Communications Company

Typeset in Times by Ellis Horwood Limited
Printed and bound in Great Britain
by Redwood Press Limited, Melksham, Wiltshire

British Library Cataloguing in Publication Data

Cluckie, I.D.
Hydrological applications of weather radar. —
(Ellis Horwood series in environmental management, science and technology)
I. Title. II. Collier, C.G. III. Series.
551.57
ISBN 0–13–441478–0

Library of Congress Cataloging-in-Publication Data

Hydrological applications of weather radar / editors, I.D. Cluckie, C.G. Collier.
p. cm. — (Ellis Horwood series in environmental management, science, and technology)
Includes bibliographical references and index.
ISBN 0–13–441478–0
1. Radar in hydrology. 2. Radar meteorology, I. Cluckie, I.D. II. Collier, C.G. III. Series.
GB656.2.R3H93   1991
551.57–dc20   91–2213
                                                                                    CIP

Contents

Editors' preface 9

Acknowledgements 11

Part 1 Raingauge–radar adjustment techniques
  1 **Calibration of weather radar data in The Netherlands** 15
       S. van den Assem
  2 **Weather radar calibration in real time: prospects for improvement** 25
       V. K. Collinge
  3 **Using and adjusting conventional radar reflectivity data for estimation
       of precipitation: past, present and future studies in Switzerland** 43
       G. Galli and J. Joss
  4 **Assessment of the adjustment of hourly radar rainfall fields by the use
       of daily raingauge totals.** 49
       T. J. Hitch
  5 **On-line calibration in HERP.** 56
       R. K. Kreuls
  6 **Local recalibration of weather radar** 65
       R. J. Moore, B. C. Watson, D. A. Jones and K. B. Black
  7 **Real-time adjustment of radar fields of precipitation by raingauges,
       allowing for orographic enhancement near the surface** 74
       C. Warner
  8 **Real-time calibration methods of radar raingauge data** 84
       T. Yamaguchi, T. Ohtsuka, K. Murata, E. Abe, F. Yoshino and S. Takemori
  9 **Real-time calibration methods of radar raingauges** 95
       F. Yoshino, E. Abe and S. Tamamoto

Part 2 Real-time radar data processing
 10 **Corrections for partial blocking of the radar beam.** 107
       T. Andersson
 11 **On the significance of radar wavelength in the estimation of snowfall** . . .117
       A. Giguere and G. L. Austin

12   **Ground clutter rejection for weather radar and weather Doppler radar** . 131
        K. Hamuzu and M. Wakabayashi

13   **The diurnal evolution of the ground return intensities and its use as a
        radar calibration procedure** 143
        O. Massambani and A. J. P. Filho

14   **Sectorized hybrid scan strategy of the NEXRAD precipitation-processing
        system** 151
        R. Shedd, J. A. Smith and M. L. Walton

Part 3   *Radar echo climatology and statistical analysis*
15   **Is radar echo climatology useful?** 163
        T. Andersson

16   **Geometrical and statistical features of Swiss radar data** 173
        R. Boesch

17   **Probable maximum flood modelling utilizing transposed
        maximized radar-derived precipitation data** 181
        I. D. Cluckie, M. L. Pessoa and P. S. Yu

18   **Variability of heavy-rainfall events in northwest England: an analysis
        of spatial structure** 192
        E. J. Stewart and N. S. Reynard

19   **Properties of echoes at first detection, resulting in multicelled rainstorms** 203
        N. Wescott and S. Changnon

Part 4   *Radar measurement accuracy including multiparameter radars*
20   **Improvement of rainfall measurements due to accurate synchronization
        of raingauges and due to advection use in calibration** 213
        B. Blanchet, A, Neuman, G. Jaquet and H. Andrieu

21   **Range and orographic corrections for use in real-time radar data
        analysis** 219
        R. Brown, G.P. Sargent and R. M. Blackall

22   **Validation of a short-range X-band system for rainfall measurement over
        an urban area** 229
        G. Delrieu, J. D. Creutin and H. Andrieu

23   **Bright-band errors in rainfall measurement: identification and
        correction using linearly polarized radar returns** 240
        S. E. Hopper, A. J. Illingworth and I. J. Caylor

24   **An integrated X-band radar system for short-range measurement of
        rain rates in HERP** 250
        A. Kammer

25   **Examination of the impact of range on the quality of daily catchment
        rainfall totals derived from the Ingham radar** 258
        M. F. Mylne and B. D. Hems

26   **Sampling errors for raingauge-derived mean areal daily and monthly
        rainfall** 267
        A. W. Seed and G. L. Austin

27   **Observations and simulation of rainfall in mountainous areas** 279
        K. Tateya, M. Nakatsugawa and T. Yamada

28 **The effect of range on the radar measurement of rainfall** 296
D. Tees and G. L. Austin

29 **An application of dual-polarization radar to radar hydrology** 305
F. Yoshino, N. Ishii, H. Mizuno and T. Ikawa

Part 5 Precipitation forecasting

30 **An advective model for probability nowcasts of accumulated
precipitation using radar** 325
T. Andersson

31 **The combined use of weather radar and mesoscale numerical model data
for short-period rainfall forecasting** 331
C. G. Collier

32 **On the evaluation of radar rainfall forecasts** 349
T. Denoeux, T. Einfalt and G. Jaquet

33 **The development of the SCOUT II.0 rainfall-forecasting method** 359
T. Einfalt, T. Denoeux and G. Jaquet

34 **Short-term rainfall forecasting using radar data and
hydrometeorological models** 368
K. P. Georgakakos and W. F. Krajewski

35 **Short-term forecasting for water level of a flash flood by radar
hyetometer** 379
T. Moriyama and M. Hirano

36 **Advanced use in rainfall prediction of a three-dimensionally scanning
radar** 391
E. M. Nakakita, M. Shiiba, S. Ikebuchi and T. Takasoa

Part 6 Hydrological forecasting

37 **Analytically derived runoff models based on rainfall point processes** . . . 411
M. Bierkens and C. E. Puente

38 **Adaptive grid-square-based geometrically distributed flood-forecasting
model** 424
S. Chander and S. Fattorelli

39 **Radar signal quantization and its influence on rainfall–runoff models** . . . 440
I. D. Cluckie, K. A. Tilford and G. W. Shepherd

40 **Never expect a perfect forecast** 452
T. Einfalt and T. Denoeux

41 **Integrating radar rainfall data into the hydrologic modelling process** . . . 459
T. L. Engdahl and H. L. McKim

42 **Reflections on rainfall information requirements for operational
rainfall–runoff modelling** 469
C. Obled

43 **Extension of lumped operational rainfall–runoff approach models to
semilumped modelling: the case of the DPFT–ERUHDIT approach** 483
J. Y. Rodriguez, D. Sempere-Torres and C. Obled

44 **The updating procedure in the MIKE 11 modelling system for real-time
forecasting** 497
M. Rungo, J. K. Refsgaard and K. Havno

45   **From radar rainfall data to hydrological data** 509
      M. Semke

46   **Modelling the time-dependent nature of the rainfall–runoff relationship**
      using on-line identification 519
      P. A. Troch, F. P. De Troch and J. Van Hyfte

47   **Hydrological relevance of radar rainfall data** 531
      H. R. Verworn

Part 7   *Operational and international experience*
48   **Wessex flood-forecasting system** 543
      C. J. Birks, A. P. Bootman, I. D. Cluckie and H. Dawei

49   **Hydrological applications of weather radar and remotely transmitted**
      data from a ground network in the Veneto region of Italy 555
      A. Capovilla, S. Chander, M. Crespi and S. Fattorelli

50   **New possibilities for precipitation estimation for river basin managers**
      in developing countries 567
      J. D. Flach, T. R. E. Chidley and A. Siyyidd

51   **Remote sensing: the environmental analysis solver** 571
      S. K. Ghosh and G. Fleming

52   **Real-time control for urban drainage systems: advantage or**
      disadvantage 580
      M. J. Green

53   **Quantitative use of radar for operational flood warning in the**
      Thames area 590
      C. M. Haggett, G. F. Merrick and C. I. Richards

54   **NEXRAD: new era in hydrometeorology in the USA** 602
      M. D. Hudlow, J. A. Smith, M. L. Walton and R. C. Shedd

55   **An application of computer-based training analysis to an operational**
      radar-based flood-warning system 613
      G. A. Kennedy

56   **COST 73: possible hydrological applications of weather radar in**
      Western Europe 623
      D. H. Newsome and C. G. Collier

57   **River basin forecasting system for Portuguese rivers** 635
      R. Oliveira and D. Ford

Editors' preface

In the last decade there has been an increasing interest amongst hydrologists in the quantitative utilization of weather radar. This has meant that the boundary between hydrology and meteorology has become more diffuse and the 'hydrometeorological' fraternity has grown in size and influence. Indeed, the historical boundaries which exist between many scientific disciplines are fast becoming eroded as environmental processes demand multidisciplinary solutions.

In recognizing the importance that weather radar offers hydrological science as a ground-based remote sensor of the precipitation process, the First International Symposium on Hydrological Applications of Weather Radar was held at the University of Salford during August 1989. The UK has been at the centre of meteorological developments in operational weather radar for many years and more recently has been in the forefront of expanding its use into the hydrological arena. The extensive collaboration that now exists between meteorologists and hydrologists was reflected in the timing of the Salford Symposium in immediately following the Fifth Scientific Assembly at Reading of the International Association of Meteorology and Atmospheric Physics.

The papers contained in the edited proceedings represent the state of the art as far as hydrological radar is concerned and define the progress evident during the 1980s whilst providing glimpses of the promising developments of the 1990s.

In the preparing of this volume the editors would wish to thank the members of the International Programme Committee for their assistance. In particular the constant help and support of members of the Water Resources Research Group at Salford is gratefully acknowledged.

Ian Cluckie and Chris Collier,
UK, 1990

Acknowledgements

International Programme Committee

Professor I. D. Cluckie, University of Salford, UK
Symposium Organizer

Dr R. A. Bailey, Severn–Trent Water Authority, UK
Mr C. Birks, Wessex Water Authority, UK
Dr J. D. Creutin, Intsitut de Mécanique, Grenoble, France
Mr C. J. Collier, Meteorological Office, UK
Mr V. K. Collinge, University of Lancaster, UK
Dr R. Fantechi, Commission of European Communities, DGXII, Belgium
Professor S. Fattorelli, University of Padova, Italy
Professor G. Fleming, University of Strathclyde, UK
Professor K. P. Georgakakos, University of Iowa, USA
Mr C. M. Haggett, Thames Water Authority, UK
Dr M. Hudlow, National Weather Service, USA
Mr R. Moore, Natural Environment Research Council (Institute of Hydrology), UK
Dr J. Rodda, World Meteorological Organization, Switzerland
Professor G. A. Schultz, University of Bochum, FRG
Ir H. Sticker, University of Wageningen, The Netherlands
Dr H.-R. Verworn, University of Hannover, FRG
Dr P. D. Walsh, North-West Water Authority, UK
Dr F. Yoshino, Public Works Research Institute, Japan

Sponsored by

University of Salford

British Hydrological Society
Commission of the European Communities
International Association of Hydrological Sciences
Institute of Hydrology
Meteorological Office
World Meteorological Organization
University of Lancaster
National Rivers Authority

Part 1
Raingauge–radar adjustment techniques

1

Calibration of weather radar data in The Netherlands

S. van den Assem
Department of Hydraulics and Catchment Hydrology, Agricultural University of
Wageningen, Nieuwe Kanaal 11, 6709 PA Wageningen, The Netherlands

ABSTRACT

A project is going on to investigate the possibilities of quantitative use of weather
radar. Radar data are combined with data of an automatic network of raingauges in
the southwestern part of the The Netherlands, which is a completely flat country site.
The results of a case study will be presented. An interpolation method is introduced
to estimate mean radar reflectivities over 15 min. Raingauge values are cumulative
over 15 min. Correction of the radar data for attenuation with a fixed Z–R relation
appears to be useless. Calibration of the radar by raingauges improves the results.
The spatial dependence of the calibration factor appears to be weak.

1. INTRODUCTION

In The Netherlands, two digitalized C-band radars are operational to detect
precipitation. One radar is situated at the airport near Amsterdam and the other at
De Bilt in the centre of the country. Both radars are used qualitatively as a tool for
analysis of the meteorological circumstances and short-term forecasting.

In 1987 a project started to investigate the possibilities for quantitative application. Accurate radar data can be very useful in rural and urban water management
(Scholma and Witter 1984, Cluckie *et al.* 1987, Collier and Knowles 1986). To obtain
reliable rainfall estimates by radar a combination is necessary with raingauges
(Collier 1986a, b). Here, use of radar is combined with data of an automatic
raingauge network in the western part of Noord-Brabant (southwestern part of The
Netherlands). This country site is completely flat. The radar data are free of ground
clutter and occultation in this area.

2. MEASUREMENT OF PRECIPITATION

2.1 Radar
The weather radar of DeBilt (Fig. 1) is a C-band radar, giving its data with a spatial
resolution of 2 km × 2 km. The bundle width is 1°. Every 15 min the atmosphere is

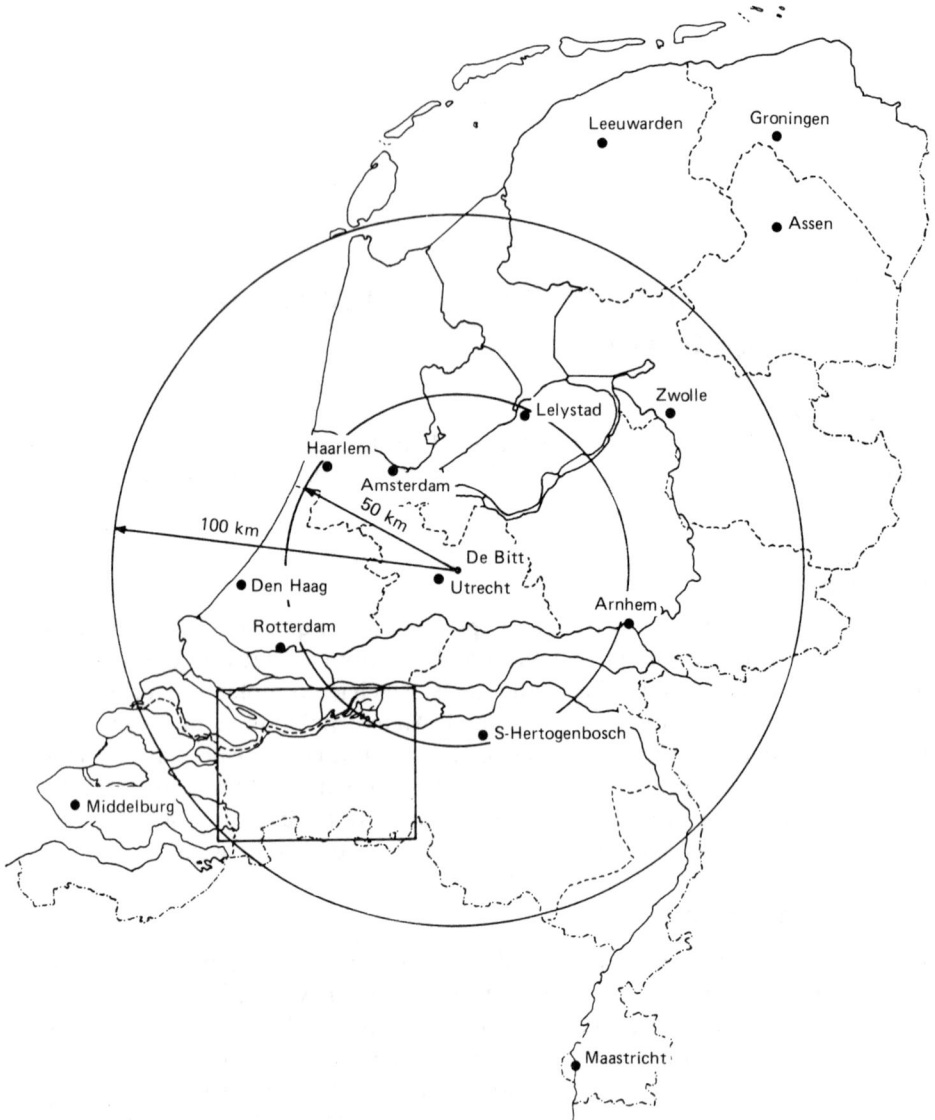

Fig. 1 — Map of The Netherlands. The radar at De Bilt is situated in the centre of the circles. A raingauge network is installed in the square area southwest of the radar.

scanned at two elevations: 0.3° and 1.7°. However, within a range of 65 km distance from the radar, only data of the highest beam are available; from 70 km, data of the lowest beam are used. Between 65 and 70 km distance from the radar the beam with the strongest radar reflection values is chosen (Table 1).

In the first instance, the rain intensities are derived from the well-known Z–R relationship $Z=200R^{1.6}$. No corrections are made for ground clutter, anomalous propagation, bright band, attenutation and variations in the Z–R relationship.

Table 1 — Distance s of the operational raingauges to the radar, height centre of the radar beam above the land surface h, and half-beam width Δh of the radar at the raingauge locations. The height of the radar beam is calculated by the so-called 'four-thirds earth approximation formula'

Location	s (km)	h (km)	Δh (km)
Zevenbergen	61.0	2.03	0.530
Baarle	72.6	0.691	0.632
Dintelsas	72.9	0.695	0.636
Seppe	73.4	0.702	0.641
Steenbergen	83.2	0.843	0.727

2.2   Raingauges

During the summer and autumn of 1988 a network of ten automatic tipping-bucket raingauges has been installed in an area of about 30 km × 45 km at a distance of 50–100 km from the radar in De Bilt. In addition, six automatic gauges are installed temporarily to investigate

(a) the distance dependence of the relation between radar estimates and raingauge values, and

(b) correlation structure of precipitation (semivariogram).

The raingauges have a resolution of 0.16 mm. Calibration of the gauges in the laboratory showed an intensity dependence (Assem 1988). At extreme intensities (greater than 2 mm/min) the tipping value is 10–15% higher than at low intensities. The raingauge data are corrected for intensity dependence. The date and time of every tipping are archieved. Continuous precipitation values during fixed time intervals can be obtained, using a linear interpolation between tippings. This method is only possible when raingauge data are not used on a real-time basis.

3.   EVENT OF 24 AND 25 SEPTEMBER 1988

3.1   Meteorological conditions

On 24 and 25 September 1988 a period occurred with more than 24 h of continuous rain in the central and southern parts of The Netherlands. The five operative raingauges in the western part of Noord-Brabant recorded more than 50 mm of rainfall during this period. The maximum 15 min values recorded at the five locations were 3–5 mm.

A frontal system approached The Netherlands in the early morning of 24 September. The cold front followed immediately after the passage of the warm front. It progressed apparently very slowly because it was located almost parallel to the westerly 500 mbar flow. Several small-scale waves were developing in the front, moving from the south of the UK to FRG. A conveyer belt was formed south of the front, laying from east to west over the middle and southern parts of The Netherlands. Small cells with very high intensities (greater than 50 mm/hr) are detected by the radar in the conveyer belt. Finally, during the evening of 25 September the front over the middle of The Netherlands retreated under the influence of a new depression from the Atlantic. Vertical soundings in De Bilt and Ukkel (Belgium) showed 0°C at 00 Greenwich Mean Time (GMT) of the 24 September at 1800 m, at 12 GMT on 25 September at 2400 m and at 00 GMT on 25 September at 2800 m height. Bright-band effects were absent during this period.

3.2 Interpolation of the radar data

The radar scans only once per 15 min. Small-scale precipitation structures passing over a location of interest will probably not be noticed correctly. In order to improve the radar rainfall estimates, we developed a new interpolation scheme, which makes use of a displacement vector, the latter is obtained as follows.

At time $t=0$ the intensity of precipitation estimated by radar at the location P is known. The displacement vector gives the position of the precipitation cells, which will arrive at the location P at the interpolation time from $t=0$, up to $t=15$, using the radar data at $t=0$. When the position of a location is not centralized in a radar pixel, an averaging procedure is used to include neighbouring radar pixels.

At time $t=15$ we can calculate backwards in time, finding which precipitation cells have passed over the location P from $t=0$ to $t=15$. When calculating backwards, the radar data at $t=15$ will be used. The corresponding pairs of cells at a certain interpolation time which are calculated forwards and backwards can be compared with each other. The mean square difference of these pairs is an indication of the accuracy of this method, using a given set of data.

A first guess of the displacement vector can be obtained from the wind velocity at 850 and 700 mbar. An iterative procedure will provide the optimal displacement vector. The amount of precipitation which has fallen at a location will be estimated by combining forward and backward interpolated data. Weighting factors will be introduced, depending on the elapsed time between the interpolation time and the time of both radar scans. The displacement vector appeared to be conservative for prolonged periods of time (up to several hours).

This implies that the use of the displacement factor allows a rainfall forecast.

Table 2 shows that the average, maximal and minimal values of non-interpolated radar data minus gauge and interpolated radar minus gauge do not differ much. However, the deviations of the non-interpolated radar minus gauge are clearly higher than the interpolated radar minus gauge. The deviations are a good indication to what extent the radar data correspond with raingauge data.

3.3 Attenuation

A C-band radar is sensitive to attenuation, due to scattering and adsorption of the microwave, mainly caused by hydrometers. These phenomena are well described by Battan (1973). An estimate of the corrected reflectivity factor Z_0 (dB) can be obtained with the relation

Table 2 — Amounts of rainfall in 15 min evaluated from non-interpolated (RN) and interpolated (RR) radar data compared with raingauge data at five locations from 0630 Universal Time, Coordinated (UTC), 24 September 1988, up to 0900 UTC, 25 September 1988: G, 15 min raingauge amount. The interpolation time step is 1 min

	Rainfall (mm) for the following locations									
	Zevenbergen		Baarle Nassau		Dintelsas		Seppe		Steenbergen	
	RN−G	RR−G	RN−G	RR−G	RN−G	RR−G	RN−G	RR−G	RN−G	RR−G
Mean	0.02	−0.08	0.00	−0.04	−0.23	−0.18	−0.21	−0.17	−0.03	−0.04
Standard deviation	0.72	0.57	0.77	0.76	0.44	0.41	0.60	0.52	0.41	0.33
Maximum	4.2	3.2	4.7	4.9	0.8	1.0	1.2	1.4	1.5	1.1
Minimum	−1.9	−2.1	−2.3	−2.3	−2.4	−2.2	−3.3	−2.5	−1.2	−1.0

$$Z_0(r) = Z(r) + 2 \sum_{x=1}^{r-1} k(x) \tag{1}$$

with Z (dB) the attenuated reflectivity factor, r (km) the range and k (dB/km) the (one-way) attenuation coefficient. The attenuation coefficient is dependent on the drop size distribution. A relation between k and the precipitation intensity R is proposed by Battan (1973) and Yoshino *et al.* (1988):

$$k = cR^d \tag{2}$$

with c and d empirical parameters.

The data from the radar in De Bilt are not yet corrected routinely for attenuation. The data are delivered in a Cartesian grid; so, in order to correct for attenuation, transformation to a polar grid must be done.

An algorithm is developed, which eliminates ground clutter at well-known locations by interpolation from neighbouring pixels, transforms the radar data to a polar grid, corrects for attenuation as given by equation (1) and retransforms the data back to the Cartesian grid. The transformation method divides all pixels into subpixels, the number of which is dependent on the refinement factor. Here, the refinement factor is put at 4, i.e. 1 radar pixel is divided into 16 subpixels. Conversion tables are used while the transformations are executed. Both transformations cause some error in the data set, owing to smoothing effects. Initially, the data in the polar grid are retransformed to the Cartesian grid, without correction for attenuation. The ratio of the variances of the data transformed twice to the original data gives an indication of the level of smoothing. Furthermore, after the polar data are corrected for attenuation and transformed to the Cartesian grid, the data are corrected for the smoothing effect. This latter correction is an approximation, because in this smoothing correction no attenuation correction effects are involved.

The attenuation algorithm is applied to the data set of 24 and 25 September 1988. The maximum transformation error never exceeded the level of ± 4 dB in a pixel, leading to maximum errors of ± 1.5 mm/h. The ratio of the variances of the twice transformed to the original data varied between 0.97 and 0.99.

Table 3 shows the bias of the attenuation-corrected radar data appears in general to be smaller than the bias of the non-corrected data. However, the standard deviation of RA−G is higher than the deviation of RR−G leading to higher mean square errors. The correction method for attenuation seems not to improve the results.

Table 3 — Amounts of rainfall in 15 min calculated by radar data which are not corrected for attenuation (RR) and attenuation-corrected radar data (RA) compared with raingauge data (G) at five locations from 0630 UTC, 24 September 1988, up to 0900 UTC, 25 September 1988. The (one-way) attenuation coefficient k is taken at $0.003R^{1.0}$

	Rainfall (mm) for the following locations									
	Zevenbergen		Baarle Nassau		Dintelsas		Seppe		Steenbergen	
	RR−G	RC−G	RR−G	RC−G	RR−G	RC−G	RR−G	RC−G	RR−G	RC−G
Mean	−0.08	0.00	−0.04	0.00	−0.18	−0.00	−0.17	−0.03	−0.04	−0.01
Standard deviation	0.57	0.41	0.76	0.41	0.41	0.30	0.52	0.54	0.33	0.35
Maximum	3.2	1.5	4.9	9.0	1.0	1.3	1.4	2.4	1.1	1.9
Minimum	−2.1	−1.6	−2.3	−2.3	−2.2	−1.1	−2.5	−2.9	−1.0	−2.0

Johnson and Brandes (1987) compared an attenuation-corrected C-band radar with an S-band radar during severe weather conditions in the USA. They concluded that correction for attenuation of a C-band radar appears to be futile, because the calibration error of the attenuated reflectivity was not restricted to extreme narrow limits. Yoshino *et al.* (1988) compared uncorrected and corrected C-band radar data with raingauges during an extreme severe storm on Japan. The Z–R relation was adjusted in order to obtain the lowest root mean square error between radar and gauge data. Correction for attenuation led to an improvement, but the results also showed that the use of an inappropriate Z–R relationship may result quickly in large errors.

The Z–R relationship was fixed to the standard values ($Z=200R^{1.6}$) in the analysis of the 24–25 September 1988 situation. A flexible Z–R relationship combined with another attenuation coefficient might have improved the results but should require an extreme amount of computing time.

3.4   Calibration by raingauges

Calibration of radar data by using raingauges improves the accuracy of the rainfall estimates (Collier 1986a, b) During 24–25 September 1988 the data from five raingauges were obtained in the southwestern part of The Netherlands (Fig. 2).

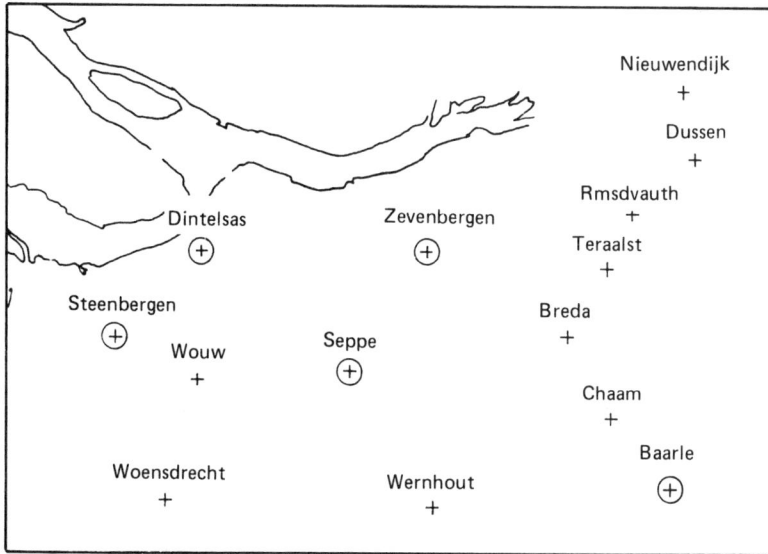

Fig. 2 — Map of the area in which the network of raingauges is located: ⊕, gauges operational, September 1988; +, gauges not (yet) operational.

Here, the calibration factor $C_0(x)$ of location x on $t=0$ is defined by

$$C_0(x) = \sum_{t=-60}^{-15} g_t(x) \Bigg/ \sum_{t=-60}^{-15} r_t(x) \tag{3}$$

with $g_t(x)$ the rain intensity at time t on location x measured by raingauge and $r_t(x)$ is the rain intensity at time t on location x measured by radar. An integration period of 45 min is introduced to avoid significant sampling errors. The most recent period of 15 min is omitted, to make it feasible to compare radar data, which are calibrated by all calibration factors. In a real-time analysis it is evident that this last quarter must be implemented.

The calibration factor found at several raingauge locations can be interpolated spatially by different techniques, such as the Thiessen method, multiquadratic surface fitting or kriging. In this case study no spatial interpolations are executed.

The radar data at the locations of the raingauges are adjusted by all five calibration factors, separately.

In Fig. 3 the mean square errors of the calibrated radar data are presented as a function of distance between the location of the radar pixel and the location where the calibration factor is derived. The lowest mean square errors are obtained at the smallest distances (0 km). This could be partly due to a bias in raingauge or radar at the location of the raingauges. The relation between the distance and mean square error appears to be weak, partly owing to the large scatter.

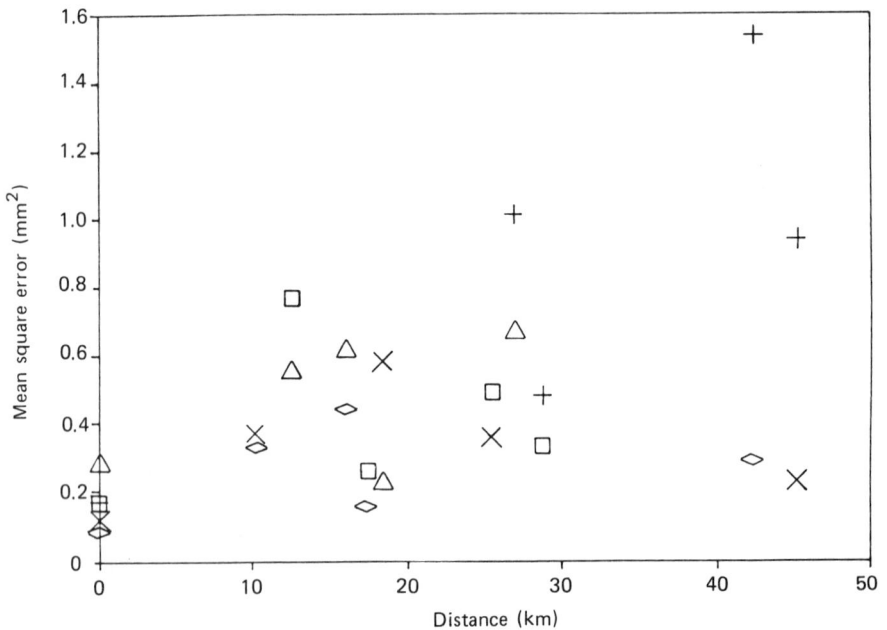

Fig. 3 — Mean square error of the radar data calibrated by calibration factors of all five raingauge locations plotted as a function of distance between the locations: □, Zevenbergen; +, Baarle Nassau; ◇, Dintelsas; △, Seppe; ×, Steenbergen.

Table 4 shows that the on-site calibration of the radar improves the results at Zevenbergen, Baarle Nassau and Dintelsas. At Seppe and Steenbergen, however, the bias is smaller after calibration, but the deviation is enlarged, leading to an almost unchanged mean square error.

4. CONCLUSIONS

(a) The accuracy of rainfall estimates obtained from radar data provided each 15 min is improved by an interpolation procedure that makes use of a displacement vector of the precipitation field.

Table 4 — Amounts of rainfall in 15 min calculated by uncalibrated radar data (RR) and radar adjusted by calibration factors of the same locations as the radar data (RC) compared with raingauge data (G) at five locations from 0630 UTC, 24 September 1988, up to 0900 UTC, 25 September 1988

	Rainfall (mm) for the following locations									
	Zevenbergen		Baarle Nassau		Dintelsas		Seppe		Steenbergen	
	RR−G	RC−G	RR−G	RC−G	RR−G	RC−G	RR−G	RC−G	RR−G	RC−G
Mean	−0.08	0.01	−0.04	0.00	−0.18	−0.00	−0.17	−0.03	−0.04	−0.01
Standard deviation	0.57	0.41	0.76	0.41	0.41	0.30	0.52	0.54	0.33	0.35
Maximum	3.2	1.5	4.9	2.0	1.0	1.3	1.4	2.4	1.1	1.9
Minimum	−2.1	−1.6	−2.3	−2.3	−2.2	−1.0	−2.5	−2.9	−2.10	−1.0

(b) The accuracy of precipitation measured by a C-band radar is not improved when the radar data are corrected for attenuation with a fixed Z–R relation. Adjustment of the Z–R relation or even the attenuation coefficient may lead to better results.

(c) Radar data calibrated by raingauges show in general an increased accuracy compared with uncalibrated radar data. The distance at which a calibration factor can still be applied will depend on the meteorological conditions. In the case study presented, a weak distance dependence was to be seen. This means that a dense network of calibration raingauges will not be necessary under these circumstances.

(d) Errors occur when areal data (radar pixels) are compared with point values (raingauges), owing to spatial variation. The radar data appear to improve in accuracy after the application of some methods; the improvement can only be interpreted statistically, when the errors caused by spatial variation are estimated.

ACKNOWLEDGEMENTS

The helpful advice of Han Stricker and Henk de Bruin, the technical support of the Department of Hydraulics and Catchment Hydrology and the data provided by the water authority Hoogheemraadschap West-Brabant, the weather service Meteo Consult and the Royal Dutch Meteorological Service KNMI is gratefully acknowledged.

REFERENCES

Assem, S. van den (1988) *Calibration of tipping bucket raingauges*, Onderzoeksverslag 85. Landbouwuniversiteit Wageningen, 18 pp. (in Dutch).

Battan, L. J. (1973) *Radar observations of the atmosphere*. University of Chicago Press, Chicago, IL.

Cluckie, I. D., *et al.* (1987) Some hydrological aspects of weather radar research in the United Kingdom. *J. Sci. Hydrol.*, **32** (3), 329–346.

Collier, C. G. (1986a) Accuracy of rainfall estimated by radar, Part I: Calibration by telemetering raingauges. *J. Hydrol.*, **83** (4), 207–223.

Collier, C. G. (1986b) Accuracy of rainfall estimated by radar, Part II: Comparison with raingauge network. *J. Hydrol.*, **83** (4), 225–235.

Collier, C. G. and Knowles, J. M. (1986) Accuracy of rainfall estimated by radar, Part III: Application for short-term flood forecasting. *J. Hydrol.*, **83** (4), 237–249.

Johnson, B. C. and Brandes, E. A. (1987) Attenuation of a 5 cm wavelength radar signal in the Lahona–Orienta Storms. *J. Atmos. Ocean. Technol.*, **4**, 512–517.

Scholma and Witter (1984).

Yoshino, F., Yoo, A. and Kouzeki, D. (1988) Accuracy improvement of radar raingauges by considering radar wave attenuation caused by heavy rainfall. *J. Res., Public Works Res. Inst.*, **26**.

2

Weather radar calibration in real time: prospects for improvement

V. K. Collinge
Institute of Environmental and Biological Sciences, University of Lancaster,
Lancaster LA1 4YQ, UK

ABSTRACT

This paper is concerned with identifying opportunities for improving the calibration of weather radars using telemetering raingauges. The basic principles are described and developments in the techniques briefly reviewed. A method for removing anomalies from radar images is described, together with research into improving calibration procedures using physical and meteorological parameters, particularly under orographic enhancement conditions. A new method of estimating the one-step-ahead assessment factor using time series analysis has been developed and gives encouraging results. The paper concludes with views on the directions in which further research is likely to be most rewarding.

1. INTRODUCTION

The rapid growth in weather radar networks throughout the world has been accompanied by progressive improvements in the hardware and data interpretation techniques. Within the UK and Ireland there is currently a network of ten radars, which is likely to increase eventually to 15. The potential value of this to the water industry in England and Wales was recognized in 1983 and as a result it became a more or less equal partner with the UK Meteorological Office in the financing of the network. A recent review of the costs to the water industry of its participation in the radar network and the resulting *potential* benefits from improved flood warnings in both urban and rural situations gave an overall benefit-to-cost ratio in the range 3.0–3.5 (Collinge 1989). The estimated improvements in flood warnings were based on assumed improvements in both rainfall forecasts and in estimates (in near real time) of the amounts of rain which has fallen, but it is clear that we are a long way off realising these benefits in practice. Part of the reason for this lies in the size of the errors in radar-derived rainfall estimates. There is, however, considerable uncertainty about the accuracy requirements. It is well known that there are substantial smoothing effects as rainfall moves through the various phases in its passage to a flood risk *zone*, and current research suggests that for this reason the accuracy requirements could be quite low (Cluckie 1987). Notwithstanding this, improving the quality of the products from weather radar, including the the accuracy of rainfall estimates, for use by the hydrologists in real-time forecasting and for other purposes remains a major challenge.

So, in common with other countries, increasing attention is now being focussed in the UK on the role of weather radars in hydrology, and particularly on using precipitation measurements and forecasts as input to hydrological models for flood-forecasting purposes. This paper reviews the development of improving radar-based rainfall estimates using calibrating raingauges, discusses the results so far obtained from ongoing research at the University of Lancaster and suggests the most promising areas for further investigation.

2. CALIBRATION TECHNIQUES

The concept of using measurements from raingauges as 'ground truth' to calibrate a weather radar dates back at least as far as 1954 (Hitschfeld and Bordan 1954). Use of the standard Z–R relationship.

$$Z = AR^b \tag{1}$$

provides a first approximation to the rainfall rates throughout the radar field, typically using values of $A = 200$ and $b = 1.6$. Data from a calibrating raingauge, together with the corresponding radar observations above it, provide the calibration factor CF:

$$\text{CF} = \frac{\text{gauge rainfall}}{\text{radar rainfall}} = \frac{R_g}{R_r} . \tag{2}$$

This factor is then used to adjust all the rainfall rates produced by the first approximation. The method is equivalent to choosing a new value of A in equation (1). This technique was used in the Dee Weather Radar Project (Harrold *et al.* 1974), where it was shown that one calibration site (which actually contained several raingauges within a small area) was sufficient to meet the accuracies required over areas within 15 km of the calibration site. Accuracy decreased with increasing distance from the calibration site, and Harrold recognized that to increase the area of quantitative radar coverage, additional calibration sites would be needed.

If several calibration sites (n) are used, then a calibration factor to be applied uniformly over the whole radar field can be derived from either

$$CF = \sum_{i=1}^{i=n} (R_g)_i \Bigg/ \sum_{i=1}^{i=n} (R_r)_i \qquad (3)$$

or

$$CF = \frac{1}{n} \sum_{i=1}^{i=n} \frac{(R_g)_i}{(R_r)_i} . \qquad (4)$$

Using equation (3) gives each observation a weight proportional to its value, whilst equation (4) gives equal weight to the ratio R_g/R_r at each site.

The results from a number of studies utilizing an average storm calibration factor are summarized by Wilson and Brandes (1979). Direct comparisons between these results are not possible because of the many differences between the studies. However, generally there is a significant reduction in the errors in radar-derived rainfall estimates when one or more calibrating gauges are used in the ways described above, but large spatial errors remain.

2.1 Spatial adjustment

If, during a rainfall event, values of R_r are obtained throughout the radar field using equation (1), the ratio R_g/R_r is found to vary spatially in a systematic way. Therefore it can be expected that, having obtained the rainfall field using equation (1), adjustment at any point can be improved by reference to the nearest calibration gauge rather than using an average calibration factor over the whole area. An early example of this approach is provided by Wilson (1970), who studied 28 storms and their rainfall totals over a 1000 mile^2 study area in Oklahoma. He showed that, by using three calibrating gauges and applying three calibration factors, each over one third of the area, better results were obtained than by using one average calibration factor derived from three gauges.

Many sophisticated schemes have been devised to combine radar observations routinely with a number of calibrating raingauges. One early example is a method devised by Brandes (1975), in which the calibration factor at each grid point is calculated by combining the factors at each of the calibrating gauges, weighting them in proportion to the inverse of the square of the distance between the grid point and the gauge. Cain and Smith (1977) used a technique known as sequential analysis, which tests the variations in $\log(R_g/R_r)$ on a probability basis and in real time. If it is significantly different from zero, an adjustment factor is applied; if not, the difference is assumed to be random. A number of methods that have been reported in the literature in the period 1977–1987 are briefly reviewed by Joss and Waldvogel (1987). More recently Moore has developed a method using a multiquadratic surface and applied this to an area of 2500 km^2 north of London covered by the Chenies radar, within which there are 20 telemetering raingauges (Moore 1988). Judged against the operational radar product which relies on five calibrating gauges, the increase in accuracy in frontal events ranges from 20 to 40%. Much of this improvement derives from the increased density of one gauge per 125 km^2. Judged against the simple average of the 20 gauges the improvement is a modest 6%.

A further approach can be identified, in which account is taken of the prevailing meteorological conditions. An example is a method developed for the calibration in real time of the Hameldon Hill radar in northwest England (Collier et al. 1983), which has been used operationally for a number of years. In this study, Collier used the inverse of the calibration factor as defined by equation (1), referred to as the assessment factor AF:

$$AF = \frac{\text{radar rainfall}}{\text{gauge rainfall}} = \frac{R_r}{R_g} \ . \tag{5}$$

An objective assessment of rainfall type was developed using a harmonic analysis of the variations in the running hourly mean assessment factors. By this means the on-site computer classifies the rainfall into one of four categories. Different geographical areas or 'domains' are defined according to the rainfall type and which take account of wind-dependent orographic effects. For each domain and rainfall type a pre-defined combination of one or more calibrating gauges is used, or in some circumstances none, in which case the rainfall is derived solely from the application of $Z = AR^b$.

3.   PROSPECTS FOR IMPROVED CALIBRATION METHODS

Research at the University of Lancaster during the past 3 years has concentrated on several approaches which have the potential for improving radar calibration. These can be considered under the following headings:

(1) improvement of the radar image before calibration;
(2) use of meteorological and physical parameters in the calibration process;
(3) use of time-series techniques in the calibration process.

The first of these, whilst not directly related to the use of calibrating gauges, has nevertheless been shown to be important for one of the radars in the UK network.

This is not an exhaustive list of opportunities, but one based on the present author's direct experience of the UK radar system. Other research directly concerned with the better use of calibration gauge data, and referred to in section 2, is directed at mathematical techniques for improving the fit of a surface representing the assessment factor or calibration factor over the area of radar coverage. For the sort of calibrating gauge density to be expected in the UK, say ten gauges over a circle of 75 km radius=1 gauge per 440 km^2, refining surface-fitting techniques are only likely to effect limited improvements.

There are other important opportunities for improving radar-derived rainfall estimates less directly concerned with the calibration process, e.g. research into corrections for bright band. Other prospects are reviewed by Joss and Waldvogel (1987) who discuss the possible advantages of dual-polarization radars for improving knowledge of the *Z–R* relationship. It is clear that much research will be needed before these advantages are quantified and even then they may be found to be quite small. These workers also discuss the advantages of developing vertical reflectivity profiles using scans at several elevations.

4.   UNIVERSITY OF LANCASTER RESEARCH

The data used for this research have been obtained from the Hameldon Hill radar in northwest England (Fig. 1), which is a Plessey C-band, 5.6 cm wavelength, 1° beam

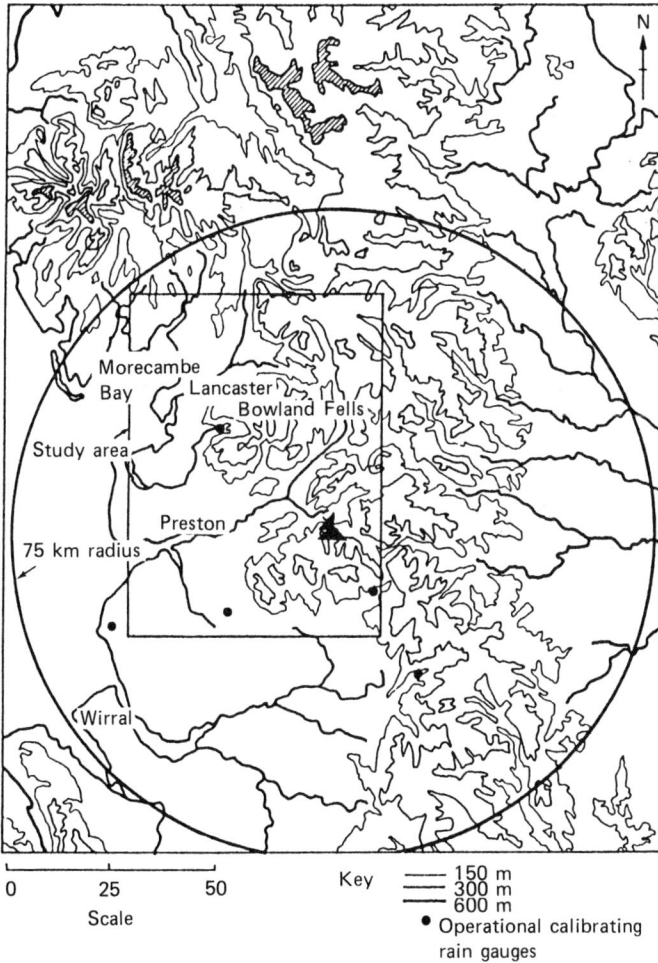

Fig. 1 — Location of Hameldon Hill radar in northwest England.

width radar. The aerial rotates at a rate of 1 rev/min, scanning each of the following the elevations in turn: 0.5°, 1.5°, 2.5°, 4.0°. Only the lower two elevations are used for estimation of surface rainfall, the higher two being used for bright-band detection and for research purposes. The 1.5° beam is used within 24 km radius to minimize clutter problems and the 0.5° beam from 24 to 210 km radius. A dedicated on-site microcomputer carries out numerous functions, including

(1) occultation correction,
(2) clutter cancellation,
(3) identification of the presence and height of the bright band,
(4) conversion of received power to rainfall rate using $Z=AR^{1.6}$ (A is usually 200),
(5) empirical range correction beyond 75 km to allow for the upper part of the beam being above the rain,
(6) attenuation correction,
(7) conversion from polar to Cartesian coordinates and
(8) calibration using telemetering raingauges.

The basic data produced are rainfall intensities in 208 levels every 5 min on a 2 km grid out to 75 km and on a 5 km grid from 75 to 210 km.

The topography of the area covered by this radar is complex (Fig. 1); within the 75 cm range there are substantial areas of high ground, particularly the central Pennines (which form an important divide between east and west) and the Bowland Fells southeast of Lancaster. There is a coastal plain to the west extending from Morecambe Bay to the Wirral peninsula. Most of the research described here has used data from a study area 60 km×80 km within the 75 km radius circle, in which a network of 28 recording raingauges was established by supplementing existing gauges owned by operational agencies (North-West Water Authority and Yorkshire Water Authority). This gave a density of one gauge per 171 km². The radar data, on a 2 km×2 km grid, was calibrated by the Meteorological Office, whose staff wrote the necessary software to remove the final-stage calibration procedure.

Software has been written for a number of data-handling tasks. Radar data are extracted from the four pixels closest to each ground truth raingauge and used to calculate the assessment factor at each of these 28 locations over any selected time interval (usually 15 min and 1 h). A contouring program has been used to produce maps of rainfall amounts based on the decalibrated radar data. Rainfall maps are also produced by combining the decalibrated radar data and the ground truth data. The latter is assumed to be correct, and assessment factors at the 28 sites are calculated. Then an interpolation routine is used to estimate the assessment factor at each pixel, these factors then being used to adjust each radar value. This can be done for any required time interval. These maps are referred to as BESTEST maps.

4.1 Improvement of radar image
Production of rainfall maps from the decalibrated radar data revealed three substantial and very obvious anomalies (Fig. 2(a)) which were equally apparent from BESTEST maps. It was clear that these anomalies would have to be removed before further work could proceed, since they covered 16% of the study area and affected

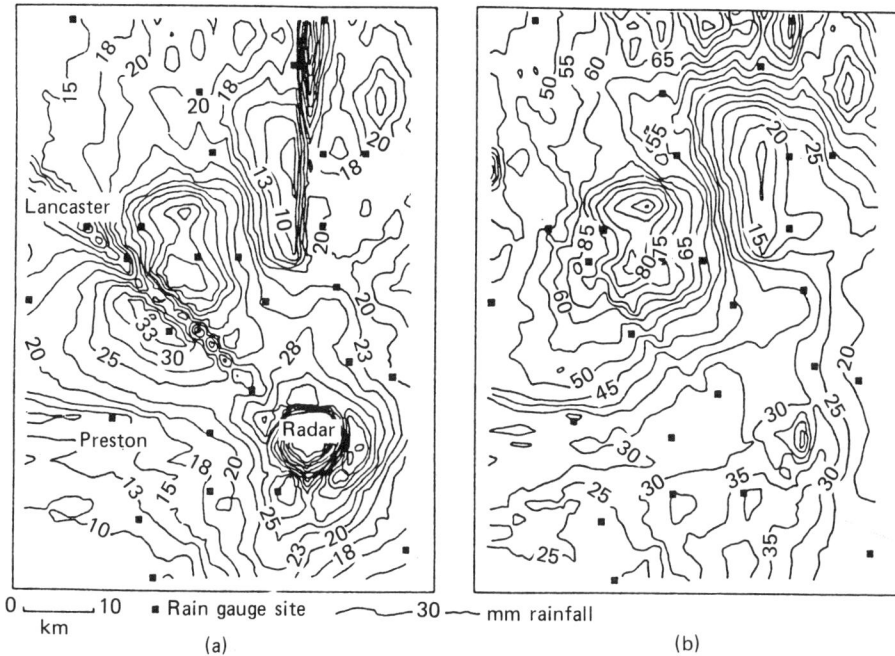

Fig. 2 — Rainfall maps from decalibrated radar data 3–4 December 1986 (a) before and (b) after removal of anomalies.

seven of the ground truth gauges including one of the operational calibrating gauges. These anomalies were found to be due to

(a) a BBC television mast close to the radar and at an azimuth of 315°,
(b) Pendle Hill (558 above sea level) which lies almost due north of the radar at a range of 13 km and which intercepts the lower part of the 0.5° beam and
(c) errors at very close range owing to inadequate time delay between the outward pulse and the return signal.

No corrections are made in the on-site computer for either (a) or (c), but occulation corrections are made for sectors where hills intercept the 0.5° beam. These corrections are largest between 352° and 001°, and between 355° and 359° over half the beam is intercepted. It is concluded that, because of the anomalies identified in this region, the occultation corrections made are not the optimum ones.

The simplest way to remove these anomalies would be to use linear interpolation across each affected area. However, this approach was rejected, because under conditions of orographic enhancement (of major interest to this research) the substantial topographic changes within these affected areas would result in non-linear changes in rainfall across them. So an alternative procedure was developed, based on the assumption that over a period of time the aggregated radar-derived rainfall in any pixel was proportional to the annual average rainfall (obtained from standard 1941–1970 maps). Radar data was aggregated for six rainfall events totalling 146 h. For the Pendle Hill anomaly, corrections were calculated on a row-by-row basis, using a pixel to the east (column X) as a reference point. Fig. 3 shows

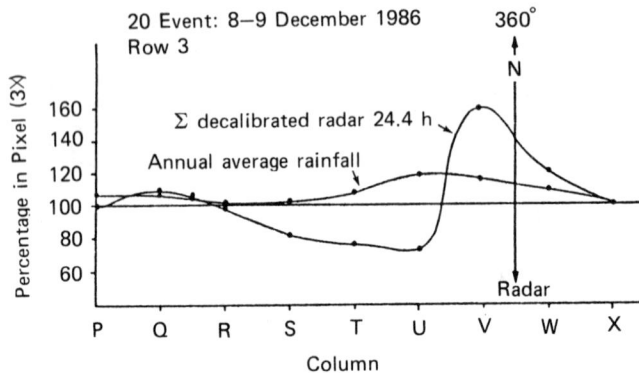

Fig. 3 — East–west section through the Pendle Hill anomaly at a range of 58 km.

the data for one event and one row, illustrating the excessive radar values in columns V and W, and low values in columns S, T and U. Once within the 24 km radius the corrections become insignificant, which is to be expected because this is where the beam elevation increases to 1.5° and occultation from Pendle Hill ceases. Similar procedures were applied to the other two anomalies. For the television mast the corrections were calculated along columns instead of along rows. The largest corrections occur within the shadow of the television mast, where radar values have to be scaled up by a factor as high as 2.7.

The results have only been assessed on a subjective basis. In all cases the improvement is very substantial, although not always quite as good as that illustrated. Other anomalies have been identified by Collier (1983) between azimuths 110° and 126° (owing to electricity power lines) and between 151° and 174° (probably owing to wrong occulation corrections). This latter anomaly has been confirmed by the present author in a recent work on the full 75 km image. Further work on this aspect is desirable but could not be undertaken by the present author with the resources available. A more objective method of testing needs to be applied. For example, Creutin has suggested plotting the ratio of the radar values at two different beam elevations.

4.2 Use of meteorological and physical parameters
A study has been made of the statistical relationship between the assessment factor at each of the 28 raingauges (using hourly data) and various parameters, concentrating on distance from radar and rate of rainfall, but also including ground altitude and beam height. A standard statistical package has been used for this work, which is

currently being extended to data from 60 gauges within the 75 km radius circle. The main results so far obtained (Table 1) show that during widespread rain, for a substantial proportion of the time, there is a decrease in assessment factor with

(i) increasing rate of rainfall and
(ii) increasing distance from the radar.

Table 1 — Regression of log (AF) on distance from radar and on rate of rainfall for study area (28 gauges) and full 75 km circle (60 gauges)

Regression of log (AF) on	Number of gauges	Time (h)	Percentage of hours when coefficient in regression equation is	
			Positive	Negative
Distance from radar	28	50	82	18
Rate of rainfall	28	50	84	16
Distance from radar	60	40	85	15
Rate of rainfall	60	40	75	25

There is supporting evidence for the first of these efects in the literature (Wilson and Brandes 1979), and for both effects from Woodley and Herndon (1970). There is also strong evidence of this range dependence from Collier (1986), whose work on the Hameldon Hill radar showed that in frontal rain conditions the assessment factor decreased substantially with increasing distance over ranges from 0 to 110 km.

The rainfall events analysed represented a range of meteorological situations, and to pursue these associations further it was decided to concentrate first on those events where significant orographic enhancement occurred.

The study area illustrated in Fig. 1 has within its regions of high ground where orographic enhancement can be substantial. For example, the Bowland Fells cover an area of approximately $800\,km^2$ and rise to an elevation of 560 m; the annual average rainfall on the highest ground is about 2000 mm in comparison with 950 mm on the plain to the west. Orographic enhancement has been identified as one of the factors expected to influence the assessment factor, and so it was decided to evaluate this effect using a simple model, illustrated in Fig.4, sections (a) and (b). It has been assumed that the radar beam is at right angles to the wind direction, that the radar is at 400 m above sea level and that the hill extends in the direction of the radar beam maintaining the same cross-section. With this configuration we have examined the changes in the assessment factor under different assumptions about the growth of the rainfall as it descends to the ground, and over a range of radar beam height (i.e. at different distances from the radar). Three models have been tested, with different assumptions about the patterns of rainfall growth.

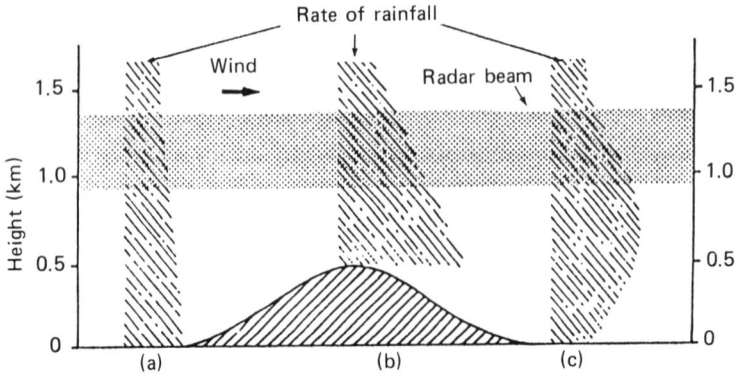

Fig. 4 — Orographic enhancement; assumed rainfall–altitude profiles (a) upwind of (b) on (c) downwind of hill.

The results presented in Figs 5 and 6 are for model 3 in which there is an assumed increase of rainfall rate upwind of the hill from zero at 2.5 km above sea level to P_0 at sea level. The orographic enhancement is assumed to give an *additional* growth of rainfall rate varying linearly from zero at 1.5 km above ground level to a value at the ground given by

$$_cP_h = \alpha h P_0 \tag{6}$$

where h (km) is the altitude of the ground and α represents the degree of orographic enhancement. Thus for example, if $\alpha = 5$, then, where $h = 400$ m, $_cP_h = 2.0\,P_0$; that is the rainfall rate is 3.0 times that at ground level upwind of the hill.

Fig. 5 shows that at all ground altitudes the assessment factor diminishes with increasing beam height. Fig. 6 presents the same data in an alternative way, showing the influence of the ground altitude on assessment factor at different beam heights. There is a fixed relationship between ground altitude and rate of rainfall (equation (6)); so values of the latter are also given in Fig. 6. Now it can be seen that generally there is an *increase* in assessment factor with increasing ground altitude (rate of rainfall) except at low ground altitudes and high beam elevations, where there is a decrease.

Clearly many different assumptions can be made about the pattern of growth of rainfall. For simplicity a linear rate of growth has been assumed, but there is no physical reason to expect this; a recent paper by Grabowski (1989) describes the physical modelling of the seeder–feeder mechanism and gives non-linear growth patterns. The adoption of a 1.5 km altitude upper limit for orographic enhancement growth was based on the work of Hill *et al.* (1981) who found that in a study in south Wales, most rainfall cases have a high proportion of enhancement occurring in the lower 1.5 km of the troposphere. Only a limited number of assumptions have been tested, but the overall conclusion is clear. Under conditions of orographic enhancement the relationship between assessment factor and beam height (or distance),

Model 3: $\alpha = 5$

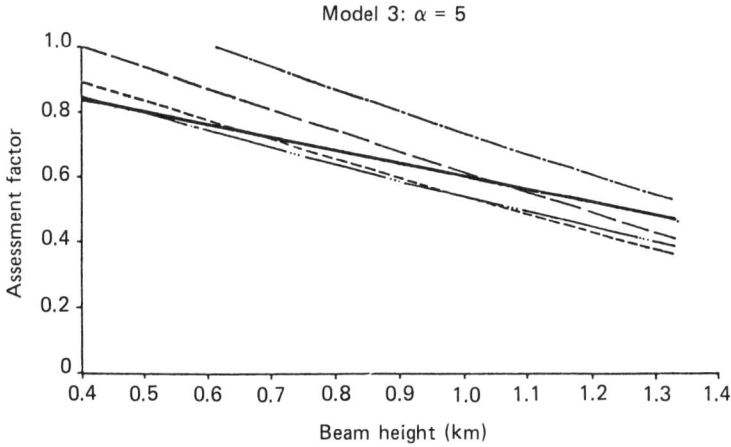

Fig. 5 — Orographic enhancement model showing the influence of beam height on assessment factor.

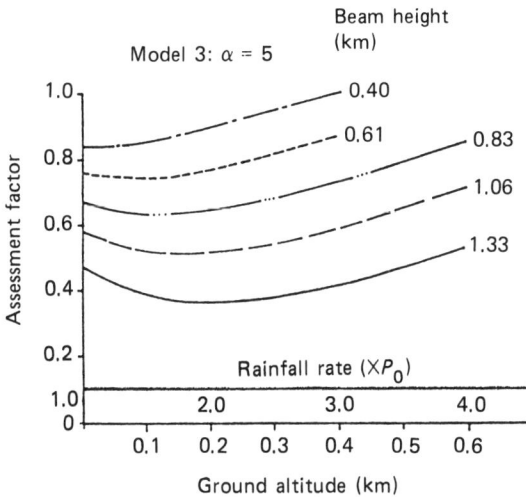

Fig. 6 — Orographic enhancement model showing the influence of ground altitude on assessment factor.

ground altitude, and degree of enhancement will not be simple, but the assessment factor should generally diminish with increasing beam height, and this effect will become more pronounced at higher orographic enhancement levels (i.e. larger values of α). In all the foregoing, no consideration has been given to conditions downwind of the hill, except in a qualitative way. Here it is assumed that evaporation at lower levels can lead to the profile shown in Fig. 4, section (c), with the consequential increase in assessment factors in the rain shadow areas.

Data from Bowland Fells have been used to test whether they conform to the general features of the model results. Fig. 7 shows the results for 29 December 1986,

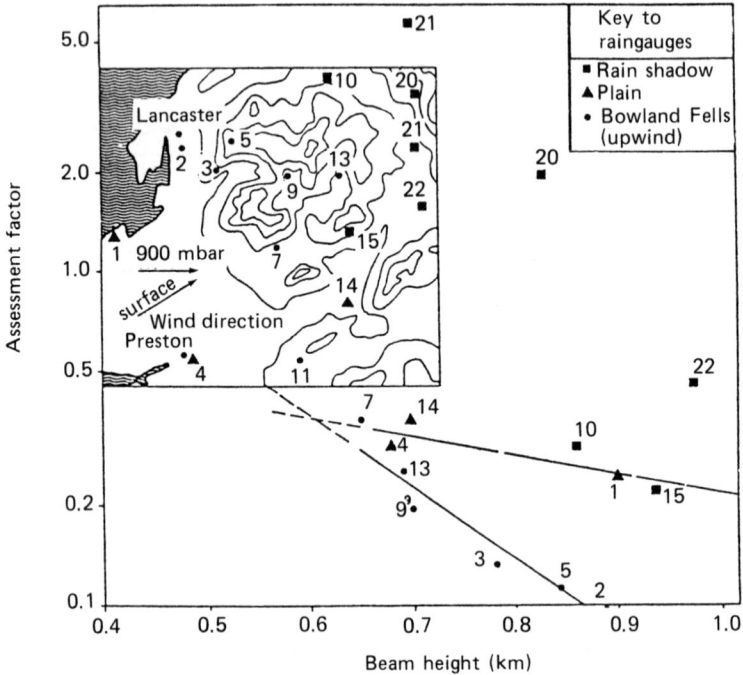

Fig. 7 — Assessment factors for Bowland Fell region on 29 December 1986 (1600–1700 hours). The insert shows the topography and wind direction for the region.

1600–1700 hours, when rainfall intensities of 0.40–1.00 mm/h on the plain were enhanced to 6 mm/h or more on the Bowland Fells, an exceptionally high level of enhancement for this area ($\alpha \approx 18$). It can be seen that

   (i) at large beam heights the assessment factors for sites subjected to orographic enhancement are much less than for the sites on the plain.

  (ii) the lines through observations from sites on the plain and through those on the upwind side of Bowland Fells cross and

 (iii) sites on the lee side of Bowland Fells in the rain shadow have very high assessment factors.

All these features are predicted by the model. Data from another hour in the same rainfall event and from several hours in another rainfall event, where the orographic enhancement was much less ($\alpha \approx 2.2$–3.5) show the same general features although with somewhat less consistency. However, the high assessment factors in shadow locations remain a prominent feature. Data from other rainfall events with the surface wind direction approximately 180°, which means that the air travelled over higher ground to the south and over a river valley before reaching Bowland Fells, show much more complicated situations. This hilly south Lancashire moorland topography creates increased surface roughness, disrupting the steady moist air inflow into the feeder cloud layer, decreasing the potential orographic enhancement maximum. Orographic enhancement is also evident over this area. Rainfall contained within the feeder cloud is partially depleted before reaching Bowland Fells, which therefore lie in a partial 'shadow' situation. This makes base rainfall level assessment, and thus the calculation of orographic enhancement over Bowland Fells, much more difficult. Work is now proceeding to quantify the enhancement and shadow effects so that assessment factors over Bowland Fells can be estimated from wind direction and speed and other relevant data. The shadow effects are proving to be the most difficult feature.

4.3 Time-series analysis

Calibration methods developed for use in real time may well be improved if they estimated the one-step-ahead rainfall by deriving the one-step-ahead assessment factor using a forecasting procedure. In general this is not done. For example, the method currently used operationally at Hameldon Hill (Collier *et al.* 1983) combines the assessment factors from up to five calibrating gauges, each being obtained by the simple averaging of data from the past hour. There are two major drawbacks to this operational technique. As already indicated, the simple averaging does not allow for variations in time of the assessment factor that are known to occur, i.e. the estimator has no 'dynamic' memory. Also the assessment factor becomes infinite when the raingauge reading is $R_g = 0$ and the radar value $R_r > 0$. It is more advantageous statistically, therefore, to formulate equation (5) in the following simple linear time series regression form:

$$(R_r)_k = (AF)_k (R_g)_k + e_k \tag{7}$$

where e_k is assumed to be a zero mean, normally distributed white-noise sequence with variance σ_e^2, and k indicates the sampling time. Note that $(AF)_k$ in (7) is simply the coefficient of the time series regression model and so it is better defined than AF in equation (5)', in the sense that no singularity occurs in the estimation process even when the R_r value is zero. Moreover, since equation (7) is formulated in a time-series context, it is possible to assume that the coefficient $(AF)_k$ is time variable and can be represented as a stochastic generalized random walk process (Young 1984) of the following form:

$$(AF)_k = (AF)_{k-1} + \eta_{k-1} \tag{8}$$

where η_k is another zero mean, normally distributed white-noise sequence with variance σ_η^2. Although such an assumption may seem unusual, it is a device which has been used successfully in the solution of other time-variable parameter systems problems, in areas such as adaptive forecasting and control.

Having defined the $(R_r)_k$ and $(R_g)_k$ relationship by equations (7) and (8), it is possible to apply Kalman filter techniques to estimate $(AF)_k$. Only one unknown parameter is normally required in such procedure, namely the noise variance ratio $NVR = \sigma_\eta^2/\sigma_c^2$. The parameter NVR effectively controls the length of significant memory of the Kalman filter estimator: a large NVR implies a short memory, while a low NVR implies a long memory. Outputs from the Kalman filter estimator based on the random-walk (NVR = 0.1) model and a more complex integrated random-walk (NVR = 0.01) model, have been produced, using the autocorrelation function and the partial autocorrelation function as simple statistical tools to evaluate the randomness of the estimated errors $\hat{e}_k$. By definition these error series should be totally random and, hence, must not produce significant 'spikes' (i.e. outside the ± 2 standard error boundaries) or structural patterns in the autocorrelation or partial autocorrelation function. It has been found that the error series are *not* totally random because there are several significant autocorrelations outside the ± 2 standard error boundaries, and there also appears to be some structural pattern in the autocorrelation function.

An alternative approach is to regress R_g data against the R_r data:

$$(R_g)_k = (CF)_k(R_r)_k + e_k \tag{9}$$

where e_k is again a white-noise series. The regression coefficient, calibration factor, is the inverse of the assessment factor. Once again, this coefficient can be modelled by one of the random-walk processes, such as the simple random walk defined as follows:

$$(CF)_k = (CF)_{k-1} + \eta_{k-1} \tag{10}$$

where η_k is again a white-noise process. The Kalman filter estimations based on these models appear to have a better-defined estimation noise sequence; the autocorrelation function and partial autocorrelation function reveal very little structural pattern in the series. The recursive filtering and smoothing estimates of the regression coefficient, calibration factor, are shown in Fig. 8 which reveal how the calibration factor varies with time at one site in this particular event on 3–4 December 1986.

A calibration method based on the foregoing (time series 1) has been developed and tested on a widespread rainfall event that occurred on the 3–4 December 1986. The method enables either the assessment factor or the calibration factor for each calibrating gauge to be modelled, and the one-step-ahead estimates, of either factor, averaged over the study area. Ten of the 28 gauges were selected to test the method, the estimated assessment factor or the calibration factor being applied to the radar observation at each test gauge to obtain the estimated rainfall. The average mean square error of the estimated one-step-ahead (15 min) rainfall was calculated for the test gauges and compared with that obtained by applying the operational calibration method. The results are given in Table 2.

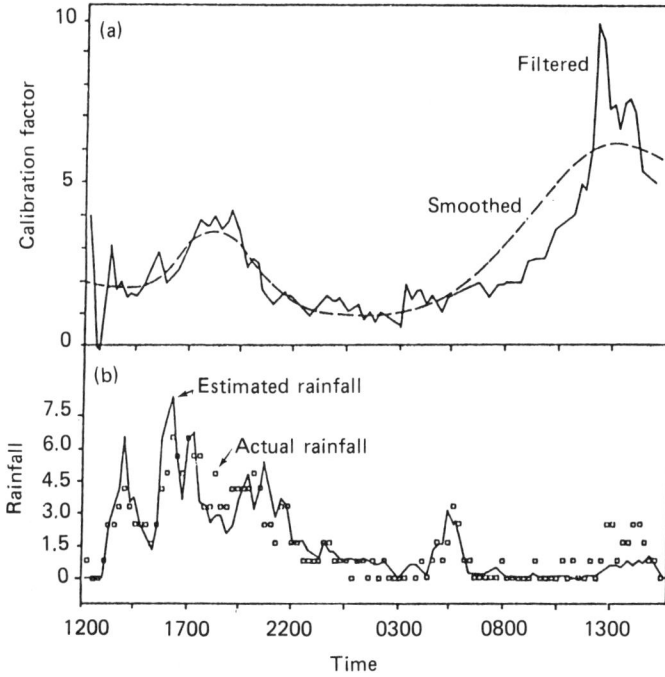

Fig. 8 — Application of time-series method 2 to one raingauge site, on 3–4 December 1986: (a) filtered and smoothed estimates of the calibration factor; (b) estimated and actual 15 min rainfall values.

Table 2 — Application of time-series method 1 to rainfall event of 3–4 December 1986

Forecasting period (10 sites) (each interval is 15 mins	Average mean square error in estimated rainfall (10 gauges)		
	Operational method	Time-series method 1 using	
		AF	CF
5.28	4.30	4.61	4.77
29–60	4.89	5.88	4.13
61–108	1.10	0.58	0.54
29–108	2.56	2.70	1.98

The following comments apply to the various forecasting periods (each interval is 15 min): 0–4, no calculations, first four data points needed to initialize method; 5–28, rain spreading across area, only by interval 28 were all five calibrating gauges recording rain; 29–60, widespread heavy rainfall; 61–108, rainfall less intense, showery.

From Table 2 it can be seen that modelling the calibration factor gave for most of the time an average mean square error lower than the operational method. These encouraging results lead to a further development, combining the time-series modelling of the calibration factor of each calibrating gauge with the knowledge that, under widespread rainfall conditions, there is usually a decline of calibration factor in assessment factor (or increase in calibration factor) with increasing distance d from the radar. This has been discussed in section 4. Estimating the one-step-ahead calibration factor at each calibrating gauge site is now followed by the regression of log (CF) against d, the regression equation then being used to estimate the calibration factors for the ten gauges, which are then applied to the radar observations to give estimated rainfall values. The process is repeated every 15 min. This method (time series 2) has been tested on three rainfall events. The results are given in Table 3, and an example of the output is illustrated in Fig. 8.

Table 3 — Application of time-series method 2 to three rainfall events

Rainfall event	Forecasting period (each interval is 15 min)	Method	Calibrating gauges	Average mean square error (10 gauges)
3–4 December 1986	29–108	Operational	5 operational	2.56
	29–108	Time series 2 (CF)	5 operational	1.85
	29–108	Time series 2 (CF)	5 alternative set	1.71
8 July 1987	5–76	Operational	4 operational[a]	3.77[b]
	5–76	Time series 2 (CF)	4 operational[a]	2.21[b]
	5–76	Time series 2 (CF)	5 alternative set	2.49[b]
31 October 1986	5–24	Operational	5 operational	5.39
	5–24	Time series 2 (CF)	5 operational	6.75
	5–24	Time series 2 (CF)	5 alternative set	3.01

[a]One operational gauge out of action.
[b]Only nine test gauges available.

With each event we have included a test using an alternative set of five calibrating gauges, not currently available for operational use as they have no telemetry facility, but chosen from the 28 ground truth gauges on a basis of geographical distribution and consistency. Table 3 shows that the time-series method 2 gave substantial reductions in the average mean square error using the five operational calibrating gauges in two events (and an increased error in the third). When using the alternative set, there are substantial reductions for all events. We have concluded that the time series techniques described here show promise for improving real-time calibration methods, although further development is needed. The choice of the noise variance ratio needs investigating and a wider range of rainfall events should be studied. It is likely that the technique could be integrated into a surface-fitting method.

5. CONCLUSIONS

The large spatial and temporal variations in assessment factors (or calibration factors) remain as formidable problems when using telemetering raingauges to calibrate weather radars, but there are clearly opportunities for improving on present methods, including the use of surface-fitting techniques and time series analysis methods, and the incorporation of meteorological and physical parameters, particularly under conditions of orographic enhancement. Some combination of these approaches is likely to provide an optimum solution but, whatever that combination is, it will be important to achieve the best possible radar image prior to calibration. It must be free as far as practicable from distortions and with all calibrating and research raingauges outside areas of doubt.

ACKNOWLEDGEMENTS

This work has been financially supported by the Natural Environment Reasearch Council, North-West Water Authority and the Commission of the European Communities. Data have been provided by the Meteorological Office, North-West Water Authority and Yorkshire Water Authority. Thanks are due to all these organizations for their help. The contribution of Professor P. C. Young and Dr Cho Ng to the time series work is gratefully acknowledged, along with assistance from James Buxton and Charles Blakeley.

REFERENCES

Brandes, E. A (1975) Optimising rainfall estimates with the aid of radar. *J. Appl. Meteorol.*, **14**, 1339–1345.

Cain, D. E. and Smith, P. L., Jr. (1977) A sequential analysis strategy for adjusting radar rainfall estimates on the basis of rain gage data in real time. *Proceedings of the Conference on Hydrometeorology, Toronto.* pp. 280–285.

Cluckie, I. D. (1987) *CEC Weather Radar and Climate Hazard Project Report.* Commission of European Communities, Brussels.

Collier, C. G. (1983) *Radar network coverage in the overlap region between Hameldon Hill and Clee Hill*, Operations and Applications Groups Report No. 72. Meteorological Office, Bracknell, Berks.

Collier, C. G. (1986) Accuracy of rainfall estimates by radar: Part 1; Calibration by telemetering raingauges. *J. Hydrol.*, **83**, 207–223.

Collier, C. G., Larke, P. R. and May, B. R. (1983) A weather radar correction procedure for real time estimation of surface rainfall. *Q. J. R. Meteorol. Soc.*, **109**, 589–608.

Collinge, V. K. (1989) *Investment in weather radar by the U.K. water industry.* British Hydrological Society.

Grabowski, W. W. (1989) On the influence of small-scale topography on precipitation. *Q. J. R. Metereol. Soc.*, **115**, 633–650.

Harrold, T. W., English, E. J. and Nicholass, C. A. (1974) The accuracy of radar-derived rainfall measurements in hilly terrain. *Q. J. R. Meteorol. Soc.*, **100**, 331–350.

Hill, F. F., Browning, K. A. and Bader, M. J. (1981) Radar and raingauge observations of orographic rain over south Wales. *Q. J. R. Meteorol. Soc.*, **107**, 643–670.

Hitschfeld, W. and Bordan, J. (1954) Errors inherent in the radar measurements of rainfall at attenuating wavelengths. *J. Appl. Meteorol.*, **11**, 58–67.

Joss, J. and Waldvogel, A. (1987) Precipitation measurement and hydrology. *Proceedings of the 40th Anniversary Conference on Radar Meteorology*.

Moore, R. J. (1988) CEC Weather Radar and Climate Hazard Project Report. Commission of European Communities, Brussels.

Wilson, J. W. (1970) Integration of radar and raingauge data for improved rainfall measurement. *J. Appl. Meteorol.*, **9**, 489–497.

Wilson, J. W. and Brandes, E. A. (1979) Radar measurement of rainfall — a summary. *Bull. Am. Meteorol.*, **60**, 1048–1058.

Woodley, W. and Herndon, A. (1970) A raingage evaluation of the Miami reflectivity–rainfall rate relation. *J. Appl. Meteorol.*, **9**, 258–263.

Young, P. C. (1984) *Recursive estimation and time-series analysis*. Springer, Berlin.

3

Using and adjusting conventional radar reflectivity data for estimation of precipitation: past, present and future studies in Switzerland

G. Galli and J. Joss
Swiss Meteorological Institute, Observatorio Ticinese, CH 6605 Locarno Monti, Switzerland

ABSTRACT

Results are presented of comparisions between estimates of precipitation by weather radar and raingauges made at the Swiss Meteorological Institute during the past 10 years. The procedure used to monitor and calibrate the radar instrumentations using microwave equipment is illustrated. Some preliminary results taken from a current study about the vertical profile of reflectivity are discussed. Except close to the radar, where the visibility of precipitation is good, this profile, together with limitations of radar to see precipitation close to the ground because of the orography and the curvature of the earth, is found to be the main source of errors for the use of radar in hydrological applications. Ways of correcting for these errors and adjusting the radar data are briefly discussed.

1. INTRODUCTION

Measurements of rain deduced from radar reflectivity data by using an empirical Z–R relation agree well with gauge measurements for the area close to the radar, but we should not forget that radar will only estimate the precipitation that it can 'see'. Reduced visibility will occur when parts of the scanned volume are lost because of occulation, e.g. behind mountains, but also in flat country at longer ranges the earth curvature will cause shielding. This fact will produce strong variability and underestimation of precipitation by radar. However, problems are especially severe in a mountainous country such as Switzerland, making the need for investigation and, if feasible, for corrections obvious.

The procedure of deducing rainfall rates from measured radar reflectivities for hydrological applications which we intend to follow, uses the following steps:

(1) making sure that the hardware is stable by means of calibration and maintenance;
(2) correcting for errors of the vertical reflectivity profile combined with shielding;
(3) taking into account enhancements of precipitation because of orographic effects;
(4) taking into account all the information about the Z–R relationship and deducing the precipitation;
(5) adjusting the estimated values with measurements from raingauges.

The method used to calibrate the radar will be described briefly. At the beginning it was applied monthly, but now only every second month. It gave for the past 10 years optimum performance involving little work.

Errors of the estimation of precipitation from conventional radar reflectivity data are then addressed and illustrated with the comparison of 1 year of data from the two Swiss radars and from the ANETZ network of 60 automatic ground-based weather stations. The spatial distribution of such errors indicates a strong dependence from the visibility of precipitation by the radars.

Improvements of accuracy may be possible by using some knowledge about the vertical distribution of the radar reflectivity. Various possibilities to obtain such data will be discussed and results of measured average profiles will be presented.

2.   CALIBRATION AND ADJUSTMENT

We should clearly distinguish a **calibration** (e.g. using a signal generator for checking the performance of microwave components and hardware parts such as receivers) from an **adjustment** (e.g. between radar and raingauge). While the first operation defines a scale and corrects for errors in the instrumentation, the second tries to estimate one variable (e.g. rain rate at the ground) from another variable (e.g. the reflectivity aloft). Both practices depend on physical laws, but, while we know how to calibrate our radar and to keep its sensitivity stable, we are not in that fortunate situation when it comes to estimating precipitation form the measured reflectivity. It remains to be shown to what level the large variations in the relationship between reflectivity aloft and precipitation at the ground may be reduced by adjusting the measured radar values. This relationship may change with time, caused by the type of precipitation, the weather situation, the height of the sampling volume, etc.

2.1   Operational calibration procedure

The operational instrumental calibration of the two Swiss weather radar stations consists in measuring the performances of the receiving chain, situated between the antenna and the image display monitor. Doing so, all the equipment except transmitter and antenna are tested. These are treated in separate steps. The calibration procedure consists in simulating the range of intensities occurring in natural precipitation distributed over the whole area around the radar. To do so, a continuous-wave signal is fed into the directional coupler, immediately after the antenna. In order to consider the natural range of intensities, one antenna. In order to consider the natural range of intensities, one antenna revolution is divided into sectors, each one having a different continuous-wave power of the signal generator. Because of the range correction, the constant power of the signal generator is equivalent to precipitation rates which are increasing with increasing range. The signal is processed as usual during normal operation and the image resulting on a display monitor analysed. The radial distances where the pixel intensity changes in level are measured and analysed to yield the stability of the receiver chain. The result is the signal-to-noise ratio of the lowest pixel intensity as a function of the introduced power level and as a function of the distance from the radar. The dynamic range of reflectivity at all ranges from the minimum to the maximum allowable distance from the radar station is therefore tested. These results verify the stability of the instrumentation and show clearly and immediately when something is wrong in the receiving chain. This procedure has been performed for about 9 years periodically once a month and gave good results for both radar stations. About a year ago, because of the high stability of the system found in the past, the period between tests has been extended to once every two months and of course after every major repair.

2.2   Adjustment of radar data

In order to know how much precipitation amounts estimated from radar would have to be adjusted to agree with the gauges, the results from 1 year of archived data were analysed. The data were extracted from the Swiss radar composite pictures, distributed in real time to some 30 users, among them six weather-forecasting centres. For the ground truth, the precipitation values measured by ANETZ, a network of about 60 automatic weather stations equipped with raingauges and evenly distributed over the whole region of Switzerland, were used. The precipitation was estimated from nine reflectivity values sampled within an array of 3×3 pixels centred around each raingauge using a fixed Z–R relationship. Daily precipitation amounts were integrated from 144 displays transmitted every 10 min. Data were selected from all days with more than 1 mm of cumulated precipitation for the period from 16 June 1984 to 10 July 1985 and daily mean values of the adjustment factor ratio of radar to ANETZ data for the precipitation amount were obtained. These values are plotted in form of isolines on a map of Switzerland, overlaid with lines of equal visibility from the radar stations, reproduced as altitudes above sea level (Fig. 1). Averaged over the whole Swiss area and over the considered time period, the radar 'sees' 40% of the precipitation captured by the network of raingauges. The ratio radar to ANETZ data varies between 11% and 91% for 68% of the population, the median value of the ratio being 40%. On average this factor decreases and its variability increases with increasing distance from the radar site; its spatial distribution reflects the strong dependence of the radar estimate from the visibility of precipitation by the radars. The main problem in Switzerland (but we expect it to be so also in other countries) is the inability of the radar to see the precipitation close enough to the terrain where the raingauge is situated. For example taking the Swiss radars Albis and La Dôle, situated on the north of the Alps and considering the region south of the Alps, the minimal visible altitude ranges from 5 to 8 km above ground at a maximum distance of 160 km from the closer radar station.

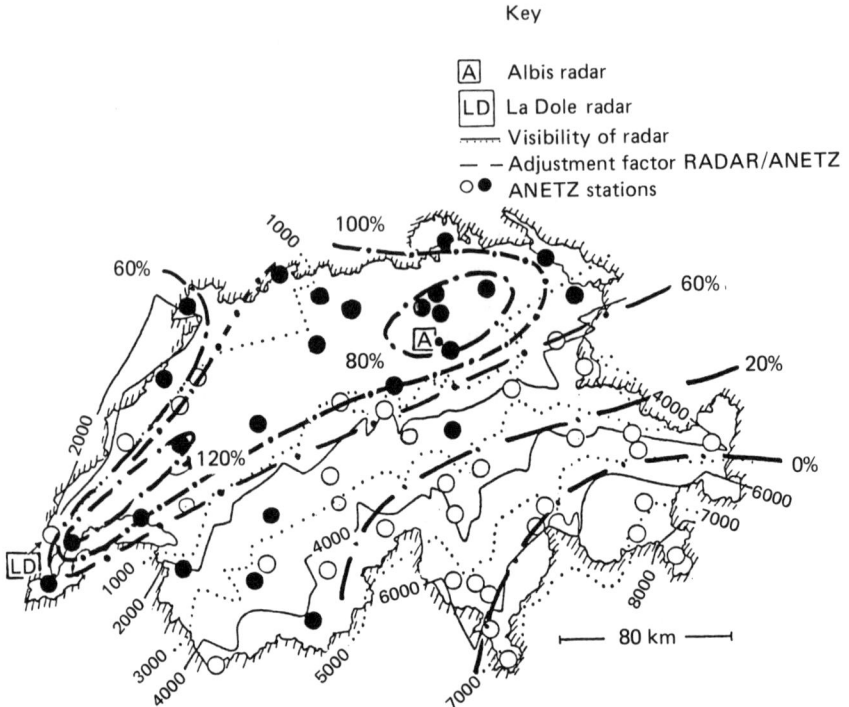

Fig. 1 — Precipitation estimated from the radar reflectivity as a function of the daily amount measured by the raingauge (16 June 1984–10 July 1985). The isolines of that fraction follow the lines of equal radar visibility labelled in metres above sea level. The locations with solid circles indicate stations with a variation of that fraction smaller than a factor of 2.

Similar results are available and being analysed now for the whole period since 1984. Preliminary results from that data set show similar behaviour to those of the 1984–1985 data and will allow us to run tests in various ways to improve the accuracy of the radar data.

3. THE VERTICAL DISTRIBUTION OF THE RADAR REFLECTIVITY

From the comparison of estimated (radar) and measured (gauge) precipitation values, we find that the main reason for the discrepancy between the two variables is caused by the inability of the radar to see precipitation close enough to the ground, owing to the orography and/or to the earth's curvature. However, knowing the vertical profile of the radar reflectivity and the minimum height where the radar sees precipitation over each point of the measurement area, we can extrapolate the radar data from a location situated vertically above the target to the ground and improve the accuracy of the precipitation estimate by radar.

Different methods can be used to estimate the vertical reflectivity profile. For example we can take data from radio sondes (height of 0° level and stability index), make use of weather-forecasting models, take a mean climatological vertical profile, with or without taking into account the type of precipitation (convective versus stratiform) or, last but not least, take the vertical profile from radar data itself, if available in volumetric form.

This is what we have tried in Switzerland; making use of the pseudo-three-dimensional data available to all users in the network, we estimated the vertical profile from these radar data. From a projection of maximum reflectivity on three planes (one horizontal and two vertical, oriented north–south and east–west respectively) we have reconstructed the volume data in an area of 100 km × 100 km around each of the two radars for a period of about 2 years. The equivalent rain rate was estimated from a fixed Z–R relationship and averaged for each volume scan, taken every 10 min in 12 horizontal planes between 1 and 12 km height. The reconstructed vertical profiles of each radar therefore consists of 12 points in altitude. Averaging 144 of these profiles per radar station and per day we obtain an average profile for each day. Taking the 10 min profile associated with the maximum precipitation intensity we obtain a maximum profile for each day.

An example for the maximum profile is shown in Fig. 2. The data were classified by season and shows an average slope of 34%/km in summer and 86%/km in winter.

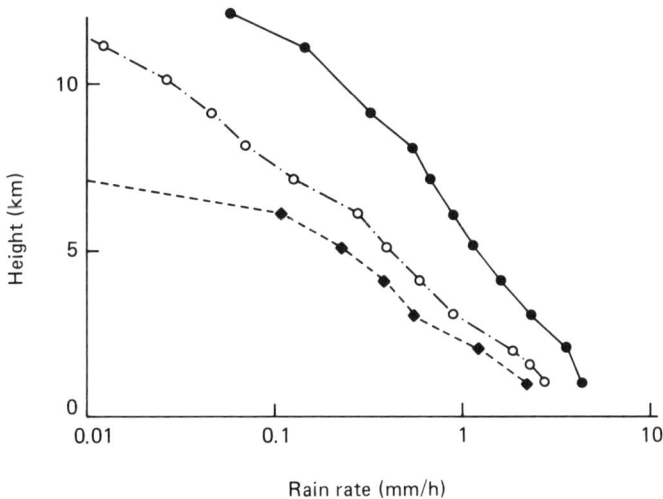

Fig. 2 — Seasonal variation in the maximum vertical reflectivity profile (23 June 1986–26 August 1988): ———●———, June–August, i.e. summer, ———○———, March–May, September–November; – –△– –, December–February, i.e. winter. The daily mean precipitation intensity estimated as a function of the altitude for three periods of the year is plotted for the radar station Albis.

Different years showed similar results. The two radars gave similar results in summer and signficantly different results in winter caused by the different visibilities from the two stations and different types of precipitation. Differences between averaged and maximum profiles are small for the gradient of the curves. The details have been given by Joss and Waldvogel (1989).

This analysis suggests an improvement by correcting radar data with seasonally averaged vertical reflectivity profiles. The good correlation between the profiles of the two radars during the summer periods may even justify a correction of the data with an estimate of the profile in quasireal time and the spatial extrapolation of the vertical profiles; the differences in winter between the results from the two stations indicate that it would be useful to consider some information about the types of precipitation, obtained for example from the spatial distribution and displacement in time of the radar reflectivity data. The reasonable correlation found between profile gradients or different altitudes justifies the extrapolation of gradient values from aloft to the ground for locations where the radar is obscured by obstacles.

As mentioned before, we shall try in future various ways to apply this correction on a pixel-by-pixel basis and then to adjust the results with data taken from raingauge measurements.

4. CONCLUSIONS

When discussing ways to improve the accuracy of the precipitation estimates, we should clearly distinguish a **calibration** (e.g. using a signal generator to check the performance of microwave components and hardware parts such as receivers) from an **adjustment** (e.g. between radar and raingauge). While the first operation defines a scale and corrects for errors in the instrumentation, the second tries to estimate one variable (e.g. rain rate on the ground) from another variable (e.g. the reflectivity aloft). Both practices depend on physical laws, but, while we know how to calibrate our radar and to keep its sensitivity stable, we are not in that fortunate situation when it comes to estimate precipitation from the measured reflectivity. It remains to be shown to what level the large variations in the relationship between reflectivity aloft and precipitation on the ground may be reduced by adjusting the measured radar values. This relationship may change with time, caused by the type of precipitation, the weather situation, the height of the sampling volume, etc. In future we hope to reduce the variations in the relationship between precipitation estimated from the reflectivity aloft and measurement at the ground by using the vertical profile of the reflectivity and adjusting the sensitivity of the radar on a pixel by pixel basis. We also may use more complicated measurement facilities such as Doppler, multiparameter or frequency agility radars, if these techniques are shown to be of benefit to our operational applications.

REFERENCES

Joss, J. and Waldvogel, A. (1989) Precipitation estimates and vertical reflectivity profile corrections. *Proceedings of the 24th Conference on Radar Meteorology, Tallahassee, FL, March* 1989. American Meteorological Society, Boston, MA.

4

Assessment of the adjustment of hourly radar rainfall fields by the use of daily raingauge totals

T. J. Hitch
Meteorological Office,
London Road, Bracknell, Berks. RG12 2SZ, UK

ABSTRACT

A technique for adjusting hourly radar rainfall fields using daily raingauge information is investigated using data from Chenies and Hameldon Hill radars. The
results suggest that the technique generally improves the agreement between hourly
radar field and hourly raingauge observations, particularly at longer ranges.

1. INTRODUCTION

The Advisory Services branch of the UK Meteorological Office provides information
for a wide variety of users (e.g. insurance companies, farmers, solicitors and water
authorities). Frequently, hourly rather than daily rainfall data are required. There
are only about 200 raingauges throughout Britain that report hourly rainfall totals to
the Meteorological Office, only 60 of which are reported each day and therefore
available to answer immediate enquiries. These hourly gauges, with an average
spacing of 40 km, cannot observe adequately the highly complex and spatially
variable hourly rainfall field.

Rainfall rates derived from weather radar, with high spatial and temporal
resolution, provide a useful source of additional information. However, radar data
are known to be influenced by a number of sources of error. For example, owing to
beam inclination, radar may overshoot rainfall and underestimate totals despite
application of a long-range correction. Also, radar tends to underestimate light
rainfalls and to overestimate heavy rainfall because a single drop size distribution is
used in the conversion from reflected radar signal to rainfall intensity.

The purpose of the work described here is to investigate the benefit of a method
of combining or 'adjusting' spatially regular hourly radar rainfall observations with
daily raingauge totals in an attempt to improve the representation of the true hourly
rainfall field.

2. ADJUSTMENT PROCEDURE

2.1 Calculation of daily adjustment factor field in PARAGON

The adjustment of daily (0900Z–0900Z) radar totals is a routine process within PARAGON (May 1988), a radar-processing and archiving system developed and operated by the Advisory Services branch. This system stores both hourly and daily 5 km×5 km radar rainfall totals operationally. These totals are derived from integrations of radar totals at 5 min intervals. The daily radar totals are adjusted using either the synoptic† or the climatological daily raingauge totals.

The adjustment procedure involves calculating an 'adjustment factor' which is the ratio of daily gauge rainfall to daily radar rainfall at each of the daily gauge locations, fitting a surface to determine an adjustment factor at each 5 km grid point and applying these to the radar observation at the grid point. The usefulness of such 'adjusted daily radar fields' has been assessed by Mylne (1988).

2.2 Adjustment of hourly radar rainfall fields

A similar method could be adopted to adjust hourly radar fields with hourly raingauge totals, but the large distances between gauges and the greater spatial variability of hourly rainfall would lead to large errors in the adjustment factor field. Instead a procedure was used in which the hourly radar observations were multiplied by the daily gauge adjustment factor fields readily available in PARAGON. Palmer *et al.* (1983) suggested the merits of such a method which may be considered as obtaining hourly rainfall totals from daily gauge totals by using the radar-derived hourly-to-daily ratio. An obvious advantage is that the hourly gauge rainfalls are not used in the adjustment process and may therefore be used as an independent ground truth to assess the radar fields. In addition, the hourly radar field adjusted by the synoptic daily raingauges could be derived in near real time before many of the hourly gauge observations are available. However, it is uncertain how representative the daily gauge adjustment factor is of the true hourly radar-to-gauge ratios.

3. THE DATA

Eight days of hourly radar rainfall data from each of Chenies and Hameldon Hill radars were adjusted by the synoptic daily gauge adjustment factor fields. All dates were chosen to avoid obvious defects in the radar data such as bright-band and anomalous propagation (ANAPROP). Either 1 h (usually with the heaviest rainfall) or a sequence of hours (those with sufficient rainfall for a sample of at least five gauges) were analysed for each day. The dates and hours studied are shown in Table 1.

4. ASSESSMENT OF THE HOURLY RADAR RAINFALL FIELDS

For the purposes of this sturdy, observations from the hourly recording gauges were regarded as the true surface rainfalls. The raw and adjusted hourly radar fields were assessed by comparison with these; the radar rainfall estimate at a particular gauge

---

† Approximately 200 synoptic daily gauge observations are available the day after a rainfall event. They are a subset of the observations from about 4000 raingauges in the climatological daily gauge network most of which will not be available until up to 3 months later.

Table 1 — Hours and dates analysed

Radar site	Date data obtained	Time data obtained (hours)	Number of hourly gauges used	
			<100 km from radar	≥100 km from radar
Chenies	23 June 1986	2100Z	13	12
	23 July 1986	1400Z	12	8
	28 July 1986	1800Z–2300Z	13 h	9 h
	3 August 1986	1000Z–1700Z	23 h	13 h
	22 August 1986	1900Z	12	5
	23 August 1986	2000Z–0200Z	12 h	7 h
	25 August 1986	2000Z	26	25
	26 August 1986	1500Z	19	4
Hameldon Hill	2 June 1986	0600Z	8	8
	9 June 1986	2000Z–0900Z	10 h	8 h
	23 June 1986	2300Z	7	10
	24 July 1986	0100Z	8	8
	30 July 1986	1000Z–2000Z	7 h	12 h
	1 August 1986	1900Z	11	16
	25 August 1986	0400Z–2000Z	8, 19	9, 19
	2 June 1987	1500Z	7	10

location for either the raw or the adjusted hourly radar field was determined by linear interpolation from the four surrounding grid-point observations. These comparisons of 'point' rainfalls with 'areal' radar estimates may introduce some additional uncertainties.

The 'log ratio' method (May, 1983) was used to compare radar and gauge rainfalls since the radar rainfall-to-gauge rainfall ratios over a rainfall field exhibit a distribution that is usually close to log-normal. The mean and standard deviation of log(radar rainfall/gauge rainfall) indicate the bias and scatter errors respectively in the radar field as an estimate of the gauge rainfall observations. This method cannot be used on occasions when either the radar or the gauge (or both) do not detect rain. The frequencies of these occasions were therefore calculated separately and are shown in Tables 2 and 3.

The mean and standard deviation of the log ratio were determined for both the raw and the adjusted hourly radar fields for gauge samples

(i) within a 100 km radius of the radar and
(ii) beyond a 100 km radius of the radar.

For each hour the relative performances of the hourly radar fields as estimates of the hourly gauge rainfalls were assessed by direct comparison between mean and standard deviation of the log ratios. The mean of the log ratio statistics were also determined over all the hourly samples with each hourly mean and standard deviation being weighted by the number of gauges in the sample.

Table 2 — Frequency of occasions when the radar and/or gauge do/does not see rain for gauges within the Chenies radar coverage: R_r, radar rainfall; R_g, gauge rainfall. The frequencies for the adjusted radar field are shown in parentheses

	$R_r=0$	$R_r>0$	$R_r>1.0$ mm
	<100 km from radar		
$R_g=0$	92 (96)	42 (41)	12 (11)
$R_g>0$	98 (106)	434 (431)	
$R_g>1.0$ mm	9 (9)		
	≥100 km from radar		
$R_g=0$	217 (213)	28 (27)	5 (4)
$R_g>0$	209 (209)	243 (241)	
$R_g>1.0$ mm	20 (19)		

Table 3 — Frequency of occasions when the radar and/or gauge do/does not see rain for gauges within the Hameldon Hill radar coverage: R_r, radar rainfall; R_g, gauge rainfall. The frequencies for the adjusted radar field are shown in parentheses

	$R_r=0$	$R_r>0$	$R_r>1.0$ mm
	<100 km from radar		
$R_g=0$	9 (9)	1 (0)	0 (0)
$R_g>0$	26 (19)	148 (147)	
$R_g>1.0$ mm	0 (0)		
	≥100 km from radar		
$R_g=0$	210 (205)	8 (8)	0 (2)
$R_g>0$	190 (205)	348 (347)	
$R_g>1.0$ mm	38 (41)		

5.   SUMMARY OF RESULTS

The results from the log ratio comparisons are summarized in Tables 4 and 5 for Chenies and Hameldon Hill radars respectively. At distances less than 100 km from Chenies radar, approximately 60% of the hours studied showed a reduction in both bias and scatter errors after the radar field was adjusted by the synoptic daily gauge observations. The percentage of occasions on which the bias error is reduced after adjustment increases to 83% for gauges beyond a 100 km radius of the radar.

    The bias error in the hourly radar field as an estimate of the hourly gauge rainfalls is reduced after adjustment on about 65% of occasions for gauges within the Hameldon Hill radar coverage. The scatter error is only reduced on 41% of the hours for gauges within 100 km of the radar site, but 59% beyond.

Table 4 — Percentage of hours on which adjustment improved the radar rainfall field for data from Chenies radar

Statistic	Percentage of hours	
	<100 km from radar	≥100 km from radar
Mean log ratio (bias error)	58% (15/26)	83% (19/23)
Standard deviation of the log ratio (scatter error)	58% (15/26)	57% (13/23)

Table 5 — Percentage of hours on which adjustment improved the radar rainfall field for data from Hameldon Hill radar

Statistic	Percentage of hours	
	<100 km from radar	≤100 km from radar
Mean log ratio (bias error)	65% (11/17)	66% (21/32)
Standard deviation of the log ratio (scatter error)	41% (7/17)	59% (19/32)

The results of averaging over all hourly samples for each radar are shown in Table 6. This global summary of the data indicates the measure of improvement introduced into the hourly radar estimates of rainfall by adjusting the radar precipitation field using the rainfall measured by the synoptic daily gauge network. The results from Chenies radar suggest that it is generally overestimating at long range for the dates studied. This may arise because the long-range calibration overcompensater for beam elevation when applied to rainfall from deep cloud.

Fig. 1 shows the hourly radar rainfall, the hourly gauge rainfall and the ratio of radar rainfall to gauge rainfall thoughout 25 August 1986 at a raingauge 60 km from the Chenies radar. The daily adjustment factor of 0.718 is a good estimate of these 'hourly adjustment factors' for the period of heavy rainfall between 1400Z and 2200Z when the majority of both radar and gauge rainfalls are greater than 2 mm. The daily adjustment factor is appropriate not only for the very light rainfalls which contribute little to the daily total; other gauges further from the radar show similar results.

6. DISCUSSION OF RESULTS

Results from the 16 dates studied indicate that adjustment of hourly radar data by synoptic daily raingauges generally improves the quality of the hourly radar rainfall

Table 6 — Summary statistics of hourly radar rainfall data samples

Radar	Radar field	Distance from radar (km)	Mean of the mean log ratios	Standard deviation of the mean log ratios	Mean of the standard deviations of the log ratio
Chenies	Raw	<100	−0.017	0.171	0.292
		⩾100	0.069	0.258	0.434
	Adjusted	<100	0.020	0.153	0.200
		⩾100	0.082	0.184	0.394
Hameldon Hill	Raw	>100	−0.174	0.084	0.244
		⩾100	−0.167	0.138	0.381
	Adjusted	<100	0.041	0.132	0.233
		⩾100	0.103	0.153	0.361

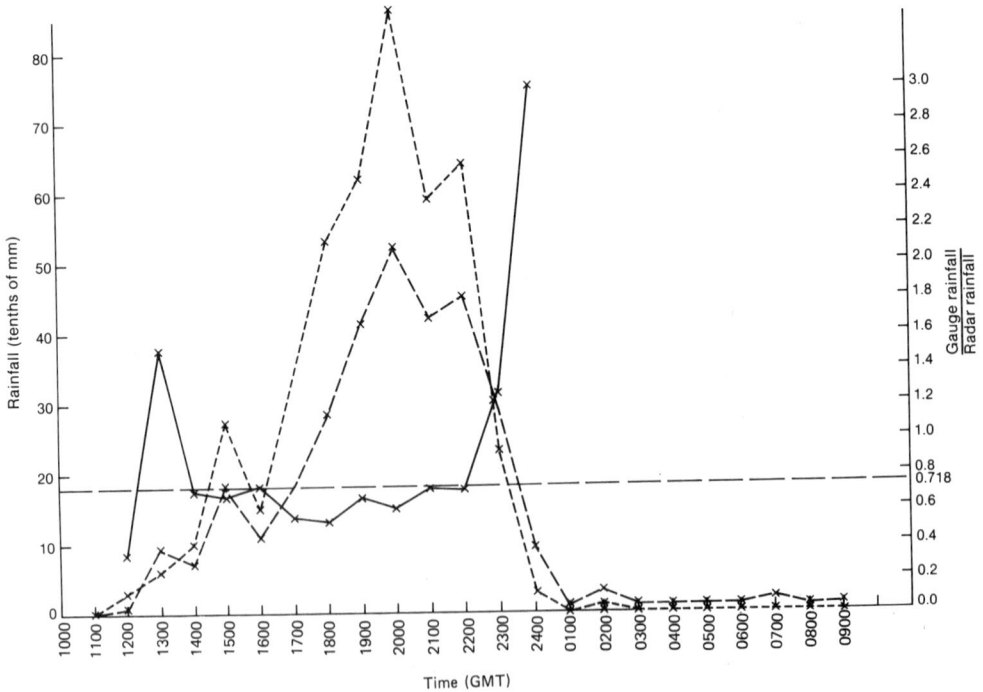

Fig. 1 — Variation in the ratio of hourly gauge rainfall to hourly radar rainfall throughout 25 August 1986 for the gauge at Bedford Meteorological Office (60 km from Chenies radar): ——, gauge rainfall; – – – –, radar rainfall; —0—, ratio of gauge rainfall to radar rainfall.

field for Chenies and Hameldon Hill radars. This is especially evident at ranges greater than 100 km where the radar data are expected to deteriorate.

Similar results were obtained after adjustment of the raw hourly radar fields by the much denser climatological daily raingauge network. However, they were not significantly better. These 'climatologically adjusted' hourly radar fields would not be available until 2–3 months after a particular rainfall event whilst the hourly radar field adjusted by synoptic daily gauges could be derived the next day, before the majority of hourly raingauge records are available.

The larger hourly rainfall totals will normally contribute most to the daily totals. The daily adjustment factor is therefore more representative of the 'hourly adjustment factor' for these heavier rainfalls than for the periods of lighter rainfall during the remainder of the day. Radar is inclined to underestimate light and to overestimate heavy rainfalls; so a daily adjustment factor strongly influenced by the heavier rainfalls throughout the day is likely to reduce further the light rainfalls. However, this is probably not a problem for most data users. The more serious case may be when prolonged light to moderate rainfall is a major contributor to the daily total so that short-duration heavy falls of greater significance may be overestimated.

7. CONCLUSIONS

Adjustment of hourly radar rainfall fields by daily gauge adjustment factors in these case studies is shown to be beneficial particularly at longer ranges. It should be borne in mind that the hours studied, chosen for their lack of major radar defects, are likely to provide the highest possible quality of raw radar data. Other hours may not give such good results for unadjusted radar and may therefore exhibit more improvement on adjustment. The success of this adjustment procedure will be dependent on the individual requirements of data users as well as the 'statistics of accuracy' of each hour. A user who needs accurate heavy-rainfall monitoring but whose requirements are not very sensitive to the accuracy of determination of light rainfall will benefit by using daily adjustment factors to adjust hourly radar.

REFERENCES

May, B. R. (1983) *Comparison of estimates and observed values — the log (ratio)*, Internal Working Document No. Met 03. Meteorological Office, Bracknell, Berks.

May, B. R. (1988) Progress in the Development of PARAGON. *Meteorol. Mag.*, **117**, 79–86.

Mylne, M. F. (1988) *Investigation of the role of radar data in the PARAGON daily rainfall archiving system: assessment of the contribution of radar data to near real-time daily rainfall estimates*, Advisory Services Technical Report No. 5. Meteorological Office, Bracknell, Berks.

Palmer, S. G., Nicholass, C. A., Lee, M. J. and Bader, M. J. (1983) The use of rainfall data from radar for hydrometeorological services. *Meteorol. Mag.*, **112**, 333–346.

5

On-line calibration in HERP

R. K. Kreuls
Radarmeteorology, Meteorologisches Institut, Universitat Bonn, W-5309 Meckenheim-Merl, Buschweg 28, FRG

ABSTRACT

The on-line calibration in HERP (Hydrological Emscher Radar Project) is carried out by hardware calibration and actual measurement of drop size distributions in order to set up $R–Z$ relationships. The relations are adapted to the needs of radar. Several methods of improvement are described and the results are verified.

1. INTRODUCTION

The best radar techniques are worthless for hydrological applications, if there is no good conversion from radar values to rain amounts. A given value of radar reflectivity Z can be related to values of rain intensity R, which vary within a range of ten to the second power. The calibration of radar values can be carried out basically in two different ways: the adjustment of raingauge amounts or the use of actual $R–Z$ relationships and hardware calibration.

2. CALIBRATION BY RAINGAUGE ADJUSTMENT

The fundamental differences between point and volume measurements are well known. Every rain event — from convective or from layer clouds — is structured in cells, which have an influence on the time and areal distribution of the rain amount at the raingauge. Therefore, areal rain computed from several gauges by one of the known methods will never be exact. In most cases it will be too low, because it is improbable that the raingauges catch the peak value of the rain. However, the longer the time of measurement lasts (months or a year), the smaller the error will become. Using raingauge networks for radar calibration causes additional errors. The problem of transferablility from rain amount at a height of some hundred metres extending some kilometres to one point at the ground is similar but even more complex. In this case the vertical fall of the raindrops and the horizontal displacement of the rain pattern are involved. Therefore, time and place compatibility between the volumes in height and at the ground are not possible. This imponderable influence can lead to errors of some hundreds per cent.

An on-line calibration needs information minute by minute in order to get optimal values. With raingauges this is hardly practicable.

3. CALIBRATION WITH CALCULATED RADAR CHARACTERISTICS AND R–Z RELATIONSHIP

The basis for the quantitative measurement of rain by radar is the radar equation (Battan, 1973). The average returned power P_r is the product of characteristics C of the radar set and parameters of the back-scattering volume (distance, r, refractive index $|K|$ and the sum of the back-scattering drops given by $\Sigma_i D_i$):

$$P_r = C\left(|K|^2 \sum_i D_i^6\right)\Big/ r^2$$

The term $\Sigma_i D_i^6$ is the radar reflectivity factor Z:

$$\sum_i D_i^6 = Z = \int N(D) D^6 \, dD$$

where $N(D)\, dD$ describes the number of back-scattering particles (of diameter D) in the unit volume. In the case of raindrops this term is the drop size distribution. From the spectra, Z (mm^6/m^3) can be computed directly. It refers to the sixth power of the drop diameters and is determined especially by the large drops.

The rain amount Q (mm), divided by time interval t (h) is the rain intensity R (mm/h). Q is created by the impact of drops with different diameters D and therefore different fall velocities $v(D)$ on a unit area:

$$R = \text{constant} \int N(D) D^3 v(D) \, dD \ .$$

The rain intensity R refers to the third power of the drop diameter. This value is mainly determined by drops of medium diameters.

Knowing the characteristics of the radar set, the radar constant can be calculated. For calibration, the relationship between R and Z is required.

4. DATA

From 1970 to 1982 we measured continuously at Mechenheim (near Bonn, FRG) 1.54 million drop size spectra using the electrodynamic drop distrometer (Joss and Waldvogel, 1967). The spectra were standardized to a 1 min measuring interval. If rain intensities R of the 1 min spectrum did not exceed 0.0001 mm/h (1.6×10^{-5} mm/min), they were eliminated. For each spectrum the values R and Z were calculated. The data were combined with rain events. Events with an amount less than 0.05 mm were rejected. The remaining 2519 rain events with 371 769 drop spectra have a total rain amount of 4250.4 mm.

5. THE *Z–R* AND THE *R–Z* RELATION

From rain event to rain event and even during an event the drop size distributions vary (and because of this the *Z–R* relations). Marshall and Palmer (1948) proposed the equation

$$Z = aR^b$$

and suggested that $a = 200$ and $b = 1.6$ as mean values. Since, then, many groups calculated this equation from their own material and a and b vary in wide ranges. In particular there are large differences between the values calculated from radar measurements and from drop size distributions. From our own radar measurements we got $Z = 170R^{1.43}$ (Breuer 1972) and from drop size distributions $Z = 200R^{1.28}$.

Two main reasons are responsible for these differences. The first is the different volume ranges analysed by the two systems. The drop distrometer collects single drops and therefore the measuring threshold is extremely low. It can be fixed by $R > 0.001$ mm/h. The radar threshold is much higher, near $Z = 10$ mm^6/m^3, which is equivalent to $R = 0.1$ mm/h. 54% of all rain minutes are situated in the range $0.001 < R < 0.1$ mm/h. For the rain amount these minutes can be neglected, but they have a large influence on the shape of curves, which determine the *Z–R* relation. The second reason for the seeming incompatibility of radar and drop spectra data is the choice of R as unconditional variable in the correlation of the logarithms of R and Z. For use in the radar equation the relation has to be rearranged: $R = aZ^b$. This seems to be permissible, because the correlation coefficient r is almost always higher than 0.9. But from our investigations we know that this may cause errors in R of up to 50%. The rain intensity R and radar reflectivity Z are calculated independently from the drop size spectra. So it is possible and necessary to use special *R–Z* relationships, derived from own regression lines, in order to avoid errors.

The first goal was to bring the configuration of the spectra data close to those of the radar, in order to set up qualified equations for quantitative rain rate measurements without considerably neglecting the rain amounts. (By consideration of the two above-mentioned points, it will become evident that *R–Z* relations from radar and from drop size distributions are congruent.) A second goal was to set up *R–Z* relationships for, for example, different atmospheric conditions, or for variability of the parameter Z, which can be used in on-line operation to calibrate the system.

6. METHOD OF CALCULATION

For each rain event the *R–Z* relation was calculated. In order to determine the quality of the relation, verifications were computed. From the Z value of each drop size distribution, the rain amount Q_{RZ} was calculated using the *R–Z* relation, which was tested. These Q_{RZ} were compared with the real rain amount Q, calculated directly. The percentage errors in the rain amount of each event were averaged. The result is the mean error F and the standard deviation σ. F represents the mean error in rain measurement over long time intervals; σ rather characterizes mean errors of single rain events.

The results were summarized for different classes of rain amount Q (of each event) in order to find the differences in the reproduction of weak and heavy rain (Table 1 and Table 2).

Table 1 — Mean percentage error F (first rows) and corresponding standard deviation σ (second rows) for given thresholds for radar reflectivity Z, rain intensity R, rain amount Q and number relevant minutes N per rain event

Q (mm)	Annual	Annual	Seasonal	Event	Event	$Z = 200R^{1.6}$
	$Z>0$ $R>0.001$ $Q>0.05$ $N>1$	$Z>10$ $R>0.1$ $Q>0.1$ $N>10$	$Z>10$ $R>0.1$ $Q>0.1$ $N>10$	$Z>0$ $R>0.001$ $Q>0.05$ $N>1$	$Z>10$ $R>0.1$ $Q>0.1$ $N>10$	
0.05–0.1	18.2	—	—	3.4	—	50.3
	61.2			24.9		60.3
0.1–0.5	12.1	18.9	16.7	3.6	3.5	29.9
	63.4	52.8	43.8	36.6	22.2	57.6
0.5–1	6.3	5.5	1.9	2.1	−1.5	15.0
	49.8	39.4	37.8	12.1	6.4	42.9
1–2	7.3	−0.0	−2.1	1.4	−2.4	9.3
	43.5	32.0	31.1	10.7	5.1	35.3
2–5	9.8	−3.5	−6.4	2.3	−2.3	5.7
	45.8	30.2	25.4	12.0	5.4	32.6
5–10	7.8	−8.1	−9.7	1.9	−2.3	0.5
	31.8	19.7	18.9	9.7	4.6	20.7
10–20	13.7	−6.2	−7.6	3.1	−1.7	2.1
	37.9	25.5	25.4	9.8	4.5	27.7
>20	16.5	−13.7	−15.1	2.1	−4.4	−6.2
	32.2	14.2	14.4	15.8	8.9	16.8
0.1–60	9.8	4.5	4.1	2.6	0.0	17.3
	53.2	42.1	37.3	25.2	14.5	47.1

7.   THE ADAPTATION OF THE R–Z RELATION TO THE NEEDS OF RADAR

First we calculated the mean annual R–Z relationship. To set this up, we did not use thresholds for radar reflectivity factor Z and the number of minutes per rain event.

Table 2 — Mean percentage error F_x (first rows) and corresponding standard deviation σ (second rows) for given classes of rain amount Q and different methods of calculation: F_{ZR} (annual), calculated from the entire data $Z = aR^b$; F_{RZ} (annual), calculated from the entire data $R = aZ^b$; F_{RZ} (summer), calculated only from the data for May to October; F_{S0}, calculated with thresholds for R and Z; $F_{Z_{max}}$, the highest Z value at any given time during the rain event determines the relationship; F_{VZ40}, the average Z and the variability of the last 10 min (separated into 40 classes) chooses the relationship to be applied; F_H, the height of the rain-generating level assigns the R–Z relationship; F_{min}, calculated on line from actual distrometer data; F_R, calculated off line distrometer data

Q (mm)	0.5–1	1–2	2–5	5–10	10–20	> 20	0.5–60
Q (%)	4.9	10.7	26.6	26.0	18.0	9.5	95.7
F_{ZR}	19.7	24.4	26.8	31.8	37.2	36.3	25.6
(annual)	50.1	49.6	49.0	44.7	43.9	42.7	48.7
F_{RZ}	6.3	7.8	9.8	7.8	13.7	16.5	8.2
(annual)	49.8	43.5	45.8	31.8	37.9	32.2	44.4
F_{RZ}	8.8	7.0	9.8	16.2	15.5	17.7	18.5
(summer)	59.7	47.3	42.7	30.8	37.2	26.7	46.7
F_{S0}	5.3	− 4.4	− 6.2	− 7.6	− 7.2	− 14.3	− 3.4
	38.9	33.0	27.5	20.4	19.7	15.6	30.7
$F_{z_{max}}$	11.7	0.8	− 2.1	− 4.6	− 5.6	− 14.1	0.9
	41.2	34.3	27.7	19.7	19.2	13.9	30.1
F_{VZ40}	5.3	− 4.9	7.9	− 10.4	− 10.3	− 18.2	− 4.7
	38.8	32.0	26.6	20.6	19.1	15.7	30.4
F_H	4.7	− 5.0	− 6.5	− 3.9	− 4.3	− 12.0	− 3.2
	36.5	31.8	25.6	21.5	19.1	12.1	29.2
F_{min}	− 1.8	− 2.4	− 3.2	− 3.7	− 3.0	− 6.0	− 2.8
	11.6	12.0	11.9	11.9	12.0	7.4	11.7
F_R	− 1.8	− 2.1	− 2.0	− 2.6	− 1.8	− 2.3	− 2.1
	5.6	4.6	5.0	4.3	5.2	2.4	4.9

The lowest value for minute rain intensity R was fixed at 0.001 mm/h. All rain events were admitted with $Q > 0.05$ mm. The result was $R = 0.0204Z^{0.7107}$. The verification was carried out and the results are given in Table 1, second column. The mean error F is 9.8% and the standard deviation σ is 53.2%. For rain events with large rain amounts ($Q > 5$ mm) σ is much better.

Step by step the four thresholds were changed and the results investigated. As a best attempt we found the following thresholds: $Z > 10$ mm^6/m^3, $R > 0.1$ mm/h; $Q > 0.1$ mm; $N > 10$. (N is the number of minutes per rain event, which meet the requirements of the Z and R thresholds. We call them 'relevant' minutes.) In Table 1, third column the effect is demonstrated. F is reduced to 4.5% and σ to 42.1%, for rain with $Q > 20$ mm even up to 14.2%. The-mean annual R–Z relation now is

$R = 0.0319Z^{0.6290}$. If the inherent R–Z relation for each rain event is computed and verified (Table 1, fifth and sixth columns) the results, of course, are much better. The adaptation of the thresholds halves the standard deviation σ for almost all Q classes.

A reduction in error is possible, if for each season of half-year periods special R–Z relations are computed. The results from verification is shown in Table 1, fourth column and the term F_{S0} in Table 2.

8. DEPENDENCE OF R–Z RELATION ON CHARACTERISTICS OF THE RAIN

The following investigations are based on the data obtained in the summer in order to find optimum values for the data, which are of special interest for hydrology. Calculations showed much higher errors in summertime than in wintertime.

8.1 R–Z relation as a function of simple parameters
As a first attempt, we tried to reduce the mean errors and standard deviations by introducing some simple information. So we tested the R–Z relationship as a function of parameters such as rain amount or rain duration or a combination from both, but this investigation achieves no progress in error reduction.

8.2 R–Z relation as a function of the structure of Z
The only value which is available from the radar itself is the reflectivity factor Z. Its spatial and temporal variations could be helpful in order to perfect the mean seasonal R–Z relation.

We assigned the Z values to six magnitude classes and computed the mean class relations. On line, every actual Z will find the relationship to which it belongs. The results were verified. The mean error F_Z is 1% lower (-2.6%), but σ is nearly constant (30.6%).

In order to distinguish convective between rain events with large drops and high Z values and advective rain with intermediate drops and from this with intermediate Z values, we chose Z_{max} as the R–Z finder. Z_{max} is the highest value of Z during a rain event. Eight classes of Z_{max} were created and Z–R relations were computed using the corresponding rain events. In the verification the highest Z value up to the actual moment determines the relation to be used. We find that $F_{Z_{max}} = 0.9\%$ and $\sigma = 30.1\%$ (Table 2, $F_{Z_{max}}$).

A promising parameter for R–Z improvement was the time variability of Z in combination with a mean Z value for a definite time interval. From the data, sliding variabilities for 10 min intervals were computed. From the resulting 70 224 intervals the mean Z values and the variabilities were computed and attached to 40 classes. The corresponding R–Z relations were determined and the accuracy was tested. Although structures in the relations were obvious, the verifications did not lead to an improvement: $F_{VZ40} = -4.7\%$; $\sigma = 30.4\%$ (Table 2, F_{VZ40}).

In order to reduce the computing time, the mean Z values and variabilities were attached to 20 classes, and in a further step to six classes. The structure in the relations became much clearer, but the verified results led to rising errors. They are even higher than the simple mean seasonal relations.

Investigations in spatial variability of radar reflectivity Z cannot be made without sufficient data from radar measurements. Studies are now going on to do this.

8.3   *R–Z* relations as a function of atmospheric parameters

Drop size distributions are a function of numerous physical parameters and meteorological conditions. Investigations proved (Kreuels 1975) that these must be meteorological quantities from the environment of the precipitation origin. We used data
from radiosonde observations in Essen. The distance between the measurements of
drop size distributions and the radiosonde data is 70 km.

In the first attempt, each rain event was attached to the height of the 0°C level
during its appearance. The heights were classified into five ranges and corresponding
R–Z relationships were set up. As a result, the mean errors F_G for all rain events
become smaller than the mean summer error F_{S0}. It is quite good with $F_G = 1.4\%$,
but σ still remains at 30.6%.

As a second parameter we looked for an indicator of the height at which the rain
was generated. We chose the upper limit of the layer where the dry and the wet
temperature differ by less than 2°C. This means saturation or nearly saturation.
Earlier investigations showed that *R–Z* equations are related to the upper 100%
humidity level. The heights were distributed into five classes and the corresponding
R–Z relationships were computed and verified. The mean error F_H (− 1.2%) and σ
become a little smaller (28.9%) (Table 2, F_H).

These were the best values that we could achieve with additional meteorological
input. (A combination of the two above-mentioned parameters did not lead to better
results.) If the places for spectra measurement and radiosonde ascent were closer,
the results might become better. In our recent project HERP the distance between
the two places is only a few kilometres and so we expected an improvement. For the
normal hydrological application, however, this method is practicable only at a few
places, because of the rareness of radiosonde stations.

9.   ON-LINE CALCULATION OF *R–Z* RELATIONS

Spectra-derived relationships at the end of a rain event showed the best results
(Table 2, F_R). The remaining errors are caused only by the calculation from the
correlation line. For on-line operation, we had to find the best method of calculation.

Following the above-obtained criterions of 'relevant' spectra, the *R* and *Z* values
of the actual drop size distributions were computed. If $Z > 10 \, \text{mm}^6/\text{m}^3$ and $R >$
0.01 mm/h, they are added to the collective spectra which have been previously
measured during the actual rain. From these the *R–Z* relation is calculated and
available to transform the radar reflectivity factors into rain intensity. This procedure
will be repeated again every subsequent minute. After verification, the mean error
F_{min} is only − 2.8% and the standard deviation σ is substantially reduced to 11.8%
(Table 2, F_{min}). This was the first time that we achieved a real improvement in σ,
which means that every rain event is described with a sufficiently small error.

During a rain event the meteorological conditions can change. Also different rain
systems can superpose on each other, which results in some change in the *R–Z*
relationship. In order to consider these factors, we tested another method of setting
up the equations. For the actual minute, the relation was calculated only from the
spectra of the last 5 mins. We also tested sliding intervals of 10, 30 and 60 min, but
after verification they all gave worse results than the first method. They are even
worse than the mean summer relations.

In Fig. 1, the results are demonstrated for practical estimation of quality of the methods. On the ordinate, the accumulated total rain amount is plotted; the abscissa gives the percentage errors. With all the different methods based on mean relationships (F_{MP} (Marshall–Palmer), F_J (F_{ZR} annual) and F_{S0}), one gets nearly the same quality of results. About 80% of the total rain amount is calculable with errors less than 30%. The last 5% of Q will have errors larger than 50%. With additional use of actual information about maximum Z values ($F_{Z_{max}}$), or time variability of Z (F_{VZ40}) or height of the rain-generating levels (F_H), one cannot get a substantial improvement of the results. Only the use of on-line (F_{min} or off-line (F_R) drop size distributions will reduce the errors. 99% of the rain amount can be calculated with the method 'F_{min}' with an error smaller than 30%.

10. ACTUAL CALIBRATION

We calibrate the hardware from time to time using a known microwave source. Therefore, in actual rain events we only calibrate the R–Z relationship. The drop distrometer is running continuously. It is situated in the area observed by the radar. Every minute the distrometer data are transmitted to the central computer, which calculates the R and Z values. In order to avoid misinterpretation from the weak side parts of the rain pattern, we start with the mean seasonal R–Z relationship. After collecting ten 'relevant' 1 min spectra (R and Z exceed given thresholds), we set up the first actual R–Z relationship. This first guess is completed in every subsequent minute by new relevant spectra which improve the relationship of the current rain event.

At the moment we run the procedure with only one drop distrometer. A second station is installed. One goal of our present investigations is to determine how representative the drop spectra are from our investigations we can conclude that every separated rain event has its 'inherent' relationship. This was demonstrated by the fact that we got the best results if we use the relationship completed minute by minute (with the history of the total rain). All other attempts, e.g. when we used relationships of the last 5, 10, 30 or 60 min, gave worse results. The R–Z relation is the result of very complex interactions of meteorological parameters, and they do not usually change within small distances within a rain pattern. (Austin (1987) stated that drop spectra are more comprehensively valid too.) However, this statement is only a conclusion; exact investigations are under way.

11. CONCLUSION

After installing the best technical solution to measuring the rain rate equipment working on the radar principle, there still remains a very large problem. Without actual knowledge of the relationship between rain rate R and radar reflectivity factor Z, the best pattern measurement cannot create the real quantitative rain structures. One reasonable way to get better results is the use of the actually measured drop size distributions for the on-line calibration. At the time, best results are achieved by using locally developed equations, which were completed every minute from actual data. In this way, we could realize a mean error of -2.8% with a standard deviation of 11.7% for this part of the measurement.

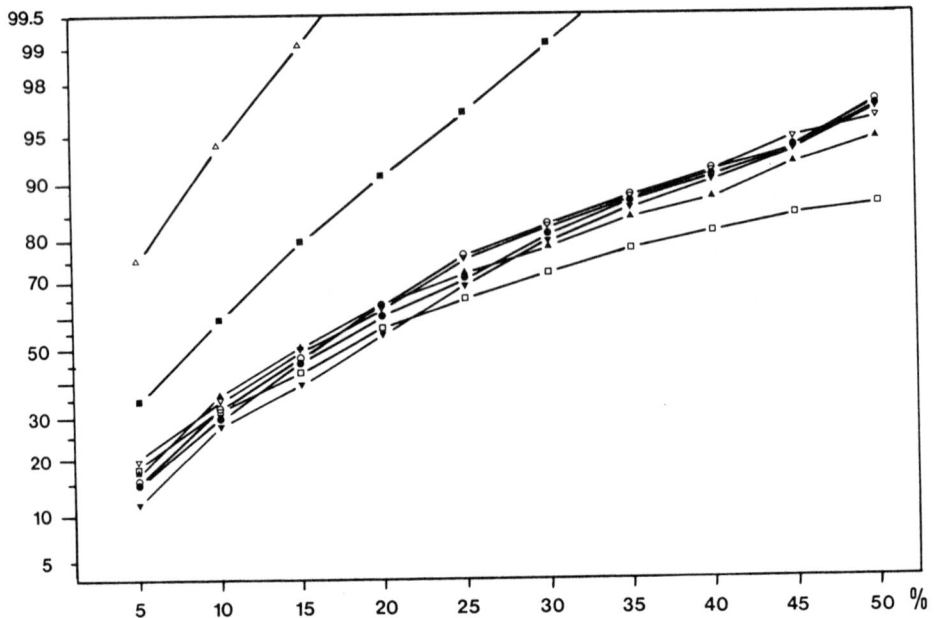

Fig. 1 — The part of the total rain amount (ordinate), calculable with given precentage error thresholds (abscissa) is demonstrated for several methods of setting up the R–Z relationships which are described in the text: $\triangle$, F_R; $\blacksquare$, F_{min}; $\bigcirc$, F_H; $\blacktriangledown$, F_{VZ40}; $\triangledown$, F_{zmax}, $\bullet$, F_{S0}; $\square$, F_J; $\blacktriangle$, F_{MP}.

REFERENCES

Austin, P. M. (1987) Relation between measured radar reflectivity and surface rainfall. *Mon. Weather Rev.* **115** (5), 1053–1071.

Battan, L. J. (1973) *Radar observation of the atmosphere.* University of Chicago Press, Chicago, IL.

Breuer, L. J. (1972) Simultaneous quantitative measurements of the rainfall rate and drop size distribution by X-band radar and drop distrometer. *Proceedings of the 15th Conference on Radar Meteorology, Urbana, Il.*, American Meteorological Society Boston, MA, pp. 167–172.

Joss, J. and Waldvogel, A. (1969) Ein Spektrograph für Niederschlagstropfen mit automatischer Austwertung. *Pure Appl. Geophys.*, **68**, 240–246.

Kreuels, R. K. (1975) Investigations and results about the relationship between some meteorological variables and radar reflectivity factors in FRG. *Proceedings 16th Conference Radar Meteorology, Houston, TX.* American Meteorological Society, Boston, MA, p. 488–491.

Marshall, J. S. and Palmer, W. Mc. K. (1948) The distribution of raindrops with size. *J. Meteorol.*, **5**, 165–166.

6

Local recalibration of weather radar

R. J. Moore, B. C. Watson, D. A. Jones and K. B. Black
Institute of Hydrology, Wallingford, Oxon., OX10 8BB, UK

ABSTRACT

The development and implementation of a procedure for local recalibration of weather radar are described. First the calibration made at the radar site, which uses a network of five calibrating raingauges, is removed to obtain the decalibrated radar field. Data from 30 raingauges are then combined with the decalibrated radar data by applying a calibration factor surface obtained by fitting a multiquadric surface to the calibration factor values calculated for each raingauge site. The procedure has been integrated into the National Rivers Authority, Thames Region, telemetry and flood-warning system running on a VAX 11/750 computer at the Thames barrier. Since 14 March 1989 the system has run operationally to support the Authority's flood-warning service to the London and the Lee Valley region. A similar surface-fitting procedure has been developed to obtain a spatial estimate of the rainfall field using data from the raingauge network alone; this serves to complement the recalibrated radar product or to replace it in the rare event that the radar malfunctions.

1. INTRODUCTION

Two factors provided the motivation for the London Weather Radar Local Calibration Study: firstly, dissatisfaction with the radar calibration procedure employed at the Chenies radar site serving the Thames region and, secondly, the prospect of increasing radar accuracy through recalibration using a reasonably dense regional network of telemetry raingauges. The 2 year study was commissioned by Thames Water Authority (now the National Rivers Authority, Thames Region) in autumn 1987 and executed by the Institute of Hydrology with the support of a Steering Commitee made up of Authority, Institute and Meteorological Office representatives. The study is now approaching completion and the prototype procedure for local recalibration of weather radar has been running operationally since 14 March 1989.

The development of the procedure is outlined in this paper. It begins with a discussion of the database of rainfall events used for technique development and evaluation, emphasizing the need for rigorous quality control of both gauge and radar data prior to analysis. The technique of multiquadric surface fitting is then outlined together with developments introduced in the study to achieve a more appropriate surface fit to the calibration factor values. Both the process of surface fitting and the evaluation of alternative procedures have led to serious consideration being given to the following: the choice of calibration factor definition, the effect of the position of the gauge within the radar grid square, the effect of the quantization error of a tipping-bucket raingauge, and the evaluation framework for assessment. Only brief consideration of each of these points is given here. A summary is given of the results which led to the recommendation of the prototype recalibration procedure. Finally, recent work concerned with fitting surfaces to raingauge data only is outlined; the results provide a valuable insight into the benefits of radar, as opposed to a raingauge network, for quantifying spatial rainfall amounts.

2. EVENT SELECTION AND DATA ASSURANCE

The first steps of the study were the selection of rainfall event data for the development and evaluation of recalibration techniques, together with quality control of these data. A database was created comprising data from 19 storm events. The radar data were 5 min values for 2 km grid squares extending to a range of 75 km from the Chenies radar installation (Fig. 1). Rainfall totals for 15 min time intervals were available from up to 30 raingauges in the London and Lee Vally area.

Forming an average rainfall value for each grid square using data from all the storm events allowed the production of Fig. 2, which reveals a number of anomalies in the radar field: radial spikes in the field to the southwest and north are due to a wooden tower and pylons blocking the radar beam respectively. Fortunately the anomalies identified are not coincident with the 30 raingauges to be used for local calibration and therefore no gauges needed to be excluded from the study.

Time series, cross-correlation and scatter plots at various time lags were used to check for time-synchronization errors between raingauge and radar data sets. In excess of 1000 plots were inspected and an inventory of times where time shifts were required was built up. Time shifts could occur twice in one day but did not follow a fixed pattern. These timing errors feature only in the archived data and arose through a fault in the raingauge telemetry archive procedure which has now been corrected. Clearly correction of time synchronization errors is very important prior to forming calibration factors from coincident gauge and radar values.

3. SURFACE-FITTING PROCEDURE

Moore (1987) reviews a number of possible approaches to merging radar and raingauge data to obtain a more accurate estimate of the rainfall field. Options range from the use of a fairly complex spatial–temporal model of storm rainfall, through extensions of linear interpolation procedures (Gandin 1965, Jones *et al*. 1979) of moderate complexity, including kriging and cokriging techniques (Matheron 1971, Journel and Huijbregts 1978, Creutin and Obled 1982), to the simple fitting of a

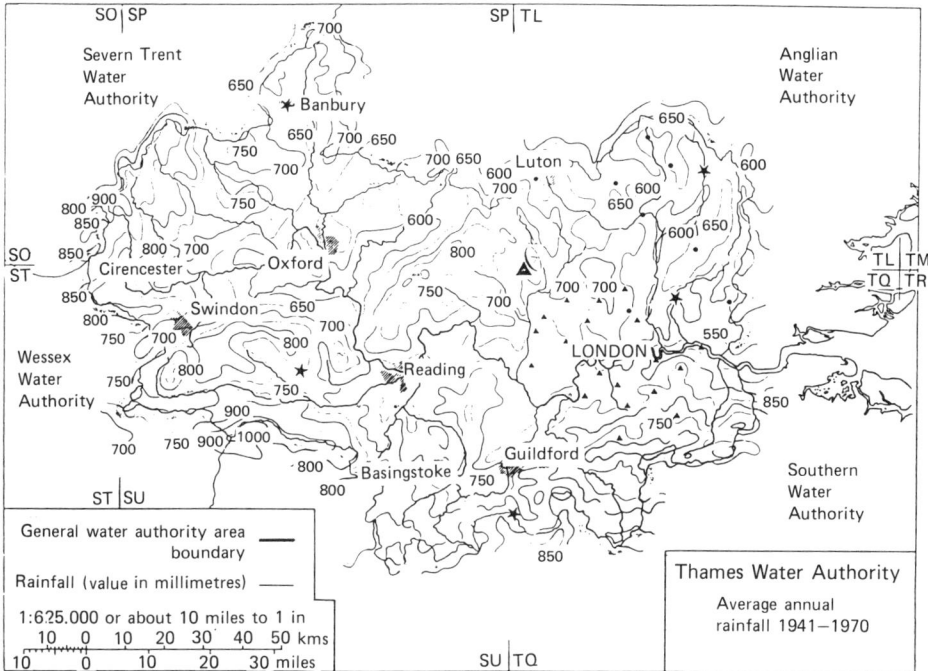

Fig. 1 — Raingauge sites and average annual rainfall for the Thames study region: △, London raingauges; ●, Lee Valley raingauges. ∗, Meteorological Office raingauges.

mathematical surface form. It is the latter approach that has been preferred in this study principally because of its speed of execution when used an an on-line calibration algorithm whilst sharing many of the attributes of linear-interpolation-type procedures. In fact, equivalence between certain surface forms and specific types of optimal linear interpolation formulae may be readily established.

The particular form of mathematical surface adopted is an extension of the multiquadric surface introduced by Hardy (1971) and expressed as follows:

$$c_i = a_0 + a_1 d_{i1} + a_2 d_{i2} + \ldots + a_n d_{in}, \qquad i = 1, 2, \ldots, n. \tag{1}$$

In this application $\{c_i, 1 = 1, 2, \ldots, n\}$ are the calibration factor values at the n raingauge sites, d_{ij} is the 'distance' between sites i and j, and $\{a_j 1 = 0, 1, 2, \ldots, n\}$ are a set of weighting coefficients.

The choice of 'distance' measure controls the form of surface fitted. A number of alternatives have been considered including the following:

(i) the simple Euclidean distance

$$d_{ij} = \sqrt{x_{ij}^2 + y_{ij}^2} \tag{2a}$$

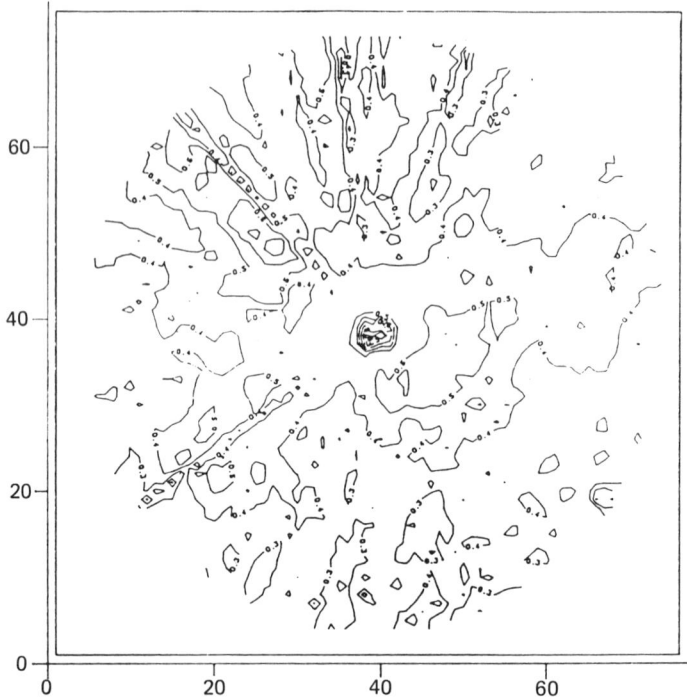

Fig. 2 — Average rainfall intensity (mm/h) field averaged across all rainfall events.

where x_{ij} and y_{ij} are the distances in the x and y coordinate directions (this is equivalent to building up the surface from a number of cones);

(ii) the smoothed Euclidean distance

$$d_{ij} = \sqrt{x_{ij}^2 + y_{ij}^2 + s^2} - s \qquad (2b)$$

where s is a smoothing parameter (in this case the surface is built up from a number of hyperboloids);

(iii) The exponential form of the Euclidean distance

$$d_{ij} = \exp\left(-\frac{\sqrt{x_{ij}^2 + y_{ij}^2}}{l}\right) \qquad (2c)$$

where the parameter l is a scaling length.

Constraints imposed when fitting the surface can be used to affect the characteristics of its behaviour. The classical method of fitting a multiquadric (Hardy 1971) forces the surface to pass exactly through the calibration factor value and no additional constraint is imposed. During the study, other fitting methods have been developed which, whilst requiring the surface to pass through the factor values, either impose a roughness minimization on the surface or constrain the surface to flatten away from the raingauge sites. A special case of the latter can be used to achieve the effect of a tendency towards no calibration with increasing distance away from the raingauge network.

In addition, variants of the above can allow the surface to depart from the calibration factor values. This is achieved in cases (2a) and (2b) by replacing the distance measure $d_{ij}=0$ when $i=j$ by the offset parameter $-K$; for case (2c), d_{ij} is changed from unity to $1+K$. The surface will then pass within a distance $a_i K$ of the calibration factor value at the ith gauge location. Both this and the roughness minimization fitting procedure allow multiquadric surfaces to be fitted to a field of calibration factor values to give a conservative calibration factor surface for use in adjusting the radar rainfall field.

4. DEFINITION OF THE CALIBRATION FACTOR

Conventionally a radar calibration factor c is defined for some fixed interval of time as the ratio of the raingauge rainfall R_g to the radar estimate R_r of rainfall for the grid square coincident with the raingauge so that $c=R_g/R_r$. The study has used some alternatives which are defined for all values of R_g and R_r (c is undefined when $R_r=0$ for the conventional definition) and which have other desirable properties for use in calibration. The ratio $(R_g+\varepsilon_g)/(R_r+\varepsilon_r)$ is used as the 'standard', where ε_r and ε_g are positive constants; reciprocal and trimmed (truncated at set lower and upper values) forms of this standard definition are also used. In addition, logarithmic forms, e.g. equation (2) in the work of Moore *et al.* (1989), are also considered.

The need for a conservative calibration factor value and the recognition that errors of quantization arise owing to the use of a tipping-bucket raingauge have led to a procedure for compensating for these errors when calculating the calibration factor values. If R_g differs from R_r by more than the intensity indicated by a single bucket tip, then Rg is adjusted towards Rr by this intensity increment prior to forming the calibration factor value; otherwise R_g is set equal to R_r.

Whilst a gauge may be coincident with a radar grid square, it may not be centrally located within it. Consequently there may be merit in forming R_r from an average of neighbouring grid squares to compensate for the non-central location of a gauge within its coincident radar grid square. Quadrant and nine-box schemes have been devised as gauge location compensation procedures in calculating R_r to obtain c.

5. ASSESSMENT FRAMEWORK

It is clear from the above that a wide range of possible options have been devised for consideration as candidates for operational implementation of a local recalibration procedure. The need for an objective choice of method led to the development of a comprehensive assessment program which became the major data-analytic tool for the study. As the basis for evaluation a procedure of removing one gauge at a time is used; this applies a calibration method to the remaining gauges and forms an error statistic from the deleted gauge value and the estimate of it obtained from the calibration method. Performance statistics are calculated by the program for both individual rainfall events and events pooled together over a selection of events; calculations can be done simultaneously for a wide choice of different calibration procedures.

The root means square error is used as the main performance statistic but the 'error' may be defined in a variety of ways. As a standard definition the simple difference $R_g - r$) is used, where r is the estimate obtained from the calibration method. However, others, e.g. some which account for raingauge quantization errors, or which use proportional errors to deflate the effect of large errors when R_g is large, are also calculated by the program.

The varieties of calibration procedure available include the options to perform time-averaging and space averaging in forming the calibration factor values. Particularly important is the ability to obtain rainfall estimates using simple alternatives to the calibration factor surface-fitting method to serve as a basis for comparison, e.g. the simple average of the raingauge values, using a constant calibration factor derived from the average of the calibration factor values, and fitting a multiquadric surface to the raingauge values only.

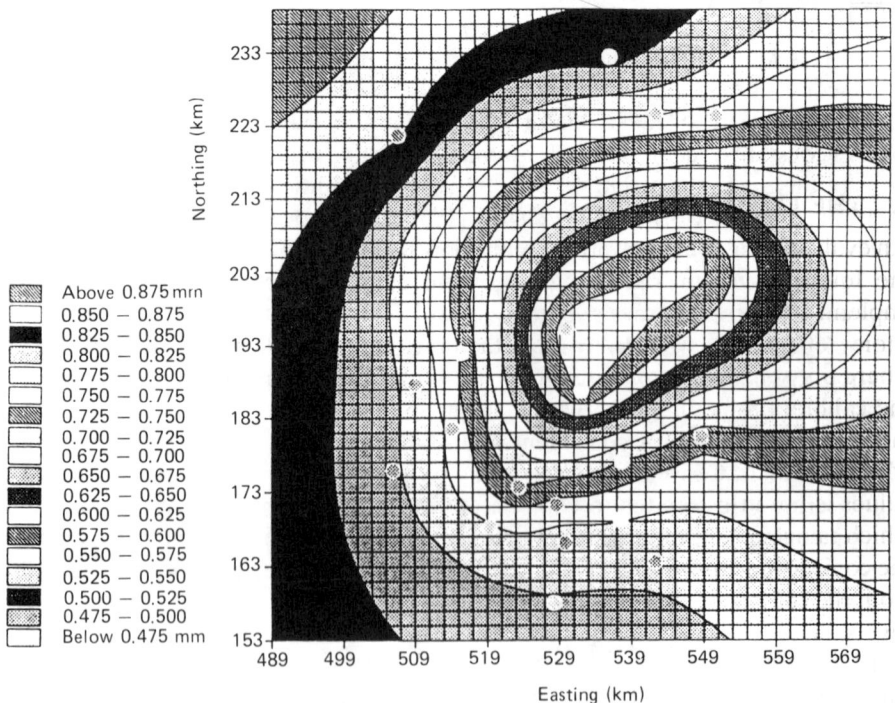

Fig. 3 — Calibration factor surface at 1915 hours, 8 May 1988; actual calibration factor values at the gauge locations are superimposed.

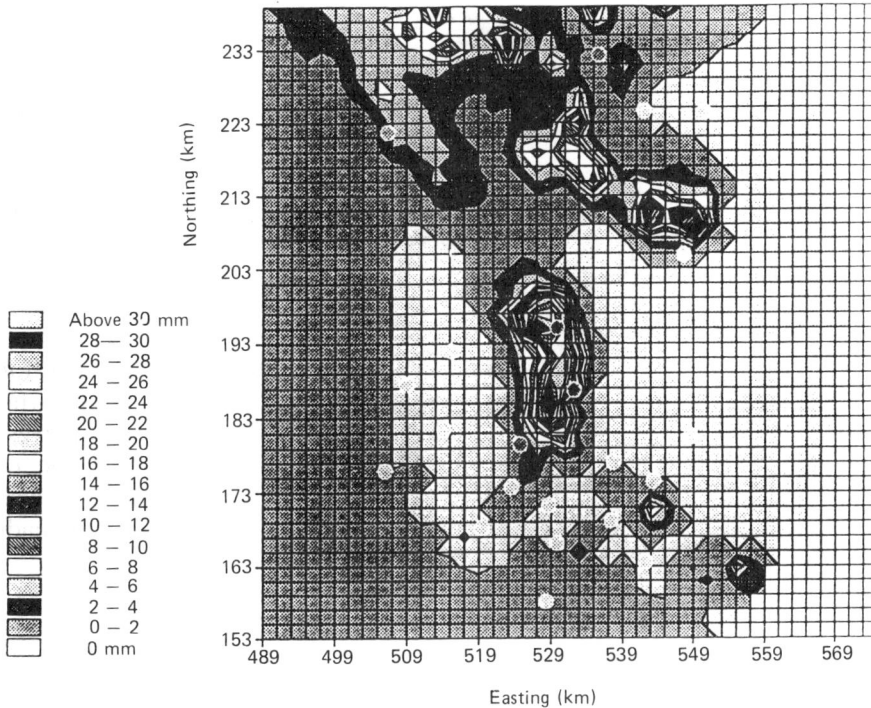

Fig. 4 — Recalibrated radar field at 1915 hours, 8 May 1988.

6. THE PROTOTYPE CALIBRATION PROCEDURE

An evaluation of the many different options for local recalibration led to adoption of the following method as the basis of the prototype calibration procedure now implemented operationally by the National Rivers Authority, Thames Region. The choice of options was arrived at by adopting the most computationally efficient where only minor benefits could be gained from other options. For the definition of the calibration factor $c=(R_g+4)(R_r+5)$ is used, where R_g and R_r refer to rainfall intensities in millimetres per hour for an interval of 15 min. No quantization adjustment is made to the raingauge values, grid-square averaging of the radar values is not invoked, and no time averaging is employed (the 15 min interval for which raingauge data are available is used). The form of multiquadric adopted is of the conic form, as represented by equation (2a), and the surface is allowed to deviate from the calibration factor values by setting the offset parameter K equal to 15 km. Fig. 3 illustrates the type of calibration factor surface obtained using this method, and Fig. 4 shows the recalibrated radar field resulting from the application of this surface to the decalibrated radar data.

This prototype calibration procedure is 23% more accurate than the calibration used at the radar site when judged at 15 min time intervals using data averaged over the 19 storm events. Operationally the calibration surface fitted every 15 min is also used to recalibrate the radar data fields at the two subsequent 5 min time steps; the

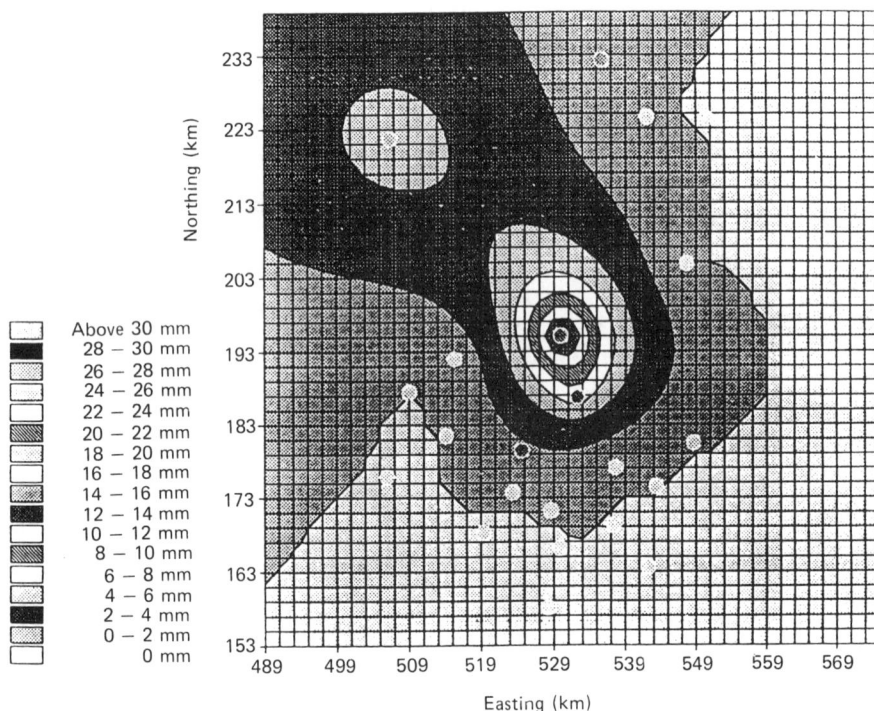

Fig. 5 — Raingauge-only rainfall field corresponding to Fig. 4.

increase in accuracy drops to 13% if the calibration factors are applied to the next 15 min period, which should be a conservative estimate of the effect of carrying forward the calibration in this way.

7. SPATIAL ESTIMATION OF RAINFALL USING RAINGAUGES ONLY

Recent work has focussed on developing a multiquadric surface-fitting procedure suitable for application to the raingauge data values only, in order to obtain an interpolated estimate of the rainfall field without using radar data. The results indicate the exponential form of the Euclidean distance as being most appropriate for application; this distance measure is also being considered in an upgrade of the prototype radar recalibration procedure. Fig. 5 shows an example of the rainfall field interpolated by this raingauge-only procedure; 22 raingauges are used. It should be compared with Fig. 4 obtained using raingauge and radar data combined using the radar recalibration procedure. The additional information on the detailed structure of this convective event that radar provides, even given a rather dense raingauge network, a rather persuasive argument for the value of weather radar.

8. CONCLUSION

The London Weather Radar Local Calibration Study has demonstrated the feasibility and value of installing local radar calibration systems which merge telemetered data from regional raingauge networks with grid-square weather radar data in real time. The procedure has been in use operationally since 14 March 1989 as part of the National Rivers Authority, Thames Region, flood-warning service to London and the Lee Valley area. Implemented on a VAX 11/750 computer at the Thames barrier, the procedure including decalibration, surface fitting and recalibration takes less than 25 s of central processor unit time. The increase in accuracy compared with at-site calibration, as suggested by off-line assessment studies, is 23% on average. Unlike the calibration made at the radar site the recalibration procedure does not introduce discontinuities in the rainfall field estimate in space and time. The computer implementation is resilient to data loss from raingauge sites. Also, recalibration can be confined to a user-prescribed window within the radar field. A further procedure for deriving a spatially continuous estimate of the rainfall field using only raingauge data is currently being integrated into the operational system. This will serve to complement the recalibrated weather radar product and to replace it when radar data reception is interrupted.

ACKNOWLEDGEMENTS

Particular thanks are due to the following members of the Steering Committee of the London Weather Radar Local Calibration Study: Chris Haggett and Peter Burrows (National Rivers Authority, Thames Region) and Chris Collier (UK Meterological Office).

REFERENCES

Creutin, J. D. and Obled, C. (1982) Objective analysis and mapping techniques for rainfall fields: an objective comparison. *Water Resour. Res.*, **18** (2), 413–431.

Gandin, L. S. (1965) *Objective analysis of meteorological fields*. Leningrad, 1963 (Transl. Israel Programme for Scientific Translation), 242 pp.

Hardy, R. L. (1971) Multiquadric equations of topography and other irregular surfaces. *J. Geophys. Res.*, **76** (8), 1905–1915.

Jones, D. A., Gurney, R. J. and O'Connell, P. E. (1979) Network design using optimal estimation procedures. *Water Resour. Res.*, **15** (6), 1801–1812.

Journel, A. G. and Huijbregts, Ch. J. (1978) *Mining geostatistics*. Academic Press, New York.

Matheron, G. (1971) The theory of regionalised variables and its applications. *Cah. Cent. Morphol. Math. Fontainebleau*, (5) 1–211.

Moore, R., J. (1987) Use of meteorological data and information in hydrological forecasting. *Proceedings of a Symposium on Education and Training in Meteorology with Emphasis on the Optimal Use of Meteorological Information and Products by all Potential users, Shinfield Park, UK, 13–18 July 1987*. World Meteorological Organization–UK Meteorological Office, 19 pp.

Moore, R. J., Watson, B. C., Jones, D. A., Black, K. B., Haggett, C. M., Crees, M. A. and Richards, C. (1989) Towards an improved system for weather radar calibration and rainfall forecasting using raingauge data from a regional tele- metry system. *New directions for surface water modelling*, IAHS Publication No. 181. International Association of Hydrological Sciences, pp 13–21.

7

Real-time adjustment of radar fields of precipitation by raingauges, allowing for orographic enhancement near the surface

C. Warner
Meteorological Office,
London Road, Bracknell, Berks. RG12 2SZ, UK

ABSTRACT

New FRONTIERS radar analysis software was introduced into the Central Forecasting Office during April 1989. The new radar analysis software reconciles calibration by raingauges with orographic growth of precipitation at low levels, partly unseen by radar. A simple procedure of truncation of high radar rainfall rates has been introduced to deal with enhanced reflections due to melting snow. An account is given of the new analysis, with emphasis on procedures of calibration. Comparisons are shown between synoptic hourly raingauge readings, radar measurements at remote sites and results from FRONTIERS.

1. INTRODUCTION

The FRONTIERS precipitation nowcasting system has recently been described by Brown (1987) and Conway and Browning (1988). As part of recent FRONTIERS research, a method of ensuring compatibility between the FRONTIERS orographic corrections and the gauge-based calibrations has been developed and incorporated into a restructured radar analysis.

2. FRONTIERS CYCLES AND CHOICES OF ANALYSES

FRONTIERS works on a half-hour period, arranged around the reception of data from the geostationary satellite METEOSAT. While the satellite data arrive every half-hour, data from the UK weather radar network arrive every quarter-hour. Incoming radar data are processed in quarter-hour cycles, alternately 'manual' and 'automatic'. The 'modified radar composite' which results from normal analysis during a manual cycle is combined with the satellite data to increase the coverage of information on the presence of rain, and a 6 h forecast is produced. The radar data arriving on the quarter-hour are processed in the same manner as the preceding data arriving on the half-hour, using parameters conventionally described as the same as last time (SALT).

The radar analysis starts with a stage of composite construction, in which all the data from a radar obviously malfunctioning can be deleted. There follows a stage of deletion of spurious echoes (Brown 1987).

In the next step, precipitation is analysed by 'regions', akin to the calibration domains treated by Collier *et al.* (1983), but without the emphasis on individual radars (unless by deliberate choice). The forecaster partitions the rainfall field over the UK into regions on the basis of his knowledge of the evolving weather situation. He may use region boundaries to isolate rainfalls all of one type, such as frontal or convective, or to distinguish areas of different parameters of orographic enhancement, or to isolate different characteristics in respect of limitation of maximum rainfall rates due to contamination by melting snow. Within a region all the precipitation is treated in the same way.

There are three different radar composites involved in the radar analysis. The site-corrected composite is derived from automatic processing at each remote site. It is the default background upon which adjusted rainfall fields are superimposed. This is the fail-safe option for the case of no activity by the forecaster during a manual cycle. A normal analysis results in two or three regions covering the precipitation of interest, and outside these the site-corrected composite.

The 'modified radar composite' is the basic field to which the FRONTIERS corrections are applied in the analysis by regions. It is derived from the site-corrected composite during input to FRONTIERS by removing remote-site gauge calibration and replacing the remote-site range corrections with the FRONTIERS moderate layer corrections. This field is gradually modified, region by region. When implementation takes place, either by choice by the forecaster or when he runs out of time allocated to the analysis, the field implemented is a combination of the modified radar composite where the region number is non-zero (where analysis has been performed) and the site-corrected composite where the region number is zero (by default).

A composite with all regions having characteristics SALT, as designated during the previous manual cycle (including calibration factors), may be chosen by the forecaster. It is the analysis which prevails in every automatic cycle. The composite with all regions SALT is a convenient starting point for modification. The forecaster may adjust the map of region boundaries by drawing or erasing; he may then opt to use region characteristics which are SALT at a touch point; all characteristics except the calibration factor then will be applied, and a new gauge calibration may be computed.

The FRONTIERS analysis proceeds alternately with manual and automatic cycles every 15 min. Every 30 min, the forecaster performs a new analysis, most conveniently by generating a composite with all regions SALT, and then modifying it.

3.   OUTLINE OF THE RADAR ANALYSIS

The principal steps in the new radar analysis are outlined in Fig. 1. A pair of leading menus — the control menu RAINGO and the auxiliary menu DRAW — allow the forecaster to begin his analysis of regions. While working on a region, the forecaster is presented with a series of other menus, eventually returning full circle to RAINGO and DRAW. When he has finished his analysis of regions, these menus allow implementation onwards to the satellite analysis. The forecaster first delineates boundaries of regions; then he chooses a region and touches inside it to begin his analysis.

Presented with the rainfall-type menus, the forecaster first assigns a rainfall type. This may be layer or convective, and the depth of the precipitation may be shallow, moderate or deep, as shown in Table 1.1 (Brown and Sargent 1988). The different types of rain require different corrections to radar rainfall rates as a function of range. The forecaster may opt for region characteristics SALT, from the previous manual cycle half an hour earlier. For this option he touches the display at the location for which the characteristics are desired, and they are read from memory.

If the rainfall type is layer, the stratification may be stable, suitable for orographic enhancement of 'seeder' rain falling into moist upslope flows (Hill 1983). Orographic parameters of humidity, wind speed and wind direction are specified by the forecaster, as in the previous FRONTIERS radar analysis. He sees the resulting orographic rain before confirming his choices and going on to the next step.

If the rainfall type is other than deep convective, the forecaster is given the opportunity to truncate the rainfall rate as a means of avoiding contamination due to melting snow. The truncation limit may be set to 16, 8, 4 or 2 mm/h.

The final step in the analysis of a region is calibration using raingauges. This is described in section 4. It concludes an analysis within a region. The programme then returns to the menus RAINGO and DRAW, from which the forecaster proceeds to analyse further regions (up to a maximum of five). When he has finished, he makes a final decision as to which composite should be used. Then he implements to the satellite analysis.

The automated choices described above are a first step towards making FRONTIERS an expert system (Conway 1989).

4. CALIBRATION USING RAINGAUGES

Calibration of the radar map of rainfall rates is accomplished individually for each region delineated by the forecaster. The accumulation of rain in a gauge over a certain time period is compared with a corresponding average of the radar measured rainfall rate in that 5 km radar pixel most nearly centred over the gauge. Several such comparisons are made, for each of the gauges in the region. Account is taken of growth of orographic rain below the radar beam close above the gauge at ground level. The several comparisons are combined into a single calibration ratio, which is then applied uniformly to the whole region. A region may include fields of view from several different radars.

Taking one gauge (identified by index i) at a time, an assessment factor $A(i)$ is formed:

$$A(i) = \frac{\text{average radar rainfall rate}}{\text{average gauge rainfall rate} - \text{amount of orographic rain not seen by radar}}$$

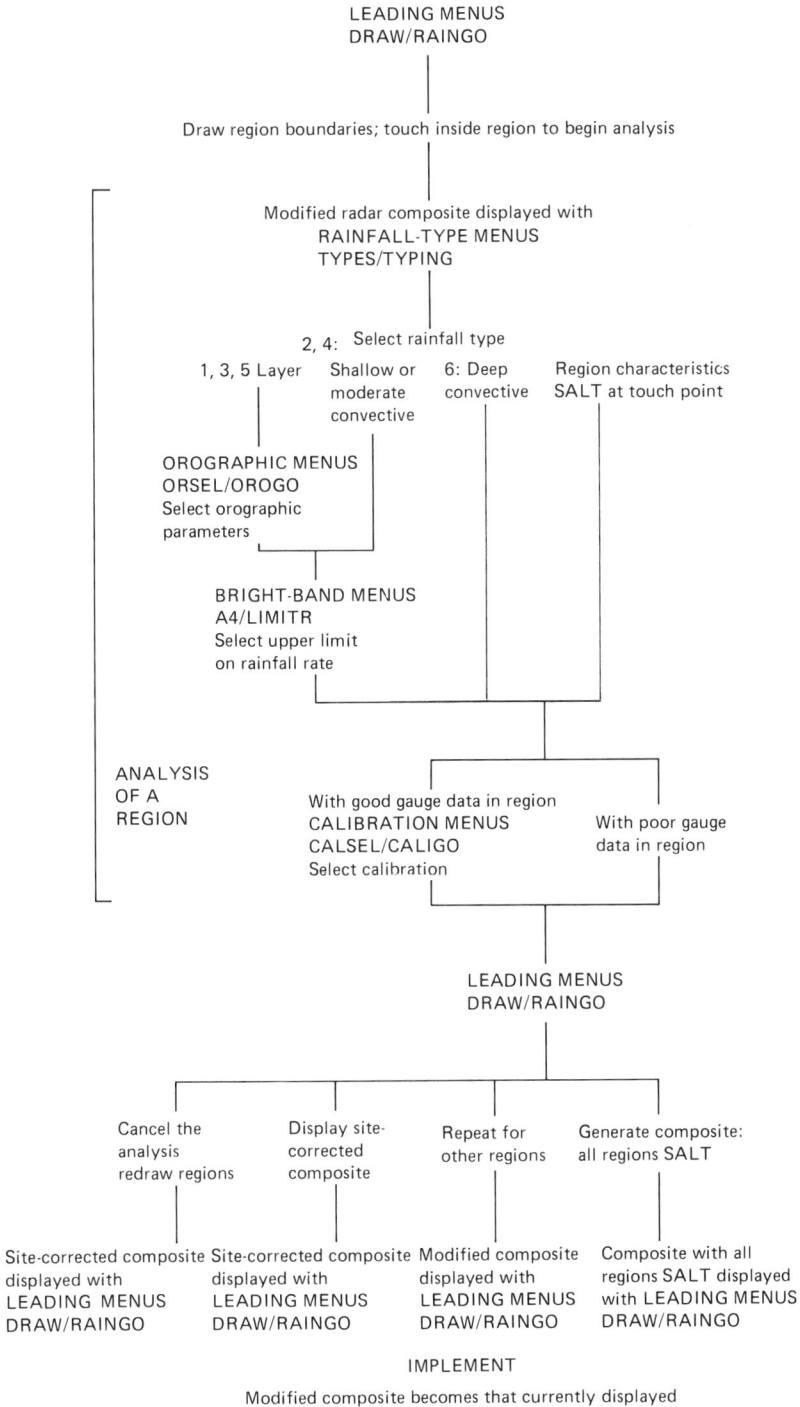

Fig. 1 — Outline of the radar analysis.

Table 1 — Rainfall types

	Layer	Convective
Shallow	Upt 3 km (10 kft)	Up to 3 km (10 kft)
Moderate	3–7 km (10–23 kft)	3–5 km (10–16 kft)
Deep	Above 7 km (23 kft)	Above 5 km (16 kft)

All those assessment factors not outside the range 0.2–5 (totalling n in number) are multiplied together, and the result is taken to the power $1/n$ to form a geometric mean assessment factor

$$A = \left(\prod_{i=1}^{n} A(i) \right)^{1/n}$$

If $n \geqslant 2$, the radar rainfall rates in the region are divided by A. If $n = 1$, A is taken as one, its default value. The use of a geometric mean follows recommendations by Koistinen and Puhakka (1986).

This calculation is not done unless there is at least one gauge which recorded two or more tips (each of 0.2 mm) during the latest hour, and there are a total of three tips (0.6 mm) or more contributed by gauges indicating one or more tips. (The units in which gauge amounts are recorded are tenths of a millimetre, and occasionally reports of half a tip or one tenth are received; these result from averaging over measurements in a closely spaced cluster of gauges.) If the calculation is not done, A is taken as one, its default value.

There are further restrictions on the calculation and use of A, designed to avoid the introduction of erroneous calibration due to mismatch of rainfall sampled by the two sensor systems. An example of such a possibility is a convective rainshaft tilting owing to shear of the wind field through which the rain falls; it could dominate the return echo in a radar pixel but mostly miss the raingauge. Sometimes frontal rain partly evaporates between the level of the base of the radar beam at long range and a gauge on the surface below. On the other hand, orographic rain falling into a gauge on a hillside might be mostly unseen by a radar far to leeward.

There is flexibility in the delineation of regions of rainfall, each to be calibrated independently. There is flexibility also in that a choice of averaging times is available, as described in the paragraphs following.

An appropriate average radar rainfall rate R_r is formed, and an appropriate average gauge rainfall rate R_g. The amount R_u of orographic rain not seen by the radar is obtained as a function of current conditions specified by the forecaster. (Consideration of the history of orographic parameters during the period of calibration is a possibility not followed here.) R_r is compared with $R_g - R_u$. Because R_g is a count in units, it may be in error by one unit, one bucket tip, 0.2 mm/h. If $R_g - R_u$) differs from R_r by no less than this unit quantization uncertainty, 0.2 mm/h, then $R_g - R_u$ is moved closer to R_u by 0.1 mm/h, or half the quantization uncertainty.

Because R_g and R_u are determined independently, their difference $R_g - R_u$ may be negative (for instance because of overspecification of orographic enhancement). If so, the gauge is ignored.

5.   CALIBRATION OVER A PERIOD OF 1 h

For comparison of radar and gauge information over the previous hour, the gauge information is straightforward; it is simply the number of tips, each of 0.2 mm, occurring over the latest hour, up to the time t (min) of the current radar image. The radar information to match this is not so straightforward. Radar rainfall rates in those 5 km pixels containing each raingauge are stored as a time series, each new value being appended to the appropriate series at the beginning of the FRONTIERS radar analysis. The radar readings in the series are values from the original of the modified radar composite; they require correction if the rainfall type is other than the standard moderate layer rainfall type. Added together are the readings for $t - 60$, $t - 45$, $t - 30$, $t - 15$ and t, with weights $\frac{1}{8}, \frac{1}{4}, \frac{1}{4}, \frac{1}{4}$, and $\frac{1}{8}$ respectively to yield an average for the hour. At the time t of the calibration, the first four of these rainfall rates are corrected with the same rainfall type correction factor and then are truncated with the same maximum rate in the event of contamination by melting snow, as is the fifth (current) value t. A more complicated but possibly better alternative would be to append new radar readings to the time series at the end rather than the beginning of the FRONTIERS rainfall analysis, after incorporation of these corrections. This matter is not likely to be problematic unless the rainfall type changes at the end of an averaging period, or substantial amounts of rain of two different types occur during an averaging period.

A calibration period of 1 h is likely to be the best choice if the rainfall is fairly evenly distributed over the hour; it is not likely to be good if most of the rain falls early rather than late in the hour. The rainfall rates at $t - 60, t - 45, t - 30, t - 15$ and t are tested in succession. Only if non-zero, are corresponding radar readings counted in the following sum or respective 'values':

$$6 + 7 + 8 + 9 + 10 \ .$$

If all five radar readings are non-zero (if all members of the sum are present), the total comes to 40. If the rainfall rate over the gauge is zero at any quarter-hour, the total amounts to less than this. If the total amounts to less than 25, the factor $A(i)$ derived from the radar–gauge comparison is not accepted. Note the following patterns of presence of rainfall which yield acceptance:

$$6 + 7 + 8 + 9 + 10$$
$$7 + 8 + 9 + 10$$
$$6 \quad + 8 + 9 + 10$$
$$6 + 7 \quad + 9 + 10$$
$$6 + 7 + 8 \quad + 10$$
$$6 + 7 + 8 + 9$$
$$8 + 9 + 10$$
$$7 \quad + 9 + 10$$

$$7 + 8 \quad + 10$$
$$6 + \quad \quad 9 + 10 \ .$$

If the current map shows no rain over the gauge, then all the others during the previous hour must do so if the calibration is to be accepted. On the other hand, calibration may be acceptable if no rain over the gauge is apparent until half an hour before the current time.

6. CALIBRATION OVER A PERIOD OF 30 min

For comparison of radar and gauge information over the previous 30 min, the gauge information is derived from the time series of accumulations of rain over the latest hour, recorded every 15 min. From this time series, the rainfall rate occurring over the latest 30 min may be derived, provided that the rainfall was zero at some time within the recent past, recorded by ten readings stretching back across 2.25 h from the current time. Let this time series be represented by the symbols $G(10)$, $G(9)$, $G(8), \ldots, G(1)$. The rainfall rate $G(30 \text{ min})$ over the latest 30 min is given by

$$G(30 \text{ min}) = G(10) - G(8) + G(6) - G(4) + \ldots \ .$$

The summation of this series is continued through $G(8)$, $G(6), \ldots$, until a zero is encountered. If no zero is encountered, the gauge is ignored. The following numerical example shows the working of this scheme. Groups of four quarter-hour amounts are seen in sum over the hour:

accumulated		observed
	$+2 + 4 + 6 + 5 \times 0.1 \text{ mm}/\frac{1}{4} =$	17×0.1 mm during hour
	$-2 - 4 - 2 - 4$	$= -12$
	$+0 + 1 + 2 + 4$	$= 7$
	$-0 - 0 - 1$	$= -1$
tenths mm/half-hour		$+11$
mm/h over the last half-hour $= 0.1 \times 22$		

If this addition yields an answer less than 2 (tenths of a millimetre fallen in the latest half-hour), a rate of 0.4 mm/h, the gauge is ignored.

The radar information to match these gauge data comes from adding together the radar readings as described above, those for $t - 30$, $t - 15$ and t with weights $\frac{1}{4}, \frac{1}{2}$ and $\frac{1}{4}$. Only if non-zero, are the corresponding radar readings counted in the following sum of respective 'values':

$$1 + 2 + 1 \ .$$

If all three radar readings are non-zero (if all members of the sum are present), the total comes to 4. If the rainfall rate over the gauge is zero at any time, the total amounts to less than this. If the total amounts to less than 3, the factor $A(i)$ derived from the radar–gauge comparison is not accepted.

7.   CALIBRATION OVER A PERIOD OF 15 min

For comparison of radar and gauge information over the previous 15 min, the gauge information is derived from the quarter-hour series of hourly accumulations of rain as described above. From this time series, the rainfall rate occurring over the latest 15 min is derived:

$$G(15 \text{ min}) = G(10) - G(9) + G(6) - G(5) + G(2) - \ldots .$$

The summation of this series is continued through $G(9)$. $G(6), \ldots$, until a zero is encountered. If no zero is encountered, the gauge is ignored. Using the same numerical example as before gives:

accumulated	observed
$+2 + 4 + 6 + 5 \times 0.1 \text{ mm}/\frac{1}{4} =$	17×0.1 mm during hour
$-4 - 2 - 4 + 6$	$= -16$
$+0 + 1 + 2 + 4$	$= 7$
$-0 - 0 - 1 - 2$	$= -3$
tenths mm/quarter-hour	$+5$

mm/h over the last quarter-hour $= 0.1 \times 20$.

In this addition yields an answer less than 2 (tenths of a millimetre fallen in the latest quarter-hour), a rate of 0.8 mm/h, the gauge is ignored.

The radar information to match this comes directly from the current, nearly instantaneous modified radar composite. (It would be possible to include in the estimate the radar data recorded 15 min earlier at the beginning of the period of averaging.)

No allowances are made for travel time of rain of 3 or 4 min between the middle of the radar beam and the gauge, or for rain drifting horizontally between radar sampling volume and gauge.

8.   PERFORMANCE OF REMOTE SITES, FRONTIERS AND RAINGAUGES

For quick-look verification studies of FRONTIERS, records from 36 surface rainfall Europe West (SREW) raingauges are recorded in the Meteorological Office's Synoptic Data Bank, immediately accessible for examination. FRONTIERS archive tapes run for 12 h each; so it is convenient to examine recordings at the 36 SREW raingauge sites over 12 h periods. Since installation of the new radar analysis on 12 April 1989, and correction on 2 May 1989 of deficiencies discovered during the first days of operations, the most notable days of rain up to the time of writing have been 24 May 1989 and 6 June 1989. Hourly integrated totals from these days are shown in Table 2 for a representative selection of gauges which accumulated large amounts of rain. The results show that on most occasions FRONTIERS and the site-

Table 2 — Hourly rain accumulations on 24 May 1989 and 6 June 1989: S, site corrected; F, FRONTIERS; G, gauge; Tr, 'trace' of rain

Hourly rain accumulation (tenths mm)

24 May 1989 GMT	Odiham			Boscombe Down			Heathrow			Nottingham		
	S	F	G	S	F	G	S	F	G	S	F	G
1300	73	77	2				197	208	26			
1400	158	166	268	1	1	Tr	29	31	18			
1500	12	13	16	128	112	270	18	19	10			
1600	15	14	2	252	204	14	36	37	16			
1700	10	8	2	7	4	6	43	44	18			
1800	41	37	8	0	0	Tr	20	21	6	18	19	34
1900	24	22	2				9	10	2	13	14	18
2000							2	2	Tr	28	29	20

Hourly rain accumulation (tenths mm)

6 June 1989 GMT	Birmingham			Gatwick			Marham			Wattisham		
	S	F	G	S	F	G	S	F	G	S	F	G
1300	1	2	Tr				5	3	20	5	4	0
1400	0	1	Tr				45	40	42	14	11	44
1500	17	57	18	1	1	Tr	2	1	12	20	13	4
1600	0	1	Tr	138	29	14	0	0	Tr	20	11	18
1700	1	1	0	173	48	30	1	1	Tr	2	1	Tr
1800	1	5	2	38	37	10	0	0	Tr	0	0	Tr
1900	1	1	Tr	17	18	40	0	0	Tr	0	0	Tr
2000	9	9	6	3	2	4				0	0	Tr

corrected composite were much the same. Calibration generally was not applied owing to mismatch in space and time between radar and gauges. Differences between FRONTIERS and the site-corrected data arise from the slightly different range corrections applied. On 6 June 1989, FRONTIERS shows enhancement over the site-corrected result for Birmingham at 1500 Greenwich Mean Time (GMT). Calibration was not applied in FRONTIERS: it was at the remote site Clee Hill, leading to reduction by a factor of 0.3 of heavy shower rainfall occurring at 1415 GMT. The results for Gatwick show FRONTIERS much closer to the gauge than the site-corrected result. In this case the site correction at the Chenies radar involved enhancements by factors of about 3, while FRONTIERS did not yield a calibration. Ways of promoting calibration in convective situations such as those recently prevalent remain to be explored.

9. CONCLUSIONS

Introduction of automated decision making into FRONTIERS has made possible improvements in calibration and treatment of melting snow. Comparison with raingauges implies that present restrictions on possibilities of calibration should be eased. It seems also that more development is in order in respect of facilitating FRONTIERS operation.

REFERENCES

Brown, R. (1987) The use of imagery in the FRONTIERS precipitation nowcasting system. *Preprints for Workshop on Satellite and Radar Imagery Interpretation, Reading, EUMET SAT.* 20–24 *July* 1987. pp. 459–472.

Brown, R. and Sargent, G. P. (1988) *New range corrections for FRONTIERS*, Nowcasting Research Group Report No. 16.

Collier, C. G., Larke, P. R. and May, B. R. (1983) A weather radar correction procedure for real-time estimation of surface rainfall. *Q. J. R. Meorol. Soc.,* **109**, 589–608.

Conway, B. J. (1989) Expert systems and weather forecasting, *Meteorol. Mag.,* **118**, 23–30.

Conway, B. J. and Browning, K. A. (1988) Weather forecasting by interactive analysis of radar and satellite imagery. *Phil. Trans. R. Soc. London, Ser. A*, **324**, 299–315.

Hill, F. F. (1983) The use of average annual rainfall to derive estimates of orographic enhancement of frontal rain over England and Wales for different wind directions. *J. Climatol.,* **3**, 113–129.

Koistingen, J. and Puhakka, T. (1986) Can we calibrate radar with raingauges? *Geophysica,* **22**, 119–129.

8

Real-time calibration methods of radar raingauge data

T. Yamaguchi[1], T. Ohtsuka[1], K. Murata[1], E. Abe[1], F. Yoshino[2] and S. Takemori[3]
[1]Research Department, Foundation of River and Basin Integration Communications, 1–3 Kojimachi, Chiyoda-ku, Tokyo, Japan. [2]Masamitsu Mizumo, Hydrology Division, Public Works Research Institute, Ministry of Construction, 1 Asahi, Tsukuba-shi Ibaraki, Japan. [3]Shin-Nihon Kisho Kaiyo Co., Ltd, Japan

ABSTRACT

The radar raingauge is an extremely effective instrument with remote sensing technology for measuring raingauge data, considering the instantaneity of the observation and the time–space continuity.

However, it is not appropriate to input values of the observation, as they are, into the various technological calculation including the runoff analysis which is at present carried out using a ground raingauge, because of the difference between radar raingauge and ground raingauge data.

We would like to report that a real-time calibration method has been developed which allows one to provide data as the direct on-line input for improving the accuracy of the radar raingauge measurement.

1. NECESSITY FOR CALIBRATION

When quantitative utilization of the radar raingauge is considered, the specialities of the required accuracy are classified as follows.

(1) The accuracy should be spatially uniform.
(2) The accuracy should be timewise uniform.
(3) Data should allow to be utilized as the ground raingauge information.

However, the radar raingauge, since it is basically a remote sensing equipment measuring only the strength of radio waves back-scattered by the rain drops in the atmosphere, has an error inherent in the instrument due to the operation principle in addition to the errors that occur when the measured values are converted into the raingauge data.

The following are the major errors.

(1) *Vertical variation in rainfall intensity*   The radar beam emitted from the radar raingauge increases its height from the ground with increasing distance. In Japan which is mountainous, the beam is emitted in a direction a little over the horizon in order to ensure that the lower portion of beam, which expands with increasing distance, avoids the mountains. The radar raingauges of the Ministry of Construction are set at 120 km for the quantitative observation range. The height of the beam centre at this distance is approximately 3000 m. Fig. 1 shows

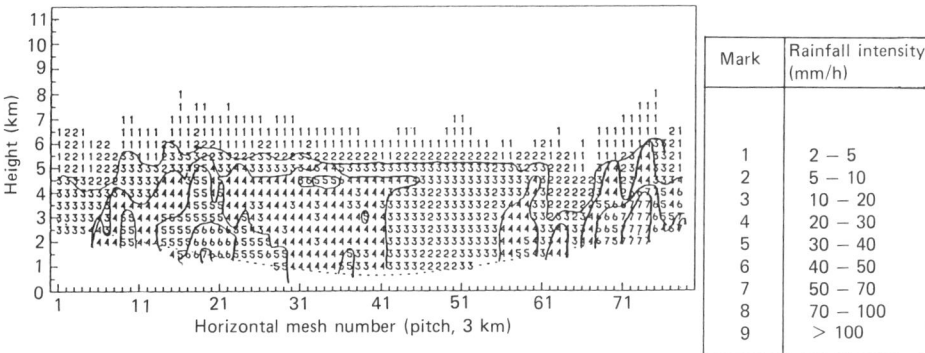

Fig. 1 — Vertical cross-section of rainfall intensity (from 2145 to 2200 hours on 16 October 1989) according to Miyama radar data.

the vertical distribution of the rainfall in a typhoon obtained by the Miyama radar raingauge of the Kinki Regional Construction Bureau with variable elevation (constant-altitude plan position indicator (CAPPI)). According to the observations, the rainfall is not distributed evenly in the vertical direction and generally tends to increase in intensity towards the ground. Summarizing the information, it is clear that in the case of a typhoon, the stronger the echo, the stronger will be the vertical gradient, indicating heavy rainfall closer to the ground. In other words, rainfall may not be observed with a beam height of 3000 m, and the observation at even lower heights gives values which deviate very much from the values observed on the ground.

(2) *Assumptions implicit in the radar equation*   The radar equation has various assumptions implicit in it. The difference between the assumptions and the actual weather phenomena results in errors.

As it is difficult to eliminate the reasons for these errors by the hardware approach of the present radar raingauge instrument, correction by software compensation is necessary.

2. FACTORS REQUIRED FOR CALIBRATION

As explained above, there are certain areas in which the radar raingauge data cannot be utilized quantitatively with the Ministry of Construction's radar raingauge which has already been installed. It is desired that the data should be effectively utilized against flood disaster, which is an annual event in Japan, in view of the nationwide observation network already established.

There are several methods conceived for compensating the radar raingauge data by the ground raingauge data. With any of the methods it is possible to obtain quantitative conformity with the ground rainfall distribution. However, in practice, a method is urgently awaited for calibrating in a time extrapolation, with on-line distribution of the information.

The ground rainfall data to be used for the calibration can be made available on line by employing a telemeter system. It is reported that, according to the present calibration technique, the larger the number of ground data, the higher will be the accuracy of the radar raingauge data. However, as it takes approximately 10–15 min to receive the data sent from the 100 or so ground rainfall observations within the 120 km radius radar raingauge quantitative observation area, it is not possible to obtain radar raingauge compensation instantaneously.

Therefore, it is required, for on-line information transfer, to calculate the compensation coefficient beforehand.

3. CALIBRATION METHOD (DYNAMIC WINDOW)

The calibration techniqe so far conceived is to calculate a curved surface for the compensation using the ratio of the ground raingauge data to the radar raingauge data as an index. However, as indicated in Fig. 2, the difference between the radar

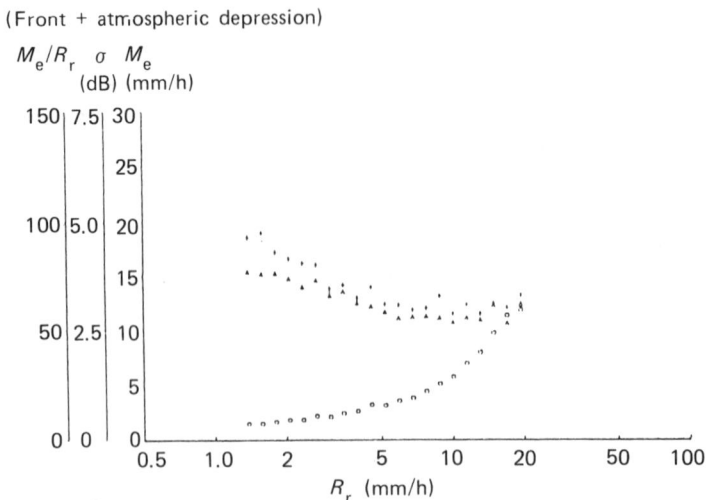

Fig. 2 — Accuracy evaluation index according to radar raingauge intensity: $\square$, M_e (mean square error); $\triangle$, σ (mean logarithmic error); +, M_e/R_r (mean relative square error).

raingauge and ground raingauge data depends upon the rainfall intensity. The difference is considered to be caused by the resolution of the ground raingauge.

A manipulation was added here by replacing the above difference with the reliability of the ground precipitation so as to equalize the reliability of the compensation coefficient F_i which formulates the curved surface of the compensation.

The procedure is as follows.

(1) Calculate the compensation coefficient F_i $(=(R_g)_i/(R_r)_i)$ for each polar coordinate mesh directly above the ground raingauge.
(2) Calculate levelled-off scales for each radar mesh. For a radar mesh j the compensation coefficient is the average of the number n_j of coefficients counted from the nearest to the mesh point. n_j is calculated, using the following formula, as the number which allows the average of the ground precipitation to gain a certain level of reliability:

$$\sum_{i=1}^{n} \left(\frac{\varepsilon_0}{C_{vi}}\right)^2 = 1 \tag{1}$$

where ε_0 is the tolerable mean error and C_{vi} is a space-variable coefficient. In practice, n is determined by adding up the $(\varepsilon_0/C_{vi})^2$ values of the order of the distance from the radar mesh j for the meshes of ground precipitation meters and by stopping the sampling when the added value exceeded unity for determining the number n.

(3) F_j is obtained as the mean value of F_i. Then the reciprocal numbers of C_{vi} of each point is used as the weight:

$$F_j = \sum_{i=1}^{n} \frac{F_i}{C_{vi}} \bigg/ \sum_{i=1}^{n} \frac{1}{C_{vi}}. \tag{2}$$

The features of the method are as follows.

(a) A constant reliability can be maintained in the compensation coefficient regardless of the rainfall intensity.
(b) When the rainfall intensity is high, a minute space variation in the compensation coefficient can be seen, as the scale of space levelling off has been reduced.
(c) Although the reliability has been reduced by the space variation of the ground precipitation meters, it can be increased by constructing patterns of the radar raingauge meter observation characteristics of the rainfall locality according to topography factors. In other words the possibility for future development is great.

4. ACCURACY VERIFICATION

The verification of the accuracy of the dynamic window was carried out using the data from the Monomiyama radar raingauge of the Tohoku Regional Construction Bureau, Ministry of Construction.

(1) *Determination of space variation coefficient and tolerable mean errors*   In order to examine this system, it is necessary to provide a space variation coefficient as a function of rainfall intensity R. This was determined using the precision rainfall observation data (Fig. 3) of the Civil Engineering Institute of the Ministry of

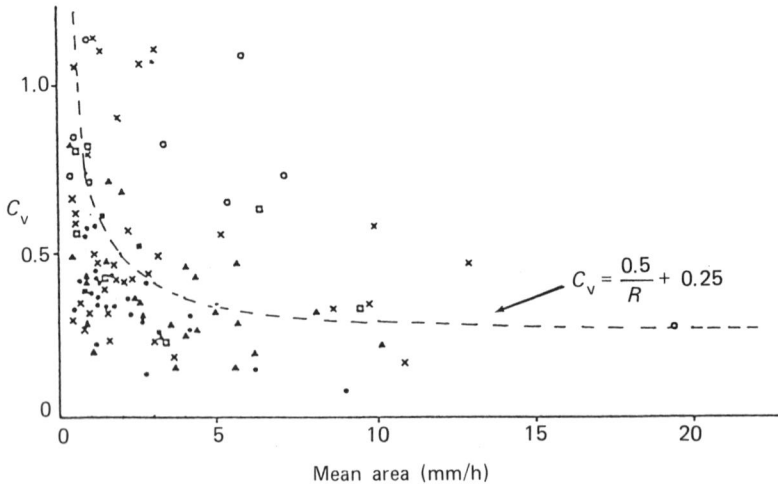

Fig. 3 — Relationship between mean area and spatial variation coefficient C_v

Construction:

$$C_v = \frac{0.5}{R} + 0.25. \tag{3}$$

Then in the quantitative observation area (radius, 120 km) of the Monomiyama radar raingauge meter, ground precipitation meters including the telemeter raingauge meters of the Ministry of Construction and the Automated Meteorological Data Acquisition System (AMeDAS) of the Weather Bureau have been distributed at a density of aproximately one set per 200 km². According to equations (1) and (2), the relationship between the tolerable mean errors to be used for the calibration of all the ground observatory data and the radii of the levelled-off area are indicated in Table 1. The tolerable mean error of 0.05 is not

Table 1 — Relationship between rainfall intensity and radii of levelled-off area (for total points)

Tolerable mean error	Radius of levelled-off area for following rainfall intensities					
	1 mm/h	2 mm/h	5 mm/h	10 mm/h	20 mm/h	50 mm/h
0.2	31	20	14	12	11	10
0.1	60	40	28	24	22	21
0.05	120	80	56	48	44	42

practical, as minor rainfalls are levelled off in the data of the entire observation range. Therefore, the tolerable levelled-off errors of 0.2 and 0.1 have been taken as the object.

(2) *Object rainfall*   Ten cases of especially large rainfalls were taken among those of front and atmospheric depression natures during 1987 and 1988. The total duration for the front and depression rainfalls were 225 h and 189 h respectively. The unit for the calculation is 1 h.

(3) *Ground precipitation parameters of the object*

Case A   the accuracy was evaluated at all the AMeDAS points, using all the telemeter points for calibration (Fig. 4).

Case B   the 21 equally distributed AMeDAS points were used for the accuracy evaluation, while all the remaining points were used for calibration (Fig. 5).

(4) *Extension of levelled-off areas to the time axis direction*   The levelled-off time was divided into three categories of 1–3 h.

Case 1 $\overline{F_i} = (F_i)_{t-1}$.
Case 2 $\overline{F_i} = \frac{1}{2}[(F_i)_{t-1} + (F_i)_{t-2}]$.
Case 3 $\overline{F_i} = \frac{1}{3}[(F_i)_{t-1} + (F_i)_{t-2} + (F_i)_{t-3}]$.

The calculated respective values of $\overline{F_i}$ have been multiplied by the radar raingauge data to calculate the difference between the ground precipitaion at t h.

(5) *Result of the accuracy verification*   The accuracy for the total of the object rainfalls is as shown in Table 2 and Fig. 6. The following are presumed.

(a) In general the compensation markedly affects the calibration.

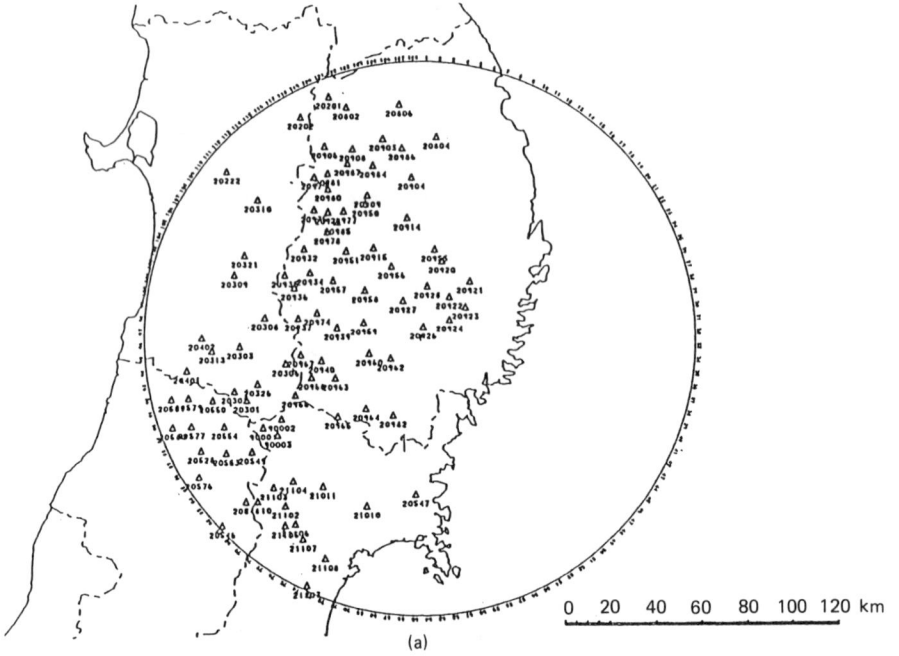

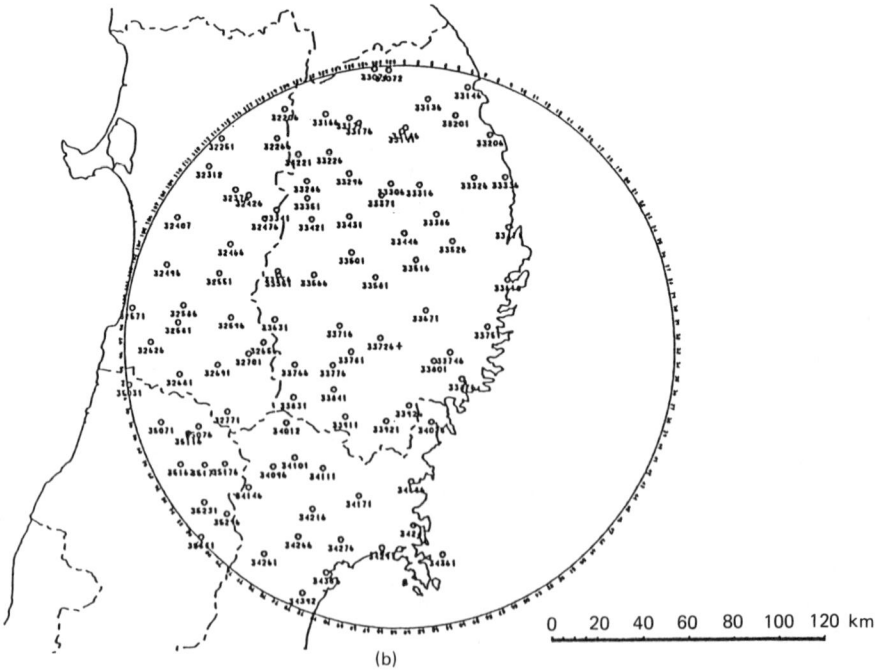

Fig. 4 — Case A: (a) telemeter observation posts ($\triangle$) used for calibration; (b) AMeDAS observation posts ($\circ$) used for evaluation.

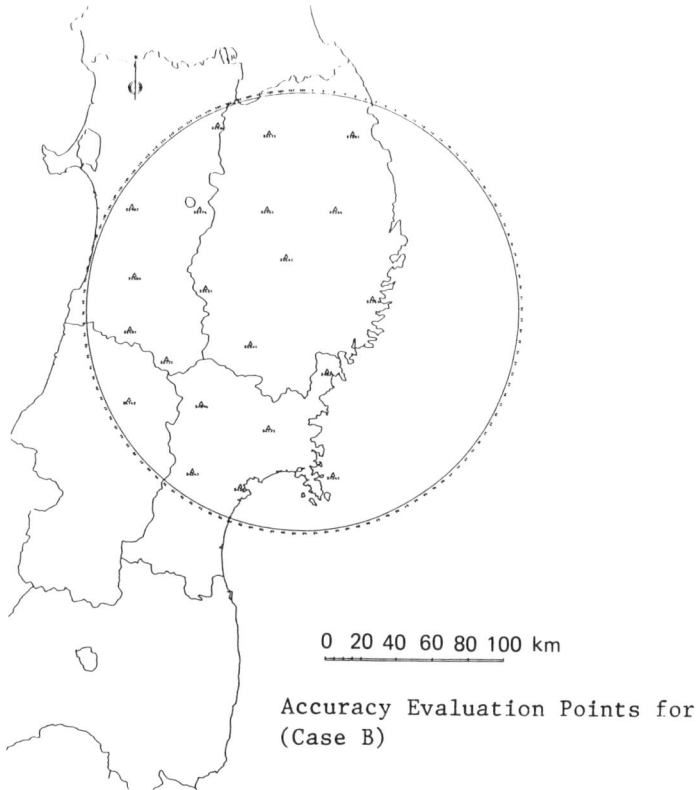

Fig. 5 — Case B: distribution of 21 AMeDAS observation posts ($\triangle$) used for evaluation.

(b) The larger the number of the ground observation posts, the higher will be the accuracy of the calibration.

(c) No evidence is observed that the accuracy is improved because of the time levelling-off.

Dispersion graphs have been prepared to investigate the compensation situation with the respective data (Fig. 7). The calibration was conducted using the compensation values of 1 h ago. Although the dispersion pattern before the compensation is preserved to a certain degree, the centre of the dispersion is on the 45° line and the correspondence between the ground precipitation and radar raingauge data is favourable. In other words, since the method presupposes that the individual ground raingauge value should have errors according to the intensity of rainfall, it is natural that not all conform to this. In order to clarify this point, a comparison has been conducted in which the ground raingauge value is close to the actual value. This means that the comparison has been carried out with the idea of reducing the error due to levelling-off (which is the essence of the method) between the ground precipitation that has been spatially and timewise levelled off using the variation coefficient and the radar raingauge

Table 2 — Accuracy of total rainfall

ε_0	Number of AMeDAS evaluation points	Without compensation	Dynamic window		
			1	2	3
		Mean square error			
0.1	21	3.80	2.78	2.85	2.88
0.2	21	3.80	3.42	3.06	3.24
0.1	96	3.80	2.96	2.98	3.03
0.2	96	3.80	3.15	3.23	3.30
		Mean logarithmic square error			
0.1	21	3.29	2.65	2.69	2.75
0.2	21	3.29	2.70	2.77	2.87
0.1	96	3.34	2.84	2.79	2.84
0.2	96	3.34	2.80	2.91	2.98
		Relative coefficient			
0.1	21	0.60	0.81	0.80	0.80
0.2	21	0.60	0.76	0.78	0.75
0.1	96	0.61	0.78	0.78	0.77
0.2	96	0.61	0.76	0.75	0.74
		Total raingauge ratio			
0.1	21	0.80	0.95	0.94	0.93
0.2	21	0.80	0.98	0.96	0.95
0.1	96	0.78	0.86	0.87	0.85
0.2	96	0.78	0.89	0.87	0.86

value of the similarly levelled-off unit (Fig. 8). Thus the compensated radar raingauge data almost agreed with the ground precipitation which is considered to be identical with the true values according to the levelling-off.

5.   CONCLUSION

A calibration method has been conceived which allows one to vary the levelled-off area according to the rainfall intensity, considering the error of ground precipitation, with the following results.

(1) Compared with the radar raingauge data without compensation, the conformity with the ground precipitation has been greatly improved.
(2) The method is extremely effective for the time extrapolation required for providing on-line data.

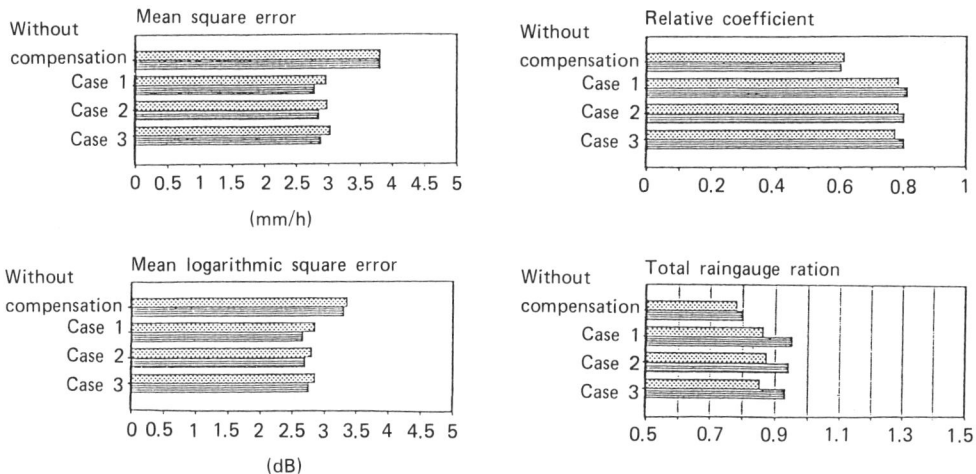

Fig. 6 — Comparison of accuracies according to systems of calibration: □, case A (calibration by telemeter only; □, case B (calibration by telemeter–AMeDAS).

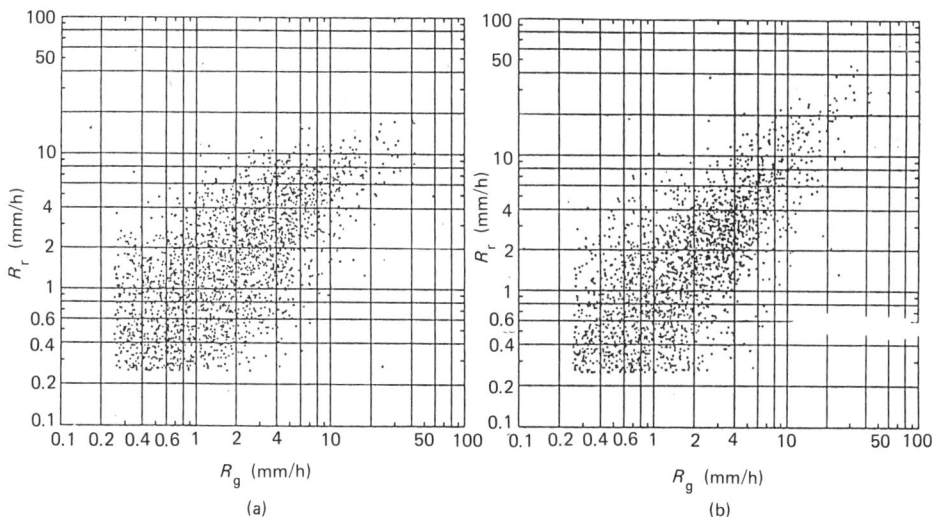

Fig. 7 — Dispersion graphs of ground precipitation and radar raingauge data (rain per total area): (a) without compensation; (b) dynamic window method.

Further study is being contemplated as follows:

(a) Comparison will be conducted with the water level or flow volume in connection with the calculation of runoff data in order to clarify the rate of accuracy improvement.

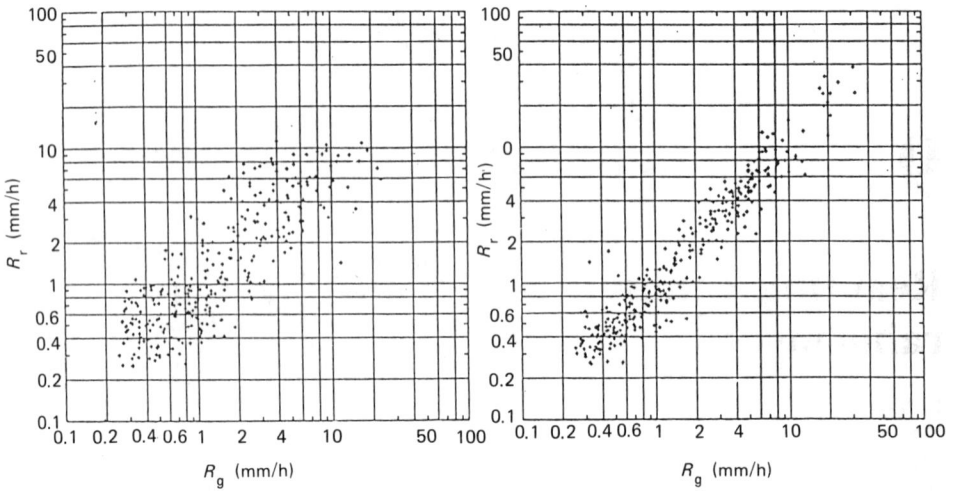

Fig. 8 — Dispersion graphs of ground precipitation and radar raingauge data (rain per total area) viewed areawise: (a) without compensation; (b) dynamic window method.

(b) The spatial accuracy of radar raingauge data and the topographical influence will be introduced, expanding the conception of the ground precipitation errors that has been the basis of the idea of the levelling-off.

9

Real-time calibration methods of radar raingauges

F. Yoshino[1], E. Abe[2] and S. Tamamoto[3]
[1]Masamitsu Mizuno, Hydrology Division, Public Works Research Institute, Ministry of Construction, 1 Asahi, Tsukuba-shi Ibaraki, Japan
[2]Research Department, Foundation of River Basin and Integrated Communications, 1–3 Kojimachi, Chiyoda-ku, Tokyo, Japan
[3]CTI Engineering, 4-9-11 Nihonbashi-Honcho, Chuo-ku, Tokyo, Japan

ABSTRACT

Radar raingauge systems have become essential in measurements of rainfall, because of their ability to clarify spatial rain distribution over large areas in short periods of time. Adjustments to the radar raingauges are needed, however, in order to analyse the rainfall on the ground, as the data are obtained while the rain is falling. Investigations were made in this study on actual rainfall results obtained using the calibration techniques and measured rainfall at ground level. As a result, it was found that the adjustment factor method produces the best results and the improvement in accuracy was approximately 20% compared with that of the system used at present.

1. INTRODUCTION

Radar raingauge systems have become an important tool in the management of rivers and roads because of their ability to detect rainfall distribution quickly over wide areas and because of increased accuracy obtained by improvements in the hardware and software. In Japan, investigations on radar raingauges began in 1941. One was installed on Mt Akagi and put into operation in 1976 by the Ministry of Construction. There are now 15 radar systems operated by the ministry. Various methods for the improvement of accuracy are still being tested in order to increase the scope of operation of these systems.

In the measurement of rainfall by radar raingauges, there is the problem that radar waves are attenuated by raindrops. If the attenuation of the waves can be accurately recorded, it can be used to improve the accuracy of the measurements.

For that purpose, it is important not only to study the relationship between the rainfall intensity and attenuation but also to estimate the rainfall intensity accurately. If the rainfall intensity is not calculated accurately, the attenuation of the waves will also be in error and the effects snowball. In particular, if the intensity of rainfall is overestimated, the attenuation of the waves will also be estimated in excess, and so rainfall intensity cannot be assessed as the intensity–attenuation data will diverge proportionally to the distance of the rainfall area from the radar site.

The radar constants B and β used in the operation are determined by a preliminary study. However, they vary in accordance with the variation in distribution of raindrops with time and space. Therefore, if the variation in radar constants B and β can be correctly estimated, excellent accuracy can be achieved. In this study, we have estimated the improvement in accuracy of real-time calibration techniques of the radar raingauge, using measured rainfall at ground level. The results are detailed below.

2. CALIBRATION TECHNIQUES

One of the most important factors of variations in the radar constants B and β with time is the change in the distribution of raindrops as they fall. A number of techniques were investigated for the estimation of this variation. In this study, it was assumed that the radar constants B and β do not vary in the observation area, and the variation with time was calibrated. β was fixed at 1.6 as it varies less than B with time. The following techniques were used for calibration of the radar constant B.

The constant B is calibrated every hour since this is the frequency of data collection and B is used to calculate the radar rainfall intensity up to the time when the next set of ground raingauge data are acquired.

2.1 Sensitivity analysis method

In this method, the radar constant B was optimized by minimizing errors, obtained in using the ground raingauge data, and the radar rainfall intensity was calculated using the constants B. In the study, $\log B$ was made to vary from 1 to 3 in 0.01 increments. The indices in the precision evaluation were, the root mean square error, the relative error and the correlation coefficient:

$$\text{root mean square error} \left(\frac{1}{mn} \sum_{i=1}^{m} \sum_{j=1}^{n} \left[(R_g ij - R_{r_{ij}} \right)^2 \right]^{1/2} \text{(mm/h)} \tag{1}$$

$$\text{relative error} \ \frac{1}{mn} \sum_{i=1}^{m} \sum_{j=1}^{n} \frac{\left| (R_r)_{ij} - (R_g)_{ij} \right|}{(R_r)_{ij}} \times 100 \ (\%) \tag{2}$$

$$\text{correlation coefficient} \ \frac{(R_r R_g - R_r^2 R_g)}{[(R_r^2 - R_r) \ (R_g^2 - R_g^2)]^{1/2}} \tag{3}$$

where m is the total number of raingauge, n is the total number of times, i is the raingauge number and j the time.

2.2 Adjustment factor method

In this method, we calibrated the radar constant B using the adjustment factor calculated as $(1/n) \sum \log(R_g/R_r)$, assuming B to be 200.

By using the assumed radar constant B_0 and the constant B_1 calibrated using the adjustment factor method, the radar rainfall intensities can be calculated from the reflectivity factor Z given in the following equations:

$$(R_r)_0 = \left(\frac{Z}{B_0}\right)^{1/\beta} \quad \text{or} \quad (R_r)_1 = \left(\frac{Z}{B_1}\right)^{1/\beta} \tag{4}$$

and the adjustment factors AF are

$$(AF)_0 = \frac{1}{n} \sum \log_{10}\left(\frac{R_g}{R_r)_{10}}\right) \quad \text{or} \quad (AF)_1 = \frac{1}{n} \sum \log_{10}\left(\frac{R_g}{(R_r)_1}\right). \tag{5}$$

Therefore the difference between these values is

$$(AF)_0 - (AG)_1 = \frac{1}{n} \sum \log_{10}\left(\frac{R_g}{(R_r)_0}\right) - \frac{1}{n} \sum \log_{10}\left(\frac{R_g}{(R_r)_{12}}\right)$$

$$= \frac{1}{\beta} \log_{10}\left(\frac{B_0}{B_1}\right). \tag{6}$$

If the radar constant B is calibrated correctly, the expected adjustment factor is zero. B_1 such that $(AF)_1 = 0$ when $(AF)_0 = k$ is calculated using the following equation:

$$B_1 = B_0 \times 10^{-\beta k}. \tag{7}$$

2.3 Sequential analysis method

Sequential analysis is a statistical examination technique used for the management of products, testing data, etc., and was presented by A. Wald in 1947. The calibration technique using sequential analysis was used by Cain and Smith (1976). The basic steps of the technique are shown in Fig. 1. It can be seen that, only when the adjustment factor calculated using the previous radar constant B does not belong to the normal distribution $N(0, \sigma^2)$, does a new radar constant need to be calculated. In this study, when the null hypothesis H_0 was rejected, we calculated a new constant B using the same method as that used in the adjustment factor method. Hence the adjustment factor method is equivalent to making calibrations every hour in the sequential analysis method.

Assumption of radar constant B

Null hypothesis	$H_0: \mu = \mu_0 = 0$
Alternative hypothesis	$H_1: \mu = \mu_1 \neq \mu_0$

Establishment of examination parameters α and β
and calculation of constants a and b
$a = (1 - \beta)/\alpha$ and $b = \beta/(1 - \alpha)$

Number of sample $m = 0$

$m = m + 1$

Calculation adjustment factor x_i

Calculation of OC function L

L ?

$b < L < a$ $L < b$

$L > a$

Rejection of hypothesis H_0	Acceptance of hypothesis H_0

Correction of radar constant B

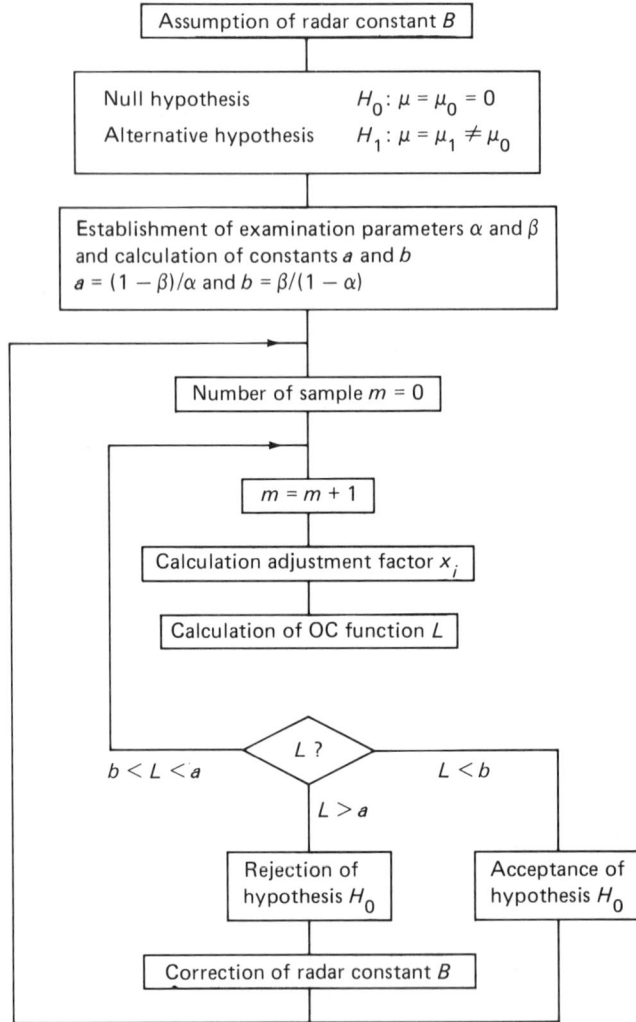

Fig. 1 — Basic steps of calibration by the sequential analysis method. The OC function is

$$L = \exp\left(\frac{m(\Delta\mu)^2}{2\sigma^2}\right) \cosh\left(\frac{\Delta\mu}{\sigma^2} \sum_{i=1}^{m} X_i - \mu_0\right) .$$

3. DATA USED FOR ANALYSIS AND INDICES OF PRECISION EVALUATION

The radar rainfall data used in this study were obtained between 1981 and 1983 from the quantitative observation area of radar raingauge (a radius of 120 km) installed at Mt Akagi. The rain classifications were six typhoons (249 h), seven fronts (372 h) and

23 thunderstorms (123 h). In this paper, the results on typhoons are included mainly because the intensities are large and because of their importance in Japan.

The ground rainfall data used for calibration were observed by the telemeter ground raingauge systems of the Ministry of Construction. The previous telemeter rainfall data and the rainfall data observed by the Automated Meteorological Data Aquisition System (AMeDAS) raingauge system of the Japan Meteorological Agency are used for precision evaluation.

The indices for precision evaluation are the root mean square error, the average error, the fitting ratio of radar to ground rainfall on the rainfall intensity ranks (used in the real-time information service of rainfall intensity and classified by 0, 2, 5, 10, 20, 30, 40, 50, 70 and 100 respectively) and the degree of improvement in these.

4.   INDICES USED IN THE SENSITIVITY ANALYSIS METHOD

The comparison of rainfall intensity and cumulative rainfall depth among the ground raingauges, the radar rainfall calculated by the existing method, and the rainfall calibrated by the sensitivity analysis method using each of the three sensitivity indexes are shown in Fig. 2. The variations in rainfall intensities calculated by the existing method correlated to those of ground rainfall intensities, but the radar rainfall R_r is generally larger than the ground rainfall R_g. On the other hand, the radar rainfall calibrated by the sensitivity analysis method using the mean square error as a sensitivity index fitted the ground rainfall favourably. The rainfall calibrated by the sensitivity analysis method using the correlation coefficient or correlative error as a sensitivity index was sometimes remarkably different from the ground rainfall results. These errors were shown especially when the rainfall intensity were large. Therefore the best index for sensitivity analysis seems to be the root mean square error.

5.   EXAMINATION PARAMETERS α, β AND δ FOR THE SEQUENTIAL ANALYSIS METHOD

The examination parameters α and β are the probability of error constants obtained from the sequential test and these are used to make judgements for sorting out the data. If the quantity α is overestimated, there is a probability of rejecting B in spite of the fact that the variation in B is within the permissible error range. On the other hand, if the quantity β is overestimated, the probability of the acceptance by mistake is likely. δ is the index of distance between the two hypotheses — the null hypothesis and an alternative hypothesis — expressed as follows:

$$\begin{array}{ll} \text{null hypothesis} & : \mu = \mu_0 \\ \text{alternative hypothesis} & : \mu = \mu_1 = \mu_0 \pm \Delta\mu = \mu_0 \pm \delta\sigma \ . \end{array} \qquad (8)$$

The smaller these parameters, the more frequently do the judgements of acceptance and rejection occur. Moreover, the frequency of judgement is related to the variation scale of the adjustment factor, and so the relationship between the quantity δ and the standard deviation σ will also directly affect the precision.

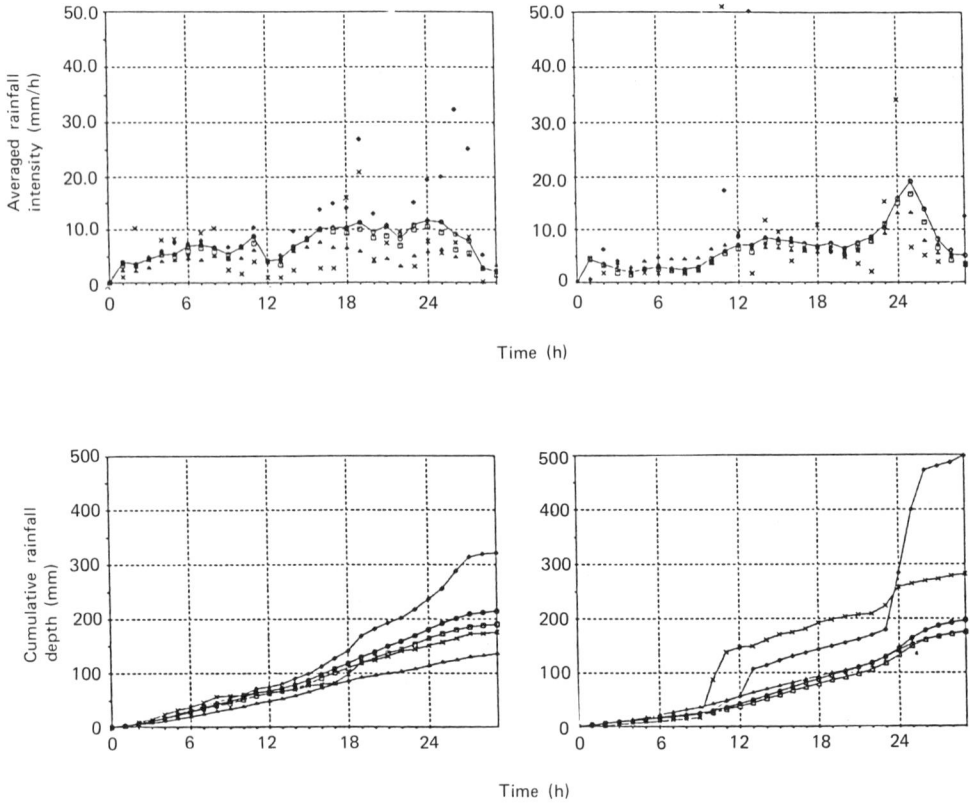

Fig. 2 — Comparison of ground rainfall (○) and radar rainfall calculated by existing method (△) or calibrated by sensitivity analysis methods with various indices (□, root mean square error; ◇, relative error; ×, correlation coefficient) for part of a typhoon.

In this study, the relationship of the errors and the parameters α, β and δ were investigated. Consequently the conditions when $\alpha = 0.2$, $\beta = 0.01$ and $\delta = 0.9$ were better in most cases. However, some of the other rain types had variable parameters and could not be assessed.

6.   THE RESULTS ON THE COMPARISON OF PRECISION EVALUATION

The indices for precision evaluation calculated from the ground rainfall and the radar rainfalls calculated by the existing method and calibrated by the sensitivity analysis method, the adjustment factor method and the sequential method are shown in Table 1. The improvements in precision evaluation of the three calibration methods compared with that of the existing method are shown in Table 2. Here, we used the root mean square error as a sensitivity index for the sensitivity analysis method and the factors $\alpha = 0.2$, $\beta = 0.01$ and $\delta = 0.9$ as parameters for the sequential analysis method.

Table 1 — Indices of precision evaluation by existing method and each calibration method (typhoon and thunderstorm). Here, the improvement degree is calculated from the following equation: improvement degree $Y = |X_1 - X_0|/X_0 \times 100$ (%) where X_0 is the index value for the existing system, X_1 is the index value for the calibrating method and Y is expressed positively when the accuracy is improved

Calibration method	Root mean square error (mm/h)	Average error (mm/h)	Fitting ratio (%)	
			With no difference	Within difference of one rank
Typhoon				
Existing	4.14	2.87	41.0%	88.4%
Sensitivity analysis	3.68	2.50	47.7	91.3
Adjustment factor	3.73	2.53	47.0	91.0
Sequential analysis	4.45	3.04	40.2	87.1
Thunderstorm				
Existing	4.70	2.27	31.2%	78.2%
Sensitivity analysis	3.53	1.53	39.7	84.1
Adjustment factor	3.48	1.51	40.5	83.9
Sequential analysis	3.41	1.48	32.1	78.8

Table 2 — Indices of precision evaluation by existing method and each calibration method (typhoon and thunderstorm). Here, the improvement degree is calculated from the following equation: improvement degree $Y = |X_1 - X_0|/X_0 \times 100$ (%) where X_0 is the index value for the existing system, X_1 is the index value for the calibrating method and Y is expressed positively when the accuracy is improved

Calibration method	Degree of movement			
	Root mean square error	Average error	Fitting ratio	
			With no difference	Within difference of one rank
Typhoon				
sensitivity analysis	11.1	12.9	16.3	3.3
adjustment factor	9.9	11.8	14.6	2.9
Sequential analysis	− 7.5	− 5.9	− 2.0	− 1.5
Thunderstorm				
Sensitivity analysis	24.9	32.6	27.2	7.5
Adjustment factor	26.0	33.5	29.8	7.3
Sequential analysis	27.4	34.8	2.9	0.7

As a result, for the sensitivity analysis method and the adjustment factor method, an improvement degree of about 10–34% was obtained, except in the fitting ratio of the rainfall intensity ranks. The improvement in the fitting ratio was small because the original system already had a good correlation of about 90%. The improvement degree in the sequential analysis method was small compared with other methods, but especially when compared with thunderstorms, improvement degree of the indices, root mean square error, and average error equal to other errors. The cumulative rainfall depth of the sensitivity analysis method and the adjustment factor method compared favourably with those of the ground rainfall data in Fig. 3. The

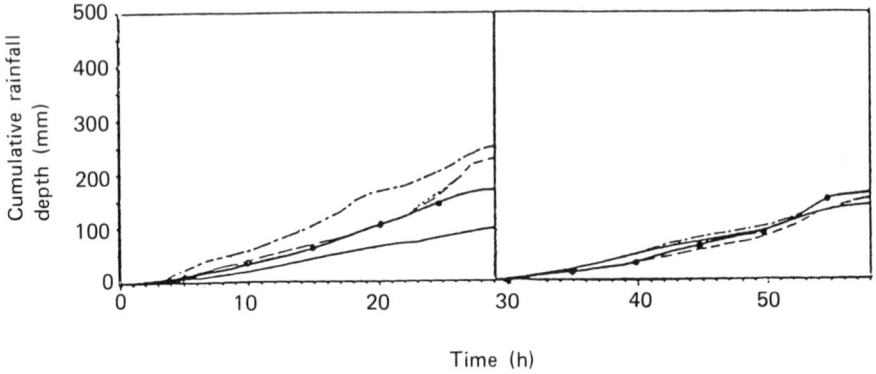

Fig. 3 — Comparison of ground rainfall (–●–) and radar rainfall calculated by the existing method (– – –) or corrected by calibration methods (– – – , sensitivity analysis method; -------, adjustment factor method; – · – , sequential analysis method) for part of a typhoon.

radar rainfall from the sensitivity analysis method or the adjustment factor method fits the ground rainfall data more closely than the radar rainfall by existing method. The rainfall from the sensitivity analysis method is nearly equal to that of the adjustment factor method. The sequential analysis method follows variations in B obtained from the sensitivity analysis method with values averaged over time. This is shown in Fig. 4.

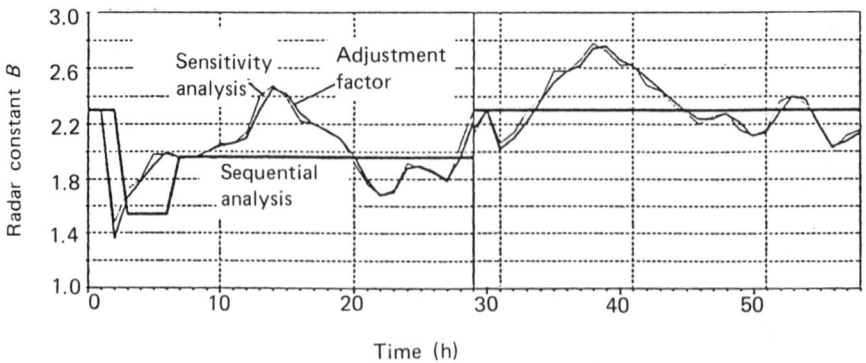

Fig. 4 — Comparison of variation of radar constant B corrected by the calibration method for part of a typhoon.

7. CONCLUSIONS

Investigations were made in this study on real-time calibration techniques of the radar raingauge system in order to suggest improvements in accuracy and applicability, for the purpose of accurate calculation of rainfall intensity by taking into account rain attenuation. This study revealed the following points.

(1) A large improvement in precision evaluation is obtained by real-time calibration of the radar constant B, compared with the accuracy of rainfall obtained by the current method.
(2) The sensitivity analysis method and the adjustment factor method are better for the real-time calibration technique than are other methods.
(3) The adjustment factor method is better than the sensitivity analysis method because more time is needed for calculation in the latter than in the former.
(4) The sequential analysis method may be a good technique for following the variations in the radar constant B due, for example, to changes in raindrop distribution, by removing apparent fluctuation in B resulting from sampling error. This results in increased accuracy for time-averaged values for precision evaluation but there are periods during which large errors occur compared with other methods because of the time delay in the resulting variations in B. Moreover, more time is needed for calculations than in the adjustment factor method.
(5) The sequential analysis method appears to be a good technique for post-analysis study in which we can feed back the calibrated constant into B calculations involving previous rainfall information.

We plan in the future to continue to apply the adjustment factor method to actual rainfalls and to investigate the real-time spatial calibration technique.

ACKNOWLEDGEMENTS

We would like to note here our gratitude towards the members of the Japan Weather Association, the Kanto Regional Construction Bureau and the Public Works Research Institute of the Ministry of Construction for their valuable advice and the data that they provided.

REFERENCES

Cain, D. E. and Smith, P. L. Jr. (1976) Operational adjustment of radar estimated rainfall with rain gage data: a statistical evaluation. *Proceedings of the 17th Conference on Radar Meteorology* 1976. American Meteorological Society, Boston, MA.

Fujiera, M. (1986) Raindrop-size distribution from individual storms. *Conference on Atmos. Sci.*, **22**, 1965.

Japan Weather Association (1987) *Primary study about introduction of rain attenuation*, Japan Weather Association, Tokyo.

Kawabata, Y. (1963) *Hydrol. Meteorol.*, March.

Public Works Research Institute (1985) *Characteristics and improvements on rain observation of radar raingauges — development of simulation system for forecasting flood flow using radar raingauge system*, PWRI No. 2078.

Public Works Research Institute (1986) *Measurements of rainfall by radar rain gauges*, PWRI Report No. 2353. Public Works Research Institute, Ministry of Construction, Tsukuba-shi Ibaraki.

Part 2

Real-time radar data processing

10

Corrections for partial blocking of the radar beam

T. Andersson
Swedish Meteorological and Hydrological Institute, Norrköping, Sweden

ABSTRACT

In order to correct for losses caused by the terrain, buildings and trees the radar horizon of the Norrköping weather radar has been measured using topographic maps and 'round-the-horizon' photographs. On the assumption of a Gaussian energy distribution in the lobe the losses are computed and corrections applied. Since the reflectivity generally decreases with increasing height and the beam overshoots most echoes at large ranges, the corrections are most effective at short ranges. Moreover the overshooting is worst where obstacles are in the way of the lower part of the beam. A method of mapping only the radar horizon using radar data is also described.

1. INTRODUCTION

The obstacles surrounding the Ericsson Doppler weather radar in Norrköping reach only fairly low elevation angles. The highest obstacle, a tree, reaches about 1.5°. The most serious blocking effects, however, comes from a grove about 20° broad and reaching an elevation angle of about 1°. To correct for losses caused by the obstacles, their azimuth and elevation angles (from the antenna) were computed in two ways.

(a) Angles related to large topographic features, such as ridges, were computed from detailed topographic maps, using a routine from Teleplan.
(b) Angles related to nearby small-scale obstacles, such as trees and buildings, were measured with a theordolite and with a 'round-the-horizon' photograph.

In (a) the resolution in azimuth was 1° and in elevation 0.01°. A Gaussian profile of power density was assumed, and corrections computed for every 1° in azimuth. Cases with uniform stratiform precipitation over the station were selected. The mean reflectivities for every 1° azimuth were then computed for three elevations angles (0.5°, 1.0° and 1.5°) for range intervals of 5 or 10 km (between ranges of 25 and 40 km). In this way the shadows cast by the obstacles are clearly seen at the lowest elevation angles. The computed corrections were then added to the measured reflectivities.

By comparing the corrected reflectivities at the three elevations used, the method was tested.

The result was, however, somewhat puzzling. The vertical reflectivity gradient was negative, i.e. the reflectivity decreased with increasing height. A high obstacle should reverse the vertical reflectivity gradient or decrease its absolute value, since energy is intercepted most at the lowest elevation angle. This effect was, however, not always observed; in a sector with high obstacles the (uncorrected) reflectivities decreased with increasing elevation angle, while in another section with low obstacles, about 180° from the first mentioned, the reflectivity decrease with increasing elevation was very small.

It is, however, possible to estimate the corrections using only radar data. This may be a better method than the mapping first tried, since obstacles such as trees have complicated shapes and are difficult to map, and it also may be difficult to get very accurate antenna elevations. In order to see how accurate the corrections computed by the radar are, those corrections were used to estimate the elevation angles of the obstacles. Comparing these estimates with the topographically measured angles shows that the mean absolute errors of the radar estimates of the angles of the obstacles' heights were somewhat below 0.15°. This should be compared with the accuracy of the radar's elevation angles; during scans, deviations of 0.1° from the normal value often occur. To this must be added possible errors in the adjustment of the antenna turntable. This adjustment is made using water levels; this probably does not permit better accuracies than about 0.1°. (Our radar is the prototype of the Ericsson radar, having an old antenna control, apparently not enduring the many starts, stops and elevation angle changes imposed by a digital radar. Later versions of the Ericsson radar have a different, more robust and accurate antenna control.)

Corrections of this type are only effective at fairly small ranges. There are geometrical as well as meteorological reasons for this. The geometrical errors are caused by the curvature of the earth and the beam divergence, both causing an increasing height of the beam with increaseing range. The meteorological reason is that the parameters studied, reflectivity or wind, vary with height. Since an obstacle stands in the way of the lower part of the beam, a measurement behind an obstacle actually refers to a higher altitude than a measurement free from obstacles. For precipitation measurements the radar is, however, best just at fairly short ranges.

2.   MAPPING OF THE RADAR HORIZON

The azimuth and elevation angles to 'large-scale' topographic features such as ridges were computed using a detailed topographic map and a routine developed by Teleplan. The resolution in azimuth was 1° and in elevation 0.01°. There are, however, several obstacles such as building and trees (some trees were cut to get a better horizon) which cannot be measured from the map. Therefore, we measured the angles also with a theodolite and prepared a 'round-the-horizon' photo. For both these methods we had to correct for parallax errors, since neither the theodolite nor the camera could be placed at the antenna focus.

The heights of buildings were obtained from the town's architect office. Using these as references for the photograph permitted rather accurate estimates of elevation and azimuth angles from the photograph. These estimates were checked against those from the theodolite. The actual accuracy is difficult to assign, especially as trees have a rather diffuse contour. In this way a map of the radar horizon was obtained.

To compute the corrections a method similar to those described by Moores and Harrold (1975), Aniol and Riedl (1979), Harju and Puhakka (1980) and Wessels (1989) was used (Fig. 1). Since one has to assume that the reflectivity is constant with height and that the targets occupy the whole pulse the efficiency of the corrections decreases with increasing range.

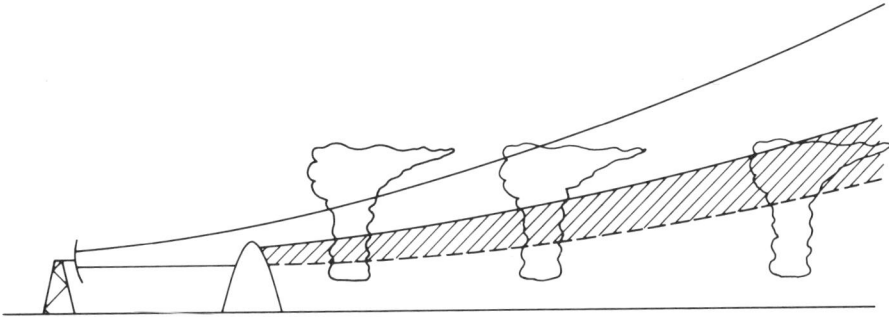

Fig. 1 — Shadowing of the beam. The corrections are most effective at small ranges.

3.   COMPUTING THE CORRECTIONS

The energy distribution within the beam was assumed to the Gaussian and described by

$$\exp(-c\theta^2) \tag{1}$$

where θ is the angle from the beam's axis and c is a constant determined by the half-power beam width of the radar. For our radar the half-power beam width is $0.425°$, giving $c = 3.837$ (deg^{-2}).

On the assumption that the obstacles are horizontal, the losses are obtained by integrating from a large angle to the actual θ of the obstacle. A table giving the losses in decibels (dB) as a function of the angle θ was prepared and used to apply the corrections.

4. ALGORITHMS FOR CORRECTION

The simplest way of correction is to apply the correction term according to the table whenever there is an echo. Denoting this correction as dBZ_corr we thus get

$$dBZ\_est = dBZ + dBZ\_corr \text{ if } dBZ > threshold. \tag{2}$$

For our radar the lowest dBZ recorded is -10; thus the threshold is -10. The Ericsson radar used by us has a function for this correction. The dBZ_corr have to be computed and written as a table.

We have also tried a slightly modified scheme, as follows.

(1) If $h \leqslant 0.5°$ (were h is the elevation angle of the obstacle), then
$$dBZ\_est = dBZ + dBZ\_ \tag{3}$$
(2) If the horizontal width of the obstacle with $h > 0.5°$ is $\ll 2°$ or less, then
$$dBZ\_est = \text{arithmetic mean of dBZ from the two surrounding azimuths.} \tag{4}$$
(3) If the obstacle with $h > 0.5°$ is broader then $2°$, then
$$dBZ\_est = dBZ\_est \text{ (next elevation).}$$

Our lowest antenna elevations were $0.5°$, $1.0°$ and $1.5°$. Thus, if an obstacle is more than $2°$ broad and reaches more than $0.5°$ above the horizon, the dBZ_est value is obtained from the scan with an antenna elevation of $1.0°$.

5. TESTS OF THE ALGORITHMS

For these tests we used the non-Doppler mode with data in polar coordinates for the antenna elevations $0.5°$, $1.0°$ and $1.5°$. The range intervals used were between 25 and 40 km. Ranges below 25 km were not used because of ground echoes, and ranges above 40 km were avoided since the reflectivity is height dependent and the beam is broadening with increasing range.

Fig. 2 shows the radar horizon and uncorrected mean reflectivities for elevations $0.5°$ and $1.5°$ from 13 scans during November 1987 to March 1988. The reflectivities at $1.5°$ are fairly even, which is to be expected since these (with very few exceptions) are not affected by the obstacles. At $0.5°$ elevation the shadowing of the obstacles is clearly seen. It is, however, surprising that in the northern section, where the obstacles are high (a ridge), the reflectivities at $0.5°$ are higher than at $1.5°$. Owing to the interception of energy at the lower elevation the reverse should be expected. The peculiar peaks at $0.5°$ elevation between azimuths $60–75°$ and $250–275°$ are caused by ground echoes. Inspection of the data showed that these ground echoes were at ranges below 30 km. Therefore seven scans using ranges below 30 km were excluded.

If the data used are good, the vertical reflectivity gradients should be reasonable, both before and after the corrections. Fig. 3 shows the mean reflectivities after corrections for the elevations $0.5°$ and $1.5°$. The $0.5°$ curve is now closer to the $1.5°$ curve. In the northern sector the reflectivity decreases most with increasing height, often at a rate of $3–4$ dBZ per $0.5°$, corresponding to $3–4$ dBZ per 300 m. The average dBZ difference between $1.0°$ and $1.5°$ elevations was 0.8 dBZ. Since the large differences in the northern sector are not caused by ground echoes, we must question the antenna elevations.

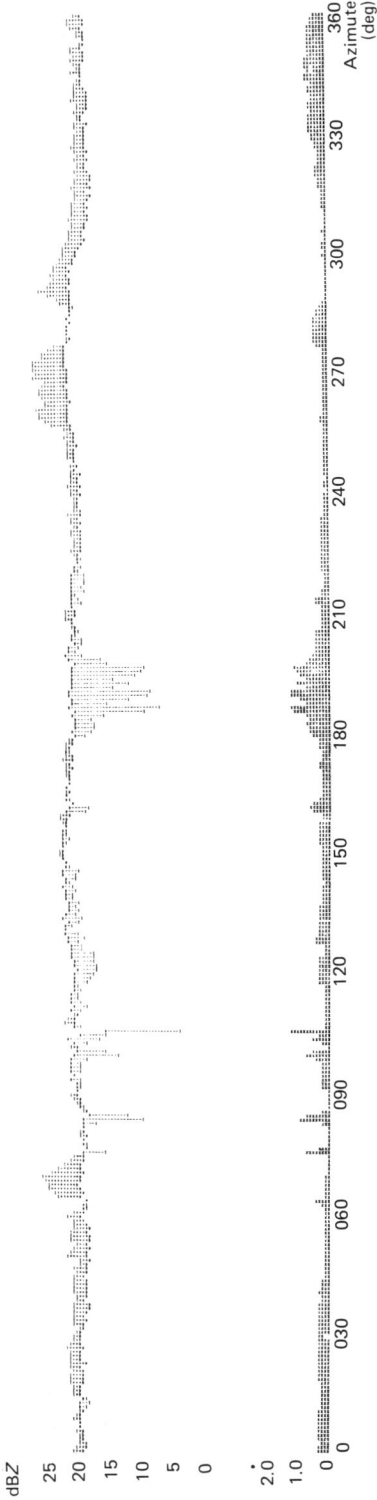

Fig. 2 — The Norrköping radar horizon (lower diagram) and uncorrected reflectivities at 0.5° and 1.5° elevations showing the means of 13 scans. 0.5° elevation is shown by a 1, and 1.5° by a 4. The figures may be impossible to read in this reproduction, but the values are connected by + if the reflectivity decreases and by − if it increases with increasing height.

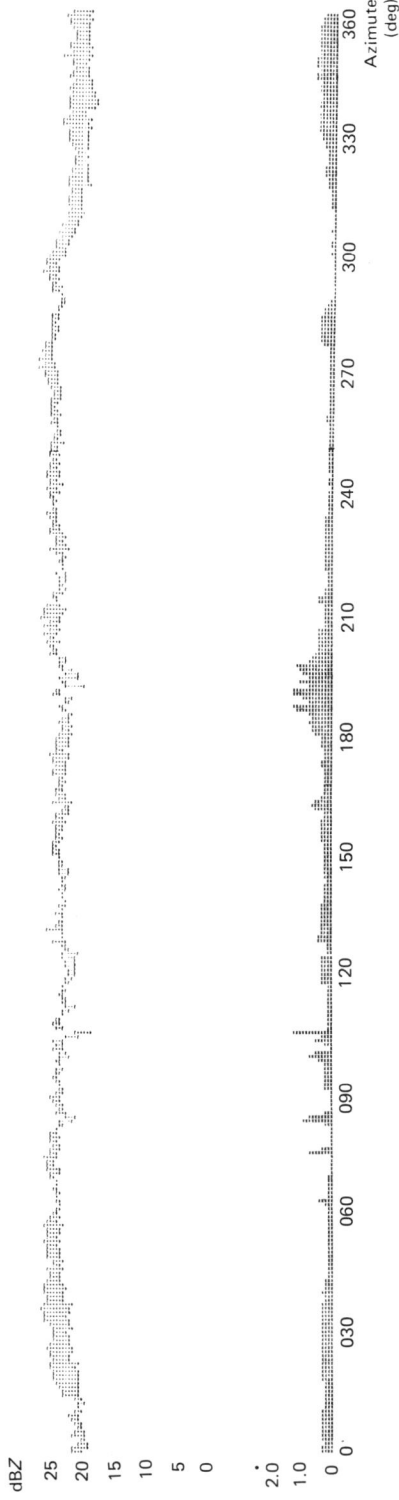

Fig. 3 — The Norrköping radar horizon and the corrected reflectivity at 0.5° elevation and the uncorrected reflectivity at 1.5° showing the means of seven scans. The same comments as made for Fig. 2 apply to Fig. 3.

6. ESTIMATING THE ELEVATIONS OF THE OBSTACLES FROM RADAR DATA

Our data contain reflectivities from one elevation (1.5°) which are not affected by the obstacles and from another elevation (1.0°) which are mostly not affected by them. From these data it is possible to estimate the 'true' vertical reflectivity gradient between these elevations. This was made excluding azimuths with obstacles reaching above 0.5°. Assuming that this gradient also is true between the 1.0° and 0.5° elevations we can estimate the 'true' reflectivity at 0.5°. Substracting this from the actually measured value gives the correction caused by the obstacles. Using the table giving the corrections as a function of the elevation angles of the obstacles we then get an estimate of them. This may be done in different ways:

$$dBZ\_est = dBZ(1.0°) + dBZ(1.0° - -1.5°) \tag{6}$$

$$dBZ\_est = dBZ(1.5°) + 2dBZ(1.0–1.5°) \tag{7}$$

dBZ_est is the estimated dBZ at the 0.5° elevation angle, dBZ(1.5°) is the measured dBZ at the 1.5° elevation angle and dBZ (1.0 − 1.5°) is the measured dBZ difference between the elevation angles 1.0 and 1.5°. Now

$$dBZ\_corr = dBZ\_est - dBZ(0.5°) \tag{8}$$

and h_est is the estimated elevation angle of the obstacle. Therefore h_est is a function of dBZ_corr.

The measured dBZ are simply the arithmetic means over a range of 5 or 10 km computed for every degree azimuth and elevation angle.

The computing differences in dBZ between elevations is not so straightforward. Of course, one difference may be computed for every degree azimuth, but these differences proved very unstable. Therefore two methods were used.

(1) Compute a mean difference over all azimuths, using only data not affected by obstacles. The restriction used was that the elevation of the obstacle, according to the map described earlier, should be less than 0.5°.
(2) Compute mean differences for smaller azimuth intervals, in order to account for possible variations in the vertical reflectivity gradient. This was made by forming nine-point running means, under the same restriction as in (1).

The dBZ values used for 1.0 and 1.5° were actually the corrected values. This simplified the work somewhat since some corrections must be added to the 1.0° elevation data when the elevations of the obstacles are high.

Equation (8) and (9) were used with dBZ (1.0°–1.5°) computed as a mean overall azimuths (method (1) above). Equation (9) was also used with a running mean dBZ(1.0°–1.5°).

What is needed for the actual corrections is dBZ_corr. If we get reliable estimates of this quantity, all that is needed is to add it to the measured reflectivities at the lowest elevation angle to correct for the losses. This correction will then work well beyond the ground echoes and further as the beam does not overshoot the precipitation. Estimating dBZ_corr does not require a detailed topographic mapping of the radar horizon; all one needs to know is which sectors must be avoided in order to get an estimate of the vertical reflectivity gradient and what is the lowest elevation angle that one can use for a scan not affected by the obstacles. Even this is possible to deduce from polar coordinate data (azimuth versus range). Thus the losses may be computed using only radar data.

7. MAGNITUDE OF THE CORRECTIONS

We obtained the following four sets of corrections:

dBZ_corr_t   correction using topographic measurements;

dBZ_corr_0   correction according to equation (8), where dBZ_est is given by equation (6), with the restriction that, if $h > 0.5°$, equation (7) is used, dBZ(1.0–1.5) is the arithmetic mean over all azimuths, excluding those with $h > 0.5$ and h is the elevation angle to the obstacle;

dBZ_corr_1   correction according to equation (8) where dBZ_est is given by equation (7) and dBZ(1.0–1.5) in as above.

dBZ_corr_2   correction according to equation (8) where dBZ_est is given by equation (7) and dBZ(1.0–1.5) is a nine-point running mean.

Some dBZ_correction were negative. This indicates a bad estimate, and the value 0 was assigned to these values.

The magnitudes of dBZ_correction according to the different methods, are shown in Table 1. If we remember that a reflectivity difference of 5 dBZ according to

Table 1 — Frequencies of corrections according to the different methods given, their arithmetic means and their mean absolute differences from method t

	Correction (dBZ)											Mean (dBZ)	Absolute difference (dBZ)
	0	1	2	3	4	5	6	7	8	9	10		
t	158	128	31	17	6	9	0	0	3	0	8	1.66	—
0	215	62	44	13	5	3	2	3	1	1	11	1.33	0.97
1	201	72	47	12	7	3	2	3	1	1	11	1.26	1.13
2	224	67	27	14	7	4	2	4	0	1	10	1.22	0.99

the Marshall–Palmer rain rate — reflectivity relations means a change in the rain rate be a factor of 2, evidently most azimuths are less shadowed than this, but the influence of the obstacles is not negligible. More than 10% of the azimuths have corrections of 3 dBZ or more.

8. ESTIMATING THE ELEVATION ANGLES OF THE OBSTACLES

Even if the elevation angles to the obstacles are not necessary, it is instructive to see how accurate the radar estimates of them are. We get three sets of estimated heights, h_est:

h_est_0 analogous to dBZ_corr_0 described earlier;
h_est_1 analogous to dBZ_corr_1 described earlier;
h_est_2 analogous to dBZ_corr_2 described earlier;

Some h_est values were below 0.00°. This indicates a bad estimate and the value 0.00° was assigned to these values.

The h_est values were then smoothed by forming three-point running means, provided that $h < 0.5°$. The reason for this restriction in that the higher obstacles are nearby buildings or trees with sharp contours that would not be smoothed. All h_values differ somewhat from the topographically mapped obstacle elevations. They generally gave lower obstacle elevations than h in the northern sector, indicating that the antenna points in too great an upward-direction there. If the antenna turntable is tilted, the differences between the topographic and estimated height angles should form a sine curve. Therefore $h–h\_est$ values were computed and a harmonic analysis performed on them.

For all the methods, h_est behaved in a similar manner. The first wave with a 360° oscillation had an amplitude of 0.14° and a phase angle of about 250°. Their equations read

$$\overline{h–h\_est\_0} = 0.034 + 0.14\sin(\text{Az}–253) \tag{10}$$

$$\overline{h–h\_est\_1} = 0.027 + 0.14\sin(\text{Az}–246) \tag{11}$$

$$\overline{h–h\_est\_2} = 0.044 + 0.14\sin(\text{Az}–263) \tag{12}$$

where Az is the azimuth angle and the bars denote the expected values. The equations indicate that the antenna turntable is tilted 0.14° upwards towards about $255° + 90° = 345°$. However, the 180° wave has nearly the same amplitude in all cases, which is difficult to explain. The subsequent waves have much smaller amplitudes.

Our intention was to check these results by

(1) repeating the test using new data and
(2) checking the antenna elevation angles against the sun.

Because of repeated breakdown of the antenna mechanism during the first half-year of 1989, no checks of this type have been possible.

9.   ACCURACY OF ELEVATION ANGLE ESTIMATES

Table 2 gives the mean value of the topographic measurements of the elevation angles and their radar estimates as well as the mean absolute deviations of the radar estimates from the topographic measurements. The mean absolute deviation is given by

$$\text{absolute deviation} = \sum_{0}^{359} \frac{|h–h\_est|}{360}.$$

Table 2 — Means of the elevation angles to obstacles according to the different methods used and the mean absolute deviations of the radar estimates from the topographic measurements

	h	h_est_1	h_est_2	h_est_3
Angle (deg)	0.252	0.217	0.224	0.207
Absolute deviations (deg)		0.135	0.145	0.129

Remembering the antenna elevation errors a random error of about 0.1° and a systematic error of about the same magnitude, one can hardly expect better results. From these tests it is, however, impossible to decide which method is the best.

10.   OPTIMIZING THE ANTENNA ELEVATION

For a weather radar one generally wants a scan with as low an elevation angle as possible. The lowest possible angle is determined by the beam width and the radar horizon. Knowing the elevation angles to the obstacles it is possible to analyse their effects. To do this we simply formed the sum of the corrections, using the topographical angle measurements, for several antenna elevations. It would of course have been equally possible to use the radar estimates of these angles, h_est. When computing the sums of the corrections, obstacles with elevation angles above 0.5° were excluded. The reason is that these obstacles are so high that measurements from their azimuths are lost. Table 3 shows that increasing the antenna elevation

Table 3 — Relative frequency of losses and their arithmetic means as a function of antenna elevation angle

Antenna elevation angle (deg)	dBZ					Mean
	0	1	2	3	4	
0.40	35	37	17	9		1.56
0.45	42	43	10	3		1.28
0.50	55	34	9	0		1.04
0.55	67	27	4	0		0.83
0.60	83	15	1	0		0.60
0.65	90	9	0	0		0.51
0.70	94	5	0	0		0.39

from 0.4° to 0.6° gives a steady decrease in the losses, the means decreasing by more than 0.2 dBZ per 0.5° increase in antenna elevation angle. Increasing the antenna elevation above 0.6 gives a continued decrease in the losses, but at only about half the earlier rate. Thus, an antenna elevation of 0.6° may be regarded as the best possible for the lowest scan, since there is also a price to be paid for increasing the elevation angle. With increasing antenna elevation the beam's height above the ground increases and at a range of 100 km an angle increase of 0.1° corresponds to a height increase of 175 m. This height increase is so small that it would probably be an advantage to change the present lowest elevation angles used for the Norrköping radar, 0.5° and 0.4° (there are several scanning schemes), to 0.6°.

11. CONCLUSIONS

Corrections for reflectivity losses caused by obstacles can be estimated using only radar data. Exact knowledge of the elevation angles of the obstacles is not necessary. Owing to geometric factors, the curvature of the earth and the beam's divergence, the corrections are most effective at fairly short ranges. These stress the importance of the exposure of the antenna. Since an obstacle stands in the way of the lower part of the beam, a measurement behind an obstacle actually refers to a higher altitude than does an unobstructed measurement. Since the wind is usually more height dependent than the reflectivity, Doppler wind measurements are even more affected by obstacles than reflectivity measurements. Although only effective at short ranges, reflectivity corrections are important since radar estimates of precipitation are best at just these ranges.

The use of radar data to estimate the corrections has the advantage that possible errors in the adjustment of the antenna turntable are automatically compensated for.

REFERENCES

Andersson, T., Magnusson, S., Lindström, B. and Karlsson, K-G. (1984) The PROMIS Doppler radar system: design and operational experience. *Proceedings of the 2nd International Symposium on Nowcasting, Norrköping*, 1984. pp. 171–176.

Aniol and Riedl (1979)

Harju, A. E. and Puhakka, T. M. (1980) A method of correcting quantitative radar measurements for partial blocking. *Preprints of the 19th Conference on Radar Meteorology, 15–18 April, 1980, Miami Beach, FL.* American Meteorological Society, Boston, MA, pp. 234–239.

Moores, W. and Harrold, T. W. (1975) Estimating the distribution and intensity of ground clutter at possible radar sites in hilly terrain. *Preprints of the 16th Conference on Radar Meteorology, 22–24 April, 1975, Houston, TX.* American Meteorological Society, Boston, MA, pp. 370–373.

Wessels, H. R. A. (1989) *Occultation diagrams for weather radars*, COST-73 Working Paper 73/WD/66. 7 pp.

11

On the significance of radar wavelength in the estimation of snowfall

A. Giguere and G. L. Austin
McGill Weather Radar Observatory, Box 198, MacDonald College,
Ste-Anne-de-Bellevue, Quebec H9X 1CO, Canada

ABSTRACT

In the measurement of snowfall by radar, the use of an X-band transmitter is expected to enhance significantly the effective range of detection of snow, and its quantitative measurement over the period of a snowstorm. The 10 m antenna of the McGill weather radar provides a beam with a width of less than 1° when an S-band transmitter is used. With an X-band transmitter, it narrows to 0.21° and is more likely to be entirely filled with precipitation. Comparisons of radar data with snowgauge data during two snowstorms, using a different transmitter in each case, show a steady diminution with increases range of the ratio R/G of the radar to the snowgauge data in both cases; the average of these ratios is much closer to unity in the X-band case, but the correlation to the ground data is poor in both cases. The factor C_f by which to multiply all radar data to bring the average of R/G to unity is 7.26 for the 'S-band' storm and 1.96 for the 'X-band' storm. A correction of 6.5 dBZ due to differences between the index of refraction for snow and that for rain, suggested by Crozier, should be taken into account to justify a correction to the data.

1. INTRODUCTION

Whether at one location on the ground or on averaging over a volume by radar, the measurement of snowfall is one of the least accurate physical measurements. The measurement of snowfall by radar is as yet at the stage where more observations are needed to justify its use and to quantify in detail the controlling physical processes. A usual way to enhance its accuracy is to calibrate radar measurement with snowgauge measurements. Among the most significant contributions in the field, Gunn and Marshall (1958), and later Sekhon and Srivastava (1971), eatablished empirical Z–R relationships between the snowfall rate and the radar reflectivity. Carlson and

Marshall (1972), using an X-band radar, measured a ratio of radar to snowgauge data and observed its diminution with increasing range. Wilson (1975), using a C-band radar, found that the radar was underestimating snowfall, even at close range. Boucher (1980) tried to design a nowcasting technique for snowfall using a radar previously calibrated with snowgauges. The intent of this study is to assess the advantage of using an X-band transmitter over an S-band transmitter in the remote estimation of snowfalls. The McGill Weather Radar Observatory is located 25 km west of Montréal downtown; its characteristics are listed in Table 1. The antenna was

Table 1 — Characteristics of the McGill weather radar (location, Ste-Anne-de-Bellevue, Québec (45°26′N 73°56′W), 25 km west of Montréal downtown area; width of the antenna, 10 m)

Characteristic	Symbol	Units	S band	X band
Peak power transmitted	Pt	kW	871	72.4
Pulse duration	τ	ms	1	1
Pulse repetition frequency		Hz	300	300
Wavelength	λ	cm	10.4	3.2
Beam width at 3 dB	θ_ϕ	deg	0.86	0.21
Gain	G	dB	44	51.1

designed to provide a beam with an angular width of less than 1° at −3 dB, when an S-band transmitter is used. If an X-band transmitter is used, the angular width narrows to less than 0.25°. A narrower beam is more likely to be entirely filled with precipitation. In the case of snow, it is crucial because it is generated at a relatively low altitude above the ground. The use of an X-band transmitter is expected to enhance significantly the effective range of detection of solid precipitation, and the quantitative measurement of the accumulation of snow over the period of a snowstorm. To prove that point, radar measurements from two snowfalls in southern Québec are compared in detail with snowgauge records from an exhaustive list of climatological and meteorological stations in the St Lawrence Valley. The first snowstorm was recorded on 20–21 November 1986, using the S-band transmitter. The second snowstorm was recorded using the X-band on 30–31 January 1987. For both events, the Z–R relationship developed by Gunn and Marshall (1958), $Z=2000R^{2.0}$, was used. This 'X versus S-band' comparison was made possible by the decision of the McGill Observatory to transmit on the X band during the winter months (from the end of November to the beginning of April), while continuing to transmit on the usual S band during spring, summer and fall.

2. DATA AND METHODOLOGY

The sensitivity of an X-band radar is significantly greater than a similar S-band system because of the greater scattering at the shorter wavelength, inversely proportional to the fourth power of the wavelength. In the radar equation the enhanced sensitivity due to the wavelength appears in the dependence of the average power P_r scattered back at the antenna on the square of the gain and on the inverse square of the wavelength. The gain is proportional to the inverse square of the wavelength, which should give a factor of $1/\lambda^4$ ($+20.4$ dB), but it is partially cancelled by the diminution in the beam width θ_ϕ, proportional to λ^2 (-10.2 dB). Including the $1/\lambda^2$ dependence ($+10.2$ dB), the overall enhancement should then be proportional to $1/\lambda^4$ ($+20.4$ dB), which would mean detecting precipitation rates lower by a factor of 10.5. However, taking the specific parameters of our modest X-band transmitter (72.4 kW) compared with our large S-band unit (871 kW), the enhancement factor is 1.7 ($+4.65$ dB) (Table 2), with a threefold narrower beam. The measured increase in the antenna gain and loss due to the beam width is actually $+14.2$ dB and -11.6 dB.

Table 2 — Value of the elements of the radar equation

	Valve (dB)		
	S band	X band	X band — S band
$10 \log (P_t)$	89.4	78.6	-10.8
$10 \log (\lambda^{-2})$	19.7	29.9	$+10.2$
$10 \log (\theta_\phi)$	-37.1	-48.7	-11.6
$10 \log (G^2)$	88	102.2	$+14.2$
$10 \log C$	71.2	71.2	0
Correction factor for Z	-60	-60	0
Correction factor for τ	-180	-180	0
Mixer noise	105.6	108	$+2.4$
Loss at the mixer	-2.25	-2.0	$+0.25$
Minimum detectable signal	-4.3	-4.3	0.0
$10 \log (1/r^2)$ (for 200 km)	-106	-106	0.0
Total	-15.75	-11.1	4.65

If snow is considered to be homogeneous mixture of air and ice, then the dielectric constant $|K|^2$ is 0.208. Crozier (1980) has shown that, if the radar is calibrated for rain, the power returned by an equal mass distribution of rain or snow will be 6.5 dB lower in the case of snow, for the same radar volume, owing to the difference in the refractive index factors of the two phases. In the case of the X band, the overall correction factor to use is 1.85 dBZ (6.5 dBZ-4.65 dBZ), which is a multiplication factor of $10^{1.85/20} \approx 1.24$ for the precipitation rate.

A snowgauge measurement is a point on the collection surface, whereas the radar scans continuously the volume of precipitation. To aid the comparison between the snowgauge records and to make the radar data less range dependent because of changing beam height, the 'constant-altitude plan position indicator' (CAPPI) technique is used to map the radar data. Fig. 1 shows radial cross-sections of the CAPPI pattern for the chosen altitude of 1.5 km. Beyond a 100 km, the CAPPI is made of the lowest elevation angle and therefore goes above the prescribed height at a certain range. That range becomes the effective maximum range of observation. Three levels for snow generation are drawn on these figures. If the generation level is at 1.5 km, we see that in both cases the beam is filled with precipitation only in its lower half; if that level is 2.0 km, then the X-band beam is completely inside the volume of precipitation up to an approximate range of 120 km, while the S-band beam, because of its larger angular width, has a significant portion of its volume above the precipitation at a range as close as 80 km. Even when the beam is entirely under the snow generation level, the inhomogeneity in the precipitation can cause the beam to be partially empty. The narrower-width use of the X band is clearly advantageous in this respect.

Ground data are retrieved from the daily records of the climatological and meteorological stations located in the province of Québec within a 200 km range of the radar, including data from the meteorological station of Ottawa International Airport. In both case studies, the stations not recording any data or being located in a known area of ground clutter (radar echoes returned from ground obstacles) were deleted from the list. For each radar volume scan, CAPPI maps at an altitude of 1.5 km were produced. The digitizer recognized 16 levels of reflectivity: anything below 16 dBZ corresponding to no precipitation, and then from 16 to 72 dBZ with an increment of 4 dBZ between each level. The data in polar coordinates were transposed onto two 128^2 Cartesian grids centred at the radar; the first one covers $(384 \text{ km})^2$, for a resolution of 9 km^2 per grid point, and the second map covers $128 \text{ km})^2$ for a resolution of 1 km^2. It was found, however, that averaging over 1 or 9 km^2 does not make a significant difference when widely spread continuous snowfall is concerned, probably because the most significant spatial variation occurs at a linear scale either smaller than 1 km or larger than 3 km. The accumulation process was simple. The time interval between two successive scans was computed in units of decimal hours, and the average precipitation was then accumulated for half the interval on each of the 128^2 points of each map, for both the preceding and the actual scan. For the first and last scans, an arbitrary chosen period of time of 10 min was given to the first preceding and the last following intervals. No attempt was made to accumulate continuously following the storm's mean motion, since both events were large-scale, slowly moving and non-convective snowfalls. The acceptability of that decision was confirmed by the smoothness of the resulting accumulation pattern.

The 20–21 November 1986 (S-band) storm left generally 20–30 cm of snow, according to the reporting ground stations. The 30–31 January 1987 (X-band) storm was less intense and left approximately half the amount of snow of the previous storm. The ground stations over which the total equivalent radar accumulation was more than 0.5 cm are the only stations whose data has been compared with the radar data; this leaves 92 stations for the 30–31 January 1987 event and 72 for the 20–21 November 1986 event. The Cartesian grid of radar data is dense enough to draw contour maps. The 0.5 cm level for the S- and X-band cases are drawn to show the effective range of detection of the radar (Fig. 2). The cutoff range of detectability of signals lower than 24 dBZ is made visible at approximately 80 km by choosing accumulation contours at 2.0 and 3.0 cm for the S band and at 3.0 and 5.0 cm for the

(a)

(b)

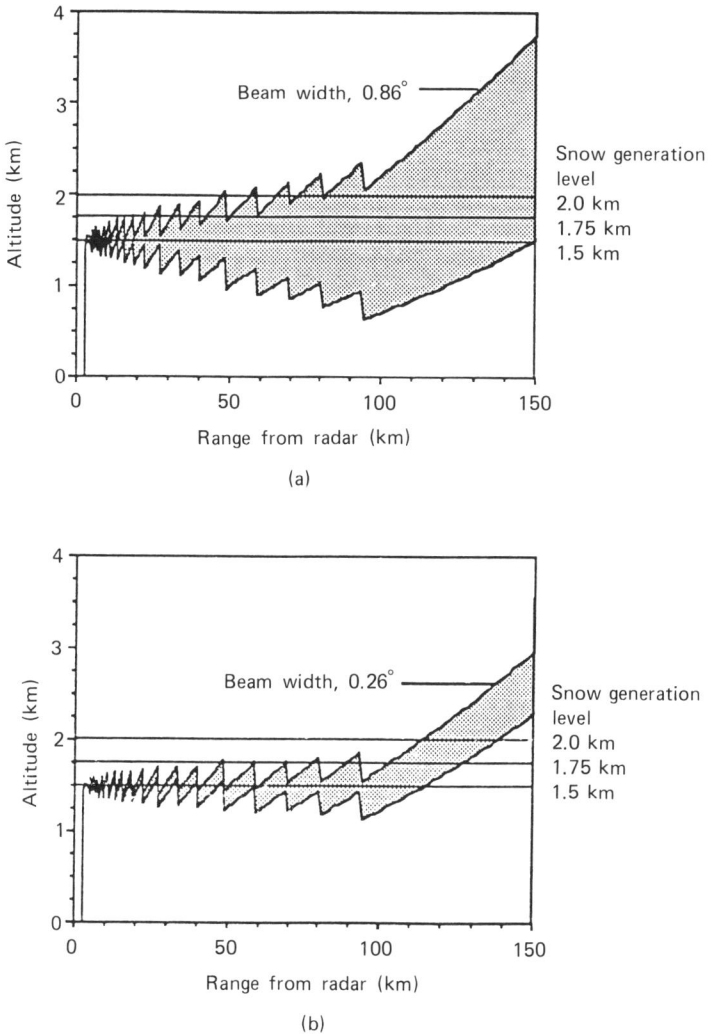

Fig. 1 — Cross-section of a 1.5 km CAPPI using (a) the S band and (b) the X band.

X band (Fig. 3). Note that the limits of a given contour level are closed and continuous, reflecting the continuity and extent of the precipitation. The radar data show more continuity than do the ground data, where two nearby stations can show dramatic differences. While the snowfall of 20 November was more intense, the radar accumulation of snow at the same range is larger on 30 January. It demonstrates that the S band's lack of sensitivity to the lighter snowfall fates has a large effect on the total 'radar' amount of snow accumulated over a long period of time, even in cases characterized by moderate to heavy snowfall rates.

A striking difference between the performances of the X and S bands is visible in comparing the accumulation history at meteorological stations, the equivalent radar

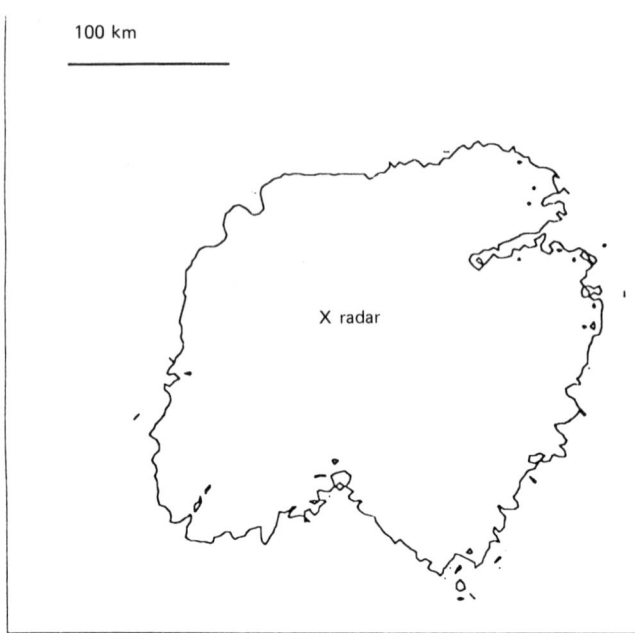

(a)

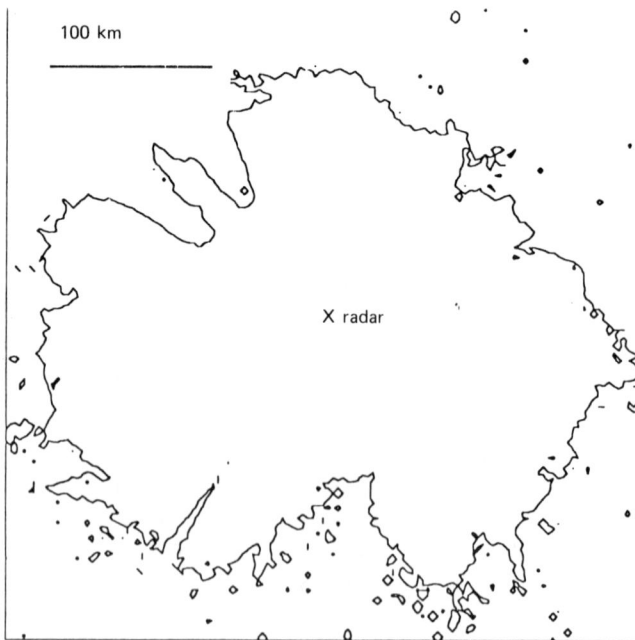

(b)

Fig. 2 — Contour of 0.5 cm accumulation over the entire storm periods, showing the effective ranges of detection: (a) 20–21 November 1986 (S band); (b) 30–31 January 1982 (X band).

(a)

(b)

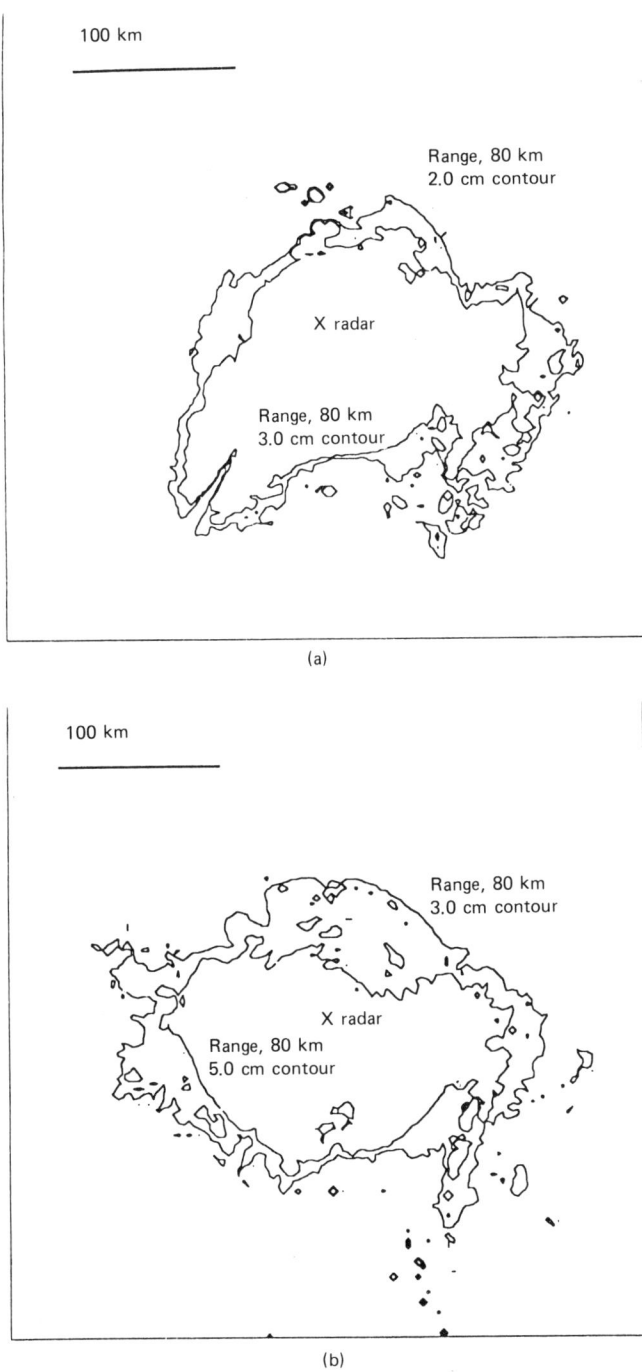

Fig. 3 — Contours of (a) 2.0 and 3.0 cm accumulation over the entire storm period 20–21 November 1986 (S band) and (b) 3.0 and 5.0 cm accumulation over the entire storm period 30–31 January 1987 (X band), showing the 24 dBZ cutoff range at 80 km.

accumulation R with the ground accumulation G and the radar-to-snowgauge ratio R/G with respect to range. The accumulation history at St-Hubert Airport meteorological station is shown in Fig. 4. On 30 January, the radar accumulation follows

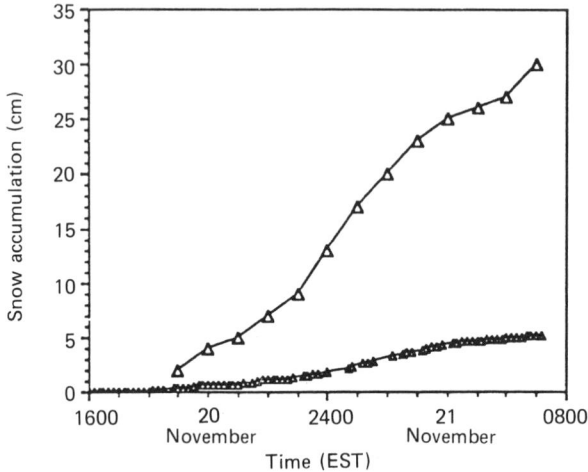

(a)

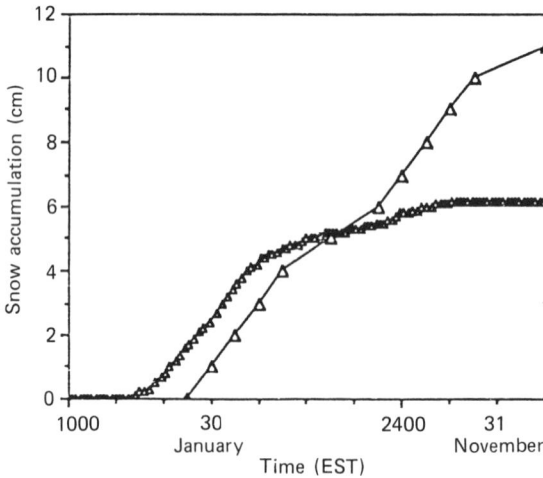

(b)

Fig. 4 — Snow accumulation —▲—, radar equivalent; —△—, ground) at St-Hubert: (a) 20–21 November 1986 (S band); (b) 30–31 January 1987 (X band).

closely the snow accumulation rate on the ground during the first 2 or 3 h of the snowfall. Then, the radar rate of accumulation levels off, particularly starting from 0100 Eastern Standard Time (EST) on 31 January. Throughout the storm, the radar accumulation is approximately half the accumulation observed on the ground. Note also that the radar accumulation starts as much as an hour and a half before the actual ground accumulation, for any event. This suggests the presence of a storm's leading edge at altitude.

The equivalent radar accumulation R is done on the 9 km^2 resolution grid (Fig. 5). The correlation of the radar measurement with the ground truth is poor in both cases. The points are widely scattered and far below the 1:1 line, but the 30–31 January X-band case is clearly closer than the 20–21 November S-band case to the 1:1 fit. A best linear fit is drawn through both sets of data. The slope through the S-band data is 5.62×10^{-2} and the correlation coefficient r is 0.032. The slope through the X-band data is 0.28 and the correlation coefficient is 0.11.

The ratio of the radar accumulation R to the ground accumulation G for each of the above stations, plotted against range, shows a steady diminution of that ratio with increasing range (Fig. 6). The range 80 km is indicated; beyond it, the radar reflectivity signals equal to or lower than 24 dBZ are not seen, and very low radar-accumulation-to-ground-accumulation ratios are predominant. A linear fit of the diminution of the ratio R/G with the range x is drawn. If the radar and ground data were perfectly correlated, the slope b would be zero and the intercept a on the R/G axis would be units. An attempt was made to use the linear fit to correct the diminution with increasing range. A corrected value R' for each radar datum was computed using the slope of the linear fit and the assumption above that perfect correlation means that the slope b' of the fit through the modified data must be zero. The value of the intercept is kept since it reflects a systematic error at all the sites, and not a dependence on the range. In both situations the fit of the data to an ideal 1:1 is greatly enhanced, with a slope $b'=0.17$ and $b'=0.47$ respectively for the S-band and the X-band data (Fig. 7). The correlation coefficients r' are 0.4 and 0.37 in the same order (one must remember that these coefficients are a measure of how well the data are correlated to the best fit and not to the ideal 1:1 fit). In both cases the dispersion of the data remains large and cannot be attributed solely to a diminution of the radar sensitivity with increasing range, different from the $1/r^2$ dependence already taken into account.

Two statistical parameters have been computed (Table 3) in an attempt to evaluate the accuracy of the radar measurements: the radar calibration factor C_f

$$C_f = \sum_i G_i \bigg/ \sum_i R_i$$

(i.e. the ratio of the sum $\sum_i G_i$ of the snowgauge amounts to the sum $\sum_i R_i$ of the radar equivalents) and the mean absolute difference AD given by

$$AD = \left(100 \sum_i \frac{|G_i - R_i|}{N}\right)\bigg/ G$$

(i.e. the sum of the point differences irrespective of their sign).

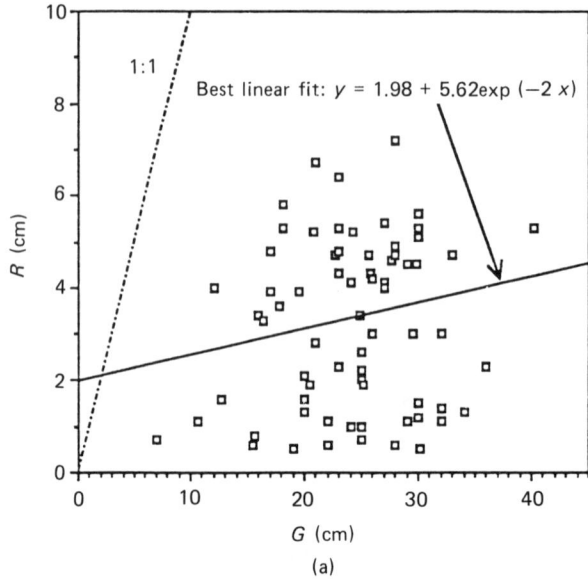

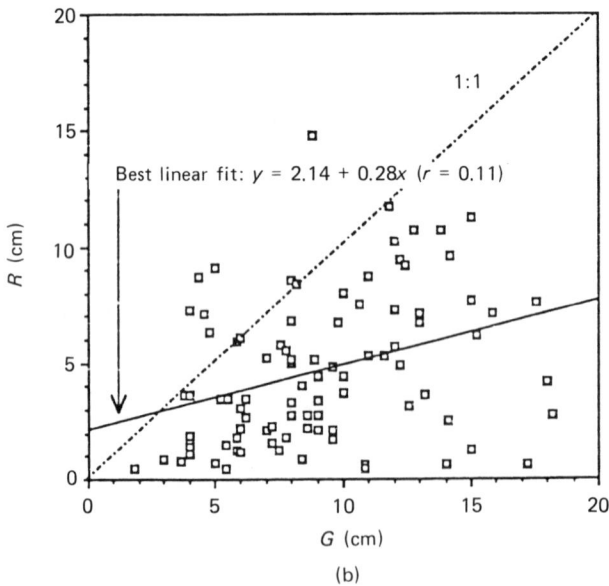

Fig. 5 — Comparison of the equivalent radar accumulation of snow *R* with ground measurement *G* at the stations where $R \geqslant 0.5$ cm: (a) 20–21 November 1986 (S band); (b) 30–31 January 1987 (X band).

AD is calculated a second time using each radar value multiplied by the radar calibration factor C_f to see what would be the reduction in AD brought by calibrating

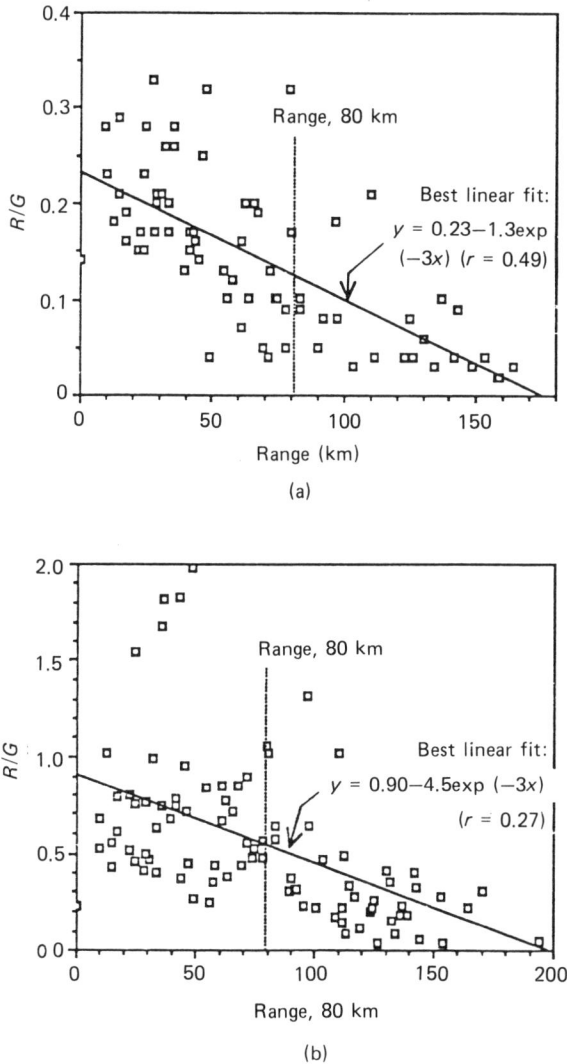

Fig. 6 — Ratio R/G with respect to distance, at the stations where $R \geqslant 0.5$ cm: (a) 20–21 November 1986 (S band); (b) 30–31 January 1987 (X band).

a priori the radar measurements. A second set C_f' and (AD') are calculated with the value of the radar data corrected with the slope of the linear fit to the ratio R/G plotted against the distance (cf. Figs 6(a) and 6(b)). C_f can be interpreted as the factor by which we would have to multiply the radar data in each case to bring to unity the mean of the ratios R/G. The underestimation of the onset of the snowfall is striking in the case of the S band, where C_f is equal to 7.26, compared with the X-band event with C_f equal to 1.96. The latter corresponds well to the factor

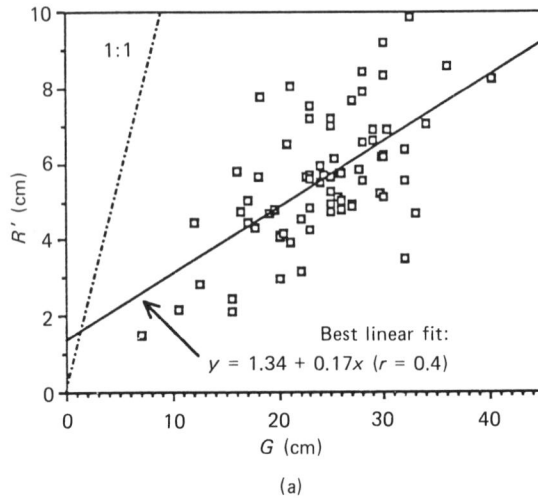

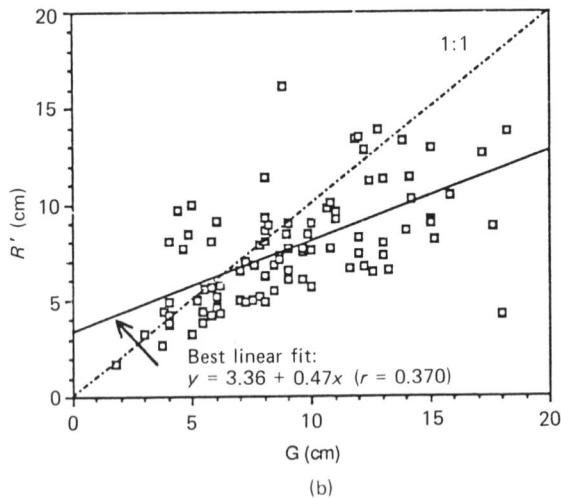

Fig. 7 — Corrected radar accumulation of snow R' versus ground accumulation G: (a) 20–21 November 1986; (b) 30–31 January 1987 (X band).

qualitatively estimated as 2 in Fig. 7(b). Weighting the radar data with respect to distance brings C_f to 4.35 for the S band and to 1.20 for the X band. The drop in the value of the mean absolute difference AD from 86.23 to 48.96 brought about by multiplying the radar data by the radar calibration factor C_f is spectacular in the case of the S band, compared with the X band (a drop of from 54.37 to only 50.87). The same remark applies for (AD'), using the corrected values: AD drops from 77.03 to 18.78 for the S band and from 27.63 to 25.67 for the X band.

Table 3 — Statistical parameters

	S band 72 stations	X band 92 stations
C_f	7.26	1.96
AD (%)	86.23	54.37
AD (C_f (%))	48.96	50.87
C_f'	4.35	1.20
(AD)' (%)	77.03	27.63
(AD)' (C_f) (%)	18.78	25.27

3. CONCLUSION

Two stratiform and continuous snowfalls have been observed at the McGill Weather Radar Observatory using two different wavelengths on the radar transmitter. As expected, the accuracy of the radar measurements of the snow accumulation is poor, on either band, and they significantly underestimate the total accumulation. For the X band, the total amount of snow on the basin is underestimated by a factor $C_f = 1.96$, after comparison with ground measurements from 92 meteorological and climatological stations; for the S band, $C_f = 7.26$ using 72 ground stations for ground truth. The mean of the point sum of the absolute differences is also found to be smaller in the first case than in the second. If the radar data are corrected by using C_f, then AD is smaller for the S band than for the X band. Another adjustment of the radar data tried was to add a correction weight according to the distance of the data from the radar site; this correction is based on a linear fit to the decrease in the ratio of the radar data to the ground data. The effective range of observation is much larger in the X-band case than for the S-band case. For lesser intensity, the X-band beam detects more snow at an equivalent range. In the area northwest of the radar site, the snowfall on 20–21 November 1986 was comparable in intensity were the snowfall on 30–31 January 1987; the former was not detected by the radar on the S band, wherever on the X band the radar gave an estimate comparable with those in other areas. It can be concluded, in the case of the January snowfall, that the level of snow generation is higher than 2 km, since the observation range extends far beyond the radius where the CAPPI looks higher than its prescribed level. The same conclusion cannot be drawn about the November snowfall, because the CAPPI rises above 1.5 km at a range where the radar accumulation decreases sharply. It could be caused not by the fact that the generation level is low but simply because the S-band beam is too large beyond 150 km to be properly filled by the precipitation. However, the main goal of this study has been reached. It has been shown that the use of an X-band transmitter does improve the remote quantitative estimation and qualitative observation of snowfalls, compared with the use of an S-band transmitter, which is generally used on weather radars. The results also emphasize the value of the very narrow beam width (0.21°) associated with the use of the large diameter (10 m) of the antenna and the X-band wavelength. This allows complete beam filling to a significantly greater range. The accumulation history over a station, for any band, suggests the presence of a leading edge at altitude of the precipitation pattern since the radar accumulation begins earlier, for every case. Appropriate vertical cross-sections through the storms are needed to investigate further these structural patterns. The combination of a correction of 6.5 dBZ, suggested by Crozier (1980), with the difference of 4.65 dB in sensitivity in favour of the X band, could bring the average ratio of radar to ground accumulation close to unity in the January case. Whether this is a coincidence or whether there is an actual physical justification in doing so must be investigated further by studying more snowfalls on the X-band transmitter.

REFERENCES

Austin, P. M. (1987) Relation between measured radar reflectivity and surface rainfall. Mon. Weather Rev. **115** (5) 1053–1071.

Battan, L. H. (1973) *Radar observation of the atmosphere*. University of Chicago Press, Chicago, Il.

Bellon, A. and Austin, G. L. (1984) Accuracy of short term radar forecasts. *J. Hydro.*, 70, 35–49.

Boucher, R. J. (1980) Snowfall rate obtained from radar reflectivity within a 50 km range. *Preprints of the* 20*th Conference on Radar Meteorology*. American Meteorological Society, Boston, MA, Paper 3C.1, pp. 271–275.

Carlson, P. E. and Marshall, J. S. (1972) Measurement of snowfall by radar. *J. App. Meteorol.* **11**, 494–499.

Crozier, C. L. (1980) *Radar measurement and display of snowfall*, Report No. APRB-108-P32. Environment Canada, Atmospheric Environment Service.

Gunn, K. L. S. and Marshall, J. S. (1958) The distribution with size of aggregated snowflakes. *J. Meteorol.* **15**, 452–461.

Marshall, J. S. (1953) Precipitation trajectories and patterns. *J. Meteorol.*, **10**, 25–29.

Marshall, J. S. and Palmer, W. M. (1948) The distribution of raindrops with size. *J. Meteorol.*, **5**, 165–166.

Sauvageot, H. (1976) Observation directe d'une cellule génératrice de traînée de précipitation-soumise à un cisaillement de vent. *J. Rech. Atmos.* **10** (2), 119–222.

Sauvageot, H. (1981) *Radarmétéorologie*. Editions Eyrolles, Paris.

Sekhon, R. S. and Srivastava, R. C. (1971) Snow size spectra and radar reflectivity. *J. Atmos. Sci.*, **27**, 299–307.

Wilson, J. W. (1975) Measurement of snowfall by radar during the IFYGL. *Preprints of the* 16*th Conference on Radar Meteorology*. American Meteorological Society, Boston, MA, pp. 508–513.

12

Ground clutter rejection for weather radar and weather Doppler radar

K. Hamuzu and M. Wakabayashi
Radar Equipment Department, Mitsubishi Electric Corporation, Amagasaki, Japan

ABSTRACT

This paper describes techniques currently used for effective rejection of ground clutters, an obstacle to the measurement of precipitation and observation of moving rainfall areas by radars. The paper also shows that the weather radars and weather Doppler radars with these techniques have produced good results by rejecting ground clutter satisfactorily.

1. INTRODUCTION

Effective rejection of unnecessary echoes such as ground clutters, etc., other than rainfall echoes is very important for the precise measurement of precipitation and sufficient observation of moving rainfall areas by radars. Large ground clutter areas existing within a radar coverage may cause serious error in precipitation measurement. For a weather Doppler radar used at a smaller elevation angle, ground clutter may obscure the rainfall-based Doppler velocity measurement, thereby making it difficult to measure wind speed near the ground surface. The rejection of ground clutter is therefore an important problem for precise measurement by weather and Doppler radars.

First, the following methods may be used to reject clutter related to the signal amplitude measurement of the echoes.

(1) Reduce the clutter signal level by using a narrow-beam antenna.
(2) Subtract the clutter signals recorded on fine days from the measured results obtained on rainy days.
(3) Reject clutter signals based on statistically analysed characteristics of their amplitude.

(4) Change the polarization of radar waves.

The above methods can be summarized as follows.

(1) This is the simplest and most effective method. However, the reduction in the clutter area means that the widths of radar pulses and radar beams must be narrowed, thereby imposing many restrictions on the system design. Nonetheless, this method is the most effective means for suppressing clutter signals. In addition, a narrow radar beam is usually used for a weather radar. For these reasons, the effects given by this method have been widely utilized.
(2) This method is effective only when the clutter level is constant. In this method, however, the measurement accuracy is affected adversely, since the actual level fluctuation is noticeable.
(3) This method rejects clutter signals by making use of a difference between the reflected signal from rainfall and ground clutters from mountains, etc. The signal reflected from rainfall is Rayleigh scattered, whereas that of the ground clutter is dispersed differently. This method, which can be implemented using relatively simple hardware and work effectively, is an excellent means of rainfall measurement.
(4) This method is based on the fact that there is a difference in reflected signals between the circularly polarized and linearly polarized waves against some targets. For the ordinary rainfall, however, reflected signals of the circularly polarized wave are attenuated remarkably. Thus, this method is unsuitable for rainfall measurement.

Secondly, in a Doppler radar used for the observation of moving rainfall area, clutter signals are rejected on the basis of the statistical features of their Doppler deviation. That is, from a viewpoint similar to that of the rainfall measurement shown in (3) above, the Doppler deviation of rainfall area, a moving target, is identified and the moving target only is extracted to reject clutter signals.

A schematic description will be given of the ground clutter rejection for weather radars and weather Doppler radars based on the concept shown in (3). Newly developed ground-clutter-rejecting systems will be introduced and the excellent results obtained described.

2. GROUND CLUTTER REJECTION FOR WEATHER RADAR

The amplitude of rainfall echo signals envelope detected through a linear intermediate-frequency receiver is Rayleigh distributed. On the assumption that the amplitude distribution between transmitted pulses is a frequency spectrum with regard to the frequency axis, besides the DC components used as the echo intensity in ordinary radar, AC components such as white noise are widespread. This means that the correlation between rainfall echo pulses to pulses can actually be considered to be null. On the other hand, although the ground clutter as a slight fluctuation, its frequency spectrum is concentrated in a lower range, showing that there is much correlation between transmitted pulses.

Let the amplitude of a reflected radar echo signal be $v(t)$, then the average value, root mean square value and dispersion will be given by the following equations:

average value (DC voltage)

$$\overline{V(t)} = \lim_{T \to \infty} \frac{1}{T} \int_0^\infty v(t)\, dt \tag{1}$$

root mean square value (all power)

$$\overline{V(t)^2} = \lim_{T \to \infty} \frac{1}{T} \int_0^\infty v(t)^2\, dt \tag{2}$$

dispersion (AC power)

$$\overline{[V(t) - \overline{V(t)}]^2} = \lim_{T \to \infty} \frac{1}{T} \int_0^\infty [v(t) - \overline{v(t)}]^2\, dt = \sigma_V^2. \tag{3}$$

If $v(t)$ is applied to a single rejecting filter consisting of a subtracter and a delay line with the delay time T of one transmission cycle, the output $V(t)$ and its dispersion σV^2 will be given by the following equations (Tatehira 1980):

$$\sigma_V^2 = \overline{V(t)^2} - \overline{V(t)}^2 \tag{4}$$
$$V(t) = \overline{v(t) - v(t - T)} = \overline{v(t)} - \overline{v(t - T)} = \tag{5}$$

and therefore

$$\sigma_V^2 = \overline{V(t)^2}$$
$$= \overline{[v(t) - v(t - T)]^2} = \overline{v(t)^2 - 2v(t)v(t - T) + v(t - T)^2}$$
$$= 2\overline{v(\tau)^2}\left(1 - \frac{\overline{v(t)\,v(t - T)}}{\overline{v(t)^2}}\right)$$
$$= 2\sigma_v^2(1 - \rho_T) \tag{6}$$

where σ_v^2 is the filter input dispersion and ρ_T is the correlation coefficient of $v(t)$ at T (dispersion normalized). This means that all the power of the filter output consists of the AC power σ_v^2 only, i.e. a function of the correlation coefficient between the filter input AC power and input signal. It will be the input AC power itself, if no correlation with the input signal exists. In other words, rainfall echoes can be picked up with ground clutter which is largely composed of DC components being rejected.

The autocorrelation $\rho(s)$ between components at a time interval s corresponding to the above-mentioned reflected radar signal $v(t)$ is given by

$$\rho(s) = \lim_{T \to \infty} \frac{1}{T} \int_0^\infty v(t) v(t+s) \, dt. \tag{7}$$

On the other hand, it can be expressed as follows in terms of a power spectrum:

$$X_T(\omega) = \int_{-\infty}^\infty v_T(t) \exp(j\omega t) \, dt \tag{8}$$

$$G_X(\omega) = \lim_{T \to \infty} \frac{1}{T} |X_T(\omega)|^2 \tag{9}$$

where $X_T(\omega)$ is the Fourier transformation partial waveform of the signal $v(t)$ within $0 < t < T$ and $G_X(\omega)$ is the mean square value of $X_T(\omega)$. There is a Fourier transformation relationship between the autocorrelation $\rho(s)$ and the power spectrum $G_X(\omega)$, i.e.

$$G_X(\omega) = \int_{-\infty}^\infty \rho(s) \exp(-j\omega s) \, dt \qquad \text{in terms of frequency} \tag{10}$$

$$\rho(s) = \frac{1}{2\pi} \int_{-\infty}^\infty G_X(\omega) \exp(j\omega s) \, d\omega \qquad \text{in terms of time base.} \tag{11}$$

Fig. 1 shows $G_X(\omega)$ and $\rho(s)$ versus correlation of reflected radar signal $v(t)$. It is obvious from the figure that the frequency spectrum is concentrated in a lower region when more correlation exists. Radar echoes reflected from the ground clutter, where there is far more correlation between transmission cycles compared with rainfall echoes, can be rejected by eliminating this lower-frequency region through a cancellor (high-pass filter).

Fig. 2 shows examples of actually measured frequency spectra of rainfall echoes and ground clutter (Aoyagi 1983). It is obvious from this figure that the frequency spectrum of rainfall echoes is wide. On the other hand, that of the ground clutter is concentrated in a lower region. This agrees with the characteristics shown in above. Table 1 shows the statistical characteristics of these signals obtained from the actually measured results (Simizu et al. 1977).

After examining the above results, we made ground-clutter-rejecting systems which mainly consist of high-pass filters. Fig. 3 illustrates the system composed of

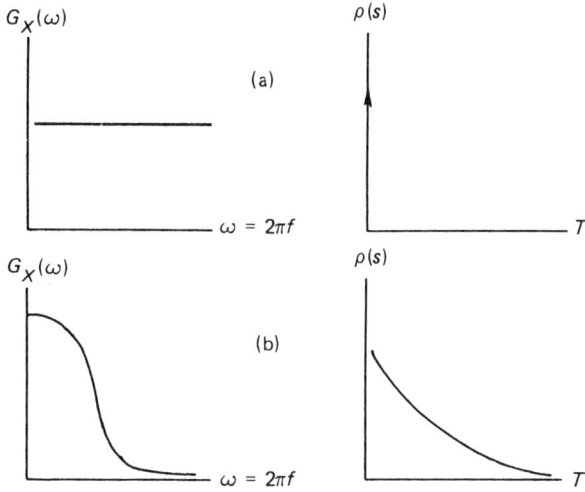

Fig. 1 — Characteristics of $G_X(\omega)$ (mean square value of the Fourier-transformed partial waveform of the radar echo signal) and $\rho(s)$ (autocorrelation of the radar echo signals): (a) more correlation; (b) less correlation.

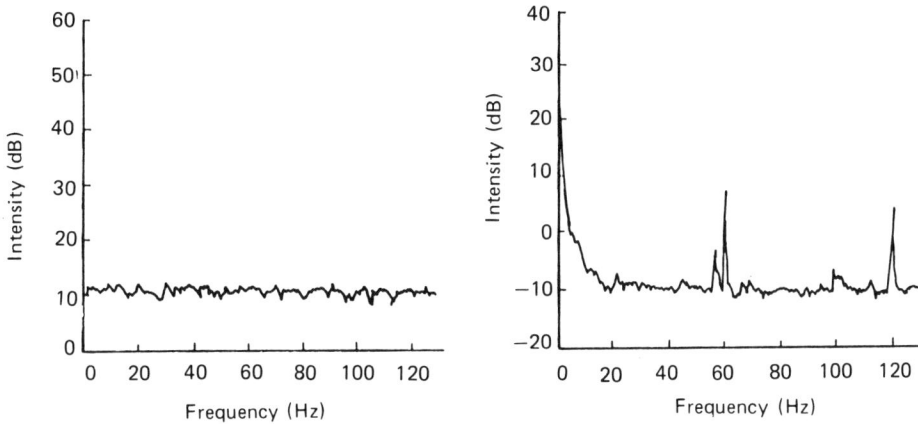

Fig. 2 — Frequency spectrum of (a) rainfall echoes and (b) ground clutter. (The amplitude distribution is with respect to the frequency axis.)

cancellers with a delay line for elimination of DC components. This canceller is called the MTI filter. For the filter, triple-feedback-type digital filters (a combination of double-feedback- and single-feedback-type filters), are used to enhance the rejection characteristics.

Table 1 — Statistical characteristics of rainfall echoes and ground clutter

	Rainfall echo	Ground clutter
AC power/all power	1/4.33	1/522.5
DC power/all power	1/1–1/4.33	1/1–1/522.5
Correlation factor between pulse to pulse	0.19	0.9974
Band width (Hz)	88.6	3.5

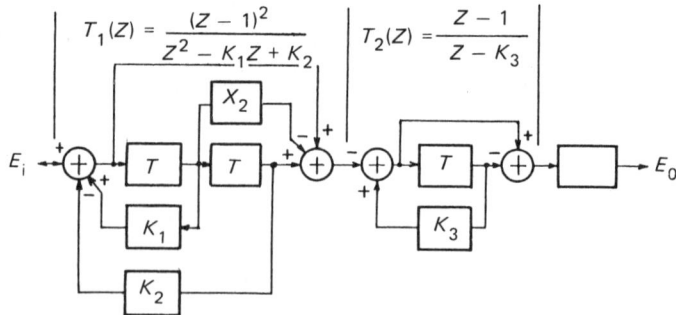

Fig. 3 — A block diagram of a ground clutter canceller for a weather radar.

3. GROUND CLUTTER REJECTION FOR WEATHER DOPPLER RADAR

Studies on frequency deviation due to radar antenna scans as well as reflected radar signals themselves must be made to implement the ground clutter rejection for a Doppler radar.

The Doppler frequency spectrum of rainfall echoes is of Gaussian distribution with the average frequency as the centre. The distribution of rainfall echo areas, which spread in space, is assumed to be constant whether the antenna scans or stops. On the other hand, the ground clutter, which is reflected from fixed and still objects, has a Doppler average velocity of 0 m/s. However, trees nodding in the wind and distortion of radar waves during propogation cause the amplitude of reflected radar signals to fluctuate, resulting in a frequency spectrum approximated by a Gaussian distribution. The ground clutter, unlike rainfall echo areas, is not distributed continuously, so that it is affected by antenna scans. In other words, the clutter is amplitude modulated with antenna patterns and appears directly in the radar received signals.

The Doppler frequency spectrum $A_r(f)$ of a rainfall echo is expressed as follows using a Gaussian distribution (Aoyagi 1986):

$$A_r(f) = A_R \exp\left(\frac{-(f - \bar{f})^2}{2\sigma_{rf}^2}\right) \tag{12}$$

where $\bar{f}$ is the average frequency and σ_{rf} the frequency dispersion. Express f, $\bar{f}$ and σ_{rf} in terms of the velocity v; then

$$v = \frac{f\lambda}{2}, \qquad \bar{v} = \frac{\bar{f}\lambda}{2}, \qquad \sigma_{rv} = \frac{\sigma_{rf}\lambda}{2} \tag{13}$$

where λ is the radar wavelength.

The Doppler frequency spectrum of ground clutter is related to antenna scans and the fluctuation of clutter amplitude. The antenna scan-caused frequency spectrum $A_{ga}(f)$ is given by

$$A_{ga}(f) = A_{GA} \exp\left(\frac{-f^2}{2\sigma_{af}^2}\right). \tag{14}$$

σ_{at}^2 is the dispersion due to antenna scans and given by

$$\sigma_{at}^2 = \left(\frac{\theta_B}{12N\sqrt{2\ln 2}}\right)^2 \tag{15}$$

where θ_B (deg) is the antenna beam width and N (rev/min) is the antenna rotation frequency; then σ_{af}^2 is expressed as follows:

$$\sigma_{af}^2 = \frac{1}{4\pi^2\sigma_{at}^2} = \left(\frac{6N\sqrt{2\ln 2}}{\pi\theta_B}\right)^2. \tag{16}$$

The ground clutter frequency spectrum including that due to clutter amplitude fluctuation is given by modulating the time series signals of clutter amplitude fluctuation with those of the fluctuation due to antenna scans. Thus, the frequency spectrum σ_{sum}^2 dispersion due to clutter amplitude fluctuation during antenna scans is as follows:

$$\sigma_{\text{sum}}^2 = \sigma_{af}^2 + \sigma_{cf}^2 \qquad (17)$$

where σ_{cf}^2 is the dispersion due to clutter amplitude fluctuation.

Fig. 4. shows an example of the actually measured ground clutter frequency

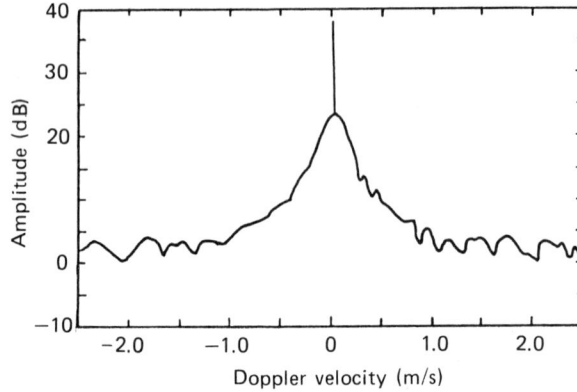

Fig. 4 — Ground clutter Doppler frequency spectrum ($\sigma = 0.29$ m/s).

spectrum (Aoyagi 1986). In this example, the velocity in the centre is 0 m/s, and the standard deviation of the fluctuation is 0.29 m/s. Fig. 5 shows the frequency spectra

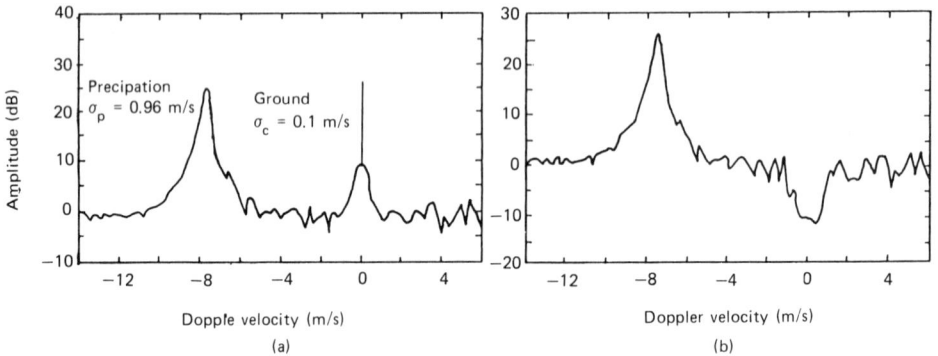

Fig. 5 — Doppler frequency spectra (a) due to rainfall echoes and ground clutter (without filter; $v = -6.3$ m/s) and (b) where ground clutter has been rejected from (a) (with filter; $v = -7.84$ m/s).

due to rainfall echoes and ground clutter measured simultaneously in a volume of radar echoes (Aoyagi 1986). In Fig. 5(b), the Doppler average velocity of rainfall echoes is -7.84 m/s, whereas in Fig. 5(a) it is reduced to -6.3 m/s owing to the ground clutter interference. It is clear from this phenomenon that the ground clutter affects rainfall measurement, although its Doppler average velocity is insignificant. In the above example, the measurement was performed with the antenna stopped. The effect caused by antenna scans is given as $\sigma_{af} = 1.5$ Hz, letting N (rotation frequency) be 1 rev/min and the antenna beam width $\theta_B = 1.5°$, or 4.3 cm/s ($\lambda = 5.7$ cm) in terms of velocity.

The frequency spectrum of ground clutter is approximated by a Gaussian distribution with 0 Hz as the centre frequency. On the other hand, it is obvious from the results described previously that rainfall echoes are approximated by a Gaussian distribution with any centre frequencies. We made a ground-clutter-rejecting system for a Doppler radar, taking these characteristics into account. This system is also so structured that it can produce the output in which a frequency spectrum of rainfall echoes which centre around a frequency of 0 Hz will be attenuated as little as possible.

Fig, 6. illustrates this system which chiefly consists of elliptical filters. There are

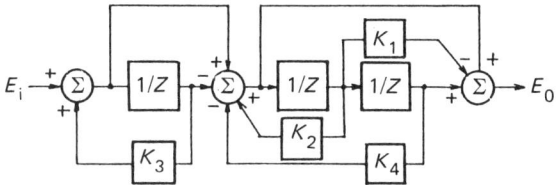

Fig. 6 — A block diagram of the clutter canceller for a weather Doppler radar:
$$\frac{E_0}{E_i} = \frac{(Z-1)(Z^2 - K_1 Z + 1)}{(Z - K_4)(Z^2 - K_2 Z + K_3)}.$$

other filters of this type, e.g. the Butterworth filter and Chebyshev filter. This system uses elliptical filters that provide the most excellent cutoff characteristics. For the elliptical filter, the sharp cutoff characteristics are obtained by allowing ripples in the pass band and cutoff area. Any of the cutoff velocities 0.5, 1.0 and 1.5 m/s can be selected.

4.   EXAMPLES OF GROUND-CLUTTER-REJECTED RESULTS

We now give examples of ground-clutter-rejected results obtained using actual radars with built-in ground-clutter-rejecting systems. Fig. 7 shows an example of the A-scope video measured with the Khepupara weather radar (S band) in Bangladesh. This figure shows that the ground clutter was rejected without affecting rainfall echoes. It is also observed from this figure that angle echoes that had happened were eliminated completely. The plan position indicator (PPI) scope picture of this example is illustrated in Fig. 8. Fig. 9 shows the result obtained from the radar echo digitizing system of the Nagoya weather radar (C band) of the Japan Meteorological Agency. Fig. 10 depicts a bright display example of the Shiratakayama radar

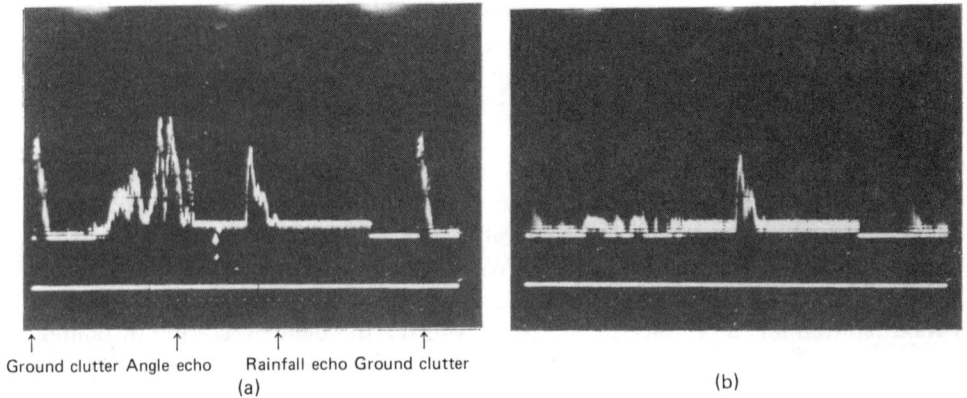

Ground clutter Angle echo   Rainfall echo Ground clutter
                        (a)   (b)

Fig. 7 — An example of the A-scope video of the Khepupara weather radar (S band) in
Bangladesh (May 1988): (a) including clutter; (b) after clutter has been rejected.

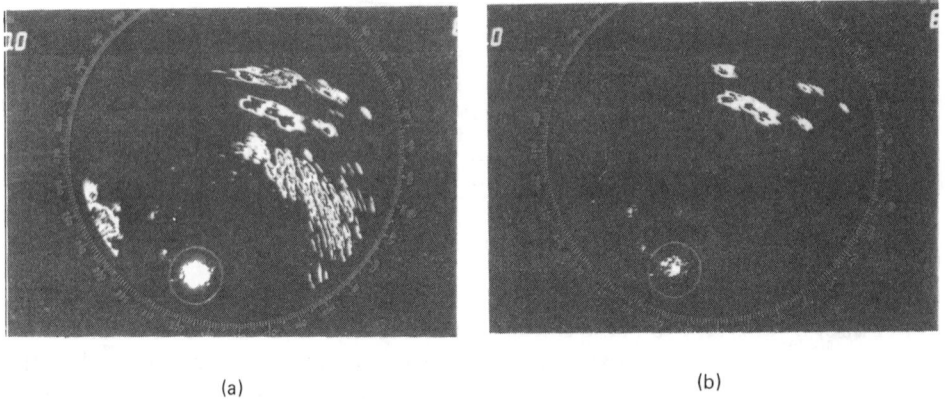

                        (a)   (b)

Fig. 8 — The PPI scope picture of the example in Fig. 7 (May 1988): (a) including clutter; (b)
after clutter has been rejected.

raingauge system (C band) of the Ministry of Construction, Japan. It is obvious from
these examples that ground clutters were rejected satisfactorily without affecting
rainfall echoes.

   Fig. 11 shows an example of the ground clutter rejection of the weather Doppler
radar (C band) of the Meteorological Research Institute, Japan. In Fig. 11(a),
showing the Doppler velocity distribution, white spots observed near the centre of
the upper left and left sections indicate the ground clutter fluctuation. Fig. 11(b)
depicts the Doppler velocity distribution after ground clutter rejection. This figure
shows clearly that the ground clutter fluctuation observed in Fig. 11(a) is eliminated.

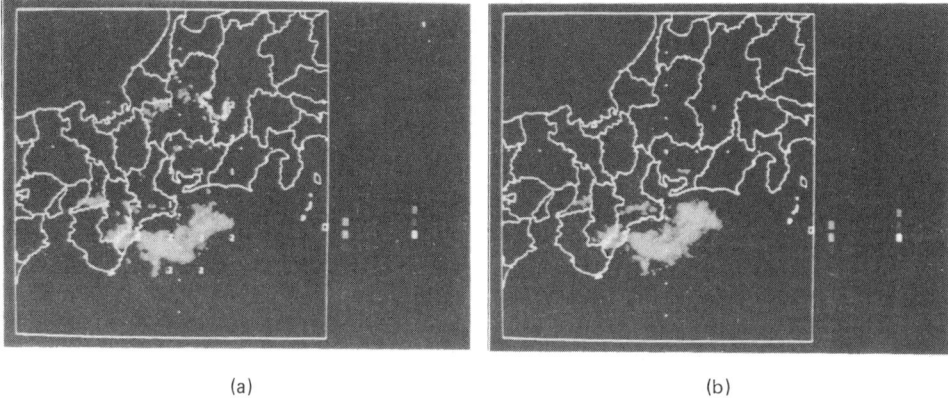

(a)   (b)

Fig. 9 — The result obtained from the radar echo digitizing system of the Nagoya weather radar (C band) of the Japan Meteorological Agency (April 1984): (a) including clutter; (b) after clutter has been rejected.

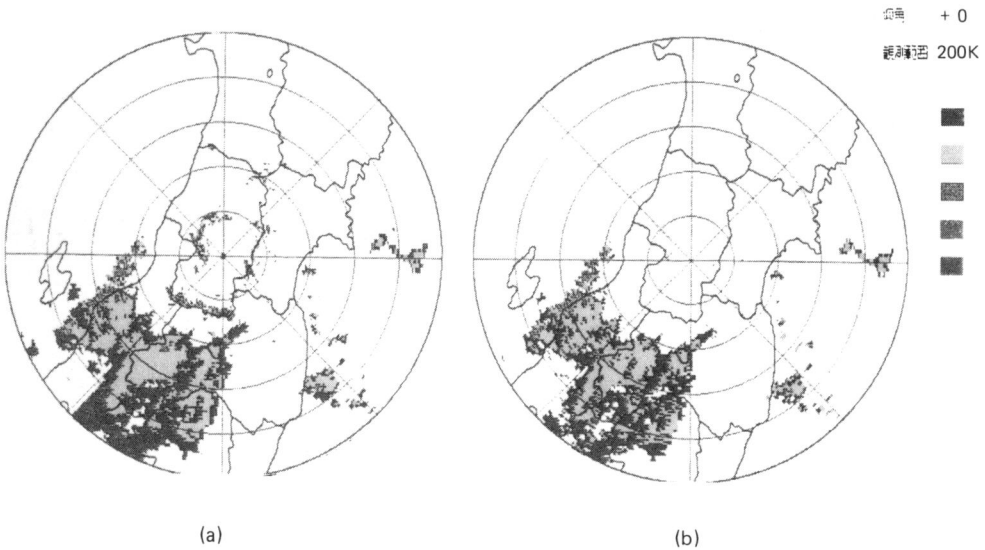

(a)   (b)

Fig. 10 — A bright display example of the Shiratakayama radar raingauge system (C band, of the Ministry of Construction, Japan (October 1988): (a) including clutter; (b) after clutter has been rejected.

5. CONCLUSION

This paper gave schematic descriptions of ground clutter rejection techniques for weather radars and weather Doppler radars. The paper also exemplified the result obtained from the ground-clutter-rejecting systems that have been made actually and used for various radars. From these examples, it is shown that ground clutter is rejected satisfactorily and rejection techniques are indispensable to the precise observation of rainfall areas.

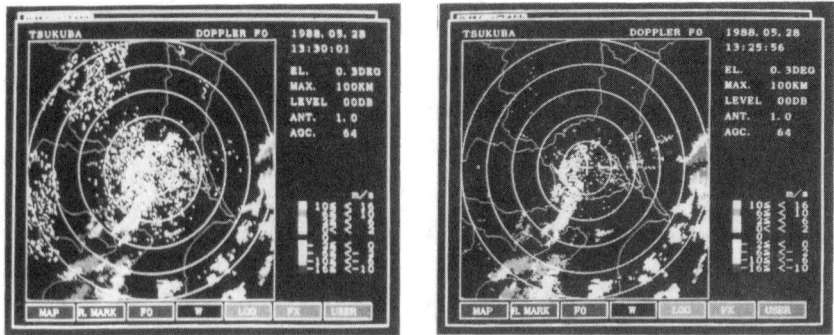

(a)   (b)

Fig. 11 — An example of ground clutter rejection of the weather Doppler radar (C band) of the Meteorological Research Institute, Japan (May 1988): (a) including clutter; (b) after clutter has been rejected.

ACKNOWLEDGEMENTS

The authors would like to express their gratitude to Dr Jiro Aoyagi of the Meteorological Research Institute, Japan, for his valuable guidance and advice on the ground clutter rejection theory throughout the work and Dr Fumio Yoshino of the Public Works Research Institute, Ministry of Construction, Japan, who also gave useful advice and the opportunity for reporting this paper. The authors also thank the personnel of the Ministry of Construction and the Meteorological Agency who offered invaluable data.

REFERENCES

Aoyagi, J. (1983) A study on the MTI weather radar system for rejecting ground clutter. *Papers Meteorol. Geophys.*, **33**, 187–243.

Aoyagi, J. (1986) *A ground clutter rejection technique for weather Doppler radars*, URSI-F Subcommittee Paper No. 305. Union Radio-Scientifique Internationale.

Simizu, N., Wakabayashi, A. and Gotoh, S. (1977) *A ground clutter rejection technique for weather radars*, Technical Report No. 11. Japan R. C., pp. 54–65.

Tatehira, R. (1980) Operational use of echo phase by conventional radar. *Papers Weather*, **27** (12), 837–842.

13

The diurnal evolution of the ground return intensities and its use as a radar calibration procedure

O. Massambani[1] and A. J. P. Filho[2]
[1]Department of Meteorology, University of São Paulo, CP 30627, São Paulo, Brazil
[2]Fundacao Centro Technologico de Hidraulica, CP 11014, São Paulo, Brazil

ABSTRACT

This paper describes an analysis of the diurnal evolution of the ground return distribution as observed by the São Paulo S-band weather radar. The 15 km constant-altitude plan position indicator maps were used to quantify the magnitude of the ground return effects. The area coverage, the average reflectivity and the maximum reflectivity were computed every 10 min, for the two major areas of ground clutter. Meteorological information was used to identify the period with the influence of the sea breeze inflow of moist air at the surface boundary layer. The time interval of the least variability of the ground return was used to monitor the receiver performance, and to correct weather data affected by a receiver frequency drift. A comparison between the correct weather radar data and the raingauge data was performed for a significative summer rainfall event in order to evaluate the effectiveness of this procedure.

1. INTRODUCTION

The São Paulo S-band weather radar has been in operational use since September 1988, monitoring one of the most important economical region of Brazil, as part of the weather radar network for São Paulo State (RADASDP Project), which has been carried out by the São Paulo Water Authority (DAEE) and the São Paulo Research Sponsoring Agency (FAPESP). The location of the surveillance area is presented in Fig. 1(a). The main objective of this radar is to monitor precipitation for nowcasting and river-flood-warning purposes.

The radar system was developed by the McGill Weather Radar Group, Canada, which allows one to monitor an area of 180 km radius and 2 km×2 km spatial resolution, every 10 min. The iterative software enables the operator to obtain real-

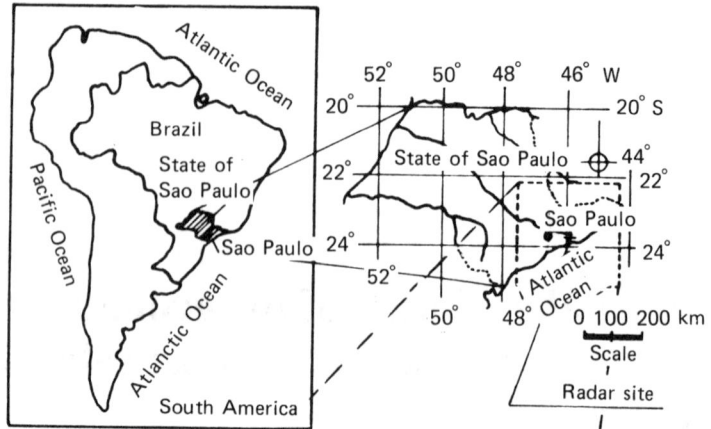

(a)

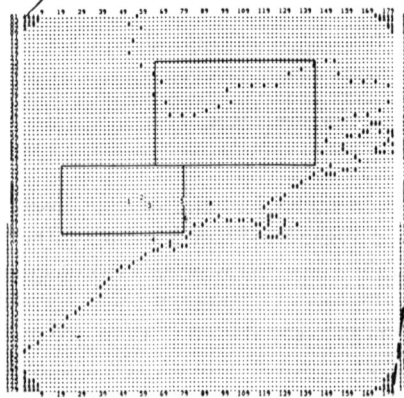

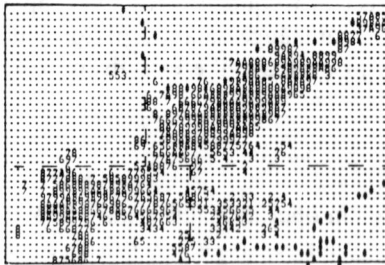

(b)

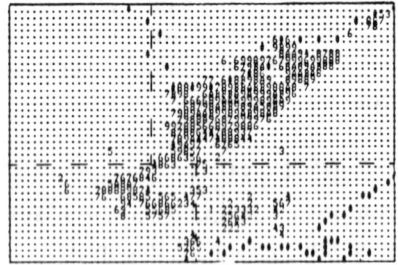

(c)

Fig. 1 — (a) The location of the São Paulo S-band weather radar, (b) a CAPPI map showing the larger area which contains the Mantiqueira ridge and (c) another CAPPI map which contains the Cantareira ridge. (b) and (c) respectively show two examples of the observed echoes at 1203 GMT and 1503 GMT on 9 September 1988.

time maps of rainfall, accumulation, echo top, vertical cross-sections, forecast, point rainfall and average accumulation for several watersheds within the area. Some of the radar characteristics are given in Table 1.

Table 1 — Radar characteristics

Peak transmitted power	650 kW
Pulse width	1.6 μs
Half-power beam width	2.0°
Wavelength	10 c
Minimum detectable signal (at 180 km range)	18.25 dBZ
Elevation range	1–30°
Number of elevations	21

The digital video integrator processor assigns a bin length of 0.5 km from 0 to 60 km of 1.0 km from 60 to 120 km, and of 2.0 km from 120 to 180 km range. The radar system is controlled by a PDP 11/73 computer and the data displayed in a colour monitor, printed and the row volume scan is stored on magnetic tape. A constant altitude plan position indicator (CAPPI) map is transferred to a second PDP computer, which houses the hydrological models.

The objective of this paper is to study the nature of the ground echoes without the presence of precipitation, and to monitor its variability in order to identify the most adequate time interval to be used as external reference of the receiver performance. The potential use of the clutter echoes to determine the stability of the radar system has also been appointed out by Reinhart (1978).

The two clutter areas chosen for this study were the Mantiqueira ridge located in the northern region and the Cantareira ridge, located in the western region of the radar site. Fig. 1 indicates the location of these areas with a mask applied, which is normally used during real-time operation. The Mantiqueira ridge is located inside the larger area and the Cantareira ridge in the smaller area.

2. DIURNAL EVOLUTION OF THE GROUND RETURN INTENSITIES

The data used for the identification of the nature of the ground returns were obtained during 7–9 September 1988. During this period the São Paulo State region was under the influence of a large-scale high-pressure system with upper-air subsidence and lower-level temperature inversion, as can be seen in Fig. 2, by the sequences of surface weather maps and by the radiosonde data observed within the radar area. Under these conditions the surface atmospheric boundary layer is most likely to be affected by a strong variation in the vertical gradient of the refractivity index, and therefore the radar beam will be bent, modifying the ground return intensities and their area coverage. In order to follow this atmospheric boundary layer evolution, the radar was operated in its normal mode during the above-mentioned period.

From the achieved 2 km×2 km pixel data, the area covered by echoes, the average reflectivity and the maximum reflectivity for the 1.5 km CAPPI, within the

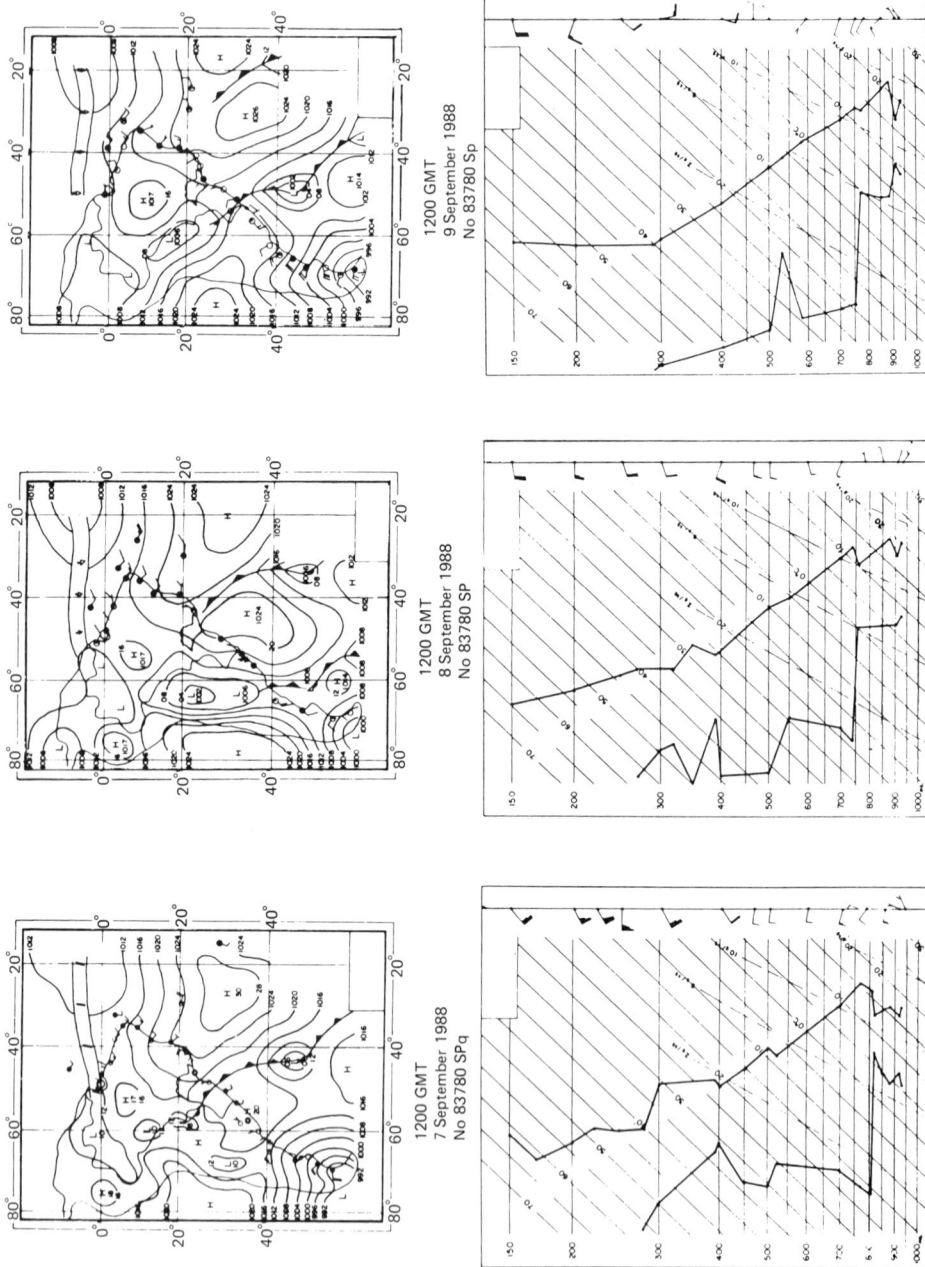

Fig. 2 — (a) Surface weather maps for 7–9 September 1988 and (b) the corresponding
radiosonde data as observed within the São Paulo S-band weather radar region.

indicated areas, were computed and are presented in Fig. 3. In this figure the relative variation in the refractivity index as measured from the surface meteorological data is also shown.

It can be observed in both areas that the ground return coverage is a minimum during the period of 1600–1700 Greenwich Mean Time (GMT). This corresponds to the period in which the atmospheric boundary layer is completely mixed owing to the turbulent heating and is connected with the upper air. At this period the refractivity index is already homogeneous, with a vertical gradient of about -40 N/km. Under this condition the beam path will follow the four thirds Earth radius curvature. Before this period the atmospheric surface layer is mixed and after this period the surface layer starts to lose heat by the radiative cooling process under clear-sky conditions. The magnitude of these effects is large, depending on the local characteristics of the ground surface. Doviak and Zrnic (1984) have discussed some of these propagation aspects.

The area chosen to monitor the Cantareira ridge, includes the Metropolitan Sãa Paulo which is a highly urbanized city, and this imposes quite a different heat store capacity compared with the Mantiqueira ridge region in which a rural kind of region prevails. A very important observation of this effect is most evident in the data of 9 September 1988, from 1200 to 1400 GMT (from 0900 to 1100 local time (LT), presented in Fig. 3. The very rapid decrease in the area coverage of ground echoes in the Cantareira region in relation to the Matiqueira region can be seen. This effect was about three times faster in the urbanized area, possibly caused by the 'urban heat island' effect. This can also be seen in the average reflectivity and the maximum-reflectivity diagrams. The computed linear correlation coefficient between the radar data and the meteorological data are indicated as 0.8 for the area coverage and average reflectivity with the refractivity index. In Fig. 3(d), this close relationship between the relative values of reflectivity and the surface refractivity data can be seen. A study of the relative importance of the moisture and the vertical wind profile on the control of the atmospheric refractivity in a tropical region has been presented by Massambani and Pereira Filho (1988).

On 7 September 1988, the sea breeze had penetrated the area at 1900 Greenwich Mean Time (GMT), as indicated by the rapid increase in the moisture in association with the southeasterly blowing wind from the Atlantic Ocean. On 9 September, the see breeze was much stronger and the effect on the clutter was much more evident. The fine structure observed in the average reflectivity graphs should be associated with the local dynamics of the atmospheric boundary layer, forced by the sea breeze circulation. A more detailed analysis is now being developed to relate these data to some environmental data for the Metropolitan São Paulo area, seeking to use the ground return analysis as a sensor of atmospheric boundary layer evolution, under clear-sky conditions.

3.   AN APPLICATION TO A RECEIVER FREQUENCY DRIFT VARIATION CASE

The rainfall observed on 20–22 December 1988 from an organized convective system in association with a cold front, monitored by the radar, showed much lower intensity than expected owing to a detected receiver frequency drift.

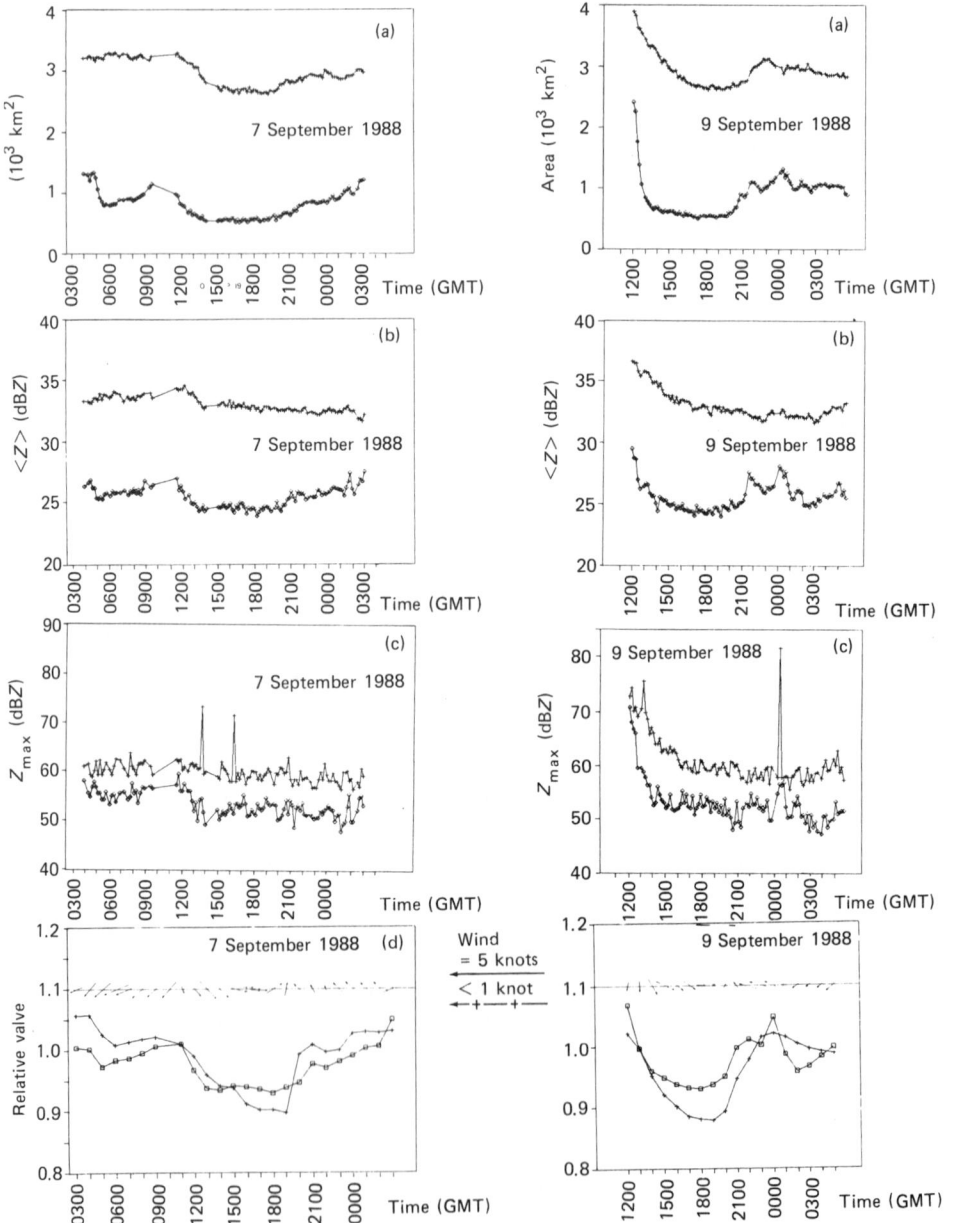

Fig. 3 — Time evolution of (a) area coverage, (b) average reflectivity and (c) maximum reflectivity as obtained during 7 and 9 September 1988, for Mantiqueira (+) and Cantareiria (○) areas with the São Paulo S-band weather radar system. (d) The relative values of the average reflectivity ($\langle \overline{z} \rangle / \langle \overline{z} \rangle$ (=26.3 dBZ) and reflectivity index $N/\overline{N}$ (=2979 N) are also shown. The time evolution of the refractivity index and the wind direction was obtained at the University of São Paulo meteorological station.

In order to correct these data, the assumption that the ground return observed during the 1600–1700 GMT period could be an indicator of the magnitude of the signal loss was applied. The mean maximum reflectivity from the Mantiqueira area for that period on 20 December was used to compare with the well-calibrated data from 7 and 9 September. The average reflectivity, and the maximum reflectivity in that area are shown in Fig. 4 for 20–22 December 1988.

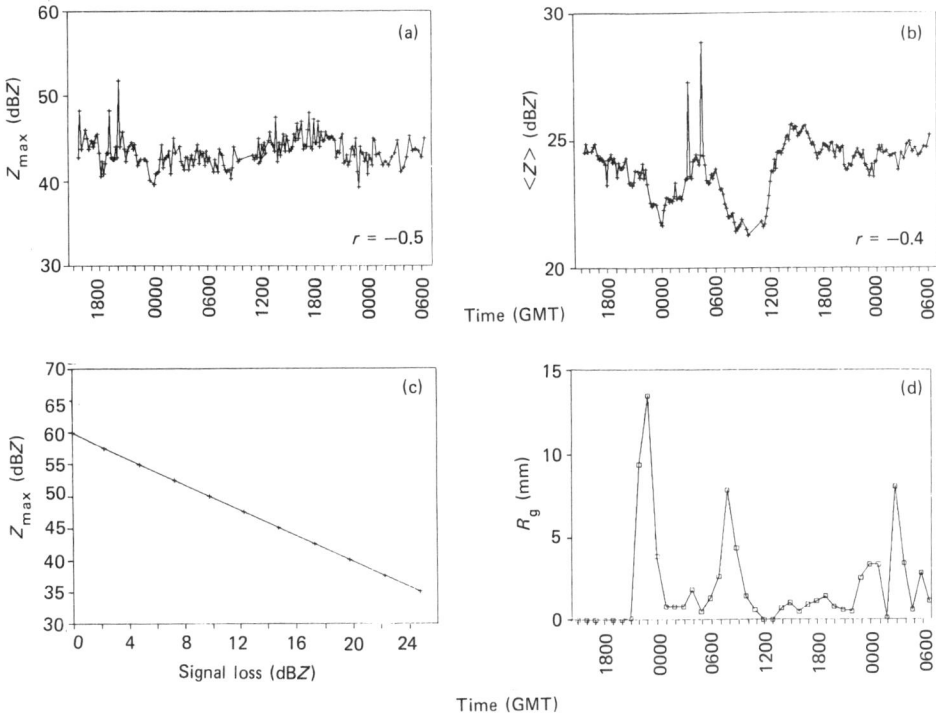

Fig. 4 — Time evolution of (a) the maximum reflectivity, (b) the average reflectivity for the Mantiqueira ridge area; (c) the signal loss which represents the correction applied to the maximum reflectivity analysis and (d) the rainfall at a raingauge located very close to the radar site.

The maximum reflectivity was found to be 16.8 dBZ lower than the September data. This value was used as a correction applied to the 3.0 km CAPPI, and a 24 h accumulation was performed and compared with six raingauge accumulated data within a 80 km range. The comparison has indicated that for a Marshall–Palmer ZR relationship, the radar data were below the gauge data, mainly owing to the loss of sensitivity below 5.4 mm/h. Another possible source of error that we have detected was a wet radome attentuation. It is shown in Fig. 4 as a rainfall evolution at a site close to the radar. These data indicate approximately the times in which the radome had been wetted. These data are negatively correlated ($r = -0.5$) with the maximum reflectivity of the Mantiqueira clutter. It is not a strong correlation, but it seems to be an indication of the possible source of error in the rainfall rate quantification.

However, the preliminary results on the use of a hydrological stochastic model to stimulate the hydrograph associated with a heavy-rainfall-rate event observed during that period show a correlation coefficient of 0.9 between the measured and the simulated hydrographs computed directly from the average reflectivity obtained for the isochrons of a very important urbanized watershed in the radar surveillance area. Further details are being studied, but a hydrograph with a peak of about 160 m^3/s was obtained, in perfect agreement with the observations. For this flood event the main contribution came from heavy rainfall rates and therefore the 5.4 mm/h loss of detectability was smoothed out by the model integration.

4.   CONCLUSIONS

It was shown that the time evolution of the area coverage by ground echoes, the average reflectivity and the maximum reflectivity values of a ground clutter area can be used to monitor variations of the local atmospheric boundary layer characteristics under clear-sky conditions.

Under the reasonable assumption that the ground surface characteristics of the Mantiqueira ridge region have not changed during the period studied, the maximum-reflectivity values during a well-known receiver calibration period could be used to monitor the receiver performance. This approach can be used as an operational indicator of the radar stability for day-to-day or year-to-year variations.

The data affected by receiver frequency drift were corrected using the relative difference of the maximum reflectivity for the Mantiqueira area. This seems to be a useful approach for recovering archived radar data.

ACKNOWLEDGEMENTS

The authors are grateful to the Radar Group of the Fundacao Centro Tecnologico de Hidraulica, and particularly to the operators for their cooperative participation in collecting this radar data. The authors acknowledge FAPESP and the DAEE for the financial support of the RADASP Project.

REFERENCES

Doviak, R. J. and Zrnic, D. S. (1984) *Doppler radar and weather observations*. Academic press, New York, 458 pp.

Massambani, O. and Perieira Filho, A. J. (1988) A importancia da umidade do ar e do perfil do vento no controle da perda de sinal em enlaces terrestres de microondas na regiao de São Paulo. *Proceedings of the 3rd Brazilian Microwave Symposium, Natal, 1988*, pp. 41–47.

Reinhart, R. E. (1978) On the use of ground return targets for radar reflectiviiy factor calibration checks. *J. Appl. Meteorol.*, 1343–1350.

14

Sectorized hybrid scan strategy of the NEXRAD precipitation-processing system

R. Shedd, J. A. Smith and M. L. Walton
Hydrologic Research Laboratory, Office of Hydrology, National Weather Service, National Oceanic and Atmospheric Administration, 8060 13th Street, Silver Spring, MD 20910, USA

ABSTRACT

In preparation for the introduction of NEXRAD in the USA, a 'sectorized hybrid scan' has been developed for enhancing precipitation processing by radar. The sectorized hybrid uses data from the four low tilts of the radar volume scan to develop a composite, or hybrid, scan. The choice of the tilt to be used at a particular location depends upon the range and altitude of the beam above the terrain. This processing has also provided for the production of regional radar coverage maps.

1. INTRODUCTION

The USA is on the verge of implementing a network of Doppler radars known as NEXRAD (Next Generation Weather Radar). A major component of the NEX-RAD software National Weather Service (NWS) Office of Hydrology. PPS converts effective reflectivity factor data into hourly accumulations of rainfall over the radar field. Details of the PPS are described within a companion paper from this symposium (Hudlow *et al.* 1989).

A significant problem that has confronted radar hydrometeorologists over the years is composting the effective reflectivity factor (hereafter in this paper simply referred to as the reflectivity) from a radar volume scan into an areal gridded field of precipitation estimates. The lower tilts are known to be more likely to be contaminated by ground clutter at the near ranges and will be more affected by anomalous propagation (AP). The higher tilts are more likely to suffer from range problems — particularly from incomplete beam filling and overshooting storm systems at further ranges. Joss and Waldvogel (1988) review a number of methods of using the vertical reflectivity profile to estimate precipitation. A reflectivity profile correction is proposed by Koistinen (1986).

This paper describes development of a 'sectorized hybrid scan' of reflectivity data that will be used within PPS. The sectorized hybrid scan is constructed to use the best data at each particular bin based on beam height, terrain effects and beam blockage. Tests using the sectorized hybrid scan have shown the procedure to provide a great improvement over other techniques to composite data. This paper also describes an automated procedure that has been developed to determine the sectorized hybrid scan. Knowing which radar tilt will be used at any particular bin also allows one to develop radar coverage maps displaying the beam elevation at any particular point.

2. BACKGROUND AND DESCRIPTION

The concept of the sectorized hybrid scan is one that has evolved over the past 15 years. Initial attempts at developing a hybrid scan are decribed in some of the documents that resulted from GATE (GARP Atlantic Tropical Experiment) in the mid-1970s (Richards and Hudlow 1977, Hudlow *et al.* 1976). The GATE processing used three tilt angles to develop the hybrid scan. The third tilt (4.0°) was used for ranges less than 16 km. The second tilt (2.0°) was used from 16 to 32 km. For the further ranges, several processing methods were examined. The most attractive method for the GATE data was termed 'bi-scan maximization'. This method uses the maximum reflectivity value from the low two tilts at a given range and azimuth. Bi-scan maximization was found to decrease the range degradation that often results from incomplete beam filling and attenuation losses.

When the PPS was being developed, it was decided to use a similar hybrid scan construction, using the four low tilts available from the NEXRAD volume coverage pattern (Ahnert *et al.* 1983). For most locations, the nominal values of the center-line axis of these tilt angles will be 0.5°, 1.45°, 2.40° and 3.35°. These four low tilts, using the NEXRAD 1° beam width, will always be contiguous. The hybrid scan, however, was developed using GATE data collected over the ocean; NEXRAD would be frequently facing problems from blockage due to mountains and other terrain features. Modifications to account for terrain features led to development of the sectorized hybrid scan.

The tilt selection procedure for the hybrid scan was a function only of range (Fig. 1); however, the procedure for the sectorized hybrid scan is a function of both range and azimuth. In areas where observations occur in the lower tilts, the sectorized hybrid scan uses a higher tilt. The sectorized hybrid array therefore will be a 230 by 360 array (to account for each 1° by 1 km bin) defining which tilt is to be used at that particular bin.

3. BENEFITS OF SECTORIZED HYBRID SCAN

The sectorized hybrid scan has a number of advantages relative to other compositing methods. The first is that the reflectivity data are collected over a more uniform altitude versus range as a result of using the increasingly higher tilts at the nearer ranges. At the far ranges (maximum range is 230 km for precipitation processing), the center line of the low tilt wil be about 5 km above the radar site.

Another significant benefit is the reduction of clutter and other radar noise. A majority of the ground clutter in the radar volume scan occurs in the lowest tilts at near ranges. Since the sectorized hybrid scan does not use data from the lowest tilts

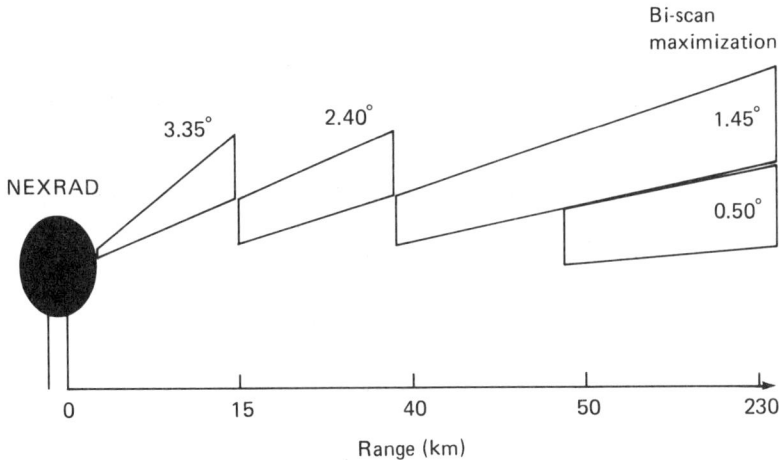

Fig. 1 — Construction of hybrid scan with no blockage present. The ranges shown are approximate and will vary by site.

close to the radar, a significant source of potential error is removed. The low tilt is also the most likely to be contaminated by AP. The pre-processing algorithm of PPS, which constructs the sectorized hybrid scan, has a tilt test to check for AP. If the percentage area reduction between the two lowest tilts for ranges between 40 and 150 km is greater than a certain threshold (currently set at 50%), the low tilt is considered to be excessively contaminated and is not used in the construction of the sectorized hybrid scan. Other tests that are performed in the pre-processing algorim include an outlier check which checks for excessively high reflectivity values and an isolated bin test. Corrections are also made for bins which may be partially obstructed by terrain or man-made obstacles.

The sectorized hybrid scan should also improve detection of orographic precipitation. Many precipitation processors simply have missing data beyond a certain range if the low tilt is blocked by terrain. By providing sufficient software intelligence, the location of these terrain features is known and the higher tilts can be used to detect precipitation occurring over the mountains. This capability will significantly increase the area of radar coverage.

Detection of bright band is another potential benefit of sectorized hybrid scan processing. Smith (1986) describes a method of detecting bright band using a similar multiple-tilt processing scheme using reflectivity data alone. Although bright-band detection is not currently a part of PPS, Smith's method is completely consistent with the capabilities of the PPS. The Office of Hydrology is in the process of determining the utility of this process and whether a change to PPS should be made in the future.

The sectorized hybrid scan and bi-scan maximization procedures are not without limitation. If a bright band is not detected and removed, it could be potentially enhanced by choosing the maximum value as well as repeated in range due to the multiple-tilt processing. Enhanced detection of virga resulting in overestimation of ground precipitation is also a possible result from bi-scan maximization. However, the potential benefits of using the sectorized hybrid outweigh these relatively minor limitations.

4.   DEFINITION PROCEDURE

A procedure has been developed to determine the tilt angle to be used for each bin
(1° azimuth by 1 km range) for the sectorized hybrid scan used in the NEXRAD PPS.
Using this information, the radar beam elevation used in the NEXRAD era to
estimate precipitation can be determined for any bin for any radar site.

The procedure uses the Defense Mapping Agency (DMA) digital terrain database. This database is a gridded (30″ by 30″) field with 10 m vertical resolution. In
performing a comparison between the elevations in the DMA database and those
determined as part of the NEXRAD site surveys, the database was found to have a
mean accuracy of 15 m relative to the elevations reported in the site surveys. For the
purpose of defining the hybrid scan and determining beam elevations, this accuracy
should be adequate. Some minor modifications to the hybrid scan may result of
higher resolution and more accurate terrain data are used.

The elevation E_{BM}, of the beam is defined as

$$E_{BM} = E_{RAD} + E_{TOWER} + H_{CA} + H_{BM} - E_G \qquad (1)$$

where E_{RAD} is the ground elevation of the radar site, E_{TOWER} is the height of the
radar tower, H_{CA} is the height from the top of the tower to the center of the radar
beam (5 m), H_{BM} is the height of the radar beam at range R and elevation angle A,
and E_G is the ground elevation for the particular bin determined from the DMA
database. H_{BM} is determined using the four-thirds effective earth radius model in the
form

$$H_{BM} = \frac{R^2 \cos^2 A}{2 \times \frac{4}{3}R_E} + R \sin A \qquad (2)$$

where R_E is the radius of the earth.

The procedure that has been developed for defining the sectorized hybrid defines
an 'optimal' elevation. The optimal elevation is a program variable currently set at
1200 m above ground. This procedure also checks for maximum elevations and
blocked azimuths and then chooses the tilt with center beam elevation closest to the
optimal elevation. The beam is also checked to assure that a minimum clearance of
the bottom of the radar beam above ground (currently set at 150 m) is maintained;
however, the fourth tilt does not need to maintain the minimum clearance in order
that data might be collected from the maximum area possible.

If a particular tilt angle is found to be blocked, it is determined whether the
blockage is more or less than 50%. If the blockage is greater than 50%, the tilt angle
is flagged as being not usuable for the remainder of that azimuth. If the blockage is
less than 50%, the tilt will not be used at the range where the blockage occurs but may
be used at further ranges. PPS includes a scheme for correcting for partially blocked
beams. Fig. 2 is an example of how the sectorized hybrid might be constructed in a
hypothetical area.

Two products, each in an alphanumeric and graphical form, are produced by this
definition procedure. The first is a display showing for each range and azimuth bin,
which of the four low tilts is to be used in constructing the sectorized hybrid scan.
Areas of missing data resulting from excessively high radar beams or total blockage

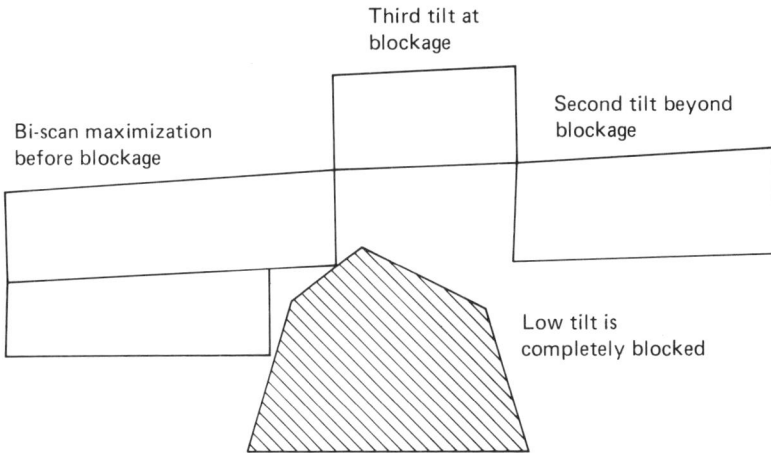

Fig. 2 — Construction of sectorized hybrid scan in the presence of blockage.

of the radar beam are also denoted. The second product is a display showing the elevation above ground for the sectorized hybrid. For display purposes, the radar beam elevation is defined from a point midway between the bottom and center line of the beam and is scaled into 750 m increments.

5.   EXAMPLES

Testing of the hybrid scan procedures has been performed for a number of sites. The first NEXRAD will be installed near Oklahoma City, OK, in the Great Plains of the USA. Obviously, in the plains where no obstructions are occurring to block the beam and only minor terrain changes to affect beam elevation, the sectorized hybrid scan is fairly uniform over the entire 360° of the radar sweep (Fig. 3), and the sectorized hybrid scan is essentially identical with a non-sectorized hybrid scan.

A significantly different situation occurs for the Tucson, AZ, NEXRAD site in the southwest USA. Tucson is located in a mountainous region and as a result the sectorized hybrid scan in significantly affected as the lower tilts are blocked most of the way around the radar (Fig. 4).

Of course, since neither of these sites has a NEXRAD yet, testing of the effects of the sectorized hybrid scan processing is not possible. However, located just outside Boulder, CO, at the base of the Rocky Mountains, is the CP-2 radar which has many characteristics similar to NEXRAD. PPS has been operational in Denver using the CP-2 radar data since May 1988. PROFS (Program for Regional Observing and Forecasting Services) attempted to implement PPS initially with a non-sectorized hybrid scan; however, the precipitation products were found to be excessively contaminated by clutter. After a sectorized hybrid scan was developed, the precipitation estimates were found to be much more useful for the forecaster as most of the mountain clutter was removed. Procedures developed at PROFS to produce the sectorized hybrid were similar to those described in this paper. The precipitation accumulation products generated using the sectorized hybrid scan as the reflectivity

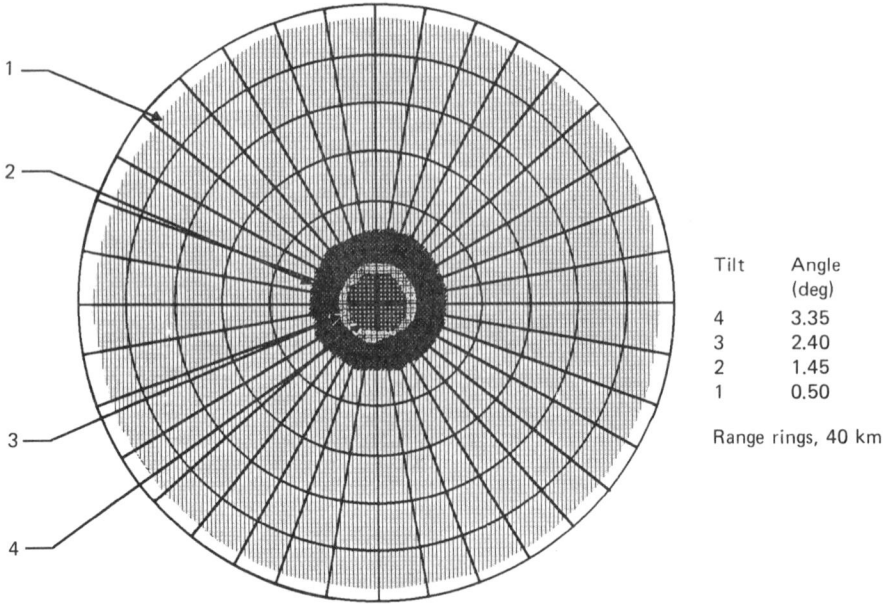

Tilt	Angle (deg)
4	3.35
3	2.40
2	1.45
1	0.50

Range rings, 40 km

Fig. 3 — Sectorized hybrid scan for Oklahoma City, OK, NEXRAD.

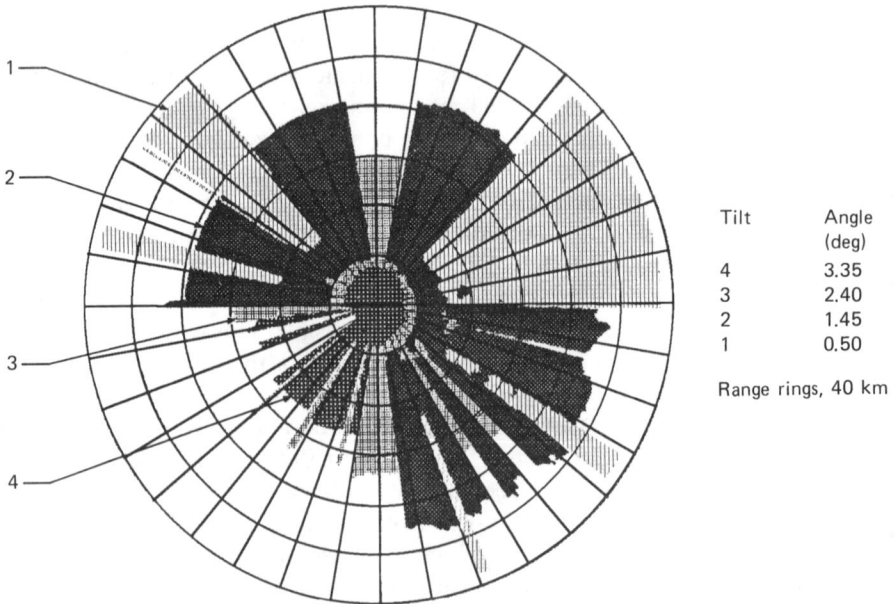

Tilt	Angle (deg)
4	3.35
3	2.40
2	1.45
1	0.50

Range rings, 40 km

Fig. 4 — Sectorized hybrid scan for Tucson, AZ, NEXRAD.

Fig. 5 — Future NEXRAD coverage for the State of Oklahoma showing the heights of point midway between the bottom and center line of the beam.

input were found to be the most-used NEXRAD-type product on the workstation located in the forecast office in Denver (Rasmussen *et al*. 1989; PROFS 1989).

The results of the sectorized hybrid definition can be used to determine the elevation of the radar beam using equation (2). The data can then be mosaicked together to show the radar coverage over a large region such as a state or river basin. Fig. 5 shows the future coverage over the State of Oklahoma resulting from the NEXRAD coverage. Four NEXRADs will be located in Oklahoma. Coverage from NEXRAD sites in adjoining states will also provide essential radar coverage over Oklahoma.

6. CONCLUSION

The sectorized hybrid scan offers significant promise for improved precipitation processing by radar. The tests that have have been performed thus far indicate that these pre-processing steps enhance the NEXRAD algorithms. The primary benefits that the sectorized hybrid and bi-scan maximization provide are in the reduction in ground clutter and AP, better altitude versus range definition, and potentially improved range performance and orographic precipitation estimates. The sectorized hybrid scan also offers the potential to be used by other NEXRAD algorithms as reflectivity input since the pre-processing that has been performed offers a much cleaner product than the base reflectivity data.

Two potential areas of limitation do exist. One is the enhanced detection of virga. However, since this phenomenon is generally relatively light precipitation, it should not seriously degrade the precipitation products. There is also some concern about azimuthal discontinuities in the mountainous regions. Fig. 4 shows the sectorized hybrid using different tilts at neighboring azimuth bins. At the far ranges, this could result in a difference of a few kilometers in height which could lead to sharp discontinuities in rainfall rate estimates. However, if the alternative is no radar coverage, these potential discontinuities would seem to be an improvement.

ACKNOWLEDGMENTS

The authors would like to acknowledge the continued support of the NEXRAD Joint System Program Office who have provided resources to support the work described in this paper. Also, Ricardo Romero assisted in preparing some of the figures for this paper.

REFERENCES

Ahnert, P. R., Hudlow, M. D., Johnson, E. R., Greene, D. R. and RosaDias, M. P. (1983) Proposed 'On-site' precipitation processing system for NEXRAD. *Preprints of the 21st Conference on Radar Meteorology*. American Meteorological Society, Boston, MA, pp. 378–385.

Hudlow, M. D., Pytlowany, P. J. and Marks, F. D. (1976) Objective analysis of GATE collected radar and rain gage data. *Preprints of the 17th Conference on Radar Meteorology*. American Meteorological Society, Boston, MA, pp. 414–421.

Hudlow, M. D., Smith, J. A., Walton, M. L., Shedd, R. C. (1991) NEXRAD: new era in hydrometeorology in the USA. *Hydrological Applications of Weather Radar*. Ed. Cluckie, I. D. and Collier, C. G. Ellis Horwood, Chichester, West Sussex, Chapter 54.

Joss, J. and Waldvogel, A. (1988) Precipitation measurement and hydrology — A review. *Proceedings of the Battan Memorial and 40th Anniversary Conference on Radar Meteorology*. American Meteorological Society, Boston, MA.

Koistingen, J. (1986) The effect of some measurement errors on radar-derived $Z.-R.$ relationships. *Proceedings of the 23rd Conference on Radar Meteorology*. American Meteorological Society, Boston, MA, pp. JP50–JP53.

PROFS (1989) PROFS Quarterly Report to NEXRAD/JSPO. Forecast Systems Laboratory, Program for Regional Observing and Forecasting Services.

Rasmussen, E. N., Smith, J. K., Pratte, J. F. and Lipschutz, R. C. (1989) Real-time precipitation accumulation estimation using the NCAR CP-2 Doppler radar. *Preprints of the 24th Conference on Radar Meteorology*. American Meteorological Society, Boston, MA, pp. 236–239.

Richards, F. and Hudlow, M. D. (1977) Use and abuse of the GATE digital radar data. *Proceedings of the 11th Technical Conference on Hurricanes and Tropical Meteorology*. American Meteorological Society, Boston, MA, pp. 216–223.

Smith, C. J. (1986) The reduction of errors caused by bright band in quantitative rainfall measurements made using radar. *J. Atom. Ocean. Technol.* **3**, 129–141.

Part 3
Radar echo climatology and statistical analysis

15

Is radar echo climatology useful?

T. Andersson
Swedish Meteorological and Hydrological Institute,
Norrköping, Sweden

ABSTRACT

This paper describes a simple experiment during the summer of 1988, studying the radar echo's frequencies and heights. The difference in echo frequency and heighy between land and sea and the effects of a lake, Vättern, on these parameters was to be studied, as well as their daily progress.

Now several digital weather radars are operating on a routine basis and if data is stored it is possible to compile radar echo climatologies. Some requirements on parameters for that purpose are discussed.

1. INTRODUCTION

Digital weather radar, as all digitized remote sounding tools, produce an enormous amount of data. Most of these are never used or even looked at by any human before being deleted. This is certainly a waste of valuable information. Several parameters are processed, some perhaps because it is technologically feasible and attractive to do so. In the worst case the user has no idea about how to use them. The situation is even worse for the scientist who feels that these data will be attractive in the future, for documenting our changing environment, but does not have resources to store them. One possibility then is to process the data in real time and only store some statistics. The difficulty then is to select the relevant statistical analysis tools. At the Swedish Meteorological and Hydrological Institute at present reflectivity data for the lowest constant altitude plan projection indicator (CAPPI) level (500 m) are stored, but only for one of our radars.

For conventional meteorological parameters, such as air temperature, the World Microbiological Organization has formulated rules for how to measure, collect, store and process them. For remote sounding parameters that only recently have been possible to measure, there are not generally accepted criteria. Even for such a parameter as radar reflectivity, which it has been possible to measure for several years, there are no accepted criteria for its climatological use. Moreover, there exist no internationally accepted criteria for the calibration of weather radars. Therefore it is doubtful whether the relfectivity measurements from different radars are comparable.

2. DATA AND AREAS USED

Three-dimensional radar data from an Ericsson C-band radar system described by Andersson *et al.* (1984) were used. Data volumes containing reflectivities for 12 CAPPI levels (0.5, 1.5, 2.5, . . ., 11.5 km) were extracted once every hour. For each data volume, 11 scans were used, with elevation angles from 0.5° to 20°. The radius of the area covered by the radar was 240 km, the horizontal resolution 2 km×2 km and the resolution in reflectivity 0.4 dBZ.

The investigation area was a north–south-oriented square with a side of 240 km, centred over the radar (Fig. 1). This area was divided into four subareas:

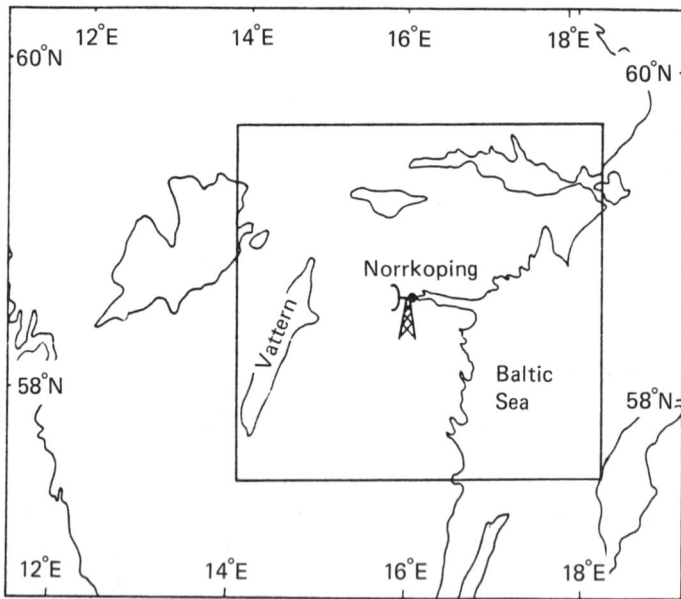

Fig. 1 — The Norrköping area. The sea area comprises the sea east of the Baltic coast.

(1) land (area, 41 336 km²);
(2) lake Vättern coast (area, 1828 km²);
(3) lake Vättern (area, 1876 km²);
(4) sea (area, 12 560 km²).

The lake Vättern coast area surrounds the lake with an east–west width of about half that of the lake and thus has an area about equal to that of the lake. The coast area should be used to compare the conditions over the lake with those over land. The reasons for using not only the land area (1) for this comparison are as follows.

(a) *The range dependence of the radar, due to the curvature of the earth and the divergence of the beam.*   Since the lake's coast area is about the same distance from the radar as the lake, the range effects should be the same for both these areas.

(b) *Parameters such as areas of high reflectivities and the heights of high relectivities.* These parameters are extreme values, and their frequency depends on the size of the area used. Moreover the precipitation climate over the large land area varies.

For each of these areas the following parameters were computed: A_{25}, the relative area of echoes with reflectivity greater than 25 dBZ at CAPPI level 1.5 km; A_{45}, the relative area of echoes with reflectivity greater than 45 dBZ at CAPPI level 1.5 km; H_{25}, the maximum height of the 25 dBZ contour; H_{25}, the maximum height of the 25 dBZ contour.

The period of investigation was 11 June–31 August 1988, during which 1796 hourly observations were collected. About 200 h were lost owing to radar or computer trouble.

The non-Doppler mode used does not suppress anomalous ground echoes. Therefore suspected observations of anomalous propagation were excluded. In this way, 76 night and early morning hours with low cloudiness, no rain according to the synoptic observations and low winds or calm were excluded.

3.   DIURNAL MARCH OF ECHO FREQUENCIES AND HEIGHTS

Fig. 2 shows the diurnal evolution of the relative area A_{25} with echoes above 25 dBZ, over the four areas for the whole period 11 June–31 August 1988 (0000–2200 Universal Time, Coordinated (UTC)). It is not surprising that the echoes over the sea have a nighttime maximum, but the same is true for the other areas. The land, lake Vättern and lake Vättern coast areas also have, as should be expected, an afternoon maximum, but it is smaller than the night maximum. The echo areas of lake Vättern and its coast are about the same, both *larger* than that of the land area. This is also surprising, since the lake during summer is a relatively cold surface which should dampen the daytime convection, resulting in reduced cloudiness and precipitation.

Fig. 3 shows the diurnal evolution of the relative area A_{45} with echoes above 45 dBZ. The land, lake Vättern coast and lake Vättern areas show the diurnal evolution to be expected, i.e. a proonounced afternoon maximum and an ill-defined night maximum. Their mean values over the day are very close. The sea area has, as also should be expected, its maximum during the night. Its mean over the day is about half that of the other areas.

From a climatological point of view, one should expect the large water surfaces to dampen the convection, especially in the early summer. Inspection of the individual months actually show that for 11–30 June 1988 the echo areas, both A_{25} and A_{45}, over

```
500 : XXXX
    : XXXX                          XX
    : XXXXXX                    XXXXXX
    : XXXXXXXX               XXXXXXXXXXXXXXX
    : XXXXXXXXXX          XXXXXXXXXXXXXXXXXXXXXX
    : XXXXXXXXXXXXXXXXXXXXXXXXXXXXXXXXXXXXXXXXXXXXXX
    : XXXXXXXXXXXXXXXXXXXXXXXXXXXXXXXXXXXXXXXXXXXXXXXXXX   XX
    : XXXXXXXXXXXXXXXXXXXXXXXXXXXXXXXXXXXXXXXXXXXXXXXXXXXXXXXX
    : XXXXXXXXXXXXXXXXXXXXXXXXXXXXXXXXXXXXXXXXXXXXXXXXXXXXXXXX
    : XXXXXXXXXXXXXXXXXXXXXXXXXXXXXXXXXXXXXXXXXXXXXXXXXXXXXXXX
  0 : XXXXXXXXXXXXXXXXXXXXXXXXXXXXXXXXXXXXXXXXXXXXXXXXXXXXXXXXX
```
Relative area with dBZ > 25; land area, mean, 385.4

```
    : XX
    : XX
500 : XXXX                          XXXX
    : XXXXXX                    XXXXXXXX
    : XXXXXXXX              XXXXXXXXXXXX
    : XXXXXXXX        XXXX  XXXXXXXXXXXXXXXXXXXX  XX
    : XXXXXXXX      XXXX    XXXXXXXXXXXXXXXXXXXX
    : XXXXXXXXXX  XXXX      XXXXXXXXXXXXXXXXXXXX     XX
    : XXXXXXXXXXXXXXXXXXXXXXXXXXXXXXXXXXXXXXXXXXXXXXXXXX  XXXXXX
    : XXXXXXXXXXXXXXXXXXXXXXXXXXXXXXXXXXXXXXXXXXXXXXXXXXXXXXXX
    : XXXXXXXXXXXXXXXXXXXXXXXXXXXXXXXXXXXXXXXXXXXXXXXXXXXXXXXX
  0 : XXXXXXXXXXXXXXXXXXXXXXXXXXXXXXXXXXXXXXXXXXXXXXXXXXXXXXXX
```
Relative area with dBZ > 25; lake Vattern coast; mean 430.2

```
    : XX
    : XXXX
500 : XXXX
    : XXXXXX                        XX
    : XXXXXXXX                 XXXXXX
    : XXXXXXXX             XXXXXXXXXXXX  XXXX
    : XXXXXXXXXX    XXXX   XXXXXXXXXXXXXXXXXXXX  XX  XX
    : XXXXXXXXXX    XXXX   XXXXXXXXXXXXXXXXXXXX     XXXXXX
    : XXXXXXXXXX XX XXXXXXXXXXXXXXXXXXXXXXXXXXXX    XXXXXX
    : XXXXXXXXXXXXXXXXXXXXXXXXXXXXXXXXXXXXXXXXXXXXXXXXXX
    : XXXXXXXXXXXXXXXXXXXXXXXXXXXXXXXXXXXXXXXXXXXXXXXXXX
    : XXXXXXXXXXXXXXXXXXXXXXXXXXXXXXXXXXXXXXXXXXXXXXXXXX
  0 : XXXXXXXXXXXXXXXXXXXXXXXXXXXXXXXXXXXXXXXXXXXXXXXXXX
```
Relative area with dBZ > 25; lake Vattern; mean, 419.4

```
    : XX
500 : XXXX
    : XXXX
    : XXXXXX
    : XXXXXXXX            XXXX
    : XXXXXXXX            XXXX        XX      XXXXXX
    : XXXXXXXXXXXX  XXXXXXXX      XXXXXX    XXXXXXX
    : XXXXXXXXXXXXXXXXXXXXXXXXXX XXXXXXX  XXXXXXXXXXXXXX
    : XXXXXXXXXXXXXXXXXXXXXXXXXXXXXXXXXXXXXXXXXXXXXXXXXX
    : XXXXXXXXXXXXXXXXXXXXXXXXXXXXXXXXXXXXXXXXXXXXXXXXXX
    : XXXXXXXXXXXXXXXXXXXXXXXXXXXXXXXXXXXXXXXXXXXXXXXXXX
  0 : XXXXXXXXXXXXXXXXXXXXXXXXXXXXXXXXXXXXXXXXXXXXXXXXXX
        0  02  04  06  08  10  12  14  16  18  20  22 UTC
```
Relative area with dBZ > 25; sea area; mean, 311.3

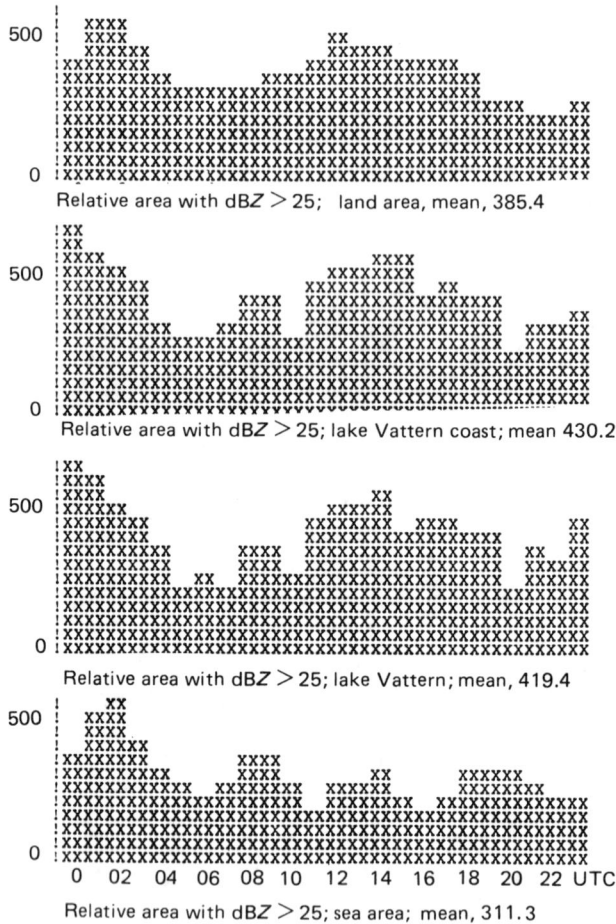

Fig. 2 — Mean relative areas with reflectivity above 25 dBZ for 11 June–31 August 1988 (unit, 0.001%).

lake Vättern and its coast are only about half that over the land. The same, however, is also true for August when the water (relative to the air) is warmest.

A range dependence of the parameters A_{25} and A_{45} should show up as decreasing values with range. The Vättern and Vättern coast areas are the most far away from the radar, and the range effect should give lower values of these parameters there than over the land and sea areas. In spite of this the Vättern and Vättern coast values are as high or higher than the land values. Therefore we must regard this feature as real, i.e. over the Vättern and its coast there have been more echoes above 25 d'bZ than over the land area, and about as many above 45 dBZ. We must, however, remember that the land area is large and contains quite different precipitation regimes, such as the Baltic coast with much less vigorous daytime convection and convective precipitation than the interior parts.

It is generally known that convective clouds tend to dissolve when approaching a lake with cool (relative to the air) water. This is also often observed on the radar. In

```
30 ¦                                      XX
   ¦                                      XX
   ¦                                    XXXXX
   ¦                                    XXXXXX
   ¦                                 XXXXXXXXXXXX
   ¦          XX                     XXXXXXXXXXXX
   ¦         XXXX                    XXXXXXXXXXXXXX
   ¦       XXXXXXXXXXXX          XXXXXXXXXXXXXXXXXX
   ¦XXXXXXXXXXXXXXXXXXXXXXXXXXXXXXXXXXXXXXXXXXXXXXXX
   ¦XXXXXXXXXXXXXXXXXXXXXXXXXXXXXXXXXXXXXXXXXXXXXXXXX  XX
 0 ¦XXXXXXXXXXXXXXXXXXXXXXXXXXXXXXXXXXXXXXXXXXXXXXXXXXXXX
   ¦XXXXXXXXXXXXXXXXXXXXXXXXXXXXXXXXXXXXXXXXXXXXXXXXXXXXX
```
Relative area with d B Z > 45; land area; mean 15.9

```
30 ¦                               XXXX      XX
   ¦                               XXXX     XXXX
   ¦                               XXXX     XXXX
   ¦                             XXXXXXX     XXXX
   ¦                           XXXXXXXXXXXXXXXXX
   ¦          XX               XXXXXXXXXXXXXXXXX
   ¦          XX               XXXXXXXXXXXXXXXXX
   ¦ XX     XXXX          XXXXXXXXXXXXXXXXXXXXX
   ¦XXXXXXX           XXXXXXXXXXXXXXXXXXXXXXXXX
   ¦XXXXXXXXXXXX  XXXXXXXXXXXXXXXXXXXXXXXXXXXXX
   ¦XXXXXXXXXXXXXXXXXXXXXXXXXXXXXXXXXXXXXXXXXXXXXXX   XX
 0 ¦XXXXXXXXXXXXXXXXXXXXXXXXXXXXXXXXXXXXXXXXXXXXXXX   XX
```
Relative area with d B Z > 45; lake Vattern coast; mean, 15.9

```
                               XX    XX
                               XX    XX
                             XXXX    XX
30 ¦                         XXXX    XX
   ¦                         XXXXXX  XX
   ¦                         XXXXXX  XX
   ¦          XX             XXXXXXX XXXX  XX
   ¦          XX             XXXXXXX XXXXXXX
   ¦          XX             XXXXXXXXXXXXXXXX        XX
   ¦ XXXXXX          XX  XXXXXXXXXXXXXXXXX     XX  XX
   ¦ XXXXXXX      XXXXXXXXXXXXXXXXXXXXXXXX       XXXXX
   ¦XXXXXXXXXXXXXXXXXXXXXXXXXXXXXXXXXXXXXXX
 0 ¦XXXXXXXXXXXXXXXXXXXXXXXXXXXXXXXXXXXXXXXXXXXXXXXXXXXX
```
Relative area with d B Z > 45; lake Vattern; mean, 16.9

```
30 ¦
   ¦
   ¦XX
   ¦XXXX
   ¦XXXX
   ¦XXXXXX         XX                XXXX   XXXX
   ¦XXXXXX        XXXXX          XXXXXXXXXXXXXXXXXXXXXX
   ¦XXXXXXXXXXXXXXXXXXXXXXXX     XXXXXXXXXXXXXXXXXXXXXX
 0 ¦XXXXXXXXXXXXXXXXXXXXXXXX     XXXXXXXXXXXXXXXXXXXXXX
     0  02  04  06  08  10   12  14  16  18   20  22 UTC
```
Relative area with d B Z > 45; sea area; mean; 8.1

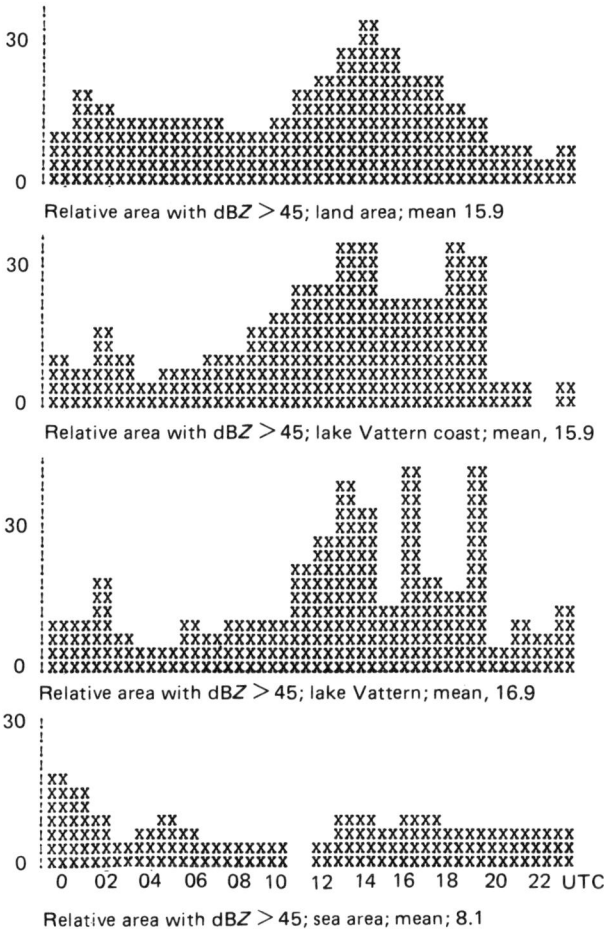

Fig. 3 — Mean relative areas with reflectivity above 45 dBZ for 11 June–31 August 1988 (unit, 0.001%).

June, isolated cumulo-nimbus echoes approaching the Vättern have a marked tendency to dissolve, and satellite pictures from convection days generally show a clear area over the lake. However, if the convection cells organize themselves into squall lines, the echoes seem to be only a little or not at all affected by the lake. Also frontal echoes generally seem unaffected by the lake. Much of the summer 1988 rainfall was frontal, with embedded cumulo-nimbi, which may explain the high echo amounts over the Vättern.

The high echo frequencies of the Vättern should result in high precipitation there, especially during July when A_{45} over the Vättern and its coast far exceeded that over the land area. The analyses of monthly precipitation indicate a rainfall minimum over the Vättern in June and August and a pronounced maximum there in July. All are quite consistent with our A_{45} areas. During summer, when the echoes generally are quite high in this region, it then should be possible to prepare echo statistics for the reflectivity (at a low CAPPI level) out to ranges of about 120 km. During other seasons, with more shallow echoes, this range is decreased.

Figs 4 and 5 show the diurnal evolution of the mean of the maximum echo height for each area. For both threshold values, 25 and 45 dBZ, the land area shows the highest echoes and the Vättern areas the lowest echoes. This is mostly caused by the different sizes of the areas to which these parameters refer. Because of the broadening of the beam the echo height should tend to decrease with increasing range. A climatology of extreme values such as echo maximum height should refer to areas of the same size.

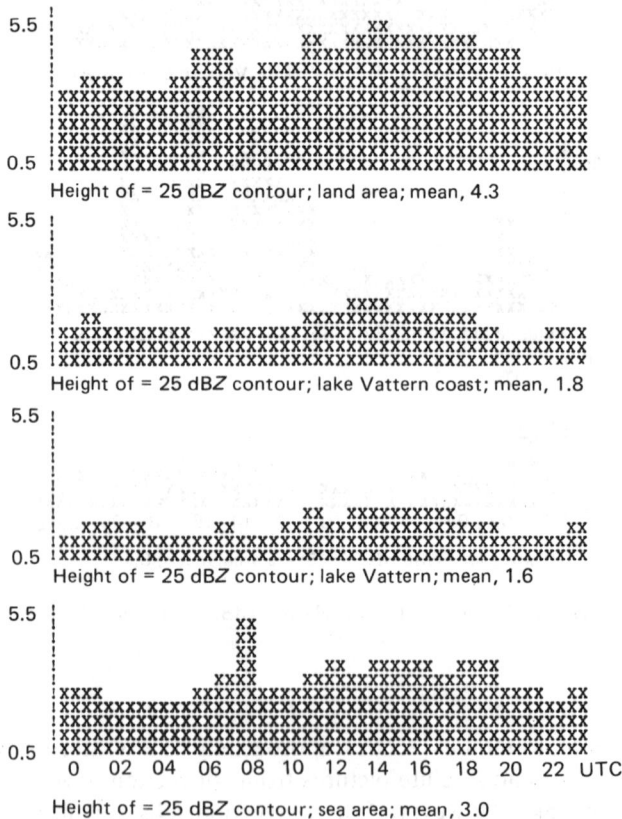

Height of = 25 dBZ contour; land area; mean, 4.3

Height of = 25 dBZ contour; lake Vattern coast; mean, 1.8

Height of = 25 dBZ contour; lake Vattern; mean, 1.6

Height of = 25 dBZ contour; sea area; mean, 3.0

Fig. 4 — Mean maximum heights of 25 dBZ contour for 11 June–31 August 1988 (unit, 1 km).

For the whole period both the echo areas and the echo-top heights generally were only somewhat larger over the lake Vättern coast area than over the lake Vättern area. Most probably a lake of this size affects different types of precipitation in different ways. While a dampening of the convection over a relatively cool water surface certainly results in lower precipitation from single cumulo-nimbus cells,

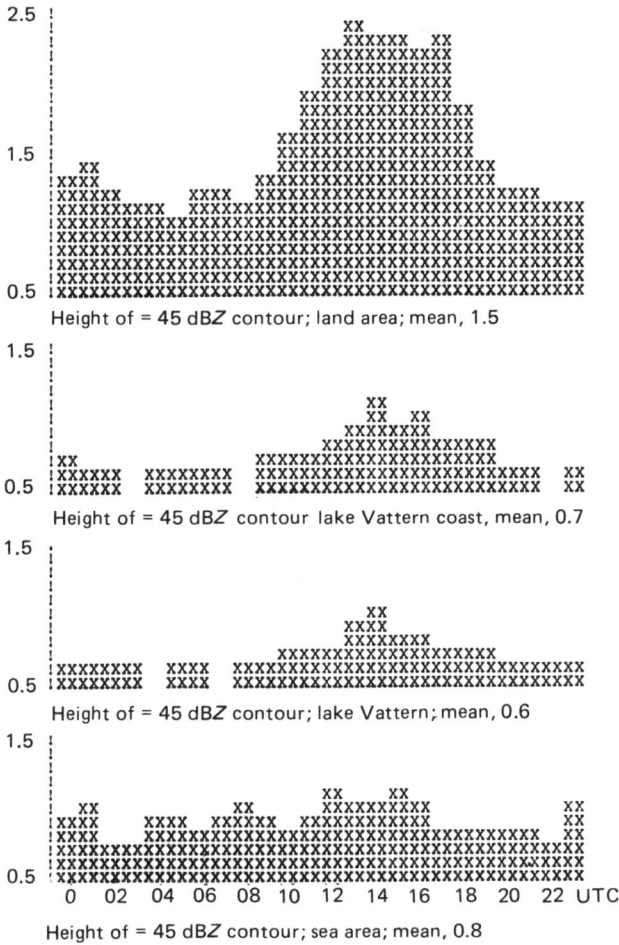

Fig. 5 — Mean maximum heights of 45 dBZ countur for 11 June–31 August 1988 (unit, 1 km).

there may be quite different effects on frontal precipitation and precipitation from organized cumulo-nimbus systems. There might have been too few pure 'convective precipitation' days this summer for the different convection patterns over the lake and land to show up in these statistics. We must also remember that the radar cannot account for any growth or decay of precipitation below the CAPPI level used, 1.5 km.

4. ECHO AREAS AND TOPS

As an example of other statistics the cumulative frequencies of area echo coverage and echo tops over the land area are shown in Figs 6 and 7. There is of course a large difference between the coverage of the areas with levels above 45 and 25 dBZ, the latter being more than ten times more frequent, and it is rare (frequency, about

0.1%) that the area with reflectivity above 45 dBZ occupies 400 km^2, i.e. 100 pixels with the resolution (2 km $\times$ 2 km) used.

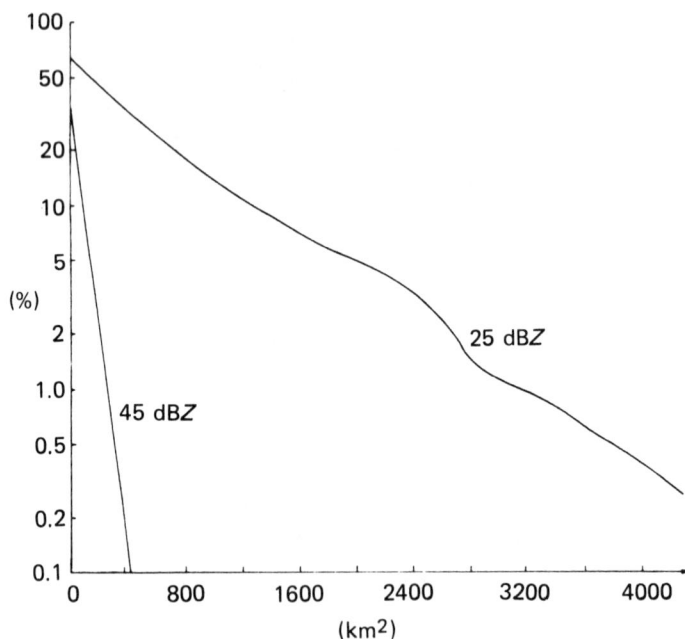

Fig. 6 — Cumulative frequencies of relative areas with reflectivities above 25 and 45 dBZ for the land area (10 334 km^2) on 11 June– 31 August 1988, 0000–2300 UTC.

A comparison between the echo coverage of lake Vättern and its coast (not illustrated here) shows that the frequencies of echoes with reflectivity *below* the limits used are somewhat larger over the lake, which is to be expected since the lake should dampen the convection. The mean echo areas were, however, as shown earlier, not smaller over the lake. This illustrates only that one summer is too short a period for climatological conclusions.

The frequencies of echoes, as well as of echo-top heights, are dependent upon the size of the area. Therefore comparison between the land and sea areas are not meaningful. To give an idea of the echo tops, Fig. 7 shows their cumulative distribution for the land area.

5.   DISCUSSION

An obvious climatological use of radar is precipitation mapping (e.g. Holtz 1983). The radar gives a resolution in space and time superior to any network of conventional gauges. Moreover it can give observations from areas where it is difficult or impossible to get conventional measurements, as over seas and mountains. The difficulties inherent in the radar technology are huge, however. The size of the volume studied is range dependent owing to the beam's divergence; its height above

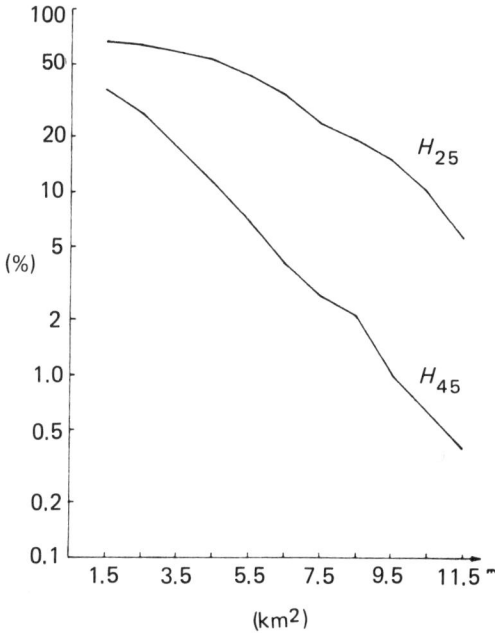

Fig. 7 — Cumulative frequencies of echo tops exceeding 1.5, 2.5, . . ., 11.5 km for reflectivities above 25 and 45 dBZ for the land area on 11 June– 31 August 1988, 0000–2300 UTC.

the earth surface increases with increasing distance owing to the curvature of the earth and there are no methods of getting reliable relations between the reflectivity measured by the radar and the rain intensity. Still the radar technique is promising.

Rogers and Yau (1982) studied the relative area occupied by radar echoes exceeding certain reflectivity thresholds at heights between 2 and 10 km over Montreal. Their investigation area had a radius of about 200 km (the nearest 12 km were excluded in order to avoid ground clutter). They give several interesting vertical profiles of reflectivity and echo coverage at different heights. For climatological purposes, however, their data are difficult to apply, since they only included cases where echoes were present at a height of 2 km in their sample.

CONCLUSIONS

Simple radar parameters, such as echo frequencies at different heights, can give valuable climatological information. These statistics should refer to areas of equal size. The optimum size must be less than the large 'land' and 'sea' areas used here. Since the radar is range-dependent, the statistics must be confined to fairly short ranges, perhaps about 60 km, if statistics are to be obtained with a reasonable height resolution and for seasons with shallow precipitation systems. An international routine for the calibration of radars is needed. If one wants to compile an echo climatology the main practical difficulty may well be to obtain funds to carry on such a project, which perhaps does not give any immediate revenues and has to go on for some years to give stable statistics.

ACKNOWLEDGEMENTS

I wish to express my gratitude to the Swedish National Research Council for supporting this work, and to Dr Hans Alexandersson for valuable discussions.

REFERENCES

Andersson, T., Magnusson, S., Lindström, B. and Karlsson, K-G. (1984) The PROMIS Doppler radar system: design and operational experience. *Proc. Second Inter. Symp. on Nowcasting*, Norrköping, 171–176.

Holtz, C. D. (1983) Radar precipitation climatology program. *Preprints*, 21*st Conf. Radar Meteorol.*, Sep 19–23, 1983, Edmonton, 390–393.

Rogers, R. R. and Yau, M. K. (1982) Areal extent and vertical structure of radar weather echoes at Montreal. *Pure and Appl. Geophys.*, **120**, 272–285.

16

Geometrical and statistical features of Swiss radar data

R. Boesch
Department of Geography, University of Zurich, Winterhurerstrasse 190, 8057
Zurich, Switzerland

ABSTRACT

One of the major research topics of the Swiss Meteorological Institute in Locarno-Monti (OTISM) deals with precipitation estimation and analysis of weather radar data. One long-term goal is to get quantitative measures of the radar data for further analysis. Further, a numerical characterization of a weather situation would be interesting for short-term forecasting and climatological purposes, as well as a tool for motion analysis of certain echo cells. To recognize radar echoes as own objects, filtering and boundary tracing algorithms from image processing have been used to vectorize the raw data into an object-oriented space. Different shape properties have been extracted from radar data and the benefits of the different descriptors have been discussed.

1. INTRODUCTION

Every 10 minutes a digital composite picture will be disseminated by OTISM. Each picture represents the ground projection of the volume data, which have been merged from two weather radars (Fig. 1).

The Swiss Meteorological Institute is interested in further studies with these data, mainly for quantitative analysis. Because of the relatively high sampling rate of 10 min, a good time resolution is available for analysis over a sequence of images. A major research topic represents motion analysis of certain echo cells and the numerical characterization of weather situations. Therefore it is necessary to transform the relatively noisy radar data into a kind of object representation. To recognize radar echoes as single objects in an image, one way is to characterize the objects by a set of shape descriptors which are reasonably invariant to changes in size, rotation and translation. The calculated 'high-level' descriptors can then be

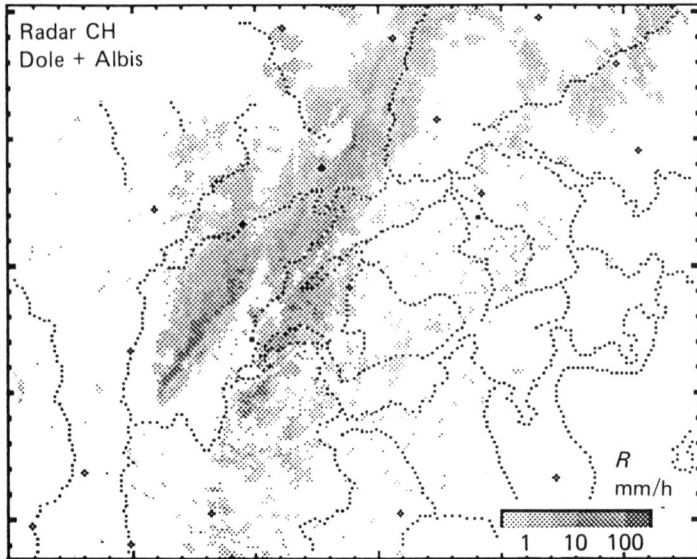

Fig. 1 — Radar composite from 8 October 1987 at 0650 hours.

used for further numerical investigations in climatology, short- and long-term forecasting.

There exist some more or less complicated descriptors for planar shapes such as

(1) simple shape descriptors, e.g. area, circularity, compactness, centre of gravity and perimeter,
(2) moments computed from the object content (mass) object and
(3) Fourier descriptors of the object boundary points.

The book by Gonzalez (1987) will provide a detailed introduction.

Most algorithms in the area of shape description either work on the boundary points of the object or require at least a first pre-processing step of the raster data, where the area of each object can be distinguished from the other objects (often known as labelling).

The crucial point in image-processing applications is commonly the edge detection process. The information in digital images is mainly contained in the edges, which represent transitions between regions of different intensities. Commonly used methods for tracing boundaries are based on the gradient information contained in the edges and work in the spatial domain. The methods of Sobel, Prewitt and Roberts are the most popular and represent spatial convolutions from the image data with an appropriate gradient mask (typically a 3×3 window). With radar echo cells of Swiss radar data, there is no reliable gradient information contained because of the small dynamic range (seven intensities) and the inherent fuzzy boundary transition within a certain cell. Further, the quantization occurs at the radar location itself and cannot be reversed later. In this paper, a simple threshold will be used to suppress the ground clutters as well as to make the edges more distinguishable from the background. This is clearly a very crude approach, but there is sufficient balance between data and processing 'quality'. Before we use the data for the tracking process, a median filter will be applied to suppress a certain noise level, as well as to smooth the boundaries of the radar echoes.

After this step, the boundary coordinates for each object will be traced and the corresponding area of each closed object are marked with a unique value (Fig. 2).

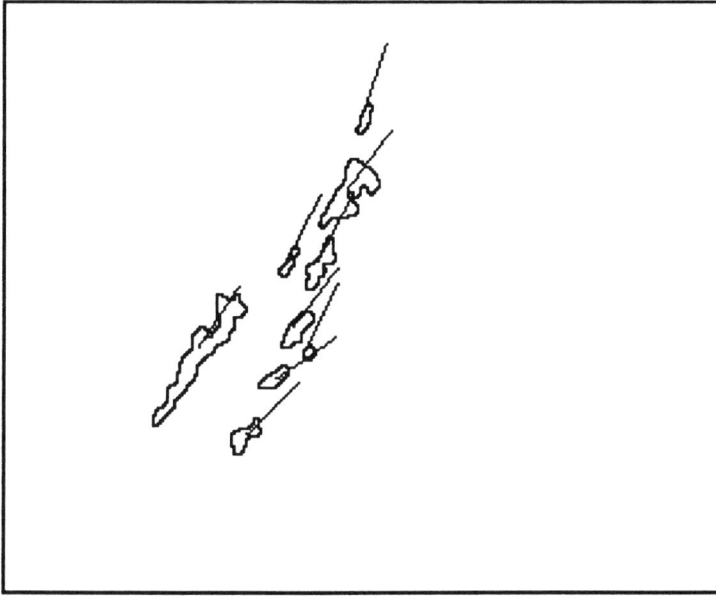

Fig. 2 — Traced radar echoes from 8 October 1987 at 0650 hours.

2. SHAPE DESCRIPTORS

Now we can calculate the different shape descriptors of all objects in a single image. For each image, about five to ten objects have been extracted (blobs below a certain minimal area have been omitted). The following formulae have been used and a comment about their applicability with radar data will be made if appropriate.

Let $f(x, y)$ be the value of a radar image cell at the location x, y (c columns and l lines).

2.1 Centre of gravity
This point will be influenced by the spectral distribution within the object. In this case, the centre can be found beside the expected location, especially in the case of large radar cells with an inhomogeneous mass distribution:

$$x_0 = \frac{1}{\text{area}} \sum_{x=1}^{c} \sum_{y=1}^{l} x f(x, y) \ , \qquad y_0 = \frac{1}{\text{area}} \sum_{x=1}^{c} \sum_{y=1}^{l} y f(x, y) \ .$$

2.2 Moments

Moments are invariant to translation, rotation and size. They are based on the centre of gravity according to the following equation:

$$M_{jk} = \sum_{x=1}^{c} \sum_{y=1}^{l} x^j y^k f(x, y) \ .$$

From this moment, the orientation θ of the major axis can be computed (which is not an invariant descriptor):

$$\theta = 0.5 \ \text{atan} \left(\frac{2M_{11}}{M_{20} - M_{02}} \right) \ .$$

There are two further descriptors based on moments: **spread** and **elongation**. Spread gives a measure of the mass distribution within the object; elongation gives a measure of the elongation of the object. When we denote the values of minimum and maximum inertia as I_{min} and I_{max}, we have

$$\text{spread} = \frac{I_{max} - I_{min}}{\text{area}^2} \ , \qquad \text{elongation} = \frac{I_{max} - I_{min}}{I_{max} + I_{min}} \ .$$

2.3 Circularity or compactness

The magnitude of the circularity can reflect the complexity of the boundary. Most commonly the following formula will be used:

$$C = \frac{4\pi \times \text{area}}{\text{perimeter}^2} \ .$$

For a circular shape, the formula takes its maximum at 1.0, which corresponds to the most compact area in relation to its boundary extension.

2.4 Fourier descriptors

Fourier descriptors of a contour with points a_k and b_k are given by

$$\theta(t) = \mu_0 \sum_{k=1}^{\infty} a_k^{\cos[kt] + b_k \sin(kt]} \ ,$$

where μ_0 is the starting point.

The boundary of an object can be seen as a function of arc length by the accumulated change in direction of the curve since the starting point. This function will be transformed in a Fourier series. The most important property of this transformation is the invariance of the harmonic amplitudes under translation, rotation, change in size and shift in the starting point (Fig. 3 gives an illustration of

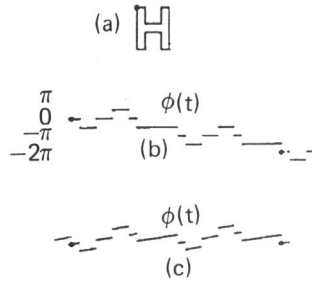

Fig. 3 — Normalized Fourier descriptors of a single character.

the invariance property.) At first glance, it seems that we have a valuable tool for matching objects in different images. If the objects are only slightly distorted from one image to another, there is a good chance for a successful matching with a generalized shape of one object for the image. The generalization of the shape with Fourier descriptors is very elegant and easy, because we only have to reduce the number of coefficients of the Fourier series and we get a smoothed shape, which still symbolizes the original shape of the object.

For distorted objects such as most radar cells are in reality, Fourier descriptors cannot be used even for simple shape matching. If we want to use Fourier descriptors, we would have to smooth the boundary so much that the characteristic form would be lost. Although the descriptors have been widely used in the field of robotics, identification of aircraft and scene analysis, they do not seem to be very useful for describing radar echo cells. Lin (1987) describes a successful use of Fourier descriptors; Zahn (1972) gives a detailed introduction into the mathematical concepts behind Fourier transformation of point data.

From all more or less significant descriptor value distributions, the variability of two interesting descriptors is shown in Figs 4 and 5 for a short sequence of images. The sequence dates from 8 October 1987 and can be characterized as a moderate advective situation (the calculated period is 90 min with a 0 min interval between the images). Among all the calculated descriptors, the orientation of maximum inertia and the location of the centre of gravity show a quite distinctive trend to the current weather situation.

The dominant angle around 60° in Fig. 4 reflects the true propagation direction of most radar cells within the radar image. From this distribution, one can also see the difficulties which arise in selecting relevant objects from a noisy aggregation of echoes. The descriptor 'elongation' now helps in discriminating objects with a

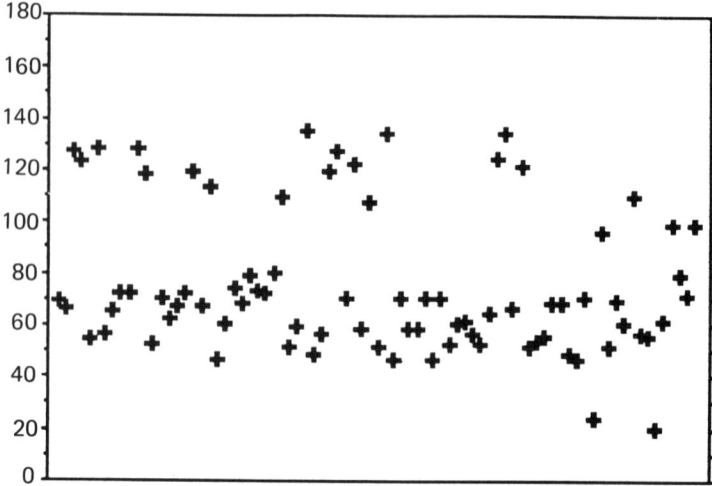

Fig. 4 — Scattergram for orientation of the maximum inertia using radar images from 0550 to 0720 hours on 8 October 1987: +, orientation of maximum inertia of a single object within an image.

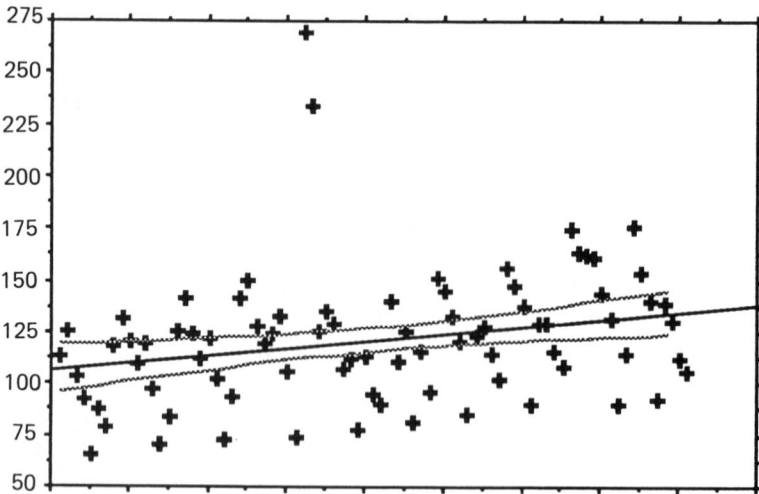

Fig. 5 — Linear regression for centre of gravity radar images from 0550 to 0720 hours on 8 October 1987: +, centre of gravity of a single object within an image.

significant orientation of the shape from objects with a small directivity or no clear orientation. If we use only truly elongated objects, some echoes above and below the 60° area in Fig. 4 can be disregarded.

Another significant descriptor can express the centre of gravity. In Fig. 5, only the x coordinate is shown; the statistical behaviour of the other coordinate is about the same. From the wavy line (90% confidence bands for the true means of the centre coordinates) it is quite obvious that the regression coefficient cannot be very high ($R=0.6$). Nevertheless the general moving direction of the whole cloud system is quite remarkable. Compared with a situation with no echoes or many very small spread-out cells, the mean centre positions can be a valuable indicator of the directivity of the whole sequence. The variance of the centre points is a measure of the width of the tracked front, as well as a divergence indicator of the cell's moving path.

In both figures, one would think that a clustering scheme in the shape descriptor space would reduce the variance. On the other hand a new emerging front within the sequence would be recognized very late in this simplification. Another improvement could be gained by the use of the spectral information contained in each identified cell. The histogram in Fig. 6 shows the distribution characteristic of Fig. 1. Some

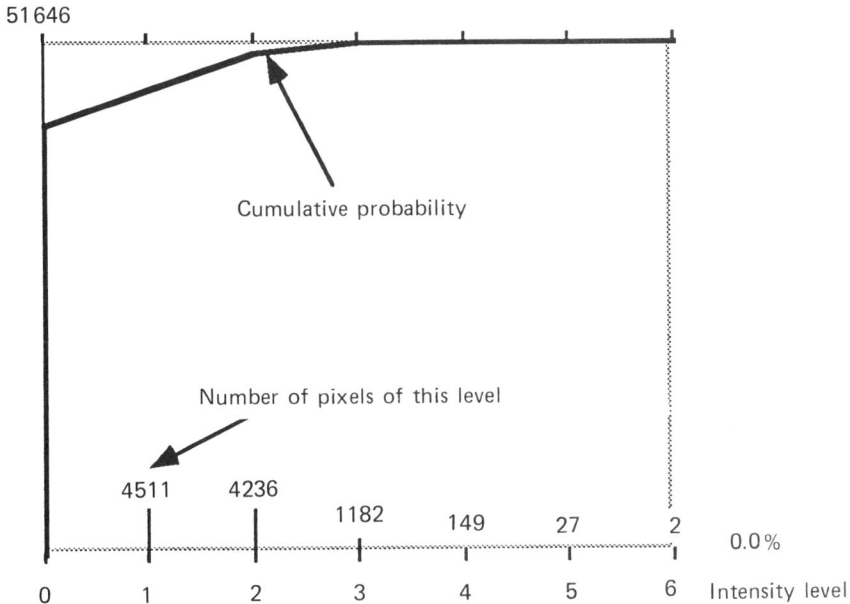

Fig. 6 — Histogram of 8 October 1987 at 0650 hours.

analysis has been made of how the intensities within a single echo are distributed. Actually this would represent a more detailed description of the interior structure than only the centre of gravity, but in most cases the size of the echo and the inherent limited data range are much too small to yield an interpretable distribution.

From the displacement of the estimated centre of gravity, we can directly calculate the average propagation speed of the whole cell aggregation. In this example, the clouds will move with an estimated speed of 40 km/h at an angle of 60°.

3. CONCLUSIONS

Different kinds of shape descriptor can lead to a descriptive transformation even for disturbed objects such as radar echo cells. Directional and location-dependent descriptors have been the primary tools for analysing the situation; other shape-oriented descriptors have been used for discriminating significant echoes from noisy echoes. A major problem that still remains is the large variance of most calculated shape descriptors. A stronger smoothing in the pre-processing stage would only lead to a flattened data range for most descriptors, which results in even more discriminating problems. However, the combined use of shape descriptors will probably aid in a quantitative description or model of certain 'well-defined' weather situations.

   This numerical description can be used for short-term forecasting as well as for climatological investigations later. A large improvement of the described analysis could be yielded by the use of the Swiss network of automatic weather stations (ANETZ), which give additional information about the current wind direction, air pressure, etc., at the station location. Although an interpolation of the ANETZ data at the cell location has to be made, the statistical independence should be a real advantage compared with the often highly correlated shape descriptors. The 10 min sampling interval of the ANETZ data minimizes the interpolation effort remarkably and the smooth behaviour of some detectors (e.g. air pressure) should also give a good estimation of the moving path of the cells.

REFERENCES

Gambotto, J. P. and Huang, T. S. (1987) Motion analysis of isolated targets in infrared image sequences. *Pattern Recognition Lett.* **5**, 357–363.

Gonzalez, R. C. and Wintz, P. (1987) *Digital image processing*, 2nd edition. Addison-Wesley, Reading, MA.

Joss, J. and Waldvogel, A. (1987) *Precipitation measurement and hydrology*, Working Report No. 145. Swiss Meteorological Institute.

Lin, C. C. and Chellappa, R. (1987) Classification of partial 2-D shapes using Fourier descriptors. *IEEE Trans. Pattern Anal. Mach. Intell.*, **PAMI-9** (5), 687–690.

Zahn, C. T. and Roskies, R. Z. (1972) Fourier descriptors for plane closed curves. *IEEE Trans. Comput.*, **C21**, 269–281.

17

Probable maximum flood modelling utilizing transposed maximized radar-derived precipitation data

I. D. Cluckie, M. L. Pessoa and P. S. Yu
Water Resources Research Group, Department of Civil Engineering, University of Salford M5 4WT, UK

ABSTRACT

The paper analyses a number of aspects related to dam design and dam safety, particularly those which are associated with the estimation of the design flood. The probable maximum precipitation was estimated by transposition and maximization of radar derived storms. Two rainfall–runoff models were used for the conversion of probable maximum precipitation to probable maximum flood: the unit hydrograph, using catchment averaged radar precipitation data, and the grid-based distributed model, taking full advantage of the rainfall spatial distribution provided by radar. The performance of the models when dealing with extreme events of precipitation and streamflow was investigated and the resulting estimates were compared. The influence of systematic errors in the radar data upon the flood estimates was also examined.

The techniques were applied to the Stocks Dam (northwest England), although both the rainfall–runoff models enable the formulation of a general simulation, which can be applied to any watershed by proper use of parameters which adequately define its physical characteristics.

The utilization of radar data for design flood estimation is considered in view of the potential of radar to improve such analyses by providing relatively high-resolution rainfall depths and rates on both temporal and spatial scales.

1. INTRODUCTION

The statistical concept of a flood with an associated return period has been used for some time as a means of mitigating against hydrologically induced failure. In recent years, however, there has been increased interest in deterministic studies for the evaluation of the maximum reasonable flood whose occurrence can be expected in a particular catchment — the probable maximum flood (PMF) — and its utilization (or the utilization of a percentage of it) as the design flood for a dam. This flood has been obtained from the estimation of the maximum theoretical depth of precipitation whose occurrence can be expected over a particular catchment, for a given duration and time of the year — the probable maximum precipitation (PMP).

These two statistical and deterministic methods are the most utilized, in the majority of the countries of the world, as the safety design criteria for dams. However, many dam disasters have occurred around the world and, on average, one third of the causes can be attributed to inadequate spillway capacity (for an up-to-date) review of 'causes of dam failures', see Pessoa (1989)). In spite of these factors, relatively little work has been carried out on more novel and perhaps more appropriate methods for the estimation of the spillway design flood.

The utilization of radar-derived distributed rainfall data in modelling the behaviour of severe storms and extreme floods (Cluckie and Pessoa 1988, Pessoa and Cluckie 1990) offers a new possibility for the provision of improved estimates of the safety design flood for dams. Radar allows the visualization of the storms as a whole, their transposition and tracking across the watershed, along the quantification of distributed rainfall amounts.

This paper analyses a number of aspects of PMF estimation from the point of view of radar-derived distributed rainfall data. The performance of two different rainfall–runoff models (one lumped and the other fully distributed) are compared, and the effect of systematic errors in the radar data upon the extreme flood estimates is examined.

2. PROBABLE MAXIMUM PRECIPITATION ESTIMATION

The PMP was estimated in this paper by transporting extreme storms — which were observed within the quantitative range of radar — over the design catchment.

Only those storms that occurred within a previously defined 'hydrometeorologically homogeneous region' were utilized. In order to define a region with such characteristics, the concept that a particular storm (to be transposed to the design catchment area) had not occurred over the design catchment *'just by mere coincidence'* was used.

Transposed storms were then maximized, and to do so three steps were followed (see Fig. 1).

(i) *'Initial time' maximization* A storm event was maximized in time firstly by selecting a particular time duration (e.g. 6 h) and an initial time of simulation ($t(0) = 0$), and thus a hydrograph was obtained. The initial time was then lagged (i.e. shifted ahead) by one 15 min radar frame and a new hydrograph was obtained. The process was repeated for the whole storm event, always considering the selected duration, and the initial time which corresponded to the largest peak flow hydrograph obtained was chosen.

(ii) *'Tracking' maximization* The actual direction the storm travelled was kept unchanged and three different storm trackings were simulated. The one which

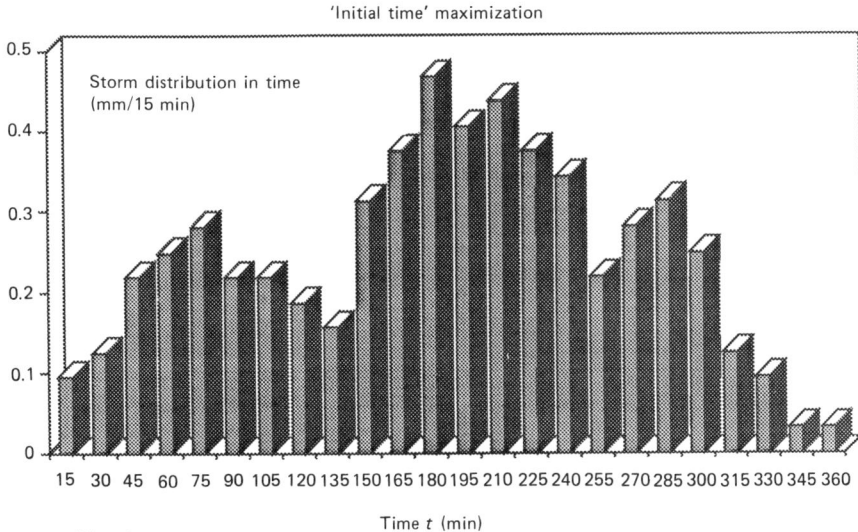

'Initial time' maximization

0.5

Storm distribution in time
(mm/15 min)

0.4

0.3

0.2

0.1

0

15   30   45   60   75   90   105 120 135 150 165 180 195 210 225 240 255 270 285 300 315 330 345 360

Time t (min)

$t(0) = 0$
$t(1) = t(0) + 15$ min
$t(2) = t(1) + 15$ min
.
.
.
$t(n) = t(n-1) + 15$ min

.... The storm initial time will be the one which produces the
most extreme peak flow hydrograph

'Tracking' maximization

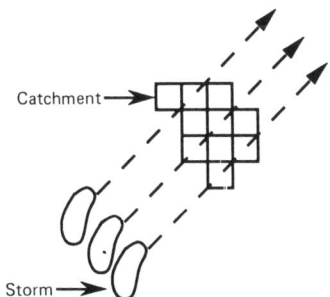

Catchment →

Storm →

The storm direction of travel is kept in the transposition.
Three different storm trackings are simulated.
The one which produces the most extreme peak hydrograph
will be considered

'Moisture' maximization

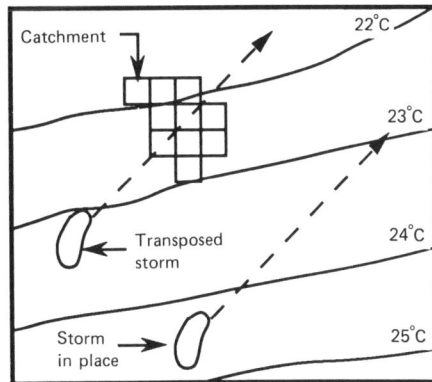

Catchment

22°C

23°C

Transposed
storm

24°C

Storm →
in place

25°C

Continuous lines indicate maximum persisting 12 h
1000 mbar dew-point temperatures for the same time
of the year that the storm occured

Fig. 1 — Storm maximization and transposition.

produced the highest peak flow hydrograph among the three simulated events was selected.

(iii) *'Moisture' maximization* Moisture maximization was carried out through the utilization of recorded maximum values of dew-point temperature (precipitable water). All radar pixels (which are, in general, squared surfaces of 2 km or 5 km side, in the UK), were multiplied by the factor PMC:

$$PMC = \frac{MRD}{MOD} \tag{1}$$

where PMC is the pixel multiplication coefficient, MRD is the maximum recorded 12 h, 1000 mbar persisting dew-point temperature (from the historical), and MOD is the maximum observed dew-point temperature for the time of the year that the storm occurred.

3. PROBABLE MAXIMUM FLOOD ESTIMATION

In this investigation the following techniques were used in order to transform the PMP estimated by storm transposition and maximization into a PMF.

(i) *Unit hydrograph* The unit hydrograph (Sherman 1932), i.e. the hydrograph resulting from one unit of rainfall excess over a specified duration, was utilized in this investigation. For the use of the unit hydrograph, which is essentially a lumped model, the radar rainfall depths (available at every 15 min) were averaged over the design catchment area. The direct runoff hydrograph due to a storm was derived by integrating the incremental rainfall excesses with the unit hydrograph. Briefly, the direct runoff resulting from the first incremental rainfall excess was calculated by multiplying it with the unit hydrograph ordinates. The direct runoff due to the second incremental rainfall excess was then calculated similarly and added to that due to the first incremental rainfall excess with the time lagging by one time interval. This procedure was continued until all incremental rainfall excesses were processed. The time increment for the rainfall excess was selected to be equal to the unit duration of the unit hydrography.

(ii) *Grid-based distributed model* The grid-based distributed model (Yu 1989) is a mathematical conceptual model which was specially devised to accept radar-derived distributed precipitation data as input. The model is based upon the concept of multilayered simulation (Fig. 2), by means of which a number of different processes of the land phase of the hydrological cycle are simulated.

(iii) *The tabular method of reservoir routing* Floods are attenuated to a certain extent when a reservoir exists to regulate the flow. The extent of the attenuation depends upon the magnitude of the flood storage with respect to the flood inflow. To account for this effect, the techniques used in this paper for both unit-hydrograph and grid-based distributed hydrological models, was that proposed by Hall and Hocking (1980) and it is called the 'tabular method of reservoir routing'. The method is based upon the equation of continuity and the storage–discharge relationship of the considered site. Further information on

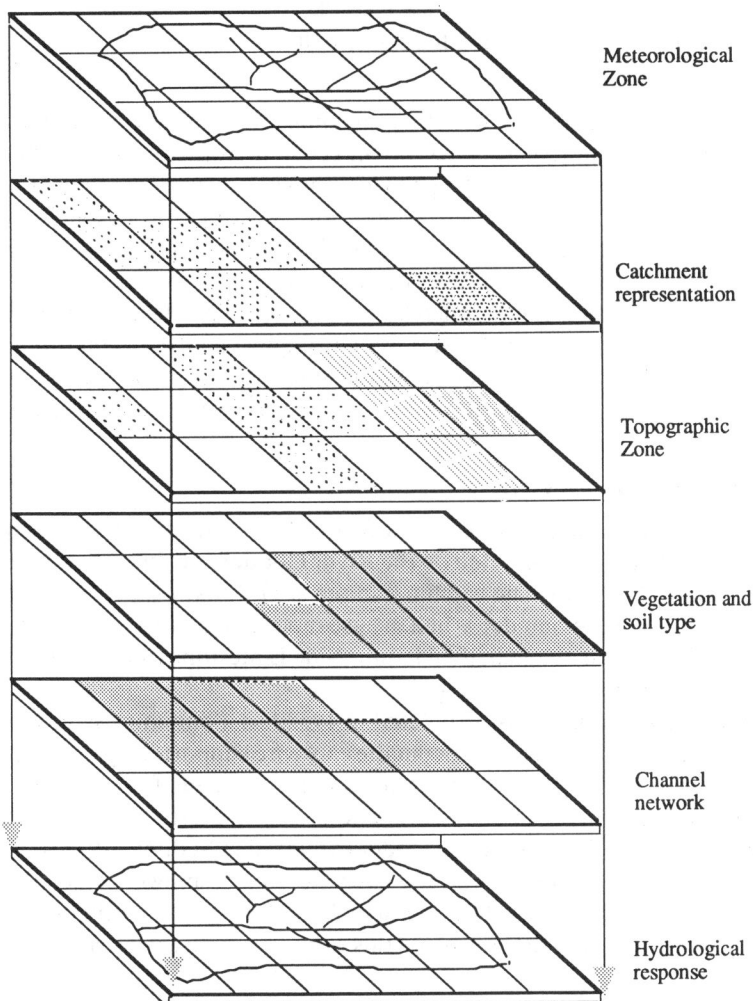

Fig. 2 — The multi-layered simulation of the grid-based distributed model.

the method may be found in the paper by Hall and Hocking (1980), who also present a complete illustrative hypothetical example.

(iv) *Antecedent moisture condition*   For the estimation of the most extreme flood that a catchment could physically produce (i.e. the PMF), the *Flood Studies Report* (Natural Environmental Research Council 1975) proposes the following method for the UH approach, which was followed in this paper.

(a) 'The unit hydrograph is made "peakier" by reducing the base length by one third and increasing the runoff ordinates by 50% to maintain the flood volume.

(b) An extreme rainfall profile is used.
(c) An allowance is made for snow melt'.

It was considered that no other maximization should be undertaken in the unit-hydrograph approach; hence no antecedent storms were considered. For the grid-based distributed model approach, however, previous storm events were considered. To do so, *the same* 'main event' was taken (i.e. its characteristics of shape, direction of travel, relative rainfall distribution between the radar pixels, etc., were all kept the same) and the absolute rainfall depth for each and every pixel was reduced by 40% (as recommended by the American Nuclear Standard N170–1976). The authors recognize that further meteorological analyses would be necessary to establish the distance in time between the antecedent and the main storm. The time gap considered in this paper to be reasonable was twice the storm duration.

4. STOCKS RESERVOIR

Stocks is a direct supply reservoir of 1.364×10^7 m^3 capacity. It has a top water area of 1.392×10^6 m^2 and is situated on the River Hodder, a tributary of the River Ribble in Yorkshire, England. The catchment area is 37.471 km^2. Some 15% of the land is afforested and the remainder is half grouse moor and half rough pasture. Half of the area is on Pendle Grits and half on Pendle Side Limestones and Bowland Shales. The Stocks catchment is shown in Fig. 3.

Stocks Dam, which was completed in 1932 has a minimum height of 33.55 m. It comprises an embankment with a puddle clay core supported by shoulders of boulder clay and protected by stone pitching on the upsteam face. The channel outlet comprises three culverts 2.44 m in diameter, on the left abutment of the embankment, discharging into the River Hodder (quoted in Seddon (1971)).

The maximum recorded outflow over the Stocks spillweir was 84.96 m^3/s which occurred during a storm on 14 December 1936. It seems likely that the reservoir was full before the onset of the storm and this would indicate a peak runoff of the order of 170.0 m^3/s.

In 1967 the Flyde Water Board commissioned an investigation into the adequacy of the Dam spillway. This study was subsequently extended to cover the modification of the structure in order to provide additional storage in the reservoir for water supply purposes.

Since the above-mentioned analyses were carried out, some new methods for spillway capacity determination have been developed involving not only non-PMF approaches (e.g. statistical procedures) but also PMF techniques.

This investigation utilizes radar-derived rainfall data in the estimation of the spillway design flood for Stocks and the results are compared with those obtained by Cluckie and Pessoa (1990), by Cluckie and Pessoa (1989) and by other previously undertaken more conventional statistically based analyses.

5. RESULTS

Cluckie and Pessoa (1990) have investigated, for Stocks, the performance of Wakeby and generalized extreme value (GEV) statistical distributions. The method of

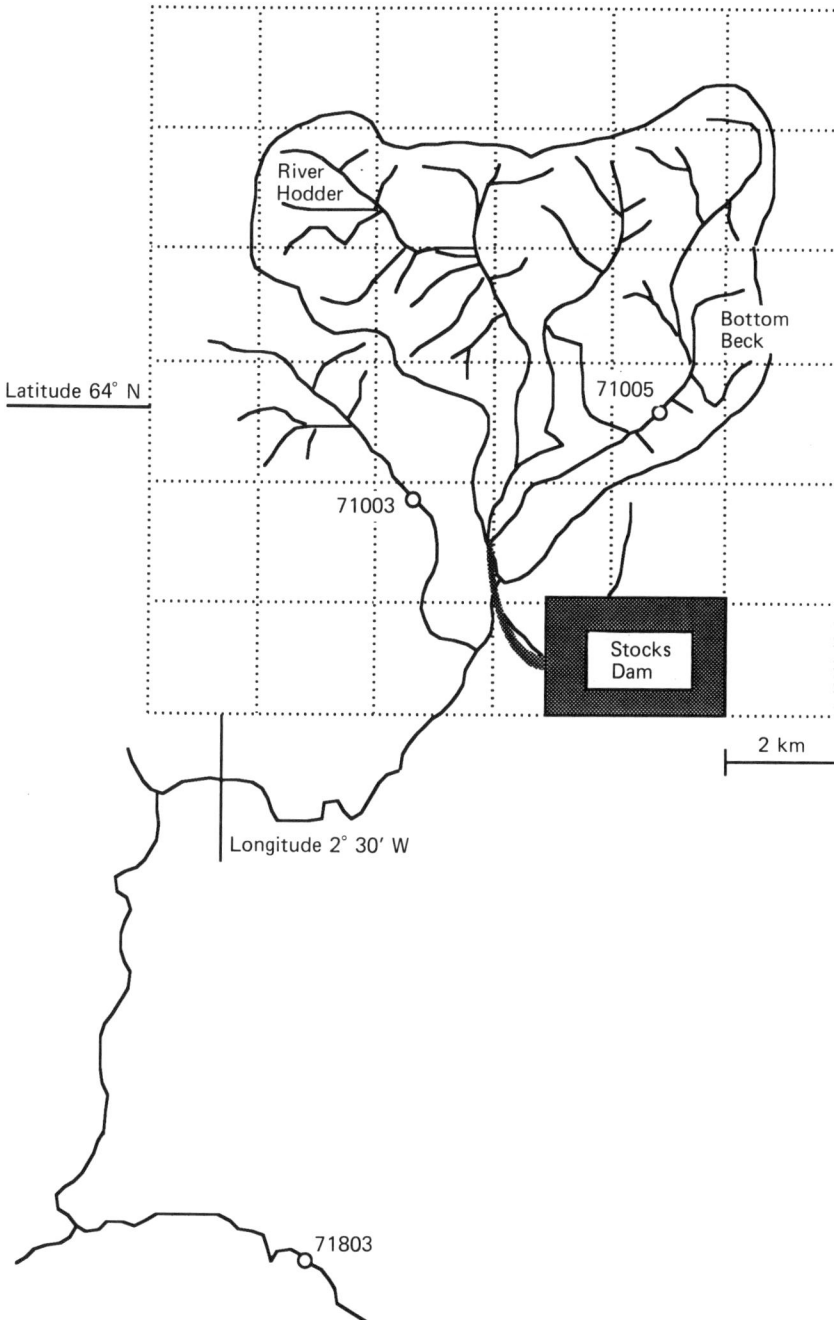

Fig. 3 — Stocks catchment area.

probability weighted moments was used for the fitting process, and the estimates were made both on an at-site and a regional basis. In the regional anaylsis, a total of 602 station years were considered (all the stations within the geographical region number 10, as defined in the *Flood Studies Report* (Natural Environment Research Council 1975). For the 10 000 year flood, the values indicated in Table 1 were found.

Table 1 — Statistical analyses for Stocks. (Adapted from Cluckie and Pessoa (1989)

Statistical analysis	10 000 year flood (m^3/s)
At-site Wakeby	250.63
At-site GEV	92.19
Regional Wakeby	147.60
Regional GEV	130.78

Pessoa (1988) applied the *Flood Studies Report* rainfall–runoff model (Natural Environment Research Council, 1975) for Stocks and found the following: for winter, 311.10 m^3/s and, for summer, 291.90 m^3/s. Binnie & Partners (1978) have also investigated the safety design flood for Stocks. They have used transposition and maximization of storms in association with the unit hydrograph and found the following: for winter, 253.20 m^3/s and, for summer, 243.90 m^3/s.

In this paper, a number of storms are investigated, which were observed within the coverage area of the Hambeldon Hill radar station, in the UK. All the storms were included in the same analysis, without considering winter and summer events separately. The floods generated by the lumped approach (unit hydrograph) and by the distributed approach are shown in Fig. 4. As may be observed from this figure, the estimated flood values for both lumped and distributed approaches were developed in order to provide the PMF.

Fig. 5 shows the sensitivity of the peak flow for both unit hydrograph and grid-based distributed model when systematic errors are introduced in the radar data. This analysis was carried out for one single-storm event (12 h duration).

6. CONCLUSIONS AND RECOMMENDATIONS

The utilization of radar distributed rainfall data in the analysis made it possible to transpose observed storm events and to maximize them not only by humidity but also in time and space, in a quite natural way. The estimates were compared with previously undertaken analyses and the methodology developed could in future be employed for the safety evaluation of such dams.

The results obtained from both lumped (unit hydrograph) and distributed (grid based distributed model) approaches were shown to be in agreement with each other, in particular for the lower storm durations. It was observed that, the greater the storm duration, the greater the difference between the lumped and the distributed estimates.

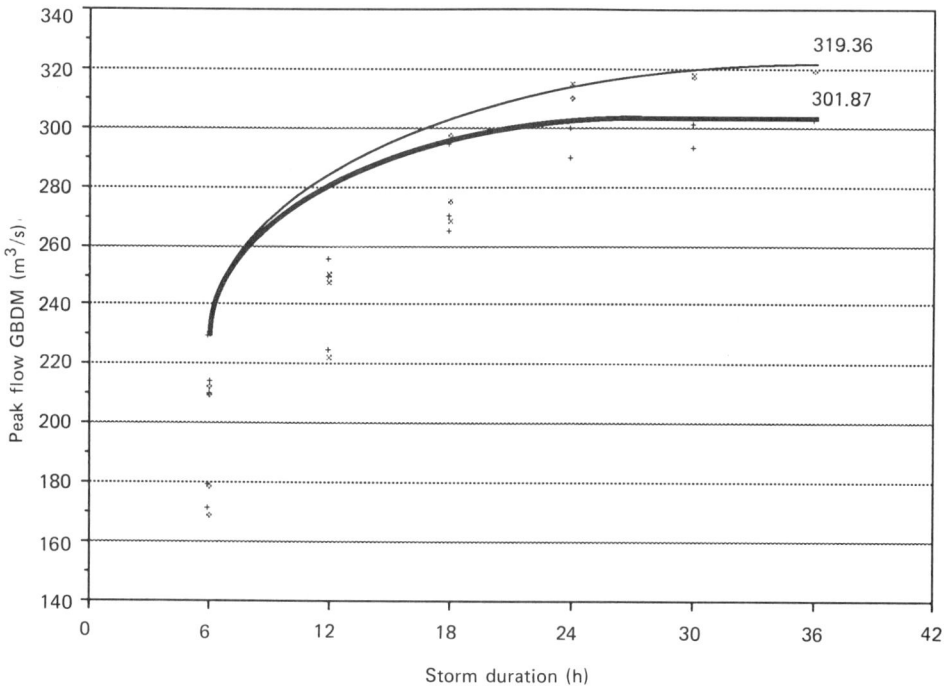

Fig. 4 — PMF from the lumped (× , ――――) and distributed approaches (+ , ■■■): GBDM, grid-based distributed model.

The distributed approach appeared to be more sensitive to systematic errors in the rainfall data, which may be explained by the fact that for the application of the unit hydrograph the radar rainfall is averaged over the catchment area, which makes the estimates more stable. Further investigation of the effects of radar calibration on extreme flood estimates are at present being carried out by the present authors.

The techniques were applied to a relatively small watershed (less than $40 \, km^2$ area). It is planned that layer catchments will be investigated, since they are more susceptible to differences in both lumped and distributed estimates, in view of the fact that the spatial variability of the storms is likely to be more significant over larger areas. The magnitude of the antecedent storm and the interval which separates it from the main storm needs to be further investigated in a meteorological context, and such an investigation is under way.

Radar has an enormous potential for hydrological investigations, and in particular for its capacity to quantify actual storms in a quasicontinuous fashion, in both time and space. Its utilization for evaluation of the design flood for dam spillways using storm transposition and maximization techniques is an additional possiblity.

7. ACKNOWLEDGEMENTS

The authors would like to thank the Conselho Nacional de Desenvolvimento Científico e Tecnológico, the Brazilian Government body which was kindly provided

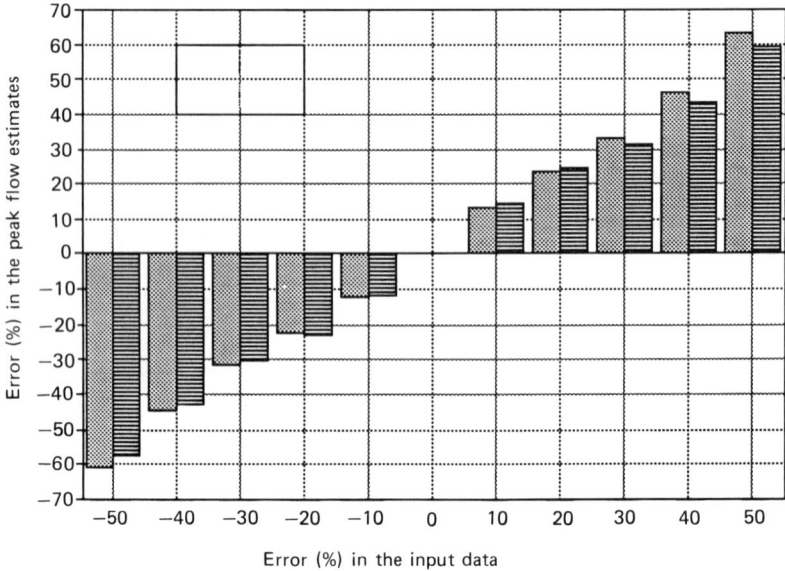

Fig. 5 — The effect of systematic error in the radar data: □, grid-based distributed model, □, unit hydrograph.

financial assistance. Thanks also go to the UK Meteorological Office, the Natural Environment Research Council (Institute of Hydrology) and the North-West Water Authority, who have assisted in providing the input data. M. L. Pessoa is currently employed by the Department of Hydrometeorology of COPEL, the Power Utility of Paraná State, Brazil, and wishes to thank the Company President, Professor F. L. S. Gomide, who has kindly allowed him leave of absence in order to undertake this project.

REFERENCES

Binnie & Partners (1978) *Report on flood studies for Stocks reservoir*, Contract Report to the North-West Water Authority. Westminster.

Cluckie, I. D. and Pessoa, M. L. (1988) Weather radar and dam safety: an evaluation. *2nd Anglo-Polish Hydrological Workshop, Birmingham, UK, 6–8th August 1988*.

Cluckie, I. D. and Pessoa, M. L. (1990) Dam safety: an evaluation of some procedures for design flood estimation. *Hydrological Sciences Journal*, (**35**) 5, **10**, 547–569.

Hall, M. J. and Hockin, D. L. (1980) *Guide to the design of storage ponds for flood control in partly urbanised catchment areas* (Draft for discussion). Technical Note No. 100. Construction Industry Research and Inst. Association, London.

National Environment Research Council (1975) *Flood studies report*, Vols I–V, Whitefriars Press, London.

Pessoa, M. L. (1989) *The hydrological utilisation of radar derived rainfall data in modelling extreme storm behaviour*, Ph.D. Thesis. University of Birmingham. Department of Civil Engineering.

Pessoa, M. L. and Cluckie, I. D. (1989) Segurança de barragem: A utilizacão de dados radar meteorológico na estimação da vazão de projeto. *40 Simpósio Luso-Brasileiro de Hidráulica e Recursos Hídricos (40 SILUSB), Tema 4, modelação Matemática em Hidráulica e Recursos Hídricos, 14–16 June 1989*, 10 Vol. LNEC, Lisboa, Portugal, pp. 159–171 (in Portuguese).

Seddon, B. T. (1971) Spillway investigations for Stocks Dam. *Proc. Inst. Civ. Eng.*, 621– 644.

Sherman, L. K. (1932) Streamflow from rainfall by unit-graph method. *Eng. News Record*, **108**, 501–505.

Yu, P-S. (1989) *Real time grid based distributed rainfall runoff model for flood forecasting with weather radar*. Ph.D. Thesis, University of Birmingham, Department of Civil Engineering, 246 pp. (including Appendix).

18

Variability of heavy-rainfall events in northwest England: an analysis of spatial structure

E. J. Stewart and N. S. Reynard
Institute of Hydrology,
Wallingford, Oxon. OX10 8BB UK

ABSTRACT

The paper describes a geostatistical analysis of the spatial structure of a number of extreme rainfall events in an area of northwest England. Radar data from the PARAGON system are used in an analysis of semivariograms. Small-scale variability is found to be a feature of frontal rainfall in the region. The results of an analysis of hourly semivarigrams suggest that only in about half of the heavy-rainfall events does the daily structure of storm rainfall totals reflect that of shorter-duration events.

1. INTRODUCTION

An understanding of the spatial variability of rainfall is required in many hydrological applications, e.g. in the definition of inputs to rainfall–runoff models. Further applications include the structuring of design storms for the assessment of flood risk and reservoir safety, for which point-to-area relationships are required, and stochastic modelling of the rainfall process itself. Until recent years, much of the research on rainfall variations was concerned with studying the spatial variability of point observations from raingauges. The major problem with this approach is the very nature of raingauges, since their phyical spacing can cause small-scale variability to be underestimated. Over the last decade, however, the situation has changed radically. The introduction of a network of weather radars in the UK has provided the hydrologist with a massive increase in rainfall information. The data are consistently at higher temporal and spatial resolutions than were previously available from raingauge networks, especially in remote areas. Furthermore, the nature of the precipitation measurements from radar differs from that of raingauges, since radar estimates represent spatial averages.

This paper demonstrates the use of radar data in an analysis of the spatial variability of storm rainfall. The methodology applied to the problem is that of semivariogram analysis, one of a series of techniques known as geostatistics. Results of an analysis of 23 heavy-rainfall events in the study region of northwest England are presented.

2. SPATIAL VARIABILITY OF EXTREME RAINFALLS

The problem of identifying the spatial characteristics of the rainfall process has been addressed in many studies. In the context of design applications, the classical approach has been the identification of statistical areal reduction factors. A statistical areal reduction factor is a value which can be applied to a point rainfall of a specified duration and return period to yield the areal rainfall of the same duration and return period. In the UK, the most widely used areal reduction factors are those given in the *Flood Studies Report* Vol. II (Natural Environment Research Council 1975). Bell (1976) reassessed the validity of these values and questioned their assumed invariance with return period. More recently, Stewart (1989) has demonstrated that radar data can be used in conjunction with raingauge data to derive areal reduction factors for short durations. The results of that work suggest that areal reduction factors calculated for northwest England for a range of durations produce lower areal rainfall values than those given in the *Flood Studies Report*. Since areal reduction factors in the *Flood Studies Report* are averages for the whole of the UK, this can be taken as an indication of differences in the extreme rainfall regime of the northwest region. In view of the dominance of frontal rainfall in the region, these results are somewhat surprising. However, they do correspond to the observation made by Dales and Reed (1989) that there is less spatial dependence in extreme rainfalls in the northwest than in the northeast of England.

The concept of an areal reduction factor is a statistical one, which gives only a broad indication of the average discrepancy between the rainfall recorded at a point and that experienced over a specified area. Therefore, in the current study an alternative analytical approach was sought, and that provided by geostatistics was adopted.

3. THE GEOSTATISTICAL APPROACH

The term geostatistics refers to a set of statistical procedures for analysing the spatial structure of random variables and for performing interpolation and areal estimation (Cooper and Istok, 1988). Until recently, the main application of geostatistical techniques has been in mining and geology, but a growing interest in the methods is currently discernible in other fields, including hydrology and hydrometeorology (for example, Lebel and Laborde, 1988; Slimani and Obled, 1986; Stewart and Reed, 1989). Geostatistics can be used wherever sample values of a spatially or temporally distributed variable can be expected to be affected by their positions and relationships with nearby values. The theoretical basis of geostatistics does not depend on the physical nature of the variable being studied. A primary advantage of geostatistical techniques over other methods is that they are tolerant of missing values. This makes them particularly applicable to the analysis of radar fields, since problems such as occulation and permanent clutter cause some data to be lost.

In this study the analysis of semivariograms of radar fields has been used to describe the spatial structure of heavy rainfall events in northwest England. A semivariogram expresses the spatial variability of a field of values in terms of intersite distance; the mean variablity between pairs of observations is plotted at regular distance intervals. In mathematical terms, the definition of an experimental semi-variogram (γ) is as follows:

$$\gamma(h) = \left[\frac{1}{2N(h)}\right] \sum_{i=1}^{N(h)} \left[z(x_{i+h}) - z(x_i)\right]^2 \tag{1}$$

where $z(x_i)$ is a measurement of a particular variable at the point x_i and $N(h)$ is the number of sample pairs separated by the distance h. The limit of distance to which the semivariogram should be calculated is usually taken as half of the maximum extent of the area sample (Clark, 1979). An ideal semivariogram tends to increase with distance, h, until it becomes approximately equal to the global variance in the field of values (the 'sill') at a distance called the 'range'. This is known as the spherical model. The distance at which the sill is reached can be used to identify the maximum extent to which observed points are similar in value to one another. Another common feature of the semivariogram is the 'nugget' effect, where γ appears to have a value greater than zero when extended back to the point where h is zero. This discontinuity indicates the existence of a random or unpredicatble component in the field of values. Several experimental semivariograms can be calculated for different directions in order to determine whether the spatial variability function is isotropic.

The analysis of semivariograms offers a way of characterizing the spatial variability of a set of observations. This study set out to identify three different characteristics of the variability of rainfall fields from radar data. These were the size of the nugget effect, and hence the extent of very small-scale variability, the range, which corresponds to the physical size of the event, and the alignment and shape of the rainfall event from directional semivariograms.

4.   ANALYSIS OF RAINFALL IN NORTHWEST ENGLAND

4.1   Study region

The study region was defined as a square of side 100 km centred on the Hameldon Hill radar station in northwest England. Details of the installation are given by Hill and Robertson (1987). The region is broadly an upland one, with altitudes ranging from sea level to over 600 m (Fig. 1). The character of the region is diverse, since it includes the urban areas of Liverpool and Manchester, as well as a section of the Pennines. A further characteristic of northwest England is that it has the highest density of upland reservoirs in the UK. Thus, a knowledge of the variability of rainfall within the region is fundamental to the setting of design criteria.

Within the study region, heavy 1 day rainfalls in the range 38–50 mm are a recurrent feature (Dales and Reed 1989). Long-term average annual rainfall (SAAR) is generally high, ranging from 660 to 2200 mm. Throughout the year the rainfall in the region is predominantly frontal. However, both the number and the strength of the depressions decrease during the summer, when heavy-rainfall events tend to originate from more localized convective cells. Frontal events are character-ized by high daily totals with steady unspectacular falls all day, while convective storms may produce only 1 or 2 h of heavy rainfall.

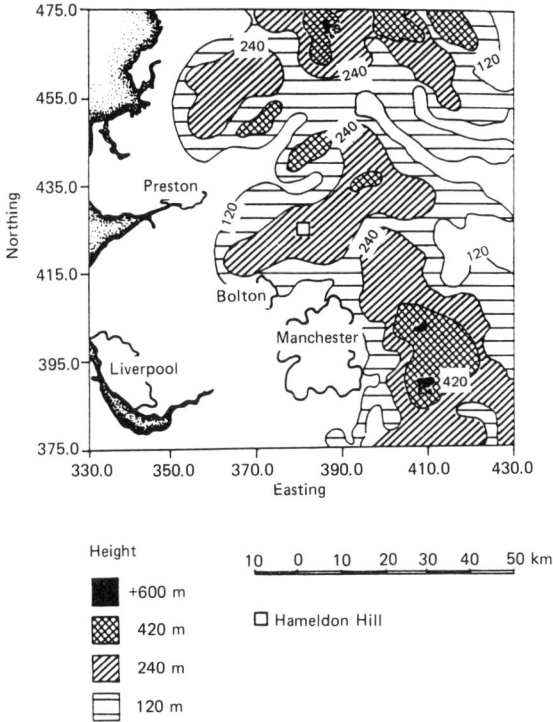

Fig. 1 — Topographical map of the study region.

4.2   Data

Radar data from Hameldon Hill, provided by the Meteorological Office's PARA-
GON data processing system (May 1988), were used in the analysis. The system
produces hourly calibrated radar measurements which are not subsequently adjusted
by ground truth data. The data consisted of hourly average rainfall depths in mm for
a grid of 400 squares of side 5 km covering the whole study area. Daily adjusted
rainfall totals were also available. The radar record was not continuous, consisting of
143 days out of a total of 185 days on which rainfall was heavy and widespread
between the years 1981 and 1987. Radar data were not available for the remaining 42
days. The heavy-rainfall days had been selected by reference to raingauge data from
the daily rainfall archive held at the Institute of Hydrology.

4.3   The requirement

The aims of the study were twofold. Firstly, the analysis was undertaken in order to
verify the results of the study of areal reduction factors in the same region (Stewart
1989). A second motive for the study was to determine whether the radar data set
could be used in an analysis of rainfall frequency for short durations. This would
allow the assessment of return periods corresponding to the maximum rainfalls of
specified subdaily durations. The main problems here are the incompleteness of the

data set, as well as the difficulty of judging its representativeness of heavy rainfalls of short duration, owing to the fact that daily rainfall was used to select the events. It was thought that, if an analysis of daily and hourly semivariograms was carried out, the similarities and differences between the two would determine the validity of selecting short-duration rainfall events on the basis of daily totals.

Accordingly, the two aspects analysed were the characteristics of the spatial structure of some of the rainfall events, and the similarities between daily and hourly rainfall fields.

4.4   Analysis of daily totals

The 23 days of heaviest rainfall were selected from the total of 143 rain days, again with reference to daily raingauge data. Rain days were selected if daily totals at more than five gauges within the study area exceeded 4% of the at-site SAAR value. This selection yielded 15 days of widespread rainfall, with the remainder showing more isolated patterns. The seasonal distribution of the rain days is given in Table 1 together with an indication of the type of event giving rise to heavy rainfall. It can be seen that 43% of the heavy-rainfall days occurred during the summer months, of which half were associated with convective storms. Table 1 emphasizes the dominance of frontal rainfall in the region.

Table 1 — Seasonal distribution of the rainfall days used in the analysis: F denotes frontal rainfall events and S denotes convective storm events

Winter (December–February)	Spring (March–May)	Summer (June–August)	Autumn (September–November)
2 January 82, F	14 March 1982, F	5 August 1981, S	17 November 81, F
8 December 1983, F	14 May 1985, F	6 August 1981, S	9 September 1983, F
3 December 1986, F	16 May 1986, F	18 June 1982, S	9 October 1983, F
29 December 1986, F		22 June 1982, S	3 September 1984, F
		25 June 1982, F	2 November 1984, F
		4 August 1982, S	3 November, 1984, F
		17 August 1982, F	
		26 July 1985, S	
		25 August 1986, F	
		18 July 1987, F	

Average directional semivariograms were constructed for the daily totals on each of the 23 days, and their characteristics were studied in conjunction with daily synoptic charts. For simplicity, each radar-derived average value was assigned to the centre point of its grid square. Two basic shapes were apparent among the average semivariograms: spherical and linear. Examples are given in Figs 2 and 3, together with contour plots of the daily adjusted radar fields. On the contour plots, blank areas denote either lack of rainfall or missing values.

An example of a spherically shaped semivariogram is given in Fig. 2(a). This indicates increasing variablity between points with increasing distance until the sill is reached at a range equal to approximately 45 km. From Fig. 2(b) it can be seen that this corresponds to the diameter of the area receiving most of the rainfall over the 1

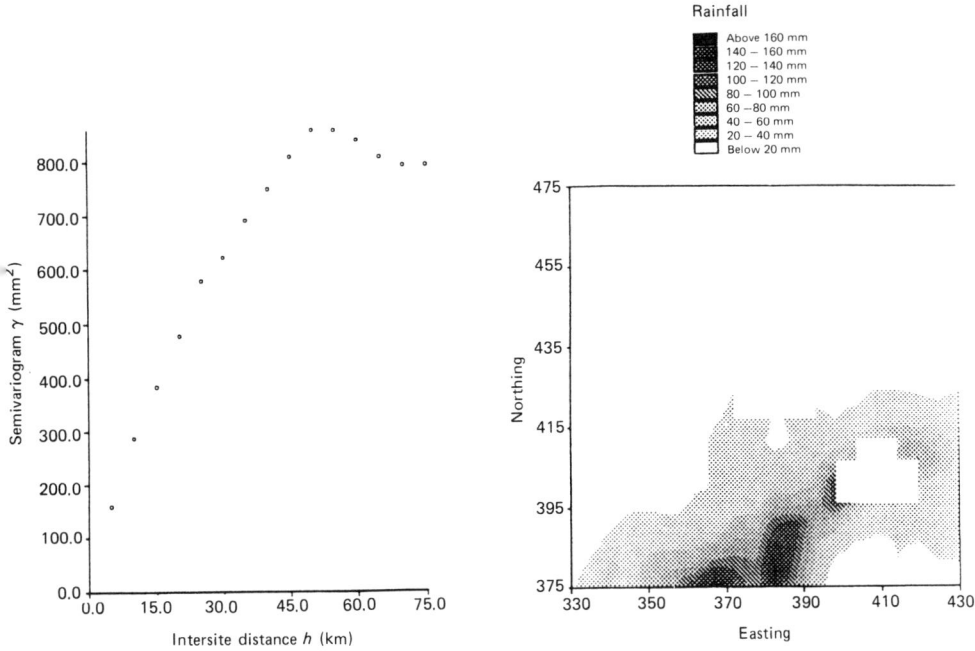

Fig. 2 — (a) Semivariogram of daily rainfall and (b) isohyetal map of daily rainfall totals on 5 August 1981.

day period. The rainfall on this day was caused by a band of thunderstorms which affected only the south and east of the study region. The semivariogram shows a small nugget value of about 45 mm². The variance of the daily totals is 717.7 mm², and thus the nugget indicates that approximately 6% of the total variability in the field is unpredictable.

A semivariogram showing a linear shape (Fig. 3(a)) indicates that the variability in the field of values increases with increasing distance to a range beyond that of the study area. This reflects the widespread heavy rainfall often associated with frontal systems. It can be seen from Fig. 3(b) that, making allowance for missing data, the rainfall covers the whole study region, although the actual totals are considerably smaller than those of the convective event on 5 August 1981. The daily semivariogram does not show a marked nugget effect.

In several cases, the daily semivariograms were found to follow a domed shape, characterized by a gradually decreasing variability after the maximum was reached. It was noted that these tended to correspond to days when the contours of daily rainfall were roughly circular and appeared to be centred on Hameldon Hill itself. These semivariograms appear to represent a special case of the spherical model, when points situated relatively far apart show only small differences in value. A number of composite semivariogram shapes were also encountered which seemed to be associated with a combination of rainfall-producing mechanisms.

Semivariograms for the five convective storm events showed spherical or dome-shaped behaviour. However, no clear tendency for frontal events to be associated with a specific semivariogram shape was identified.

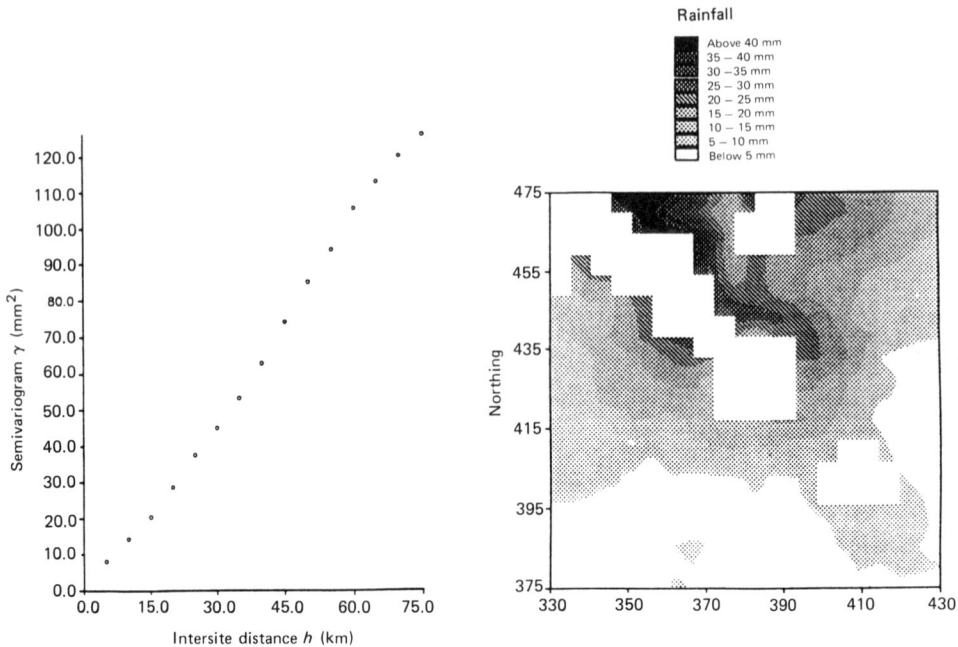

Fig. 3 — (a) Semivariogram of daily rainfall and (b) isohyetal map of daily rainfall totals on 2 January 1982.

Semivariograms for four different directions were produced for each of the 23 rain days. These were found to indicate the alignment of the areas of maximum daily rainfall. An example is given in Fig. 4, in which the semivariograms in the northwest–southeast and north–south directions are broadly similar to the daily average (Fig. 2(a)) and indicate that the spatial variability is well defined. Those in the west–east and northeast–southwest directions show that the variability is generally increasing beyond the boundaries of the study region.

4.5   Comparison of daily and hourly semivariograms

Eleven of the 23 rain days were selected for the comparison of daily and hourly semivariograms. They were selected on the basis of having few missing data and being representative of the whole data set. Two basic types of rain day could be distinguished: days on which most of the rain fell within a few hours, and days on which the rainfall was almost continuous throughout the day. In the case of the former, the hourly semivariograms were very similar in form to those for the daily totals. An example of the semivariogram of the hour of heaviest rainfall on the rain day of 5 August 1981 (defined as the period from 0900 Greenwich Mean Time (GMT) on 5 August to 0900 GMT on 6 August), together with a plot of the rainfall totals for the same hour, is given in Fig. 5. It can be seen that the hourly semivariogram basically follows the same shape as the daily one (Fig. 2(a)) but shows a discontinuity at about 25 km which reflects the lateral extent of the rainfall cell (Fig.

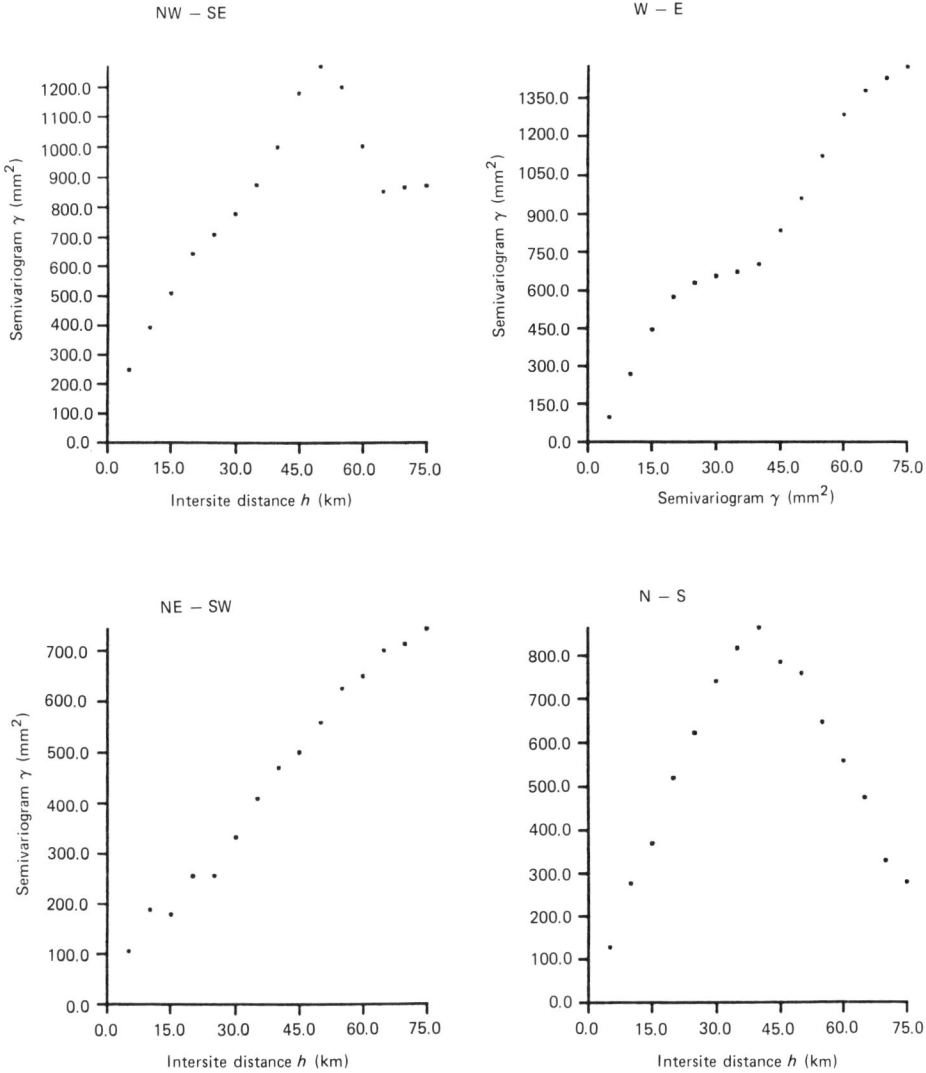

Fig. 4 — Directional semivariograms of daily rainfall on 5 August 1981.

5(b)). In the case of heavy rainfall of longer duration, much more variability was apparent between hourly and daily semivariograms. An example is given in Fig. 6 for 14 March 1982. The synoptic chart for this day indicates a generally westerly airflow, with a waving cold front slightly to the south of the study region. This produced a showery day. The semivariogram of daily totals (Fig. 6(a)) reflects the widespread rainfall, indicating increasing variability with distance to a range beyond that of the study region. The semivariogram for the hour ending 1500 GMT (Fig. 6(b)) demonstrates a different spatial structure, with maximum variability reached at a

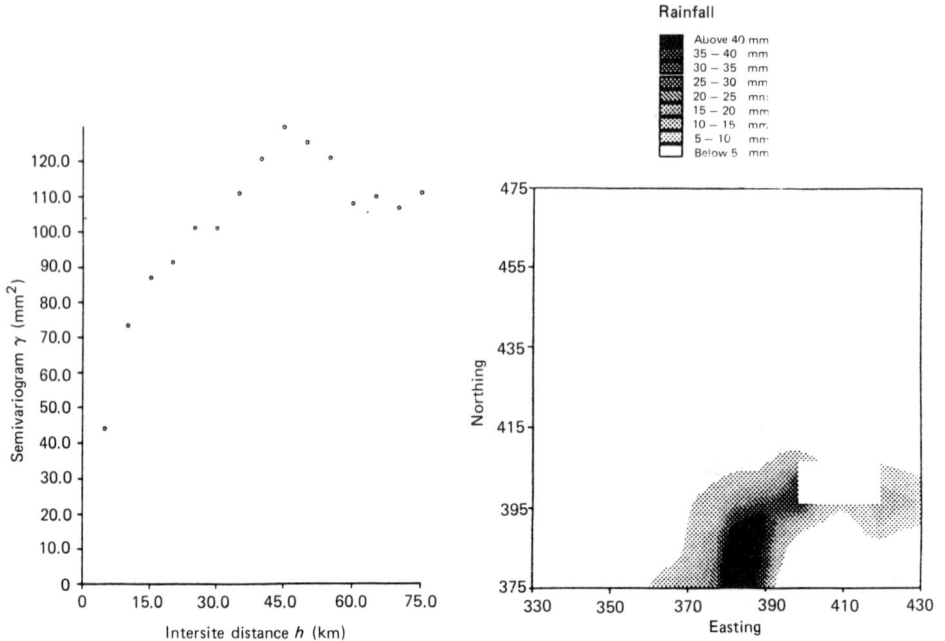

Fig. 5 — (a) Semivariogram of rainfall for the hour ending 0200 GMT on 6 August 1981 and (b) isohyetal map of rainfall.

range of 45 km. The rainfall totals for this hour show the existence of a fairly intense shower centred to the southeast of Hameldon Hill.

5. DISCUSSION

Examination of the daily semivariograms revealed that, although two different types of rainfall (convective and frontal) affect northwest England, corresponding types of spatial structure are not easily distinguished. All the summer storms studied showed spherical or dome-shaped semivariograms of daily rainfall, values of the range and nugget being related to the physical size of the event and the extent of very small-scale variability respectively. However, spherical and dome-shaped semivariograms were also found to characterize 13 out of 18 days of frontal rainfall, when high daily totals were locally concentrated into circular or elliptical patterns. On such days, the front itself seems to have passed quickly over the study area, so that for the majority of the time the region was affected by a showery airstream, often causing localized downpours. Orographic enhancement is also a feature of the rainfall regime of much of the region. This tended to be characterized by an initially spherical semivario-gram, indicating localized showers over the Pennines, which showed subsequent linear behaviour, thus reflecting general rainfall of lower intensity over the whole region. The daily semivariograms showed a linear shape, reflecting widespread moderate rainfall in only 5 of the 18 days of frontal rainfall. This indicates that frontal rainfall in northwest England is often characterized by small-scale variability, and is

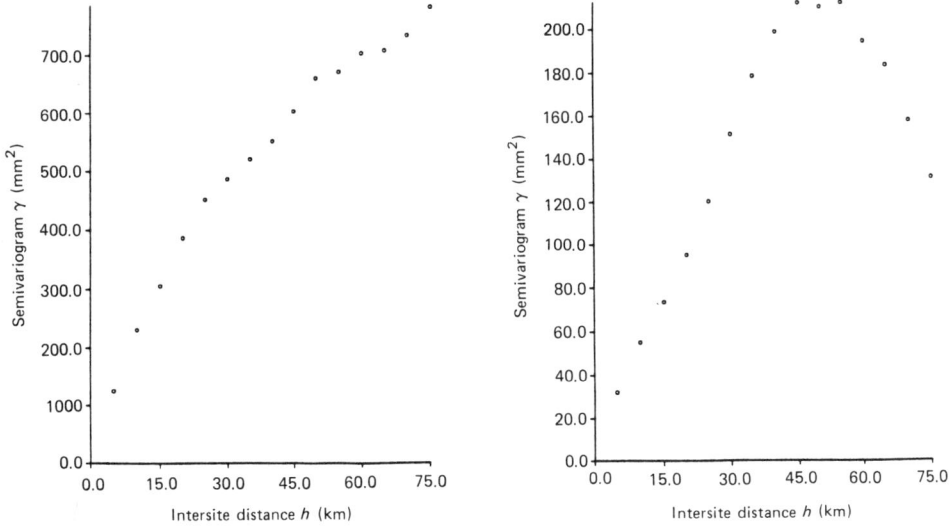

Fig. 6 — (a) Semivariograms of (a) daily rainfall and (b) rainfall for the hour ending 1500 GMT
on 14 March 1982.

consistent with the results of the study of areal reduction factors in the same region
(Stewart 1989).

The apparent similarities in the spatial structure of different types of rainfall may
be a function of the resolution of the radar data and the fact that the analysis has been
concerned with variability over a large region. In a recent study of small areas within
the same region, Shepherd *et al.* (1988) described a method of identifying rainfall
type from weather radar and discussed its application to real-time calibration. The
method is derived from space–time correlation surfaces and thus is similar to an
analysis of directional semivariograms but requires data at higher temporal and
spatial resolutions.

The results of the analysis of hourly semivariograms suggest that only in about
half of the heavy-rainfall events within the region does the daily structure of storm
rainfall totals reflect that of shorter-duration events. This has implications for the use
of the data set in design storm analysis for short durations.

6. CONCLUSIONS

The use of radar data for spatial rainfall studies has been illustrated. The analysis of
semivariograms has revealed a high degree of small-scale variability within frontal
events in northwest England. It has not been possible to characterize rainfall type on
the basis of semivariogram structure, although this may reflect the spatial and
temporal resolution of the data. The spatial structure of daily rainfall has been found
to correspond to that of hourly totals in about half of the rain days studied. Thus, in
order to draw conclusions about the spatial structure of hourly rainfall from daily
totals, further meteorological information is required.

ACKNOWLEDGEMENTS

Work on the variation of extreme rainfall events in upland areas was commissioned by the UK Department of the Environment (Contract No. PECD7/7/190). Strategic research on rainfall estimation is also supported by the Ministry of Agriculture, Fisheries and Food.

REFERENCES

Bell, F. C. (1976) *The areal reduction factor in rainfall frequency estimation*, Report No. 35. Institute of Hydrology, Wallingford, Oxon.

Clark, I. (1979) *Practical geostatistics*. Applied Science, Barking, Essex.

Cooper, R. M. and Istok, J. D. (1988) Geostatistics applied to groundwater contamination. I: Methodology. *J. Environ. Eng.*, **114** (2), 270–286.

Dales, M. Y. and Reed, D. W. (1989) *Regional flood and storm hazard assessment*, Report No. 102. Institute of Hydrology, Wallingford, Oxon.

Hill, G. and Robertson, R. B. (1987) The establishment and operation of an unmanned weather radar. In: V. Collinge and C. Kirby (eds), *Weather radar and flood forecasting* Wiley, Chichester, West Sussex, pp. 55–59.

Lebel, T. and Laborde, J. P. (1988) A geostatistical approach for areal rainfall statistics assessment. *Stoch. Hydrol. Hydraul.*, **2**, 245–261.

May, B. R. (1988) Progress in the development of PARAGON. *Meteorol. Mag.*, **117**, 79–86.

Natural Environment Research Council (1975) *Flood Studies Report*, Vols I–V. Whitefriars Press, London.

Shepherd, G. W., Cluckie, I. D., Collier, C. G., Yu, S. and James, P. K. (1988) The identification of rainfall type from weather radar data. *Meteorol Mag.*, **117**, 180–186.

Slimani, M. and Obled, C. (1986) Regionalization of extreme rainfall parameters through kriging and correlation with topography descriptors. *Proceedings of an International Symposium on Food Frequency and Risk Analyses, Baton Rouge, LA, May 1986.*

Stewart E. J. (1989) Areal reduction factors for design storm construction: joint use of raingauge and radar data. *Proceedings of the symposium on New Directions for Surface Water Modelling, May 1989 Baltimore, MD*, IAHS Publication No. 181. International Association of Hydrological Sciences.

Stewart, E. J. and Reed, D. W. (1989) Spatial structure in point rainfall: a geostatistical approach. *Proceedings of the British Hydrological Society Symposium, Sheffield, September 1989.* British Hydrological Society.

19

Properties of echoes at first detection, resulting in multicelled rainstorms

N. Wescott and S. Changnon
Climate and Meteorology Section, Illinois State Water Survey, Champaign, 61820, USA

ABSTRACT

The initial characteristics of 144 radar echoes have been examined for two summer days when convective storms were observed. It was found that, for a given rain period, the distribution of height, area and reflectivity of the echoes which subsequently merged were shifted towards higher values than for those that dissipated without becoming joined to another echo at the 20 dBZ level. Between-day differences and results regarding the first echo location with respect to other echoes also are discussed.

1. INTRODUCTION

It is a common feature of convective cloud systems that they be composed of a number of cloud elements. The manner in which clouds aggregate together has been the subject of a number of observational studies (Changnon 1976, Cunning *et al*. 1982, Foote and Frank 1983, Ackerman and Westcott 1984, Cunning and DeMaria 1986, Ackerman and Kennedy 1989, Westcott and Kennedy 1989) and cloud modeling efforts (Orville *et al*. 1980, Tao and Simpson 1984, 1989) (see the review by Westcott (1984)). These studies have shown that there are a variety of factors resulting in the initiation, growth and subsequent aggregation of clouds. Mesoscale convergence and convective scale outflows generated by downdrafts have been found to be important in the initiation of new cells (Simpson 1980, Fankhauser,1982, Westcott and Kennedy, 1989). The strength of the dynamic forcing and the local thermodynamic environment in which a cloud develops have a large impact on its rate of growth. Differential cloud motion, the horizontal expansion of cloud cells, new cell growth between existing clouds, and natural seeding by adjacent clouds have been observed to be important in the aggregation of cloud cells.

Many of these cloud investigations have looked at the evolution of only a few cases or have examined the general rainfall, properties of differing sized cloud systems. A new study has been undertaken using three-dimensional radar reflectivity data to describe the aggregation of echoes for a population of echoes on several days when different environmental conditions exist.? This preliminary study concentrates on the ecoes at first detection to determine whether clouds which join together have different initial characteristics which result in a higher likelihood of their aggregating together, than those that do not join.

2. DATA ANALYSIS

The data used in this study were obtained by the ISWS/NSF CHILL radar on two summer days in 1986. The data were processed by a program which compressed the data into six two-dimensional Cartesian grids of maximum echo top height, minimum echo height and height of the maximum reflectivity, as well as the reflectivity at the top of the echo and at the echo base, and the maximum reflectivity of the echo, for each x, y grid-point location. A 121 km×121 km grid was used, with a horizontal resolution of 1.25 km, and the radar located at the center of the grid. The minimum reflectivity recorded on the first case day 25 July 1986) was 20 dBZ, and on the second day (26 August 1986), 14 dBZ.

First echoes were identified and tracked in time first by computer and then checked by a radar analyst. The echo had to contain at least two 20 dBZ grid points and had to exist for at least two consecutive time periods to be included in the sample. Care was taken to insure that only echoes topped by the radar antenna were included in the sample. Echoes that moved into the analysis area also were excluded. The echo cores were tracked to the time when they either dissipated or merged with an adjacent echo core. In order for a echo to be considered to have joined with a neighboring echo, it had to be linked at more than one grid point and the bridge had to be present for at least two consecutive volumes.

3. DESCRIPTION OF THE CASE DAYS

The environment conditions under which convection developed on the two days were similar in several respects. First the trigger mechanism for both rain periods was an approaching cold front. The two periods were characterized by a warm cloud bases, ample moisture and an unstable atmosphere. Additionally, the vertical shear of the wind was weak (10–18 m/s) and concentrated below the cloud base. Other conditions prevailed which influenced the growth of the convective cells.

3.1 25 July 1986, 1533–1901 CDT

During this afternoon, convection formed along a cold front which moved through central Illinois at about 17700 Central Daylight Time (CDT). The 1634 CDT sounding launched from the radar site (CMI) indicated unstable conditions, as well as the presence of the front. A strong inversion was observed at about 500 mbar; the air was conditionally unstable both below and above the inversion. Ample moisture was indicated. The cloud base was estimated as 2.1 km with a temperature of about 17°C. Two major lines developed and moved eastward at about 13 m/s through the area, one about 60 km to the north of the CHILL and one about 50 km to the south of the CHILL. Both were oriented from the WSW to the ENE, generally parallel to the mean cloud layer winds. The individual cells were moving to the ENE at about 12 m/s, while the system as a whole moved to the east at about 13 m/s.

3.2   26 August 1986, 0800–1230 CDT

On this morning, thunderstorms were forming in a warm air mass about 250 km ahead of a strong, slowly moving cold front. The 0700 CDT sounding at Salem, IL (160 km to the SSW of CMI), showed conditionally unstable conditions to 500 mbar above a shallow surface inversion, and substantial moisture to 450 mbar (6.5 km mean sea level). Visual aircraft reports indicated a variable cloud base with multiple layers of clouds. The cloud-base temperature and height were estimated at 19°C, at about 1–1.5 km mean sea level. Surface heating was inhibited by a cirrus overcast and an analysis of surface winds indicated only weak convergence in the area.

The area of convection had originated in southern Iowa near midnight and moved southeastward into the study area by morning. The area consisted of a large number of small line and several large lines, typically oriented SSW to NNE, more perpendicular to the main cloud layer winds, and to the cloud layer vertical shear vector. The storms moved to the east at about 19 m/s and the individual cells moved to the northeast, also at about 19 m/s. The winds were light (2.7 m/s) and from the south at the surface, veered to the WSW at 14 m/s by 1.5 km and remained steady aloft. By 1145 CDT, the clouds were moving out of the study area and were rapidly diminishing in intensity and areal extent.

4.   RESULTS

During the afternoon of 25 July 1986, only 36 echoes were found during the $3\frac{1}{2}$ h period. The echoes in general tended to be larger and longer lived than those during the morning period. Of the 36 first echoes, 27 (75%) joined with another echo. During the morning period of 26 August 1986, 108 first echoes were observed and, of these, 78 (72%) became aggregated with another. About 90% of the dissipating echoes had lifetimes of less than 30 min on both days. Of those that merged, 35–40% did so within 5 min on both days and 75–80% joined within 20 min.

4.1   First-echo studies

The first echoes were stratified in a number of ways. First, they were examined to determine whether any obvious differences could be observed in their initial character that might allow us to predict which of the echoes would merge and which would dissipate. The first-echo area, maximum reflectivity and height were examined. In general, while both subsets of data covered a range of values, the echoes which subsequently merged were first observed at higher levels, tended to be larger and have stronger reflectivities. The median value and range of values which included more than 50% of the sample are presented in Table 1.

The indication that the echoes later merge tend to be larger and taller when initially observed suggests that they are more vigorous and thus may grow large enough or remain active long enough to aggregate with another echo eventually. Differences between the two days also were observed. The merging echoes during

Table 1 — Median and range of values which include more than 50% of the sample for echo characteristics at first detection: maximum reflectivity, area, maximum height, height of the maximum reflectivity and minimum height of the echo

| | | Afternoon 25 July 1986 | | Morning 26 August 1986 | |
| | | Dissipate, | Merge, | Dissipate, | Merge, |
		Sample of 9	Sample of 27	Sample of 30	Sample of 78
Maximum Z	Median	25.0–27.5	25.0–27.5	25.0–27.5	27.5–30.0
(dBZ)	Range	22.5–30.0	22.5–30.0	20.5–27.5	22.5–30.0
Area (km^2)	Median	3–5	6–10	3–5	3–5
	Range	3–5	3–10	3–5	3–10
Maximum	Median	4.5	7.5	5.5	5.5
height	Range	3.5–4.5	7.5–10.5	4.5–5.5	5.5–6.5
(km)					
Height of	Median	3.5	6.5	4.5	5.5
maximum	range	3.5–6.5	4.5–7.5	3.5–5.5	4.5–6.5
Z (km)					
Minimum	Median	2.5	4.5	1.5	1.5
height (km)	Range	1.5–3.5	3.5–6.5	1.5	1.5

the afternoon period tended to be larger and form higher than the echoes which dissipated on this day and than both the merging and the dissipating cores of the morning case. The difference between days could be attributed to stronger forcing and to the presence of strong surface heating on the afternoon of 25 July.

After observing the differences between the initial characters of the first echoes, an attempt was made to determine whether the initial location of the first echo might have a bearing on whether or not they subsequently merge. It was hypothesized that those that merged initially formed closer to an adjacent echo, perhaps between two nearby echoes or in a preferred direction from the parent storm. During the morning period, for both groups of echoes, those that dissipated and those that merged respectively, 76–82% formed within 5 cm and 93% formed within 10 km of another echo. During the afternoon, the echoes tended to form further from nearby echoes. About 60% formed within 5 km and 90% of the merged echoes formed within 15 km of a neighbor. Thus, the merged echoes in this sample do not appear to form closer to a neighbor. The between-day difference in distance of formation is larger than the difference between merging and dissipating echoes.

The data were then examined to determine whether the echoes that merged appeared to form in a preferred location in comparision with those that dissipated. Byers and Braham (1949), in looking at echoes on three summer days in Ohio, found that the most likely location for new echo formation on two of the days was between two existing echoes. Similar results were found here. On 26 August, 33 (31%) of the first echoes were located less than 5 km away from and between two neighboring echoes. During the afternoon of 25 July, however, only five (14%) of the echoes were between two neighboring echoes, reflecting the wider cell spacing. Of this group of echoes, only 60% subsequently merged, compared with 77% for all other echoes. During the morning period, when echoes forming between two closely spaced neighbors were more common, 79% of these subsequently merged, as opposed to 69% for all other echoes. These between-neighbor echoes on 26 August typically merged more rapidly than the other merging first echoes but were similar in size and height to the other echoes.

Other investigators have observed clouds to form on the right flank of an existing cloud system (Browning and Ludlum 1962, Dennis *et al.* 1970, Maurwitz 1972). Typically these storms studied developed under conditions of moderate to strong shear and were severe in nature. Again, as observed by Byers and Braham (1949), echoes formed in all directions with respect to a 'PARENT ECHO'. However, a preferred direction was found on both days. Here we found that on 25 July the preferred direction of first-echo location was on the right flank and included ten (28%) of the first echoes; 90% of these echoes joined. On 26 August, the preferred direction of formation was on the right flank as well, but this included only 17 (19%) of the first echoes. Of these, 81% subsequently merged, about 10% more than of those forming elsewhere.

At the time of first-echo detection, little difference was observed in the size and height of the echoes which formed in a preferred location, In contrast with the echoes which formed between two existing nearby echoes, these echoes tended to form further away from the parent echo with which they would merge and took somewhat longer to join with the adjacent echo. On 25 July, the south west echoes formed approximately 10 km further from adjacent echoes and took more than 15 min longer to merge. During the morning case this difference was on the order of 5 min and 5 km. This again reflects the larger characteristic spacing between echoes found during the afternoon period.

4.2   Echo cores already merged at the time of first detection
Another difference between the two rain periods that is not reflected in the first-echo analysis was the multicelled nature of the merged storms. On both days, adjacent cores were observed that had already merged at the 20 dbZ level. This occurred in 48 cases on 26 August, and in 40 cases on 25 July, Thus, on 25 July, more adjacent cores formed already attached than there were first echoes while, on 26 August, only half as many adjacent cores formed that were merged. Generally, it was the echoes that were merging that developed the adjacent cores.

The first echoes and joined cores that developed during the morning case were generally distinctive through much of their history and especially during their growth stage. This characteristic is reminiscent of the strongly evolving multicelled storms described by Foote and Frank (1983). On 25 July, often the already-joined cores were first observed as protrusions from the parent echo and only as they matured did they become distinguishable. This is more typical of 'weakly evolving' storms.

Thus, on 25 July 1986, the echoes which subsequently merged were first observed at a higher altitde than during the morning period. These echoes also were more distant (by 5 km in the mean) from already-existing echoes. Additionally, the multicelled storms showed evidence of weak evolution, i.e. formation at adjacent echo protrusions which later appeared as reflectivity cores. These characteristics suggest that the echo cores resulted from stronger forcing than for the morning case as their subsequent history confirmed.

5. SUMMARY AND FUTURE WORK

This paper has mainly dealt with the first-echo characteristics of echo cores which subsequently join together. They were found to form at a higher elevation (by 1 km), with slightly stronger reflectivities ($\approx$2.5 dBZ) and slightly larger areas ($\approx$5 km^2) than those which dissipated without joining. While a preferred location was found for first-echo formation, echoes were observed to form in all quadrants with respect to the mean translation direction of the parent cloud. Little difference was found in the area, height and reflectivity characteristics of these clouds when examined by location. The echoes on both days as well as those which formed between two close neighbors during the morning case were 10–30% more likely to join to another.

On the two days studied thus far, clouds of varying sizes (3–850 km^2) are found to join. The majority of the merging events (56–65%) occurred between an individual echo and another echo that had already merged. Only a quarter to a third of the single echoes joined with another single echo. About 10% of the mergers occurred between two echo complexes. The relative frequency of the way in which the echoes bridge together will be made on the basis of the relative age and size of the merging pair. Additionally, the subsequent growth history of the aggregated systems will be examined.

ACKNOWLEDGMENTS

The authors would like to thank Nancy Westbrook for many laborious hours tracking echoes. The work was supported by the National Science Foundation under grant No. ATM87-15893 and by the National Oceanic and Atmospheric Administration Federal-State Program under Grant No. COMM-NA88RAH08107.

REFERENCES

Ackerman, B. and Kennedy, P. C. (1989) Doppler velocity fields during the merger of two small echoes. *Preprint of the 24th Conference on Radar Meteorology*, American Meteorological Society, Boston, MA, pp. 73–76.

Ackerman, B. and Westcott, N. E. (1986) Midwestern convective clouds: A review. *J. Weather Modif.*, **18**, April.

Ackerman, B. and Westcott, N. E. (1984) The morphology of merging clouds. *Preprint of the 9th International Conference on Cloud Physics*, Vol. II. pp. 403–406.

Browning, K. A. and Ludlum, F. H. (1962) Airflow in convective storms. *Q. J. R. Meteorol. Soc.*, **88**, 117–135.

Byers, H. R. and Braham, R. R. (1949) *The thunderstorm*, Report of the Thunderstorm Project. U.S. Government Printing Office, Washington, DC., 287 pp.

Changnon, S. A., Jr., (1976) Effects of urban areas and echo mergining on radar echo behavior. *J. Appl. Meteorol.*, **15**, 561–570.

Cheng, L. and Rogers, D. C. (1988) Hailfalls and hailstorm feeder clouds — an Alberta case study. *J. Atmos. Sci.*, **45** (23), 3533–3545.

Cunning, J. R., Holle, R. L., Gannon, P. T. and Watson, A. T. (1982) Convective evolution and merger in the FACE experimental area: Mesoscale convection and boundary layer interaction. *J. Appl. Meteorol.,* **21** (7), 953–977.

Cunning, J. R., Poor, H. W. and DeMaria, M. (1986) An investigation of the development of cumulonimbus systems over south Florida. Part II: In-cloud structure. *Mon. Weather Rev.,* **114** (1), 25–39.

Dennis, A. S., Schock, C. A. and Koscielski, A. (1970) Characteristics of hailstorms of western South Dakota. *J. Appl. Meteorol.,* **9**, 127–135.

Fankhauser, J. C. (1982) The 22 June 1976 case study: large-scale influences, radar echo structure and mesoscale scale circulations. *Hailstorms of the central high plains*, Vol. 2. Colorado University Press, Boulder, CO, pp. 1–35.

Foote, G. B. and Frank, H. W. (1983) Case study of a hailstorm in Colorado. Part III: Airflow from triple Doppler measurements. *J. Atmos. Sci.,* **40**, 686–707.

Heymsfield, G. M. (1981) Evolution of downdrafts and rotation in an Illinois thunderstorm. *Mon. Weather Rev.,* **109** (9), 1969–1988.

Maurwitz, J. D. (1972) The structure and motion of severe hailstorms. Part II: Multicell storms. *J. Appl. Meteorol.,* **11**, 180–188.

Orville, H. D., Kuo, Y. H., Farley, R. D. and Hwang, C. S. (1980) Numerical simulation of cloud interactions. *J. Rech. Atmos.,* **14**, 499–516.

Simpson, J. (1980) Downdrafts as linkages in dynamic cumulus seeding effects. *J. Appl. Meteorol.,* **19**, 477–487.

Tao, W.-K. and Simpson, J. (1984) Cloud interactions and merging: numerical simulations. *J. Atmos. Sci.,* **41**, 2901–2917.

Tao, W.-K. and Simpson, J. (1989) A further study of cumulus interactions and mergers: 3-dimensional simulations with trajectory analysis. *J. Appl. Meteorol.,* in press.

Westcott, N. E. and Kennedy, P. C. (1989) Cloud development and merger in an Illinois thunderstorm observed by Doppler radar. *J. Atmos. Sci.,* **46** (1), 117–131.

Westcott, N. E. (1984) A historical perspective on cloud mergers. *Bull. Am. Meteorol. Soc.,* **65**, 219–226.

Part 4

Radar measurement accuracy including multiparameter radars

20

Improvement of rainfall measurements due to accurate synchronization of raingauges and due to advection use in calibration

B. Blanchet, A. Neuman, G. Jaquet and H. Andrieu
Cergrene, Ecole Nationale des Ponts et Chaussées, La Courtine, 93167 Noisy-le-Grand Cédex, France

ABSTRACT

This paper deals with the reliability and calibration of radar rainfall measurements. Four rainfall events, which have been recorded in the Paris suburban area, are calibrated with synchronized raingauge data. The results are improved by integration of the radar measurements with the advection of the echoes.

1. INTRODUCTION

Radar rainfall measurement has been studied for almost 25 years by numerous scientists. Although the techniques are well explored, the reliability of radar measurements is still under research.

Several types of errors can occur:

(1) errors due to the measurement of reflectivity;
(2) errors due to the transformation of reflectivity to intensity;
(3) differences between the rainfall intensity in high altitudes and on the ground;
(4) errors due to fast echo movement;
(5) anomalous propagation and ground clutter.

This paper deals with aspects (2) and (4), which concern the calibration of radar measurement.

2. RADAR RAINFALL MEASUREMENT AND CALIBRATION

The reflectivity of the radar beam can under certain assumptions (relatively homogeneous rainfall, small drop size, etc.) be expressed by

$$Z = A \, R^b$$

where Z (dBZ) is the radar reflectivity, R (mm/h) is the rainfall intensity, and A and b are parameters. The parameters A and B are dependent on the characteristics of precipitation (drop size, temperature, etc.) The most commonly used values are $A=486$ and $b=1.37$ for convective rainfall (the Jones relationship), and $A=200$ and $b=1.6$ for stratiform rainfall (the Marshall–Palmer relationship).

Since the rainfall characteristics can change rapidly within an event, it is inconvenient to use fixed values. A better technique is to calibrate the roughly calculated intensities by point raingauge measurements. This means changing the parameters A and/or b with respect to the actual situation.

3.   EXPERIENCE IN THE SEINE-SAINT-DENIS COUNTY

In Seine-Saint-Denis county, in the suburban area of Paris, radar rainfall measurements are held since 1982. The area has a surface of about 300 km^2 and is situated 20–50 km away from the radar location. The radar measurement interval is 2 min, and the spatial resolution 800 m$\times$800 m. For calibration purposes a network of 20 raingauges with an average grid of 4 km is used. The measurement conditions are especially good for the ranges are rather short and the temporal and spatial resolution is high.

Four runoff-producing rainfall events, two convective rainfalls and two large stratiform events were recorded in summer 1982 and have been under research earlier (Andrieu 1986, Jacquet *et al.* 1987).

Andrieu tested and compared several calibration techniques (mean correction factor, the Brandes method, interpolation of the A parameter, and kriging). This research has shown that none of these methods leads to significantly better results than the others. This has been confirmed by recent publications (Dalezios 1988). Therefore the technique with the minimal computational requirement (calculation time and computer memory needs), i.e. the calibration with a mean correction factor, was applied in Seine-Saint-Denis.

The mean error, calculated at 5 min time intervals between calibrated radar data and raingauge measurement, is between 30% and 90%.

4.   SYNCHRONIZATION OF RAINGAUGE AND RADAR MEASUREMENT

In the measurement period of 1982 the raingauges were not connected to the central control unit. This caused certain time laps in the recordings of the different instruments because of:

(1) incorrect time calibration of raingauge recording devices,
(2) incorrect paper insertion in raingauge recording devices,
(3) non-constant speed of recording for climatic reasons and
(4) incorrect digitalization of raingauge data.

Different techniques were used to check asynchronization of data:

(a) maximizing the correlation coefficient between radar and raingauge data integrated over 5 min;
(b) maximizing the correlation coefficient between the calculated increments of the intensities between $t-5$ and t;
(c) minimizing the deviation between the two data series.

Where the different techniques led to similar results, the elapsed time was adopted (Agostini-Blanchet 1988) and was checked with the recorded paper. Constant-speed deviation was found by comparing weekly maintenance visiting hours as recorded on the paper and recorded hours, but initial deviations due to poor paper insertion were also observed after using this constant speed.

For each event a synchronization was necessary for about half of the raingauges (Fig. 1). The time lap was mostly small, between 5 and 10 min. In a few cases the lap was greater, for one gauge a time deviation of 35 min was observed.

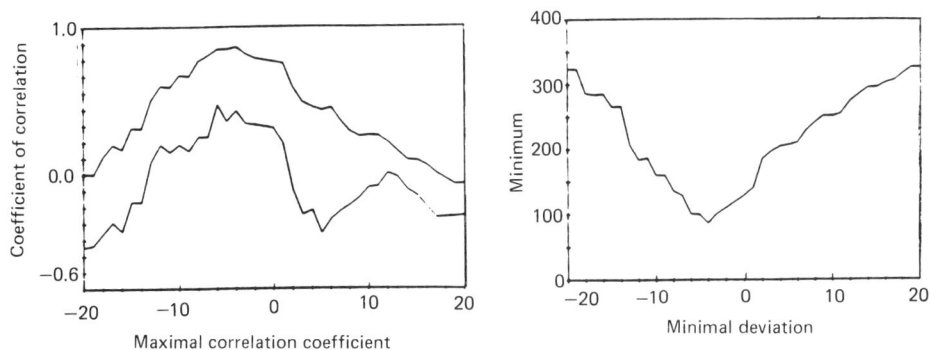

Fig. 1 — Resynchronization of raingauge data.

The synchronization improved the results of calibration significantly. The deviation on 5 min intervals was decreased from 0.8 to 0.6 for the showers and from 0.45 to 0.3 for the extended rainfall events (Fig. 2). The deviation was even decreased for storms where high intensities lasted for a very short time.

5. INTEGRATION WITH ADVECTION OF THE ECHOES

As opposed to the raingauge, the radar gives a time-punctual measurement. Therefore an integration in time is necessary to calculate rainfall volumes (Fig. 3). Andrieu (1986) used an average of two radar measurements (three on special occasions). For fast movement of echoes this is inconvenient, since even for rather slow advections of about 40 km/h more than one grid point of 800 m×800 m passes a certain point in an interval of 2 min. To take this into account, an integration within the radar images was made, taking into consideration the displacement vectors.

The integration was made for time steps of 5 min. Within an interval of 5 min the image with the median values was adopted. The resulting values were calibrated with the synchronized raingauge data.

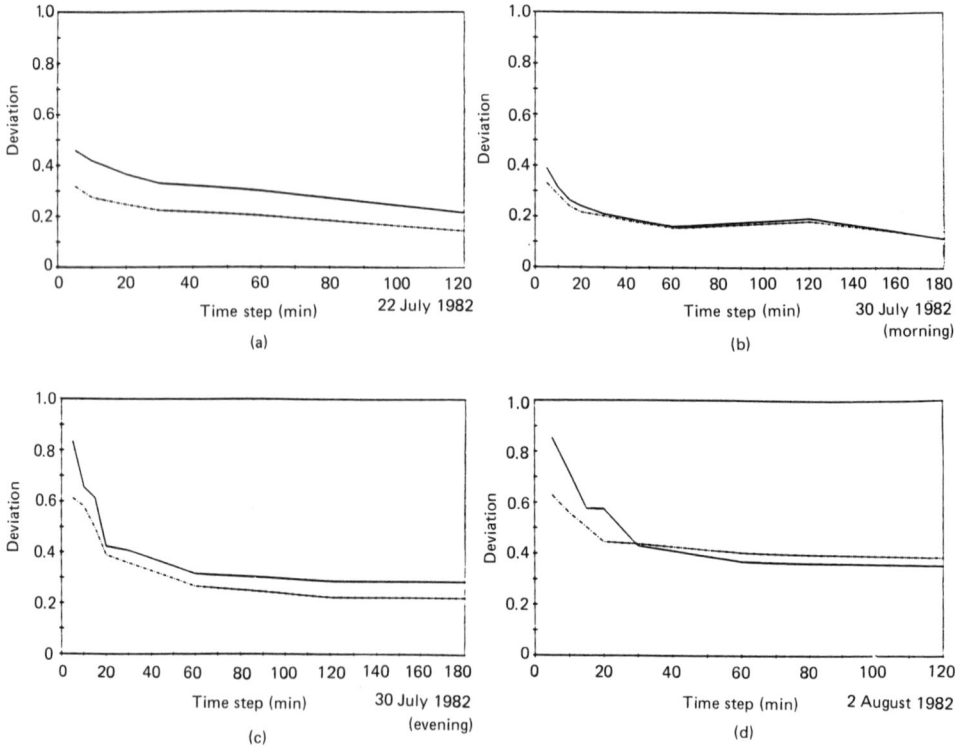

Fig. 2 — Deviation between radar and raingauge measurements for (a), (b) stratiform events and (c), (d) showery events: ——, original data; — · — · —, synchronized data.

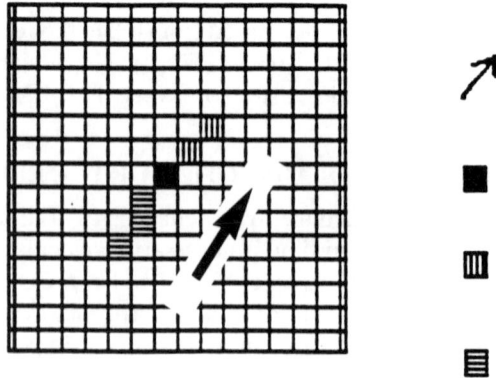

Fig. 3 — Principle of the integration method ($t_1 < t_i < t_2$): →, displacement vector; ■, pixel over gauge; ▨, pixels that have passed the gauge between t_i and t_1; ▨, pixels that will pass the gauge between t_i and t_2.

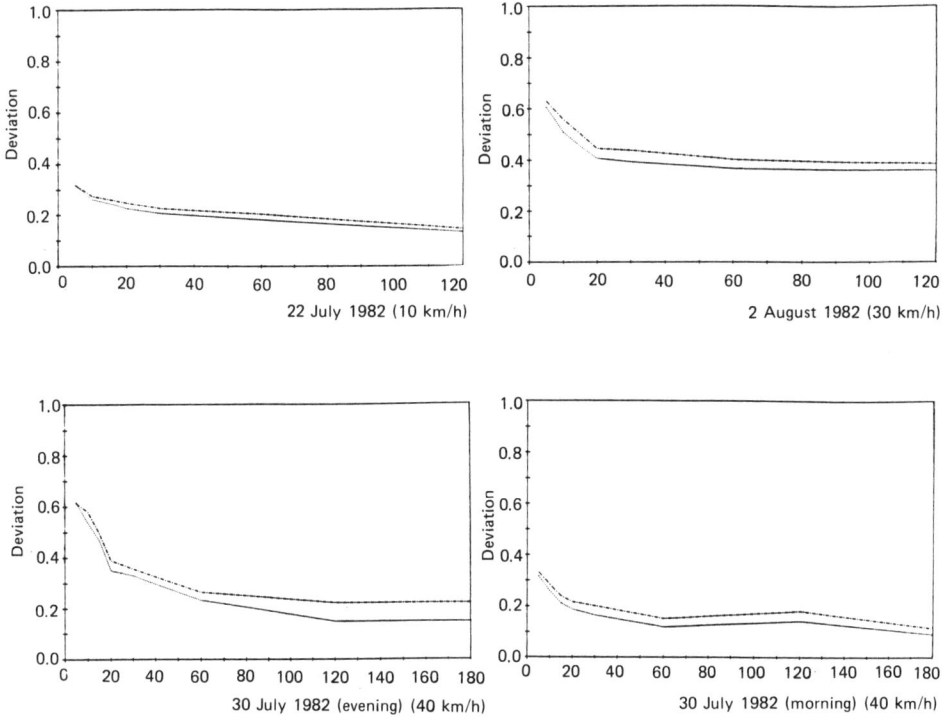

Fig. 4 — Deviation between integrated and radar and raingauge data: $-\cdot-\cdot-$, interpolated data; ———, integrated data.

The calibration factor CF was calculated every 5 min using the rainfall amounts of the last 15 min. For the calculation of the deviation between radar and raingauge data a different factor was calculated for every gauge:

$$(CF)_i(t) = \sum_{j \neq i = 1}^{p} \sum_{t' = t-15}^{2} (IP)_j(t') \bigg/ \sum^{j \neq i = 1p} \sum_{t' = t-15}^{t} (IR)_j(t') \ .$$

The deviation between calibrated radar and raingauge data is then calculated as

$$\mathrm{dev}(\Delta t) = \sum_{T=1}^{N} \sum_{i=1}^{p} \left| \sum_{t=T}^{T+\Delta t} (IP)_i(t) - \sum_{t=T}^{T+\Delta t} (IR)_i(t)^*(CF)_i(t) \right| \bigg/ \sum_{i=1}^{p} \sum_{t=T}^{T+\Delta} (IP)_i(t)$$

where $(IP)_i(t)$ is the intensity of rainfall measured by gauge i at time step t, $(IR)_i(t)$ is the intensity of rainfall measured by the radar over the site of gauge i at time step t, N is the number of time steps of Δt min and p is the number of gauges.

With the integration technique a reduction in this error of 10–20% for the slow-moving events (less than 25 km/h) and by 20–40% for the faster-moving events (40 km/h) was achieved. For faster displacement an even higher difference is expected. These results have been acknowledged by work done by McGill University (19**) where a similar technique for 2 km×2 km constant-altitude plan position indicator (CAPPI) radar maps were developed.

6. CONCLUSION

Continuing our research in radar rainfall measuring, two types of principal error were considered in this paper: errors due to incorrect data handling and errors due to rapid echo movement.

An improvement of the mean error between radar and raingauge measurement was achieved by careful synchronization of raingauge data. Furthermore it could be shown that it is essential to take echo movement into account, especially when advection speeds exceeds or equal 40 km/h.

These results prove the importance of using advection to calibrate and estimate rainfall. It has proved efficient where most static techniques, even the optimal best-fit type (e.g. kriging), proved inefficient.

Results should still be improved, but the first advance shows the need for better tracking procedures and we are currently working on this.

REFERENCES

Agostini-Blanchet, B. (1988) *Incertitudes liées à la mesure de la pluie'*, Rapport. DEA Techniques et Gestion de l'Environnement, Ecole Nationale des Ponts et Chaussées, Paris.

Andrieu, H. (1986) *Interprétation de mesures du radar rodin de trappes pour la conaissance en temps réel des précipitations en Seine–St Denis et Val-de-Marne*, Thèse de Doctor-Ingénieur. Ecole Nationale des Ponts et Chaussées, Paris.

Dalezois, N. R. (1988) Objective rainfall evaluation in radar hydrology. *J. Water Resour. Plan. Manage,* **114** (5).

Jacquet, G., Andrieu, H. and Denoeux, T. (1987) About rainfall measurement. *Proceedings of the 4th International Conference on Urban Storm Drainage.* IAHR Lausanne.

McGill University (1985) *User's guide for the PPS system*, Internal Report. Stormy Weather Group, McGill University, Montreal.

21

Range and orographic corrections for use in real-time radar data analysis

R. Brown, G. P. Sargent and R. M. Blackall
Meteorological Office, London Road, Bracknell, Berks. RG12 252, UK

ABSTRACT

This paper describes a method of calculating range and orographic corrections to be applied to real-time radar data. The range corrections allow for the effect on the measured rainfall rate of incomplete filling of the beam and the growth of precipitation beneath the beam at long range. The orographic corrections are increments to be added to the observed rainfall rate to allow for the low-level growth of precipitation by the seeder–feeder mechanism. Example corrections are shown and compared with the observed radar performance.

1. INTRODUCTION

The FRONTIERS precipitation nowcasting system allows for additional quality control to be applied to the radar data from the UK weather radar network, beyond that applied automatically by the radar site software. Because the system is operated by a forecaster, a more subtle choice of corrections is possible than can be applied automatically, since the forecaster can use his knowledge of the current synoptic situation. The purpose of this paper is to describe the calculation of two of the correction fields offered to the forecaster — the range-dependent and orographic corrections — since this has not been published before. Further details of the FRONTIERS system and its operation, including descriptions of other corrections applied to the radar data, can be found in the papers by Brown (1987) and Conway and Browning (1988).

For a full understanding of the form of the corrections it is important to note that the UK radar data are used in the form plan position indicators (PPIs), obtained using as an elevation angle as possible, mainly 0.5°. Such an observational strategy is in line with the emphasis on obtaining an accurate estimate of the surface rainfall rate, as required for hydrological use. However, the errors in the radar data then become more range dependent than in the constant-altitude plan position indicator (CAPPI) mode, also used commonly, where the reflectivities are used from a fixed height (typically 3 km).

Sources of error in radar data, which led to reduced accuracy in extrapolation forecasts, were listed by Browning *et al.* (1982) in a report on a pilot nowcasting study which informed the development of FRONTIERS. A common problem was the underestimation of surface rainfall rate at long range where the beam reaches an altitude of several kilometres and starts to overshoot the top of the precipitation. Growth of precipitation beneath the beam also contributes to the underestimation. To compensate for this, a fixed-range correction is applied at the radar sites, derived from a climatological comparison with raingauges. Typically the rainfall rates are multiplied by a factor of up to 5 at the maximum range of 210 km. However, on occasions when the precipitation develops predominantly low down in the atmosphere, larger corrections are required, this being particularly true when the freezing level is low. On the other hand, when deep convection of a high freezing level is present, the climatological corrections produce unrealistically high rainfall rates at long range. Within FRONTIERS the fixed-range correction is replaced by a choice of six correction fields, calculated from prescribed reflectivity profiles.

Another deficiency in the radar data is caused by the growth of precipitation beneath the radar beam over hills and mountains due to the seeder–feeder mechanism. Such enhancement occurs when a strong, moist, low-level flow, often ahead of a cold front, is lifted over hills, leading to the formation of capping clouds 1 or 3 km deep. Pre-existing precipitation from the associated front scavenges the droplets in the capping cloud, producing enhancements in rainfall rate of several millimetres per hour over that found upwind of the hills. Further details of the seeder–feeder mechanism can be found in the observational study of Hill *et al.* (1981) and the numerical simulations reported by Carruthers and Choularton (1983) and Robichaud and Austin (1988).

Stored within the FRONTIERS system are climatological tables of orographic enhancements derived from daily gauge data. These are increments to be added to the observed radar rainfall rate, rather than the usual multiplicative correction factors. Although the enhancement occurs typically in the lowest 1500 m of the atmosphere, close to a radar most of the enhancement will be detected because the beam will still be close to the surface. Progressively less of the enhancement is detected with increasing range from the radar. To allow for this, the FRONTIERS system contains tables of the percentage of the orographic enhancement seen by each radar in each pixel (subsequently referred to as the 'percentage-seen' tables). We concentrate in the paper on the method of calculating the percentage-seen table but for completeness summarize here the method of obtaining the enhancement fields, complete details of which can be found in the paper by Hill (1983). To obtain sufficiently high spatial resolution it was necessary to use daily gauge data. Cases were chosen where the rainfall pattern was dominated by orographic enhancement. Mean hourly rainfall rates were derived from the daily totals using a smoothed rainfall duration field derived from recording gauges and synoptic observations. The enhancement field was obtained by subtracting the downwind coastal rainfall rate from that over the hills. Cases within specified wind speed and direction categories were averaged to produce the final tables.

2.   BASIC ELEMENTS OF THE CALCULATION

Many elements of the calculation are common to both the range corrections and the percentage-seen fields and will be described only once. The basic difference lies in the form of the reflectivity profile assumed, since the range corrections account for the development of the precipitation down to about 1500 m whilst the percentage-seen fields apply beneath this.

2.1   Beam power profile

The power profile from the beam centre ($\theta = 0$ is parameterized by the expression

$$P(\theta) = \left(\frac{\sin(k\theta)}{k\theta}\right)^2 \tag{1}$$

where k is adjusted for the actual beam width. For a $1°$ beam width (to the half-power point), $k = 159.46$ for θ in degrees. In the calculations, the square of (1) is used to allow for the two-way response of the radar.

2.2   Radar infilling map

Although the majority of the radar data are obtained using a beam elevation of $0.5°$ ($0°$ for Clee), out to a range of 25–50 km, depending upon the radar, it is necessary to use data from a higher-elevation scan to avoid permanent ground clutter. A machineable data set exists for each radar, called the infilling map, which indicates the elevation of the beam from which data are used at every pixel in the single-site field. During the calculation of the correction fields, results are obtained for all the elevations used in the final radar image, then the appropriate answer is selected for each pixel by consulting the infilling map.

2.3   High-resolution topographic data set

A machineable data set of the UK topography with a horizontal resolution of 500 m was used to calculate the effects of beam occultation. Over Ireland only a 5 km resolution data set was available but extra data points were inserted manually to enhance the resolution in areas where the beam was likely to be obstructed. Man-made obstructions were not accounted for, except in the case of the Upavon radar where they are a dominant feature.

3.   THEORY OF THE RANGE CORRECTIONS

The radar beam, the top and bottom of which have elevation angles α and β respectively, encounters a layer of precipitation in which the reflectivity Z may be constant or decrease with height and which may or may not extend to the top of the beam. The range correction is defined by the ratio of the measured reflectivity to the 'surface' reflectivity Z_0. The measured reflectivity is taken to be the mean reflectivity across the beam weighted by the beam power profile, written $\overline{Z}$. In general at a given range r,

$$\overline{Z} = \int_\alpha^\beta Z(\theta)f(\theta)\,\mathrm{d}\theta \tag{2}$$

where Z has been transformed from a function of height to a function of a generalized elevation angle θ and $f(\theta)\,d\theta$ is the fraction of the beam power in the range from θ to $\theta + d\theta$, i.e.

$$f(\theta)\,d\theta = (P(\theta)\,d\theta)\Big/ \int_\alpha^\beta P(\theta)\,d\theta \qquad (3)$$

where $P(\theta)\,d\theta$ is the relative beam power in the range from θ to $\theta + d\theta$ from (1). The fractional power return relative to the surface value is $F(r) = \overline{Z}/Z_0$ at range r. Since within FRONTIERS the radar reflectivity is converted to rainfall rate (R using a Z–R relationship of the form

$$Z = AR^{1.6} \qquad (4)$$

$F(r)$ is converted to a fraction of the surface rainfall rate by taking the root of 1.6. Subsequent division of the observed radar rainfall rate by $F(r)$ gives the range-corrected value.

In general the computation of $F(r)$ allows for three effects:

 (i) a reflectivity gradient between the level of the beam and the surface;
 (ii) precipitation not filling the beam in the vertical;
(iii) the bottom of the beam occluded by hills.

The latter two reduce the angular limits of the integration in (2) to be less than the beam width.

Equation (2) is integrated using 27 angular increments in the vertical increasing to 453 beyond 100 km range, on a radial horizontal grid of resolution 0.75 km by 0.5°. The $\overline{Z}$ values are interpolated onto the 84 × 84 Cartesian grid with 5 km resolution used by the single-site radar data, before converting to rainfall rate equivalent.

4.   CHOICE OF REFLECTIVITY PROFILES FOR THE RANGE CORRECTIONS

The six range correction fields within FRONTIERS are divided into three for stratiform and three for convective precipitation. For each type, three depths of precipitation are assumed, described as shallow, moderate and deep. The division into stratiform and convective, besides allowing the use of differently shaped reflectivity profiles, allows different values of A to be used in the Z–R relationship (200 for stratiform and 300 for convective cases). This reduces the convective rainfall rates by about 30% to allow for the excess large drops found in convective compared with stratiform precipitation.

The stratiform reflectivity profiles are assumed to result from widespread slow ascent and their shape has been based upon published profiles measured by radar (e.g. Hamilton 1964, Dissanayke *et al.* 1983, Gori 1983, Stewart *et al.* 1984. Below heights of 1.0 km (shallow), 1.5 km (moderate) and 3 km (deep) the reflectivity is assumed constant at the surface value. The published profiles all show peaks caused by bright-band effects which have been ignored. Above the constant layer the reflectivity decreases roughly exponentially but with tops set at 5, 6 and 8.5 km for the three depths.

Whilst the published stratiform profiles exhibit very similar shapes, much greater temporal and spatial variations occur in convective precipitation. A phenomenon which could not be parametrized is the development of a reflectivity maximum aloft caused by precipitation suspended in the updraft, which subsequently descends to the surface as the hydrometeors grow too large to be held aloft, or the updraft decays, e.g. given by the profiles Hamilton (1963) and Illingworth *et al.* (1987). Therefore we account mainly for the finite depth of the precipitation by assuming a constant reflectivity from the surface to 2, 3 and 6 km for the shallow, moderate and deep convective profiles, with the reflectivity decreasing to zero over approximately 1.5 km above this. The unequal spacing reflects the fact that the corrections required vary much more rapidly with precipitation depth for shallow precipitation. For both the stratiform and the convective cases the reflectivity profiles have been designed to produce approximately equispaced corrections, rather than have equispaced precipitation depths.

5. EXAMPLES OF THE RANGE CORRECTIONS FIELDS

Examples of the range correction fields for moderately deep stratiform precipitation are shown in Fig. 1 for the Upavon, Clee, Hameldon and London radars. The range correction factor reaches 2 at the inner edge of the first black strip, 3 at the outer edge, 4 at the inner edge of the second strip, etc. The picture is slightly confused to the east of Upavon because in some places the correction factor jumps from 3 to 5 within one pixel so that there is not a complete black strip.

The shape of the correction fields is dominated by beam occulation effects. The radar with the best horizon is Clee, although the influence of the Welsh mountains is apparent to the west. In fact, Clee is the only radar which operates with a 0° beam elevation — hence the small corrections required. Hameldon has a good visibility to the west but is obstructed to the east and southeast and also due north. Upavon is such a poor site that a mixture of 1.0° and 1.6° beam elevations is used to the north and east to maximum range.

The range corrections for moderately deep stratiform precipitation attain a maximum value of about 5, except where the occulation of the beam is severe, which agrees with the values applied at the radar sites, deduced from comparisons with gauges. To validate the predicted shape of the correction fields, especially the effect of topography, 3 months' worth of radar data has been processed to record the number of times precipitation was detected in each pixel. The results show that precipitation was detected less frequently with increasing range, especially in directions where the beam is occluded. This is illustrated by the boundaries in Fig. 2, outside which precipitation was detected on less than 5% of occasions. Agreement with the shape of the range corrections fields in Fig. 1 is generally excellent. However, in Fig. 2 the effect of local man-made obstructions is apparent in a few directions, e.g. northeast of Clee, west of Hameldon, and this is not reproduced in Fig. 1. The performance of the London radar also seems worse than predicted to the northeast.

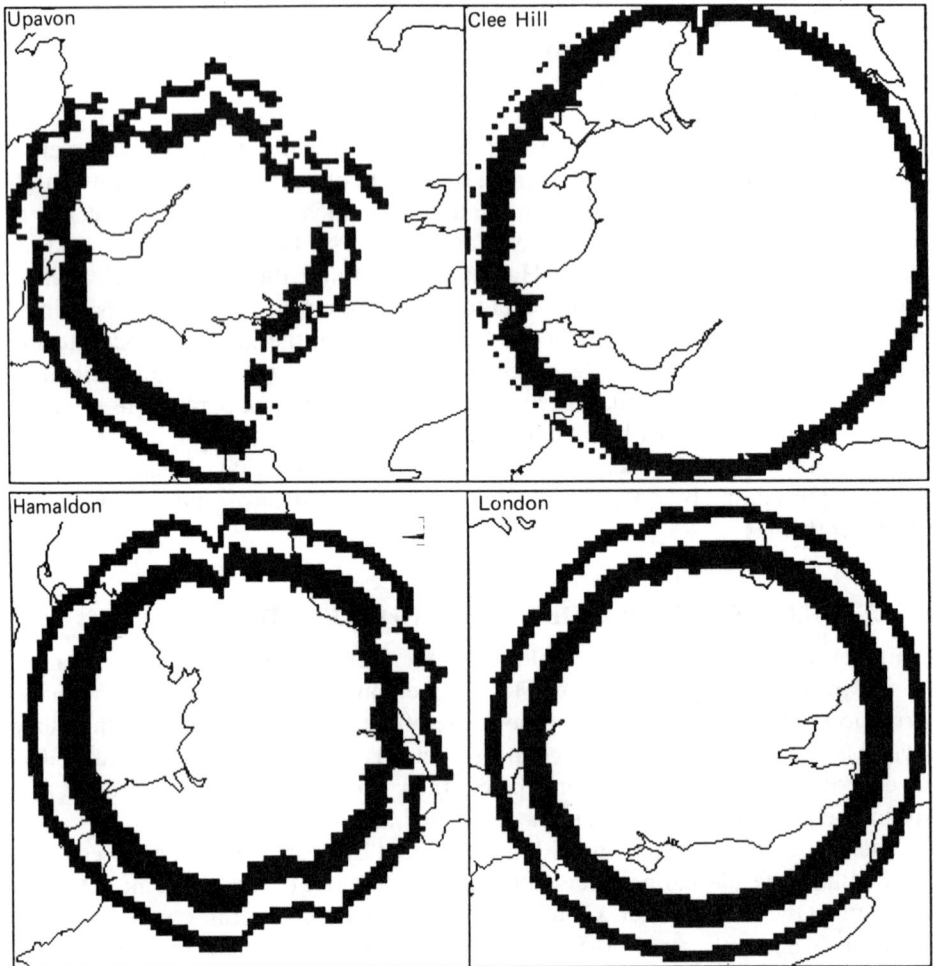

Fig. 1 — Computed range corrections fields, assuming moderately deep stratiform precipitation, for the Upavon, Clee Hill, Hameldon Hill and London (Chenies) radars, which are at the centre of the boxes. The range correction reaches a factor of 2 at the inner edge of the first black strip, 3 at the outer edge, 4 at the inner edge of the second black strip, etc.

6. CALCULATION OF THE PERCENTAGE-SEEN TABLES

An idealized reflectivity profile exhibiting low-level enhancement is shown in Fig. 3. For the percentage-seen calculations the top of the layer in which orographic enhancement occurs was assumed to be at 1500 m, the typical value observed by Hill *et al.* (1981). Their profiles (from which the bright-band peak had been removed) showed that the reflectivity decreased with increasing height through the seeder could, as illustrated by the broken curve in Fig. 3. However, it has not been found possible to formulate the percentage-seen calculation to take this into account. Therefore the reflectivity above 1500 m was assumed constant, taking a value Z_c, the reflectivity at the top of the feeder cloud, as shown in Fig. 3. The mean reflectivity $\bar{Z}$

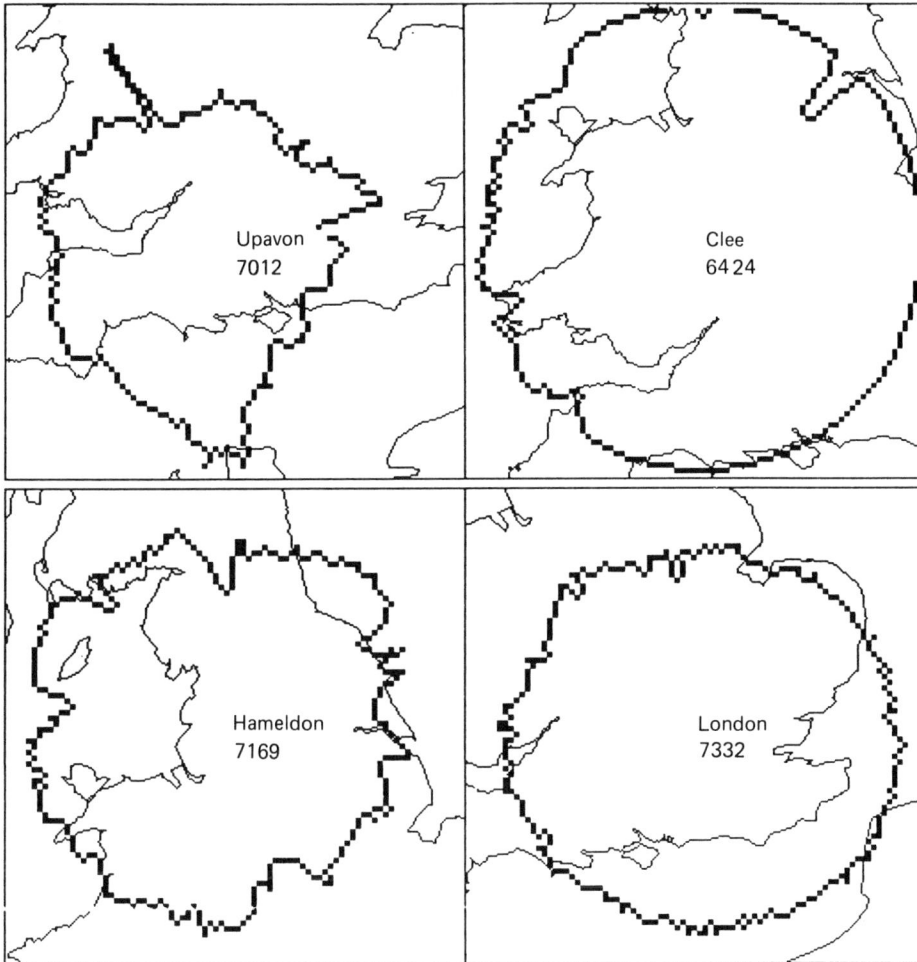

Fig. 2 — Observed radar visibilities for the same radars as Fig. 1 determined from 3 months' data. The actual number of fields used for each radar is annotated. The boundary separates regions where precipitation was detected on less than and more than 5% of occasions. The spike northwest of Upavon was caused by persistent noise.

was calculated from equation (1), just as for the range corrections. The fraction f_Z of the orographic enhancement seen in terms of reflectivity is given by

$$f_Z = \frac{\overline{Z} - Z_c}{Z_0 - Z_c}. \tag{5}$$

It can be shown using simple algebra that f_Z is independent of Z_c which is taken as zero. Unfortunately because of the non-linear relationship between Z and R, shown by equation (4), the fraction (f_R) of the enhancement seen expressed in terms of rainfall rate, is not independent of the rainfall rate (R_c) at the top of the feeder cloud. It can be shown that f_R is a function of f_Z and R_c/O_G, where O_G is the enhancement

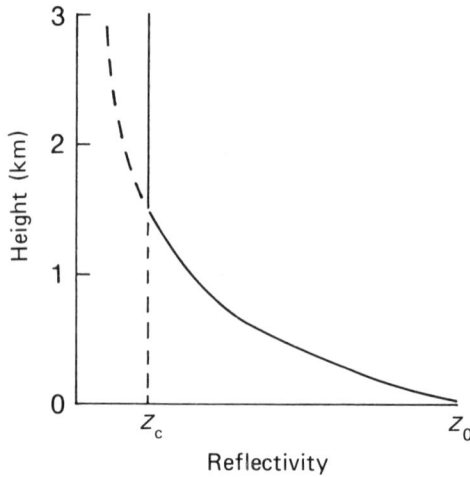

Fig. 3 — Idealized reflectivity profile for an orographic enhancement case (neglecting bright-band effects). Z_0 is the 'surface' reflectivity and Z_c the reflectivity at the top of the feeder cloud. Above 1.5 km, the assumed top of the feeder cloud, observations indicate decreasing reflectivity as shown by the broken line. For the percentage-seen calculations the reflectivity is taken to be constant above 1.5 km, as shown by the full curve.

(i.e. $O_G = R_0 - R_c$) (Warner 1987). Luckily f_R depends mainly on f_Z and only varies slowly with R_c/O_G.

The reflectivity profile used for the final calculations was based upon the rainfall rate profiles given by, Hill et al. (1981, Fig. 13), which were obtained during orographic enhancement episodes over the Gower Peninsula in south Wales. These were converted to reflectivity using equation (4) and then averaged. A value of 0.35 was used for R_c/O_G based upon Fig. 15 of Hill et al. (1981). The integration was performed using the same grid and methodology as for the range corrections, with one important exception. For the range correction calculation the reflectivity profile was always based at sea level but for the percentage-seen calculation it was based at the local surface.

7.   SENSITIVITY TO THE ASSUMED REFLECTIVITY PROFILE

The percentage-seen fields are extremely difficult to verify. It is clear from visual examination that topographical effects dominate, especially beam occultation. The results presented in Figs 1 and 2 suggest that such effects will be modelled reasonably. Thus the largest uncertainty arises from variations in the reflectivity profile through the feeder cloud and the depth of the latter. To obtain an estimate of this uncertainty, the percentage-seen calculation for the Hameldon radar has been performed for the seven mean reflectivity profiles presented in Fig. 13 of Hill et al. (1981). The results are presented in Fig. 4 as a function of range looking due north of Hameldon. The percentage-seen value have been converted to rainfall rates in this figure using the FRONTIERS orographic enhancements for a WSW, 55 kt wind, which are shown as the 'total enhancement' curve in Fig. 4(a). The amount of enhancement seen by the radar is shown in Fig. 4(a) and that undetected in Fig. 4(b). Three curves are shown in each figure derived from the maximum percentage seen in

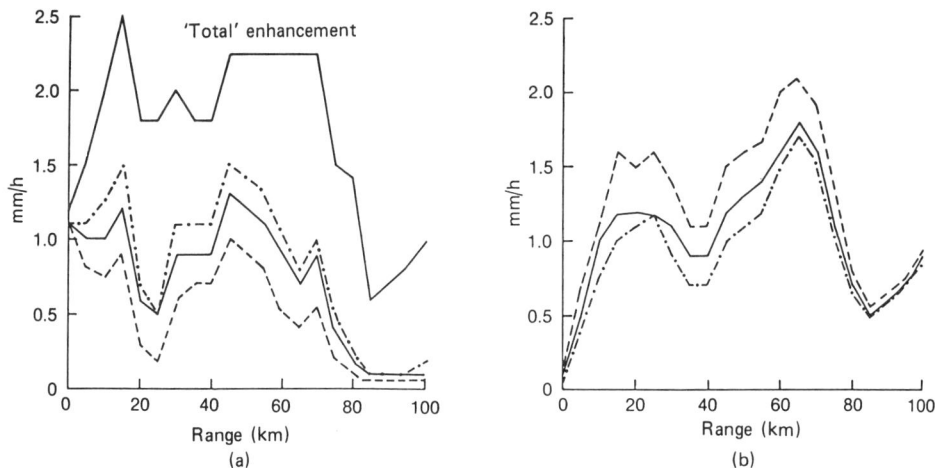

Fig. 4 — The predicted orographic enhancement (a) 'seen' and (b) 'not seen' looking north from the Hameldon radar, assuming a WSW, 55 kt wind at 900 m. The total orographic enhancement is shown as the top curve in (a). The calculations were performed for the seven profiles given in Hill et al. (1981, Fig. 13). Three estimates are shown: (i) using the maximum percentage-seen value (·–·–·–); (ii) using the minimum percentage-seen value excluding case 2 of Hill et al. (———); (iii) using the percentage-seen calculated from case 2 (– – –).

each pixel, the minimum percentage seen (excluding case 2 of Hill et al.) and for case 2. The minimum in the enhancement seen at a range of 25 km is caused by the transition from 1° to 0.5° beam elevation data. The results for all profiles except case 2 lie between the maximum and minimum curves. Case 2 appears to the exceptional because most of the enhancement occurred below 500 m.

8. CONCLUSIONS

Theoretical range corrections based upon prescribed reflectivity profiles have been developed and are applied during real-time quality control of the radar data. Corrections for the growth of precipitation beneath the radar beam by the seeder–feeder mechanisms of orographic enhancement have also been developed and applied. Encouraging agreement between the predicted azimuthal variations in the range correction field and the observed radar visibility field was found. However, further quantitative evaluation of the corrections is required.

ACKNOWLEDGEMENTS

We wish to acknowledge the help given by Mr F. H. Hill in formulating the percentage-seen calculation and by Mrs C. Banks who produced the original range correction software.

REFERENCES

Brown, R. (1987) The use of imagery in the FRONTIERS precipitation nowcasting system. *Proceedings of a Workshop on Satellite and Radar Imagery Interpretation, Reading, 1987, EUMETSAT.* pp. 459–472.

Browning, K. A., Collier, C. G., Larke, P. R., Menmuir, P., Monk, G. A. and Owens, R. G. (1982) On the forecasting of frontal rain using a weather radar network. *Mon. Weather Rev.*, **110**, 534–552.

Carruthers, D. J. and Choularton, T. W. (1983) A model of the seeder–feeder mechanism of orographic rain including stratification and wind-drift effects. *Q. J. R. Meteorol. Soc.*, **109**, 575–588.

Conway, B. J. and Browning, K. A. (1988) Weather forecasting by interactive analysis of radar and satellite imagery. *Philos. Trans. R. Soc. London*, **324**, 299–315.

Dissanayke, A. W., Chandra, M. and Watson, P. A. (1983) 'Backscattering and differential scattering characteristics of the melting layer. *Proceedings of the Union Radio-Scientific Internationale Symposium on Wave Propagation and Remote Sensing*. European Space Agency, pp. 363–370.

Gori, E. G. (1983) Simultaneous measurements of rainfall parameters by means of a vertical radar and distrometers along a mountain slope. *Proceedings of the 21st Conference on Radar Meteorology*. American Meteorological Society, Boston, MA, pp. 693–698.

Hamilton, P. M. (1964) *Precipitation profiles for the total radar coverage*, Scientific Report No. MW37). Stormy Weather Group, McGill University, Montreal.

Hill, F. F. (1983) The use of average annual rainfall to derive estimates of orographic enhancement of frontal rain over England and Wales for different wind directions. *J. Climatol.*, **3**, 113–129.

Hill, F. F., Browning, K. A. and Bader, M. J. (1981) Radar and raingauge observations of orographic rain over Wales. *Q. J. R. Meteorol. Soc.*, **107**, 643–670.

Illingworth, A. J., Goddard, J. W. F. and Cherry, S. M. (1987) Polarisation radar studies of precipitation development in convective storms. *Q. J. R. Meteorol. Soc.*, **113**, 469–489.

Robichaud, A. J. and Austin, G. L. (1988) On the modelling of warm orographic rain by the seeder–feeder mechanism. *Q. J. R. Meteorol. Soc.*, **114**, 967–988.

Stewart, R. E., Marwitz, J. D. and Pace, J. C. (1984) Characteristics through the melting layer of stratiform clouds. *J. Atmos. Sci.*, **41**, 3227–3237.

Warner, C. (1989) *On FRONTIERS calibration and orographic corrections*, Met. O 24 Technical Note No. 37. Meteorological Office, Bracknell, Berks.

22

Validation of a short-range X-band system for rainfall measurement over an urban area

G. Delrieu[1], J. D. Creutin[2] and H. Andrieu[2]
[1]Institut de Mécanique de Grenoble, Domaine Universitiare, BP 53 X 38041, Grenoble Cédex, France
[2]Laboratoire Central des Ponts et Chaussees, BP 59 44340, Bougenais, France

ABSTRACT

The results of the specially designed Grenoble 88 experiment have been analysed to assess the performance of a short-range X-band radar setup for rainfall measurement in urban hydrology applications. Range attenuation by precipitation is shown to be significant for the observed events, but other factors such as electronic calibration uncertainties and 'on-site' attenuation effects also seem to explain a large part of the discrepancies between radar and raingauge measurements.

An iterative procedure for the range attenuation correction is then described and the need for a good pre-estimate of reflectivity data prior to the correction is stressed. Compared with a simple adjustment of reflectivity data using a suitable additive correction factor, the method involving an initial additive adjustment followed by the range attenuation correction proves more effective in significantly reducing the bias and in increasing the correlation between radar and raingauge rainfall rates integrated over 6 and 24 min time steps.

1. INTRODUCTION

The use of low-cost short-range weather radar could be an effective solution for urban or mountainous areas subject to severe hydrological problems and where adequate coverage cannot be provided by synoptic radar networks. The 'Grenoble 88 experiment' was specially designed to assess the quantitative performance of such a radar setup, the LARS 88, recently developed by Enterprise Electronics Corporation.

The cost of this system was reduced mainly by choosing the following characteristics: X-band radar, 25 kW peak power and 1 m antenna size. The low peak power leads to an important limitation on the maximum useful range for quantitative measurement (about 30–50 km), while the short antenna size leads to a degradation of the angular resolution of the system (beam width, about 2.5°). These choices are not crippling provided that the radar location is optimized with respect to the area to be covered. Therefore, the main drawback appears to be related to the choice of the 3.2 cm wavelength which is known to be severely attenuated by rainfall.

In this paper, after a brief presentation of the experiment, we try to quantify the degree of attenuation at the concerned spatial scales and for the observed precipitation. Then, a correction procedure is described and its effectiveness assessed using raingauge data.

2. GRENOBLE 88 EXPERIMENT

Table 1 gives the main characteristics of the LARS 88 radar set-up. Special attention was paid to the electronic calibration of the system, but some uncertainties remain owing to the lack of actual measurements of the antenna parameters (losses in the waveguides, antenna diagram and gain). An interface with a microcomputer was developed in order to collect polar reflectivity data coded over 256 levels.

Table 1 — Main technical characteristics of the LARS 88 radar system

Wavelength	3.2 cm
Peak power	25 kW
Antenna diameter	1 m
Gain	36 dB
Half-power beam width	2.5°
Pulse repetition frequency[a]	100 Hz
Pulse duration[a]	1 μs
Antenna rotation speed[a]	2 rev/min
Displayed data[a, b] Maximum range	Selectable: 15, 30, 60, 120 km
Spatial resolution of Cartesian grid	125 m × 125 m for 15 km maximum range
	250 m × 250 m for 30 km maximum range
Intensity levels	6

[a]Parameters for 'intensity' mode; 'Doppler' mode giving radial velocity fields may also be available but the interest of this additional information for hydrology is beyond the scope of this paper.
[b]The characteristics of the reflectivity data actually used in this study are presented in section 2.

For convenience, the radar was set up at the Institute. This choice led us to install the validation raingauge network in the Grésivaudan Valley (Fig. 1) which is

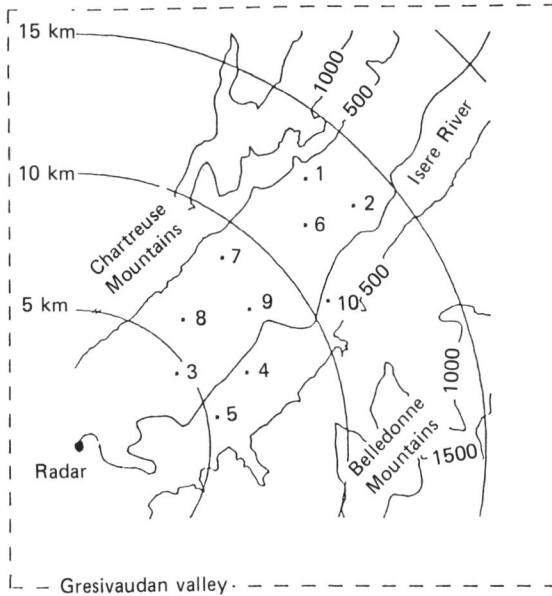

Fig. 1 — Location of the radar and the raingauge network for Grenoble 88 experiment.

virtually the only part of the very mountainous Grenoble area not affected by ground clutters. The ten tipping-bucket raingauge (sensitivity 1/10 mm; time resolution, 1 min) were spread over an area of about 50 km^2 at distance of 4.3 to 13.4 km from the radar. Radar data were collected in plan portion indicator (PPI) mode (site 3°) with a frequency of one image every 3 min. Reflectivities were first converted to rainfall rates using the Marshall–Palmer relationship and then averaged on a Cartesian grid with a 500 m × 500 m mesh. Great attention was paid to the synchronization of all the measurement devices, and we chose a working time step of 6 min in order to limit the uncertainty on raingauge data due to the tipping-bucket discretization.

During the two month measurement period (June–July 1988), ten significant rainfall events were measured. The maximum rainfall intensities observed at ground level were about 40–50 mm/h over 6 min with a maximum of 72 mm/h) and 10–20 mm/h over 1 h, and the maximum totals for a given event were about 20–30 mm.

3. EVIDENCE OF ATTENUATION

Fig. 2 gives the evolution of the ratio M_G/M_R for the different events. M_G represents the mean of the 6 min intensities measured by the set of the ten available raingauges, while M_R represents the mean of corresponding raw radar data, uncorrected for attenuation. M_G/M_R varies from 1.9 to 9.6, indicating a very strong and highly fluctuating underestimation of the radar.

Fig. 3 shows, for the three main events, the evolution of $M_G(d)/M_R(d)$ and $M_G(d)$ where d represents the distance to the radar installation. Note that, whatever the

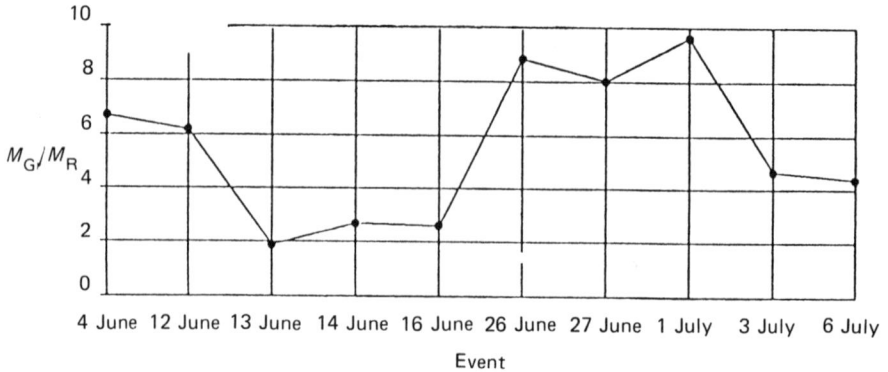

Fig. 2 — Fluctuations of the ratio MG/Mr for the different observed rainfall events.

$M_G(d)$ profile, the $M_G(d)/M_R(d)$ profile tends to increase with increasing range. This can be attributed to the effects of attenuation by rainfall.

Fitting a regression line through the $M_G(d)/M_R(d)$ profiles, the value $M_G(0)/M_R(0)$ extrapolated for $d = 0$ can be used to compute the following criterion:

$$C = \frac{M_G/M_R - M_G(0)/M_R(0)}{M_G/M_R}$$

C can be considered as a rough estimate of radar underestimation due to attenuation. For the three given examples, C equals respectively 0.11, 0.36 and 0.41, indicating that other major underestimation factors are present.

First of all, the uncertainties of the electronic calibration may explain a significant part of the remaining underestimation. However, when rainfall of medium to strong intensity was observed at the radar site itself, some momentary 'blinding' of the radar was revealed by the vanishing of the returned echoes (including strong ground clutters) at ranges greater than 5 km. This phenomenon can probably be explained by the presence of water on the antenna and also by the presence of raindrops in the vicinity of the antenna which may alter the transmitted wave. These effects will be referred to as 'on-site attenuation effects' and we believe that they have to be clearly distinguished from the 'range attenuation effects' shown in Fig. 3. We shall focus now on correction possibilities for the latter on-site attenuation effects which are at present under study at the Institute.

4. CORRECTION PROCEDURE FOR RANGE ATTENUATION EFFECTS

This correction first relies on the choice of an empirical relationship between the attenuation K (dB/km) and the rainfall rate R (mm/h). As an initial approximation, we have used in this work the K–R relationship proposed for $\lambda = 32$ cm and $T = 18°C$ by Gunn and East (1954):

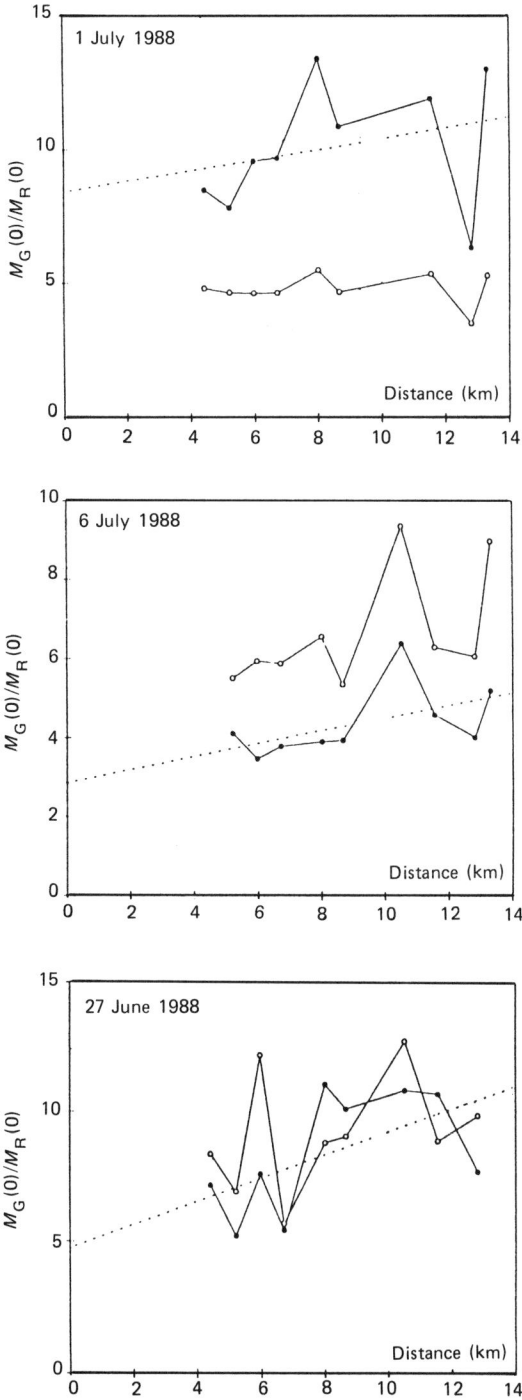

Fig. 3 — $M_G(d)/M_r(d)$ (●) and $M_G(d)$ (mm/h) (○) profiles for the three main events.

$$K = 0.0074 R^{1.31} \ .$$

A step-by-step correction scheme can then be proposed:

$$Z_c(k) = Z_r(k) + 2p \sum_{i=1}^{k-1} K_c(i)$$

where $Z_c(k)$ represents the corrected reflectivity for the kth range bin, $Z_r(k)$ the raw reflectivity, $K_c(i)$ (dB/km) the attenuation deduced from the *corrected* reflectivity $Z_c(i)$ through the use of adequate $K - R$ and $Z - R$ relationships. p (km) represents the range bin depth.

Sims *et al.* (1964) noted that this 'full' correction procedure may lead to unrealistic corrected values, especially when reflectivities are initially overestimated (miscalibration of the radar). They therefore adopted a 'minimal' correction:

$$Z_c(k) = Z_r(k) + 2p \sum_{i=1}^{k-1} K_r(i)$$

where $K_r(i)$ is deduced from the *raw* reflectivity $Z_r(i)$. This procedure was shown to be effective in reducing the production of excessive corrected values, but it leads *de facto* to an undercorrection of attenuation.

An iterative alternative procedure was therefore proposed by Hildebrand (1978). For iteration j, it can be written as follows:

$$(Z_c)_j(k) = Z_r(k) + 2p \sum_{i=1}^{k-1} K_j(i)$$

where $K_j(i)$ is calculated from $(Z_c)_{j-1}(i)$ and $K_j(i) = K_r(i)$.

For each iteration, the total correction over the full range is computed as

$$K_T(j) = 2p \sum_{i=1}^{N_{bin}} K_j(i) \ .$$

The procedure is stopped when $K_T(j) - K_T(j-1)$ is less than a given threshold, e.g. 1 dB. When the initial reflectivity profile is overestimated or when the K–R relationship is unsuitable, the procedure may diverge. In this case, the calculations are reinitialized and stopped after a small empirically defined number of iterations.

After numerous simulations, we have chosen the latter correction scheme for its ability to provide the maximum correction when reflectivity data is pre-estimated well, and to detect and limit the production of erroneous data in the case of initial reflectivity data overestimation.

5. VALIDATION OF THE RANGE ATTENUATION CORRECTION PROCEDURE

In the present context, an initial adjustment of the raw reflectivity data is necessary to take into account the uncertainties of the electronic calibration and the on-site attenuation effects. Indeed, if the range attenuation correction procedure is applied to the strongly underestimated reflectivity data, it will have almost no effect.

The value $M_G(0)/M_R(0)$ (see section 3) can be used to compute an additive correction factor ΔZ_1 for raw reflectivity data prior to the application of the range attenuation correction procedure:

$$\Delta Z_1 = 10b \, \log \left(\frac{M_G(0)}{M_R(0)} \right)$$

where b is the exponent of the Z–R relationship.

Thus, the radar processing procedure can be summarized as follows:

(i)   additive correction of polar raw reflectivity data using ΔZ_1 (ΔZ_1 varies from storm to storm but remains constant in time and space for a given event);

(ii) application of the range attenuation correction procedure;

(iii) normal end of process, i.e. Z–R conversion, polar-to-Cartesian conversion and time integration.

In order to asssess the effectiveness of the range attenuation correction procedure, a simpler processing method can be used:

(i)   additive correction of polar raw reflectivity data using an alternative correction factor ΔZ_2, defined as

$$\Delta Z_2 = 10b \, \log \left(\frac{M_G}{M_R} \right)$$

(ii) normal end of process.

These methods will be referred to subsequently as method 1 and method 2 and their performance will be evaluated with respect to raingauge data using the following criteria:

(i) M_G/M_{CR}, the ratio of raingauge mean intensity to corrected radar mean intensity for various events;
(ii) ρ, the correlation coefficient between corresponding raingauge and corrected radar intensities.

The results obtained for the set of the three main events are presented in Table 2,

Table 2 — Validation of the range attenuation correction procedure for the set of the three main events (27 June 1988, 1 July 1988 and 6 July 1988). For the meaning of M_G/M_{CR}, ρ_1 and ρ_2, see section 5. The values in parentheses indicate the number of radar–raingauge pairs taken into account in the calculation of the corresponding criterion

	Method 1: uniform adjustment and attenuation	Methd 2: simple uniform adjustment
	Time step, 6 min	
M_G/M_{CR}	1.04 (588)	1.13 (588)
ρ_1	0.782 (588)	0.738 (588)
ρ_2	0.635 (372)	0.585 (374)
	Time step, 24 min	
$M_G M_{CR}$	1.05 (142)	1.14 (142)
ρ_1	0.872 (142)	0.821 (142)
ρ_2	0.678 (105)	0.628 (105)

for 6 and 24 min integration time steps. The M_G/M_{CR} criterion denotes a significant remaining underestimation of the radar values corrected by simple uniform adjustment (method 2). The corresponding scattergrams (Fig. 4) show that the adjustment is correct for low values but insufficient for high values, indicating the non-linear effect of attenuation. The M_G/M_{CR} criterion is significantly closer to unity for method 1.

The comparison of the correlation coefficients calculated for the whole data set (ρ_1, Table 2) shows that the use of the attenuation correction procedure provides an appreciable improvement (by at least 0.05 whatever the integration time step). However, the assymetry of the radar and raingauge distributions is obvious in the scattergrams presented in Fig. 4. A large number of low values, generally well estimated by the two devices, may have a predominant weight in the correlation coefficient estimate. In order to test the robustness of these statistics, the two sets of data were logarithmically transformed and truncated at a 1 mm/h threshold. In this way, almost normally distributed variables are obtained. The corresponding correlation coefficients (ρ_2, Table 2) are 0.15–0.20 less than those computed on the original values. This result is, in fact, more consistent with the behaviour of the medium values as shown on the scattergrams (Fig. 4). In any case, the improvement of

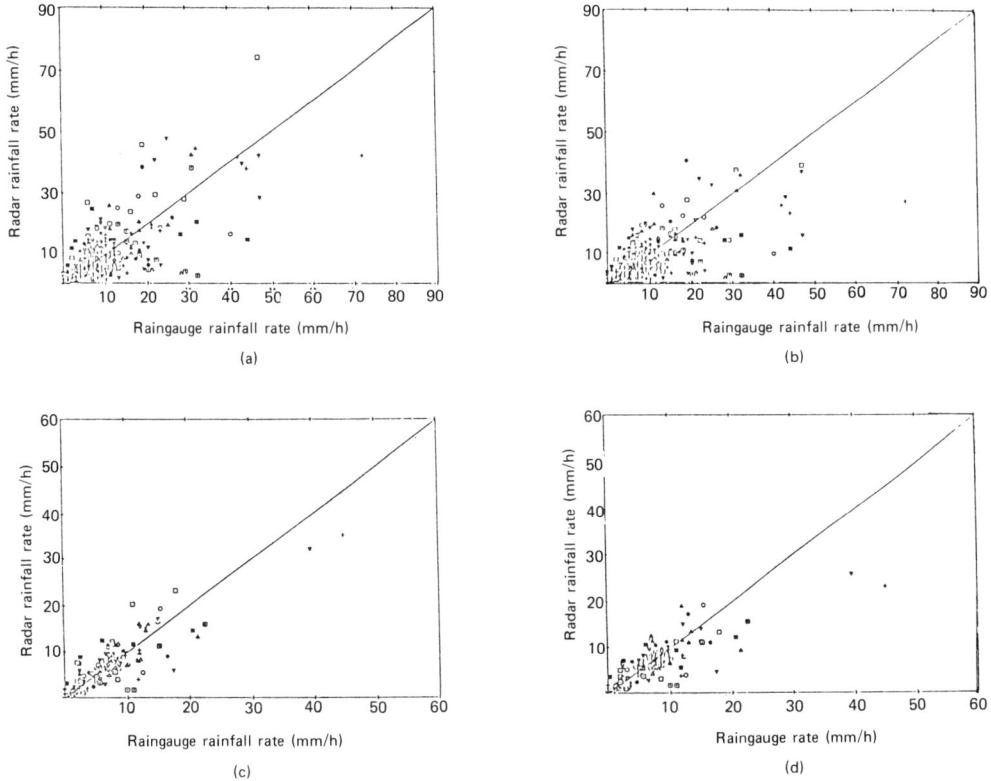

Fig. 4 — Scattergrams of raingauge intensities versus corrected radar intensities: (a) method 1, integration time step of 6 min; (b) method 2, integration time step of 6 min; (c) method 1, integration time step of 24 min; (d) method 2, integration time step of 24 min.

roughly 0.05 related to the application of the attenuation correction procedure is confirmed by this more detailed correlation analysis.

In order to provide a more visual illustration of the results obtained, Fig. 5 shows a sequence of 6 mm corrected radar and raingauge maps for the 27 June 1988 event. Raingauge maps were obtained using spline interpolation. The on-site attenuation effects mentioned in section 3 are obvious for the first time step; a value of 15 mm/h (not mapped on the raingauge image) was observed at the radar site and the cell which was clearly developing in the middle of the raingauge network is not seen by the radar. At the next time step, the rain ceased at the radar site and the two sets of maps subsequently show very similar patterns with respect to the relatively low spatial resolution of the raingauge network.

6. CONCLUSION

Future work is necessary in order to confirm these encouraging initial results.

The radar system stability and the quality of the electronic calibration appear to be extremely important factors. Further investigation would benefit from an automatic calibration control procedure based on the use of radar data only (known

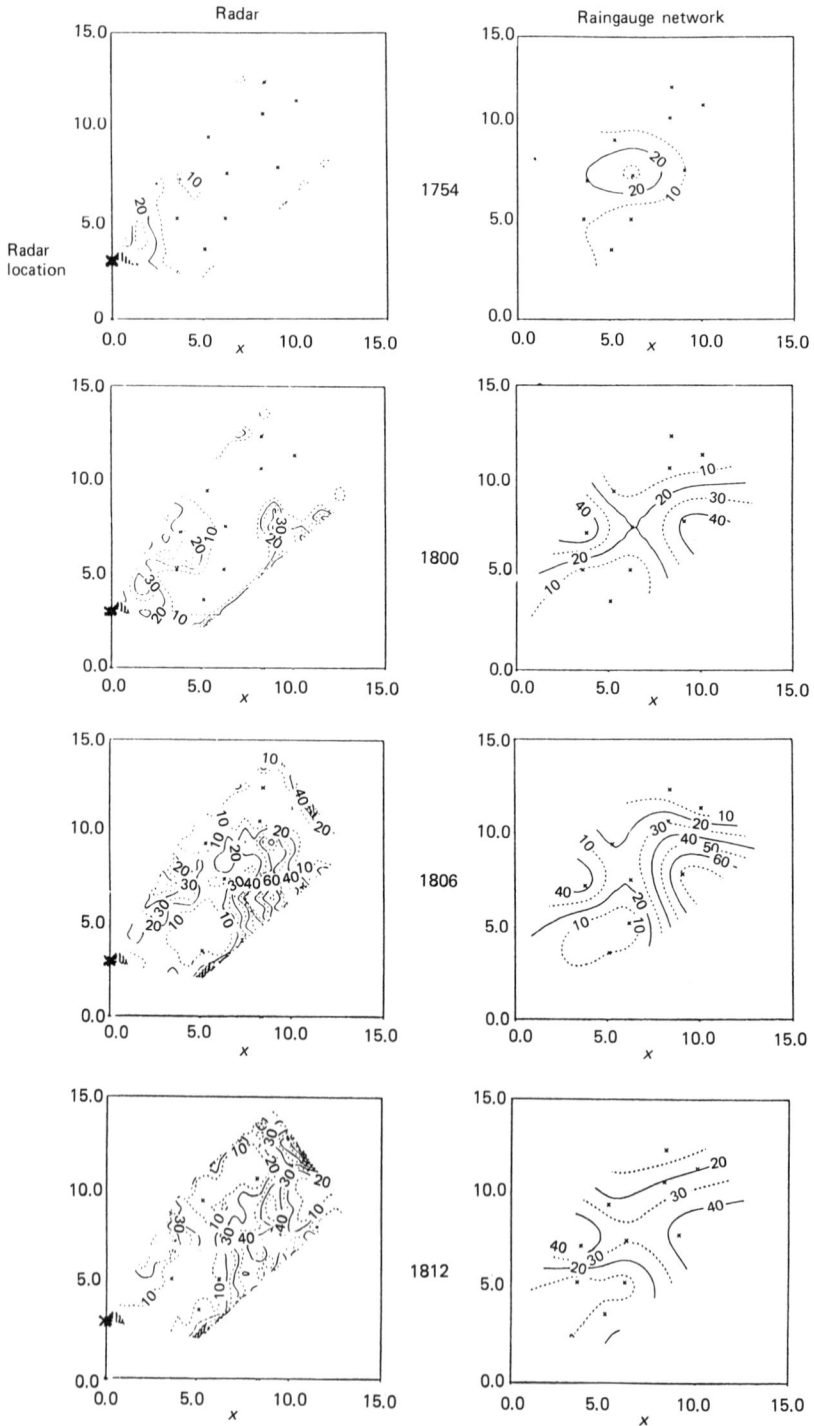

Fig. 5 — 6 min radar and raingauge maps for 27 June 1988 event.

reflectively target) to obtain well-pre-estimated reflectivity data, the attenuation correction procedure being very sensitive to this factor.

The determination of specific $K–R$ and $Z–R$ relationships could also introduce some improvements. Raindrop size distributions obtained from various hydrological experiments in the South of France could be used for this purpose.

Technical solutions have to be found for the on-site attenuation problem. Various experiments and numerical simulations are in progress to obtain an order-of-magnitude estimate of the different identified sources of attentuation and to predict the effectiveness of the proposed solutions (radome, etc.).

Finally, we must note that the data collected during the limited duration of the Grenoble 88 experiment are not sufficient to provide reliable statistics concerning the range at which the probability of no-rain detection owing to range attenuation and weak peak power becomes prohibitive. Data collected in the Cevennes region over 3 years with a 10 cm weather radar system could serve as 'true reflectivity' fields to assess this very important point.

ACKNOWLEDGEMENTS

The authors gratefully acknowledge Enterprise Electronics Corporation, Electricité de France, the Laboratoire d'Electronique, Microonde et Optique Guidée and the Centre National de la Recherche Scientifique for their important contributions to the Grenoble 88 experiment. We are also indebted to Ch. Masnou, A. Nonga, A. Mérigeon, M. Rauhoff, J. M. Taunier, R. Laty and J. P. Gaudet for their part in the development and operation of the radar system and the raingauge network during the experiment.

REFERENCES

Gunn, K. L. S. and East, T. W. R. (1954) The microwave properties of precipitation particles. *Q. J. Roy. Meteorol. Soc.*, **80**, 522–545.
Hildebrand, P. H. (1978) Iterative correction for attenuation of 5 cm radar in rain. *J. App. Meteorol.*, **17**, 508–514.
Sims, A. L., Mueller, E. A., Stout, G. E. and Larson, T. E. (1964) *Investigation of the quantitative determination of point and areal precipitation of radar echo measurements*, 9th Technical Report. US Army Electronics Research and Development Laboratory, Fort Monmouth, NJ.

23

Bright-band errors in rainfall measurement: identification and correction using linearly polarized radar returns

S. E. Hopper, A. J. Illingworth and I. J. Caylor
Department of Physics, University of Manchester Institute of Science and Technology,
Manchester M60 1QD, UK

ABSTRACT

A principal source of error in rainfall rates derived from the radar reflectivity Z is caused by the enhanced radar return due to melting snowflakes in the bright band. We present observations made by the narrow beam-width S-band Chilbolton radar which involve transmission of horizontally and vertically polarized radiation and reception of the co-polar and cross-polar return signals. The linear depolarization ratio (LDR) is defined as the ratio of the cross-polar to the co-polar return. The high values of Z in the bright band are accompanied by values of LDR above -18 dB; in echoes where no bright band is present, the values of LDR are everywhere below -20 dB.

1. INTRODUCTION

The conventional radar reflectivity Z is proportional to ND^6, where N is the concentration of particles of diameter D, summed over all sizes. Z is usually expressed in units relative to the signal from a 1 mm raindrop per cubic metre, even though from Z alone it is not possible to distinguish rain from ice and the reflectivity of a raindrop is about 7 dB higher than the equivalent mass of ice. Neither can Z be used to differentiate between the various forms of frozen hydrometers (snow, hail, hailstones, etc.), or to measure the sizes and concentrations of raindrops. Z is usually converted into a rain rate R using an empirical relation of the form

$$Z = 284R^{1.47} \tag{1}$$

which, for a given R, is equivalent to assuming a constant size distribution of raindrops. Some of the problems in estimating rainfall using equation (1) are demonstrated in Fig. 1, which is a vertical section of Z through stratiform rainfall observed by the 0.25° beam-width Chilbolton radar.

The most notable feature in Fig. 1 is the layer of enhanced reflectivity or 'bright band' at about 1.1 km altitude. At a range of 45 km the value of Z reaches 43 dBZ, leading (via equation (1)) to an estimated rain rate of about 18 mm/h. Values below 150 m are affected by obscuration of the radar beam, but from 200 to 800 m, in the rain, the value of Z is only 28 dBZ, equivalent to a rainfall rate of only 1.7 mm/h. The bright band leads to an overestimate of R by a factor of 10.

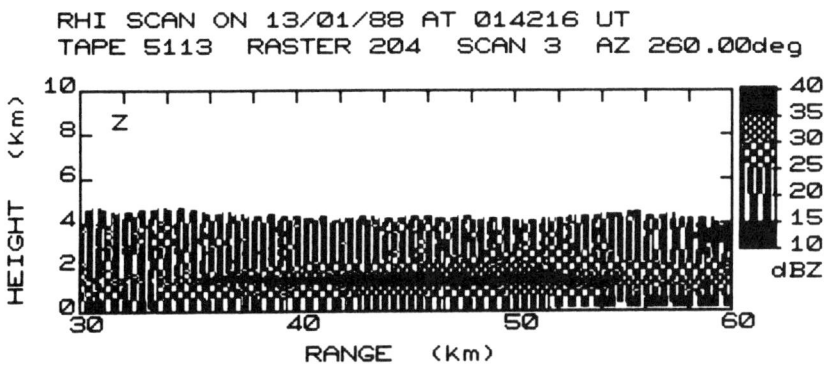

Fig. 1 — An example of the structure of the radar reflectivity Z for stratiform precipitation with a well-defined bright band.

The enhanced return in the bright band is caused by large low-density wet snowflakes which reflect microwaves as if they were giant raindrops. Dry snowflakes have a lower return because of their low dielectric constant. Below the bright band, Z falls owing to two factors: when the snowflakes melt completely, they collapse to smaller raindrops and, secondly, the raindrop concentration falls as the terminal velocity increases.

Other problems are evident form Fig. 1. The Z values at altitudes greater than 2.5 km are lower than in the rain, much of the growth and aggregation of the ice occurring at lower levels. Consequently, the higher-elevation scans of a 1° beam-width radar will underestimate rainfall at larger distances (Joss and Waldvogel 1989), while sampling of the rain alone with the lower-elevation scans without ground clutter and/or obscuration is difficult, because, in the UK, the melting layer is usually below 2 km.

Ground-based raingauges can provide localized real-time information for correcting bright band errors (Smith 1986). Most radar networks scan in plan position indicator (PPI) mode to obtain a complete spatial coverage, and in truly stratiform rain the bright band should be recognizable as a concentric ring of enhanced reflectivity centred on the radar. Collier (1986) has suggested an automatic means of identifying the bright band by comparing the range of the maximum values of Z at a particular azimuth for two different beam elevations. However, in practice the height of the bright band can change, the precipitation is never stratiform and, in the UK at least, quite vigorous showers often have bright bands. In this paper we demonstrate a means of uniquely identifying the bright band by analysing the cross-polar radar return.

2.   THE CHILBOLTON POLARIZATION RADAR

The Chilbolton radar operates at S band (10 cm) and, with a 25 m dish, is the largest steerable meteorological radar in the world, having a beam width of only 0.25°. Earlier reports (Cherry and Goddard 1982, Hall *et al.* 1984, Illingworth *et al.* 1987) have considered the implementation and interpretation of differential reflectivity Z_{DR}, we now analyse observations made in 1988 of a new parameter, the linear depolarization ratio LDR.

The differential reflectivity Z_{DR} provides an estimate of mean hydrometeor shape. It is defined as

$$Z_{DR} = 10 \log (Z_H/Z_V) \qquad (2)$$

where Z_H and Z_V are the radar reflectivites measured at horizontal and vertical polarizations respectively. For small raindrops or tumbling ice particles, Z_H and Z_V are equal and Z_{DR} is zero. The theoretical values of Z_{DR} for oblate particles with their minor axes aligned in the vertical are plotted in Fig. 2; Z_{DR} increases with greater oblateness and higher dielectric constant. In heavier rain, Z_{DR} is positive and reflects the mean shape (and hence the size) of the raindrops. Z_{DR} for ice is more complex (Illingsworth *et al.* 1986). Because of the low dielectric constant, dry snowflakes have a Z_{DR} value close to zero, but wet snowflakes can have high positive values. Graupel tends to tumble and to be associated with a zero Z_{DR} value.

The linear depolarization ratio LDR is a measure of the hydrometeor fall mode and appears to be an excellent indicator of wet ice. It is defined as

$$LDR \equiv 10 \log (Z_{VH}/Z_H) \qquad (3)$$

where Z_{VH} is the (horizontal) cross-polar return from a vertically polarized transmitted pulse, and Z_H (as in equation (2)) is the co-polar (horizontal) return for horizontally polarized transmission. A cross-polar return occurs only when oblate hydrometeors fall with their major or minor axis at an angle to the vertical. Computations of LDR for tumbling oblate spheroids (Fig. 3) are consistent with the Chilbolton observations (Illingworth and Caylor 1989). Snowflakes have such a low dielectric constant that, even if they are very oblate, their LDR is below the antenna limit of −32 dB; oblate dry hail or graupel could have a value up to −20 dB if the axial ratio were as low as 0.5, but LDR values above −20 dB can only realistically occur for wet tumbling ice particles. Such high values are restricted to the bright band. Raindrops give rise to a very low cross-polar return.

3.   COMPARISON OF ECHOES WITH AND WITHOUT BRIGHT BANDS

An RHI scan through widespread stratiform precipitation is displayed in Fig. 4, where the reflectivity does not exceed 30 dBZ in the rain but reaches 40 dBZ at 2 km

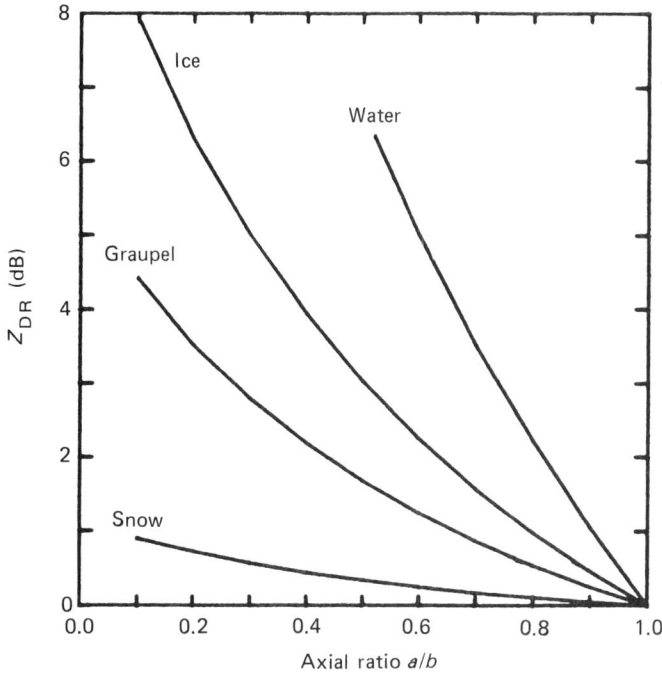

Fig. 2 — Z_{DR} values as a function of axial ratio for various precipitation particles, assuming that the major axis is horizontal.

altitude in the bright band. At this height the oblate melting snowflakes give a clearly visible bright band in Z_{DR} with values reaching 2 dB. However, automatic recognition of the Z_{DR} bright band can be difficult. Z_{DR} values in the rain at 10–20 km range reach 0.5 dB, and in heavier rain they can be much higher. We also note in Fig. 4 the positive Z_{DR} values above 2 km altitude, this low-Z region presumably containing aligned high-density ice crystals.

It is much easier to identify the Z bright band from the LDR data. The maximum values of Z coincide with the peak LDR of about -15 dB, which is consistent with wet tumbling snowflakes having an axial ratio of about 0.5 (Fig. 3). In contrast, LDR values in the rain are near the antenna limit of -32 dB and reach about -27 dB in the low-Z ice region above 3 km where the Z_{DR} indicated high-density crystals. Fig. 4 also shows that ground clutter results in LDR values above -10 dB near to the ground. Because LDR involves measuring the low-power cross-polar return, it is much more susceptible to ground clutter than is Z or Z_{DR}.

Fig. 5 illustrates an example of a heavy shower with a horizontal extent of about 25 km. The Z values in the rain are higher than in Fig. 4, but the enhancement of Z in the bright band is over 10 dB. Again the presence of the bright band can be most easily identified by the values of LDR which exceed -16 dB.

A vigorous shower with no bright band is depicted in Fig. 6 and it is clear that the polarization parameters have a quite different character: the LDR values are much lower than for the bright-band case in Figs 4 and 5. We believe that the data in Fig. 6

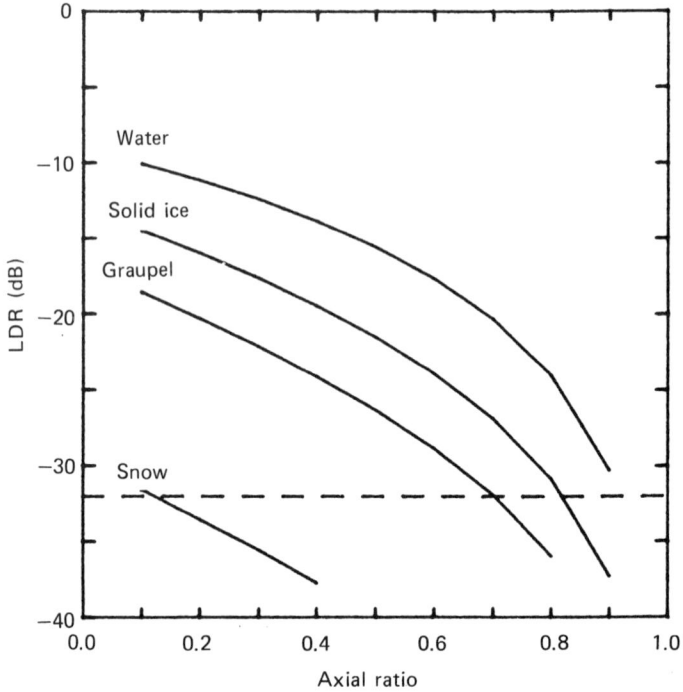

Fig. 3 — LDR values for randomly tumbling particles as a function of axial ratio.

indicate the presence of graupel. The dry tumbling graupel gives negligible LDR, but melting occurs at about 2 km altitude and LDR rises to about −25 dB. This weak 'LDR graupel bright band' is consistent (Fig. 3) with tumbling wet ice with an axial ratio of 0.8. In the heavy rain the LDR is just detectable and is explicable in terms of a canting angle of about 5°. Values of Z_{DR} are low for the tumbling dry ice but rise monotonically as the graupel melts and assumes the equilibrium shape of the large raindrops. This vertical profile in Z_{DR} should be contrasted with the bright-band case in Figs 4 and 5, where a maximum in Z_{DR} is caused by the low-density oblate wet snowflakes, which subsequently, on complete melting, collapse to more spherical raindrops.

It should be emphasized that, in the UK at least, the presence of a bright band is not restricted to stratiform clouds with low Z. In some showers, Z values can reach 50 dBZ in the bright band, while others, with no bright band, have lower peak values of Z.

4. LINEAR DEPOLARIZATION RATIO STATISTICS

In order to test our hypothesis that the peak value of LDR in a vertical profile is related to the increased reflectivity in the bright band, the results from scans on 11 different days in 1988 are summarized in Fig. 7. The altitude of the maximum value in LDR (which is found in all types of cloud) is used to fix the melting level. The

RHI scan on 29 November 1988 at 120130 UT
tape 5148 Raster 58 Scan 1 AZ 285.00 deg

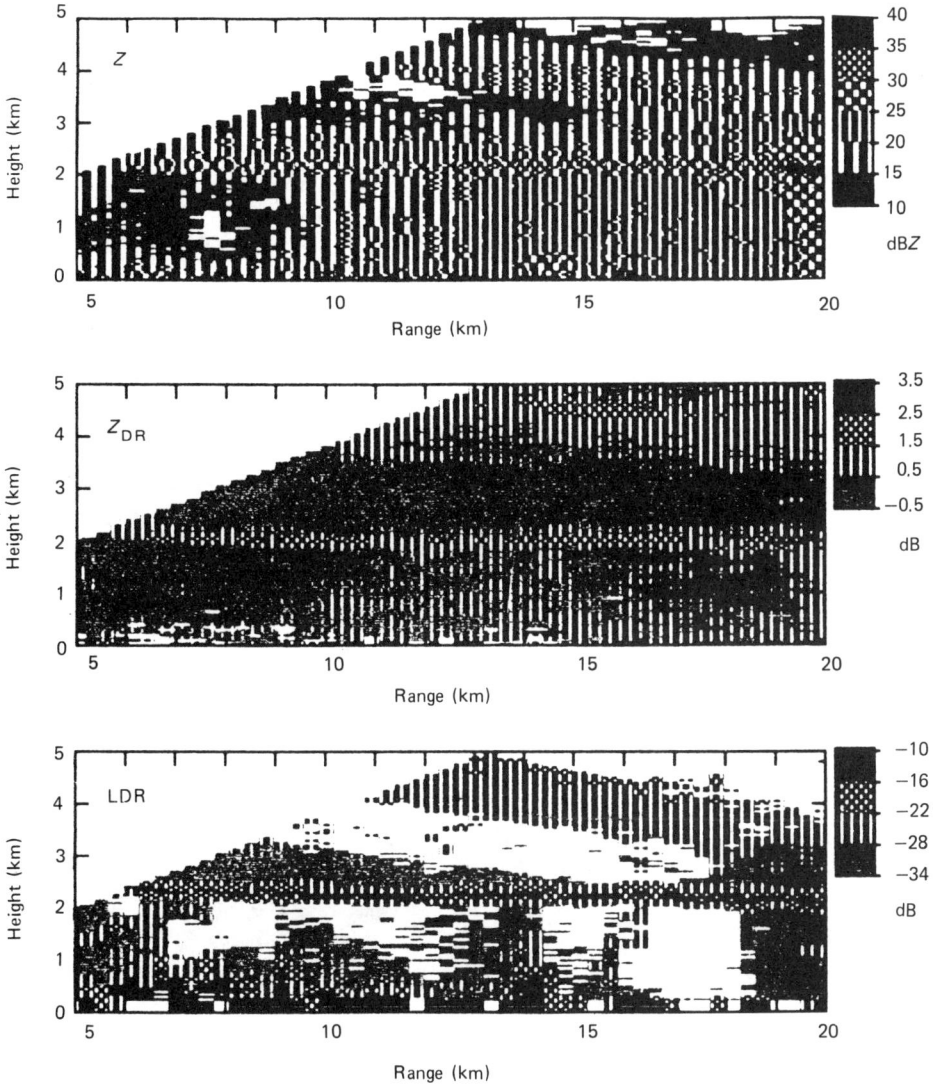

Fig. 4 — A typical RHI scan through stratiform cloud on 29 November 1988 with a bright band for Z, Z_{DR} and LDR.

enhancement ΔZ of Z is then estimated by comparing the Z value at the melting level with the Z value in the rain 500 m below. Histograms of the enhancement of the reflectivity in the bright band are plotted for each 4 dB increment in LDR. For most vertical profiles the peak LDR values are in the range from -14 to -18 dB and in these cases the enhancement of Z is, on average, about 10 dB. Less common, in this UK sample, are the peak values of LDR in the range from -18 to -22 dB and from

RHI scan on 29 May 1988 at 155541 UT
tape 7150 Raster 52 Scan 2 AZ 255.00 deg

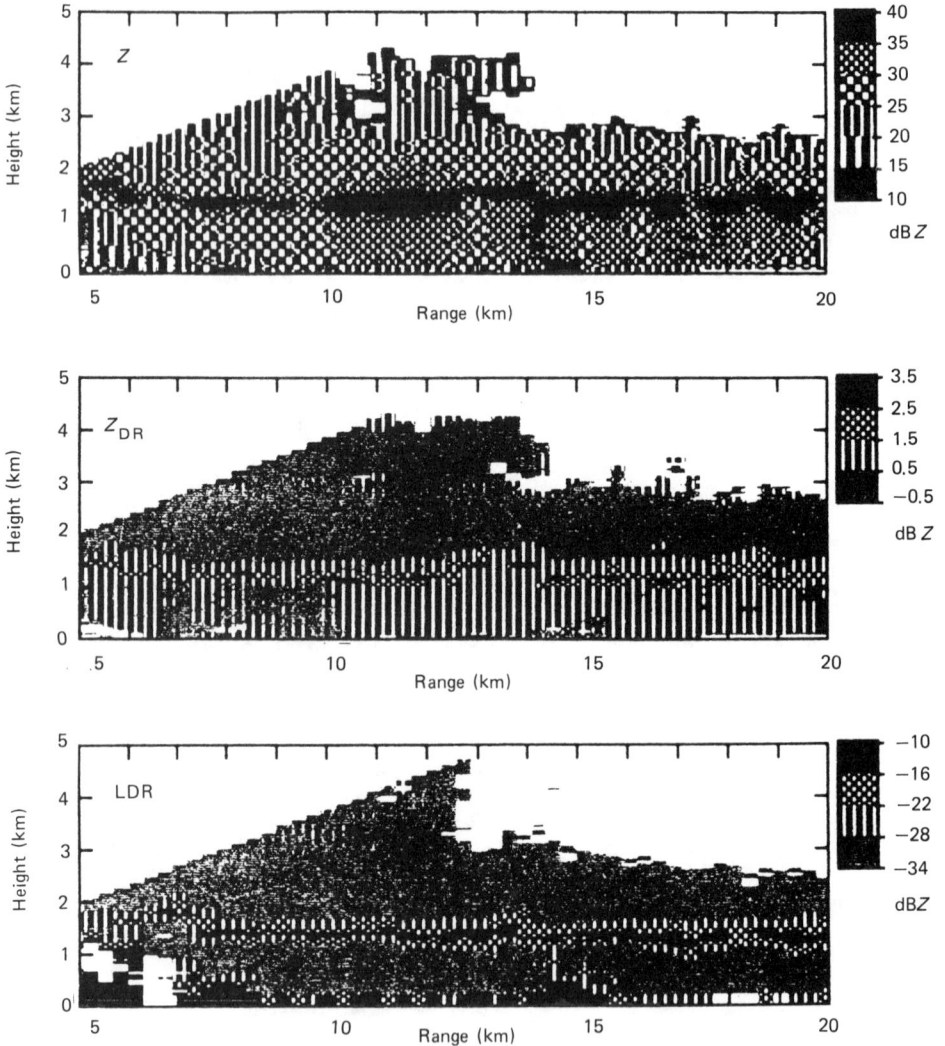

Fig. 5 — An RHI scan Z, Z_{DR} and LDR through a vigorous shower on 29 May 1988 with a bright band.

−22 to −26 dB where the Z enhancement is essentially zero and no bright band is present.

It should be stressed that in 1988 there were no observed cases of deep vigorous convection. Measurements of LDR at 3 cm (Herzegh and Jameson 1989) suggest that hail in wet growth can give high LDR values, although the measurements at this wavelength are affected by propagation problems. For the observations discussed in this paper the depth of the bright band is greater than the beam width of the

RHI scan on 13 July 1988 at 155423 UT
tape 5131 Raster 22 Scan 1 AZ 253.00 deg

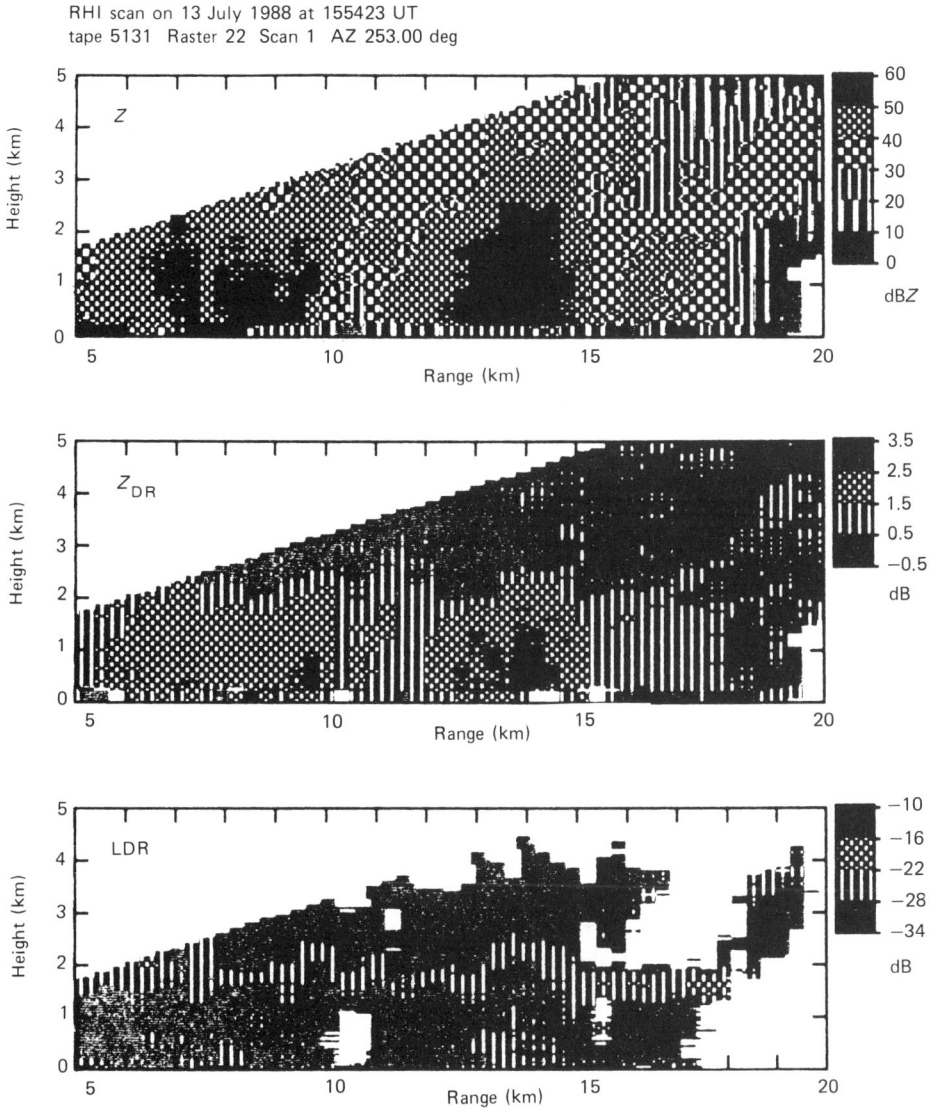

Fig. 6 — A typical RHI scan through a convective cloud on 13 July 1988 with no bright band for Z, Z_{DR} and L_{DR}.

Chilboltin radar. In future we shall analyse the effect on the LDR measurements for a 1° beam-width radar which is only partially filled by the bright band.

5. DISCUSSION AND CONCLUSION

Several factors need to be considered if the LDR technique is to be implemented on conventional C-band radars. The LDR measurement has the advantage that no fast

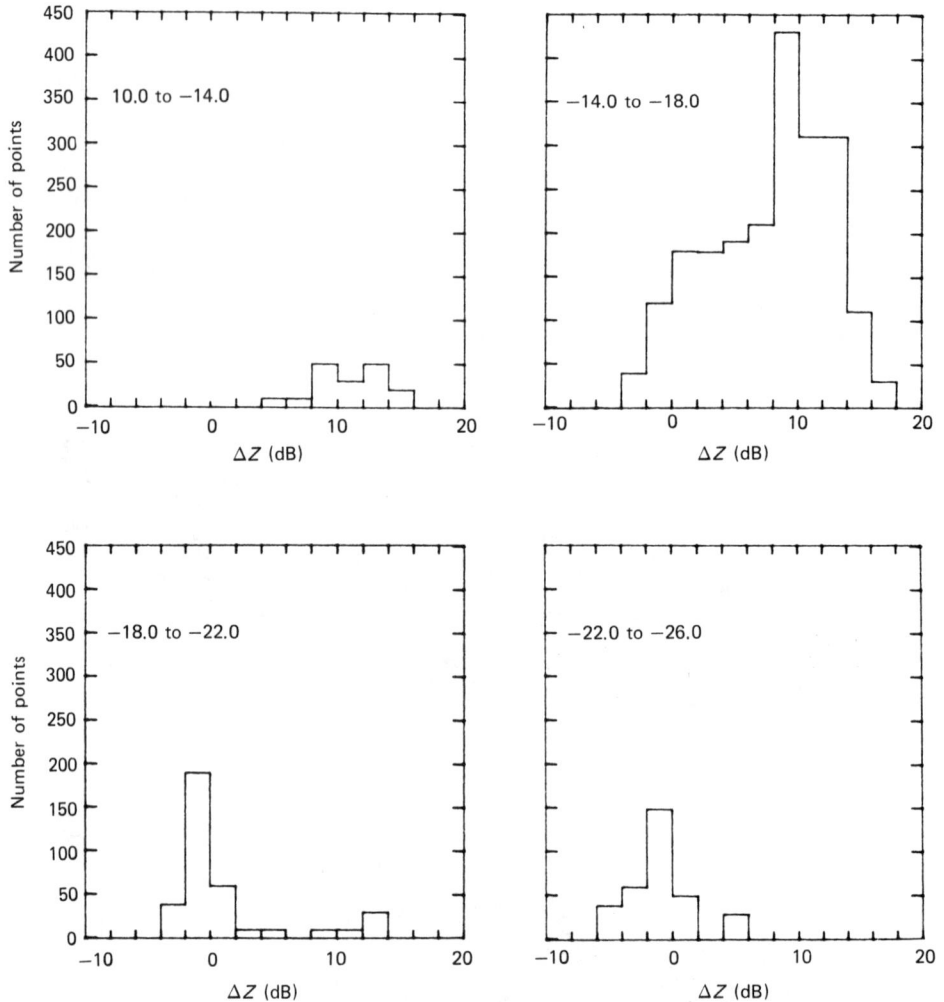

Fig. 7 — Histograms of the enhancement ΔZ of the reflectivity in the bright band as a function of LDR in the bright band.

switching of the transmitted signal is needed. However, the cross-polar signal power is very low, and a signal-to-noise ratio of more than 20 dB is required if an LDR of down to −20 dB is to be detected. We also note that the low-power cross-polar signal is much more sensitive than Z and Z_{DR} to ground clutter contamination. Finally, and most importantly, the LDR signal is affected by propagation and attenuation problems at shorter wavelengths. Correction algorithms may be possible at C band.

These observations suggest that high values of LDR are associated with the presence of melting snowflakes and can be used to identify the bright band. This parameter may also be of use in identifying anomalous propagation and ground clutter, both of which should have high values of LDR.

ACKNOWLEDGEMENTS

This work was supported by US Air Force Office of Scientific Research (Grant 89-0121), Natural Environment Research Council (Grant GR3/5896) and the Meteorological Office. We also thank John Goddard of RAL who implemented the LDR parameter on the Chilbolton radar.

REFERENCES

Cherry, S. M. and Goddard, J. W. F. (1982) *Union Radio-Scientifique Internationale Symposium on Multiple Parameter Radar Measurement of Precipitation, Bournemouth, UK.*

Collier, C. G. (1986) *J. Hydrol.*, **83**, 207–223.

Hall, M. P. M., Goddard, J. W. F. and Cherry, S. M. (1984) *Radio Sci.*, **19**, 132–140.

Herzegh, P. H. and Jameson, A. R. (1989) *Preprints of the 24th Conference in Radar Meteorology*. American Metorological Society, Boston, MA, pp. 315–317.

Illingworth, A. J. and Caylor, I. J. (1989) *Preprints of the 24th Conference in Radar Meteorology*. American Meteorological Society, Boston, MA, pp. 323–327.

Illingworth, A. J., Goddard, J. W. F. and Cherry, S. M. (1987) *Q. J. R. Meteorol. Soc.*, **113**, 469–489.

Joss, J. and Waldwogel, A. (1989) *Preprints of the 24th Conference in Radar Meteorology*. American Meteorological Society, Boston, MA, 682–688.

Smith, C. J. (1986) *J. Atmos. Ocean. Technol.* **3**, 129–141.

24

An integrated X-band radar system for short-range measurement of rain rates in HERP

A. Kammer
Meteorological Institute, University of Bonn, Auf dem Huegel 20, W-5300 Bonn 1, FRG

ABSTRACT

Corresponding to hydrological requirements, a radar system, as part of an intended local radar network, was developed during the last 3 years. This system allows the measurement of rain rates with a resolution of 128 intensity classes for areas of the size 600 m × 600 m updated every minute. Based on the self-limitation in range of about 40 km, a small X-band radar can be used. Radar-oriented tasks are done by a microprocessor system (MC6809), whereas all further products are calculated by a PDP11/73, including range and attenuation corrections. A special look-up table accelerates the transformation from polar to Cartesian pixels and, at the same time, suppresses ground clutter efficiently (clutter map method, based on the original polar pixels).

1. INTRODUCTION

Within a still-running 6 year field experiment HERP (Hydrological Emscher Radar Project) the foundations for a low-cost radar system for the quantitative determination of rain rates was developed. For urban hydrology, high resolution in time and space is needed, because of the short reaction times of the hydrological system. Additionally, the accuracy of rain data is important for the quality of the runoff simulation (Verworn 1991, Semke 1991). The relation between radar-measured reflectivity factors Z and the corresponding rain intensities R varies from rain type to rain type. Errors introduced by using a fixed $R–Z$ relation (e.g. the Marshall–Palmer relation) should be avoided. Therefore, an on-line calibration by ground-based measurement of actual $R–Z$ relations, calculated from drop spectra, is provided (Kreuels 1991).

During the past 3 years the prototype of a hydrological radar, as part of an aspired local net, was developed and installed at the Regional Weather Forecast Center

Essen, FRG, about 10 km south of the river Emscher. In the following, the hardware structure and signal processing will be presented.

2. SYSTEM DESIGN ASPECTS

Corresponding to the requirements in urban hydrology, a radar system was designed, which allows the determination of rain rates

(1) with a resolution of 128 intensity classes (7 bit),
(2) for areas of the size $600 \text{ m} \times 600 \text{ m} = 0.36 \text{ km}^2$,
(3) within a range of about 40 km and
(4) updated every minute.

The intensity classes 0–26 are spaced linearly $(0.0, \ldots, 2.6 \text{ mm/h})$, whereas higher intensities $(2.7, \ldots, 275 \text{ mm/h})$ are represented by the logarithmic classes 27–127. This classification is sufficient, because the relative difference between neighbouring classes is smaller than 5%.

The size of the integration areas of $600 \text{ m} \times 600 \text{ m}$ was chosen with respect to the resolution in range of the (given) radar set of 300 m. The video signal is digitized at intervals of also 300 m. Therefore, at least two independent samples belong to one Cartesian pixel.

The reliability of radar-measured rain rates (or reflectivity factors) decreases with increasing range r. There are two reasons for this.

(1) The increase in the pulse volume with r^2, which increases the risk of inhomogeneous filling by raindrops;
(2) The increase in beam height because of the earth's curvature and the necessary elevation angle of the antenna; the probability that a measurement is within the falling rain, below the cloud base, decreases, and the risk of bright-band effects increases.

In both cases the range of validity of the radar equation for rain is limited (Battan 1973). In Herp the antenna has a half-power beam width of 1.1°. The elevation angle is fixed at 1.5°. On the assumption of a tolerable beam height above ground of 1 km, the range is limited to about 40 km.

Because of the self-limitation in range, it is possible to use a small (low-cost) X-band microwave sensor. The necessary correction of the received power, corresponding to attenuation by rain, is not a problem in general. Only in extreme situations, e.g. heavy rain at the radar site, can the radar sight be reduced for a short time in a way so that there is nothing left to be corrected. With regard to the intended local network, this information loss will be compensated by a second radar system with momentarily better conditions.

3. SIGNAL PROCESSING

The hardware structure of the complete system is shown in Fig. 4 at the end of this paper. It is divided into three main parts:

(1) radar components (antenna, transmitter, receiver, etc.);

(2) radar processor (radar control, signal analogue-to-digital conversion and signal pre-processing);
(3) master processor (radar meteorological calculations and communication).

3.1   Radar receiver

The radar receiver has to process very different received powers, from -14 dBM (heavy rain; 150 mm/h; range, 450 m) down to -94 dBm (drizzle; 0.1 mm/h; range, 38 550 m). This dynamic power of 80 dB consists of the dynamic power of rain (41 dB) and the range dependence $(1/r^2)$ of the received power (39 dB). With an overall dynamic power of 85 dB and a minimum detectable signal of -99 dBm, the receiver is able to process the incoming signal without reducing the range dependence by an STC circuit. To be on the safe side, the use of a radio-frequency STC followed by a low-noise radio-frequency pre-amplifier between a transmit–receive limiter and mixer should be taken into consideration (see Fig. 4). Based on the logarithmic transfer function of the receiver, the amplitude of the resultant video signal in volts is proportional to the received power in dBm.

3.2   Signal pre-processing

All the tasks close to the radar set are done by the radar processor. The erasable programmable read-only memory resident program, developed in assembler language,

(a) controls the radar set,
(b) drives the antenna,
(c) digitizes the video signal and
(d) pre-processes the digitized data.

Within a sector of 90° (129 azimuth positions; 315–45°; increment, 0.7°) the video signal is digitized in 128 range steps (600 m$\leqslant r \leqslant$38 700 m; increment, 300 m) and integrated over 16 pulses (Fig. 1).

3.3   Radar meteorological calculations

All further calculations, up to rain intensities, are coded in Fortran 77 and run on a PDP11/73 under RSX-11M. A summary is given in Fig. 2.

On-line calibration of the radar system is provided by the use of actual R–Z relations (Kreuls 1991). During a rain event the drop spectrum is measured with a distrometer, and pairs of R and Z values are calculated from this every minute. After about 10 relevant minutes the actual R–Z relation can be used instead of a fixed one.

Attenuation by rain is calculated with the formula of Eissing (1976):

$$A \ (\text{dB/km}) = 0.0238 \ [R \ (\text{mm/h})]^{1.063}$$

two-way attenuation for $\lambda = 3.2$ cm.

In a former unpublished study of the present author, ten formulae were tested with a data set of 90 rain events for a catchment close to Bonn FRG. Only two relations — those of Eissing (1976) and of Greene et al. (1966) — could satisfy the data for this event. Most of the formulae overestimated the attenuation by rain at higher intensities.

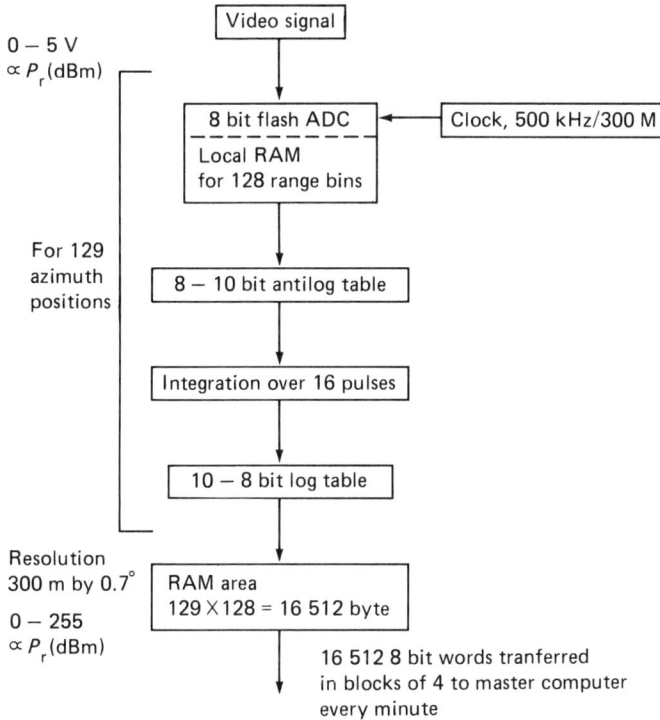

Fig. 1 — Signal pre-processing: ADC, analogue-to-digital converter; RAM, random-access memory.

Clutter removal and polar-to-Cartesian transformation are done using the same look-up table. The two-dimensional table (129×128 values) is organized in the same manner as the original radar data. A positive number indicates the Cartesian pixel that the polar value belongs to. If the number is negated, the corresponding radar value is superimposed by clutter and, therefore, it will not be taken into account. In contrast with the weather radar system in the UK (Collier and James 1986) or the Swiss radar system (Joss 1981), where the data are averaged in space before clutter removal, the clutter identification here operates on the basic polar data set, which enlarges the efficiency of the clutter map method.

4. SYSTEM ACCURACY

On the assumption of well-calibrated hardware, the standard deviation of the radar-measured reflectivity factor Z depends on the number k of independent samples, which belong to a Cartesian pixel. In the case of a logarithmic rceiver, the standard deviation is given by the formula (Sirmans and Doviak 1973)

$$\sigma(k) = \frac{5.57\,dB}{(\sqrt{k})}\,.$$

Amplitude, 300 m by 0.7°	Radar data, 16512 byte	Digitized video signal, 129 azimuth positions by 128 range bins
Received power P_r, 300 m by 0.7°	Amplitude to P_r d (dBm)	Calibration function in form of a look-up table
	Clutter removal	By a stored clutter map
Reflectivity factor Z, 300 m by 0.7°	P_r (dBm) to Z (dBZ)	Radar equation including $1/r^2$ correction
Rain rate R, 300 m by 0.7°	Z (dBZ) to R (dBR)	R–Z relation fixed or actual
	Linearization R (dBR) to R (mm/h)	By a look-up table
	Attenuation to correction R (mm/h) to A (dB/300 m)	Attenuation by rain for λ = 3.2 cm
Rain rate R, 600 m by 600 m	Polar to cartesian and integration	By a look-up table of size 129 ×128
	Interpolation	Only for cartesian pixels without hits
Rain rate R, 600 m by 600 m, 128 intensity classes	R (mm/h) to 128 intensity classes	

2290 7 bit words transfered
for hydrological calculations
every minute

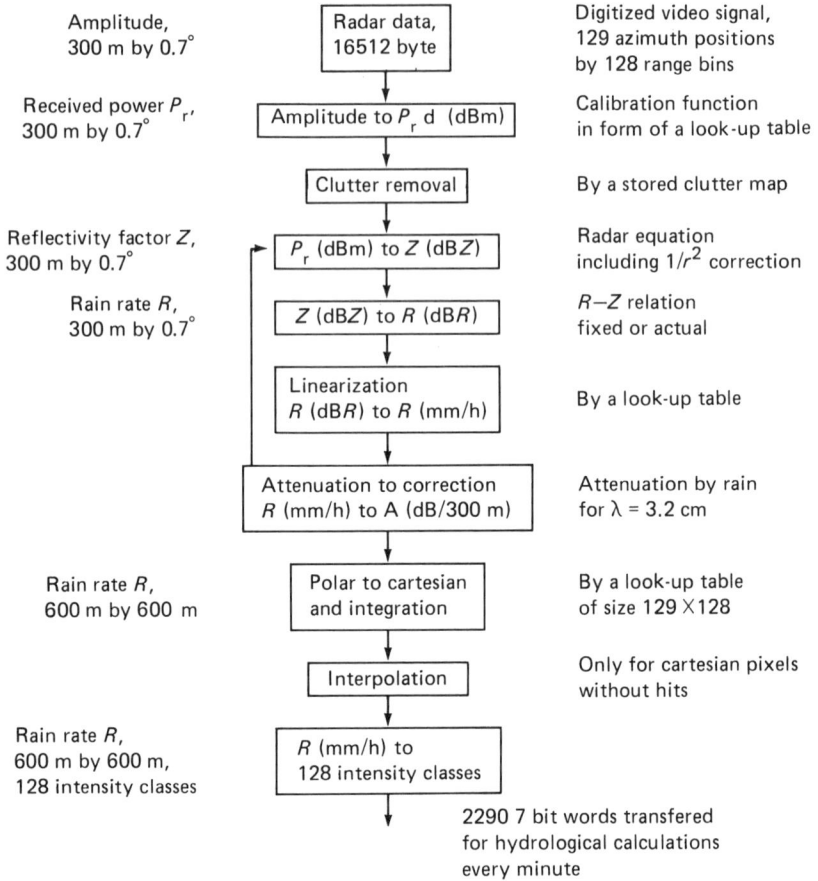

Fig. 2 — Radar data processing.

At a wavelength of 3.2 cm, pulse-to-pulse samples are statistically independent, if the pulse repetition frequency is lower than 312 Hz. The Emscher radar has a pulse repetition frequency of 245 Hz. Therefore, after pre-processing, every polar pixel (300 m; 0.7°) corresponds to the arithmetical mean of 16 independent samples. Changing to Cartesian pixels (600 m × 600 m), a variable number of polar values are combined with respect to range (186 for the nearest and 2 for the farthest). Therefore, within a range of 450 m up to 38 550 m the standard deviation varies from 0.1 to 0.98 dB, corresponding to a relative standard error of 2.3–25.3%. Fig. 3 describes the range dependence of system accuracy, including the case when several Cartesian pixels are combined to subcatchments.

It should be noted that additional errors occur by calculating the rain rate R from a fixed R–Z relation. Therefore, on-line calibration by using rain-type-dependent relations is provided. Moreover, the accuracy decreases by increasing the clutter presence, because the number of independent samples (i.e. samples not superimposed by clutter) decreases.

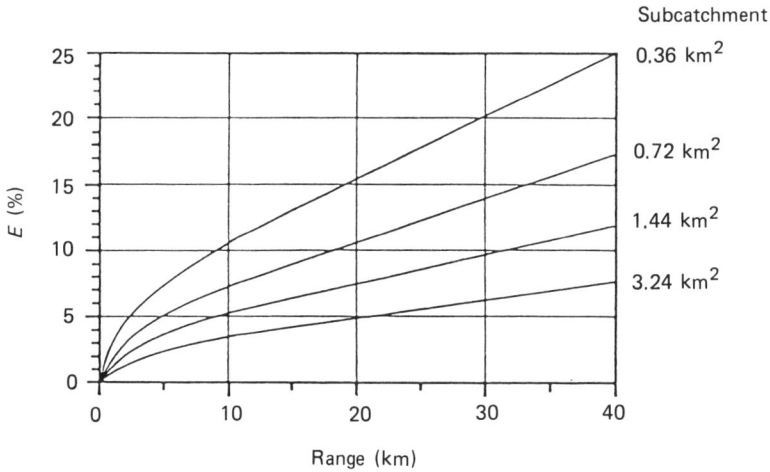

Fig. 3 — Relative standard error E of the determined reflectivity factor Z (mm^6/m^3) versus range.

5. FINAL REMARKS

A local radar network, based on simple and low-cost X-band components, as described above, cannot be compared with existing or planned C- or S-band networks of the national weather services. It is a microwave sensor for rain rates that has to try conclusions in costs and accuracy with an on-line network of a comparable number of raingauges. Especially in urban hydrology, remote sensing often is the only alternative for nowcasting rain rates with high resolution in space, time and intensity levels.

ACKNOWLEDGEMENTS

The project is financed by the Minister of Research and Technology of the FRG. Furthermore, thanks are due to the FRG radar manufacturer Gematronik GmbH for much technical support, and to the Regional Weather Forecast Center Essen, FRG, for the friendly accommodation and assistance.

REFERENCES

Battan, L. J. (1973) *Radar observation of the atmosphere.* University of Chicago Press, Chicago, Il.

Collier, C. G. and James, P. K. (1986) On the development of an integrated weather radar data processing system. *Proceedings of the 23rd Conference on Weather Radar, Snowmass, CO, 1986.* American Meteorological Society, Boston, MA, pp. JP95–JP98.

Eissing, R. (1976) *Streuung und Daempfung elektromagnetischer Wellen an Niederschlaegen in cm/mm-Wellenlaengenbereichen,* Berichte No. 33. Institute of Radiometeorology and Maritime Meteorology, University of Hamburg.

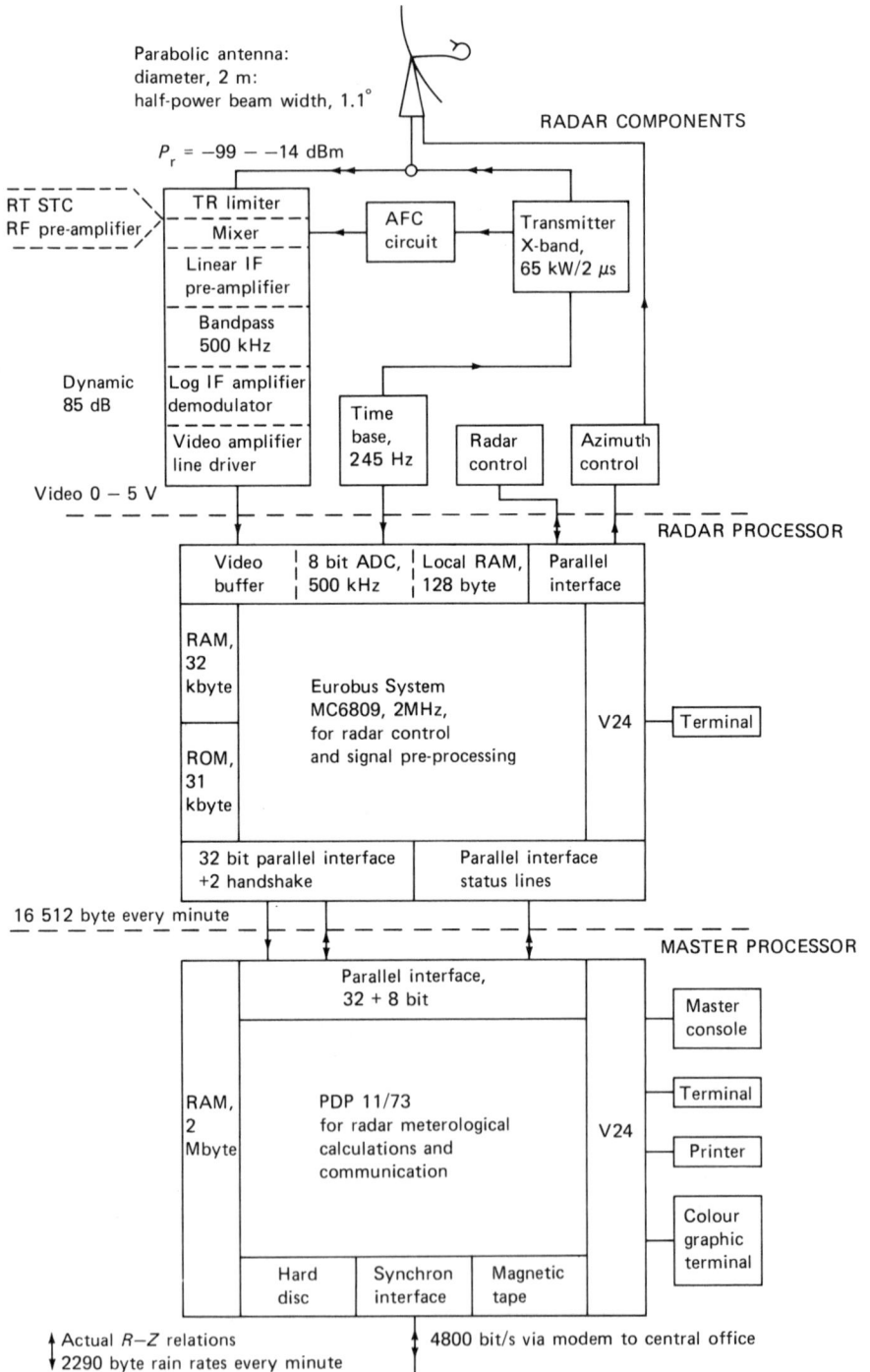

Fig. 4 — Hardware structure of the radar system in HERP: RF STC, radio–frequency – – – ;
TR, transmit–receive; IF, intermediate frequency, AFC, automatic frequency control; ADC,
analogue-to-digital converter; RAM, random-access memory; ROM, read-only memory.

Greene, D. R., Clark, R. A. and Moyer, V. E., (1966) 3.2 cm attenuation related to rainfall rate and liquid water content *Proceedings of the 12th Conference on Weather Radar, Norman, OK, 1966.* American Metorological Society, Boston, MA, pp. 250–253.

Joss, J. (1981) Digital radar information in the Swiss Metorological Institute. *Proceedings of the 20th Conference on Weather Radar,* Boston, MA, 1981. American Meteorological Society, Boston, MA, pp. 194–199.

Kreuels, R. K. (1991) On-line calibration in HERP. *Hydrological Applications of Weather Radar.* Ed. Cluckie, I. D. and Collier, C. G. Ellis Horwood, Chichester, West Sussex, Chapter 5.

Semke, M. (1991) From radar rainfall to hydrological data. *Hydrological Applications of Weather Radar.* Ed. Cluckie, I. D. and Collier, C. G. Ellis Horwood, Chichester, West Sussex, Chapter 45.

Sirmans, D. and Doviak, R. J. (1973) *Meteorological radar signal intensity estimation,* Technical Memorandum No. ERL NSSL 64. National Oceanic and Atmospheric Administrations, Norman, OK.

Verworn, H. R. (1991) Hydrological relevance of radar rainfall data. *Hydrological Applications of Weather Radar.* Ed. Cluckie, I. D. and Collier, C. G. Ellis Horwood, Chichester, West Sussex, Chapter 47.

25

Examination of the impact of range on the quality of daily catchment rainfall totals derived from the Ingham radar

M. F. Mylne and B. D. Hems
Meteorological Office, London Road, Bracknell, Berks. RG12 2SZ, UK

ABSTRACT

The aims of the Long Range Calibration Study (LORCS) Group in assessing and improving the quality of data from Ingham radar are outlined. Subcatchment daily rainfalls estimated from radar are compared with estimates derived from raingauges. A variety of assessment techniques are described. The results of these comparisons for October–December 1988 are presented.

1. INTRODUCTION

Regular transmission of data from Ingham radar began on 13 September 1988. This is the seventh radar in the UK network. The radar is owned by a consortium of Anglian, Severn–Trent and Yorkshire Water Authorities and the UK Meteorological Office. A study group with representaatives from each of the consortium members has been set up to investigate the accuracy of the data. In this paper we examine the accuracy of daily catchment total rainfall calculated from the radar output data stream. Hourly and daily catchment totals are used in both flood forecasting and water management. For rapid-response catchments an option of 15 min radar data is also available. Previously catchment totals used by water authorities have been obtained by a variety of surface-fitting techniques to data from a few recording and daily raingauges. If they are sufficiently accurate, the radar data can provide real-time rates and daily totals derived from higher spatial resolution input observations than is possible with gauge networks.

Many of the hydrological applications have to be met using data from beyond the nominal limit (75 km) of quantitative radar coverage. At this range the radar beam can overshoot low-level precipitation or beam attenuation can cause underestimation of rainfall. The particular aims of the Lincoln weather radar project Long Range Calibrations Study (LORCS) Group, are

(i) to summarize the relative accuracy of catchment totals derived from operational gauges and radar,
(ii) to establish the frequency with which certain accuracy requirements are met, and
(iii) to determine the impact of distance from the aerial on the quality of catchment totals from the radar and to assess whether long-range calibration can be improved.

In this paper the sources of data available for computation, the techniques used and the results from the first few months of data are described. The project is expected to continue for 18–24 months and the investigation methods will evolve as more results are accumulated.

2.   SOURCES OF DATA FOR THIS COMPARISON

The principal comparisons summarized in this paper are between daily catchment totals from data and raingauge data for 32 catchments of variable size within 210 km of the Ingham aerial. The location of the aerial and catchments are shown in Fig. 1.

2.1   Radar

Both 2 km and 5 km square averages are derived from the basic radar data. The averages are stored as grid-point values at the centre of 2 km and 5 km squares respectively, on the site tapes at Ingham. The tapes are then collected and sent to the Meteorological Office for processing. In real time, average hourly and daily rainfall rates for each catchment are derived using on-site processing software and transmitted to water authorities. Catchment totals are later sent to Bracknell. In this study only the 5 km, daily data are used.

2.2   Sparse gauge network

A national network of raingauges with average spacing of approximately 40 km reports daily precipitation totals (0900Z–0900Z) to the Meteorological Office at the end of each rainfall day. Within the study area of this investigation, 16 gauges are available, six of which lie within study catchments. To derive catchment totals from gauges, first the daily total is expressed as a percentage of annual average rainfall at each gauge location; next a surface is fitted to these percentage values; at each 5 km grid point the surface value is multiplied by the annual average rainfall; finally the catchment daily total is the average from each of the grid-point values within a catchment. There are about 37 additional gauges used by water authorities to produce catchment totals operationally. These will be incorporated into the project at a later date.

2.3   Dense gauge network

A national network of raingauges with average spacing of approximately 8 km reports daily rainfall totals which are available to the Meteorological Office for use

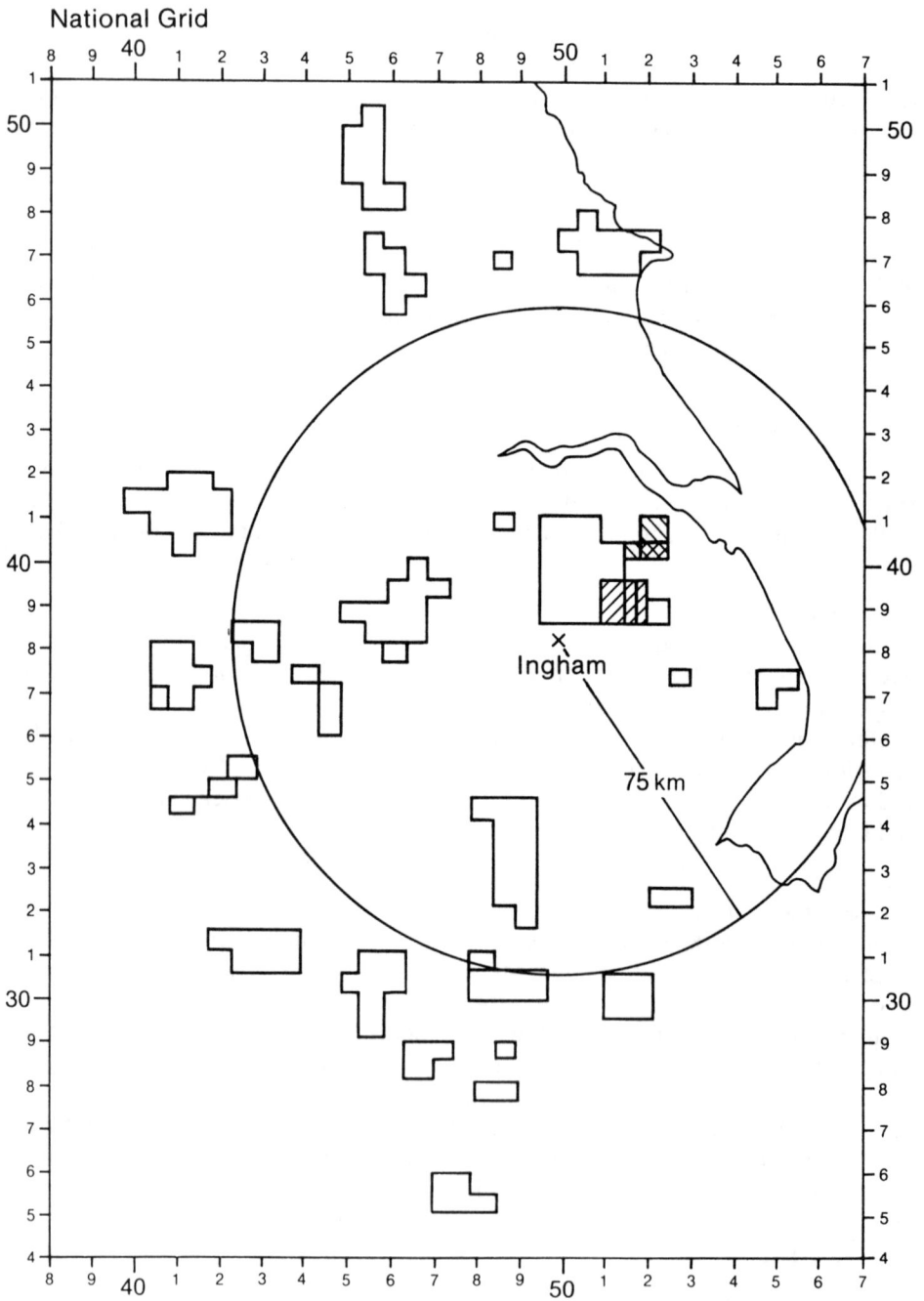

Fig. 1 — Location of radar and catchments used in this study.

about 3 months after the event. Catchment totals are derived in the same way as described for the sparse network. About 460 gauges contribute to the rainfall surface generated in the study area. 88 lie within study catchments.

3.   ASSESSMENT TECHNIQUES

All comparisons are between radar catchment rainfall totals and dense gauge catchment totals unless otherwise stated.

3.1   Distinction between wet and dry days
A threshold of 2 mm was chosen to distinguish between wet and dry days. A gross appraisal of the performance of the radar is made calculating the percentage of occasions when the radar and gauge catchment totals agree whether rainfall was above or below 2 mm. The percentage of occasions when the radar total is wet and the gauge total dry and vice versa provides an indication of the occurrence of clutter (ground echo) due to anomalous propagation (ANAPROP) or beam overshoot, etc.

3.2   Magnitude of R/G
R/G is the ratio of the radar daily catchment rainfall to the gauge catchment total. Based on the requirements stated in UKON documents 7 and 5 — internal Meteorological Office policy documents for raingauge networks radar observation networks and respectively (Meteorological Office 1986, 1989), it was decided to test the percentage of occasions for which R/G lies in the ranges 0.5–2.0 and 0.625–1.6. The results are presented only for cases when both radar and gauge totals exceed 0.5 mm as high accuracy is unnecessary for many purposes when rainfall amounts are low.

3.3   Magnitude of difference between R and G
A range of accuracy requirements were suggested by water authority members of the LORCS Group. The most stringent required that a radar catchment total should be within 10% or 1 mm (whichever is larger) of the 'truth'. For the purpose of this preliminary investigation the rainfall totals derived from the dense gauge network are the assumed 'truth'. The representativeness of these totals for hydrological application depends not only on the density of the gauges but also in the quality of the observation and the position of the gauge relative to hydrologically important areas of the catchment.

3.4   Percentage of occasions on which the radar total is closer than the sparse
gauge to the dense gauge catchment total
An important step in assessing the usefulness of the Ingham radar data is to compare both them, and the gauge data similar to that which would be available to the water authorities in near real time, with the best estimate of the truth available. In this investigation the relative merits of the radar and sparse gauge catchment totals were assessed against the dense gauge catchment totals. The sparse network has many fewer and less strategically located gauges than are available to the water authorities

in near real time. Follow-up studies will be performed to investigate the effect of using these latter gauges.

4.   RESULTS

The assessment statistics are calculated each month for each catchment as the data are received. At the time of writing the results have been proceed for October–December 1988. In addition to individual catchment statistics, statistics averaged over all catchments within annuli of 0– < 75, 75–100 and > 100 radius from the radar site have been derived as a first step in assessing the range dependence of radar performance.

The period for which data are currently available coincides with the beginning of the period of little rainfall in the winter of 1988–1989. Except in October the sample of days with more than 0.5 mm rainfall in each month was generally less than 10. The small sample size should be borne in mind when considering these preliminary results.

The percentage of occasions on which there is agreement between the radar catchment totals and gauge totals that it was either wet (2 mm or more) or dry (less than 2 mm) for each annuli and month are shown in Table 1. In all months, agreement is better than 80% and there is no notable influence or range on the quality of observations. Similar success rates at distinguishing between wet and dry conditions have been found in case studies of Hameldon Hill and Chenies radar data (Mylne 1988). Certain catchments had totals with a tendency for radar totals to be 'wet' when the gauge value was 'dry'. This probably indicates susceptibility to ANAPROP.

Table 1 — Percentage of occasions on which there is agreement between the radar and the dense gauge catchment totals whether daily rainfall is greater than or less than 2 mm

Month	Catchment distance from radar		
	< 75 km	75–100 km	> 100 km
October 88	390/459 = 84.9%	296/346 = 85.5%	109/120 = 90.8%
November 88	395/480 = 82.3%	304/360 = 84.5%	110/120 = 91.7%
December 88	384/432 = 88.9%	275/324 = 84.9%	91/108 = 84.3%
Season average	1169/1371 = 85.3%	875/1030 = 85.0%	310/348 = 89.1%

The number of occasions when both radar and gauges totals exceeded 0.5 mm and the ratio R/G lay in the range 0.625–1.6 for each month and range annuli are shown in Table 2. The same statistics for occasions when R/G lay within 0.5–2.0 are given in Table 3. The sample size in December is very small and the quality of these data is poor at all ranges; less than 30% of ratios lie within 0.625–1.6 and more than 60% outside the often-quoted acceptable ratio range 0.5–2.0.

Table 2 — Number of occasions on which both radar and dense gauge catchment totals exceed 0.5 mm and the ratio R/G lies within the range 0.625–1.6

Month	Catchment distance from radar		
	< 75 km	75–100 km	> 100 km
October 88	95/156 = 60.8%	69/145 = 47.5%	28/47 = 59.6%
November 88	59/107 = 55.1%	39/75 = 52.0%	15/26 = 57.7%
December 88	8/66 = 12.1%	12/45 = 26.6%	2/15 = 13.3%
Season average	162/329 = 49.2%	102/2650 = 45.3%	45/88 = 51.1%

Table 3 — Number of occasions on which both radar and dense gauge catchment totals exceed 0.5 mm and the ratio R/G lies within the range 0.5–2.0

Month	Catchment distance from radar		
	< 75 km	75–100 km	> 100 km
October 88	123/156 = 78.8%	98/145 = 67.6%	32/47 = 68.1%
November 88	80/107 = 74.8%	58/75 = 77.3%	20/26 = 76.9%
December 88	15/66 = 22.7%	18/45 = 40.0%	6/15 = 40.0%
Season average	218/329 = 66.3%	174/265 = 65.6%	58/88 = 65.9%

The seasonal, in particlar the October, statistics are more encouraging but still do not meet the standards required by the LORCS consortium based on UKON documents. In October there is evidence of deterioration in the quality of estimate with range over the closest two annuli. 60.8% of ratios lie within 0.625–1.6 for catchments closer than 75 km to the radar but only 47.5% for the 75–100 km catchments (Table 2). The closer catchments also had more occasions of ratios within 0.5–2.0. Similar but less extreme relationships are reflected in the seasonal totals. The apparent improvement in quality beyond 100 km is being investigated. It is thought that this may arise largely because these catchments are located in positions that are less prone to clutter and ANAPROP than are the intermediate catchments. The seasonal performance in this study is slightly worse than the values of 50% within

0.625–1.6 and 28% outside 0.5–2.0 identified in a previous radar–gauge comparison for 14 Oklahoma thunderstorms within 45–100 km of a radar (Wilson and Brandes 1979). However, in both October and November the present study had better results. This is expected as radar totals in general are too large in heavy-rainfall conditions when a fixed drop size distribution is assumed in the signal-to-rainfall conversion.

A similar pattern is revealed by the percentage 'acceptable' statistic (Table 4). For October and November, approximately 50% of radar catchment totals are within 1 mm or 10% (whichever is larger) of the dense gauge catchments total. As many as 62% are acceptable for the near catchments in October but this falls to only 35.2% for the lower totals in December. Seasonally the near and far catchment totals are 'acceptable' on a slightly higher percentage of occasions than the intermediate catchments.

Table 4 — Number of occasions on which radar catchment totals are within 1 mm or 10% (whichever is larger) of the dense gauge catchment total

Month	Catchment distance from radar		
	< 75 km	75–100 km	> 100 km
October 88	135/218 = 62.0%	96/190 = 50.5%	31/58 = 53.4%
November 88	70/138 = 50.7%	51/107 = 47.7%	18/35 = 51.4%
December 88	37/105 = 35.2%	26/78 = 33.3%	11/24 = 45.8%
Season average	242/461 = 52.5%	173/375 = 46.1%	60/117 = 51.3%

The final statistic, the number of occasions on which the radar catchment totals are closer than the sparse gauge totals to the 'truth' (Table 5), reveals little range

Table 5 — Number of occasions on which radar totals are closer than sparse gauge totals to the dense gauge catchment totals

Month	Catchment distance from radar		
	< 75 km	75–100 km	> 100 km
October 88	79/130 = 60.8%	60/133 = 45.1%	23/41 = 56.1%
November 88	31/93 = 33.3%	28/68 = 41.2%	11/23 = 47.8%
December 88	14/59 = 23.7%	3/15 = 20.0%	8/44 = 18.2%
Season average	124/282 = 43.9%	91/216 = 42.1%	42/108 = 38.9%

dependence. In general, radar was better on less than half of the occasions. Only in October was radar better on more occasions than the sparse gauges (60.8% for the close catchments and 56.1% for the four catchments beyond 100 km). The October results are similar to those from grid-point comparisons on wet case study days for grid points within 75 km of Hameldon Hill and Chenies radars (Mylne 1988). Delrieu *et al.* (1988), from a study of 11 rainfall events in Montreal, also concluded that for a daily rainfall event a gauge network of similar density to the sparse network provided results equivalent to those of radar, except in the case of events with high spatial variability.

Seasonally most statistics show that best performace was achieved by catchments within 75 km of the radar. Poorest (but only marginally so) results were mostly from the 75–100 km catchments. Examination of the performance of the radar in individual catchments has shown that certain catchments persistently perform better or worse than the majority of catchments in the same distance annuli. Poor performance may be associated with susceptibility to ANAPROP, bright band or clutter which will mask the impact of range.

5. SUMMARY AND FUTURE WORK

Simple statistics have been used in the initial stages of this project to provide a rapid appraisal of the quality of radar from the Ingham aerial. Ingham data were not calibrated by raingauges until mid-February 1989; so the quality of radar data is expected to improve. From the 3 months of data available the sample of wet days is small and the catchment totals do not yet reach the accuracy standards identified by the consortium that R/G values should lie in the range 0.625–1.6 on 68% of occasions, within 0.5–2.0 on 95% of occasions and in general errors be less than 1 mm or 10% (whichever is larger).

The annuli average values showed only slight deterioration in data quality with increasing range out to 130 km. Hitch and Hems (1988), using daily-radar-to-daily-guage ratios at gauge locations, similarly found little evidence of deterioration in the quality of Clee Hill data until ranges in excess of 130 km. Persistent poor performance of radar in individual catchments masks the impact of range. The causes of these errors are being investigated to isolate the influence of distance from the radar.

It is encouraging that, in the wettest month, more than 50% of radar catchment totals were within 1 mm or 10% (whichever is larger) of the assumed 'truth'. There is clearly good agreement between gauge and radar on the distinction between wet and dry conditions. Future comparisons will be concentrated on wet conditions. When sufficient data are available, performance for different daily intensity ranges will be examined on seasonal and annual times scales.

Contour maps of annual mean R/G values are providing useful tools in identifying range and geographical influence on the quality of data from other aerials (Hems and Brownscombe 1991). Similar maps are to be produced for the Ingham aerial to aid in the realization of the aims of the LORCS Study Group, i.e. to appraise the quality of data from this site and to compare, especially at long ranges, the performance with that of neighbouring radars.

ACKNOWLEDGEMENTS

This work is supported by the Lincoln Weather Radar Consortium comprised of the UK Meteorological Office and from the Anglian Water, Severn–Trent Water and Yorkshire Water Units of the National Rivers Authority.

REFERENCES

Delrieu, G., Bellon, A. and Creutin, J. D. (1988) Estimation de lames d'eau spatial a l'aide de donnees de pluviometres et de radar meteorologique — application au pas de temps journalier dans la region de Montreal. *J. Hydrol.* **98**, 315–344.

Hems, B. D. and Brownscombe, J. L. (1991) Comparison of radar precipitation measurements with a dense network of daily rainguages — application to radar networking. *Proceedings of COST 73 International Seminar on Weather Radar Networking*, Brussels, 5–8 September, 1989. In press.

Hitch, T. J. and Hems, B. D. (1988) A comparison of radar and gauge measurement of rainfall over Wales in October 1987. *Meteorol. Mag.*, **117**, 276–279.

Meteorological Office (1986) *UKON 7 United Kingdom Observational Networks — weather radar*, Meteorological Working Document. Meteorological Office, Bracknell, Berks.

Meteorological Office (1989) *UKON 5 United Kingdom Observational Networks — rainfall stations* (1988), Working Document. Meteorological Office, Bracknell, Berks.

Mylne, M. F. (1988) *Investigation of the role of radar data in the PARAGON daily rainfall archiving system: assessment of the contribution of radar data to near real-time daily rainfall estimates*, Advisory Services Technical Report No. 5, Unpublished Advisory Services Branch Memorandum.

Wilson, J. W. and Brandes, E. A. (1979) Radar measurements of rainfall — a summary. *Bull. Am. Meteorol. Soc.*, **60**, 1048–1058.

26

Sampling errors for raingauge-derived mean areal daily and monthly rainfall

A. W. Seed[1] and G. L. Austin[2]
[1]Hydrological Research Institute, Department of Water Affairs, P.Bag X313, Pretoria, South Africa
[2]McGill Weather Radar Observatory, Box 198, MacDonald College, Ste-Anne-de-Bellvue, Quebec H9X 1CO, Canada

ABSTRACT

Radar data from two geographical locations are used to simulate the mean standard error in using a sparse raingauge network to estimate daily and monthly mean areal convective rainfall over areas ranging from 45 000 to 180 000 km^2. It was found that a network with a regular configuration gave somewhat less variable errors than the uniform random raingauge network, although the mean errors were very similar. The difference became more pronounced for the very sparse networks. The mean standard error for a particular network and rainfield was found to be a function of the number of gauges in the network, the raining fraction of the area and the ratio of the standard deviation over the mean of the non-zero portion of the rainfield. A simple three-parameter relationship was proposed to relate the mean standard error, expressed as percentage of the mean areal rainfall, to these variables. It was found that a single relationship was able to explain 63% of the variability in the estimated mean standard estimation error, combining data from both regions. Finally, the domain over which the relationship is able to make reasonable predictions is discussed, the principal constraint being that the raining fraction of the area should not exceed 0.5 for networks with more than 200 raingauges.

1. INTRODUCTION

In spite of the heavy emphasis that the modern literature places on various exotic rainfall-measuring systems, the old-fashioned raingauge still provides the bulk of the rainfall data to practising hydrologists and climatologists throughout the world and will inevitably have a major role to play in the TRMM ground truth strategy. This is simply due to the length of record that exists for the raingauge compared with radar for example. A fundamental question that has to be asked, therefore, is how well can a raingauge network with a particular geometrical configuration measure mean areal rainfall over fairly large areas. The accuracy of raingauge-derived mean areal rainfall needs to be understood before gauge data can be used as ground truth satellite and radar rainfall-measuring systems.

It is surprising that we are as yet unable to measure rainfall with perfect accuracy, even at a point. Rainfall measured by a raingauge is strongly affected by small-scale wind effects and local turbulence around the lip of the gauge. The local topography surrounding the gauge, particularly the slope and aspect also affect the gauge measurement (e.g. Rodda 1971). Rodda compared the annual rainfall measured by pit gauges and standard gauges and found that the standard gauge measured up to 30% less rainfall in some parts of Britain. Once the gauge design and siting guidelines have been established, the best that we can hope for is some relative measure of rainfall. However, it is the spatial variability of the rainfield being sampled that limits the accuracy of the gauge-measured mean areal rainfall.

The accuracy with which a gauge network can measure rainfall depends on the variability of the rainfield and the geometric organization of the network. In areas where physiographic variables, distance from the sea, altitude and rain-shadow effects, for example, influence the rainfield, the network configuration should be analysed in the space of these variables rather than the more usual Cartesian space. Seed (1987) used multidimensional cluster analysis to identify homogeneous physio-graphic regions when mapping convective rainfall in a physiographically complex region of Natal, South Africa. The gauge network as it exists in Natal has the classic problem that is the mountainous areas that have the highest rainfall and river runoff and therefore are hydrologically the most significant areas, but also they have virtually no raingauges owing to the practical problems involved in siting and maintaining the gauges.

This study uses a large quantity of radar data from both Florida and South Africa to determine empirically the measurement error for daily and monthly rainfall accumulations over a large area as a function of the network organization and density. In particular the two following questions will be addressed.

(a) To what extent do the errors in estimating mean areal rainfall over large areas depend on the network configuration as a random or rectangular array?
(b) What are the network and rainfield characteristics that influence the estimation error?

2. METHOD

Florida radar data collected at Patrick Air Force Base for the period 8–30 August 1987 have been processed into a single database consisting of approximately 1000 5 min rainfall maps. An entire summer of convective rainfall from Nelspruit. South Africa, has also been processed into a data base of 5 min rainfall maps. These data were checked for radar artefacts and the clean data were accumulated to form maps of daily rainfall. For the purposes of this analysis, these data sets were assumed to have the same statistical properties in both space and time as the real rainfields occurring in their respective climatological regions. Various random and regular 'gauge networks' were generated for a given network density; each gauge was assigned the value of the map pixel at the gauge position. These 'gauge measurements' were then used to recreate the rainfields and the original and estimated fields were compared. The mean standard error for the mean areal rainfall over the entire area ($180\,000\,\mathrm{km}^2$) was calculated for each day and expressed as a percentage of the actual mean areal rainfall. These statistics were then accumulated and used to calculate the median, maximum and minimum error for the days analysed.

3. SELECTION OF THE RAINGAUGE INTERPOLATION SCHEME

A large variety of methods are available when interpolating from a random scatter of data points in an area onto a regular grid. Perhaps the first of such methods was published by Thiessen (1911) and Thiessen polygons are still widely used in hydrology today. There are basically two types of interpolation scheme: local estimation where only the known points in a restricted neighbourhood are used to interpolate onto an unmeasured point, and global estimation techniques where the entire set of points is used, usually by means of least-squares regression.

Trend surface and multiple-regression techniques are commonly applied to annual or mean monthly and mean annual rainfall where the rainfield is non-zero at all points in the map area. Examples of such methods include Hutchinson (1968) who mapped mean annual precipitation in New Zealand and Storr and Ferguson (1973) who mapped mean annual precipitation in British Columbia, Canada. All these regression methods are fairly sensitive to the spatial distribution of the raingauges, particularly near the edges of the map area; see Whitten (1975) for a discussion on this problem.

Local estimation techniques use the known values within a small neighbourhood around the point of interpolation. There are a great number of operational schemes in use, particularly in the field of computer-generated contour mapping. McLain (1974) lists a number of schemes that were common at that time. The underlying assumption of these techniques is that data are more likely to be useful if they were measured near the point of interpolation. Delfiner and Delhomme (1975, p. 96) made the following interesting comment with regard to distance weighting schemes: 'Clearly, no general rule can be derived from experiment on particular datas and point configurations. Consequently, the choice of a distance weighting function is more or less a matter of personal belief, of tradition or of confidence in the device of "influential authorities".'

Although distance weighting schemes suffer from a certain arbitrariness in the selection of parameters, Ripley (1980) was able to show that, in order for the interpolated surface to be smooth in the neighbourhood of the data points, the derivative of the weighting function must tend to zero as the distance to the point tends to zero and that the function should decay at a rate faster than the inverse square of the distance. Distance-weighting schemes also do not cope well with clustered data although *ad hoc* solutions can be used to remedy the situation.

Optimal interpolation techniques minimize the variance of the interpolation error are another major class of local interpolators. This class of interpolation techniques include the various flavours of kriging, and methods proposed by Gandin

(1965) and Ripley (1980). These techniques do in fact outperform most of the other interpolation techniques; see Creutin and Obled (1982) and Tabios and Salas (1985) for comparative studies. However, problems are experienced when the rainfield has zero rain rates, the 'hole effect' in kriging parlance, which requires special treatment (e.g. Creutin and Barancourt 1988) and they are far more expensive in computer time than the other techniques.

The validity of any interpolation scheme has to be seen against the extreme variability in the short-duration accumulated rainfields. The observed existence of extreme gradients and a generally discontinuous behaviour with rain often falling over less than 10% of the area leads to the conclusion that any interpolation technique will not show great accuracy in these cases. It is only recently that the meteorological community has started to deal with the extreme intermittency of rainfall and cloud fields compared with the more traditional variables of temperature, pressure and wind. The underlying cause for this extreme variability is the drastic non-linearity involved in a cloud and rain formation. The response from a hydrological point of view is to exercise extreme caution about the likely accuracy of any interpolation scheme, including those of great mathematical complexity.

A casual examination of a rainfield interpolated by means of Thiessen polygons will be sufficient to convince one that the technique does not produce aesthetically pleasing rainfields. The comparative analysis between regular and random gauge networks will be done using Thiessen polygons, in the interests of reducing the computer time. Thereafter, a distance-weighting technique will be used to estimate the measurement error variance as a function of network density and averaging area. Finally, a relationship using raining area, number of gauges, and variability of the rain field will be used to predict the mean standard error in mean areal daily rainfall estimates.

4.   RANDOM VERSUS REGULAR RAINGAUGE NETWORKS

Since this study was exploratory in nature, and the computer time requirements were substantial if the entire data set were to be analyzed, seven 24 h accumulations were chosen as the 'truth' for the study. The maps included 1 day of intense convective rainfall, both with a small rain area which was expected to provide the largest errors. The mean standard error MSE is given by

$$\text{MSE} = \sqrt{\frac{1}{n} \sum_{i=1}^{n} (\hat{y}_i - y_i)2}$$

where $\hat{y}_i$ is the estimated mean areal rainfall using the ith gauge network and y_i is the actual mean areal rainfall for that day. MSE was calculated for each of the 7 days in the data set using nine raingauge networks. Fig. 1 shows the maximum, median and minimum MSE as a function of network density using random and regular network configuration. The random network gave a larger median and maximum MSE for the 7 days analysed. In particular, the maximum MSE for the random network configuration increased more quickly than the maximum MSE for the regular network as the network density was decreased.

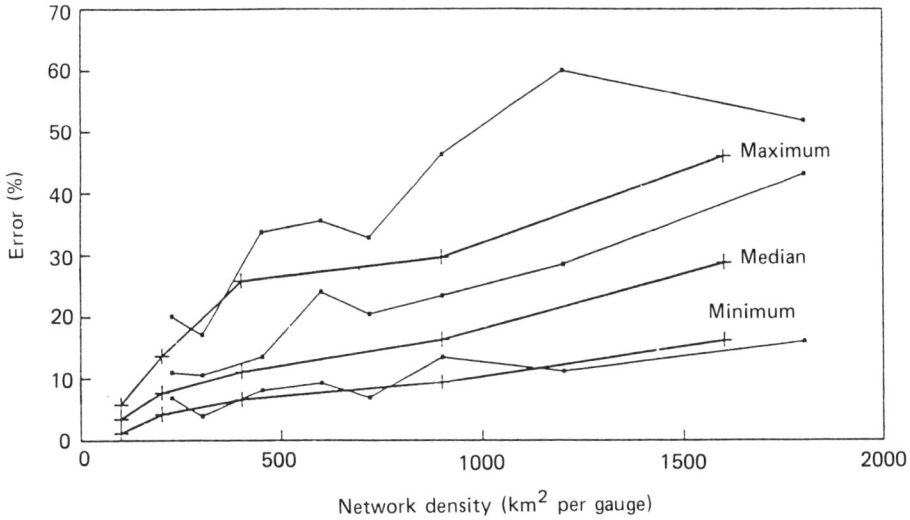

Fig. 1 — Errors in the estimation of mean areal daily rainfall over 180 000 km² using Thiessen
polygons with regular (+) and random (●) raingauge networks.

The errors for each of the nine networks for each of the 7 days were normalized by
means of $e = [(\hat{y}_i - y_i)/y_i] \times 100$. The cumulative distribution of e for the 900 km² per
gauge density is shown in Fig. 2. From this figure it is clear that the random network

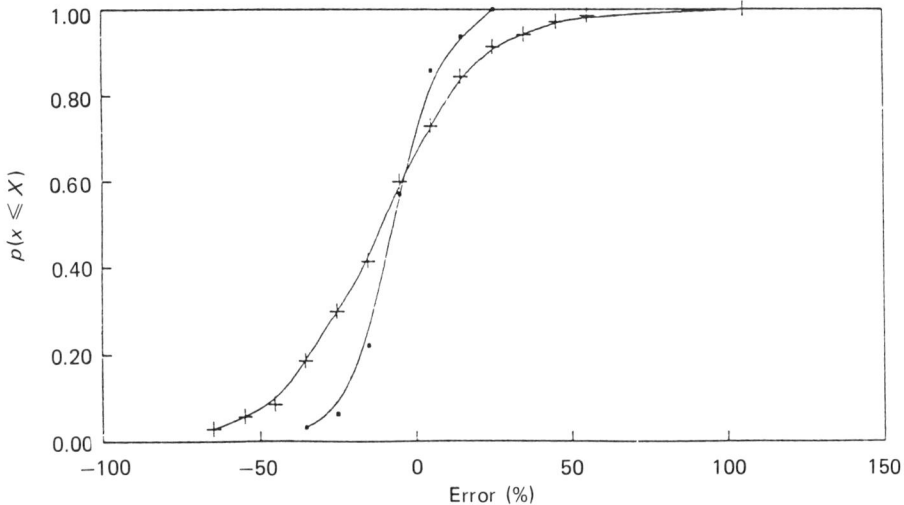

Fig. 2 — Cumulative distribution of normalized errors in the estimation of mean areal rainfall
over 180 000 km² using Thiessen polygons with regular (●) and random (+) raingauge
networks.

configuration produces longer tails, particularly on the overestimation tail. The underestimation tail is constrained by − 100 by construction; the worst underestimation possible is to measure none of the rain that fell on that day.

It is interesting to note that the rate at which the maximum MSE decreases with increasing gauge density is noticeably faster after 400 km^2 per gauge. The probability that the raingauge network makes a drastic error rapidly decrease once the network density exceeds 400 km^2. Therefore, although the rate at which MSE decreases with increasing gauge density is depressingly slow, the dense network at least has a fairly constant error from one day to the next.

An examination of MSE for the various days and gauge densities revealed that the days with a large rain area in general had a low error and those with small rain areas had high errors. Fig. 3 shows a plot of error versus rain area for network

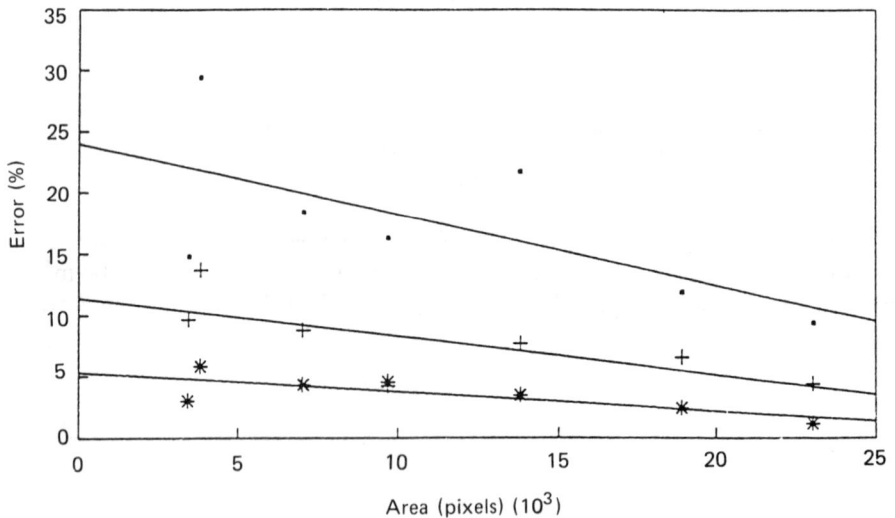

Fig. 3 — Estimation error as a function of rain area for various network densities: *, 100 km^2; + , 200 km^2; ●, 900 km^2.

densities of 100, 200 and 900 km^2 per gauge using a regular network design and Thiessen polygon interpolation. The dependence of error on rain area is striking. Therefore, the raingauge network would tend to have lower estimation errors on the days of heavy mean areal rainfall since rain area alone is able to account for most of the mean areal rainfall variance (Rosenfelt et al. 1988). The dense network was less sensitive to the rain area since the network was able to sample even sparse rainfields with good probability.

5.   MEAN STANDARD ERROR VERSUS NETWORK DENSITY

The distance-weighting function

$$w(d_i) = \exp\left(-\frac{d_i}{a}\right)$$

where $w(d_i)$ is the weight from the ith gauge a distance d_i away and a is the average nearest-neighbour distance for the network was used to generate daily and monthly rainfall maps for the Florida and Nelspruit data. The Nelspruit data had a maximum range of 120 km; so the areal mean rainfall for 44000 km^2 was calculated using 16 days for Nelspruit, and 7 days for Florida, and nine different gauge networks. The entire simulation took 6 h of central processing unit time on a super-minicomputer. Figs 4 and 5 show the error as a function of gauge network density for daily and

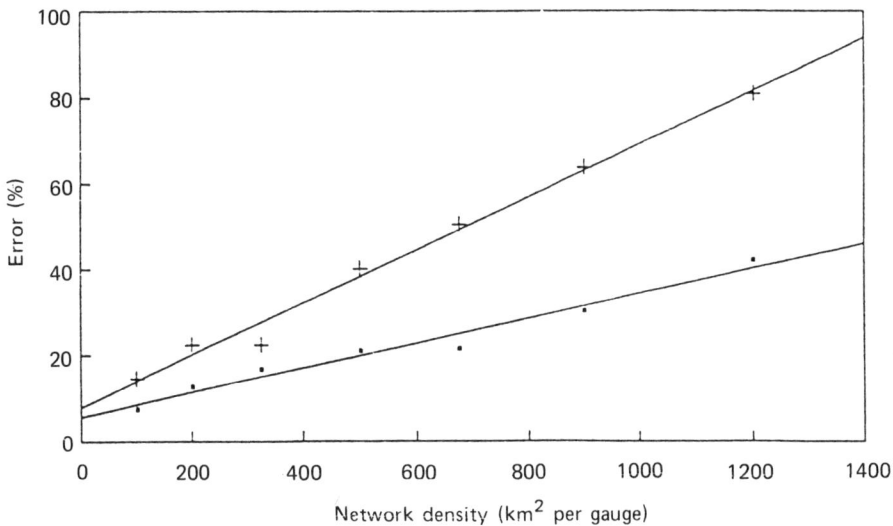

Fig. 4 — Errors in the estimation of mean areal daily rainfall over 45000 km^2 using distance-weighting interpolation: ●, Florida; +, Nelspruit.

monthly data, respectively. From these figures it is apparent that Nelspruit has interpolation errors that are approximately a factor of 2 higher for monthly rainfall and a factor of 4 higher for daily rainfall. The ratio of standard deviation over the mean 2 km mean rainfall was 1.5 for Florida and 2.47 for Nelspruit; the mean number of raining pixels per day was 2500 for Florida and 3000 for Nelspruit. It seems than that the raingauge network error is quite sensitive to the relative variability of the rainfield — not an unexpected result.

6.   MEAN STANDARD ERROR VERSUS AREAL AVERAGING

The Florida data were used to determine the error for monthly mean areal rainfall over 45000, 100000 and 180000 km^2 as a function of network density. For sparse

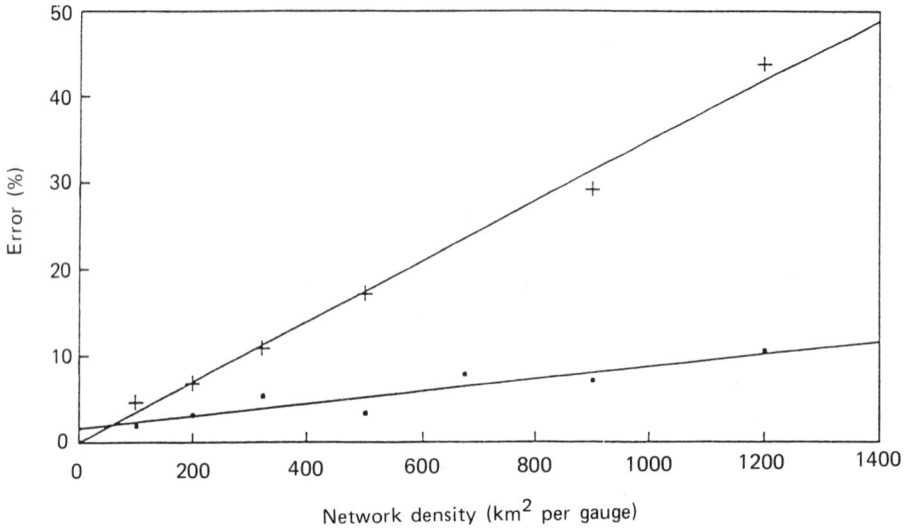

Fig. 5 — Errors in the estimation of mean areal monthly rainfall over $45\,000$ km^2 using distance-weighting interpolation: ●, Florida; +, Nelsruit.

networks, the correlation between any two raingauges is likely to be slight, and therefore the measurement error, expressed as the mean standard error, would be expected to be inversely proportional to the square root of the number of gauges, independent of the averaging area. Fig. 6 shows a log–log plot of measurement error

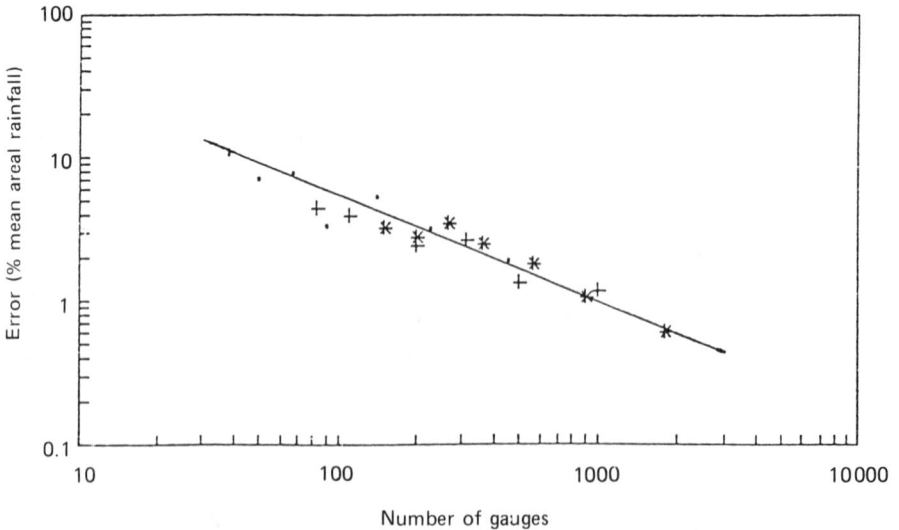

Fig. 6 — Estimation error for mean areal monthly rainfall over $45\,000$ km^2 (●), $100\,000$ km^2 (+) and $180\,000$ km^2 (*) as a function of the network density.

versus number of gauges for the three averaging areas. Unfortunately, the maximum range of the radar at Nelspruit limited the coverage area to 44 000 km^2 and therefore it was not possible to repeat the analysis for those data. While the results from an analysis using only one monthly mean areal rainfall map can hardly be considered conclusive, Fig. 6 suggests that error is indeed independent of averaging area, provided that the network is sparse in the sense that the mean correlation between any two gauges is small. It should also be noted that the gauges in the various networks were deployed over the entire area.

7. MEAN STANDARD ESTIMATION ERROR FOR DAILY MEAN AREAL RAINFALL

From the above analysis, it is apparent that the estimation error for mean areal rainfall over large areas and sparse networks, the most common kind of network, is a function of the number of gauges, the raining area and the variability of the non-zero fraction of the rainfield. A function of the form

$$E = (a + bV)N^{-0.5} cA$$

where E is the mean standard error expressed as a percentage of the areal mean rainfall, V is the standard deviation divided by the mean, A is the raining area divided by the total area, N is the number of gauges in the network, and a, b and c are empirical constants to be estimated from the data would seem a reasonable first guess, assuming that N and V have no influence on c.

A least-squares fit was undertaken for the 16 days of Nelspruit data and 7 days of Florida data first separately and then using the combined data set. Table 1 gives summary of the three sets of parameters and Fig. 7 plots the estimated against predicted E using the combined data set parameters.

Table 1 — Summary of the model fit to the Nelspruit, Florida and combined data sets

	a	b	c	r^2
Nelspruit	461.6	25.0	44.5	0.50
Florida	416.5	55.8	44.6	0.66
Combined	411.7	42.3*	39.1	0.63

*Some authorities argue this parameter should be exactly 42 (Adams 1980).

From Table 1 it can be seen that the parameters a and c were very similar for all three data sets. The second parameter, b, was possibly badly estimated in the Florida case since the 7 days used in the analysis all had very similar, relatively low variability. Since the regression was based on estimates of the mean standard error, which itself has a large variance, the regression was able to estimate E surprisingly well in explaining 63% of the variance of E. The model has the interesting property

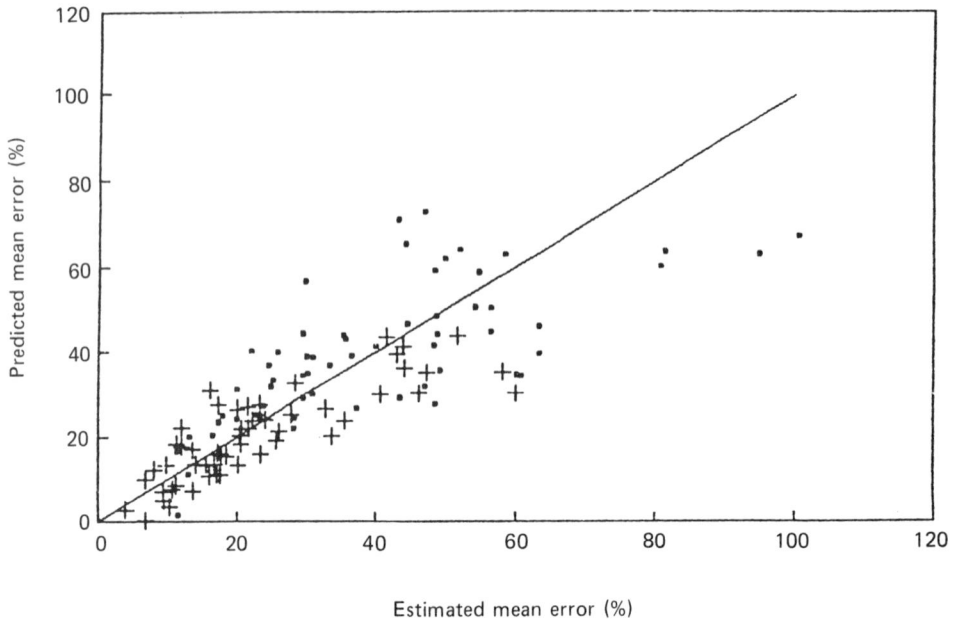

Fig. 7 — Predicted versus estimated mean standard error: ● Nelspruit; +, Florida.

of predicting negative errors when the raining fraction exceeds 0.5, the relative variability of the field is low, 1.4 say, and N is of the order 200 greater. Clearly, the error function is not able to make predictions in this domain. The lower limit for the raining fraction A appears to be somewhat less than 0.05, the lower limit for A in the combined data set. The data set to hand only had 2 days with A greater than 0.5, making it impossible to explore this domain more thoroughly.

Of course, the actual measurement error for a given rainfield and network can rather easily be very much higher than the mean measurement error, the variance of E increasing as the mean increase. A complete description of E must therefore include an estimate of the variance or preferably a description of the probability distribution.

8. CONCLUSIONS

In conclusion then, it has been found that the regular networks were somewhat better than the uniform random networks, insofar as the variance of the estimation error tended to be lower for regular networks, particularly for the more sparse networks. As the number of gauges increased, the difference between the two configurations, as expected, diminished.

The mean standard estimation error is independent of the averaging area if the gauge density is expressed as the number of gauges in the network and not in square kilometres per gauge. This only holds true when the density of the network is less than one gauge per 15 km, in which case the interstation correlations can be ignored. Since this is of the order of the correlation length for convective rainfields, one could speculate that this result, suitably scaled by the correlation length, can be applied to shorter rainfall accumulations over smaller areas.

After the number of gauges in the network, the raining fraction of the area covered by the network was found to affect the network measurement error significantly. In general, the small rain areas gave the largest mean measurement error; once again this effect is greatest for the more sparse networks. A simple model to predict the mean standard measurement error given the variability of the rainfield, number of gauges in the network and the raining fraction of the area covered by the network was proposed. The model was able to explain 63% of the variance in a combined data set using data from South Africa and Florida data separately. The nature of the model is such that it is unable to predict measurement errors for raining fractions that exceed about 0.5, except when the number of raingauges is less than about 200.

REFERENCES

Adams, D. (1980) *Hitch-hiker's Guide to the Galaxy*. Pan, London.
Creutin, J. D. and Barancourt, C. (1988) Pattern and variability analysis of heavy rainfall fields in a mountainous Mediterranean Region. *Conference on Mesoscale Precipitation*: *Analysis Simulation and Forecasting*, Cambridge, MA, 13–16 September 1988.
Creutin, J. D. and Obled, C. (1982) Objective analysis and mapping techniques for rainfall fields: an objective comparison. *Water Resour. Res.*, 413–431.
Delfiner, P. and Delhomme, J. P. (1975) Optimum interpolation by kriging. In: J. C. Davis and M. J. McCullagh (eds). *Display and analysis of spatial data*, Wiley, Chichester, pp. 96–112.
Gandin, L. S. (1965) *Objective analysis of meteorological fields*, translated from Russian by R. Hardin. Israel Program for Scientific Translation, Jerusalem, p. 242.
Hutchinson, F. (1968) An analysis of the effect of topography on rainfall in the Taieri catchment area, Otago, Earth Sci., **2**, 51–68.
Lebel, T., Bastin, G., Obled C. and Creutin, J. D. (1987) On the accuracy of areal rainfall estimation: a case study. *Water Resour. Res.*, **23** (11), 2123–2134.
McLain, D. H. (1974) Drawing contours from arbitrary data points. *Comput. J.*, **17**, 318–324.
Ripley, B. (1980) *Spatial statistics*. Wiley, London, pp. 44–75.
Rodda, J. C. (1971) *The precipitation measurement paradox the instrument accuracy problem*, WMO Report No. 316. World Meteorological Organization, Geneva, 42 pp.
Rosenfelt, D., Atlas, D. and Short, D. A. (1988) The estimation of convective rainfall by area integrals, Part II: The height area rainfall threshold (HART) method. *Conference on Mesoscale Precipitation*: *Analysis, Simulation and Forecasting, Cambridge, MA, 13–16 September* 1988.
Seed, A. W. (1987)*The techniques of rainfall mapping*. Unpublished M.Sc. Thesis. Department of Agricultural Engineering, University of Natal, South Africa.
Storr, and Ferguson (1973) The distribution of precipitation in some mountainous Canadian watersheds. *Proceedings of a Symposium on Mountainous Areas*,

WMO Report No. 326. World Microbiological Organization, Geneva, pp. 244–263.

Tabios III, G. Q. and Salas, T. D. (1985). A comparative analysis of techniques for spatial interpolation of precipitation. *Water Resour. Bull.*, **21** (3), 365–380.

Thiessen, A. H. (1911) Precipitation averages for large areas. *Mon. Weather Rev.*, July, 1082–1084.

Whitten, E. H. T. (1975) Practical use of trend surface analysis in the geological sciences. In: J. C. Davis and M. J. McCullagh (eds), *Display and analysis of spatial data*, Wiley, Chichester, pp.82–296.

27

Observations and simulation of rainfall in mountainous areas

K. Tateya,
Foundation of Hokkaido River Disaster Prevention Research Centre,
Sapporo, Japan
M. Nakatsugawa,
Civil Engineering Research Institute, Hokkaido Development Bureau,
Sapporo, Japan
T. Yamada,
Department of Civil Engineering, Hokkaido University,
Sapporo, Japan

ABSTRACT

In this paper, the results of the field observation of rainfall carried out by the present authors are shown and the effects of height or direction of the mountain on rainfall are discussed. With these investigation, we calculate the wind field in mountainous areas by using the digital maps and spectral analysis. Combining the calculated wind field with the Kessler model for rainfall, we simulate the rainfall field in the mountainous areas. Comparing the observed data with the calculated rainfall pattern, we confirm that the above-mentioned calculation can precisely simulate the actual orographic effect on rainfall fields. In addition, an attempt is made to transform the three-dimensional model to the two-dimensional model, to reduce the computational burden for practical applications.

1. INTRODUCTION

The observation and forecast of heavy rainfall are very important for the purpose of disaster prevention. The Japanese Ministry of Construction intends to cover the whole country by 22 radar stations. Such a radar network system will supply information on temporal and spatial distributions of rainfall. The simulation model, coupled with the observed data from radar, is of practical use to forcast the rainfall accurately. Many researchers have suggested that occurrences of heavy rainfall are often connected with the orographic conditions. However, it is extremely difficult to verify the orographic effects, because there are few raingauge stations in mountainous areas. Therefore, it is necessary to develop a model which can estimate and forecast heavy rainfall in such areas. Previous studies of orographic meteorology have been carried out by many other authors. In the present paper, the results of the field observations by the present authors are shown from the topographical point of view. On the other hand, we calculate the rainfall field by the numerical method and obtain some knowledge of the characteristics of orographic rainfall.

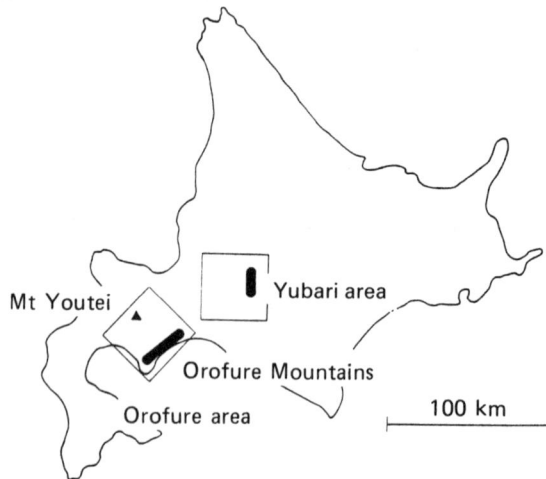

Fig. 1 — Map of Hokkaido Island.

2. FIELD OBSERVATION OF OROGRAPHIC RAINFALL
2.1 Previous studies by Kikuchi and co-workers

Takeda and Kikuchi (1978), Kikuchi *et al.* (1978) and Konno and Kikuchi (1981) carried out the observations of rainfall amounts using a special raingauge network along the slope in the Orofure Mountains. The Orofure Mountains are located in the southwest part of Hokkaido Island, Japan, as shown in Fig. 1 and is famous for occurrences of heavy rainfall. From the results of their investigations, they have suggested that, when the humid southeasterly wind from the Pacific Ocean prevails on the wall-like Orofure Mountains, the strong ascending draft generates along the southeast slope of the ridge and such a condition is important as a direct trigger of heavy rainfall in this area. Furthermore, the orographic rainfall is classified into 'continuous type' generated by a humid air mass along the slope and 'convective type' influenced by the large disturbance. A series of their observations indicate that the former shows a more remarkable orographic feature than the latter does.

2.2   Field observation carried out by the authors

Fig. 1 also shows the area around Mt Yubari which is located in the central part of Hokkaido Island and Fig. 2 gives this area in detail. There is the mountain range stretching from the south to the north in this area. We installed ten raingauges (full squares in Fig. 2) along the west slope of Mt Yubari in cooperation with the Ishikari River Local Head Office, Hokkaido Development Bureau, and we carried out special observation of rainfall from 20 July 1988 to 15 October 1988. Our study includes several data obtained from eight existing raingauges (full triangles in Fig. 2). During our observations, we fortunately observed the heavy rainfall on 25 and 26 August 1988. The synoptic situations in this period showed that the humid airflow from tropical cyclone in the Pacific Ocean and the cold airflow from the Eurasian Continent prevailed on Hokkaido and the stationary front grew remarkably. The maximum rainfall intensity observed along the west slope of Mt Yubari was about 30 mm/h and the meteorological chart just showed that the wind changed from an easterly direction to a westerly direction in connection with the front movement in the morning on 26 August. Figs 3 and 4 illustrate examples of the results obtained. Fig. 3 shows the behaviour of the total rainfall amounts against the elevation in the whole observation period (20 July–15 October 1988), while Fig. 4 shows the amounts during a week (25–31 August 1988) which include the event of the above-mentioned heavy rainfall. These examples show that the total amount of rainfall on the slope of the mountain increases linearly with increasing elevation. Furthermore, it follows from Figs 3 and 4 that the rainfall at the highest site is nearly twice that at the lowest site (1000 m difference between the two sites). The rainfall pattern is not necessarily regular for the cases where convective unstableness might take place under the large disturbance. However, the behaviour of the accumulated rainfall amount against elevation shows clearly the typical orographic effect during the weekly or monthly periods.

3.   MODELLING OF THE WIND FIELD IN A MOUNTAINOUS AREA

3.1   Calculation method

In previous studies, the Scorer equation has often been used to calculate the airflow over the mountains. In this paper, the typical orographic effect on the wind field is considered, assuming that the atmosphere is saturated and has an undisturbed and steady uniform flow. Additionally, it is assumed that the airflow which contributes to rainfall generation has a larger inertia than topographical fluctuations, and the Coriolis effect is assumed to be negligible in the scale from the geodynamical point of view. According to the above assumptions, we adopt the potential flow for modelling the wind field. Fig. 5 shows the flowchart for calculation of the wind field. Any contour line on the map of a basin is transformed to the digital map data, using the two-dimensional discrete fast Fourier transform (FFT) for these digital map data. The three components of wind speed can be obtained as the solutions of the three-dimensional Laplace equation

$$\frac{\partial^2 \phi}{\partial x^2} + \frac{\partial^2 \phi}{\partial y^2} + \frac{\partial^2 \phi}{\partial z^2} = 0 \qquad (1)$$

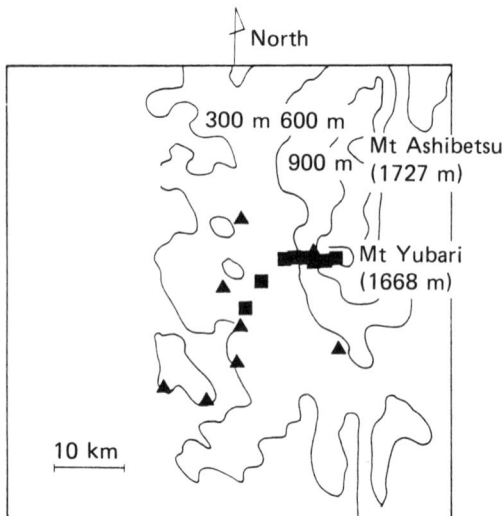

Fig. 2 — Location of raingauges installed in the Yubari area: ▲, existing raingauges; ■, newly installed raingauges.

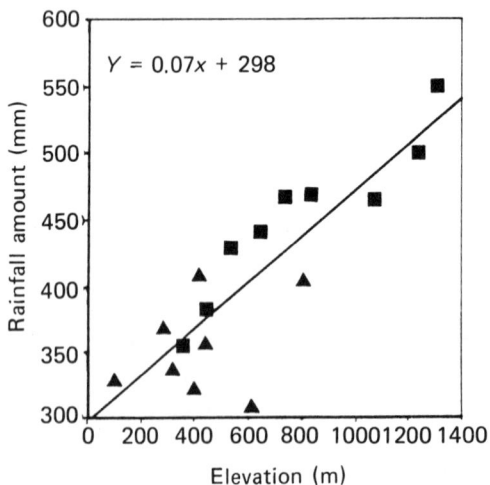

Fig. 3 — Observed rainfall amount plotted against elevation (20 July–15 October 1988): ▲, existing raingauges; ■, newly installed raingauges.

by a semianalytical method as follows.

The velocity potential is

$$\phi = U_x + \phi(z) \qquad \exp(jkx)\,\exp(jly) \tag{2}$$

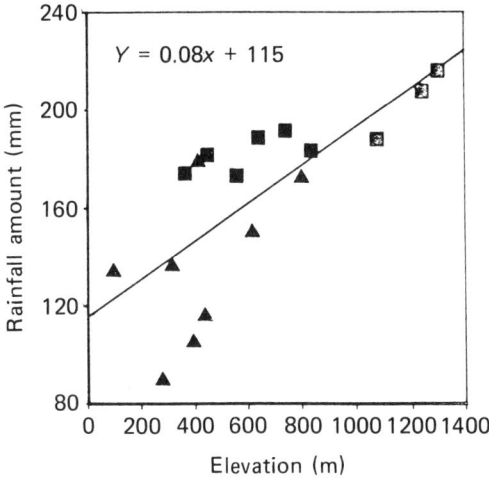

Fig. 4 — Observed rainfall amount plotted against elevation (25–31 August 1988): ▲, existing raingauges; ■, newly installed raingauges.

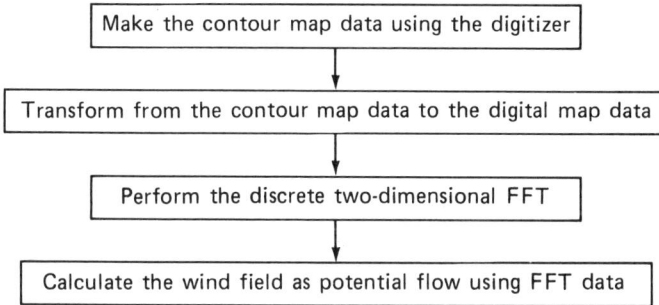

Fig. 5 — Flow chart for calculation of the wind field.

where u(m/s) is the wind speed of the main stream, $\phi(z)$ (m/s) is the velocity potential in the z direction, and k and l are the horizontal wavenumbers ($\beta = \sqrt{k^2 + l^2}$). The topographic function is

$$\eta = -h + a \quad \exp(jkx) \quad \exp(jly) \tag{3}$$

where h (m) is the altitude of the upper boundary. The boundary conditions are

$$\frac{\partial \phi}{\partial z} = 0 \quad \text{at } z = 0 \tag{4}$$

$$\frac{\partial\phi}{\partial z} = U\frac{\partial\eta}{\partial x} \quad \text{at } z = -h . \tag{5}$$

The solutions for the wind speed are

$$u = U + U\frac{1}{N^2}\sum\sum F(p, q)\frac{K^2 \cosh(\beta z)}{\beta \sinh(\beta h)} \exp\left(\frac{2\pi j}{N} mp\right) \exp\left(\frac{2\pi j}{N} nq\right) \tag{6}$$

$$v = U\frac{1}{N^2}\sum\sum F(p, q)\frac{Kl \cosh(\beta z)}{\beta \sinh(\beta h)} \exp\left(\frac{2\pi j}{N} mp\right) \exp\left(\frac{2\pi j}{N} nq\right) \tag{7}$$

$$w = -U\frac{1}{N^2}\sum\sum F(p, q)\frac{Kj \sinh(\beta z)}{\beta \sinh(\beta h)} \exp\left(\frac{2\pi j}{N} mp\right) \exp\left(\frac{2\pi j}{N} nq\right) \tag{8}$$

where u (m/s), v (m/s) and w (m/s) are the wind speeds in the x, y and z directions respectively, N is the number of FFT terms and $F(p, q)$ is the FFT component. The coordinate system for the calculation is depicted in Fig. 6.

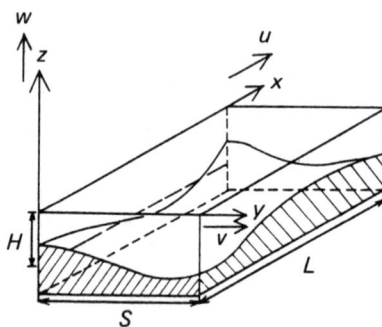

Fig. 6 — Coordinate system for calculation.

3.2. Topographical effects on wind field

To understand how the mountain shapes exert influence on the ascending wind along the slope, we calculated the wind field shown in the Orofure area and the Yubari area of Fig. 1. The computational domain covers 10 km in altitude and 63 km in horizontal square area. The grid sizes are 1 km in the horizontal directions x and y and 0.2 km in the vertical direction z. The Orofure area includes the Orofure Mountains with an elevation of 1200–1300 m which lies southwesterly along the Pacific Ocean and Mt Youtei with the height of 1893 m which has an isolated peak. In the Yubari area, there is a mountain range at an elevation of 1600–1700 m stretching from south to north. The dotted parts in Figs 7 and 8 show the areas where the strong ascending wind speeds which are more than 2% of the main stream at an altitude of 5000 m are calculated. In the case of the southeasterly wind in the Orofure area, the ascending

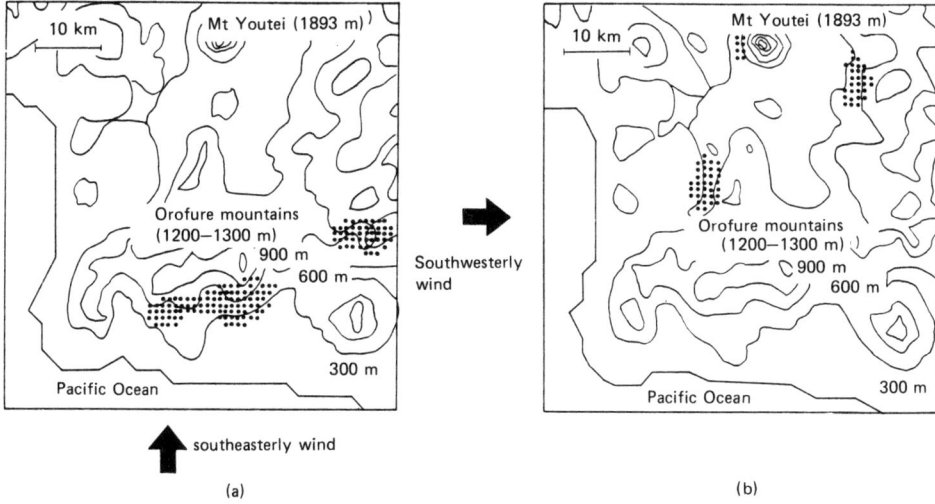

Fig. 7 — The area of strong ascending wind (the Orofure area): (a) southeasterly wind; (b) southwesterly wind.

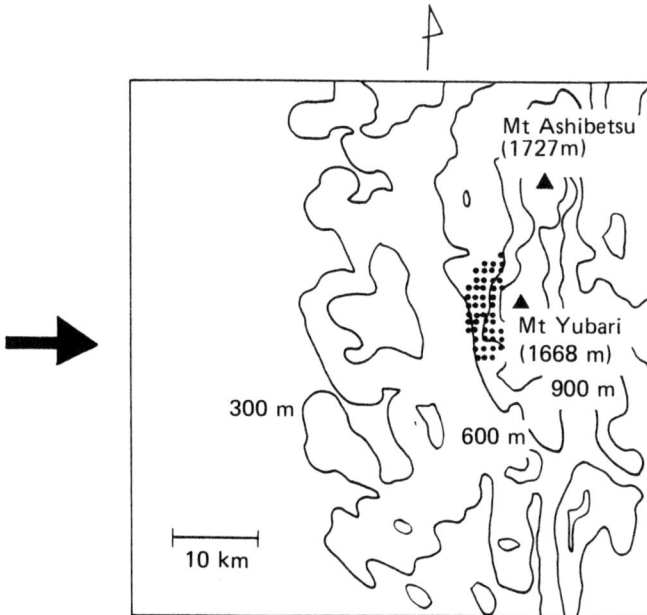

Fig. 8 — The area of strong ascending wind (The Yubari area).

wind is strong at the upper height above the southeast slope of the Orofure Mountains, which is caused by the existence of a wall-like slope, as shown in Fig. 7(a). This figure also shows that, around an isolated mountain such as Mt Youtei, the ascending wind is not so strong, because of the so-called evading effect of the wind in spite of a higher altitude. Fig. 7(b) shows that the southwesterly wind cannot form a wider and stronger ascending wind field than the southeasterly wind which attacks the wall of the mountain range. Fig. 8 clearly shows that the ascending wind field is formed above the west slope of Mt Yubari where we installed some raingauges.

4. ANALYSIS OF OROGRAPHIC RAINFALL FIELD

4.1 Model of rainfall

In this study, the three-dimensional Kessler model is adopted to calculate the rainfall field, combined with the calculated wind field. Fig. 9 is a schematic description of the

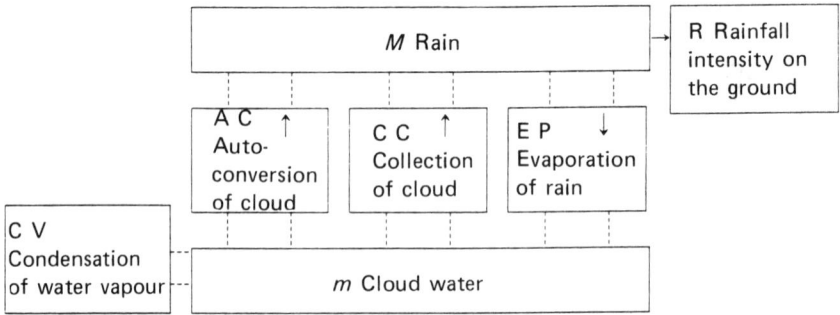

Fig. 9 — Rainfall mechanism in the Kessler model.

rainfall mechanism in the Kessler model where the liquid of water is deparated by cloud water and raindrops which are defined by the size of water drops. The fundamental conservation equations for cloud water and raindrops are as follows:

$$\frac{\partial m}{\partial t} = -u\frac{\partial m}{\partial x} - v\frac{\partial m}{\partial y} - w\frac{\partial m}{\partial z} - AC - CC + EP + CV \tag{9}$$

$$\frac{\partial M}{\partial t} = -u\frac{\partial M}{\partial x} - v\frac{\partial M}{\partial y} - (w+V)\frac{\partial M}{\partial z} + AC - CC - EP \tag{10}$$

where

$$AC = \begin{cases} K_1(m-a), & m > a \\ 0, & m < a \end{cases} \tag{11}$$

$$CC = 6.96 \times 10^{-4} E(\text{No})^{1/8} m M^{7/8} \exp(KZ/2) \tag{12}$$

$$EP = 1.93 \times 10^{-6} (\text{No})^{7/20} m M^{13/20} \tag{13}$$

$$CV = w(A + Bz) \tag{14}$$

$$V = -38.3(\text{No})^{-1/8} M^{1/8} \exp(kz/2) \tag{15}$$

$$R = 138 (\text{No})^{-1/8} M^{9/8} \tag{16}$$

m is the cloud water content ($\lg^{-3}$), M is the rain water content, u(m/s), v(m/s) and w are the wind speeds in the x, y and z directions respectively, AC ($/g^3$ s) is the autoconversion of cloud, CC ($/g^3$ s) is the collection of cloud, EP ($/g^3$ s) is the evaporation of rain, CV ($/g^3$ s) is the condensation of water vapour, V (m/s) is the fall speed of raindrops, R (mm h) is the rainfall rate, K_1 ($= 10^{-3}$/s) is a constant, a ($= 0.5/g^3$) is the threshold value of autoconversion, E ($= 1$) is the collection efficiency, No ($= 10^7/n^4$) is the Marshall–Palmer constant, k ($= 10^{-4}/n^1$) is the lapse rate of air density, z is the altitude (m), $A(= 3 \times 10^{-3}/g^4)$ is a constant and B ($= -3 \times 10^{-7}/g^5$) is a constant. Equations (9) and (10) are solved numerically by a finite-difference method using the same three-dimensional computational grid as in the wind calculation. These calculations are performed under some assumptions, such as a steady wind field, the Marshall–Palmer distribution in microphysical parameters and a constant terminal velocity as the fall speed of raindrops. Saturated air is assumed as the initial condition ($m = 0$) everywhere. Convective unstableness induced from thermodynamical processes should be considered to explain the rainfall mechanism fully. However, in this study, we investigate the basic mechanism of rainfall caused by the topographical effects and deal with the stratiform rainfall known as the so-called warm rainfall. Therefore, the storm rainfall as caused by convective unstableness is not considered.

4.2   Basic characteristics of calculated rainfall using the Kessler model
In order to understand the characteristics of rainfall simulated by the Kessler model, some numerical experiments are carried out for topography of simple shape. The results obtained are summarized as follows.

(1) *Steady characteristics of rainfall*   The calculated rainfall amount and its spatial distribution reach a steady state in 30 or 40 min. When we decrease the threshold value of autoconversion (a in equation (11)), the raindrops from the cloud water are quickly formed and the rainfall field reaches steady state promptly.

(2) *Advection effects*   Fig. 10 shows the calculated results for some cases of uniform wind speed around a concentric circular peaked mountain. It is recognized from this figure that, the larger the main stream wind speed, the stronger the rainfall intensity is, because the ascending wind speed increases. However, when the horizontal advection velocity increases excessively, the rain band shifts to the lee side and the rainfall intensity decreases beyond some threshold value.

(3) *Condensation of cloud water from water vapour*   To investigate the effect of condensation on rainfall, some different ascending wind speed in the condensa- tion term (w in equation (14)) are used in the calculation. Fig. 11 shows the

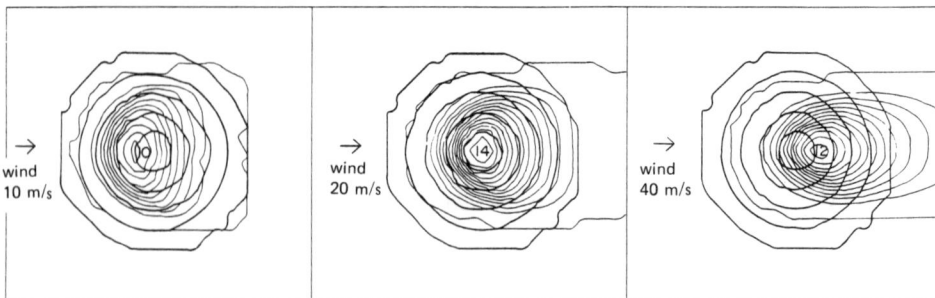

Fig. 10 — The rain band calculated with different main stream wind speeds: ———, rainfall intensity (mm/h); ———, contour lines.

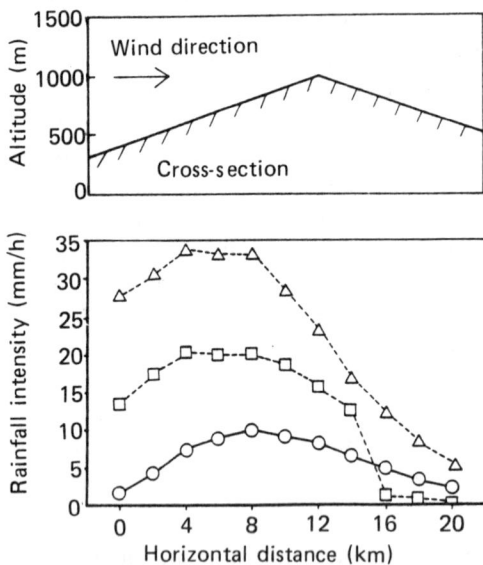

Fig. 11 — Rainfall intensity calculated with different ascending winds in the condensation term (CV) (along the line through the summit in a concentric circular mountain): ○, calculated by potential theory; □, 2W; △, 3W.

distribution of the rainfall intensity calculated along the line of wind direction passing through the summit of the concentric circular peaked mountain. The result clearly shows that the rainfall intensity is linearly proportional to the ascending wind speed in the condensation term.

(4) *Mountain shape*   To investigate the topographic effect on rainfall, especially mountain shapes, two typical cases are considered: the concentric circular peaked shape and the two-dimensional wall-like shape with the same elevation. Fig. 12 shows the calculated rainfall intensity along the line of wind direction

passing through the summit (the same cross-sectional shape). From this figure, it is recognized that the ascending wind with or without the evading effects directly influence the rainfall intensity. Such topographic effects can be thought of as an important factor in the actual rainfall field.

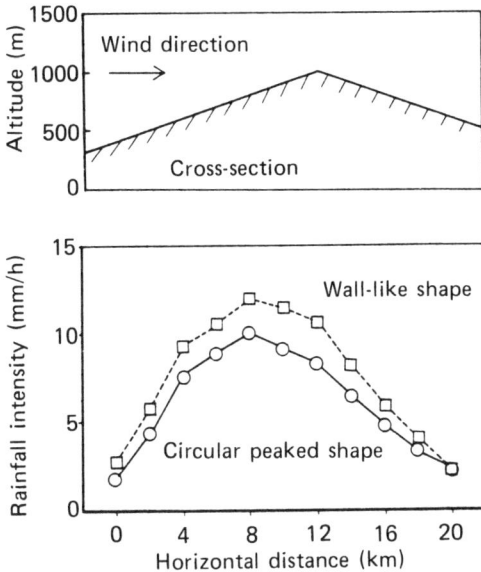

Fig. 12 — Rainfall intensity calculated with different topographic shapes (along the line through the summit).

4.3   Results calculated with actual topography

Fig. 13(a) shows the contour lines of rainfall intensity calculated in the Orofure area where the southeasterly wind is prevailing. The wind speed of main stream is 10 m/s and saturated air is given as the initial condition. The calculated result shows the same tendency as the observation results performed by Kikuchi and co-workers, i.e. a typical rain band is formed along the southeast slope of the Orofure Mountains. The simulated rainfall intensity at the centre of the rain band is about 10 mm/h. However, the actual heavy rainfall recorded in this area contained the rainfall intensity higher than 50 mm/h. Therefore, the calculated rainfall amount is under-estimated, because the above-mentioned simulation considers only the stratiform rainfall due to the typical orographic effect. In the case of the southwesterly wind, a rainfall intensity only half that for the southeasterly wind is calculated, as shown in Fig. 13(b). These results clearly show the effect of mountain shapes on rainfall production. Fig. 14 shows the rainfall contour lines calculated in the Yubari area where the westerly winds are predominant. The result illustrates that the typical rain band is formed by the presence of the wall-like mountain range as described in the previous example. Furthermore, Fig. 15 shows the change in the rainfall intensity with the elevation at the nearest computational grid point where the raingauge shown in Fig. 2 was installed. It can be recognized from Fig. 15 that the simulation model reproduces the observed fact that the rainfall amount increases linearly with increasing elevation (see Figs 3 and 4).

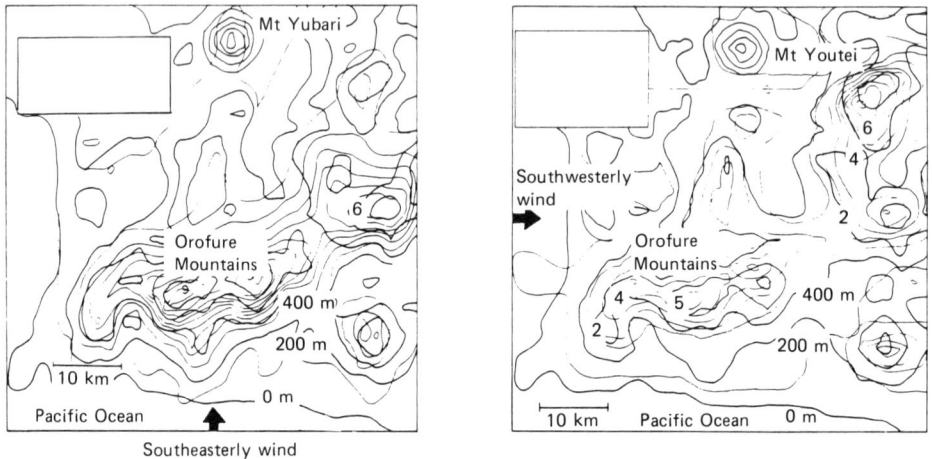

Fig. 13 — Calculated rainfall intensity (The Orofure area) with (a) a southeasterly wind and (b) a southwesterly wind: ——, rainfall intensity (mm/h); ——, contour lines.

4.4   Effects of topographic scale on rainfall

As mentioned previously, the present calculation method utilizes the FFT data of the digital map to obtain the wind field numerically. Fig. 16 shows a spectrum of the FFT components for the Orofure area. In order to understand the scale effects of topography on rainfall, we removed high frequency terms larger than the tenth term of FFT, which correspond to scales less than 5 km in the actual topography according to the sampling theorem. We also carried out a series of recalculations using the removed topographic components. The calculated rainfall field under the southeasterly wind in the Orofure area is shown in Fig. 17. The result obtained demonstrates the same tendency of rain bands as that in Fig. 13(a). It is understood from this figure that the scales less than 5 km have no significant effects on orographic rainfall pattern.

4.5   Two-dimensional modelling of rainfall

Calibration and computational efforts are too stringent in the practical application of a three-dimensional rainfall model. For the purpose of real-time rainfall forecasting using radar information, the two-dimensional model is of practical use to reduce computer storage and time. Hence an attempt is made to transform the three-dimensional model to a two-dimensional model by integrating and averaging

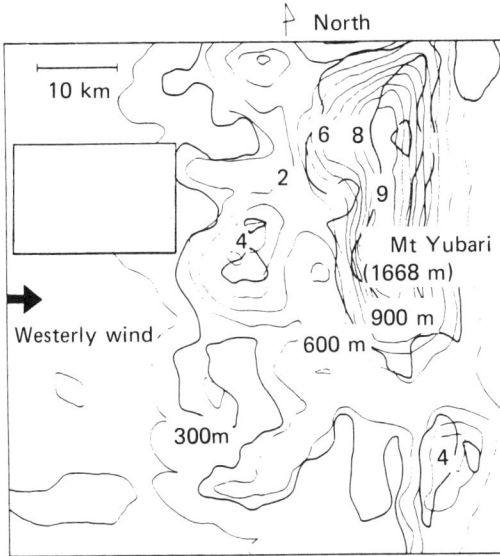

Fig. 14 — Calculated rainfall intensity (the Yubari area); ———, rainfall intensity; ———, contour lines.

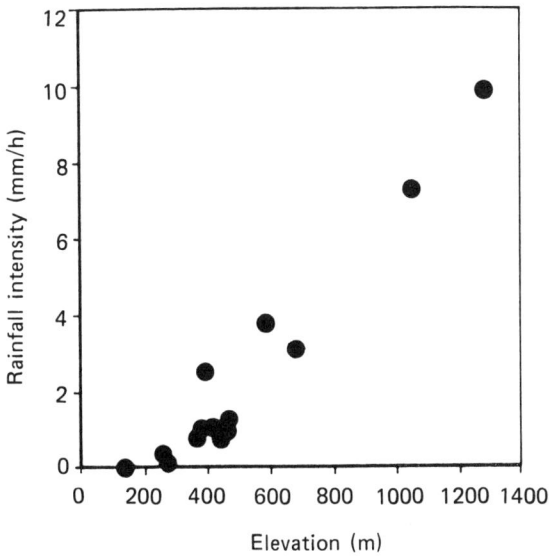

Fig. 15 — Calculated rainfall intensity against elevation (the Yubari area).

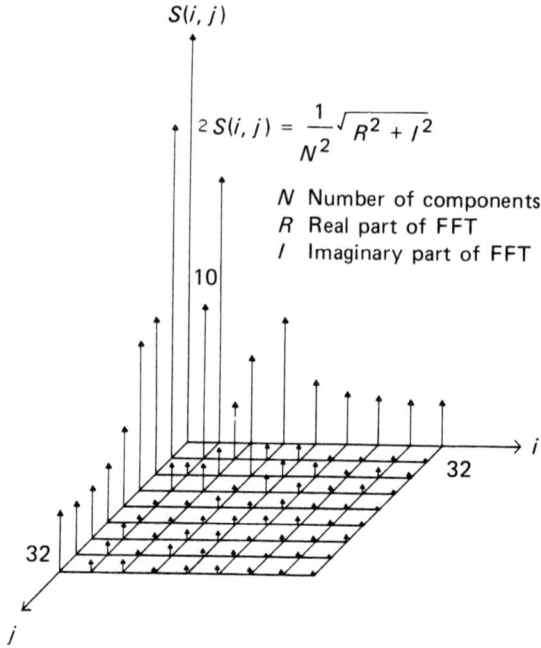

Fig. 16 — FFT components for the actual topography (the Orofure area).

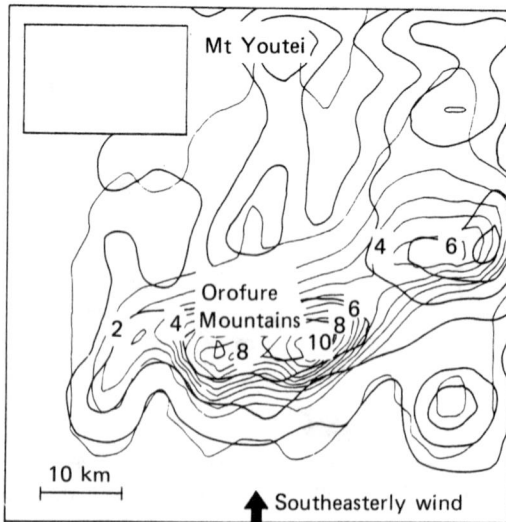

Fig. 17 — Calculated rainfall intensity (high-frequency components of FFT are removed) (the Orofure area).

relevant variables and parameters in the vertical direction. In this operation, the basic equations are averaged vertically, assuming that cloud and rainfall amounts are uniform in each vertical column.

$$\frac{\partial \overline{m}}{\partial t} = -\overline{u}\frac{\partial \overline{m}}{\partial x} - \overline{v}\frac{\partial \overline{m}}{\partial y} - \frac{\overline{u}\,\overline{m}}{H-h}\frac{\partial h}{\partial x} - \frac{\overline{v}\,\overline{m}}{H-h}\frac{\partial h}{\partial y}$$

$$- \overline{AC} - \overline{CC} + \overline{EP} + \overline{CV}$$

$$\frac{\partial \overline{M}}{\partial t} = -\overline{u}\frac{\partial \overline{M}}{\partial x} - \overline{v}\frac{\partial \overline{M}}{\partial y} - \frac{\overline{V}\,\overline{M}}{H-h} - \frac{\overline{u}\,\overline{M}}{H-h}\frac{\partial h}{\partial x} - \frac{\overline{v}\,\overline{m}}{H-h}\frac{\partial h}{\partial y}$$

$$+ \overline{AC} - \overline{CC} + \overline{EP}$$

where

$$\overline{AC} = \begin{cases} K_1(\overline{m}-a), & \overline{m}>a \\ 0, & \overline{m}<a \end{cases} \tag{19}$$

$$\overline{CC} = 6.96 \times 10^{-4} E(No)^{1/8}\overline{mM}^{7/8} \tag{20}$$

$$\overline{EP} = 1.93 \times 10^{-6}(No)^{7/20}\overline{mM}^{13/20} \tag{21}$$

$$\overline{CV} = \overline{w}\left(A + \frac{B}{2}(H+h)\right)\Big/(H-h) \tag{22}$$

$$\overline{V} = -38.3\,(No)^{-1/8}\overline{M}^{1/8} \tag{23}$$

$\overline{m}$ $(/g^3)$ is the vertical averaged cloud water content, $\overline{M}$ $(/g^3)$ is the vertical averaged rain water content, $\overline{u}$ (m/s), $\overline{v}$ (m/s) and $\overline{w}$ (m/s) are the vertical averaged wind speeds in x, y and z directions, $\overline{AC}$ $(/g^3\,s)$ is the vertical averaged autoconversion of cloud, $\overline{CC}$ $(/g^3\,s)$ is the vertical averaged collection of cloud, $\overline{EP}$ $(/g^3\,s)$ is the vertical averaged evaporation of rain, $\overline{CV}$ $(/g^3\,s)$ is the vertical averaged condensation of water vapour, $\overline{V}$ (m/s^1) is the vertical averaged fall speed of raindrops, H (m) is the altitude of upper boundary layer, h is the elevation of the ground surface (m), K_1 $(=10^{-3}/s)$ is a constant, a $(=0.5/g^3)$ is the threshold value of autoconversion, E $(=1)$ is the collection efficiency, (No) $(=10^7/m^4)$ is the Marshall–Palmer constant, A $(=3 \times 10^{-3}/g^4)$ is a constant and B $(=-3 \times 10^{-7}/g^5)$ is a constant.

In this section, we confirm that the simplified model is able to simulate rainfall field as easily as the three-dimensional model. Equations (17) and (18) are solved numerically using the finite-difference method, and a series of calculations for rainfall are performed under the same conditions as for the three-dimensional model. The wind components in the equations are obtained by averaging the solutions of the three-dimensional potential flow vertically.

First, the rainfall fields are calculated on the wall-like and concentric circular peaked shapes of the mountain as done in the numerical experiments using the three-dimensional model. The rainfall pattern in the former case has a similar tendency to the results obtained from use of the three-dimensional method. However, in the latter case, the simulated rain bands shift to the lee side in comparison with the three-dimensional simulation, because the horizontal advection effect is emphasized excessively. Accordingly, the two-dimensional model fails to produce the evading effect of wind around a mountain. The vertical wind component influenced by the topographic effect is one of the significant factors in rainfall fields. Therefore, the ascending wind velocity must be estimated precisely. Numerical experiments of rainfall for the actual topography are conducted in the case of southeasterly wind in the Orofure area as shown in Fig. 18. The synoptic rainfall pattern shows the same tendency as the result of the three-dimensional simulation (compare Fig. 13(a)). Even the simplified version can reproduce the heavy rainfall along the wall-like slope of the Orofure Mountains. It follows from these calculations that the two-dimensional model proposed herein offers considerable promise for practical use.

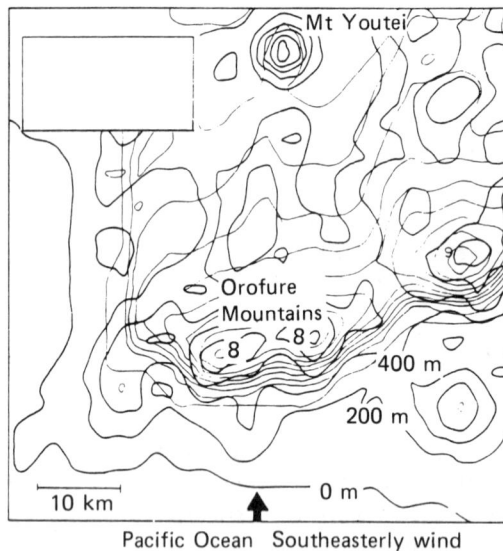

Fig. 18 — Calculated rainfall intensity using two-dimensional model (The Orofure area): ———— rainfall intensity (mm/h); ———, contour lines.

5. CONCLUSIONS

On the basis of the field observations and analysis for rainfall field in mountainous areas, the results are briefly summarized as follows.

(1) The observed data clearly show that rainfall amounts on the slope of the mountain increase linearly with increasing elevation. Such orographic effects of rainfall can also be reproduced by the simulation results.

(2) When the humid wind blows to the wall-like mountains, the heavy rainfall is observed along the slope which faces the wind direction.
(3) When the ascending wind in the condensation term is stronger, the amount of rainfall increases linearly. On the other hand, when the horizontal advection effect is in excess, the rainfall intensity decreases.
(4) No significant effects of the distribution and the amount of rainfall are discerned, even when high-frequency components of FFT (i.e. when the topographic scale in the horizontal directions is less than 5 km) are neglected.
(5) The simplified model can operationally simulate the same synoptic rainfall field as easily as the three-dimensional model.

ACKNOWLEDGEMENTS

The authors wish to express their great appreciation to Professor Fujita and Dr Hoshi who gave us many invaluable suggestions and comments. Thanks are also due to the flood-forecasting section, the Ishikari River Local Head Office, Hokkaido Development Bureau, which cooperated in the field observations and offered many data and materials.

REFERENCES

Kessler, E. (1969) Models of microphysical parameters and processes. *Meteorol. Monogr.* (10) 26–31.
Kikuchi, K., Harimaya, T. and Horie, N. (1981) Analyses of the properties of local heavy rainfall in the southwestern part of Hokkaido Island in the latter part of August, 1980. *Geophys. Bull. Hokkaido Univ.*, **40**, 55–77 (in Japanese).
Konno, T. and Kikuchi, K. (1981) Properties of local heavy rainfall on the southeast slope of Orofure mountain range in the Iburi District, Hokkaido, Japan (1). *Geophys. Bull. Hokkaido Univ.*, **39**, 1–18 (in Japanese).
Nakatsugawa, M., Yamada, T., Naitou, O. and Mizushima, T. (1989) Simulation of wind field and rainfall in catchment area. *Proceedings of the 33rd Japanese Conference on Hydraulics.* pp. 109–114 (in Japanese).
Takeda, E. and Kikuchi, K. (1978) Local heavy rainfalls in Hokkaido Island, Japan (1). *Geophys. Bull. Hokkaido Univ.*, **37**, 19–29 (in Japanese).
Yamada, T. and Watabe, G. (1987) Numerical simulation of orographic rainfall. *Proc. Jpn. Soc. Civ. Eng.*, **42**, 98–99 (in Japanese).
Yamada, T. and Watanabe, H. (1988) Analyses of wind field in mountainous areas. *Proc. Jpn. Soc. Civ. Eng.*, **43**, 68–69 (in Japanese).

28

The effect of range on the radar measurement of rainfall

D. Tees and G. L. Austin
McGill Weather Radar Observatory, Box 198, MacDonald College, Ste-Anne-de-Bellevue, Quebec H9X 1CO, Canada

ABSTRACT

This paper describes a simulation of radar range-dependent beam-filling and beam height errors using high-resolution near-range volume scans from the McGill radar as input data. An effort is made to quantify the errors associated with different systems currently used to combine radar and raingauges. It is found that using individual gauge-to-radar ratios to calibrate radar can result in serious unnecessary correction. Cumulative probability methods employing range-dependent $Z-R$ relationships also tend to result in serious overestimates of rainfall at long range. The best method is to find the physical factors affecting the comparison if the raingauges are significantly different from the radar values.

1. INTRODUCTION

In the past 30 years, a great deal of research has been done on the use of radar in the measurement of rainfall. As yet, however, there is still no consensus on the accuracy of the estimates that radars produce. Up until recently the radar community has focussed its attention on the $Z-R$ relationship. Wilson and Brandes (1979), for example, conclude that variation in the $Z-R$ relationship is the primary cause of the difference between radar and raingauges. Accordingly, many workers both before and after this paper was published have tried to develop dual-wavelength (Ulbrich and Atlas 1975) and dual-polarization radars (Seliga and Bringi 1976). These provide the observer with two parameters instead of the normal one, allowing for a determination of the drop size distribution, and hence a $Z-R$ relationship. However, the usefulness of two-parameter radars in reducing the radar 'error' depends crucially on whether the conclusion of Wilson and Brandes above is correct: are $Z-R$ relationships the major factor in the difference between radar and raingauges?

The answer would appear to be No. Results have been presented by Zawadzki (1984) which suggest that the drop size distribution is not the major factor. Although this is a count against the direct drop size distribution measurements, there are a number of empirical methods which use raingauges to correct the radar directly, perhaps resulting in more accurate estimates of radar. In some form or other this is certainly necessary. At McGill, long-term radar accumulations are verified against rainfall as measured by raingauges.

One popular method, described by Collier (1986), is to fit radar data to a small number of telemetering raingauges. By using these as ground truth, one may compare the radar pixel immediately above and get a correction factor for that pixel and (hopefully) other nearby pixels. This method can have disastrous effects on the correction factors, however, if, because of sampling effects, one of the measuring systems records rain while the other gets nothing. In addition, the comparison may be complicated by sub-bin variability which the radar smooths out, but which registers on the gauge. Surprisingly, it has been reported (Collier 1986, Wilson and Brandes 1979), that calibrations of this kind by a few gauges can reduce the total error between radar data and other gauges.

Another method, which has been suggested by Calheiros and Zawadzki (1987), is based on mapping the cumulative probability of rainfall in raingauges to the cumulative probability of radar. Here, a radar intensity is supposed to correspond to the rainfall with the same cumulative probability. Since the distributions change with range, this method produces range-dependent Z–R relationships. Their procedure attempts to match the radar to the climatology of the gauges. The method is only intended to work 'on the average' since the comparison of any particular day with the climatology is doubtful. It has the advantage that simultaneous radar and raingauge data are not necessary.

The main problem with the approach is that, because of sampling differences, the probability that rain occurs over a small area may not be the same as that over a larger area; for example, for a high rainfall rate to show up in a gauge, it needs simply to exist at the right place. For it to show up on a radar, however, it must occur over a volume the size of a radar pixel. Because of smoothing, it is quite possible that these high rates may never show up on a radar, thus making the comparison of probabilities past a certain point impossible. It also may map spuriously rainfall at high elevations into rainfall at low elevations where the statistics are no doubt different. Finally, one may quarrel with the whole idea of a relationship which is content with being correct only 'on average'.

The problem with these empirical calibration methods is that gauges and radar measure the rain-field differently. Radar produces an estimate of the mean rainfall in a volume illuminated by its beam. This beam is typically small (say about 1 km across), but it might well be larger than the scale size of significant variations in the rain field. It is certainly much larger than the gauge catchment area. A network of raingauges, on the other hand, produces an estimate of the rain that fell into an array of 8 in buckets. It is quite possible for a rainfall extremum of small horizontal extent to register on a gauge but not to be seen by a radar because the echo does not occur over an area the size of a radar bin. Thus, even if both systems measure the rainfall correctly, some discrepancies in point comparison can be expected owing to these scale effects. Austin (1987) recognizes this and makes the assumption that the radar–raingauge comparison variation is mostly due to sampling differences. After looking at 374 radar–raingauge comparisons over 20 storms, she suggests a number of corrections which should be made to the radar data depending on the type of storm and the physical cause of any bias seen. Thus one should consider all the available information from both radar and raingauges before making corrections.

One of the physical factors affecting the radar–raingauge discrepancy is the range dependence of radar errors. There are two reasons why radar error should be range dependent: first, because of a finite radar beam width, radar resolution is much better at close range than at extreme or even midrange (volume can change by a factor of 100 between the nearest and farthest ranges); second, because of the curvature of the earth, a radar beam cuts a storm higher up at long range. Because of these range effects, there is a need for an examination of

(1) the effect of range on radar estimates of rainfall;
(2) the effect of range-dependent sampling differences on radar–raingauge comparisons;
(3) the effect of (2) on the different radar–raingauge combination schemes.

This chapter will consider all these questions.

2. SIMULATION

The problem with all attempts to examine remote sensing errors is the lack of a dense 'correct' rain field. Were there such a parameter, then it could be degraded and compared with the initial field. One way to proceed is to assume that some wrong field is in fact 'correct' enough for it to be a plausible 'correct' field. The choice used here follows that of Damant *et al.* (1983) in a similar context (finding the errors in the Thiessen polygon method of rainfall interpolation). This method uses radar data at the highest available resolution and proposes that even though this is not the actual field that fell on the date it was recorded, it is a possible field that could have fallen on some hypothetical day. If this possibility is admitted, then it should be possible to use this field as the input to a simulation of any facet of the radar–raingauge comparison problem. This method requires only the assumption that the subbin variability does not introduce any other significant effects. This choice is certainly less drastic than the assumption that one would have to make if one were to take, say, a sparse field of raingauges. In addition, the radar gives a spatial picture of the rainfall in both area and height. The input data for the simulation will use close-range polar radar data which have excellent resolution.

The simulation used to study range dependence uses high-resolution data from close to the McGill radar (for the parameters of this radar, see Table 1). A full three-dimensional volume scan is used (21 elevation angles). These data are then rotated around a curved earth (of radius $\frac{4}{3}R_{earth}$) to simulate the effects of normal atmospheric refraction. It is then rebinned using radar's binning program as the radar would have seen it. The power corresponding to the dBZ values on the tape are averaged and converted back to dBZ. The transformation has been done for seven convective storms that occurred in the Montreal area between 1979 and 1985. For

Table 1 — McGill radar parameters

Wavelength	10.4 cm
Peak transmitted power	1 MW
Pulse length	1 μs
Pulse repetition frequency	300 Hz
Antenna size (diameter)	10 m
Minimum detectable signal	0.5 mm/h at 150 km
Beam width	0.86°

more details on the simulation, see Tees (1989). In addition to the main simulation, a number of others have been done to look at specific features. For example, since it turns out that the increasing height of the beam tends to dominate the curved-earth scenario, a simulation has been done on a flat earth where the beam elevation is not such a problem. The flat-earth simulation provides an opportunity to examine beam-filling errors which, as the results show, tend to be swamped in the real curved-earth simulation. In addition, it is possible to do the simulation with different radar parameters. It is thus possible to use different elevation programs or beams widths and see what happens to the statistics of the field. Some of the statistics which were computed are shown in Figs 1 to 6.

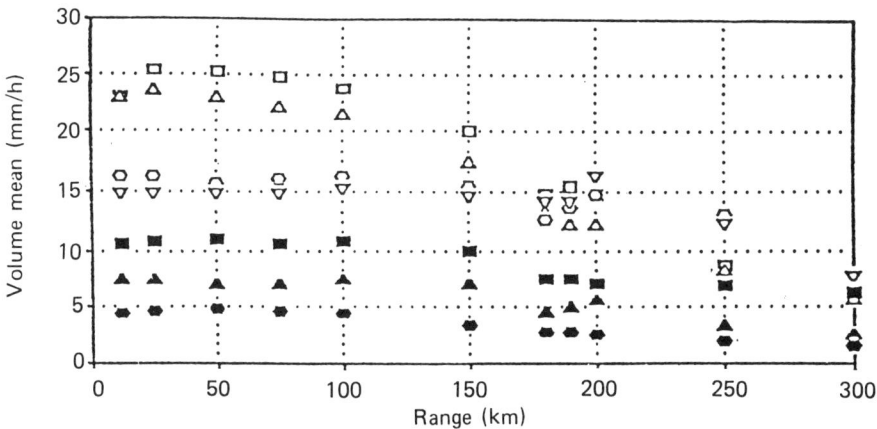

Fig. 1 — Graph of whole volumes scan mean versus range for the 'curved-earth' simulation. The different symbols show the behaviour of the seven different rain fields used in the study.

The first is the mean rainfall of the transformed volume scan (Fig. 1). This mean has been computed using $Z = 200R^{1.6}$ to turn dBZ to rainfall. Beyond 150 km the curved-earth simulation shows a drop in field mean indicative of a beam elevation. The flat-earth simulation (Fig. 2) shows no such drop. Instead it shows a slow falloff up to 150 km indicative of a beam-filling problem. The two sets of curves are almost identical up to 100 km, however. Thus is seems that, closer than 100 km, the radar

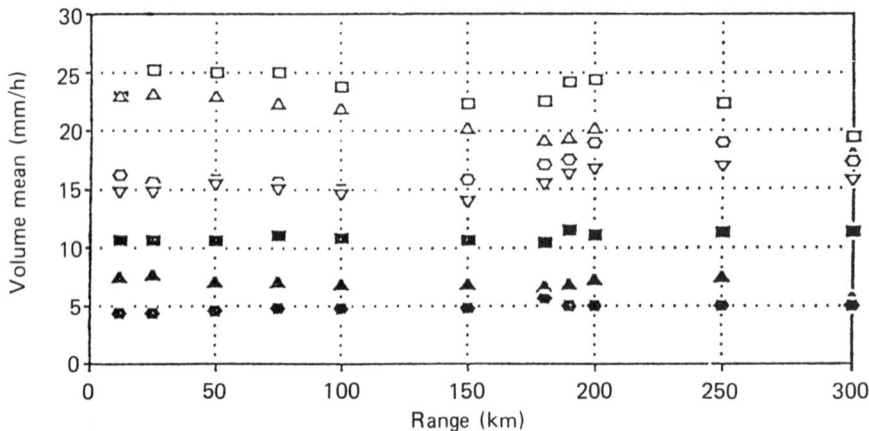

Fig. 2 — Same as Fig. 1 for the 'flat-earth' simulation.

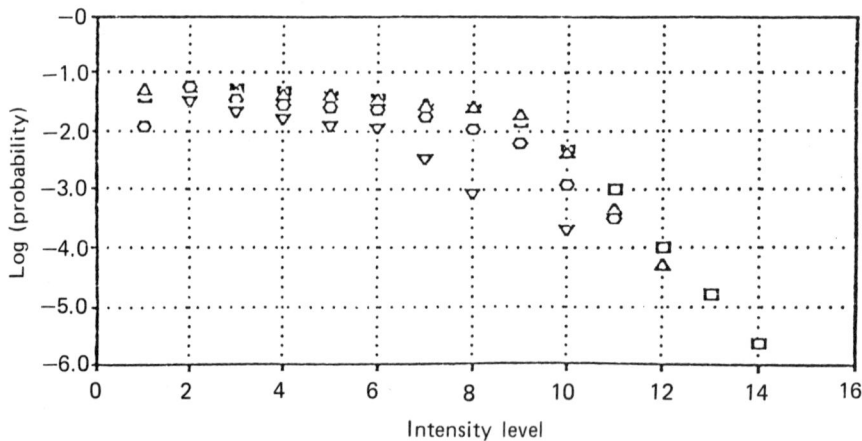

Fig. 3 — Logarithm of intensity distribution versus intensive level for the ranges 12 km (□),
100 km (△), 200 km (○) and 300 km (▽).

error (which is slight) is caused mostly by beam filling, while beyond this range, there
are significant errors due to the change in statistics with height.

Fig. 3 shows the intensity distribution, averaged over the seven fields used, at
different ranges. These curves shows that the high reflectivities disappear at long
ranges. Using these curves, empirical range dependent Z–R relationships can be
calculated using the method of Calheiros and Zawadzki (the distribution at the
closest range has been used as the distributions for the raingauges). Fig. 4 shows the
mean rainfall calculated using the relationship plotted against range for the fields
used in this study. It is found that, at long ranges especially, rain rates tend to be

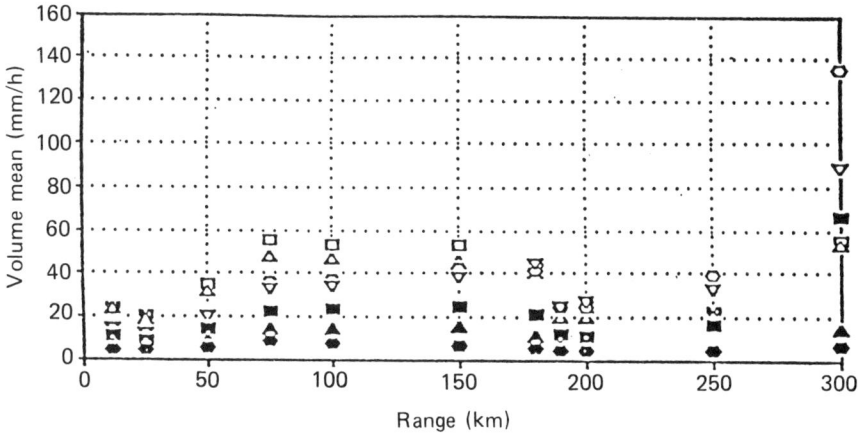

Fig. 4 — Graph of field mean rainfall versus range using a simulated Calheiros–Zawadzki cumulative probability method. This is based on the average distribution of intensities of the seven storms (shown in Fig. 3).

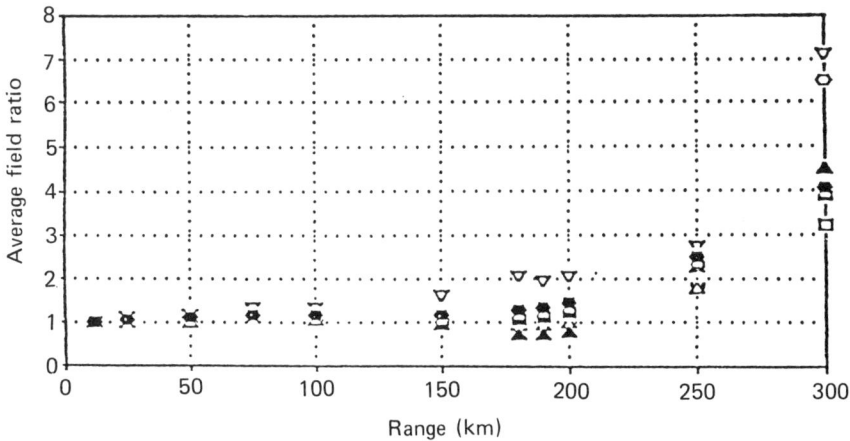

Fig. 5 — Average ratio of G/R versus range for the 'curved-earth' simulation.

drastically overestimated. The reason for this is that at long ranges when one is sampling the field high up, the assumption that one can map the rainfall at the ground into the rainfall aloft is certainly wrong. The statistics of the field at the different elevations are simply different. If one is to related these high altitude returns to the climatological raingauge rainfall, it would be necessary to know the relation between rainfall and height. Only if knowledge of the rainfall aloft can lead to estimates of the rainfall at more reasonable heights can the method succeed.

The average field ratio (Fig. 5) finds the average ratio of each initial field point ('raingauges') to the corresponding transformed field point (radar). Beyond 150 km

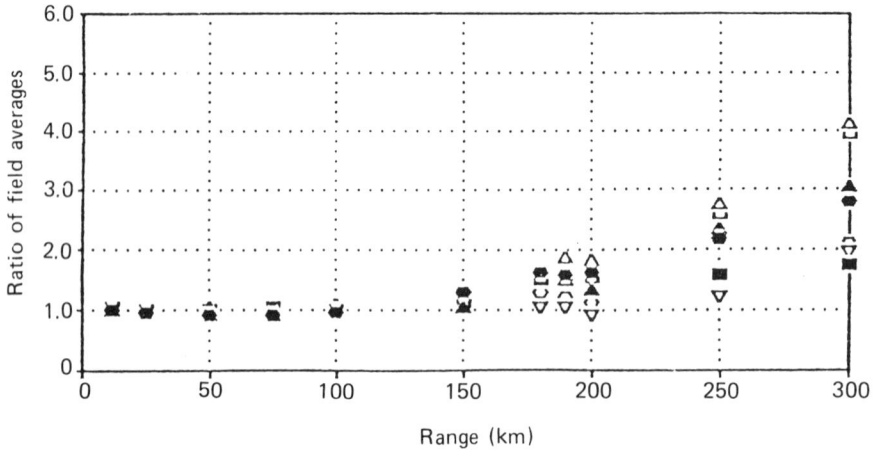

Fig. 6. — Ratio of the transformed field to the initial field for the 'curved-earth' simulation.

this statistic rises very quickly from 1 to values of 3–7. Even the lighter rain fields used here can show large average field ratios at middle ranges; thus it seems reasonable to assume that this conclusion holds for all fields which are at least somewhat variable on the small scale. Another ratio statistic based on the calibration procedure used at the McGill radar is shown in Fig. 6. The ratio of field averages finds the ratio of the transformed field mean to the initial field mean. This static also rises at long range, but the values range from 2 to 4 depending on the field. These are much more modest than those of the average field ratio. From this it is clear that the method of Collier (1986) (using individual gauge-to-radar ratios to correct the radar) can easily lead to large erroneous 'corrections' in heavy convective rain.

Another application of the program developed here involves seeing what kind of radar setup is needed to achieve a desired accuracy. One wishes to know within what range one needs to know the rainfall with a certain error. Several configurations have been studied. The results for various programs are shown in Fig. 7.

(1) *Plan position indicator (PPI) radar*   The two PPI simulations shown in Fig. 7 were done using both a low (0.9°) and a relatively high (2.5°) elevation angle. Both had disastrous effects on the field mean outside a certain range band. The high elevation angle was virtually useless beyond 75 km. The low elevation angle was seriously low at close ranges and also at long ranges. For middle ranges between 50 and 150 km it gives values which are not so bad. The other curve on the graph shows the range behaviour of a radar at Maniwaki which has four elevation angles. The curve is much closer to that of the McGill radar. In general, single-elevation-angle radars are unable to give height information and are limited range. Even a modest number of elevation angles improves the estimates of field mean, and the difference between radar and 'raingauges'.

(2) *Constant-altitude plan position indicator* (CAPPI) radar   The simulations show that this setup is accurate up to about 150 km where the beam begins to rise above the area of interest.

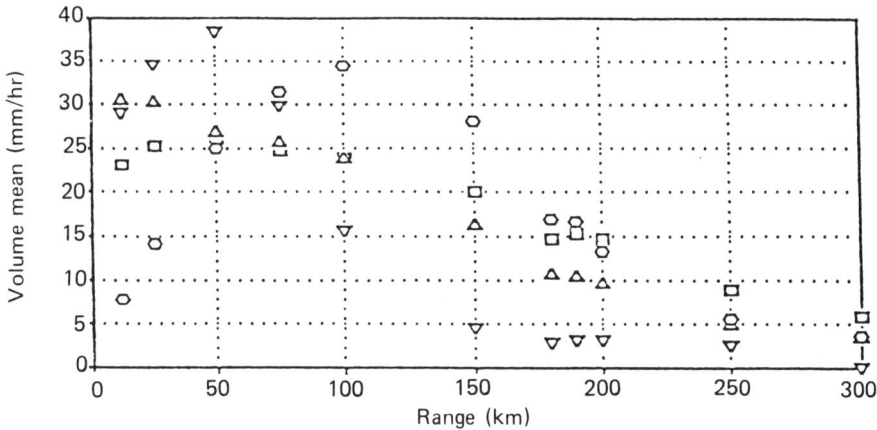

Fig. 7 — Field mean using a high-intensity rain field observed using four different elevation programs: □, that used by the McGill radar, △, the elevation program of the Maniwaki radar (four elevation angles); ○, 0.9° PPI; ▽, 2.5° PPI.

3. CONCLUSIONS

The major conclusion to be drawn here is that one does not need to postulate erroneous $Z–R$ relationships to get significant differences between radar and gauges of the kind that have been reported for example by Wilson and Brandes (1979). Any attempt to rely on the ratio of a single gauge to radar comparison may well go astray by as much as 150% in heavy convective rain at 150 km. Any correction to the raw radar data should be made on the basis of the ratio of field means which follows most closely the behaviour of the radar mean as a function of range. Even this is not likely to work on an instantaneous basis, however. A better plan would be the restriction of radar range. For accurate estimates, it is best to restrict the range of a 10 cm radar to about 150 km if multiple elevation angles are used, and to a certain range band if only a single angle is available (in case of the latter, an even better procedure would be to convert to some form of multiple-elevation scanning, which flattens the range behaviour of radar estimates of the statistics). At longer ranges, since one is only seeing the tops of the storm, correction must be based on a knowledge of the height structure of the weather in the area. Any such correction is likely to be of low precision.

REFERENCES

Austin, P. M. (1987) Relation between measured radar reflectivity and surface rainfall. *Mon. Weather Rev.* **115**, 1053–1070.

Calheiros, R. V. and Zawadski, I. (1987) Reflectivity–rain rate relationships for radar hydrology in Brazil. *J. Climate Appl. Meteorol.*, **26**, 118–132.

Collier, C. G. (1986) Accuracy of rainfall estimates by radar, Part I: calibration by telemetering raingauges. *J. Hydrol.*, **83**, 207–223.

Damant, C. G., Austin, G. L., Bellon, A. and Broughton, R. S. (1983) Errors in the Thiessen technique for estimating areal rain amounts using weather radar data. *J. Hydrol.*, **62**, 81–94.

Seliga, T. A. and Bringi, V. N. (1976) Potential use of radar differential reflectivity measurements at orthogonal polarizations for measuring precipitation. *J. Appl. Meteorol.*, **15**, 69–74.

Tees, D. (1989) *The effect of range on the radar measurement of rainfall: a simulation*, M.Sc. Thesis, McGill University, Ste-Anne-de-Bellvue.

Ulbrich, C. W. and Atlas, D. (1975) The use of radar reflectivity and microwave attenuation to obtain improved measurements of precipitation parameters. *Preprints of the 16th Conference on Radar Meteorology, Houston, TX, 1975.* American Meteorological Society, Boston, MA, pp. 496–503.

Wilson, J. W. and Brandes, E. A. (1970) Radar–rain gage precipitation measurements: a summary. *Bull. Am. Meteorol. Soc.*, **60**, 1048–1058.

Zawadzki, I. (1984) Factors affecting the precision of radar measurements of rain. *Preprints of the 22nd Conference on Radar Meteorology, Zurich, 1984.* American Meteorological Society, Boston, MA, pp. 251–256.

29

An application of dual-polarization radar to radar hydrology

F. Yoshino, N. Ishii, H. Mizuno and T. Ikawa
Masamitsu Mizuno, Hydrology Division, Public Works Research Institute,
Ministry of Construction, 1 Asahi, Tsukuba-shi Ibaraki, Japan

ABSTRACT

We have studied observation characteristics by dual-polarization Doppler radar (DND radar). The following interesting results were obtained.

(a) The radar observables Z_H and Z_{DR} are nearly coincident with the results observed by distrometers for a storm.
(b) The drop size distribution is most appropriately expressed by a gamma distribution $N(D) = D^m N_0 \exp(-\Lambda D)$ using $m=2$.
(c) Bright band is probably discriminated by the values of Z_H and Z_{DR}, and by using their variational tendency.
(d) The examination of the statistics of the Z_{DR} signal shows that the variance of the Z_{DR} estimate is a function of both the sample pair correlation and the Doppler spectrum width.
(e) It is concluded that the resolution ability 0.2 dB of Z_{DR} may be attained by a time integration of more than 300 pulses of signals. This means that some averaging in range has to be considered for the resolution ability $0.1 Z_{DR}$.

1. INTRODUCTION

Radars have long been used as tools for the understanding of the formation, growth and dissipation of clouds and precipitation. However, many hydrologists have recently been studying the observational characteristics of weather radars. The studies include the improvement of their accuracy for estimating rainfall rate by raingauge calibration methods, the short-term precipitation forecast and runoff forecast using radar, distributed runoff models using areal precipitation values, etc. Therefore these studies may be called radar hydrology.

However, the most important problem seems to arise from the principle of measurement of rainfall rate by radar. Based on the conventional radar data, the rainfall rate is expreseed by empirical $Z–R$ relations. The $Z–R$ relations depend usually upon storm characteristics such as their causes, hydrometeor types, etc., and their constants included in the $Z–R$ relation vary in a wide range. Principally, the precipitation intensity is expressed by the product of drop size distribution in the air and the falling velocity of the drops.

Therefore, it is necessary to study the methods of observing drop size distribution and falling velocity of raindrops. It has recently been said that dual-polarization radar enables one to observe the deformed ratio of raindrops by the differential reflectivity factor and therefore gives the drop size distribution in the air. Dual-polarization radar measurements are an efficient tool to improve the accuracy of the estimates of meteorological parameters such as the hydrometeor type, the rainfall rate and the location of bright band.

We have studied the observation characteristics by dual-polarization Doppler radar (DND radar). The following interesting results were obtained. These studies may reveal the specification of the next-generation radar raingauge in Japan.

2. SPECIFICATION OF DND RADAR

DND radar is a dual-polarization Doppler radar. The dual-polarization radar can estimate not only the type and shape of hydrometeor but also the raindrop size distribution using the differential reflectivity Z_{DR}. According to the research by Seliga *et al.* (1984) and Goddard and Cherry (1984), the polarization radar can measure the rainfall intensity more accurately than conventional radar can. There are more than 15 dual-polarization radars in the world.

Doppler radar can measure the velocity and direction of the horizontal and vertical wind in the sky. The minimum resolution of the wind velocity of DND radar is 0.125 m/s. When the angle of elevation is 20°, the radar can measure the velocity and direction of the wind in the sky every 170 m altitude.

Table 1 shows the main specifications of the DND radar. The radar enables one to measure snowfall intensity from 0.2 to 29 mm/h within the range 20 km from the radar and to measure rainfall intensity from 0.3 to 108 mm/h.

The antenna type is a centre-fed paraboloid. Z_{DR} should be measured with 0.1 dB error to measure the precipitation intensity within 10% error by dual-polarization radar data (Sachidananda and Zrnić 1986). Therefore the feed type must be a rear feed. The antenna pattern of the horizontal polarization coincides with that of the vertical polarization. Polarization control is attained with a ferrite switch. The radar can transmit horizontal and vertical polarization electromagnetic waves every 9 pulses (about $\frac{1}{30}$ s) or each pulse alternately. In order to measure snowfall accurately, the minimum detectable signal is determined as -107.5 dBm. The minimum resolution is about 1% of the precipitation intensity.

3. STATISTICAL ANALYSIS FOR THE VARIATION OF RECEIVED SIGNALS Z_H AND Z_{DR}

There are two characteristics of meteorological scatterers that must be considered when designing the modulation and signal processing for meteorological radars: the

Table 1 — Specification of DND radar

Parameter	Dual-polarization radar	Doppler radar
Frequency	5280 MHz	5280 MHz
Peak power	75 kW	75 kW
Antenna diameter	2 m	2 m
Antenna rotation	6 rev/min	1 rev/min
Beam width	1.8°	1.8°
Gain	36.6 dB	36.6 dB
Side-lobe level	23.6 dB	23.6 dB
Pulse repetition frequency	280 pulses/s	1120 pulses/s
Pulse width	2.0 µs	0.5 µs
Minimum detectable signal	−107.5 dBm	−107.5 dBm
Polarization radiated	LIN-H, LIN-V	LIN-H
Clutter canceller	MTI and nothing	Nothing

scatterers are, in general, distributed throughout the radar resolution volume and the back-scattered signal (the phasor sum of the individual scatterers) has a limited correlation time because the scatterers are moving relative to each other.

One of the problems with Z_{DR} measurement has been its relatively long acquisition time. The feasibility of Z_{DR} measurements depends on the radar hardware and on the signal-processing techniques used. In practice, it is desirable to have Z_{DR} estimates with the same number of samples as required for Z_H and mean velocity estimation. Therefore, it is important to consider the effect of averaging non-independent sample pairs in estimating the acquisition time required for accurate Z_{DR} measurement. The effect of data averaging on the accuracy improvement mainly depends on the scanning method of the radar and sampling method of data (simultaneous or alternate sampling). The theoretical results were compared with the data obtained by DND radar observation.

3.1 Signal-processing algorithms
The signal correlation depends on the radar wavelength and the reflectivity-weighted distribution of the radial velocities of the scatterers (the Doppler spectrum width). For a Gaussian input signal power spectrum, the autocorrelation function ρ (m) of the signal is usually modelled as a function of spectrum width and time lag $\tau = mT_S$ (Zrnić, 1979):

$$|\rho(\tau)|^2 = \exp\left(-\frac{16\pi^2\tau^2\sigma_V^2}{\lambda^2}\right) \tag{1}$$

where λ is the radar wavelength and T_S is the pulse repetition time.

Sachidananda and Zrnić (1985) have examined the properties of the signal fluctuations and obtained the variance of a square law estimator of Z_{DR} for alternate sampling:

$$\text{Var}(Z_{DR}) = 2\,Z_{DR}^2 \sum_{m=-(M-1)}^{M-1} \frac{M - |m|}{M^2}\,[|\rho\,(2m)|^2 -$$
$$|\rho(2m+1)|^2|\rho_{HV}\,(0)|^2] \tag{2}$$

where M is the number of averaged sample pairs and ρ_{HV} is correlation between horizontal polarization signal P_H and vertical polarization signal P_V.

Chandrasekar *et al.* (1985) have also obtained a similar result for a logarithmic receiver:

$$\text{Var}(Z_{DR}) = \frac{37.72\,\pi^2}{M^2\,6}\left(\sum_{m=-(M-1)}^{M-1}(M-|m|)[\rho_{HH}^{\log}(2m) - \rho_{HV}^{\log}(2m+1)]\right) \tag{3}$$

where

$$\rho^{\log} = \frac{6}{\pi^2}\sum_{m=1}^{\infty}\frac{|\rho|^2 m}{m^2}$$

and

$$\rho_{HV}^{\log}(m) \approx \rho_{HH}^{\log}(m)\rho_{HV}(0).$$

DND radar has the ability to employ a polarization switching interval for each pulse and every nine pulses. Therefore the effect of averaging every nine pulses on the accuracy of Z_{DR} estimates is also analysed.

Schidananda and Zrnic' (1985) have also obtained the covariance $\text{Cov}(P_1, P_2)$ between two received powers P_1 and P_2 after the integration of M pulses when noise is omitted:

$$\text{Cov}(P_1, P_2) = \frac{P_1 P_2}{M^2}\sum_{m=-(M-1)}^{M-1}(M-|m|)(|\rho_{12}(m)|^2). \tag{4}$$

As the signals P_1 and P_2 are considered to be the same polarization signals with a consecutive small time difference, the autocorrelation function $C(\tau)$ is expressed by

$$C(\tau) = \frac{P^2}{M^2}\sum_{m=-(M-1)}^{M-1}(M-|m|)(|\rho(M+\tau)|^2). \tag{5}$$

As the variance of received power P_M integrated for M pulses equals $C(0)$, the autocorrelation $\rho_M(\tau)$ for the M-pulse averaging signal is expressed as

$$\rho_M(\tau) = \frac{\displaystyle\sum_{m=-(M-1)}^{M-1} (M - |m|)(|\rho(m+\tau)|^2)}{\displaystyle\sum_{m=-(M-1)}^{M-1} (M - |m|)(|\rho(m)|^2)}. \tag{6}$$

Therefore, the variance of Z_{DR}, which is obtained after the integration of every M pulse, is expressed by the convertion of Z_{DR}^2 to $Z_{DR}^2 (P_M^2/P^2)$ and $\rho(\tau)$ to $\rho_M(\tau)$ in equation (2) or (3).

3.2 Performance of the algorithms

The standard error of the estimates of Z_H, Z_{DR} and σ_V are analysed through the above-mentioned alogrithms.

Fig. 1 shows the spectrum of received power from rain echoes obtained by DND radar (the antenna scan was fixed at the azimuth direction and an elevation angle of 5°). The observed power spectrum may be simulated by Gaussian distribution ($\sigma_V = 0.654$ m/s). This result confirms the assumption of the above-mentioned theoretical considerations.

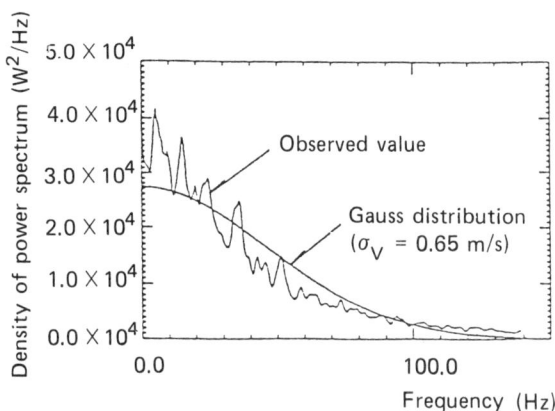

Fig. 1 — Power spectrum of received signal where σ_V is the Doppler spectrum width.

Fig. 2 gives the variation in the autocorrelation with the pulse width, for the Doppler spectrum width σ_V, expressed by equation (1). The figure shows that the autocorrelation rapidly increases for a Doppler spectrum width of 2.0 m/s. Also it is shown from equation (1) that the autocorrelation decreases as the wavelength decreases.

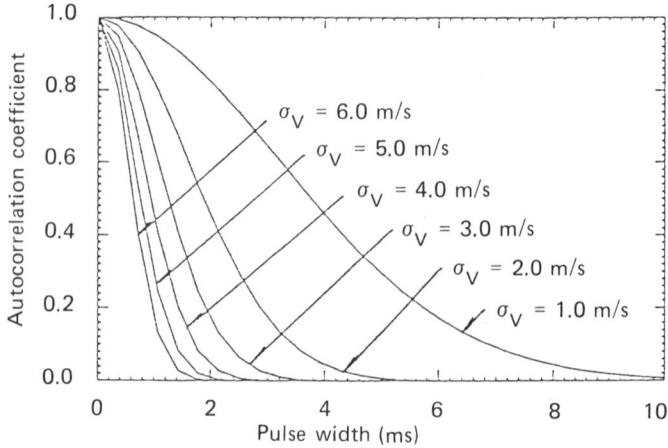

Fig. 2 — The relation between pulse width and autocorrelation coefficient.

Fig. 3 shows the variation in $\sigma_{Z_{DR}}$ with the pulse repetition frequency for alternate sampling of each pulse, every four pulses and nine pulses of polarization switching interval (square law estimator; Doppler spectrum width, 2.0 m/s; correlation coefficient $\rho_{HV} = 0.995$).

Fig. 3 also gives the variation in $\sigma_{Z_{DR}}$ with number of sample pairs for the alternate-sampling scheme. These results show that, as the pulse repetition frequency increases, $\sigma_{Z_{DR}}$ decreases for the alternate-sampling scheme of each pulse and that $\sigma_{Z_{DR}}$ for a pulse repetition frequency of 500 Hz is approximately halved for a pulse repetition frequency of 300 Hz. However, the figure shows that $\sigma_{Z_{DR}}$ for the alternate sampling of every nine pulses becomes larger than that for each pulse. Also the result shows that $\sigma_{Z_{DR}}$ decreases as the wavelength increases.

Fig. 4 shows the variation in $\sigma_{Z_{DR}}$ with number of sample pairs actually observed by the alternate-sampling scheme at each pulse. The white-noise effects have been neglected. The spectrum width is shown as a parameter and the pulse repetition time T_S is set at $\frac{1}{280}$ s. The result shows that, for the logarithmic amplifier, $\sigma_{Z_{DR}}$ becomes 0.2 dB, when about 300 sample pairs are used for the Z_{DR} estimation. It is pointed out that the values of $\sigma_{Z_{DR}}$ for logarithmic receivers are higher than those for square law receivers. This suggests that a square law receiver is preferable to a logarithmic receiver in the context of Z_{DR} estimation. These results are the same as obtained by Chandrasekar *et al.* (1985).

The rain rate R is fairly sensitive to errors in Z_H and Z_{DR}. Concerning the report by Sachidananda and Zrnić (1985), it is also pointed out that an uncertainty of 1 dB in Z_H and 0.2 dB in Z_{DR} can give rise to as much as a 40% standard error in the rain rate and that, to keep the standard error in rate less than 25%, Z_H and Z_{DR} have to be estimated to an accuracy of 1.0 dB and 0.1 dB respectively.

The examination of the statistics of the Z_{DR} signal shows that the estimate of the variance of Z_{DR} is a function of both the sample pair correlation and the Doppler spectrum width. Therefore, the accuracy of 0.1 dB for the Z_{DR} estimate is only attained for Doppler spectrum width less than 0.5 m/s. This means that it is very

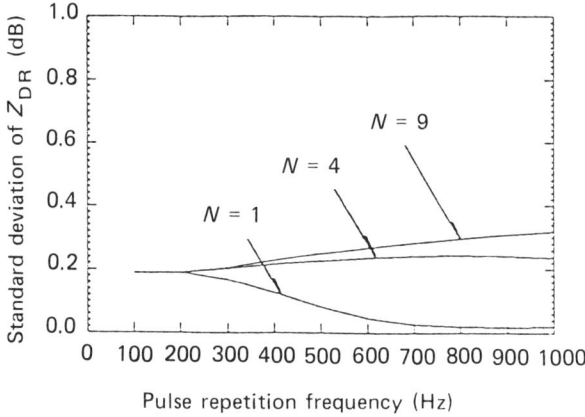

Fig. 3 — The relation between pulse repetition frequency and standard deviation of Z_{DR} where N is the switching interval of polarization (number of pulses).

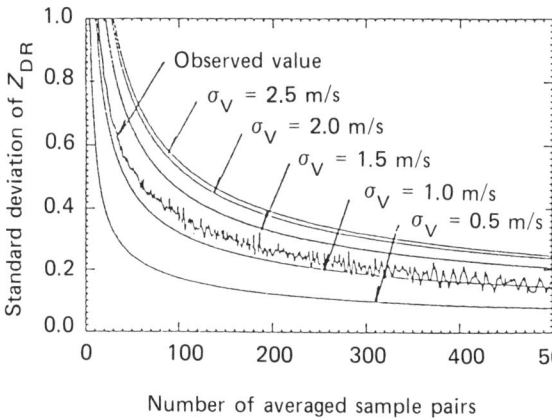

Fig. 4 — The relation between averaged sample pairs and their standard deviation of Z_{DR} for the logarithmic amplifier.

difficult to estimate Z_{DR} values with a resolution less than 0.1 dB. Although some averaging in range is available, the number of samples must be small if decent range resolution is to be preserved.

4. THE ASSESSMENT OF THE ABILITY OF DUAL-POLARIZATION RADAR (CO-POLAR LINEAR) TO PREDICT RAINFALL RATE BY SIMULTANEOUS DISTROMETER MEASUREMENTS

4.1 Theoretical background

The radar reflectivities for horizontally and vertically polarized waves, Z_H and Z_V respectively, are expressed as

$$Z_{H,V} = \frac{10^6 \lambda^4}{\pi^2 |K|^2} \int_0^{D_{max}} \sigma_{H,V}(D) N(D) \, dD \quad mm^6/m^3 \tag{7}$$

where $\sigma_H(D)$ and $\sigma_V(D)$ are the back-scattering cross-sections for horizontally and vertically polarized microwaves of an oblate raindrop whose volume is equal to that of a corresponding spheroidal raindrop of diameter D (cm). In addition, λ (cm) is the radar wavelength, $|K|^2 = 0.93$ is the refractivity factor for liquid water, $N(D)$ (/m^3 cm) is the number of drops per unit volume per unit size interval, and the integration is carried out over all diameters $0 \leqslant D \leqslant D_{max}$ where D_{max} (cm) is the maximum raindrop diameter.

From these definitions the differential reflectivity factor is found from

$$Z_{DR} = 10 \log \left(\frac{Z_H}{Z_V} \right) dB. \tag{8}$$

The drop size distribution is usually expressed as a gamma distribution:

$$N(D) = N_0 D^m \exp(-\Lambda D)$$

where N_0 (/m^3 cm), m and Λ (/cm) are parameters.

By using these relations, the differential reflectivity is expressed as follows:

$$Z_{DR} = 10 \log \left(\int_0^{D_{max}} \sigma_H(D) D^m \exp(-\Lambda D) \, dD \middle/ \int_0^{D_{max}} \sigma_V(D) D^m \exp (-\Lambda D) \, dD \right). \tag{10}$$

Therefore Λ is determined from equation (10), when the scattering cross-sections $\sigma_H(D)$ and $\sigma_V(D)$ were calculated with Gan's extension of Rayleigh scattering from spherical drops applied to oblate spheroids as described by Van de Hulst (1975), using values of drop deformation a/b (where a and b are the respective semiminor and semimajor axes of the oblate drop) for various values of D as measured by Pruppacher and Pitter (1971). At this stage, the m values were assumed as 0, 2, 4 and 6. N_0 is determined from equations (7) and (9).

The rainfall intensity is estimated from the relation

$$R = 3.6 \times 10^{-3} \frac{4}{3} \pi \int_0^{D_{max}} \left(\frac{D}{2} \right)^3 N(D) V(D) \, dD \tag{11}$$

where R (mm/h) is the rainfall intensity and $V(D)$ is the falling velocity of raindrops. In this analysis, the falling velocity expressed by the Best equation

$$V(D) = 958.0\left\{1 - \exp\left[-\left(\frac{D}{0.0885}\right)^{1.147}\right]\right\} \text{cm/s} \qquad (12)$$

is applied for the diameter D (mm) of raindrops. Therefore the ratio of Z_H to R is expressed by

$$\frac{Z_H}{R} = \left(10^6 \frac{\lambda^4}{\pi^5} \left|\frac{e+2}{e-1}\right|^2 \int_0^{D_{max}} \sigma_H(D) D^m \exp(-\Lambda D) \, dD\right) \Big/$$

$$3.6 \times 10^{-3} \tfrac{4}{3}\pi \int_0^{D_{max}} \left(\frac{D}{2}\right)^3 V(D) D^m \exp(-\Lambda D) \, dD). \qquad (13)$$

4.2 Radar and distrometer measurements

Simultaneous measurements by DND radar and four distrometers were conducted in the weak rainstorm on 5 October 1988 at Kurume in northern Kyushu. The rainfall intensity less than 20 mm/h for each 5 min were observed on the ground. The storm was so weak that it might be thought not appropriate for the dual-polarization analysis.

The echoes detected by the receiver were digitized and averaged for each nine pulses before recording on magnetic tape. The power received was first averaged in range over two pulse volumes to give a total radar sampling volume 500 m long. The power recorded in each of 80 of these range gates 500 m long was then averaged for 256 sample pairs on each polarization. The sampling errors when measuring Z_H and Z_{DR} were estimated to have standard deviations of typically 1.0 dB and 0.2 dB respectively.

The comparative but not absolute calibration of Z_H has been conducted with references to standard values of B and β (which were accepted for the case of weak-storm events) and theoretically estimated values for Z_H and Z_V from distrometer measurements. The bias error of Z_H and Z_V between radar and distrometer data were estimated to be 0.245 dB and -0.005 dB respectively. These discrepancies were adjusted for the Z_{DR} estimates.

Ground observations were done by four distrometers (manufactured by the Distromet Company, Switzerland). They were located 0, 8.7, 12.1 and 18.2 km from the DND radar site. The number $n(D)$ of drops falling onto the distrometer's sensor (area, 50 cm^2), over a period of 1 min was recorded according to size in 20 channels, corrresponding to a range of drop diameters D between 0.3 and 6 mm. The size distribution measured by the distrometer was for drops falling onto an area in a given time and was converted to the volumetric size distribution $N(D)$ that would be measured by the radar, using

$$N(D)\,\delta D = \frac{n(D)}{V(D)\,\mathrm{DT}\,A}$$

where $N(D)\,\delta D$ is the number of drops per cubic metre in the inteval δD of the distrometer channel which has an arithmetic mean diameter D (mm), $V(D)$ (m/s) is the terminal velocity of drops of diameter D, DT(s) is the measurement duration and A (mm^2) is the area of the sensor.

4.3   Comparisons of radar and distrometer data for Z_H and Z_{DR}

The data for this experiment were collected with the antenna stationary and pointing at an elevation of 3° to observe the rain vertically above the distrometers. The data selected from the range gate corresponding to the location of distrometers were subsequently averaged over 20 s periods for comparison with the data collection period of the distrometer. This averaging is expected to reduce the standard errors of measuring Z_{DR} to 0.2 dB. For the computation, the maximum drop size was determined as 6.0 mm and the m value is assumed to be 2.

An example of 200 min of comparison data is shown in Fig. 5. Distrometer values of Z_H and Z_{DR} were calculated for each 1 min sample by using equations (7) and (10). Radar measurements of Z_H and Z_{DR} are shown at 20 s intervals, i.e. before final averaging to 1 min samples. From the figure, it is understood that Z_H values show fairly good correspondence between radar and distrometer observed values (Fig. 6), but Z_{DR} values show larger scattering between these two values (Fig. 6). However, the time variation of Z_{DR} values are sell observed. The difference might arise from the difference of spaces observed by radar and distrometer, but the observed rainfall intensity might also affect the results because this technique is thought to be appropriate for an intensity of more than 20 mm/h.

The values of cross-correlation between the radar and distrometer time series profiles of Z_H were calculated for various time offsets, and a maximum cross-correlation (of 0.896) was found when an offset of 360 s was added to each radar time. This offset time has been applied in Fig. 5 and in all subsequent comparisons between ground-based and radar measurements.

Fig. 6. shows all available comparisons of distrometer and radar Z_H and Z_{DR} values. The correlation of Z_{DR} values is 0.671 and standard deviation is 0.214 dB. The comparison of radar and distrometer Z_{DR} values has shown less correlation than expected. As the Z_{DR} values are almost less than 1.5 dB, an exact conclusion will be obtained after more intense storms have been observed.

4.4   Comparison of estimated dropsize distribution by DND radar and distrometer measurements

The drop size distribution parameters N_0 and Λ are estimated by the above-mentioned methods. Fig. 7 shows an example of a 1 min differential drop size distribution, where the distrometer data are compared with the radar estimated drop size distribution. In this example, fairly good correlation is obtained.

Fig. 8 shows the correlation of the mean drop size observed by distrometer with that estimated by radar data. Fairly good correlation between these two values exists.

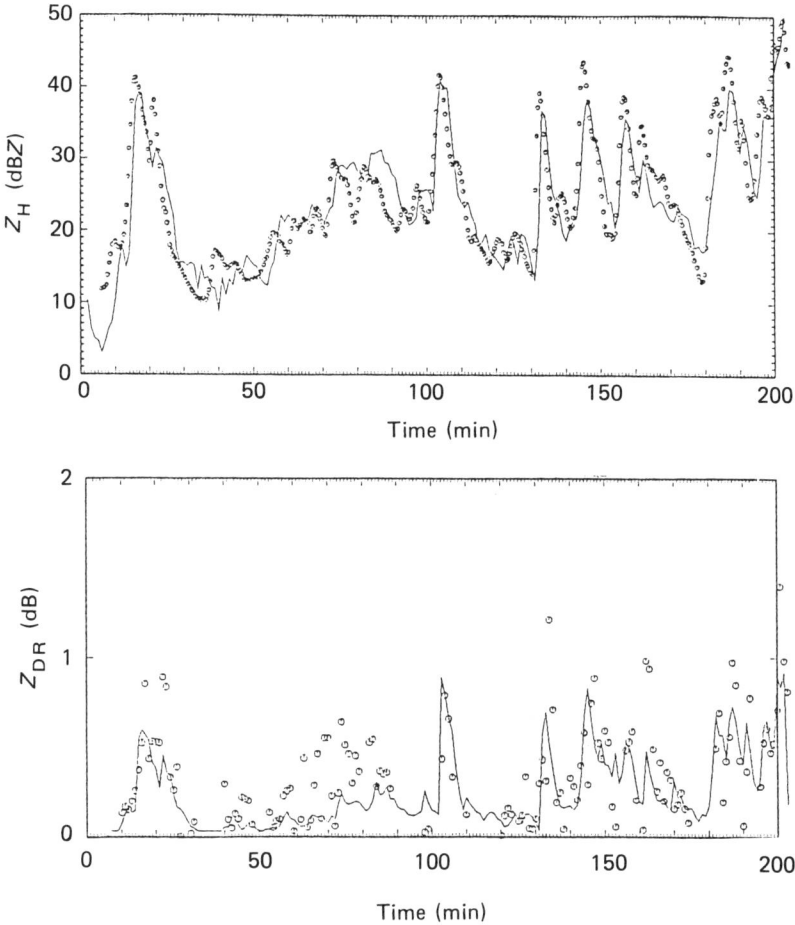

Fig. 5 — Time sequence of radar and distrometer data collected on 5 October 1988: •, radar, 20 s samples; —, distrometer, 1 min samples.

Fig. 9 shows the correlation diagram between radar observed values and distrometer observed values for $Z_H > 30$ dB. Good correlations are recognized between these two measurements. As the Z_{DR} values are almost less then 1.5 dB, an exact conclusion will also be obtained after more intense storms have been observed.

Fig. 10 also shows a correlation diagram of rainfall rate with distrometer measurements and radar estimates. The correlation coefficient is 0.712, but a relatively large scattering of data is shown in Fig. 10.

5.   IDENTIFICATION OF HYDROMETEORS BY DUAL-POLARIZATION RADAR

In the exact prediction of the rainfall rate, an essential step is the identification of hydrometeors both near the ground and aloft. It was pointed out that joint measurements of Z_H and Z_{DR} can provide extra information about the types of

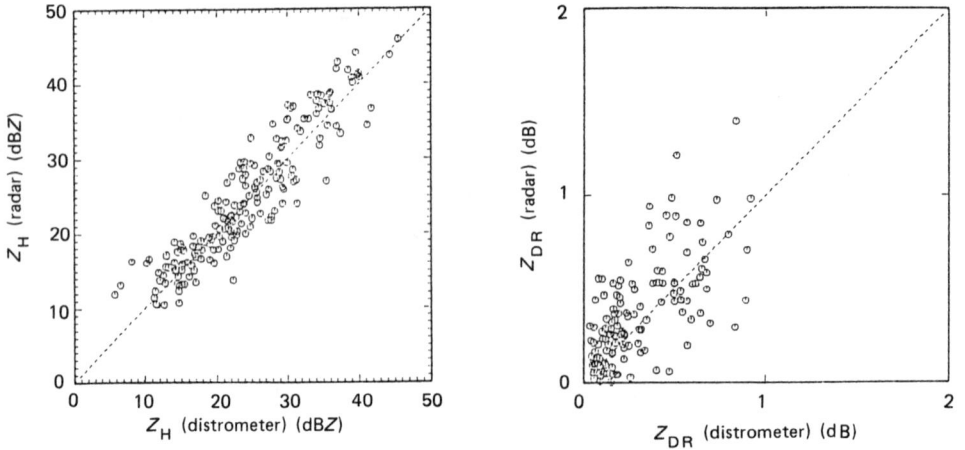

Fig. 6 — Comparison of distrometer and radar values of Z_H and Z_{DR}.

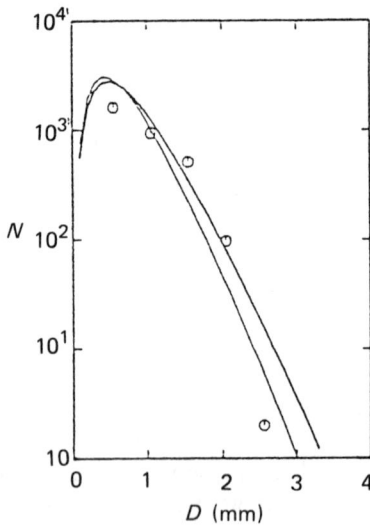

Fig. 7 — An example of comparison between drop size distribution estimated by radar (——)
and value observed by distrometer (○—○).

precipitation particle. This section suggests some results concerning the identification criteria, using bright-band observation data.

Fig. 11 shows an example of the results observed for bright band layers, together
with radiosonde temperature values. Observation was conducted on 11 January
1989. The radiosonde data were observed at Wazima, about 90 km remote from the
radar site. A well-known criterion to separate rain from ice-phase particles in stratus

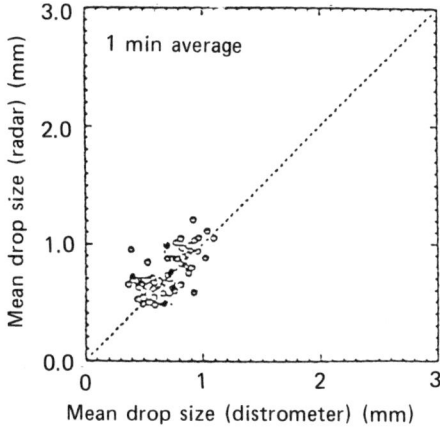

Fig. 8 — Comparison of mean drop size between radar estimate and distrometer measurement.

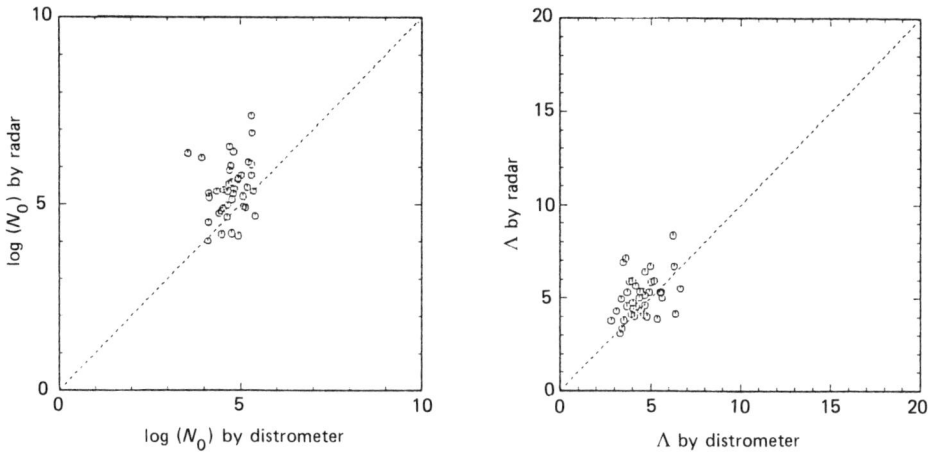

Fig. 9 — Relation between radar and distrometer measurements of N_0 and Λ (data for $Z_H > 30$ dB).

clouds is the location of the melting region which has been identified as a region of high Z_H, typically 300 m below the 0°C isotherm (Battan 1973). Fig. 11 shows the 0°C isotherm at a height of 1500 m above sea level and the highest Z_H value at 1200 m. It is pointed out that a clear feature could be seen in Z_{DR}, which often shows a sharp peak just below the bright band, corresponding to a region where highly oblate melted snowflakes are found just before collapse.

From Fig. 11, it is understood that the clear melting layer (Z_H values of about 40 dB) is formed at a height of around 1.3 km above the ground. At about 300 m below this height the Z_{DR} value has a peak of about 1.1 dB. It was said that dry

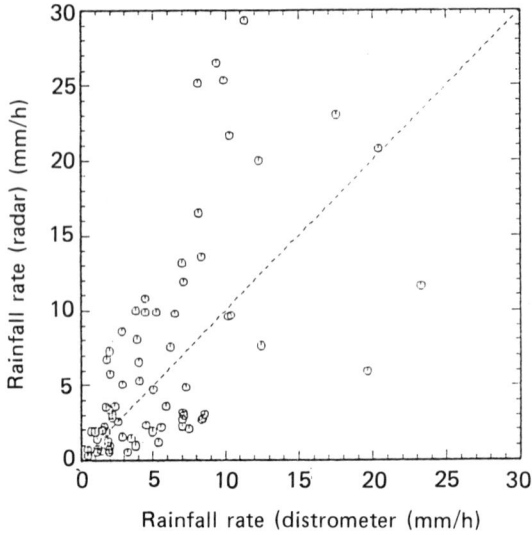

Fig. 10 — Comparison of distrometer and radar values of rainfall rate.

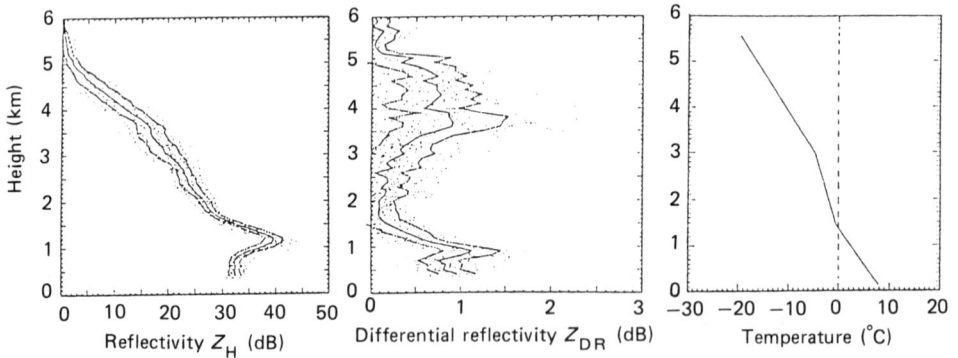

Fig. 11 — Variation in Z_H and Z_{DR} with height. The data were assembled from scanned radar data during about 20 s. Also shown is a temperature profile from a radiosonde ascent at Wazima, 90 km northwest of the radar site (observed from 2111 to 2138 hours on 11 January 1989).

snowflakes above the bright band have relatively low Z_H compared with rain. The low refractive index of snow, which is a low-density ice–air mixture, means that Z_{DR} is always very low, 0.5 dB or less. This coincides with the result of Hall *et al.* (1984).

At a height of about 3.5 km, there is another high Z_{DR} region, but the reflectivity Z_H is not so high. Above this layer, the Z_H and Z_{DR} characteristics suggest a combination of increasing preferred orientation and an increase in particle dielectric constant due to a density increase.

As snowflakes melt to become raindrops, the fall speeds increase by a factor of about 4, and so Z_H reduces by 6 dB from this cause alone. In the example shown in Fig. 11, these general considerations about Z_H and Z_{DR} help to distinguish hydrometeor types.

A likely explanation of the bright band may be offered by reference to Fig. 12,

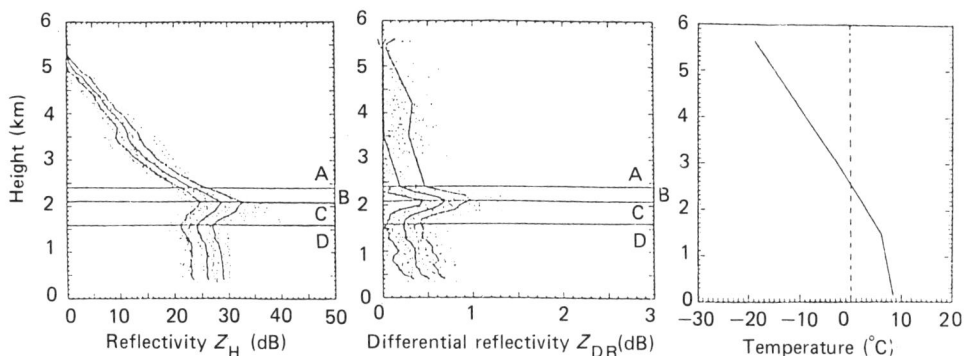

Fig. 12 — Same as Fig. 11 (observed from 0953 to 1020 hours on 19 January 1989).

which shows values of Z_H and Z_{DR}, together with radiosonde temperature values, as a function of height. Although there is no high Z_{DR} value above the bright-band layer in the Fig. 12, the general features of the bright band are the same as in Fig. 11. The various height regions (A–D) can be considered in terms of physical processes as follows.

Region A Z_H increases by 20 dB, suggesting crystal growth by riming or accretion. At the same time, Z_{DR} does not vary widely ranging from -1.0 to 1.0 dB.

Region B The 0°C layer is located at a height of about 2400 m above sea level. Both Z_H and Z_{DR} increase, because of the increased dielectric constant of the water-coated ice particles. Evidently, the particles will have had some degree of preferred horizontal orientation in the region above. This orientation is made clear once the particles become water coated. Z_{DR} takes a positive value, ranging from 0 to 1.0 dB.

Region C Both Z_H and Z_{DR} are now decreasing; the smaller snowflakes have now melted completely and their fall speeds have increased to values appropriate for raindrops, causing a decrease in number density and a decrease in reflectivity.

Region D No further significant changes in Z_H or Z_{DR} occur in the rain, which has an intensity of 1.5 mm/h.

Although even the small amount of radar data presented in this section illustrates many meteorological effects, the use of Z_{DR} in interpreting developments of hydrometeor types in precipitation cells may hold great promise.

6. CONCLUSION

This paper has described the observation characteristics of dual-polarization radar, especially the signal-processing method, the assessment of rainfall measurement simultaneously using four ground-based distrometers and its ability of identification of hydrometeors.

An examination of the statistics of Z_{DR} measurement using the alternate-sampling method showed that the variance $\sigma_{Z_{DR}}$ was expressed by the function of correlation betweeen the horizontal and vertical polarization signals, namely the function of the spectrum width, the pulse repetition frequency, the pulse width and also the number of averaged sample pairs. Therefore, the accuracy of 0.1 dB for the Z_{DR} estimate may only be attained for a Doppler spectrum width of less than 0.5 m/s.

The field observation has compared dual-polarization radar measurements of various rain-related parameters with simultaneous measurements by ground-based distrometers. The comparison problem has been separated into two components:

(1) the estimation of reflectivity and differential reflectivity, which required no assumption to be made concerning the drop size distribution;
(2) the parameters described for drop size distribution.

Good correlation between radar and distrometer measurements of Z_H and Z_{DR} has been obtained, with a small but significant difference in Z_{DR} values. The drop size distribution is most appropriately expressed by the gamma distribution $N(D) = D^m N_0 \exp(-\Lambda D)$ using $m = 2$.

Although even the small amount of radar data on the identification of hydrometeor types in this paper illustrates many meteorological effects, the main points may be summarized by reference to Figs 11 and 12. The dual-polarization radar parameter Z_{DR} clearly distinguishes the melting layer from regions of rainfall. Rain has a wide range of positive Z_{DR} generally correlated with Z_H. The melting layer has generally higher values of Z_{DR} than weak rainfall. $\sigma_{Z_{DR}}$ is also larger in this region than in the rain below.

This study clearly indicates that Z_{DR} may have important applications in the detection of certain types of ice phase hydrometeors, and therefore more exact estimation of the rainfall rate for the hydrologist. Future studies are required to deal with more data from field observation and related hypotheses should be clarified by them. These studies may reveal the specification of the next-generation radar raingauge for operational hydrology.

ACKNOWLEDGEMENTS

The authors would like to express grateful thanks to members of the River Bureau and Telecommunication Division of Ministry of Construction and also to Dr T. Oguti, Radio Research Laboratory, for his kind assistance in using his computer program. This work was performed under the cooperation study with Toshiba Electric Company, Mitsubishi Electric Company and Japan Radio Company Ltd.

REFERENCES

Battan, L. J. (1973) *Radar observation of the atmosphere*, University of Chicago Press, Chicago, IL.

Chandrasekar, V., Bringi, V. N. and Brockwell, P. J. (1985) Statistical properties of dual-polarized radar signal, *Proceedings of the 23rd Conference on Radar Meteorology, 1985*. American Meteorological Society, Boston, MA.

Goddard, J. W. F. and Cherry, S. M. (1984) Quantitative precipitation measurement with dual linear polarization radar. *Proceedings of the 22nd Conference on Radar Meteorology, 1984*. American Meteorological Society, Boston, MA.

Hall, M. P. M., Goddard, J. W. F. and Cherry, S. M. (1984) Identification of hydrometeors and other target by dual-polarization radar. *Radar Radio Sci.*, **19** (1), 1984.

Pruppacher, H. R. and Pitter, R. L. (1971) A semi-empirical determination of the shape of cloud and raindrops. *J. Atmos. Sci.*, **28**.

Sachidananda, M. and Zrnić, D. S. (1985) Z_{DR} measurement considerations for a fast scan capability radar. *Radio Sci.*, **20** (4).

Sachidananda, M. and Zrnić, D. S. (1986) Differential propagation phase shift and rainfall rate estimation. *Radio Sci.*, **21** (2).

Seliga, T. A., Aydin, K. and Direskeneli, H. (1984) Comparison of distrometer-derived rainfall and radar parameter with differential reflectivity radar measurement during MAYPOLE'83. *Proceedings of the 22nd Conference of Radar Meteorology, 1984*. American Meteorological Society, Boston, MA.

Van de Hulst, H. C. (1957) *Light scattering by small particles*. Wiley, New York, pp. 85–102.

Zrnić, D. S (1979) Estimation of spectral moments for weather echoes. *IEEE Trans. Geosci. Electron.*, **GE-17**, 113–128.

Part 5
Precipitation forecasting

30

An advective model for probability nowcasts of accumulated precipitation using radar

T. Andersson
Swedish Meteorological and Hydrological Institute, Noorköping, Sweden

ABSTRACT

The present model selects a source region upwind of the forecast spot. All pixels within the source region are considered to have the same probability of hitting the forecast spot. A pixel hitting the forecast spot is assumed to precipitate there a short time (about 10 min). A drawing is performed, and a frequency distribution of accumulated precipitation during the first time step of the forecast is obtained. A second drawing gives the frequency distribution of accumulated precipitation during the first to second time step, a third that during the first to third time step, and so on, until the end of the forecast period is reached.

1. INTRODUCTION

At present the main operational use of weather radar is subjective nowcasting of precipitation. Even forecasts of accumulated precipitation are often made subjectively, although there are numerical methods (Carpenter and Owens 1981, Austin *et al*. 1986, Walton and Johnson 1986). These models give categorical answers and are mainly advective.

The main difficulty is the rapid changes with time in the reflectivity (rainfall) pattern even for non-convective precipitation. Since at present there are no operational models capable of forecasting the development of precipitating cells on this scale, it seems natural to develop a probabilistic model. As far as I know, only one such model has been described (Zuckerberg 1976). This model builds upon Poisson distribution and only gives a yes–no answer.

In the present work an advective model, giving the probability of accumulated spot precipitation, is developed. An area (the source area) upwind of the forecasting spot is advected over it (we have used the 850 hPa forecasted wind but plan to use a wind measured by the radar itself). The frequency distribution of reflectivities within this area is then computed. The class limits are selected so that there is a doubling of the rain rate for every class increment. The horizontal resolution is 2 km. Using a pre-defined rain rate–reflectivity relation the distribution of accumulated precipitation during a small time interval, the time step, is computed. In the source region the reflectivities are assumed to be randomly distributed. During the advection all pixels within the source region have equal probabilities of hitting the forecast spot. For every time step a drawing is made, and we get the frequency distribution of accumulated precipitation for the first time step, the first to the second time step, the first to the third time step, etc., up to the first to the last time step. The last mentioned is the forecasted probability distribution of accumulated precipitation.

2.   METHOD

2.1   Geometry

The geometry is shown in Fig. 1. A source area upwind of the forecasting spot is selected. As steering wind the 850 hPa wind according to the latest forecast from the

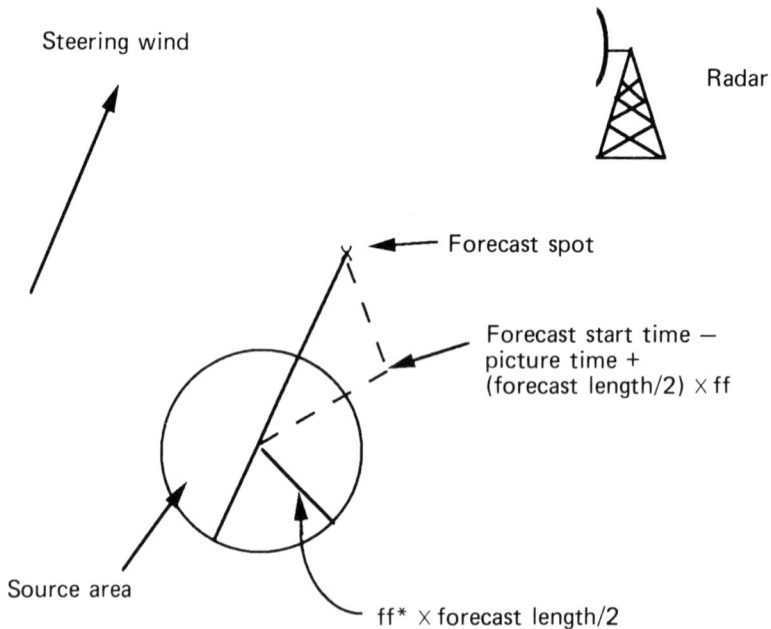

Fig. 1 — Definition of the source area. The area has a minimum radius; see text.

Swedish limited-area model is selected. Since the radar has Doppler capability, it is, however, possible to use the actual wind according to the radar. The wind at a proper level may be obtained from manual interpretation of the radial winds on the plan projection indicator (PPI), or from automatic interpretations as velocity azimuth display or uniform wind technique (Persson and Andersson 1987).

If we denote the wind speed as ff, the radius of the source area is ff×forecast length/2 or, if this is below a minimum value, the minimum value (selected as 20 km). To produce a forecast, at least half the source region should be within the range of the radar.

2.2 Frequency distributions

The equivalent radar reflectivity factor, hereafter called the reflectivity and given in dBZ, is the main parameter. Its frequency distribution within the source area is computed. For practical reasons a minimum dBZ value has to be selected. Reflectivities below this value are considered to give rain rate zero. The class width is selected so that the rainfall rate increases with a factor of 2 for every class increment. The class width is then dependent upon the reflectivity–rain rate relation used. With the Marshall–Palmer relation.

$$Z = 200R^{1.6} \tag{1}$$

the class width becomes 4.8 dBZ (where Z (mm^6/m^3) is the reflectivity and R (mm/h) the rain rate).

We have chosen eight classes, the class number 0 giving rain rate equal to 0 and class number 7 a rain rate of 64 mm/h (class midpoint). Thus we have the frequency distribution

f_rate cl

where cl is the class number, and each cl corresponds to a rain rate.

We can also let cl correspond to an accumulation during a time interval, the time step. The accumulated rain is then

$$cl\_acc = R \times time\ step \tag{2}$$

and we get the frequency distribution

f_rate(cl_acc).

At the forecast spot the frequency distribution of accumulated rain is

f_acc(t, acc)

where t is the number of time steps.

In practice we work with this as a one-dimensional array, since only the result from the last time step is saved. The resolution for acc. is 0.01 mm.

2.3 Computing the expected frequencies of accumulated precipitation

At the start $(t = 0)$

$$f\_acc(0, 0) = 1.0 \text{ and } f\_acc(0, x) = 0.0$$

where $x > 0$, i.e. there is then no accumulated rain.

After a drawing

$$f\_acc(t + 1, \text{acc} + \text{cl\_acc}) = \sum_{cl=0}^{7} \sum_{0}^{max} f\_acc(t, \text{acc}) \times f\_rate(\text{cl\_acc}) \tag{3}$$

and we must remember that different combinations of acc and cl_acc may result in the same acc + cl_acc.

This is then repeated until the end of the forecast period is reached, when the left side of equation (3) is the forecast.

3. FORECASTS TESTED AND THEIR DISPLAY

Several forecast lengths and times of forecast start may be selected and several parameters may be displayed. We have chosen to test forecasts for 1 h accumulated precipitation, starting at the radar image time. We intend to use other start times (+ 1 and + 2 h) but as yet we have not verified such data. For display we have chosen the decentile probabilities for different rain amounts (unit, 0.1 mm) as shown in Table 1. The forecast spots used (Fig. 2) have recording raingauges (resolution, 0.1 mm) so that the forecasts may be verified.

Table 1 — Forecast probabilities of 1 h accumulated precipitation (unit, 0.1 mm) for ten stations (Fig. 2). VER gives the actually measured precipitation. 999 means no measurement (Norrköping; 25th February 1989, 1300; forecast length, 1.0 h; advective wind, 180°, 35 knots; radius, 32 km)

	P_{700}	P_{708}	P_{719}	P_{721}	P_{722}	P_{723}	P_{724}	P_{729}	P_{732}	P_{733}
100	0	0	0	0	0	0	0	0	0	0
90	8	6	1	0	0	0	0	0	4	0
80	11	9	1	0	0	0	0	1	6	1
70	13	9	3	0	0	0	1	3	8	1
60	14	11	4	0	0	0	1	3	9	1
50	16	13	4	0	0	0	3	4	11	3
40	18	14	6	0	0	0	3	6	14	3
30	21	16	8	0	0	0	4	9	16	4
20	23	18	9	0	0	0	6	13	19	4
10	27	22	11	0	0	1	8	16	24	6
VER	999	9	14	0	0	1	4	1	999	6

4. VERIFICATION

Hitherto, forecasts for 19 occasions during three rain events have been made. We have verified them as categorical forecasts, using the 50% probability accumulated

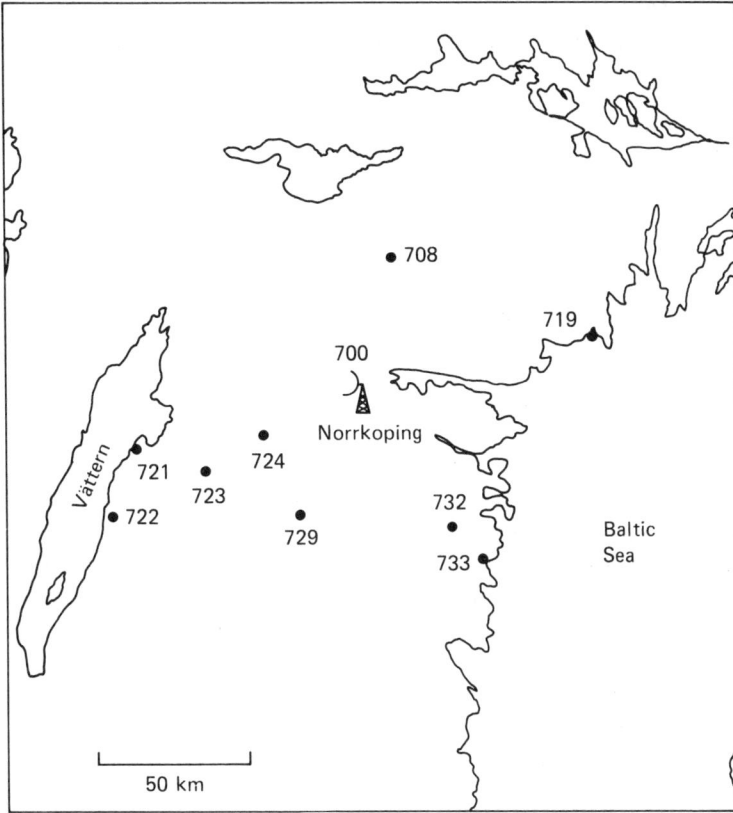

Fig. 2 — Positions of the radar and stations.

sum as the forecast and the limit rain–no rain as 0.1 mm. Thus, in Table 1, a value of 0 means no rain and a value of 1 or more means rain.

The results are shown in Table 2. Some verification parameters are given in Table 3.

Table 2 — Contingency table for forecasts of 1 h precipitation during three rain events (22–27 February 1989) (start time equal to radar image time)

		Forecast	
		Yes	No
O	Yes	74	25
B			
S	No	9	43

Table 3 — Verification parameterrs for forecasts of 1 h
precipitation (start time equal to radar image time)

'Correct', rain–no rain	77%
Probability of detection	75%
False alarm ratio	11%
Critical success index	69%

After having obtained a larger sample, most important other start times and heavier showers, we intend to verify using the Brier score, which is more suitable for probability forecasts.

REFERENCES

Austin, G. L., Kilambi, A., Bellon, A., Leoutsarakos, N., Hausner, A., Trueman, L. and Ivanich, M. (1986) Rapid II, an operational highspeed interactive analysis and display system for intensity radar data processing. *Preprints of the 23rd Conference on Radar Meteorology*, Vol. 3, *Snowmass, Co. 1986*. American Meteorological Society, Boston MA, MP. pp. JP79–JP82.

Carpenter, K. M. and Owen, R. G. (1981) Use of radar network data for forecasting rain. *Proceedings of the COST 72 Workshop–Seminar on Weather Radar, 9–11 March 1981*. European Centre for Medium Range Weather Forecasts and Meteorological Office, Radar Research Laboratory, Bracknell, Berks., pp. 159–182.

Persson, P. O. G. and Andersson, T. (1987) *Automoatic wind field interpretation of Doppler radar radial wind components*, PROMIS Report No. 6, 72 pp. Swedish Meteorological and Hydrological Institute, S-60176 Norrköping.

Walton, M. L. and Johnson, E. R. (1986) An improved precipitation projection procedure for the NEXRAD flash-flood potential system. *Preprints of the 23rd Conference Radar Meteorology*, Vol. 3, *Snowmass, CO, 1986*. American Meteorological Society, Boston, MA, pp. JP62–JP65.

Zuckerberg, F. L. (1976) An application of the binomial distribution to radar observations for short-range precipitation forecasting. *Preprints of the 6th Conference on Weather Forecasting and Analysis, 1976*. pp. 228–231.

31

The combined use of weather radar and mesoscale numerical model data for short-period rainfall forecasting

C. G. Collier
Meteorological Office, London Road, Bracknell, Berks. RG12 2SZ, UK

ABSTRACT

For about a year, very-short-period (up to 6 h ahead) forecasts of rainfall in the UK have been made routinely (but not operationally) using linear extrapolation of the motion of radar echoes. The system within which this is done is known as FRONTIERS. Forecasts for 3–18 h ahead are being made using mesoscale numerical weather prediction models. There is a growing understanding that extrapolation forecasts will benefit from the use of mesoscale model wind fields, and the model forecasts will benefit from incorporation of radar information in the model data assimilation procedure. In this paper we identify likely problems with these approaches and outline how these quite different methods of forecasting rainfall are being implemented in the Meteorological Office.

1. INTRODUCTION

Rainfall results in surface runoff via streams and rivers to lakes or the sea, or surface infiltration. Evaporation from open water returns moisture to the atmosphere. Water which infiltrates the earth's surface may be returned to the atmosphere by evapotranspiration, may be stored as soil moisture for a short time or may enter long-term storage in the water table. This circulation is the hydrological cycle.

The hydrological cycle is not constant, and changes in rainfall produce corresponding changes in surface runoff and infiltration. Meteorologists attempt to understand and forecast rainfall such that hydrologists may have the necessary information they require for water management. We shall be concerned in this paper with forecasting rainfall over periods up to about 24 h ahead, although discussion will concentrate on forecasting up to 12 h ahead. Hence the main concern is for rainfall forecasting for use in hyrological flow forecasting, particularly for high flows. Other non-hydrological uses of such forecasts will not be discussed here.

Jamieson and WIlkinson (1972) have pointed out that, if hydrological forecasts are based only upon current measurements of rainfall, then the implicit assumption made is that no further rainfall will occur after the time that a forecast is made. Such an assumption is clearly possible if forecasts are being made during a rainfall event, and therefore there is an important requirement for rainfall forecasts (see also Central Water Planning Unit 1977). This point has been reinforced by the results of a number of hydrological simulations presented by Labadie *et al.* (1981) and Cluckie and Owens (1987), and observations of a severe flood in London by Haggett (1988) in which two severe rainfall events occurred within hours of each other.

The rainfall measurement accuracy required for hydrological forecasting will depend upon the amount of rainfall. Nemec (1986) suggests accuracies in total storm precipitation amount of 2 mm below 40 mm and 5% above 40 mm with a reporting interval of 6 h and a distinction made between rainfall and solid precipitation. This is a little more stringent at lower totals than the requirement suggested by Salomonson *et al.* (1975) of 3–8 mm for a 40 km^2 river basin. The reporting interval is dependent upon the speed of response of the river to rainfall. The storm total could be generated over periods ranging from an hour or so (thunderstorms) to greater than 1 day (frontal rainfall and tropical storms). This accuracy may not be necessary for all hydrological models, as procedures are now being developed which involve error feedback mechanisms designed to cope with less accurate data (Simpson *et al.* 1980, Cluckie *et al.* 1987).

2. TYPES OF WEATHER FORECASTING

What approaches then are likely to meet this varied requirement for rainfall forecasts?

The method adopted in weather forecasting and the degree of detail achievable depends on the length of the forecast period. Browning (1980) has subdivided short-term forecasting as shown in Fig. 1. The standard approach to forecasting involves the use of supercomputers to solve equations representing the dynamics and thermodynamics of the atmosphere by means of numerical weather prediction (NWP) models. Conventional NWP models, although they may be enhanced by statistical interpretation of the model output, represent explicitly only the larger-scale features of the weather and are best suited to providing forecasts of a general nature for periods beyond 12 h ahead. So-called mesoscale NWP models, representing features of the weather on smaller scales of tens to hundreds of kilometres, are coming into operation, and these provide greater detail for the forecast period a few hours to 18 h ahead. While NWP models have the major advantage that they can be in principle predict the development of new weather systems, they suffer (even the mesoscale models) from an inability to represent variability in cloud, rain and associated parameters on the smallest scales of interest. Thus for forecasts up to a few hours ahead a better forecast is often achievable simply by observing the detailed distribution and movement of weather patterns, particularly rainfall, and assuming that they will continue to travel without change over the very short period concerned. This approach — a full description of the weather now together with extrapolation up to 2 h ahead — is known as nowcasting. Since extrapolation is the basis of nowcasting systems, the quality of the resulting forecasts depends upon the time ahead for which

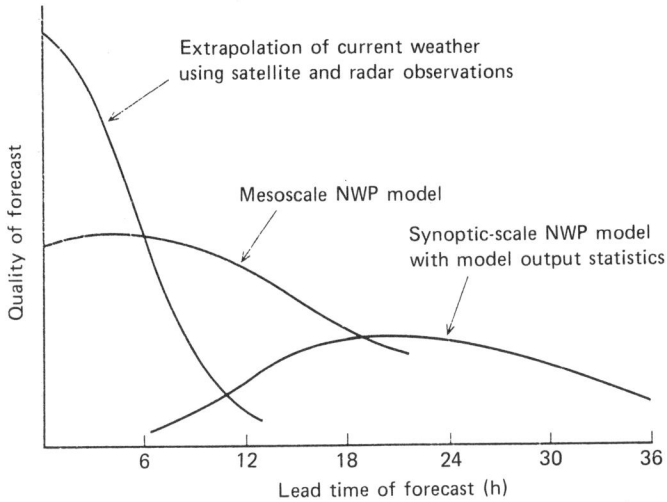

Fig. 1 — The quality of weather forecasts, defined as the product of the accuracy and detail achievable, shown as a function of lead time for three different forecasting methods. The figure is highly schematic and the stage at which the quality of one technique becomes superior to another not only will change over the years with the development of the different methods but also will depend on the particular phenomenon being forecast. (From Browning (1980).)

linear extrapolation is valid. This varies for different weather systems as discussed by Zipser (1983) (see also Doswell 1986). For example, linear extrapolation may be very useful for forecasting the passage of frontal systems many hours ahead but for individual showers may only be useful for up to 20 min or so.

Where linear extrapolation is no longer capable of producing a good forecast because of development or decay, a non-linear forecasting procedure must be applied. However, in the case of a detailed site-specific forecast of intense local phenomena the application of a non-linear model can produce a worse result than linear techniques; it is one thing to predict that conditions are ripe for a new development to occur but quite another to predict precisely when and where they will be triggered, unless there is some well-defined topographical forcing for example.

Although several types of non-linear forecasting procedure have been developed, including purely statistical techniques, the most reliable and generally applicable are procedures based upon numerical models of the atmosphere. Such models may enhance the use of weather radar data in two ways, namely

(a) by providing data with which to aid the extrapolation of radar echoes (wind fields, likely development and precipitation depth and type) and
(b) by using the radar data as an integral part of the model data assimilation procedure defining the initial humidity field. In this paper we shall discuss both these approaches.

3. CURRENT PERFORMANCE OF FORECASTING PROCEDURES

It is difficult to assess objectively the relative performance of the several different methods of extrapolating echo movement, as most of the techniques claim to be successful in a limited sense and have been used in different weather situations. However, Elvander (1976) (see also Alaka *et al.* 1977) has described a series of experiments using three different techniques: a cross-correlation method, tracking individual echoes using a linear least-square extrapolation of the motion of the echo centroid, and a technique involving the tracking of individual echoes by considering the entire echo complex.

It was concluded by Elvander (1976) that the simple cross-correlation model was the most effective when used with zero-tilt (i.e. low-altitude) reflectivity data, but the linear least-squares interpolation of echo centroids was the most effective method when the data on the vertically integrated liquid water content were used. The data used in the experiments were representative only of convective rainfall; no stratiform rainfall cases were considered.

These conclusions were based upon forecasts up to 90 min ahead, using instantaneous pictures at both 10 and 30 min intervals. Forecasts made using input data at 10 min intervals were usually about 10–40% more accurate than forecasts made using data at 30 min intervals.

Hill *et al.* (1977) have demonstrated that a cross-correlation procedure does provide quite successful forecasts up to 6 h ahead for one case of frontal rainfall using data smoothed over a grid length of 20 km. However, both Browning *et al.* (1974) and Hill and Browning (1979) have presented data showing significant differential motion of mesoscale precipitation areas within frontal systems which this technique is not suited to coping with.

The improvement that can be achieved using subjective techniques has been examined by Browning *et al.* (1982), who compared forecasts for a 400 km^2 area made using data from four radars in the UK. The forecasts derived during a total of 29 frontal events between November 1979 and June 1980 were prepared using both a totally objective echo centroid technique (Collier 1981) and subjective linear extrapolation. Fig. 2 shows the percentage error, regardless of sign, in the forecast hourly rainfall for the 400 km^2 area centred on Malvern (located in the middle of the radar network), plotted as a function of lead time. Various errors in the forecasts were identified and partially corrected such that the subjective forecasts were considerably better than the objective forecasts. However, comparable data for totally objective cross-correlation forecasts up to 3 h ahead assessed by Bellon and Austin (1984) are also plotted in Fig. 2 and show that objective procedures may approach the performance of subjective techniques.

The performance of subjective forecasting procedures in operational use during frontal precipitation has recently been confirmed by Brown *et al.* (1988). Satellite data are used to extend the area of coverage provided by the radar network with a concomitant increase in forecast accuracy. The effect of the satellite data depends upon the type of rainfall and so far has only been used to produce extended estimates of rain–no rain areas. Further work is needed to examine the performance achieved operationally in convective rainfall for which showers are of short duration and random organization. When the convection is organized on the mesoscale, it might be hoped that a performance approaching that in frontal rainfall can be achieved.

So far we have discussed the performance of linear extrapolation procedures used with radar data. However, attempts to develop site-specific models for forecasting precipitation using knowledge of orography and atmospheric stability have met with

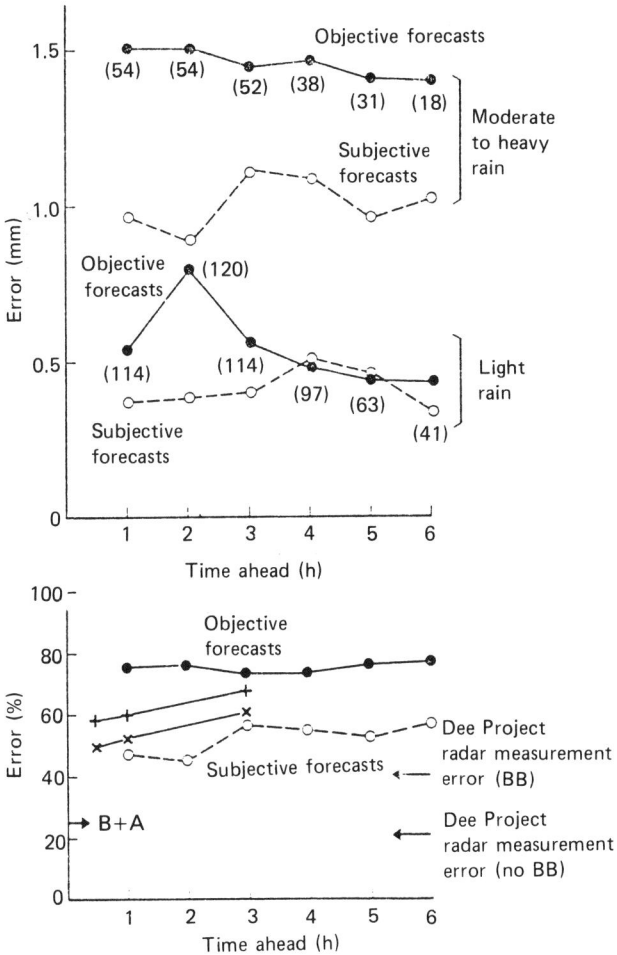

Fig. 2 — (a) The mean error regardless of sign in the forecast hourly rainfall for the 20 km square (400 km²) centred on Malvern, plotted as a function of the lead time of the forecast for light rain (trace, 1 mm/h) and moderate to heavy rain (more than 1 mm/h)): ———, results of objective forecasts (without manual modification); – – –, results of subjective forecast; the numbers in parentheses denote the number of cases. (b) The percentage error for the cases of moderate to heavy rain: ———, results of objective forecasts (without manual modification); – – –, results of subjective forecast; ←, radar measurement errors found in the Dee Weather Radar Project; ×———×, performance of the cross-correlation technique described by Bellon and Austin (1984) for a 600 km² area; +———+, performance of the cross-correlation technique described by Bellon and Austin (1984) for a single point (gauge site); →, B+A, error of the radar estimates of rainfall derived by Bellon and Austin; BB, bright band. Partly from Browning *et al.* (1982).)

some success (Collier 1975, 1977). More sophisticated models incorporating cloud physics have been linked to mesoscale dynamical models to provide rainfall forecasts for many hours ahead. Fig. 3 shows a comparison of daily rainfall forecasts over north Wales made with an operational model and the same model used to drive the detailed orographic model of Bell (1978).

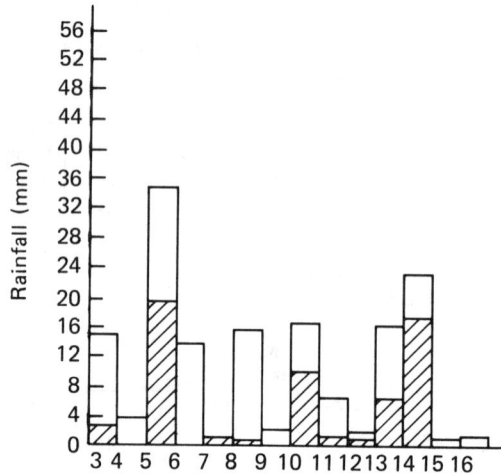

(a)   Day number (October)

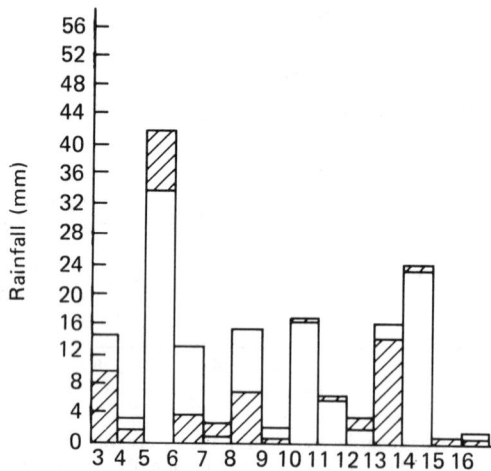

(b)   Day number (October)

Fig. 3 — Comparison of forecast (▨) and observed (from raingauges (▢) rainfall in an area of
south Wales: (a) daily rainfall forecast using operational numerical model with ten levels (▨);
(b) daily rainfall forecast using orographic model (▨). Forecast and estimated actual values of
rainfall are superimposed with common zero on the axis; the arrangement in the vertical dotted
and blank areas depends on which value is the greater. (From Bell (1978).)

In recent years this work has followed to approaches. Rainfall has been modelled
statistically, and the resulting description fitted to existing raingauge observations
(e.g. Sivapalan and Wood 1987). A degree of success is claimed, but there remains
doubt as to the general operational reliability. In parallel with this work the
development of site-specific models for rainfall estimation from meteorological
parameters has continued (e.g. Georgakakos 1986, 1987). Fig. 4 shows the root

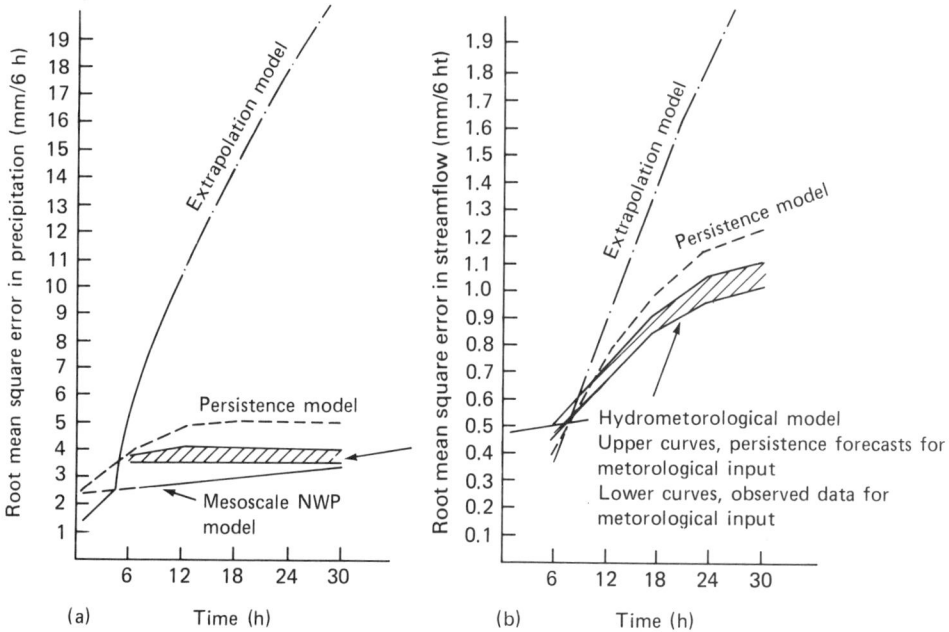

Fig. 4 — (a) Comparison of least-squares performance in precipitation-forecasting models and for various forecast lead times. (b) Comaprison of least-squares performance in streamflow forecasting for three flow models and for various forecast lead times. (Partly from Georgakokes (1986).)

mean square errors in precipitation and streamflow forecasts as a function of lead time achieved to date using this type of model. Note the poor performance of precipitation field extrapolation beyond 6 h. However, extrapolation outperforms persistence and the complex model for streamflow predictions at 6 h ahead and probably at lead times less than 6 h ahead. The current performance of numerical weather prediction models has been added to Fig. 4. This type of forecasting now outperforms other techniques beyond about 6 h ahead (e.g. Gyakum and Samuels 1987). However, up to 3 h ahead, extrapolation forecasts are usually better. Further work is needed to ascertain the best mix of extrapolation and numerical forecasts up to 6 h ahead.

4.   USING NUMERICAL MODEL DATA TO IMPROVE EXTRAPOLATION FORECASTS

It is clear that wind fields obtained from numerical models, particularly mesoscale models having grid lengths of 10–20 km, can provide the means to move radar echoes in quite different and yet spatially consistent directions, and with different speeds. Unfortunately, the echo motion is associated with the wind velocity at a particular height which may vary depending upon the meteorological situation. Fig. 5 shows an example of echo motion compared with motion defined from a mesoscale numerical model. Therefore it is necessary to obtain the depth and type of precipitation from

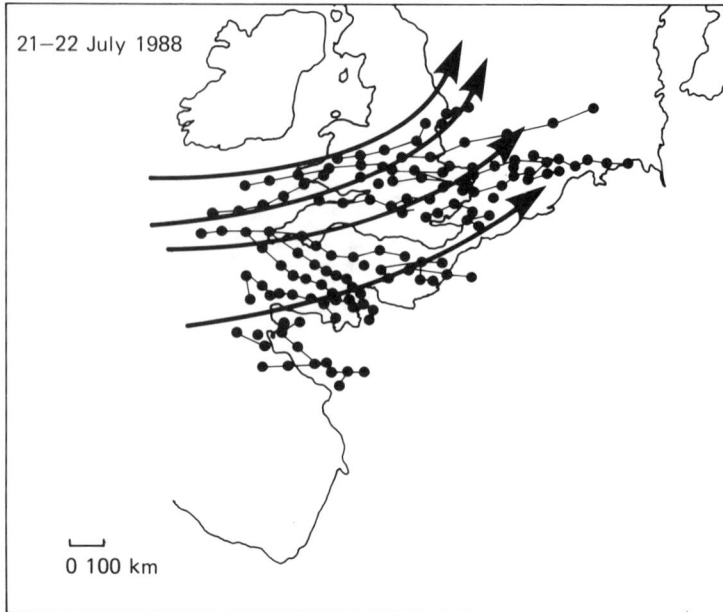

Fig. 5 — Examples of radar echo tracks over northwest Europe derived from radar data from UK. The Netherlands, France and Switzerland. The direction of motion is shown by arrows, and the centroid positions at hourly intervals are indicated by dots. The lines without dots are streamlines at a height of 1010 m forecast by the UK mesoscale model at 1200 Universal Time, Coordinated (UTC), on 21 July 1988.

the model in order to assess the steering level for the radar echoes. It may be necessary to use winds from different heights in different parts of the radar coverage, e.g. across a front. In addition, the errors in the model winds, whilst usually small, could be larger than those derived from echo-matching techniques. Hence it is not obvious that the use of model winds will always produce a better forecast than other methods do, and it may be necessary to modify the wind field subjectively before it is applied to the radar field. More research is needed to clarify the impact of such a technique.

Development of a rainfall pattern can, in principle, also be deduced from numerical forecast models. Methods of recognizing and extrapolating echo development are difficult to implement reliably and therefore must only have limited impact operationally. Mesoscale models may indicate how a rainfall field is likely to develop or decay. Such information can be used by forecasters to modify forecasts derived from radar data subjectively. This could be one of the most effective ways of using numerical model output to aid radar extrapolation procedures.

The blending of numerical model data and objective pattern recognition algorithms must be done in a way which ensures the production of weather forecasts of reproducible accuracy. The use of artificial intelligence techniques, such as expert systems, promises to allow the computer to perform some of the humans high-level judgemental tasks and to lead to the development of forecasting systems which have the required reliability. Work is under way in the UK Meterological Office, as elsewhere, to produce such a system for operational use in mid-1990.

5. THE ASSIMILATION OF RADAR DATA INTO NUMERICAL WEATHER PREDICTION MODELS

So far we have concentrated on the extrapolation of radar echoes, to produce very-short-period forecasts of rainfall. However, radar data may also contribute to the initial data assimilation procedures used with NWP models. Rainfall analyses derived from radar data may improve numerical forecasts for periods very much greater than a few hours ahead. Much of this work is in the early stages of development, but the indications are that significant benefits are likely to result from the mutual interaction of radar and numerical forecast data. Because physical parametrization is complex, the initial data used in integrations must be specified in real time accurately; otherwise the model will deviate rapidly from the observed atmospheric evolution. Likewise, model boundary conditions must also be specified accurately, particularly the lower boundary condition.

Zhang and Fritsch (1986) and Golding (1987) emphasize the importance of both topographic and thermal forcing to the accuracy of predictions and note that correct representation of such forcing requires the accurate specification of the moisture distribution in the atmosphere. This is because the primary response of the atmosphere appears to occur through condensation of water, and the consequent release of latent heat. Support for this comes from numerical experiments reported by Mills (1983), Diallo and Frank (1986), Danard (1987) and Bell and Hammon (1989). Thus a detailed specification of the three-dimensional moisture distribution is required. Given this distribution, the distribution of condensation and consequent latent heating is determined, and applied over a period of integration at the start of the forecast. Recent experiments by Wright and Golding (1989) demonstrate the effect that radar data may have on mesoscale model (grid length, 15 km) forecasts.

The way in which precipitation data are absorbed into the data assimilation is critical, and this is an active area of research. Krishnamurti *et al.* (1988), in developing an assimilation scheme for a global *spectral* model aimed at reducing the spin-up time (the time that a model takes to reach a climatological steady state), proposed the application of a reverse Kuo cumulus parametrization algorithm to providee a modified humidity field. The specific humidity is defined as

$$
q_m(q) = \left(R \Big/ -\frac{1}{g} \int_{p_T}^{p_B} \omega \frac{\delta q}{\delta p} \right) q(p) + \left(\frac{1}{g} \int_{p_T}^{p_B} q \, dp \Big/ \frac{1}{g} \int_{p_T}^{p_B} dp \right) \times
$$
$$
\left[1 - R \Big/ \left(-\frac{1}{g} \int_{p_T}^{p_B} \omega \frac{\delta q}{\delta p} \, dp \right) \right] \tag{1}
$$

where $q(p)$ and $q_m(p)$ are the original and modified values of specific humidity at pressure level p, w is the p vertical velocity, and p_B and p_T denote the bottom and top pressure of a cumulus column.

This relationship has the following properties.

(1) The conservation of moisture gives

$$
\frac{1}{g} \int_{p_T}^{p_B} q_m \, dp \equiv \frac{1}{g} \int_{p_T}^{p_B} q \, dp. \tag{2}
$$

(2) It agrees with the observed rainfall;

$$-\frac{1}{g} \int_{P_T}^{P_B} \omega 1 \frac{\delta q}{\delta p} \delta p \equiv R.$$ (3)

(3) It is consistent with the Kuo scheme (Krishnamuti *et al*. 1984).

The modification to the humidity field is applied only to the regions where

$$-\frac{1}{g} \int_{P_T}^{P_B} \omega 1 \frac{\delta q}{\delta p} \delta p > 0.$$ (4)

Also the modified humidity $q_m(p) \leqslant q_s$, the saturated specific humidity.

It was found that, if a prediction experiment was started some 2 days prior to the initial time, then the spin-up time was considerably reduced. There is a clear need for continuous data assimilation.

A simpler approach has been adopted by Bell and Hammon (1989) for a σ-level model (pressure coordinates). The model relative humidity fields were compared with diagnosed cloud fields. Each model level was amended directly as required to make them compatible with the available observations. It was necessary to estimate the cloud base and top subjectively, and usually only one layer of cloud was catered for. Where cloud had to be removed, rather arbitrary values of 50–70% relative humidities were assigned. Where cloud was present, the model was assigned a value of relative humidity above the threshold used by the radiation scheme according to the same criteria used by that scheme, namely

$$Q = \frac{(U - U_{crit})^2}{(1 - U_{crit})^2}$$ (5)

where Q is the cloud fraction, and U is the relative humidity (where $U > U_{crit}$ and $U_{crit} = 85\%$ threshold relative humidity).

Fig. 6 shows one example of forecasts made using radar data as part of the data assimilation. The effects can be large. However, the extent to which errors in the estimates of rainfall can be assimilated without seriously degrading the forecast accuracy remains to be investigated. For example equation (3) indicates that agreement with the modified humidity field is always forced. This may not be good, and it may be necessary to use some quality procedure to weight the input data according to their quality.

6. COPING WITH RAINFALL FORECAST ERRORS IN HYDROLOGICAL FORECAST MODELS

As soon as digital radar data became available in near real time the prospect of accurate, very detailed precipitation information over wide areas stimulated hydrologists to input these data to flow-forecasting models. Some early assessments (Anderl *et al*. 1976, Barge *et al*. 1979) were optimistic, and hydrograph simulations were discribed showing considerable improvement over hydrographs derived using

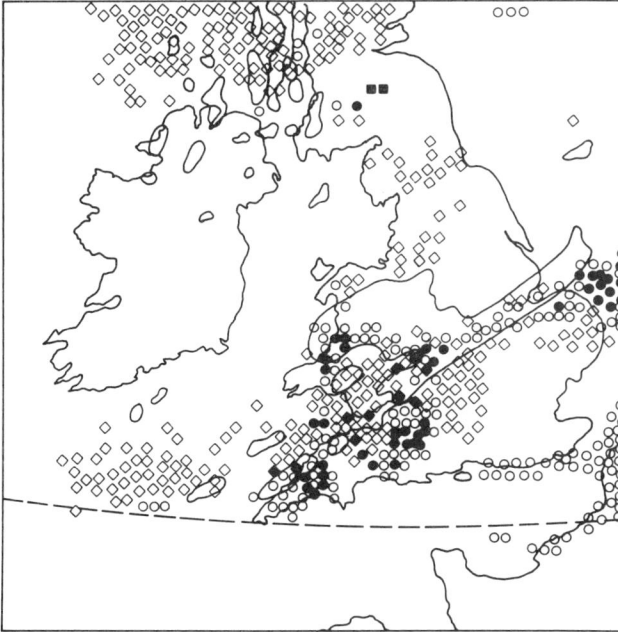

Fig. 6 — 6 h forecast precipitation for 1200 Greenwich Mean Time (GMT) on 15 February 1989: ·, rain, more than 0.05 mm/h; ○ rain, more than 0.1 mm/h, ● rain, more than 0.5 mm/h; ×, snow (rain equivalent) 0.05 mm/h; ■, snow (rain equivalent), more than 0.5 mm/h; ◇, showers (local rate) more than 0.4 mm/h; ◆, showers (local rate), more than 2 mm/h. The observed radar echoes are outlined. (After Wright and Golding (1989).)

raingauge data alone. However, others (Gorrie and Kouwen 1977) noted that generally very little improvement was evident using radar data in uniform rain, although a clear advantage was evident in convective rainfall. The effects of errors in radar measurements of precipitation were highlighted by Collier and Knowles (1986), and Fig. 7 shows examples of good and bad simulations.

Flow forecasting in urban basins should provide the opportunity to demonstrate the benefit of the detail in the precipitation field provided by radar data. Austin and Austin (1974) noted that the speed and direction of storm travel in an urban basin are as important as the total rainfall accumulation, although Roberts (1987) stressed the difficulty of using radar data which were not adjusted accurately in real time. On occasions, therefore, in these situations also radar data were found to provide good hydrograph simulations, but on other occasions quite bad simulations.

The explanation for the differences in performance measured using radar data becomes easier to understand when one considers changes in runoff brought about by changes in precipitation. In a wet river basin in the USA a 10% change in precipitation produces 10–30% change in runoff (Nemec 1985). Clearly then, if radar data or any other data are inaccurate and are used to derive the precipitation input to a model, the error may translate to a substantially larger percentage runoff error. Since the mean errors in real-time radar estimates of rainfall over river basins are around 20–30% (Collier 1986a), it is not surprising that the use of radar data may

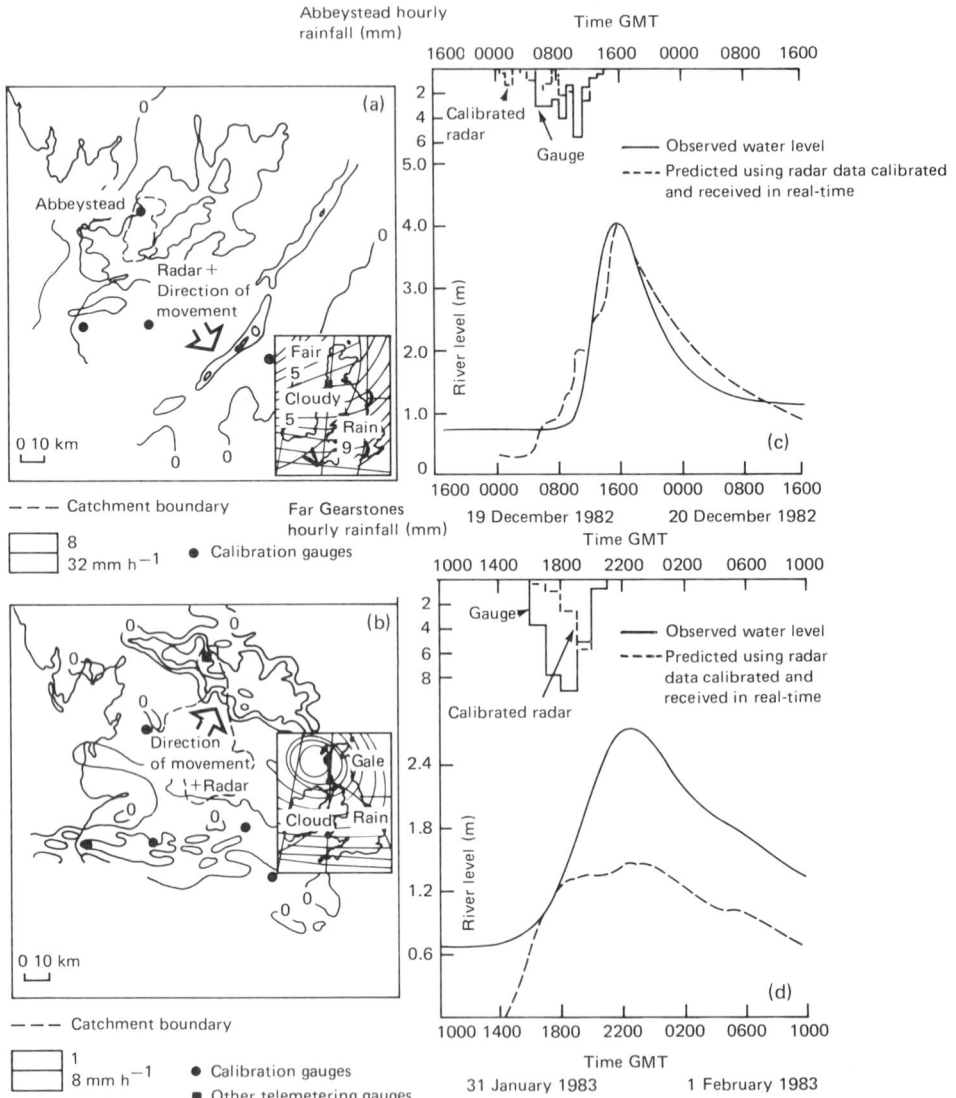

Fig. 7 — (a) The instantaneous rainfall at 1232 GMT on the 19 December 1982 in northwest England. The catchment boundary of the River Wyre is shown, and the location of the radar site with the locations of the telemetering raingauges used to calibrate the radar in real time. The synoptic situation close to this time is also shown. A cold front is located over northwest England. (b) The hourly rainfall totals given by the Abbeystead gauge and the radar, and river hydrographs as observed at the bottom of the catchment, and as predicted using real-time-gauge-adjusted radar data. (c), (d) As (a), (b) for the Ribble catchment at 1819 GMT on 31 January 1983. A warm front has moved over northern England, leaving the area within the warm sector of a depression. The gauge data in (b) are for the Far Gearstones gauge, indicated by a cross in (a). (From Collier and Knowles (1986).)

sometimes result in inaccurate hydrograph simulations. What is often not appreciated is that the use of sparse telemetering raingauge networks may have an even greater detrimental effect (Collier 1986b).

The problem of inaccurate input is compounded when applying precipitation forecasts which are likely to have even larger errors. Cluckie and Owens (1987) present the results of streamflow forecasts made using several short-period (up to 6 h ahead) rainfall-forecasting scenarios including one based upon the extrapolation of radar data. Examples are shown in Fig. 8. For most of the forecasts the extrapolation technique outperforms the other scenarios, but on one occasion the forecast is wildly inaccurate.

Collier and Knowles (1986) point out that, only if ways can be found for dealing with errors induced by errors in radar estimates of precipitation, will radar attain its potential in hydrological forecasting. Fortunately, self-correcting hydrological models are being developed for operational use. These models monitor flow simulation accuracy and use measured error characteristics to improve model performance (e.g. Kitanidis and Bras 1980a,b, Simpson *et al.* 1980, Moore 1982).

7. PROSPECTS FOR THE FUTURE

If radar data are to realize their full potential within very-short-period rainfall-forecasting systems, then this author believes that it will be necessary both to use mesoscale numerical model data to aid extrapolation of radar echoes and to employ the radar data within the model data assimilation procedures. This will require the operational availability of numerical products (wind field, and depth of rainfall fields) derived using radar and satellite data.

In the UK a mesoscale model, having a 15 km grid, will become available operationally during 1990. Work is under way to develop a method for using model data with the FRONTIERS system. This involves the deployment of a front-end microprocessor within which radar echo motion derived from a cross-correlation pattern matching technique will be compared with the model wind fields at various levels (Fig. 9). The model wind field so selected, which may use winds from different model levels in different areas, will be used to extrapolate the radar echo motion. This selection procedure may involve a simple form of expert system.

Whilst it is expected that this work will lead to more reliable forecasts of rainfall, a number of problems will remain. Perhaps most important of these will be the forecasting of convective rainfall particularly from thunderstorms. Many countries regard this problem as the most challenging in mesoscale meteorology, and several countries, e.g. the USA and China as well as the UK, have specified programmes to use expert system technology as a basis for future developments. Doppler weather radar giving measurements of wind speed and turbulence spectrum width offer new possibilities in combination with both numerical model and conventional meteorological data. The challenge is to use the data that Doppler radars can provide in ways which are useful to the operational forecaster. The UK are installing during 1991 a prototype low-power Plessey Doppler radar at the Wardon Hill site in central southern England. The aim is to try to meet this operational challenge.

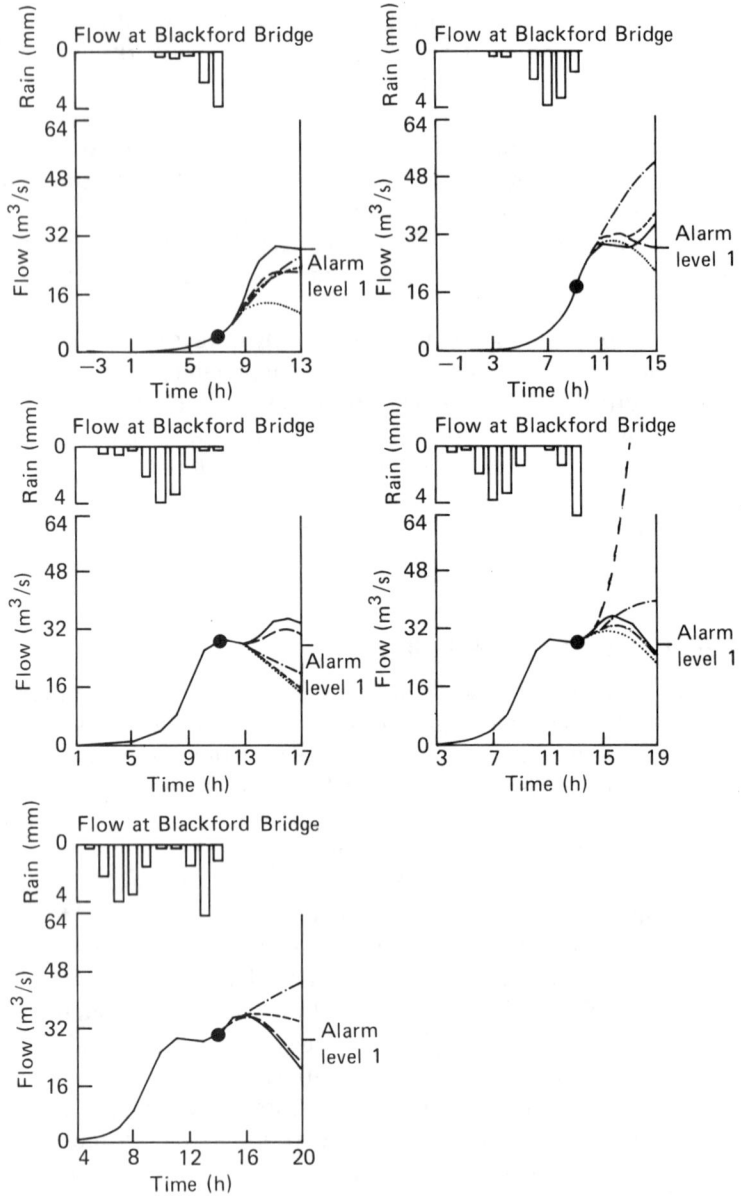

Fig. 8 — Forecast sequence for 29 January 1985 showing various future rainfall scenarios. FRONTIERS is the operational forecast system using radar network data being developed in the UK Meteorological Office: ●, time of forecast; ———, actual flow forecasts; – – –, no more rain; - - -, perfect foresight; —·—, average past rain; — — —, FRONTIERS. (From Cluckie and Owens (1987).)

Finally, in recent years there has been an increasing awareness of the need for the development of fully integrated systems linking operational meteorological forecasting models to hydrological models (Georgakakos and Kavvas 1987). In addition, it is

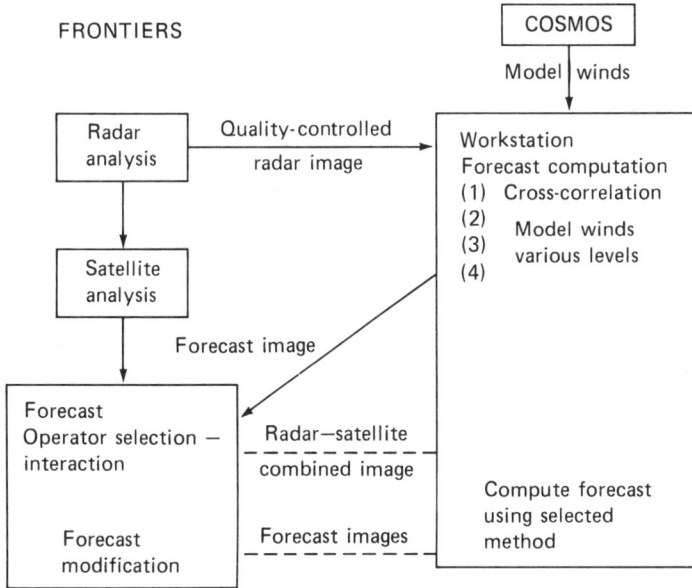

Fig. 9 — Use of workstation to automate the computation of the velocity field in the forecast system of the FRONTEIRS system. Mesoscale model winds are obtained from the large mainframe computer system COSMOS via a local area network. These winds are used with echo-motion vectors computed from pattern matching procedures.

recognized that reliable flood predictions are only useful if they can be disseminated rapidly to warn those who may be at risk from floods. The development of communications systems and warning procedures must proceed at the same pace as technological advances.

REFERENCES

Alaka, M. A., Charba, J. P. and Elvander, R. C. (1977) *Thunderstorm prediction for use in air traffic control*, Report No. FRA-RD-77-40. US Department of Transportation, Federal Aviation Administration, Washington, DC, 32 pp.

Anderl, B., Attmannspacher, W. and Schultz, G. A. (1976) Accuracy of reservoir inflow forecast based on radar rainfall measurements. *Water Resour. Res.*, **12** (2), 217–223.

Austin, G. L. and Austin, L. B. (1974) The use of radar in urban hydrology. *J. Hydrol.* **22,** 131–142.

Barge, B. L., Humphries, R. G., Mah, S. J. and Kuhnke, W. K. (1979) Rainfall measurements by weather radar: applications to hydrology. *Water Resour. Res.*, **15** (6), 1380–1386.

Bell, R. S. (1978) The forecasting of orographically enhanced rainfall accumulations using 10-level model data. *Meteorol. Mag.*, **107**, 113–124.

Bell, R. S. and Hammon, O. (1989) The sensitivity of fine-mesh rainfall forecasts to changes in the initial moisture fields. *Meteorol. Mag.*

Bellon, A. and Austin, G. L. (1984) The accuracy of short-term rainfall forecasts. *J. Hydrol.*, **70,** 35–49.

Brown, R., Sargent, G. P. and Blackall, R. M. (1988). *An assessment of FRON-TIERS forecasts for the winter of 1987–88*, Internal Working Paper No. SCWSPF 17/10 May. Meteorological Office, Bracknell, Berks.

Browning, K. A. (1980) Local weather forecasting. *Proc. R. Soc. London, Ser. A*, **371**, 179–211.

Browning, K. A., Collier, C. G., Larke, P. R., Menmuir, P., Monk, G. A. and Owens, R. G. (1982) On the forecasting of frontal rain using a weather radar network. *Mon. Weather. Rev.*, **110,** 534–552.

Browning, K. A., Hill, F. F. and Pardoe, C. W. (1974) Structure and mechanism of precipitation and the effect of orography in a wintertime warm sector. *Q. J. R. Meteorol. Soc.*, **100,** 309–330.

Central Water Planning Unit (1977) *Dee Weather Radar and Real Time Hydrological Forecasting Project*, Report by the Steering Committee. Central Water Planning Unit, Reading, Berks. 172 pp.

Cluckie, I. D., Ede, P. F., Owens, M. D., Baily, A. C. and Collier, C. G. (1987) Some hydrological aspects of weather radar research in the United Kingdom. *Hydrol. Sci. J.* **32** (319) 329–346.

Cluckie, I. D. and Owens, M. D. (1987) *Real-time rainfall runoff models and use of weather radar information*. In: U. K. Collinge and C. Kirby (eds), *Weather radar and flood forecasting*. Wiley, Chichester, West Sussex, pp. 171–190.

Collier, C. G. (1975) A representation of the effects of topography on surface rainfall within moving baroclinic disturbances. *Q. J. R. Meteorol. Soc.*, **101,** 407–422.

Collier, C. G. (1977) The effect of model grid length and orographic rainfall 'efficiency' on computed surface rainfall. *Q. J. R. Meteorol. Soc.*, **103,** 247–253.

Collier, C. G. (1981) Objective rainfall forecasting using data from the United Kingdom weather radar network. *Proc. IAMAP Symp*. Hamburg, 25–28 August, Special Publication No. ESA SP-165, 201–206.

Collier, C. G. (1986a) Accuracy of rainfall estimates by radar, Part I: Calibration by telemetering raingauges. *J. Hydrol.*, **83,** 207–223.

Collier, C. G. (1986b) Accuracy of rainfall estimates by radar, Part II: Comparison with raingauge network. *J. Hydrol.*, **83,** 225–235.

Collier, C. G. and Knowles, J. M. (1986) Accuracy of rainfall estimates by radar, Part III: Application for short-term flood forecasting. *J. Hydrol.*, **83,** 237–249.

Danard, M. (1987) On the use of satellite estimates of precipitation in initial analyses for numerical weather prediction. *Atmos.-Ocean.*, **23** (1), 23–42.

Diallo, N. T. and Frank, W. M. (1986) Effects of enhanced initial moisture fields on simulated rainfall over west Africa and the east Atlantic. *Mon. Weather Rev.*, **114,** 1811–1821.

Doswell, C. A. III. (1986) Short-range forecasting. In: P. Ray (ed.) *Mesoscale meteorology*. American Meteorological Society, Boston, MA, Chapter 29, 689–719.

Elvander, R. C. (1976) An evaluation of the relative performance of three weather radar echo forecasting techniques. *Preprints of the 17th Conference on Radar Meteorology, Seattle, WA, 1976.* American Meteorological Society, Boston, MA, 526–532.

Georgakakos, K. P. (1986) A generalized stochastic hydrometeorological model for flood and flash-flood forecasting. 2. Case studies. *Water Resour. Res.,* **22** (13), 2096–2106.

Georgakakos, K. P. (1987) Precipitation analysis, modeling, and prediction in hydrology. *Rev. Geophys.,* **25** (2), 163–178.

Georgakakos, K. P. and Kavvas, M. L. (1987) Precipitation analysis, modeling, and prediction in hydrology. *Rev. Geophys.,* **25** (2), 163–178.

Golding, B. W. (1987) Strategies for using mesoscale data in an operational mesoscale model. *Preprints of the Workshop on Satellite and Radar Imagery Interpretation, Reading, Berks., 20–24 July 1987, EUMETSAT.* 341–364.

Gorrie, J. E. and Kouwen, N. (1977) Hydrological applications of calibrated radar precipitation measurements, *Reprints of the 2nd Conference on Hydrometeorology, Toronto, Ontario, 25–27 October 1977.* American Meteorological Society, Boston, MA, 272–279.

Gyakum, J. R. and Samuels, K. J. (1987) An evaluation of quantitative and probability-of-precipitation forecasts during the 1984–85 warm and cold seasons. *Weather Forecast.* **2,** 158–168.

Haggett, C. M. (1988) Thunderstorms over north-west London — 8 May 1988. *Weather,* **43** (7), 266–267.

Hill, F. F. and Browning, K. A. (1979) Persistence and orographic modulation of mesoscale precipitation areas in a potentially unstable warm sector. *Q. J. R. Meteorol. Soc.,* **105,** 57–70.

Hill, F. F., Whyte, K. W. and Browning, K. A. (1977). The contribution of a weather radar network to forecasting frontal precipitation; a case study. *Meteorol. Mag.,* **106,** 69–69.

Jamieson, D. G. and Wilkinson, J. C. (1972) River Dee research program, 3: A short-term control strategy for multipurpose systems. *Water Resour. Res.,* **8** (4), 911–920.

Kitanidis, P. K. and Bras, R. L. (1980a) Real-time forecasting with a conceptual hydrologic model, 1, Analysis of uncertainty. *Water Resour. Res.,* **16** (6), 1025–1033.

Kitanidis, P. K. and Bras, R. L. (1980b) Real-time forecasting with a conceptual hydrologic model, 2, Applications and results. *Water Resour. Res.,* **16** (6), 1034–1044.

Krishnamurti, T. N., Bedi, H. S., Heckley, W. and Ingles, K. (1988) Reduction of the spinup time for evaporation and precipitation in a spectral model. *Mon. Weather Rev.,* **116,** 907–920.

Krishnamurti, T. N., Ingles, K., Cocke, S., Pasch, R. and Kitade, T. (1984) Details of low latitude medium range numerical weather prediction using a global spectral model II. Effect of orography and physical initialization. *J. Meteorol. Soc. Jpn.,* **62,** 613–649.

Labadie, J. W., Lazaro, R. C. and Morrow, D. M. (1981) Worth of short-term forecasting for combined sewer overflow control. *Water Resour. Res.,* **17** (5), 1489–1497.

Mills, G. A. (1983) The sensitivity of a numerical prognosis to moisture detail in the initial state. *Aust. Meteorol. Mag.,* **31,** 111–119.

Moore, R. J. (1982) Transfer functions, noise predictors and the forecasting of flood events in real-time. In: V. P. Singh (ed.), *Statistical analysis of rainfall and runoff.* Water Resource Publishers, CO, 229–250.

Nemec, J. (1985) Water resource systems and climate change. In: J. C. Rodda (ed.), *Facets of hydrology,* Vol. II. Wiley, Chichester, West Sussex, 131–152.

Nemec, J. (1986) *Hydrological forecasting. Design and operation of hydrological forecasting systems.* Reidel, Dordrecht, 239 pp.

Roberts, G. K. (1987) The use of radar rainfall data in urban drainage models in Manchester. *Public Health Eng.,* **14** (6), 61–64.

Salomonson, V. V., Ambaruch, R., Rangor, A. and Ormsby, J. P. (1975) Remote sensing requirements as suggested by watershed model sensitivity analyses. *Proceedings of the International Symposium on Remote Sensing of Environment,* Vol. 2, Ann Arbor, MI, *6–10 October 1975,* Environmental Research Institute, MI, pp. 1273–1284.

Simpson, R. J., Wood, T. R. and Hamlin, M. J. (1980) Simple self-correcting models for forecasting flows on small basins in real time. *Proceedings of the Oxford Symposium on Hydrological Forecasting,* IAHS Publication No. 129. International Association of Hydrological Sciences, pp. 433–444.

Sivapalan, M. and Wood, E. F. (1987) A multidimensional model of nonstationary space-time rainfall at the catchment scale. *Water Resour. Res.,* **23** (7), 1289–1299.

Wright, B. J. and Golding, B. W. (1989) The impact of radar and satellite imagery in a mesoscale NWP system. *Preprints of the COST-73 International Seminar on Weather Radar Newtorking, 5–8 September 1989.* Commission of the European Communities, Brussels.

Zhang, D.-L. and Fritsch, J. M. (1986) A case study of the sensitivity of a numerical simulation of mesoscale convective system to varying initial conditions. *Mon. Weather Rev.,* **114,** 2418–2431.

Zipser, E. J. (1983) Nowcasting and very-short-range forecasting. In: *National STORM program: scientific and technical bases and major objectives.* United Corporation of Atmospheric Research, Boulder, CO, 6.1–6.30.

32

On the evaluation of radar rainfall forecasts

T. Denoeux[1], T. Einfalt[2] and G. Jaquet[2]
[1]Cergrene Ecole Nationale des Ponts et Chausées, La Courtine, 93167 Noisy-le-Grand Cédex, France
[2]RHEA 1 rue Albert Einstein, Champs-sur-Marne, 77436 Marne le Vallee Cédex 2, France

ABSTRACT

In order to compare different forecasting techniques, or to determine their reliability under various meteorological conditions, a quality criterion has to be used. In this paper, criteria reported in the literature on radar rainfall forecasting are reviewed and compared using a set of 149 forecasts. The results of this comparison demonstrate the necessity to define a quality criterion in connection with the use of the forecasts: this is done in the case of the real-time control of the sewer network in the Seine–Saint-Denis county, near Paris.

1. INTRODUCTION

Forecasts of such dynamic phenomena as meteorological ones cannot generally be expected to be perfect (Einfalt and Denoeux 1991). In the case of very-short-term rainfall forecasts obtained from the interpretation of radar data, error sources fall into three main categories.

(1) *Measurement errors*   Rainfall-forecasting methods use radar rainfall measurements in the form of plan position indicator (PPI) or constant-altitude plan position indicator (CAPPI) images. The different types of measurement errors (e.g. Wilson and Brandes 1979) account for part of the total forecast error.
(2) *Errors in the estimation of rainfall field evolution*   Radar images are used in forecasting to estimate the displacement of rainfall areas (Collier 1976, Austin 1985, Einfalt 1988). This estimation is always based on some assumptions, e.g. the uniformity of displacement for cross-correlation methods (Austin and Bellon 1974), or a slow variation in the characteristics of rainfall areas for pattern

recognition methods (Einfalt 1988, Einfalt *et al.* 1991). The validity of these assumptions is not guaranteed under all meteorological conditions (Denoeux 1989).

(3) *Extrapolation errors*   Once the displacement of rainfall areas has been estimated, the next step is its extrapolation. Phenomena such as cells growth or decay, non-linearity of cells trajectories, variation in speed, birth of new rainfall areas, etc., have not, up to now, successfully been taken into account (e.g. Tsonis and Austin 1981.)

From this brief analysis, it is evident that the forecast skill depends not only on the forecasting technique but also on the radar and signal-processing hardware and software and, above all, on some characteristics of the meteorological situation, such as atmospheric stability, size and duration of rainfall areas, etc.

For that reason, one often has to compare the results obtained by different forecasting methods (Elvander 1976, Ciccione and Pircher 1984, Carpenter and Owens 1981, Tsonis and Austin 1981, Einfalt 1988), or the results obtained by one method under various meteorological conditions (Austin and Bellon 1974, Denoeux *et al.* 1989, Denouex 1989). To do such comparisons a quality criterion has to be selected, which is the problem addressed in this paper.

2.   REVIEW OF QUALITY CRITERIA USED FOR THE EVALUATION OF RADAR RAINFALL FORECASTS

Radar rainfall-forecasting techniques have the ability to provide the user with a variety of information, which include a forecasted displacement between times t_0 and $t_0 + \Delta t$, a forecasted rainfall intensity field at time $t_0 + \Delta t$, and forecasted hyetographs during $[t_0, t_0 + \Delta t]$, on some zones of interest. As a consequence of this diversity, researchers have used many different quality criteria to evaluate rainfall forecasts. These criteria can be classified according to the type of result on which they are based: displacements, rainfall fields or hyetographs.

2.1   Criteria based on displacement
The absolute difference (ΔV) and relative difference ($\Delta V/V$) between observed displacement V_O and forecasted displacement V_F, between times t_0 and $t_{0+\Delta t}$, are good examples of criteria in this category (Austin and Bellon 1974):

$$\Delta V = \|V_O - V_F\| , \qquad \frac{\Delta V}{V} = \frac{\Delta V}{\|V_O\|} .$$

Austin and Bellon found, for 32 forecasts from 30 min to 120 min ahead, a mean $\Delta V/V$ of 12.8%, the values ranging 0 to 49%.

2.2   Criteria based on rainfall fields
These criteria were mainly used by meteorologists (Bellon and Austin 1978), but also by some hydrologists (Einfalt 1988). Most frequently used is the critical success index CSI defined as

$$\mathrm{CSI} = \frac{a}{a+b+c}$$

where a, b and c are the numbers of pixels in the areas indicated in Fig. 1.

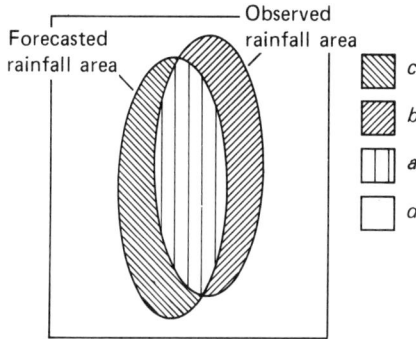

Fig. 1 — Definition of a, b, c and d for the calculation of CSI and RI.

Ciccione and Pircher (1981) used an index based on the same idea, called the Rousseau index RI and defined as

$$\mathrm{RI} = \frac{4ad - (b+c)^2}{(2a+b+c)(2d+b+c)}$$

where a, b and c have the same meaning as above, and d is the number of pixels where rainfall was neither forecasted nor observed.

These criteria equal 0 for very poor forecasts, and 1 for perfect forecasts.

2.3 Criteria based on hyetographs

Criteria in this third category have been used to evaluate rainfall forecasting techniques in the context of hydrological applications, e.g. by Huff *et al.* (1980), Carpenter and Owens (1981), Tsonis and Austin (1981), Damant *et al.* (1983) and Bellon and Austin (1984). All these workers used the average of absolute (ΔH difference) and relative difference ($\Delta H/H$) between observed rainfall depth $(H_\mathrm{O})_i$ and forecasted rainfall depth $(H_\mathrm{F})_i$ on a set of n areas:

$$\Delta H = \frac{1}{n} \sum_{i=1}^{i=n} \left| (H_\mathrm{O})_i - (H_\mathrm{F})_i \right|$$

$$\frac{\Delta H}{H} = \frac{1}{n} \sum_{i=1}^{i=n} \frac{\left| (H_\mathrm{O})_i - (H_\mathrm{F})_i \right|}{(H_\mathrm{F})_i} \ .$$

As an example of results with $\Delta H/H$, Huff *et al.* (1980) and Bellon and Austin (1984) reported, from experiments in the Chicago and Montreal regions respectively, mean relative errors of the order of 60%, for forecasts 1 h ahead (raingauges providing the reference values).

3.   COMPARISON OF MOST FREQUENTLY USED QUALITY CRITERIA

Before addressing the problem of a quality criterion selection, one should first consider the following questions.

(1) If a forecast f_1 is better than a forecast f_2, according to some criterion, what is the probability that it is worse according to another criterion?
(2) Does the result of a comparison between forecasting techniques depend on the choice of a quality criterion?

As an attempt to answer these questions, we selected a set of 149 30 min periods, taken from various hydrologically significant rainfall events, and for which PPI images from the Trappes radar (located near Paris), with the following characteristics, were available: space resolution, $1.6 \, km \times 1.6 \, km$; one image every 15 min; 16 reflectivity levels. For each of these periods, forecasts 30 min ahead were simulated by two methods: firstly a simple cross-correlation method (CROS) and secondly, the SCOUT method, based on advanced pattern recognition techniques (Einfalt and Denoeux 1987a, b, Einfalt 1988).

The forecasts were then evaluated by the six criteria defined in section 2: ΔV, $\Delta V/V$, $- CSI$, $- RI$, ΔH, $\Delta H/H$ (the signs of CSI and RI were changed so that all criteria are decreasing functions of forecast quality) For all these evaluations, radar measurements served as a reference.

In the first step, we calculated, for each pair of criteria (Q_1, Q_2), the percentage P of pairs (f_1, f_2) of forecasts (performed by CROS), verifying that

$$[Q_1(f_1) \geqslant Q_1(f_2) \text{ and } Q_2(f_1) < Q_2(f_2)]$$

or

$$[Q_1(f_1) < Q_1(f_2) \text{ and } Q_2(f_1) \geqslant Q_2(f_2)] \ .$$

Thus, P, represents the percentage of cases when criteria Q_1 and Q_2 gave different relative evaluations of forecast quality.

As indicated in Table 1, P is relatively small for two criteria in the same category (from 13% to 25%), but it is very large (from 40% to 60%) for two criteria in different categories.

In the second step, we tried to find out to what extent the results of a comparison between the forecasting skills of CROS and SCOUT methods could be influenced by the choice of quality criterion. The averages of CSI, RI, ΔH and $\Delta H/H$ were therefore calculated for the 149 forecasts performed by CROS on the one hand, and for the 149 forecasts performed by SCOUT on the other hand.

Table 1 — Percentage of cases when two criteria gave different relative evaluations of pairs of forecasts

	$-$ CSI	$-$ RI	ΔH	$\Delta H/H$	ΔV	$\Delta V/V$
$-$ CSI	0	13	42	34	59	51
$-$ RI		0	47	39	56	54
ΔH			0	20	52	45
$\Delta H/H$				0	55	46
ΔV					0	25
$\Delta V/V$						0

The results of these calculations, reported in Table 2,, show that CROS and SCOUT can be regarded as equivalent according to CSI and RI (with a slight superiority of CROS), whereas SCOUT is definitely better than CROS according to ΔH and $\Delta H/H$.

From these results, it can be concluded that the quality criteria most frequently used to evaluate radar rainfall forecasts cannot be considered as equivalent, whether used to compare two forecasting techniques, or from the viewpoint of the reliability of one method in different situations. Therefore, the importance of a criterion selection must be emphasized. This problem is tackled in section 4.

4. AN EXAMPLE OF A RAINFALL FORECAST QUALITY CRITERION SELECTION IN RELATION TO THE REAL-TIME CONTROL OF A SEWER SYSTEM

As we have already mentioned, the choice of ΔH and $\Delta H/H$ seemed reasonable to researchers working on the hydrological applications of weather radar. However, this choice is based on the following three underlying hypotheses, which cannot be expected to be valid in all situations.

(1) The time distribution of rainfall does not need to be forecast precisely (therefore, only rainfall depths are considered in the forecast evaluation, instead of the full hyetograph).
(2) Underestimations and overestimations have the same importance (so differences between observed and forecasted rainfall depths can be calculated regardless of sign).
(3) The average is the best parameter to characterize the error distribution on the n considered drainage catchments (as a consequence, a forecast resulting in a 50% error on two catchments is equivalent to a forecast with 0% error on one catchment, and 100% on the other catchment).

As long as these hypotheses have not systematically been verified, the ΔH and $\Delta H/H$ criteria cannot be guaranteed to be coherent with the user's view of forecast quality, i.e. one cannot be sure that, if a forecast f_1 is better than a forecast f_2 according to ΔH or $\Delta H/H$, it would be preferred by the user.

Table 2 — Mean values of CSI, RI, ΔH and $\Delta H/H$
for 149 forecasts performed by CROS and SCOUT

	CROS	SCOUT
CSI	0.42	0.41
RI	0.47	0.44
ΔH (mm)	0.81	0.71
$\Delta H/H$ (%)	108.4	52.4

While this kind of verification may not be possible for all types of situation, it is possible when the forecasts are used to make decisions of actions, e.g. in the context of the real-time control of a sewer system. The *objective* evaluation of forecast quality, provided by such criteria as ΔH or $\Delta H/H$, can then be compared to the user's *subjective* evaluation, based on the consequences of the actions taken in accordance with the forecasts. A quality criterion Q can then be considered as *relevant* to that application, if, for every pair of forecasts (f_1, f_2), the following proposition is true.

'f_1 is better than f_2, according to Q, if and only if the consequences of the actions taken after f_1, are preferable, from the user's viewpoint, to the consequences of the actions taken after f_2'.

This general idea was applied to the case of the real-time control of the sewer network in the Seine–Saint-Denis county, near Paris (Frerot *et al.* 1985, Denoeux *et al.* 1987).

The equipment controlled in real time on this network are two retention basins, with volumes of 95 000 and 65 000 m^3. Control strategies are optimized according to forecasts of inflow from four main drainage catchments. The objective function F results from the aggregation of the three following objectives, in decreasing order of importance:

(1) overflow reduction;
(2) preservation of water quality in one of the two retention basins, used as a recreational area;
(3) limitation of gate operations.

This function, built together by researchers and sewer engineers, can be viewed as an approximation of the user's utility function; it can therefore be used to compare the consequences of control strategies, resulting from different rainfall forecasts.

In order to verify the coherence of $\Delta H/H$ with F, four situations, corresponding to two hydraulically significant rainfall events, were considered. For each situation, 200 rainfall forecasts were generated at random, independently of the four drainage catchment areas. The values of $\Delta H/H$ could then be compared with the value of F, obtained from the simulation, using the observed rainfall values, of the control strategies generated in accordance with the forecasts. The results of this comparison, reported in Fig. 2 for one of the four situations, reveals an absence of correlation

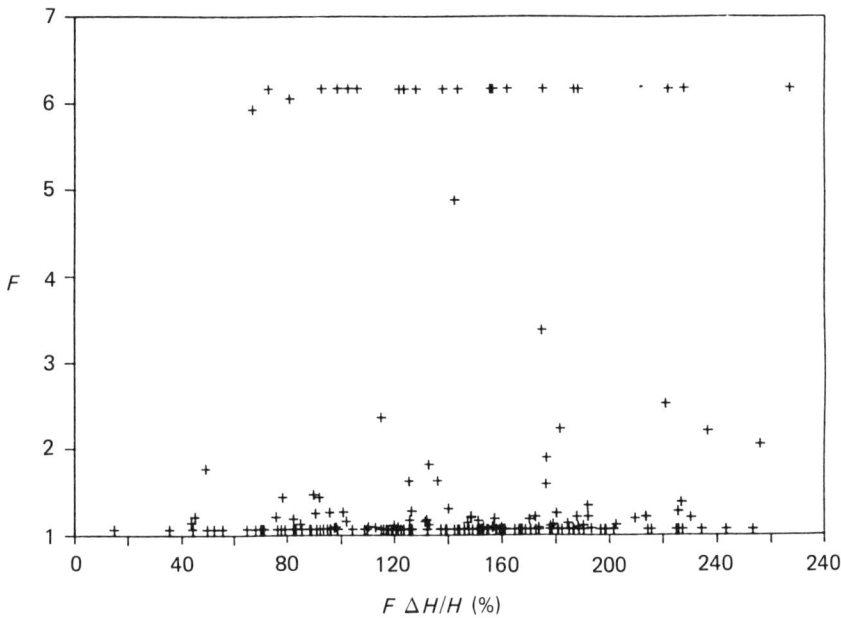

Fig. 2 — Plot of $\Delta H/H$ versus objective function F for 200 randomly generated rainfall forecasts.

between $\Delta H/H$ and F; forecasts with high values of $\Delta H/H$ do not necessarily lead to 'bad' control strategies. The analysis performed on the other situations gave similar results.

After a detailed analysis of the problem, this rather surprising phenomenon could be explained by the fact that only overestimations greater than 150% on the two catchments upstream, and underestimations greater than 50% on the two catchments downstream, were associated with bad control strategies. Thus, hypotheses (2) and (3) mentioned above are not valid for this application of rainfall forecasts.

These findings led to the definition of a new quality criterion, equal to the number of catchments upstream where an overestimation greater than 150% was made, plus the number of catchments downstream where an underestimation greater than 50% was made. This new *specific* criterion, named NMP (Nombre de Mauvaises, Prévisions, i.e. number of bad forecasts), varies between 0 for good forecasts, and 4 for very bad forecasts.

Fig. 3 represents, for the same situation as in Fig. 2, the average of $F \pm$ one standard deviation, corresponding to each value of NMP. In this situation (as well as in the three others), the new criterion appears much better correlated than $\Delta H/H$ with the objective function F; therefore, it can be regarded as more relevant to this application of the forecasts.

5. CONCLUSIONS

The results presented in this paper show the limits of an evaluation of rainfall forecasts with general criteria, such as CSI and ΔH. Comparisons of forecasting

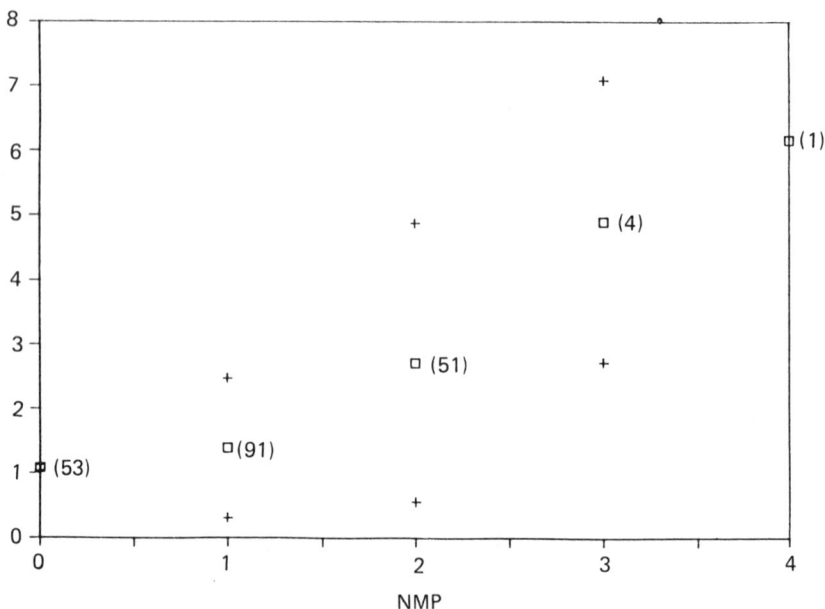

Fig. 3 — Average of objective function F, $\pm$ one standard deviation for each NMP value (the number of cases are indicated in parentheses).

techniques, or of the performance of one technique in many situations, can lead to conclusions that may be dependent on the quality criterion used.

However, it is possible, when the forecasts are used to make decisions, to verify the coherence between a forecast quality criterion, and the efficiency of the decisions made on the basis of the forecasts. This general approach, applied to the real-time control of a sewer network, led to the definition of a new *specific* quality criterion, to be used in subsequent studies on the application of rainfall forecasting in this context.

REFERENCES

Austin, G. L. (1985) Application of pattern recognition and extrapolation techniques to forecasting. *ESA J.*, **9**, 147–155.

Austin, G. L. and Bellon, A. The use of digital weather radar records for short-term precipitation forecasting. *Q. J.R. Meteorol. Soc.*, **100**, 658–664.

Bellon, A. and Austin, G. L. (1978) The evaluation of two years of real-time operation of a short-term precipitation forecasting procedure (SHARP). *J. Appl. Meteorol.*, **17** (12), 1778–1787.

Bellon, A. and Austin, G. L. (1984) The accuracy of short-term radar rainfall forecasts. *J. Hydrol.*, **70**, 35–49.

Carpenter, K. M. and Owens, R. G. (1981) Use of radar data for forecasting rain. *Proceedings of the COST 72 Workshop–Seminar on Weather Radar*. pp. 161–181.

Ciccione, M. and Pircher, V. (1984) Preliminary assessment of very short-term forecasting of rain from single radar data. *Proceedings of the Nowcasting Symposium, Norrköping, 1984.*

Collier, C. G. (1976) *Objective forecasting using radar data: a review*, Research Report No. 9. Meteorological Office Research Laboratory, Malvern, Worcs., 16 pp.

Damant, C., Austin, G. L., Bellon, A., Osseyrane, M. and Nguyen, N. (1983) Radar rain forecasting for wastewater control. *J. Hydraul. Div., Am. Sci. Civ. Eng.*, **109** (2), 293–297.

Denoeux, T. (1989) *Fiabilité de la prévision de pluie en hydrologie urbaine*, Ph.D. Thesis, Ecole Nationale des Ponts et Chaussées, Paris.

Denoeux, T., Einfalt, T. and Jacquet, G. (1987) Introduction of a weather radar as an operational tool. *Proceedings of the 4th Conference on Urban Storm Drainage, Topics in Urban Storm Water Quality, Planning and Management, Lausanne, 31 August–4 September 1987.* International Association of Hydrological Sciences, pp. 289–290.

Denoeux, T., Einfalt, T. and Jacquet, G. (1989) Reliability of radar rainfall forecasts *Proceedings of the International Conference on Topical Problems in Urban Drainage and in Industrial Plant Systems, Strbské Pleso, 25–28 April, 1989.* Dom Techniky CSVTS, Bratislava, pp. 73–75.

Einfalt, T. (1988) *Recherche d'une méthode optimale de prévision de pluie par radar en hydologie urbaine*, Ph.D. Thesis. Ecole Nationale des Ponts et Chaussées, Paris, 189 pp.

Einfalt, T. and Denoeux, T. (1987a) Radar rainfall forecasting for real time control of a sewer system. *Proceedings of the 4th Conference on Urban Storm Drainage, Topics in Urban Drainage Hydraulics and Hydrology, Lausanne, 31 August–4 September 1987.* International Association of Hydrological Sciences, pp. 47–48.

Einfalt, T. and Denoeux, T. (1987b) Utilisation d'images radar en prévision de pluie. *Reconnaissance des Formes et Intelligence Artificielle, Actes du VIe Association Française pour la Cybernetique, Economique et Technique Congrés, Antibes, 16–20 November 1987.* Dunod ,Paris, pp 467–472.

Einfalt, T. and Denoeux, T. (1991) Never expect a perfect forecast. *Hydrological Applications of Weather Radar.* Ed. Cluckie, I. D. and Collier, C. G. Ellis Horwood, Chichester, West Sussex, Chapter 40.

Einfalt, T., Denoeux, T. and Jacquet, G. (1991) The development of the SCOUT II.0 rainfall forecasting method. *Hydrological Applications of Weather Radar.* Ed. Cluckie, I. D. and Collier, C. G. Ellis Horwood, Chichester, West Sussex, Chapter 33.

Elvander, R. C. (1976) An evaluation of the relative performance of three radar echo forecasting techniques. *Proceedings of the 17th Conference on Radar Meteorology, Seattle, WA, 1976.* American Meteorological Society, Boston, MA, pp. 526–532.

Frérot, A., Jacquet, G. and Delattre, J. M. (1986) Elaboration des consignes optimales de gestion d'un réseau d'assainissement. *Eau et Informatique, Actes du Colloque, Paris, May 1986.* Ecole des Ponts et Chausées, Paris.

Huff, F. A., Changnon, S. A. and Vogel, J. L. (1980) Convective rain monitoring and forecasting system for an urban area. *Proceedings of the 19th Conference on Radar Meteorology, Miami, FL, 1980.* American Meteorological Society, Boston, MA, pp. 56–61.

Tsonis, A. A. and Austin, G. L. (1981) An evaluation of extrapolation techniques for the short-term prediction of rain amounts. *Atmos. Ocean,* **19** (1), 54–65.

Wilson, J. W. and Brandes, E. A. (1979) Radar measurement of rainfall — a summary, *Bull. Am. Meteorol. Soc.,* **60** (9), 1048–1058.

33

The development of the SCOUT II.0 rainfall-forecasting method

T. Einfalt[1], T. Denoeux[2] and G. Jaquet[1]
[1] RHEA, Rue Albert Einstein, Champs sur Marne, 77436 Marne la Vallee Cédex 2, France
[2] Cergrene Ecole Nationale des Ponts et Chausées, La Courtine, 93167 Noisy-le-Grand Cedex, France

ABSTRACT

This paper describes the variety of choices in the development of the SCOUT II.0 radar rainfall-forecasting method. A review of the crucial steps of the so-called structured approach shows the methodological improvements that have been introduced in the areas of feature selection and echo matching. A main contributor to this development has been the orientation towards the needs of a hydrological user.

1. INTRODUCTION

During the last 25 years, a variety of studies on different aspects of radar rainfall measurement, analysis and forecasting, performed on different radar systems, have contributed to an increasing of knowledge on various topics concerning radar rainfall forecasting.

Wilson (1966) analysed radar data in order to select invariant features which might exhibit the characteristics of the observed rainfall event. Wilson and Brandes (1979) and Zawadski (1984) worked on measurement quality and potential factors for measurement uncertainties. Calibration techniques for radar measurements in hilly regions have been developed in the UK (Collier 1986).

2. DEVELOPMENT OF FORECASTING TECHNIQUES

From the comparison of black-and-white paper photographs (e.g. Wilson 1966) to digital data representation on a colour graphics screen, two basic radar image forecasting approaches have emerged, both with appropriate significance.

(1) A 'global' approach, based on statistical comparison of the image as a whole (e.g. Bellon and Austin 1978).   This approach consists of a search of the translation of an image $I(t)$ versus an image $I(t + dt)$, yielding the maximal cross-correlation coefficient of the two images (or other index of similarity). Under the assumption of a linear radar echo movement and a negligible change in the echo's shape, the forecast is a simple extrapolation of the obtained translation value, uniformly applied to the whole image $I(t + dt)$.

(2) A 'structured' approach, based on a simple kind of human perception modelling, treating image features (e.g. Einfalt and Schilling 1984).   The most distinctive features of each radar echo on a radar image are calculated, e.g. centre of gravity, size and mean intensity. The comparison of only these features allows the recognition of the echoes of image $I(t)$ on the image $I(t + dt)$ so that for each echo an individual displacement vector can be obtained.

The global approach is characterized by

(a) being simple and
(b) being reliable for many meteorological situations

while the structured approach

(a) allows for different movement directions of echoes,
(b) allows a recognition of ground echoes,
(c) is more intuitive and
(d) issues more information of the situation to the user.

Already in 1978, Collier (1978) had reviewed these two types of forecasting approach, yielding no significantly different performance for them. In the meantime there have been developed forecasting techniques for a multilayer rainfall forecast (Bjerkaas and Forsyth 1980) (structured approach), simplified techniques for microcomputer use (Einfalt and Schilling 1984) (structured approach), in the UK a radar- and satellite-based measurement and forecasting tool (Browning 1979) (global approach), and in Japan work on a growth and decay forecast (Yoshino and Kozeki 1985) (structured approach).

However, even today it must be recognized that both approaches do not differ very much in their statistical peformance indices (Denoeux *et al.* 1991), but for local hydrological problems (Jaquet and Einfalt 1985) the structured approach seems to offer more possibilites.

3.   UPGRADING A SIMPLE STRUCTURED METHOD

A structured method usually consists of four steps (Fig. 1):

```
┌─────────────────────────────────────┐
│          Echo definition            │
└─────────────────────────────────────┘
                   │
┌─────────────────────────────────────┐
│        Echo characterization        │
└─────────────────────────────────────┘
                   │
┌─────────────────────────────────────┐
│           Echo matching             │
└─────────────────────────────────────┘
                   │
┌─────────────────────────────────────┐
│             Forecast                │
└─────────────────────────────────────┘
```

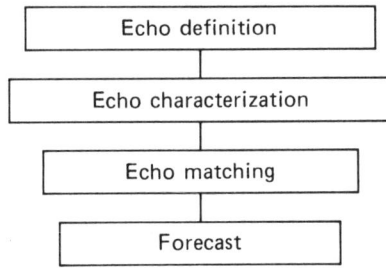

Fig. 1 — The four steps of the structured approach.

(1) the echo definition (the decision on which point belongs to which echo — an echo being a set of uniformly treated pixels on a radar image);
(2) the feature calculation;
(3) The echo matching (recognition) using only the calculated features;
(4) the forecast by extrapolating the movement vectors obtained in step (3).

The approach relies very much on the selection of the echo features (step (2)) and on the echo recognition technique (step (3)). The main problems in this context are

(a) important changes in shape and size of echoes,
(b) splitting or merging of echoes,
(c) image boundary effects, and
(d) the choice of the parameters on which the recognition process is based.

4.   THE ECHO DEFINITION PROCESS

Originally, this definition meant an echo determination for the whole sequence of the forecast, and, so this step was of highest importance, but the introduction of 'artificial echoes' (Einfalt *et al.* 1989) now allows one to unify neighboring echoes which have not been regarded as one echo by the echo definition process.

There is a large number of possibilities to cluster rainfall pixels. However, the first fundamental question is, 'what will be considered as a 'rainfall pixel'?'.

Usually, it is convenient and intuitive to select an intensity threshold for this purpose. Different techniques have been proposed, e.g. Blackmer *et al.* (1973) have worked on a fixed threshold level, adding the directly neighbouring pixels if they were not more than one level below (see Fig. 2 and Fig. 3 for an example with a threshold of 4). Wolf *et al.* (1977) used a simple connectivity definition of directly or diagonally touching pixels (Fig. 4). Only directly touching pixels (Fig. 5) were used by Einfalt and Schilling (1984). There are other solutions which do not work in this way (e.g. Crane 1976), but cluster pixels which are surrounding a recognized local centre of intense rainfall.

2	2	3	4	3	3	4
3	5	4	2	2	2	3
2	4	2	1	1	2	4
2	2	1	1	0	0	1

Fig. 2 — Echo definition (Blackmer *et al.* 1973), initial state.

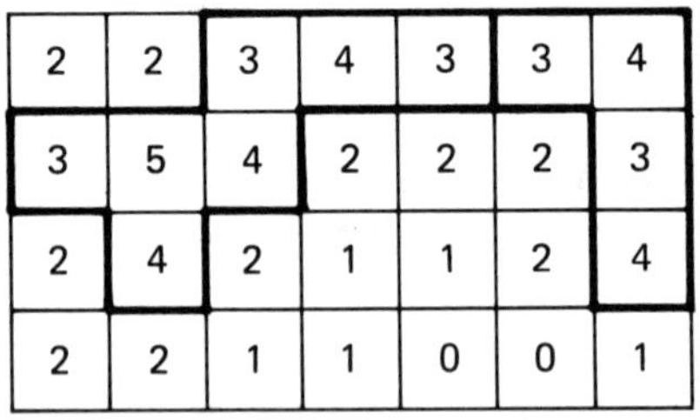

2	2	3	4	3	3	4
3	5	4	2	2	2	3
2	4	2	1	1	2	4
2	2	1	1	0	0	1

Fig. 3 — Echo definition (Blackmer *et al.* 1973), final state.

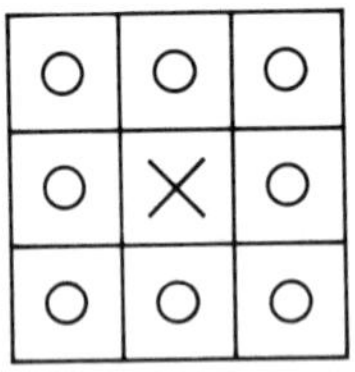

Fig. 4 — Echo definition (Wolf *et al* 1977): ×, actual pixel; ○, neighbouring pixel.

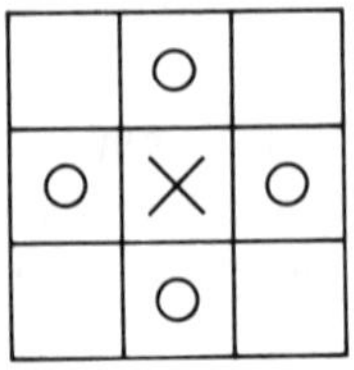

Fig. 5 — Echo definition (Einfalt and Schilling 1984): ×, actual pixel; ○, neighbouring pixel.

On the assumption that a threshold has been set for a whole image, a decision has to be made which connectivity definition is to be applied for clustering. Intuitively, that of Blackmer *et al.* is as valid as those of Wolf *et al.* or Einfalt and Schilling. So a decision has to be made, based on the underlying model concept or according to the calculation time needs of an application.

5.   FEATURE SELECTION

There is a considerable number of features which have been used for echo tracking. Only two of them, size and centroid, are present in nearly all studies. The other features, be it from features, historical data or statistical analysis, have been chosen to very different degrees. In order to describe the echo shape in an optimal manner, the following have been used:

(1) the Fourier transform (Blackmer *et al.* 1973, Östlund 1973);
(2) the maximal echo diameter (Wolf *et al.* 1977; Östlund 1973);
(3) the moments of intertia on the X and Y axes (Wolf *et al.* 1977);
(4) the elongation and orientation (Einfalt 1988).

For statistical and historical analysis the following have been selected:

(a) the mass of the echo (Einfalt 1988, Östlund 1973);
(b) the average intensity (the mass divided by the size) (Wilson 1966, Wolf *et al.* 1977).
(c) the previous displacement vector (Blackmer *et al.* 1973, Einfalt 1988, Wolf *et al.* 1977);
(d) the number of recognitions (Blackmer *et al.* 1973);
(e) the intensity variance (Wilson 1966);
(f) the internal structure of an echo (Wilson 1966);
(g) the intensity distribution (Einfalt 1988).

6.   ECHO MATCHING

This step has the goal of recognizing echoes which have been analysed on one radar image on the following image, only using the calculated echo features.

As there is a wide variety of features that can be used and no objective value for rejection or acceptance of a potential partner exists, there is no commonly agreed systematic approach to the matching process. All the researchers either extremely simplify their matching rule (e.g. Ciccione and Pircher, 1984) or use intuitive heuristic methods which tend to yield satisfactory results (Wolf *et al.* 1977, Bjerkaas and Forsyth 1980, Yoshino and Kozeki 1985, Einfalt 1988).

The matching criteria are based on a minimal squared distance for obtaining a common displacement vector, or a minimal shape difference for obtaining a common displacement vector, or a maximal value for a quality criterion derived from several features for obtaining individual vectors or a sufficiently large population of matched echoes by individual form feature comparison.

7. THE FORECASTING STEP

The extrapolation of the obtained displacement vectors is either done in a uniform manner using a mean vector, or individually echo by echo. Whether the forecasting vector is the obtained vector from the last two images or whether it includes further historical information is principally the only decision to be taken, but it can be important for certain situations.

A relatively new idea is to add a module inside the forecasting step which calculates user-oriented information such as rainfall volume accumulated over a certain time. However, it seems to be vital for the acceptance of every operational forecasting tool that it is guided by the user's needs.

8. QUALITY CONTROL

The variety of quality criteria are nearly as large as the possibilities for selecting matching criteria. For a description of application-oriented and image-oriented criteria, see Denoeux *et al.* (1991).

9. CHOICES FOR THE SCOUT II.0 FORECASTING METHOD

The SCOUT II.0 methods is an urban hydrology oriented forecasting tool (SCOUT is an acronym for second-moments cloud tracking). A thorough examination of the advantages and disadvantages of the proposed forecasting techniques of nearly three decades has led to the definition of the forecasting method SCOUT II.0 which overcomes some of the typical problems of 'structured' approaches while preserving the usual benefits as follows.

(1) Splitting and merging of echoes can be recognized.
(2) The image boundary does not necessarily split echoes into several pieces.
(3) The crucial selection of the echo features for echo recognition has been optimized.
(4) The echo recognition process has been reformulated in favour of a more intuitive technique.

Since the urban hydrologist is mostly concerned with heavy rainfall, a threshold for echo definition is fixed after a statistical intensity analysis of the whole image, leading to a selection of the most important rainfall areas. The clustering procedure is simply that in Fig. 5.

The radar echoes are characterized by the feature in Table 1.

Every echo, can additionally, be a member of a so-called 'artificial echo' which is the union of two echoes which are sufficiently close. By this technique (Einfalt *et al.* 1989), splitting and merging of echoes can be recognized and, consequently, the image boundary does not necessarily split echoes into several pieces.

The matching process is based on a heuristic analysis of the echo movement. First, the general movement is used to get a rough idea of the overall displacement direction and speed. Initially, wind data from regular meteorological observations are used for this purpose. After this, the individual movement of each echo determines the region where a potential partner has to be found. On the basis of

Table 1 — Features used for characterizing the echoes

n	Size, i.e. number of pixels
m	Mass, i.e. rainfall volume $= (1/n)\sum\limits_{i\in e} R(i)$
	where $R(i)$ is the measured rainfall intensity at pixel i
c	Centroid, i.e. the centre of gravity
a	Orientation (angle of the principal axis of inertia)
E	Elongation (quotient of the principal moments of inertia)
D	Intensity distribution
n_0	Previous size
υ	Previous movement vector
σ_υ	Root mean square variation in previous movement vectors
n_p	Number of previous recognitions of the echo

elongation, orientation and intensity distribution, the similarity hypothesis is accepted or rejected for each echo in this region, provided that the echoes do not differ too much in size.

For each of the selected echoes, the forecasting vector is calculated as the mean of the previous displacement and the actual displacement.

This information is now used to calculate the rainfall depth for the following hour, using forecasted images for every 5 min. The rainfall depth information can be processed in several ways:

(1) to issue a flood warning;
(2) to calculate a rainfall volume for a subcatchment or a catchment;
(3) to be used on a pixel-by-pixel basis (e.g. radar–raingauge comparison).

An additional feature of the SCOUT II.0 forecasting method is the possibility of a forecasting quality assessment which is only based on imagery information. Using echo number and echo number variation, the forecasting quality can be estimated even without running the forecasting program. By this means, the user can decide whether a forecast should be calculated, depending on whether it is sensible for him to use this tool or not.

10. CONCLUSION

The development of rainfall analysis and rainfall-forecasting techniques has led to a considerable variety of methodologies for forecasting purposes. A number of shortcomings of forecasting methods could be eliminated or reduced by the SCOUT II.0 method.

The selection of informative echo features has been oriented towards a combination of form features, statistical data and historical information which represent the echoes and their differences in a reliable and simple manner. The matching process has been designed for a quick and intuitive recognition of the echoes, including 'artificial echoes'. Thus, this method is able to recognize splitting and merging of echoes.

Additionally, the development of the method has very much been oriented by the users' needs, resulting in a variable-intensity threshold definition for the method, a forecast of catchment-oriented information instead of radar images and the capability to assess the forecasting quality only by imagery features.

ACKNOWLEDGEMENTS

This work has been conducted at the Cergrene Research Centre and at the real-time control centre of the Siene-St-Denis county.

REFERENCES

Bellon, A. and Austin, G. L. (1978) The evaluation of two years of real-time operation of a short-term precipitation forecasting procedure (SHARP). *J. Appl. Meteorolog.*, **17** (12), 1778–1787.

Bjerkaas, C. J. and Forsyth, D. E. (1980) *An automated real-time storm analysis and storm tracking program.* US Air Force Geophysics Laboratory.

Blackmer, R. H., Duda, R. O. and Reboh, E. (1973) *Application of pattern recognition techniques to digitized weather radar data*, Report No. 36072. Stanford Research Institute, Menlo Park, CA.

Browning, K. A. (1979) The FRONTIERS plan: a strategy for using radar and satellite imagery for very-short-range precipitation forecasting. *Meteorol. Mag.* **108**.

Ciccione, M. and Pircher, V. (1984) Preliminary assessment of very short-term forecasting of rain from single radar data. *Proceedings of the Nowcasting Symposium, Norrköping, 1984.*

Collier, C. G. (1978) *Objective forecasting using radar data: a review.* Research Report No. 9, Meteorological Office, Research Laboratory, Malvern, Worcs.

Collier, C. G. (1986) Accuracy of rainfall estimates by radar, Part I: Calibration by telemetering raingauges. *J. Hydrol.* **83**, 207–223.

Crane, R. K. (1976) Rain cell detection and tracking. *Proceedings of the 17th Conf. on Radar Meteorology, Seattle, WA, 1976.* American Meteorological Society, Boston, MA.

Denoeux, T., Einfalt, T. and Jaquer, G. (1991) On the evaluation of radar rainfall forecasts. *Hydrological Applications of Weather Radar.* Ed. Cluckie, I. D. and Collier, C. G. Ellis Horwood, Chichester, West Sussex, Chapter 32.

Denoeux, T. (1989) *Fiabilité de la prévision de pluie par radar en hydrologie urbaine*, Ph.D. Thesis. Cergrene Ecole Nationale des Ponts et Chausées, Paris.

Denoeux, T., Einfalt, T. and Jacquet, G. (1987). Introduction of a weather radar as an operational tool. *Proceedings of the 4th International Conference on Urban Storm Drainage, Lausanne, 1987.*

Einfalt, T. (1988) *Recherche d'une méthode optimale de prévision de pluie par radar en hydrologie urbaine*, Ph.D. Thesis, Cergrene Ecole Nationale des Ponts et Chausées, Paris.

Einfalt, T. and Schilling, W. (1984) SCOUT — a storm tracking procedure for a microcomputer. *Proceedings of the 3rd International Conference on Urban Storm Drainage,* Göteborg, 1984.

Jaquet, G. and Einfalt, T. (1985) Radar rainfall forecasting for Paris suburban area. Poster session of the Symposium on Radar Rainfall and Flood Warning Symposium, Lancaster, 1985.

Östlund, S. S. (1973) *Computer software for rainfall analysis and echo tracking of digitized radar data,* Technical Memorandum ERL WMPO-15. National Oceanic and Atmospheric Administration, Boulder, CO.

Wilson, J. W. (1966) *Movement and predictability of radar echoes,* ESSA Technical Memorandum No. NSSL-28. Norman, OK.

Wilson, J. W. and Brandes, E. A. (1979) Radar measurement of rainfall — a summary. *Bull. Am. Meteorol. Soc.,* **60** (9).

Wolf, D. E., Hall, D. J. and Endlich, R. M. (1977) Experiments in automatic cloud tracking, using SMS–GOES data. *J. Appl. Meteorol.,* **16**, 1219–1230.

Yoshino, F. and Kozeki, D. (1985) Study on short-term forecasting of rainfall using radar rain-gauge. *Proceedings of the Weather Radar and Flood Warning Symposium, Lancaster, 16–18 September 1985.*

Zawadski, I. (1984) Factors affecting the provision of radar measurements of ran. *Proceedings of the 22nd Conference on Radar Meteorology, Zürich, September, 1984.*

34

Short-term rainfall forecasting using radar data and hydrometeorological models

K. P. Georgakakos and W. F. Krajewski
Department of Civil and Environmental Engineering and Iowa Institute of
Hydraulic Research, The University of Iowa, Iowa City, IA 52242-1585, USA

ABSTRACT

A physically based dynamic model of rainfall prediction suitable for use with hydrologic prediction models has been enhanced with the capability to utilize radar rainfall and reflectivity data. A state estimator was designed for the updating of the model state from real-time observations of reflectivity and rainfall and for the generation of prediction-uncertainty measures. Results from initial tests of the formulation are presented using the 10 cm RADAP II radar at Oklahoma City, OK, USA. The results show the utility of radar rainfall and radar reflectivity measurements in the real-time rainfall forecasting.

1. INTRODUCTION

In recent years, quantitative precipitation prediction models suitable for use in real-time hydrology have appeared in the hydrologic literature (Georgakakos and Bras 1984a, b, Georgakakos 1984). These models are based on an expression of the convertion of liquid-water mass over the hydrologic basin of interest. Adiabatic and pseudoadiabatic processes are utilized for the determination of the condensation source term in the conservation equation. Cloud microphysical parameterizations are used for the calculation of cloud precipitation rates. Evaporation of cloud drops in the subcloud layer of unsaturated air is also modelled. Those models predict hourly precipitation rates given input in the form of surface air temperature, pressure and dew-point temperature. The models have been formulated in state-space form and state estimators have been designed to process mean areal precipitation in real time for firstly state updating and secondly determination of forecast uncertainty. The models were developed in an effort to improve short-term precipitation predictions on the scale of small- and medium-size hydrologic basins (100–1000 km^2). They have already been coupled to hydrologic models (Georgakakos 1986a, b) to form integrated hydrometeorological forecast systems for the real-time prediction of floods and flash floods (Georgakakos 1987).

It is the purpose of this paper to propose using a precipitation model of the type described above as a physically based integrator of real-time hydrometeorological data from remote and on-site sensors for the short-term (hourly) prediction of rainfall. In particular, we propose a formulation that utilizes volume-scan radar reflectivity data together with surface meteorological data. It is expected that the present formulation would produce improved results compared with the previous formulations because high radar tilts are utilized to obtain a direct observation of the cloud- and rain-water mass. Such an observation is particularly useful in cases of light rain over the domain of interest at the beginning of storms for the initialization of the model state. In cases of saturated lower troposphere with a temperature inversion such an additional observation can help to correct the erroneous overprediction of the model state by the condensation component that utilizes only surface input data (see Georgakakos (1984b) for a discussion). In addition at long radar ranges in the absence of raingauge data, even the lowest-radar-tilt observations are not low enough to provide reliable surface rainfall estimates to be used for updating. Such observations, however, are direct observations of the model state (cloud- and rain-water mass) and *can be used* for updating. In fact, in such cases updating is only possible with the enhanced formulation proposed herein. The utility of the radar data and the importance of the reliable estimation of radar measurement errors for the real-time prediction of rainfall and resultant flows is indicated by way of a sensitivity analysis that utilized data from the RADAP II radar in Oklahoma City, OK.

2. MODEL FORMULATION

Consider the domain over a small watershed. It is of interest to predict rainfall over the domain with short forecast lead time (of the order of an hour). The spatially lumped rainfall prediction model used is given by the following equations in state space form:

$$\frac{dX(t)}{dt} = h(\mathbf{u}(t))X(t) + f(\mathbf{u}(t)) + w(t) \tag{1}$$

or

$$z(t_i) = \Delta t \, \phi(\mathbf{u}(t_i))X(t_i) + \upsilon(t_i) \, , \qquad i = 1, 2, \ldots \tag{2}$$

where equation (1) describes the dynamics of the scalar state variable $X(t)$ and equation (2) describes the relationship between the model state and the rainfall observation at time t_i over the domain of interest (mean areal rainfall). The following nomenclature was used: $X(t)$ is the model state condensed liquid-water equivalent over the domain of interest at time t, $z(t_i)$ is the observed mean areal rainfall accumulated over the interval (t_{i-1}, t_i), $h(\cdot), f(\cdot)$, and $\phi(\cdot)$ are time-varying functions dependent on the model input variables, $\mathbf{u}(t)$ is the vector of the three model input variables (surface air temperature, pressure and dew-point temperature over the domain of interest) at time t, $\Delta t = t_i - t_{i-1}$, $w(t)$ is a time-continuous random error process that simulates random errors in the model structure, input time series and model-parameter values, and $v(t_i)$ is a time-discrete random error sequence that simulates random errors in the observations of mean areal rainfall. The functions h, f and ϕ, as they have been defined by Georgakakos and Bras (1984a, b) and by Georgakakos (1986a), are used in the following development. The term $h(\mathbf{u}(t))X(t))$ represents the cloud output in terms of liquid-water equivalent at time t, and $f(\mathbf{u}(t))$ represents the condensation input rate (mass divided by the product of time and area). Thus, equation (1) is a statement of mass conservation for rain and cloud water. The term $\phi(\mathbf{u}(t_i))X(t_i)$ represents the model predicted instantaneous mean areal rainfall amount over the domain of interest of time t_i. Equation (2) represents an approximation to the integral equation that defines rainfall accumulations over Δt as functions of the state of the time-continuous model. Georgakakos (1986c), however, has shown that for small Δt (less than 3 h) such an approximation is not detrimental to the model predictions. Since only hourly predictions would be analyzed in this paper, equation (2) is used as the observation equation of the state-space model form. Discussion on the probabilistic character of $w(t)$ and $v(t_i)$ is reserved for section 3.

Supporting relationships are given next (see also Georgakakos and Bras 1984a, b). The convective updraft velocity is given by

$$v = \varepsilon_1 \sqrt{c_p(T_m - T_s)} \tag{3}$$

The cloud-top pressure P_t is given by

$$\frac{p_t - p_1}{p_0 - p_1} = \frac{1}{1 + v} . \tag{4}$$

The hydrometeor size distribution is

$$n(D) = N_o \exp \left(-\frac{D}{\varepsilon_4} \right) . \tag{5}$$

ε_1 and ε_4 are parameters to be determined from observed data and are believed to be storm invariant for a particular hydroclimatic regime (Georgakakos 1984), T_m is the air temperature at a characteristic pressure level inside the storm cloud, T_s is the ambient air temperature at the same pressure level, v is the updraft velocity, p_t is the cloud-top pressure, p_1 is the lower bound for the cloud-top pressure, p_0 is the upper bound for the cloud-top pressure, D represents the diameter of the hydrometeors, $n(D)$ is the hydrometeor size distribution characterizing cloud and rain water, and N_0 is a parameter of the size distribution characterizing small sizes. Georgakakos and Bras (1984a, b) derived an expression for N_0 as a function of the model state X. It is noted that the size distribution can be time varying.

Given reflectivity factor measurements Z_i^j ($i = 1, \ldots, n$) at various tilts of the radar antenna angle Φ_j with $j = 1, \ldots, m$, assumed to be taken at instances t_1 through t_n during the time interval Δt, the following reflectivity factor versus size distribution relationship exists (e.g. Burgess and Ray 1986):

$$Z_i^j = \int_0^\infty D^6\, n_i^j(D)\, \mathrm{d}D \tag{6}$$

where the size distribution $n(D)$ of equation (5) has been localized as $n_i^j(D)$ for the time instant t_i and the tilt angle Φ_j. Localization of the distribution is through localization of its parameters N_0 and ε_4. Because of the absence of upper-air meteorological data for the temporal scales (hourly), utilization of reflectivity factor for each tilt angle is not recommended. Instead, as a compromise, we propose to use an average reflectivity factor $Z(t_i)$, defined by

$$Z(t_i) = \frac{1}{m} \sum_{j=1}^m Z_i^j \tag{7}$$

as a characteristic measure of nonzero reflectivity factors over the height of the storm clouds and for time instant t_i. In addition, we assume that an 'equivalent' size distribution $n_i'(d)$, which is uniform over the height of the storm clouds exist at time t_i such that

$$Z(t_i) = \int_0^\infty D^6 n_i'(D)\, \mathrm{d}D \; . \tag{8}$$

It is noted that because of the high power of D the use of ∞ as an upper bound might introduce appreciable errors in the computations for relatively flat size distributions. Also, for diameters of hydrometeors near zero, no return signal (and hence no reflectivity factor) is expected. However, for simplicity we retain the previous equation with the understanding that it might be improved.

It also holds that, for time instant t_i,

$$X(t_i) = Z_{\mathrm{cl}}(\mathbf{u}(t_i))\, \frac{\pi}{6}\, \rho_w \int_0^\infty D^3 n_i(D)\, \mathrm{d}D \tag{9}$$

where ρ_w represents the density of liquid water, and $Z_{cl}(\mathbf{u}(t_i))$ is the average depth of the storm clouds over the hydrologic area of interest and for time t_i. The size distribution $n_i(D)$ is an average size distribution over the depth of the clouds over the domain of interest. It is noted that the two distributions $n'_i(D)$ and $n_i(D)$ need not be the same since the functional form of the right-hand side of equations (8) and (9) differs. It is hypothesized that the difference between the two size distributions lies in the average hydrometeor size ε_4, with $\varepsilon'_4 = \alpha \varepsilon_4$, and ε'_4 corresponds to $n'_i(D)$. Given that the larger hydrometeors produce the bulk of the radar return signal, it is expected that the parameter α will satisfy $\alpha > 1$.

The last two equations (after substitution of the exponential size distribution relationships for $n_i(D)$ and $n'_i(D)$) have four unknowns: X, Z, N_0, ε_4. One, then, can eliminate two of them by solving the equations for those two unknowns. Since $X(t_i)$ is the model state variable and $Z(t_i)$ is observed, it is convenient to solve for the parameters of the hydrometeor size distribution. This way, if reflectivity factor measurements are available, they could be used in a second observation equation within the model state-space form. The second observation equation would be

$$Z(t_i) = \phi'(\mathbf{u}(t_i))X(t_i) + \upsilon'(t_i) \tag{10}$$

with

$$\phi'(\mathbf{u}(t_i)) = \frac{720\varepsilon_4^4\alpha^7}{\pi\rho_w Z_{cl}(\mathbf{u}(t_i))} \tag{11}$$

and with $\upsilon'(t_i)$ simulating random measurement errors in reflectivity factor observations.

3. MODELS OF UNCERTAINTY

Equations (1), (2) and (10) represent the mathematical state-space form of a spatially lumped rainfall prediction model that utilizes real-time observations of rainfall and vertically averaged reflectivity factor. It is noted that both the dynamics equation (1) and the observation equations (2) and (10) are linear in the state variable X. As such, they are suitable for use with state estimators for the sequential processing of the observations and real-time prediction (e.g. Kalman filters as described in Gelb (1974)). Crucial for the successful design of a state estimator is the determination of the statistical properties of the error terms in the state-space form (i.e. w, υ and υ'). Those terms summarize the uncertainty in the dynamics and observation equations.

The model-error noise $w(t)$ simulates random (possibly non-stationary) errors in the model structure, input-variable observations or predictions, and model-parameter estimates. For the rainfall prediction model at hand, dominant structural errors are those associated with the determination of mass of condensate based on only surface observations. Input errors are due to sensor errors in the measurement of air temperature, pressure and dew-point temperature (or humidity), and, for the case of large watersheds, to the determination of mean areal quantities from a few point measurements of those variables. Parameter estimation errors represent an important component of model errors particularly in cases when the parameters are location dependent and no historical data sets are available for model calibration and the definition of reliable parameter estimates. It is noted that the representation of model errors via the additive random process $w(t)$ is successful as long as the second-moment characteristics of the errors match those of $w(t)$. Rajaram and Georgakakos (1989) derive expressions for the second-moment properties of $w(t)$ that adequately represent input and parameter errors. In cases when structural errors are dominant, then direct derivation of the second-moment properties of $w(t)$ is impractical and adaptive methods should be used for the determination of those properties (e.g. Jazwinski 1970). Georgakakos (1984) has presented an adaptive procedure that resulted in robust estimates of the second-moment properties of $w(t)$ for the case of the rainfall prediction model, utilizing only the rainfall observation equation for updating. In this work we parametrize the errors in the input meteorological variables is described by Rajaram and Georgakakos (1989). The error process $w(t)$ was assumed to be an independent Gaussian white-noise sequence with mean zero and variance parameter (units of spectral density) $Q(t)$.

The random sequence $v(t_i)$ represents errors in the observations of rainfall over the domain of interest. For small domains, the observation error is mainly due to sensor errors. For relatively large domains the error should incorporate additional errors generated in the determination of mean areal rainfall from several point raingauge measurements and radar data (e.g. Krajewski 1987). When radar data are used for the determination of rainfall, several factors contribute errors to the observations (e.g. Doviak and Zrnic, 1984). Most important among those for short and medium ranges are the definition of the Z–R relationship (R denotes rainfall rate), anomalous radar beam propagation due to temperature inversions and strong humidity gradients, and the presence of a bright band due to the melting of falling ice crystals in the storm clouds. For long ranges, partial beam filling and evaporation of hydrometeors below the lowest radar tilt angle become significant contributors to radar observation errors while bright-band-induced errors become less important (e.g. Collier, 1987). For the purposes of this analysis, $v(t_i)$ is assumed to be an independent Gaussian white sequence with mean zero and variance $R(t_i)$.

Errors in the observations of vertically averaged reflectivity factor are represented by the random sequence $v(t_i)$. Such errors are mainly due to the difference between the reflectivity factor and the effective reflectivity factor in cases when the Rayleigh approximation is not valid (e.g. Doviak and Zrnik 1984), anomalous propagation and bright-band effects are discussed in the previous paragraph, and partial beam filling for long radar ranges. It is noted that use of a vertically averaged reflectivity factor as an observation implies equation (8) and consequently the existence of a characteristic size distribution representative of the cloud column hydrometeors that is exponential. Such an assumption may lead to structural errors in the observation equation (10) which have to be included as part of $v'(t_i)$. The sequence $v'(t_i)$ was assumed to be an independent Gaussian white random sequence with zero mean and variance equal to $R'(t_i)$. The sequences $v(t_i)$ and $v'(t_i)$ were assumed to be independent in this analysis. Such an assumption is expected to hold in cases when errors in $v(t_i)$ are mainly due to errors in the Z–R relationship rather than in the measurement of reflectivity.

Once the probabilistic character of the noise sequences w, v, and v' has been defined, a state estimator for a system with continuous-time dynamics and with discrete-time observations can be designed. The state estimator performs

(1) updating of model-predicted state estimates from observations $z(t_i)$ and $Z(t_i)$, or just $z(t_i)$, or just $Z(t_i)$, and
(2) computation of the variance of the mean areal rainfall predictions in real time.

4. INITIAL TESTS AND RESULTS

Reflectivity factor data were obtained from the Oklahoma City RADAP II radar for selected convective storms in the summer of 1985. The data had been archived during routine operation of the Oklahoma City radar. The radar is a 10 cm radar with a base tilt angle of 0.5° and with tilt angle increments of 2° from 2° up to 22°. The radar takes a base elevation scan every 10 min. Scans for higher tilt angles were not available continuously, thus making the computation of the observations of the vertically integrated reflectivity factor problematic at short ranges. The radar data was provided by the Hydrologic Research Laboratory of the US National Weather Service. No quality control of the data other than checks for transmission errors was performed. Corresponding surface meteorological data for the first-order meteorological stations in the area were obtained from the US National Climate Data Center in Asheville, NC. Initial tests of the proposed methodology with the data available at the time of writing are reported in the following.

Fig. 1 presents the hourly predictions of the spatially lumped rainfall prediction model and the corresponding observations for each of four updating scenarios: no updating (deterministic runs), updating from only reflectivity factor observations, updating from only rainfall observations, and updating from both reflectivity factor and rainfall observations. The deterministic rainfall model parameters were calibrated for this particular storm. The parameters took the values $\varepsilon_1 = 0.001\,65$, $\varepsilon_4 = 50\ \mu$m, $\alpha = 1.8$, $p_1 = 200$ mbar and $p_0 = 900$ mbar. The rainfall data were obtained from base-scan radar reflectivity data using: $Z = 200R^{1.6}$, where Z (mm^6/m^3) is reflectivity factor and R (mm/h) is rainfall. For the runs with updating, standing error of 3 K was assumed for both the surface temperature and the surface dew-point temperature data and a value of 0.054^2 mm^2/h was used for $Q(t)$ which was assumed to be constant in time. When updating from reflectivity factor data was used, a value of 10^3 mm^{12}/m^6 was used for $R'(t)$. When updating from rainfall data was used, a value of 1 mm/h was assigned to $R(t)$. It can be seen that the addition of the reflectivity factor data improves both the deterministic predictions and the predictions obtained when updating from rainfall data is used.

Table 1 presents the values of the least-squares performance indices: coefficients of efficiency, persistence and extrapolation. The coefficient of efficiency is a measure of the rainfall variance reduction by the model predictions. The coefficient of persistence compares the performance of the rainfall prediction model with the performance of a simple persistence scheme that gives the current rainfall observation as the next prediction. The coefficient of extrapolation compares the performance of the rainfall prediction model with the performance of a linear extrapolation

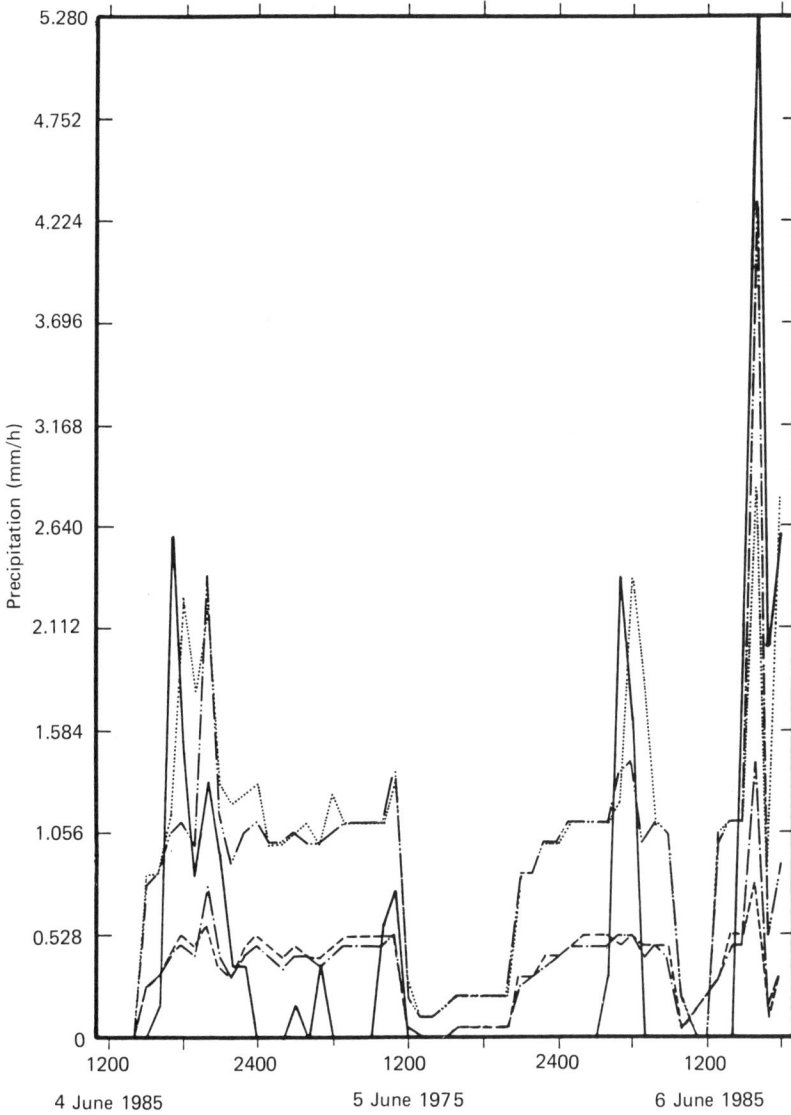

Fig. 1 — Hourly rainfall predictions and corresponding observations for the 4–6 June 1985
storm for various updating scenarios: − −, observation; − − −, deterministic; − · −, updating
from reflectivity data; . . . ., updating from main data; − ·· −, updating from both reflectivity
data and rain data. Predictions are with calibrated model parameters.

scheme that predicts rainfall using the current and previous rainfall observations.
Those performance indices have been used by Georgakakos and Bras (1984b) and
Georgakakos (1986b) to assess the performance of spatially lumped rainfall predic-
tion models. Table 1 corroborates the results displayed in Fig. 1.

Fig. 2 presents analogous (to Fig. 1) results obtained for the case when the values
of the parameters of the rainfall model were far from optimum. In particular, the

Table 1 — Values of the least-squares performance indices for the 4–6 June 1985, storm when best estimates of model parameters have been used

Updatiang scenario	Coefficient of efficiency	Coefficient of persistence	Coefficient of extrapolation
No updating	0.104	− 0.120	0.517
Reflectivity	0.214	0.018	0.577
Rainfall	0.120	− 0.100	0.526
Rainfall and reflectivity	0.268	0.086	0.606

value of p_0 was changed to 700 mbar. The values of the rest of the model parameters and the values of the state estimator parameters remained the same as in the previous case examined. The same trend is observed here. For this case, even though the deterministic model overpredicts the rainfall accumulations considerably, it can be seen that substantial improvement is obtained by updating from reflectivity data. Table 2 presents the corresponding values of the least-squares performance indices. It is noted that, in both cases examined (results in Figs 1 and 2) and for all updating scenarios, the value of the standard deviation of the normalized prediction residuals ranged from 0.7 to 0.8 while the value of the lag-one autocorrelation coefficient of the same series ranged from 0.1 to 0.5, indicating near-optimal state estimator performance.

5. PROSPECT

Encouraging initial results have been obtained using radar reflectivity data with a spatially lumped dynamical rainfall model to predict rainfall on hydrologic scales. It is recommended that a spatially distributed rainfall prediction model of the type presented by Lee and Georgakakos (1989) be enhanced with the capability to utilize radar reflectivity data and extensive tests to be done to assess fully the utility of such data in real-time rainfall and flood prediction. Utilization of a spatially distributed rainfall model would allow for maximum utilization of the spatial radar data and for a more appropriate definition of the second-moment properties of the radar observation errors. Such efforts are considered essential in view of the upcoming deployment of NEXRAD Doppler radars in the USA.

ACKNOWLEDGEMENTS

This work was supported by the USA Cold Regions Research and Engineering Laboratory, Corps of Engineers, US Department of the Army, under Contract DACA89-87-K-0004. Additional support was provided by the US National Science Foundation under the Presidential Young Investigator Award CES-8657526. The authors wish to acknowledge the assistance of Dr J. Seo of the Hydrologic Research Laboratory of the US National Weather Service in the acquisition and preparation of the RADAP II radar data.

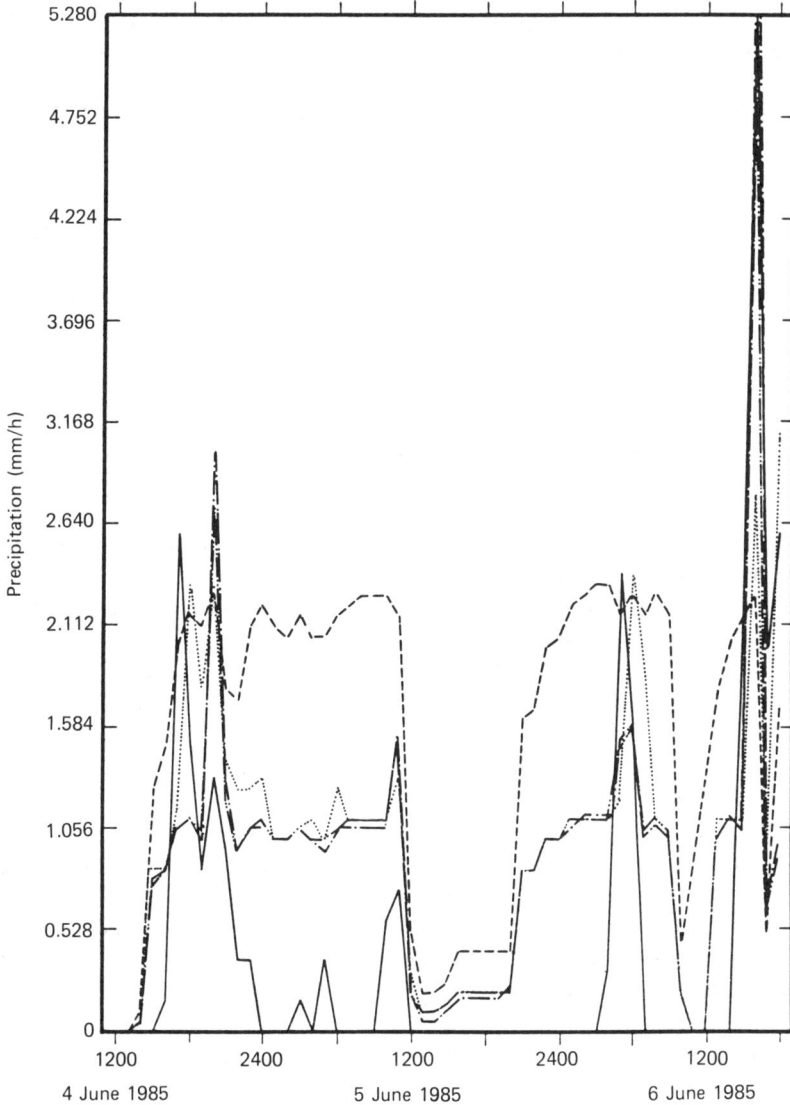

Fig. 2 — Hourly rainfall predictions and corresponding observations for the 4–6 June 1985 storm for various updating scenarios: − −, observation; − − −, deterministic; − · −, updating from reflectivity data; . . . ., updating from main data; − ·· −, updating from both reflectivity data and rain data. Predictions are with sub-optimal parameter values.

REFERENCES

Burgess, D. and Ray, P. S. (1986) Principles of radar. In: P. S. Ray (ed.) *Mesoscale meteorology and forecasting*. American Meteorological Society, Boston, MA, pp. 85–117.

Collier, C. G. (1987) Accuracy of real-time radar measurements. In: V. Collinge and C. Kirby (eds), *Weather radar and flood forecasting*. Wiley, New York, pp. 71–95.

Table 2 — Values of the least-squares performance indices for the 4–6 June 1985, storm when suboptimal values of the model parameters have been used

Updating scenario	Coefficient of efficiency	Coefficient of persistence	Coefficient of extrapolation
No updating	− 1.510	− 2.137	− 0.352
Reflectivity	0.254	0.068	0.598
Rainfall	0.103	− 0.121	0.517
Rainfall and reflectivity	0.274	0.093	0.609

Doviak, R. J. and Zrnic, D. S. (1984) *Doppler radar and weather observations.* Academic Press, New York, pp. 179–202.

Gelb, A. (ed.) (1974) *Applied optimal estimation.* MIT Press, Cambridge, MA, 374 pp.

Georgakakos, K. P. (1984) Model-error adaptive parameter determination of a conceptual rainfall prediction model. *Proceedings of the 16th IEEE 1984 Southeastern Symposium on System Theory.* IEEE Computer Society Press, Silver Spring, ML, pp. 111–115.

Georgakakos, K. P. (1986a) A generalized stochastic hydrometeorological model for flood and flash-flood forecasting, 1, Formulation. *Water Resour. Res.,* **22** (13), 2083–2096.

Georgakakos, K. P. (1986b) A generalized stochastic hydrometeorological model for flood and flash-flood forecasting, 2, Case Studies. *Water Resour. Res.,* **22** (13), 2097–2106.

Georgakakos, K. P. (1986c) State estimation of a scalar dynamic precipitation model from time-aggregate observations. *Water Resour. Res.* **22** (5), 744–748.

Georgakakos, K. P. and Bras, R. L. (1984a) A hydrologically useful station precipitation model, 1, Formulation. *Water Resour. Res.,* **20** (11), 1585–1596.

Georgakakos, K. P. and Bras, R. L. (1984b) A hydrologically useful station precipitation model, 2, Case Studies. *Water Resour. Res.,* **20** (11), 1597–1610.

Georgakakos, K. P. (1987) Real time flash flood prediction. *J. Geophys. Res.,* **92** (D8), 9615–9629.

Jazwinski, A. H. (1970) *Stochastic processes and filtering theory.* Academic Press, New York, pp. 311–318.

Krajewski, W. F. (1987) Co-kriging radar-rainfall and rain gage data. *J. Geophys. Res.,* **92** (D8), 9571–9580.

Lee, T. H. and Georgakakos, K. P. (1989) A two-dimensional stochastic-dynamical quantitative precipitation forecasting model. *J. Geophys. Res.* (In press).

Rajaram, H. and Georgakakos, K. P. (1989) Recursive parameter estimation of hydrologic models. *Water Resour. Res.,* **25** (23), 281–294.

35

Short-term forecasting for water level of a flash flood by radar hyetometer

T. Moriyama and M. Hirano
Department of Civil Engineering Hydraulics, Kyushu University, Fukuoka 812, Japan

ABSTRACT

Recently, weather radar has been fully utilized and detailed real-time information over a wide range of area has become obtainable with high time–space resolution, but the accuracy of the radar hyetometer is not always high enough to use for flood forecasting.

In this paper, a forecasting method by using radar data is proposed, in which a parameter for the $Z–R$ relationship of the radar hyetometer is included in the system parameters and identified together by Kalman filtering. The application of this method to the Onga river basin shows that the forecasting of water stages by using radar data is possible with the same accuracy of that by rainfall data on the ground.

1. INTRODUCTION

Recently, rainfall observation systems using radar and telemeters are being fully utilized along the river basins throughout Japan. Flood forecasts using the on-line information from these systems will predict and provide effective ways to lessen or prevent damage due to floods. However, the forecasting method for floods has yet to be fully developed.

The longer the forecast made, the more time is allowed to prepare for a forthcoming flood. However, longer lead times in forecasts will require rainfall predictions which are not always accurate enough to use for flood forecast. In this situation, the short-term forecast without rainfall prediction may also be useful in making a more accurate forecast. In the Kyushu District in Japan, rivers are steep and have small catchment areas. Even the largest river, the Chikugo, has a basin of only 2860 km^2 at the mouth of the river. The lead times of forecasts required for warning and protective actions are considered to be 3 h or so in this district.

Recently, we have proposed a dynamic model for prediction of flood stages. In the model, the equations of continuity and momentum for unsteady flow were applied to a river flow and solved by the kinematic wave theory. The unit hydrograph was introduced for estimation of the lateral inflow from residual slopes. From the viewpoint that the stage data are of as major importance as rainfall data, a system model for the short-term forecasting of the flood stages was derived. By feeding on-line data regarding the rainfall and water levels into the model, the operating forecast of which are the unit hydrographs, the constants for kinematic wave theory and a parameter for the Z–R relationship of the radar hyetometer.

To examine the applicability of this system, we apply it to the Onga River in Kyushu District.

2. DEVELOPMENT OF FORECASTING MODEL

The equations of continuity and momentum for unsteady flow in an open channel are

$$\frac{\partial A}{\partial t} + \frac{\partial Q}{\partial x} = q_* \tag{1}$$

$$A = KQ^P \tag{2}$$

where A is the cross-sectional area, Q is the flow discharge, t is the time, x is the downstream distance and q_* is the lateral inflow.

Equations (1) and (2) are solved by the kinematic wave theory as

$$A_2(t) - A_1(t - \tau_{2,1}) = \int_{t-\tau_{21}}^{t} q_* dt - \int_{t-\tau_{2,2}}^{t} A^P \frac{\partial K}{\partial x} dt \tag{3}$$

where $\tau_{2,1}$ is the time which the flood wave takes to travel from the upstream site to the downstream site, and the subscripts 1 and 2 denote the upstream and the downstream sites respectively. The second term of the right-hand side of equation (3) can be approximated as

$$\int_{t-\tau_{2,1}}^{t} A^P \frac{\partial K}{\partial x} dt = \int_{0}^{L_{21}} \frac{A}{KP} \frac{\partial K}{\partial x} dt$$

$$\approx \frac{L_{21}}{2} \left(\frac{A_2}{K_2 P_2} \frac{\partial K_2}{\partial x} + \frac{A_1}{K_1 P_1} \frac{\partial K_1}{\partial x} \right)$$

$$= C_1 A_1(t - \tau_{21}) + C_2 A_2(t) \tag{4}$$

where L_{21} is the length between the upstream and the downstream sites respectively, and $C = (L/2KP)(\partial K/\partial x)$. In the same way as mentioned above, the cross-sectional area at the moment $t + I$ can be given by

$$A_2(t+I) - A_1(t+I - \tau_{21}) = \int_{t+1+\tau_{21}}^{t+1} q_* dt - C_1 A_1(t+I - \tau_{21}) - C_2 A_2(t+I). \tag{5}$$

Subtracting equation (3) from equation (5), one obtains

$$A_2(t+I) - A_2(t) = k[A_1(t+I-\tau_{21}) - A_1(t-\tau_{21})] + \left(\int_{t+I+\tau_{21}}^{t+I} q_* \, dt - \int_{t-\tau_{21}}^{t} q_* \, dt \right) \tag{6}$$

where $k = (1 - C_1/(1 + C_2)$ and $m = 1/(1 + C_2)$. Equation (6) is the fundamental computing equation of the cross-sectional area at the forecasting time I. The cross-sectional areas can easily be transformed into the water stages.

The lateral inflow q_* can be expressed by

$$\int_{t-\tau_{21}}^{t} q_* \, dt = \tau_{21} \, q_*(t - \tau_{s21}) \tag{7}$$

where τ_{s21} is the appropriate value of time between 0 to τ_{s21}. By the unit hydrograph, equation (7) is given by

$$q_*(t - \tau_*) = \int_0^{t-\tau_s} lfu(\tau)r(t - \tau_s - \tau) \, d\tau \tag{8}$$

where l is the length of the slope, f is the runoff coefficient, $u(\tau)$ is the instantaneous unit hydrograph, r is the rainfall intensity and τ_s is the lag time which the rainfall takes to flow into the river. Substitution of equation (8) into equation (6) yields

$$A_2(t+I) - A_2(t) = k[A_1(t+I-\tau_{s21}) - A_1(t-\tau_{21})]$$
$$+ \left(\int_0^{t+l-\tau_s} U(\tau)r(t - \tau_s - \tau) \, d\tau - \right.$$
$$\left. \int_0^{t-\tau_s} U(\tau)r(t - \tau_s - \tau) \, d\tau \right) \tag{9}$$

where $U(\tau) = m\tau_{21}lfu(\tau)$.

The prediction of the water stage with a lead time I is possible when the rainfall data up to the moment $t + I - \tau_{21}$ are given. Therefore, if the lead time taken in the forecast is equal to or less than lag time τ_s, i.e. $I \leqslant \tau_s$, the prediction of rainfall is not necessary while, in the case when $I > \tau_s$, prediction of rainfall with a lead time of $I - \tau_s$ is required.

2.1   Equation for a river with a tributary
If a river has a tributary with a gauging station as shown in Fig. 1, then equation (3) is arranged for the confluence as

$$A_t(t) = \frac{1 - C_{11}}{1 + C_t} A_{11}(t - \tau_{11}) + \frac{1 - C_{12}}{1 + C_t} A_{12}(t - \tau_{12})$$
$$+ \frac{1}{1 + C_t} \left(\int_{t-\tau_{11}}^{t} q_{*11} \, dt + \int_{t-\tau_{12}}^{t} q_{*12} \, dt \right) \tag{10}$$

where A_t, A_{11} and A_{12} are the cross-sectional areas at the confluence, the upper gauging station (station 1) and the station of the tributary (station 2) respectively, τ_{11}

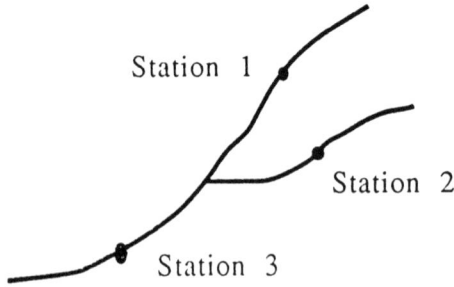

Fig. 1 — Schematic sketch of a river with a tributary.

and τ_{12} are the travelling times of flood waves from the stations 1 and 2 to the confluence respectively, and subscripts 11, 12 and t denote the station 1, station 2 and the confluence respectively. For the confluence and the downstream station, equation (3) is reformed as

$$A_3(t) = \frac{1 - C_t}{1 + C_3} A_t(t - \tau_3) + \frac{1}{1 + C_3} \int_{t-\tau_3}^{t} q_{*3} \, dt \tag{11}$$

where A_3 is the cross-sectional area at the downstream site (station 3), τ_3 is the travel time from the confluence to the downstream site and q_{*3} is lateral inflow between the confluence and station 3. By substitution of equation (10) into equation (11), we obtain

$$A_3(t) = k_1 A_{11}(t - \tau_1) + k_2 A_{12}(t - \tau_2)$$
$$+ m_1 \int_{t-\tau_1}^{t-\tau_3} q_{*11} \, dt + m_2 \int_{t-\tau_2}^{t-\tau_3} q_{*12} \, dt + m_2 \int_{t-\tau_3}^{t} q_{*3} \, dt \tag{12}$$

where

$$\tau_1 = \tau_{11} + \tau_3, \quad \tau_2 = \tau_{12} + \tau_3$$
$$k_1 = \frac{(1 - C_1)(1 - C_{11})}{(1 + C_3)(1 + C_t)}, \quad k_2 = \frac{(1 - C_t)(1 - C_{12})}{(1 + C_3)(1 + C_t)},$$
$$m_1 = \frac{(1 - C_{11})}{(1 + C_3)(1 + C_t)}, \quad m_2 = \frac{(1 - C_{12})}{(1 + C_3)(1 + C_t)},$$
$$m_3 = \frac{1}{(1 + C_3)(1 + C_t)}.$$

In the same way as mentioned above, the computing equation for forecasting time I is given as

$$A_3(t + I) - A_3(t) = k_1[A_{11}(t + I - \tau_1) - A_{11}(t - \tau_1)]$$
$$+ k_2[A_{12}(t + I - \tau_2) - A_{12}(t - \tau_2)]$$
$$+ m_1\left(\int_{t+I-\tau_1}^{t+I-\tau_3} q_{*11} \, dt - \int_{t-\tau_1}^{t-\tau_3} q_{*11} \, dt\right)$$

$$+ m_2 \left(\int_{t+I-\tau_2}^{t+I-\tau_3} q_{*12}\, dt - \int_{t-\tau_2}^{t-\tau_3} q_{*12}\, dt \right)$$

$$+ m_3 \left(\int_{t+I-\tau_3}^{t+I-} q_{*3}\, dt - \int_{t-\tau_3}^{t} q_{*3}\, dt \right) \tag{13}$$

If a river has two tributaries with a gauging station as shown in Fig. 2, then one

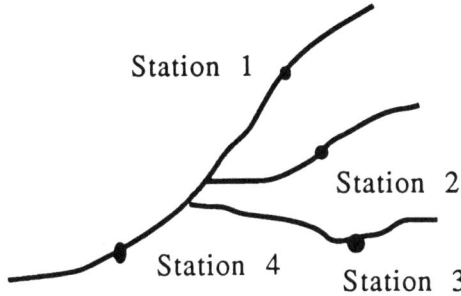

Fig. 2 — Schematic sketch of a river with two tributaries.

obtains in the same way as follows:

$$
\begin{aligned}
A_4(t+I) - A_4(t) &= k_1[A_{11}(t+I-\tau_1) - A_{11}(t-\tau_1)] \\
&+ k_2[A_{12}(t+I-\tau_2) - A_{12}(t-\tau_2)] \\
&= k_3[A_{13}(t+I-\tau_3) - A_{13}(t-\tau_3)]
\end{aligned}
$$

$$+ m_1 \left(\int_{t+I-\tau_1}^{t+I-\tau_4} q_{*11}\, dt - \int_{t-\tau_1}^{t-\tau_4} q_{*11}\, dt \right)$$

$$+ m_2 \left(\int_{t+I-\tau_2}^{t+I-\tau_4} q_{*12}\, dt - \int_{t-\tau_2}^{t-\tau_4} q_{*12}\, dt \right)$$

$$+ m_3 \left(\int_{t+I-\tau_3}^{t+I-\tau_4} q_{*13}\, dt - \int_{t-\tau_3}^{t-\tau_4} q_{*13}\, dt \right)$$

$$+ m_4 \left(\int_{t+I-\tau_4}^{t+I} q_{*4}\, dt - \int_{t-\tau_4}^{t} q_{*4}\, dt \right)$$

$$\tag{14}$$

$$\tau_1 = \tau_{11} + \tau_4, \quad \tau_2 = \tau_{12} + \tau_4, \quad \tau_3 = \tau_{13} + \tau_4$$

$$k_1 = \frac{(1-C_t)(1-C_{11})}{(1+C_4)(1+C_t)}, \quad k_2 = \frac{(1-C)(1-C_{12})}{(1+C_4)(1+C_t)}$$

$$m_1 = \frac{1-C_{12}}{(1+C_4)(1+C_1)}, \quad m_2 = \frac{1-C_{12}}{(1+C_3)(1+C_t)}$$

$$m_3 = \frac{1-C_{13}}{(1+C_4)(1+C_t)}, \quad m_4 = \frac{1}{(1+C_4)(1+C_t)}$$

where A_t, A_4, A_{11}, A_{12} and A_{13} are the cross-sectional areas at the confluence, the lower gauging station (station 4), the upper gauging station (station 1) and the stations of the tributaries (station 2 and 3) respectively T_{11}, T_{12} and T_{13} are the travelling times of flood waves from the stations 1, 2 and 3 to the confluence respectively, and subscripts 11, 12, 13, 4 and t denote station 1, station 2, station 3, station 4 and the confluence respectively.

2.2   Estimation of lateral inflow using a raingauge on the ground and a radar hyetometer

The first term of the right-hand side of equation (9) is written in the form of discrete time with unit interval as

$$A_2(t+I) - A_2(t) = k[A_1(t+I-\tau_{21}) - A_1(t-\tau_{21})]$$

$$+ \sum_{i=0}^{t-\tau_2} U(i)R(i) \tag{15}$$

where

$$U(i) = flmu(i),$$

$$R(i) = \sum_{j=0}^{t} [r(t+I+\tau_s-i-j) - r(t+\tau_s-i-j)]$$

and r is the rainfall intensity using the data from the raingauge on the ground. The system parameters to be identified are k and $U(i)$.

If the river has a tributary with a gauging station as in Fig. 1, then equation (13) becomes

$$A_3(t+I) - A_3(t) = \sum_{i=1}^{3} k_i[A_i(t+I-\tau_i) - A_i(t-\tau_i)]$$

$$+ \sum_{j=1}^{3} \sum_{i=0}^{t-\tau_{sj}} U_j(i)R_j(i). \tag{16}$$

If the river has two tributaries with a gauging station as in Fig. 2, then equation (14) becomes

$$A_4(t+I) - A_4(t) = \sum_{i=1}^{4} k_i[A_i(t+I-\tau_i) - A_i(t-\tau_i)]$$

$$+ \sum_{j=1}^{4} \sum_{i=0}^{t-\tau_{sj}} U_j(i)R_j(i). \tag{17}$$

Next, let us consider using data from the radar hyetometer. By applying the equation $z = Br^\beta$ which expresses the relation between the reflectivity factor z (mm^6/m^3) and rainfall intensity r (mm/h), the second term of the right-hand side of equation (14) can be expressed by

$$\sum_{i=0}^{t-\tau_s} U(i)Z(i) \tag{18}$$

where

$$U(i) = m\tau fl \frac{B_0}{b} \frac{1}{\beta} u(i)$$

$$Z(i) = \sum_{j=0}^{\tau_{21}} \{[z(t + I - \tau_{21} + \tau_s - i + j)]^{1/\beta} - [z(t - \tau_{21} + \tau_s - i + j)]^{1/\beta}\}$$

where β and B_0 are the standard values which are used to estimate the rainfall on a radar hyetometer, and z is the reflectivity factor observed by the radar hyetometer. The system parameters to be identified are k and $U(i)$. If the river has a tributary with a gauging station as in Fig. 1, then equations (13) and (18) become

$$A_3(t + I) - A_3(t) = \sum_{i=1}^{3} k_i[A_i(t + I - \tau_i) - A_i(t - \tau_i)]$$

$$+ \sum_{j=1}^{3} \sum_{i=0}^{t - \tau_{sj}} U_j(i)Z_j(i). \tag{19}$$

If the river has two tributaries with a gauging station as in Fig. 2, then equation (14) becomes

$$A_4(t + I) - A_4(t) = \sum_{i=1}^{4} \{k_i[A_i(t + I - \tau_i) - A_i(t - \tau_i)]\}$$

$$+ \sum_{j=1}^{4} \sum_{i=0}^{t - \tau_{sj}} U_j(i)R_j(i). \tag{20}$$

3.   APPLICATION

3.1   The applied basin
The model is applied to the Onga River whose basin is illustrated in Fig. 3. There are eight water level stations and 13 raingauges on the ground. The total area of this basin is 926 km^2.

3.2   The rainfall data
The rainfall data used in this study are from the raingauges on the ground and a radar hyetometer. Both kinds of data are observed by the Ministry of Construction. This radar site is located on Mt Shakadake in Kyushu island and the coverage area is the north part of Kyushu island as shown in Fig. 4. The whole catchment area of the Onga River is covered by this radar. The data observed by this radar are recorded on magnetic tapes every 5 min, and spatial resolution is 3 km for the radius direction and $2.8125°$ ($= 360°/128$) for the horizontal angle. The rainfall data are accumulated for an amount time (1 h in this case) on each catchment area in the Onga River.

The standard values of the parameters in the Z–R relationship are taken as $\beta = 1.58$ and $B_0 = 224.4$, which are used by the Ministry of Construction for this radar

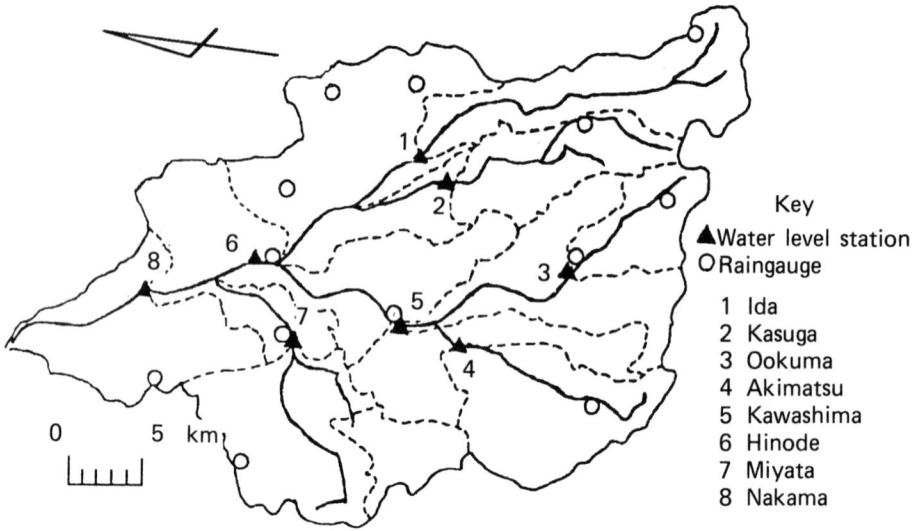

Fig. 3 — Catchment area of the Onga River.

Key
▲ Water level station
○ Raingauge

1 Ida
2 Kasuga
3 Ookuma
4 Akimatsu
5 Kawashima
6 Hinode
7 Miyata
8 Nakama

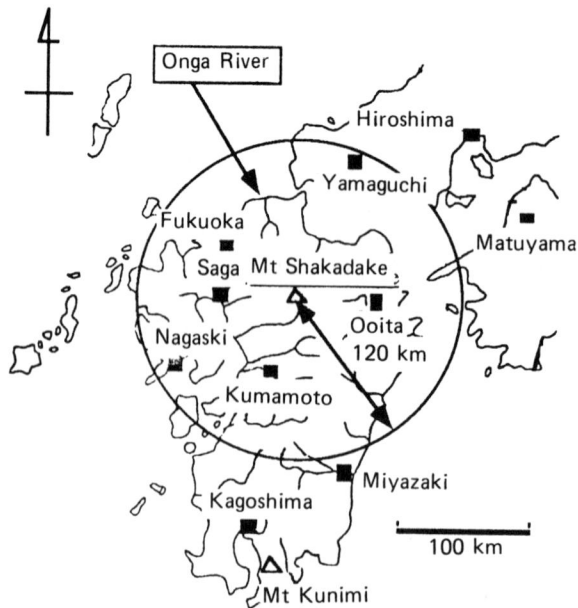

Fig. 4 — The coverage area of radar hyetometer.

hyetometer. As shown in Fig. 5, however, these values of rainfall intensity are underestimated compared with the rainfall intensity from the raingauge on the ground.

3.3 The water level data
The water level is recorded every 1 h at eight stations.

3.4 Identification of the system parameters
To identify the system parameters, we used the Kalman filtering method as follows.
The state equation and measurement function are written as

$$X(j+1) = X(j) + w(j) \qquad (21)$$

and

$$Y(j+1) = M(j+1) + v(j) \qquad (22)$$

where X is the state vector, j is the discrete time whose interval is unit, w is the system noise, Y is the measurement vector, M is the measurement matrix and v is the measurement noise.

For example, applying the Kalman filtering to equation (16) which corresponds to the measurement function, the terms in equations (21) and (22) are expressed as

$$Y(j+1) = A_2(t+I) - A_2(t)$$
$$M(j+1) = [A_1(t+I-\tau_{21}) - A_1(t-\tau_{21}),\ R(1),\ R(2),\ ...]$$
$$X(j+1) = [k,\ U(0),\ U(1),\ U(2),\ ...]. \qquad (23)$$

3.5 The application of equations
The water level at Nakama station is predicted by the procedure as follows.

(1) The water level is translated to the cross-sectional area.
(2) The cross-sectional area at Kawashima station is computed from equation (16) or (19) using the cross-sectional area at the upper stations at Chikuma and Akimatsu.
(3) The cross-sectional area at Hinode station is computed from equations (17) or (20) using the cross-sectional area at the upper stations at Kawashima, Ida and Kasuga.
(4) The cross-sectional area at Nakama station is computed from equation (16) or (19) using the cross-sectional area at the upper station at Hinode and Miyata.
(5) The computed value of the cross-sectional area is translated to the water level.

Here, the water level which is forecasted by using the data from the radar hyetometer is compared with the water level computed by using the data from the raingauge on the ground. An example of the former is shown in Fig. 6 and the latter in Fig. 7. Each example has a lead time of 3 h.

In Figs 6 and 7, the predicted values of the stage with a lead time of 3 h are compared with the observed values at Nakama station. In these computations the values of the lag times of rainfall for each basin were determined so as to avoid rainfall predictions. Although the radar gives smaller values of rainfall than the

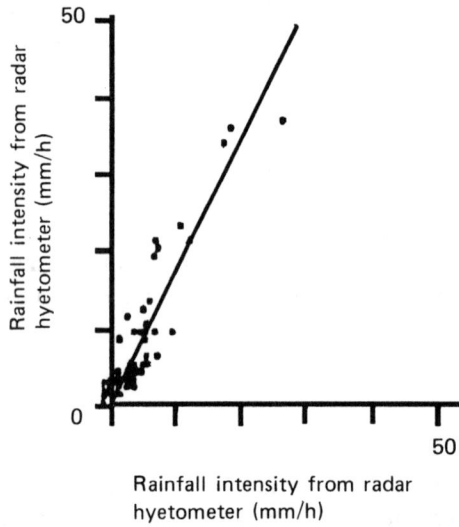

Fig. 5 — Rainfall intensity of ground raingauge versus radar hyetometer.

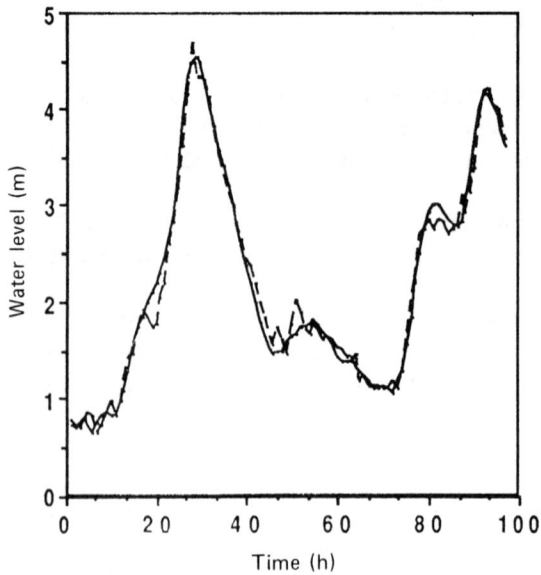

Fig. 6 — 3 h prediction at Nakama station (radar Hyetometer): ———, observed, – – –, calculated.

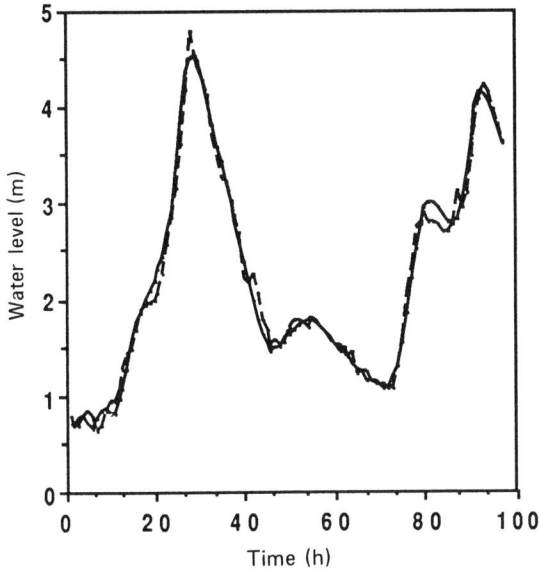

Fig. 7 — 3 h prediction at Nakama station (raingauge on the ground): ————, observed; – – –, calculated.

ground raingauges do the forecasting values obtained by using radar data have the same accuracy as those obtained by ground rainfall data.

4. CONCLUSION

A dynamic model for real-time forecasting of the water stages has been developed from the basic equations of unsteady flow. The inputs into the model are the water stages of the upper stages and data from the raingauges on the ground or radar hyetometer. The Kalman filtering technique is used to identify the system parameters of the model. The model was applied to the Onga River by using the data from the raingauges on the ground and from the radar hyetometer. In both cases, the forecasting water levels with a lead time of 3 h showed close agreement with the observed data. The model was applied to the Onga River by using the data both from the raingauges on the ground and from the radar hyetometer. In both cases, the forecasted water levels with a lead time of 3 h showed close agreement with the observed values, in spite of the fact that there is significant difference between both the rainfall data.

ACKNOWLEDGEMENTS

The rainfall data from the raingauges on the ground and the radar hyetometer and the water levels were observed by the Ministry of Construction.

REFERENCES

Alekhin, Yu. M. (1964) *Short-range forecasting of lowland-river runoff*. Israel Program for Scientific Translations, Jerusalem.

Hirano, M., Moriyama, T., Matsui, M., Nakayama, H. and Matuo, K. (1986) Real-time forecasting for water stages of a flash flood. *Proceedings of the 5th Congress of APD*. International Association of Hydrological Sciences.

36

Advanced use in rainfall prediction of a three-dimensionally scanning radar

E. Nakakita, M. Shiiba, S. Ikebuchi and T. Takasoa
Disaster Prevention Research Institute, Kyoto University,
Gokasho, Uji, Kyoto 606,
Japan

ABSTRACT

A computation method of the rainfall distribution to be utilized for the short-term rainfall prediction based on the analysis of a heavy-rainfall event which arose during the Baiu season in Japan, and a method of extracting information required to utilize this computation method in real time from the three-dimensionally scanning radar are presented. The former method of rainfall distribution is based on the idea that the rainfall occurs because of the interaction between the moving modelled instability field and the given stream field of the water vapour under the influence of the topography, and the results were a good representation of the time variation in the shape of the rain band which moved across the radar observation area. On the other hand, the latter method consists in estimating the conversion rate from water vapour to liquid water by use of the information from the three-dimensionally scanning radar, and, in terms of the water budget, the estimated values of the conversion rate are consistent with the water vapour flux which is estimated by the computation method.

1. INTRODUCTION

Many methods of short-term rainfall prediction have been proposed in Japan (e.g. Tatehira and Makino 1974, Ohkura *et al.* 1983, Shiiba, *et al.* 1984) and other countries (e.g. Wilk and Gray 1970, Austin and Bellon 1974). these methods, however, are essentially an extrapolation method of the pattern of the rainfall distribution detected by the radar in the form of plan position indicator (PPI) or constant-altitude plan position indicator (CAPPI) and are only applicable to a lead time of 1 or 2 h because variations of the rainfall distribution are too complicated under the influence of the topography, in particular, in the mountainous country Japan, to be represented by extrapolation of a two-dimensional distribution pattern. In particular, it is important for flood forecasts and real-time dam operations but is very difficult, owing to the influence of topography, to predict time variations in heavy rainfall from an organized convective cell system according to typhoons or fronts.

To cope with this and in particular to extend the lead time of prediction of the heavy rainfall influenced by typography, a meteorologically based prediction method by use of the information from radar should be developed. Therefore, information about the water vapour, the source of precipitation, is indispensable to the development. Furthermore, to realize this, utilization of information about three-dimensional distribution of rainfall is required.

From this point of view, we have been investigating how to utilize a three-dimensionally scanning radar (in the rest of this paper, *three-dimensional radar* refers to the radar of this type) for short-term rainfall prediction based on meteorology. First, we started the investigation with the visualization of three-dimensionally spread rainfall distributions and their time variation using computer colour graphics, in order to investigate how fine a resolution the three-dimensional radar that we use has (Nakakita *et al*. 1987, 1988a). Next, using many graphic screens together with meteorological data such as surface wind and upper observations, we investigated the features of a heavy-rainfall event and determined the fundamental relations between the time variation in the rainfall, the inflow pattern of the water vapour in the lower atmospheric layer and the topography, on a meso-β scale (Nakakita *et al*. 1987, 1988c). Furthermore, to confirm the estimation of the features, we carried out some numerical experiments based on a mesoscale dynamic model (Nakakita *et al*. 1988c).

This paper includes not only a summary of these investigations but also the computation method of the rainfall distribution to be utilized for the short-term rainfall prediction based on the idea that the rainfall occurs because of the interaction between the moving modelled instability field and the given stream field of the water vapour under the influence of the topography, and a method of extracting information required to utilize this computation method in real time from the three-dimensionally scanning radar (Nakakita *et al*. 1988b).

2. FEATURES OF A HEAVY-RAINFALL EVENT DETECTED BY THREE-DIMENSIONAL RADAR

The three-dimensional radar that we used is the Miyama radar installed by the Ministry of Construction in 1981, which is located in the central part of the Kinki District in Japan as Fig. 1 shows. The observation domain of the radar is the inside of the circular cylinder which has a height of about 15 km and has a circular horizontal section with a radius of 120 km (Fig. 2). The beam scanning procedure in routine work is presented in Fig. 3. The beam scans only azimuthally; hence the elevation angle varies discretely. It takes 5 min to scan all the observation domain. Therefore, data are collected every 5 min. The wavelength is 5 cm, and the received power is averaged over the region with a length of 3 km in the beam direction and a width of $(360/128)°$ in the azimuthal direction. In our analysis, grid points are arranged in a

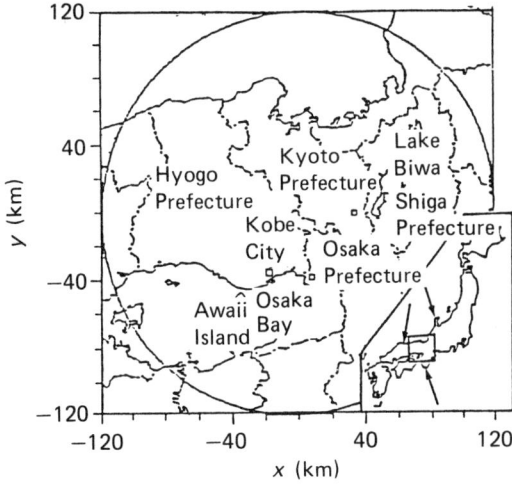

Fig. 1 — Locations of the Kinki District and the Miyama radar.

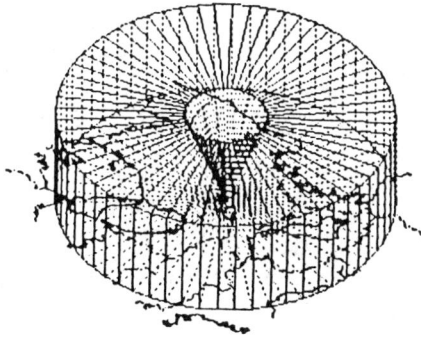

Fig. 2 — Observation domain of the Miyama radar.

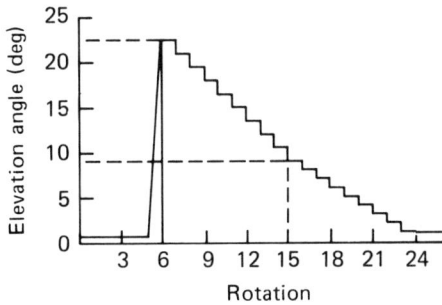

Fig. 3 — Beam-scanning procedure of the Miyama radar.

Cartesian coordinate system over the Kinki District with a horizontal interval of 3 km×3 km and a vertical interval of 1 km, and the original data observation are transformed into the values on these grid points.

Having described the characteristics of the Miyama radar, let us now turn to the features of a heavy-rainfall event which arose along a Baiu front in July 1986. Baiu front is a stationary front which appears near or south of the Japanese Archipelago from early June to mid-July. This air pattern continues for about 40 days, and this period is called the Baiu season or 'Tsuyu', the rainy season. If a moist tongue of air comes into the stationary front from the south, it brings localized heavy rainfall to western Japan.

Fig. 4 shows the GMS infrared image at 0100 Japanese Standard Time (JST) on 22 July 1986. The band of clouds extending from China to northeast in the Baiu front, and the Kinki District is under the warm air which spreads to the south of the front.

Fig. 4 — GMS infrared image at 0100 JST on 22 July 1986.

Furthermore, over the Kinki District, organized convective clouds can be detected. Fig. 5 shows the distribution of rainfall intensity at 2.5 km height with the surface wind. The rainfall intensity was estimated from the radar reflective factor by use of the size distribution of diameter of raindrops proposed by Marshall and Palmer (1948). The surface wind was estimated by linear interpolation of the data observed by Automated Meteorological Data Acquisition System (AMeDAS), which is managed by the Japan Meteorological Agency. The observation points of AMeDAS are arranged on a meso-β scale, and the observation height is from about 10 m to several metres. In addition, observation is done every hour.

The curved rain band extending from northwest to southeast corresponds to the organized convective clouds in Fig. 4, although the rain band is located slightly westward owing to the movement of the system towards the east. In this rain band,

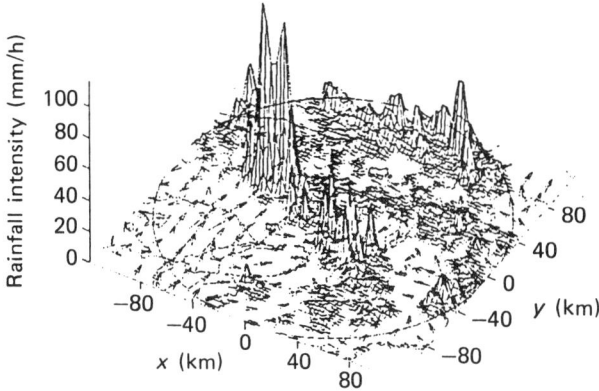

Fig. 5 — Rainfall distribution at 2.5 km height estimated from radar data at 2300 JST on 21 July 1986.

two heavy-rainfall areas are found., One of them can be seen over the centre of the Hyogo Prefecture, and the other over the southern part of Kyoto Prefecture (see also Fig. 1). The central part of Hyogo Prefecture is the place where the air mass laden with moisture from the south strikes the mountain ranges in the lower atmospheric layer. In contrast, the air mass, diverted from the current which flows into the Hyogo Prefecture over Awaji Island to the east, flows into the southern part of Kyoto Prefecture through Osaka Bay. This pattern of the current on a meso-β scale in the lower atmosphere continued for more than 10 h. From this fact and the comparison between this pattern of the current and the topography around Kinki District shown in Fig. 6, one can guess that this pattern was mainly produced by the influence of the topography under the synoptic scale conditions of the atmosphere and therefore was stable for several hours. Moreover, this guess was confirmed by a three-dimensional

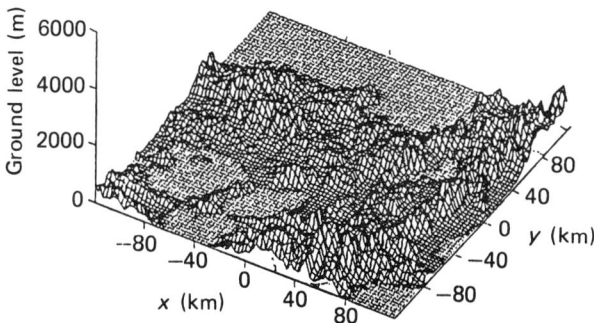

Fig. 6 — Topography around the Kinki District.

numerical experiment based on a mesoscale dynamic model by use of data only on the topography and meteorological upper observations (Nakakita *et al*. 1988c).

Having described the features of the surface wind, let us return to the rainfall distribution. Fig. 7 shows sequential rainfall distribution at 2.5 km height from 2230

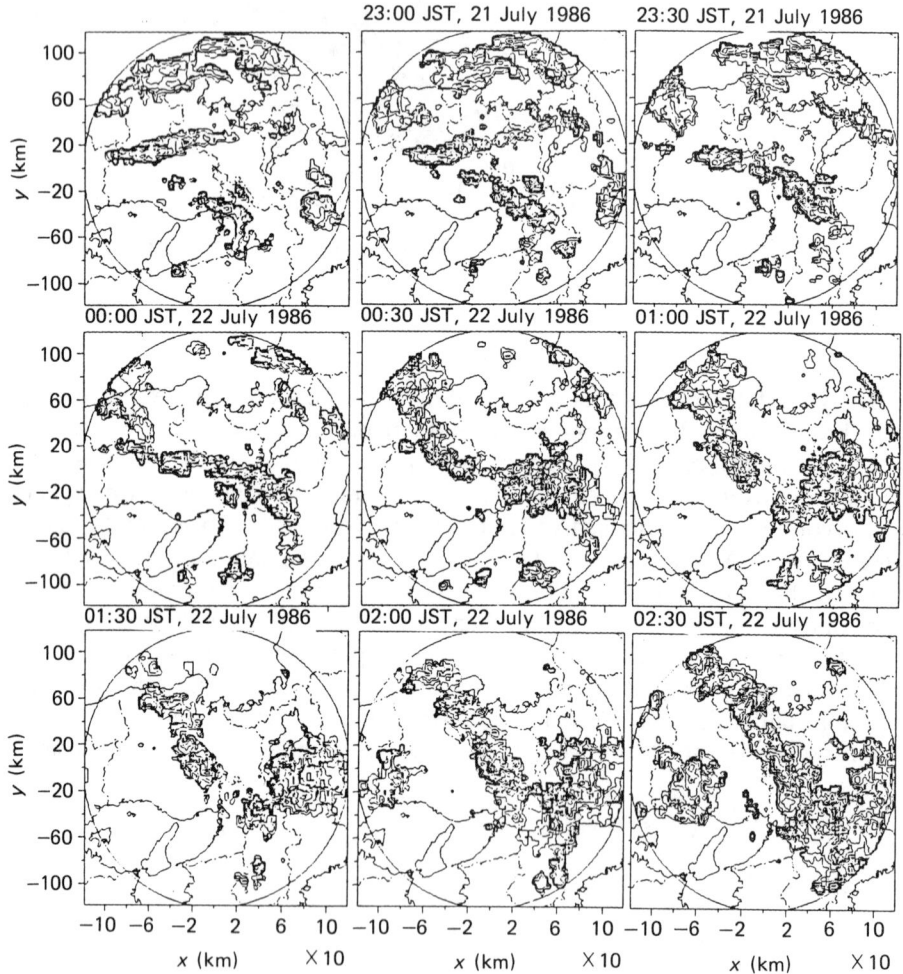

Fig. 7 — Rainfall distribution at 2.5 km height estimated from radar data. The contours correspond to 1, 2, 4, 8, 16, 32 and 64 mm/h.

JST on 21 July, 1986 to 0230 JST on 22 July 1986 for every half-hour. Contours correspond to intensities of 1, 2, 4, 8, 16, 32, 64 mm/h. In terms of the rain band which extends from northwest to southeast and moves towards the east, the shape varies from a curved type to a straight line as it moves towards the east under the influence of the topography. In particular, over the central part of Hyogo Prefecture, where the wet wind from the south is lifted by the mountains, the rainfall intensity of the rain band which moves from west to east, increases greatly. Furthermore the rainfall area over this region seems to stop its movement. From the three-dimensional rainfall distribution, it is found that the heavy rainfall is caused from severe storms of multicell type located over the mountain ridge (Nakakita *et al.* 1987, 1988a). On the other hand, over the current of the wet wind which flows into the southern part of Kyoto Prefecture from Osaka Bay, convective cells triggered around mountains which are located to the west or the northwest of Kobe City grow stronger as they move towards the east and bring localized heavy rainfall over the mountain ranges around the southern part of Kyoto Prefecture.

From these investigation we concluded that the information about the topography and the current pattern in the lower atmosphere in a meso-β scale can be used to predict where heavy rainfall will occur in the radar observation region, and because of their stationarity they are useful for a meteorologically based short-term rainfall prediction; furthermore, the variation in rainfall distribution can be represented by the idea that the rainfall occurs because of the interaction between a moving modelled instability field and a given stream field of the water vapour under the influence of the topography.

3.   BASIC EQUATIONS

In our methods, first the three-dimensional wind field is determined under the condition that the Coirolis force, the pressure gradient force on a synoptic scale and the vertical shear stress balance. Other values are calculated by the use of this wind field. The basic equations used in this method are transformed from the Cartesian coordinate system (x, y, z) into a terrain-following coordinate system (x, y, s) by the transformation

$$s = \frac{z - h(x, y)}{H - h(x, y)}, \tag{1}$$

which was adopted by Colton (1976), where H is the elevation of the top grid point in the model and $h(x, y)$ the terrain elevation. The basic equations are as follows. The equation of continuity is

$$\frac{\partial}{\partial x}(\rho_0 u) + \frac{\partial}{\partial y}(\rho_0 v) + \frac{\partial}{\partial s}(\rho_0 \omega) = \frac{1}{H - h}\left(\rho_0 u \frac{\partial h}{\partial x} + \rho_0 v \frac{\partial h}{\partial y}\right) \tag{2}$$

The horizontal momentum equations are

$$f(v - v_{g0}) + \frac{1}{\rho_0 (H - h)^2} \frac{\partial}{\partial s}\left(\rho_0 K \frac{\partial u}{\partial s}\right) = 0,$$

$$-f(u - u_{g0}) + \frac{1}{\rho_0 (H - h)^2} \frac{\partial}{\partial s}\left(\rho_0 K \frac{\partial v}{\partial s}\right) = 0 . \tag{3}$$

The thermodynamic equation is

$$\frac{\partial \theta}{\partial t} + u\frac{\partial \theta}{\partial x} + v\frac{\partial \theta}{\partial y} + \omega\frac{\partial \theta}{\partial s} = S_t,$$ (4)

The water vapour conservation equation is

$$\frac{\partial m_v}{\partial t} + u\frac{\partial m_v}{\partial x} + v\frac{\partial m_v}{\partial y} + \omega\frac{\partial m_v}{\partial s} = -S_m .$$ (5)

The liquid-water conservation equation is

$$\frac{\partial m_l}{\partial t} + u\frac{\partial m_l}{\partial x} + v\frac{\partial m_l}{\partial s} = S_m + \frac{\rho_w}{\rho_0}\frac{\partial R}{\partial z}$$ (6)

$$R = \rho_0 w_t m_l .$$ (7)

In the above,

$$\omega = \frac{\partial s}{\partial t} + u\frac{\partial s}{\partial x} + v\frac{\partial s}{\partial y} + w\frac{\partial s}{\partial z}$$ (8)

and (u, v, w) is the wind velocity, (u_g, v_g) is the geostrophic wind velocity, ρ, and ρ_w are the density of the air and the density of the water, θ is the potential temperature, m_v is the water vapour mixing ratio and m_t is the liquid-water mixing ratio. The subscript θ indicates the variables of the synoptic scale. Moreover,

$$K = \begin{cases} 10 \ m/s & z - h > d \\ 0.35(z - h)u_* , & z - h < d \end{cases}$$ (9)

and the value of the friction velocity u_* should be selected so that $0.35du_* = 10$ m/s. The value of d will be mentioned later. On the other hand, w_t is the relative fall velocity of water particles and is computed by the empirical equation obtained by Ogura and Takahasi (1971) for the first trial, as was done by Colton (1976).

4.   ESTIMATION OF VALUES ON A SYNOPTIC SCALE

The synoptic values of ρ and (u_g, v_g) are estimated from TTAA data of the upper observation over three points around the Kinki District (the positions are indicated by the arrows in Fig. 1) based on the relations

$$u_g = -\frac{1}{f}\frac{\partial \phi}{\partial y}, \quad v_g = \frac{1}{f}\frac{\partial \phi}{\partial x}, \quad \frac{\partial \phi}{\partial p} = -\frac{1}{\rho} = -\frac{RT}{p} \tag{10}$$

where ϕ is the geopotential height, p, the pressure, T the temperature and R the gas constant, and the hydrostatic assumption is used. From these relations, the assumption

$$\phi = [A_x(\ln p)^2 + B_x \ln p + C_x]x + [A_y(\ln p)^2 + B_y \ln p + C_y]y + \\ + A_c(\ln p)^2 + B_c \ln p + C_c \tag{11}$$

leads to

$$u_g = \frac{-[A_y(\ln p)^2 + B_y \ln p + C_y\}}{f}, \quad v_g = \frac{A_x(\ln p)^2 + B_x \ln p + C_x]}{f} \tag{12}$$

$$T = \frac{-[2(A_x x + A_y y + A_c)\ln p + B_x x + B_y y + B_c]}{R}.$$

Because all variables in equations (11) and (12) are given by TTAA data, by the method of least squares of the residuals between the left- and right-hand sides of these four equations one can estimate the values of A_x, B_x, C_x, A_y, B_y, C_y and C_c, and, therefore, the synoptic values of ρ and (u_g, v_g) (Nakakita *et al.* 1988c, d).

5.   ESTIMATION OF THREE-DIMENSIONAL WIND FIELD

The three-dimensional wind field is estimated by the following procedure by the use of equations (2) and (3). These equations are approximated to difference equations by the central-difference scheme. Moreover, grid points are arranged such that

$$s = 0, 5, 10, 50, 100, 200, 400, 600, 800, 1000, 1200, 1400, 1600, 1800, 2000, \\ 3000, 4000, 5000, 6000, 7000, 8000, 9000, 10000, 11000 \, (\times 1/11\,000)$$

in the vertical direction so that the vertical profiles of horizontal wind should be smooth when H equals 11 km. With top and bottom boundary conditions, simultaneous linear equations are independently obtained for every vertical line and can be easily solved.

The boundary conditions that we use are

$$(u, v) = (u_b(x, y), v_b(x, y)) \, (s = 10/11\,000), \\ (u, v) = (u_{g0}(x, y, 1), v_{g0}(x, y, 1)) \, (s = 1). \tag{13}$$

Because, as mentioned above, the patterns of the surface wind on a meso-β scale observed by AMeDAS are stable and can be easily used for the short-term prediction of rainfall, we estimate the values of the bottom boundary by the use of AMeDAS data. The procedure of computing $(u_b(x, y), v_b(x, y))$ is as follows.

(1) For the grid point over the land, the linearly interpolated vectors from those of AMeDAS are stored.

(2) Over the sea, the solutions at $s = 10/10\,000$ of the simultaneous linear equations obtained under the condition that $u = v = 0$ at $s = 0$ are stored, and the value of d is selected so that the order of velocity obtained by this procedure equals that of the wind velocity observed by AMeDAS.

(3) These stored vectors are smoothed by the averaging over 36 km × 36 km area, and these averaged vectors are used as $(u_b(x, y), v_b(x, y))$.

6. MODELLING OF THE MOVING INSTABILITY FIELD AND PROCEDURE OF RAINFALL COMPUTATION

We take a mesodisturbance as a moving field that has a high ability to convert water vapour to liquid water and call this field the instability field. Furthermore, over the mountainous region, the approach of the precipitation field should be applied not to the rainfall distribution itself but to the instability field because, as mentioned above, the rainfall distribution and its variation are too complicated to be represented by any stochastic models of precipitation fields, when they are influenced by the topography. In other words, the instability field which can be applied to the stochastic models should be defined. Many kinds of instability are defined in meteorology and our aim is to utilize the concept of the convective instability. For the first trial, however, we defined the instability field as the distribution of index α which is the percentage of the amount of water vapour that converts to liquid water with respect to the amount of supersaturated water vapour.

For the first trial, we define the distribution of index α such that

$$\alpha(r) = \frac{a}{\sqrt{2\pi}\sigma} \exp\left[-\frac{1}{2}\left(\frac{r}{\sigma}\right)^2 \right] \tag{14}$$

around a moving centre line of instability field, considering an application to the stochastic precipitation field. In this equation, r is the horizontal distance from the moving centre line. The parameter a is defined such that $\alpha(0) = 2$ when $\sigma = 10$ km. Therefore, there is region where $\alpha > 1$. Although the three-dimensional wind field which can be estimated by the method described in section 5 does not include any convection, because the value of α is greater than 1, we can represent the effect of convection on the rainfall as follows when convection exists, the air mass laden with moisture can be lifted from a lower layer to a higher layer where the pressure and the temperature are lower and, therefore, the saturation water vapour is lower and the amount of water vapour that converts to liquid water is larger than those in the lower layer. Other parameters to be determined are the velocity and angle ϕ of the central line to the x axis.

The rainfall distribution is calculated using equations (4)–(7). These equations also approximate to the difference equations of the forward upstream scheme. S_t and

S_m are calculated indirectly by the procedure described below, in order to express the interaction between the inflow water vapour and the instability field. The procedure is as follows.

(1) Time-updated values θ^*, m_v^*, m_l^* and m_s^* are computed under the condition that S_t, $S_m = 0$ in equations (4)–(6), where m, is the saturation mixing ratio.
(2) According to the definition of the instability field, updated values are corrected by the following procedure based on the method of Asai (1965):

$$\delta m = \begin{cases} \alpha(m_v^* - m_s^*), & (m_v^* > m_s^*) \\ m_v^* - m_s^*, & (m_v^* < m_s^*) \end{cases} \tag{15}$$

$$\delta m^* = \delta m \bigg/ \left[1 + \frac{L^2}{C_p R_v} \left(\frac{1000}{p} \right)^{2(R_d/C_p)} \frac{m_s^*}{\theta^{*2}} \right],$$

$$\delta m^{**} = \begin{cases} \min(m_v^*, \delta m^*), & \delta m^* > 0 \\ -\min(m_l^*, -\delta m^*), & \delta m^* < 0 \end{cases} \tag{16}$$

$$\theta = \theta^* + \frac{L}{C_p} \left(\frac{1000}{p} \right)^{(R_d/C_p)} \delta m^{**}, \quad m_v = m_v^* - \delta m^{**}, \quad m_l = m_l^* + \delta m^{**} \tag{17}$$

where L is the latent heat of condensation, C_p is the specific heat at constant pressure, R_v and R_d are the gas constants for the water vapour and the dry air, and p (mbar) is the pressure.

The boundary conditions are needed only for the region in the boundary where the wind flows into the domain because we use the upstream scheme. In the south and north, the initial values are given. On the other hand, for another boundary, the condition that the partial derivatives normal to the boundary equal to zero is used.

7.   METHOD OF ESTIMATING THE CONVERSION RATE OF WATER VAPOUR USING THREE-DIMENSIONAL RADAR

Finally, we present a method of estimating the conversion rate from water vapour to liquid water by using the information from the three-dimensional radar. We have developed this method in order to make it possible to estimate the parameters of the instability field in real time.

The basic equation is

$$\frac{\partial m_l}{\partial t} + u \frac{\partial m_l}{\partial x} + v \frac{\partial m_l}{\partial y} + w \frac{\partial m_l}{\partial z} = \frac{Q}{\rho_0} + \frac{\rho_w}{\rho_0} \frac{\partial R}{\partial z}, \quad m_l = \frac{M}{\rho_0} \tag{18}$$

which is equivalent to equation (6), but is an expression in the Cartesian coordinate system, where M is the liquid water content. The unknown variable Q is the conversion rate from water vapour to liquid water. (u, v, w) and ρ_0 are estimated from the methods described above. On the other hand, M and R are estimated from the radar reflective factor Z by the use of the size distributions proposed by Marshall and Palmer (1948)and Gunn and Marshall (1958). Below the height of the bright band, the Marshall–Palmer distribution is employed and a value of 0.93 is used for $|K\pi|^2$, which is the conventional notation for the complex refractive index in the radar equation. Below the height of the bright band, the Gunn–Marshall distribution is employed and a value if 0.197 is used for $|K|^2$. Furthermore, at the height of the bright band, R and M are calculated by interpolation using the values at the grid points just above and below the height, in order to reduce the estimation error of R and M caused by the appearance of the bright band.

Equation (18) approximated to the difference equation by the central scheme. Moreover, in this method, $\Delta x = \Delta y = 3 \, km$, $\Delta z = 1$ km and $\Delta t = 5$ min. In contrast, the values of R and M are smoothed by the averaging over 15 km $\times$ 15 km in horizontal space and 20 min in time before substitution into the equation.

In addition, we can also estimate both Q and w simultaneously using the method of least squares under the assumption that Q and w are uniform over the grid points around the grid point at which we want to estimate the values of Q and w. By this method we can estimate the value of w on a meso-β scale from the viewpoint of the water balance, although the wind field which we can estimate by the method described in section 5 is that of nearly synoptic scale, and therefore the order of w is much smaller than that of the horizontal wind.

8. RESULTS AND DISCUSSION

Fig. 8 shows the vertical profile of the geostrophic wind estimated from TTAA data

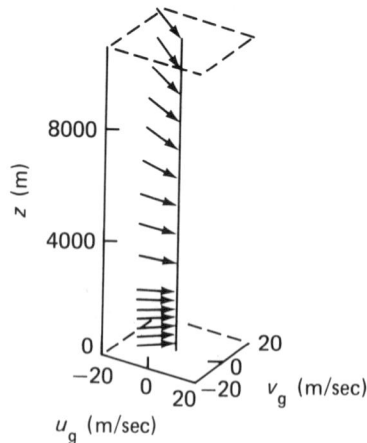

Fig. 8 — Vertical profile of the estimated geopostrophic wind in synoptic scale.

observed at 2100 JST on 21 July 1986. In contrast, Fig. 9 shows the distribution, of (u_b, v_b), which is the bottom boundary condition of the estimation of three-dimensional wind field. Although, in this trial, for the calculation of this distribution,

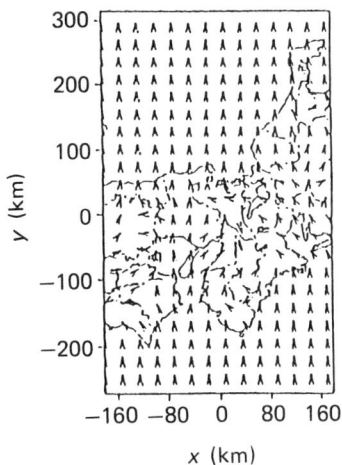

Fig. 9 — Horizontal wind used for the bottom boundary condition.

averaging was done for the data from AMeDAS from 1900 JST on 21 July 1986 to 0500 JST on 22 July 1986, the current pattern on a meso-β scale described in section 2 can be clearly detected. In addition, this pattern can be found until 200 m height above the surface the estimated three-dimensional wind field when $d = 30$ m.

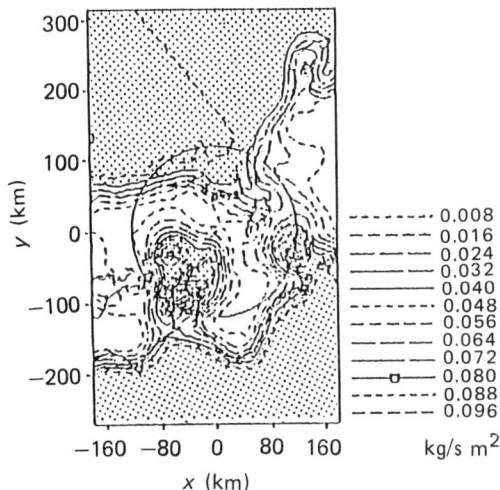

Fig. 10 — Estimated distribution of the quantity of the water vapour flux at 10 m height (time, 360 min).

Moreover, Fig. 10 shows the distribution of the quantity of the water vapour flux at 10 m height above the surface. The computation was done using equation (5), in which the right-hand side, i.e. S_m, equals to zero. We determined the initial values based on the upper observation so that the relative humidity is 98% at 0 m height, 90% at 3000 m and 50% at the top boundary. In addition, $\Delta x = \Delta y = 9$ km and $\Delta t = 15$ s.

On the other hand, in this trial, we defined the centre line of the instability field as shown in Fig. 11, corresponding to the rain band extending over Kyoto Prefecture at

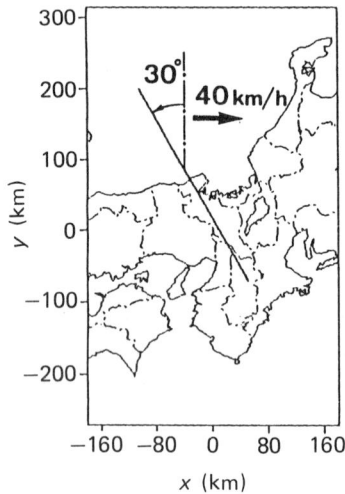

Fig. 11 — Characteristics of the centre line used to computation.

0200 JST on 22 July 1987 (see Fig. 7). The moving speed was roughly estimated from Fig. 7. The computation started when the east-side edge of the line was at the west boundary and was carried out while the line was on the area shown in Fig. 10. Initially, m_1 was zero over the domain. Moreover, to obtain the initial values of m_v, computation by use of equation (5), in which $S_m = 0$, was done for 2 h with the same time interval described above. Furthermore, as the initial values of θ, we used the estimated values on the synoptic scale. The successive rainfall distributions at the surface computed by the method described in section 5, when $\sigma = 10$ km in equation (14), are presented in Fig. 12. This result provides a good representation of the time variation in the shape of the rain band which varied from a curved type to a straight line and therefore is ample proof that the rainfall was much influenced by the mountains. In addition, this time variation cannot be represented by any methods based on extrapolation. Furthermore, over the area near Kobe city, the rainfall is calculated from 220 to 330 min. This is proof that convective cells can be triggered around this area as mentioned in section 2. On the other hand, Fig. 13 shows the results when $\sigma = 30$ km. The represented shape of the rain band is very like that of the radar observations at 0000 or 0030 JST on 22 July 1986.

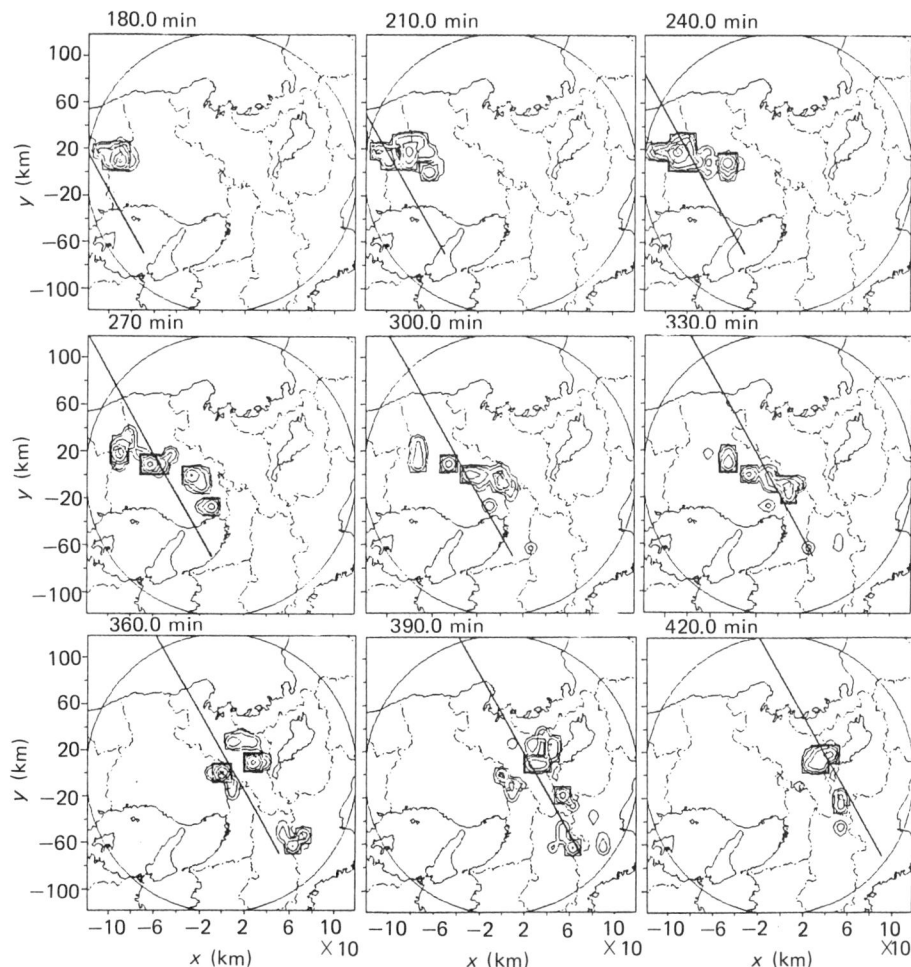

Fig. 12 — Computed rainfall distribution at the surface when $\sigma = 10$ km. The contours correspond to 1, 2, 4, 8, 16, 32 and 64 mm/h. The moving line is the centre of the instability field.

Finally, we represent the results of estimating the conversion rate of water vapour using the three-dimensional radar. Fig. 14 shows the estimated distribution of the conversion rate Q at 3.5 km height for 2300 JST on 21 July 1986. The contours correspond to 1×10^{-7}, 2×10^{-7}, 4×10^{-7} and 8×10^{-7} kg/m^3 s, and the full lines represent positive areas. No negative area exists in this figure. In particular, large positive values are estimated around the area of heavy rainfall described in section 2 (see Fig. 5). In addition, the order of the estimated conversion rate is consistent with the order of the water vapour flux, considering the extent of the heavy-rainfall area. In contrast, Fig. 15 shows the comparison between the time series of range-averaged rainfall intensity at 1.5 km height and those of the integral of the estimated condensation rate from 2 to 4 km height. This figure means that about half the liquid water that falls through 1.5 km height is produced in this vertical range and therefore

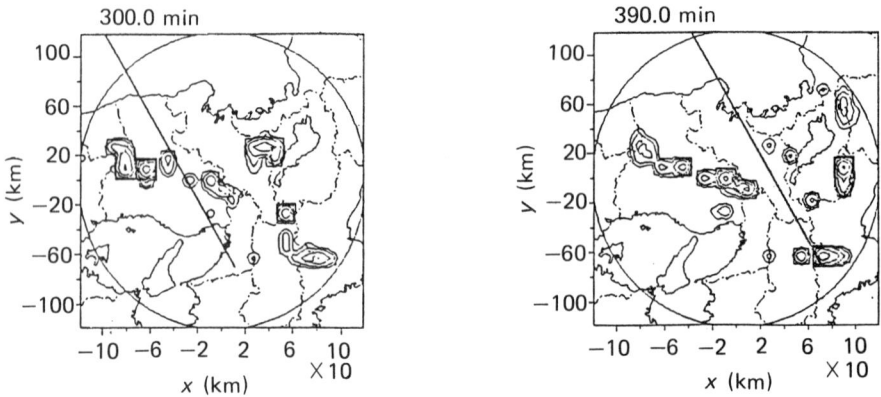

Fig. 13 — Computed rainfall distributions at the surface when σ = 30 km. The contours correspond to 1, 2, 4, 8, 16, 32 and 64 mm/h.

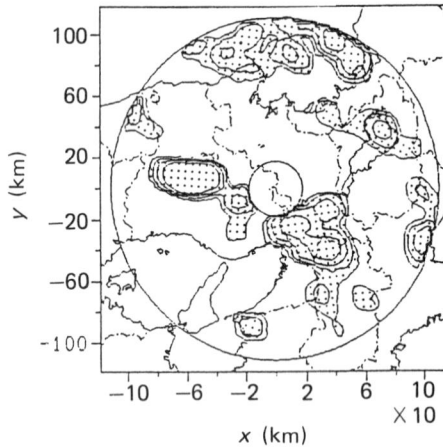

Fig. 14 — Estimated distribution of the conversion rate at 3.5 km height from water vapour to liquid water at 2300 JST on 21 July 1986. The contours correspond to 1×10^{-7}, 2×10^{-7}, 4×10^{-7} and 8×10^{-7} kg/m³ s.

supports our policy that the current pattern of the wind or the water vapour in the lower atmosphere is important information. Furthermore, Fig. 16 shows the distribution of the estimated vertical wind. Over the heavy-rainfall area where severe storms exist, a strong upward wind is estimated.

9. CONCLUSION

Both the computation method of the rainfall distribution based on the concept of instability field and the method of estimating the condensation rate from the

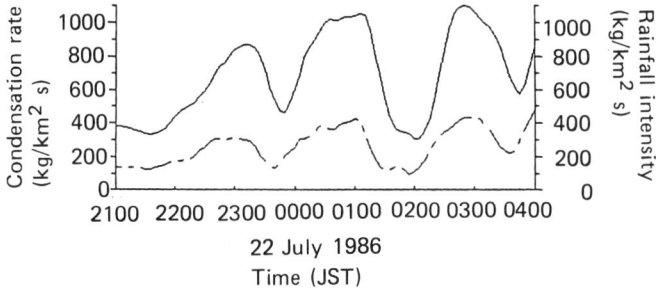

Fig. 15 — Comparison between the time series of the range-averaged rainfall intensity (——) at 1.5 km height and those of the integral of the estimated condensation rate (— · —) from 2 to 4 km height.

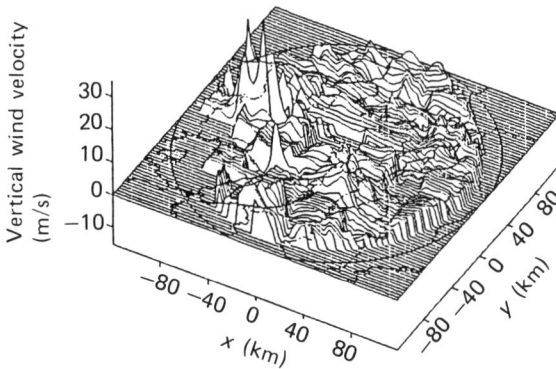

Fig. 16 — Distribution of vertical wind velocity at 3.5 km height estimated from three-dimensional radar at 2300 JST on 21 July 1986, from the viewpoint of the water balance.

information of three-dimensional radar gave effective results on the application of these methods to short-term rainfall prediction. In the next stage, we are going to unite these two methods as a real-time prediction method.

ACKNOWLEDGEMENTS

We are grateful to the Yodo River Dams Control Office in the Ministry of Construction for providing the radar data used in this research, and to Mr K. Yamaura for his help in computation.

REFERENCES

Asai, T. (1965) A numerical study of the air-mass transformation over the Japan sea in winter. *J. Meteorol. Soc. Jpn.*, **43**, 1–15.

Austin, G. L. and Bellon, A. (1974) The use of digital weather record for short-term precipitation forecasting. *Qt. J. R. Meteorol. Soc.*, **100**, 658–664.

Colton, D. E. (1976) Numerical simulation of orographically induced precipitation distribution for use in hydrologic analysis. *J. Appl. Meteorol.*, **15**, 1241–1251.

Gunn, K. L. S. and Marshall, J. S. ((1958) The distribution with size of aggregate snowflakes. *J. Meteorol.*, **15**, 452–466.

Marshall, J. S. and Palmer, W. M. K. (1948) The distribution of raindrops with size. *J. Meteorol.*, **5**, 186–192.

Nakakita, E., Shiiba, M., Ikebuchi, S. and Takasao, T. (1988a) Visualization of the information from three-dimensionally scanning radar raingauge. *Proc. Jpn. Soc. Cis. Eng.*, **393** (II-9), 161–169 (in Japanese).

Nakakita, E., Shiiba, M., Ikebuchi, S. and Takasao, T. (1988b) Fundamental Study for making better use of a three-dimensionally scanning radar raingauge (II). *Ann. Disaster Prevention Res. Inst., Kyoto Univ.*, **31**B(2), 231–240 (in Japanese).

Nakakita, E., Tsutsui, M., Ikebuchi, S. and Takasao, T. (1987) Fundamental study for making better use of a three-dimensionally scanning radar raingauge. *Ann. Disaster Prevention Res. Inst., Kyoto Univ.*, **30**B(2) 265–282.

Nakakita, E., Tsutsui, M., Ikebuchi, S. and Takasao, T. (1988c) Analysis of rainfall distribution based on mesoscale dynamic model. *Ann. Disaster Prevention Res. Inst., Kyoto Univ.*, **31B** (2), 209–229 (in Japanese).

Nakakita, E., Tsutsui, M., Ikebuchi, S. and Takasao, T. (1988d) Analysis of rainfall distribution based on mesoscale dynamic models. *Proceedings of the 32nd Japanese Conference on Hydraulics*. Japanese Society of Civil Engineers, Tokyo, pp. 12–18 (in Japanese).

Ogura, Y. and Takahashi, T. (1971) Numerical simulation of the life cycle of thunder storm cell. *Mon. Weather Rev.*, **99**, 895–911.

Ohkura, H., Ishizaki, K., Nakao, H. and Morimoto, R., (1983) Short-term precipitation forecasting by radar raingauge, *Proceedings of the 27th Japanese Conference on Hydraulics*. Japanese Society of Civil Engineers, Tokyo, pp. 349–354 (in Japanese).

Shiiba, M., Takasao, T., and Nakakita, E. (1984) Investigation of short-term rainfall prediction method by a translation model. *Proceedings of the 28th Japanese Conference on Hydraulics*. Japanese Society of Civil Engineers, Tokyo, pp. 423–428 (in Japanese).

Tatehira, R. and Makino, Y. (1974) Use of digitized echo patterns for rainfall forecasting. *J. Meteorol. Res., Jpn. Meteorol. Agency*, **26**, 187–199 (in Japanese).

Wilk, K. E. and Gray, K. C. (1994) Processing and analysis techniques used with the NSSL weather radar system. *Proceedings of the 14th Conference on Radar Meteorology, 1974*. American Meteorological Society, Boston, MA, pp. 369–374.

Part 6
Hydrological forecasting

37

Analytically derived runoff models based on rainfall point processes

M. Bierkens[1] and C. E. Puente[2]
[1] Department of Hydraulics and Catchment Hydrology, Agricultural University of Wageningen, Nieuwe Kanaal 11, 6708 PA Wageningen, The Netherlands
[2] Department of Land, Air and Water Resources, University of California, Davis, CA 95616, USA

ABSTRACT

This paper presents four stochastic and lumped-in-space rainfall–runoff models. Rectangular pulses arriving according to some stochastic process represent rainfall while runoff is modeled by routing these pulses through linear reservoirs. Closed-form expressions for the mean and second-order properties of the rainfall and runoff processes are respectively reviewed and introduced. The reservoir coefficients are shown to be of major importance in the statistical behaviour of runoff, mainly influencing the scale on which the process varies. An outline is given of the statistical conditions which rainfall and runoff data must initially satisfy in order for the models to be applicable.

1. INTRODUCTION

Common models that describe hydrological processes are not consistent across spatial and/or temporal scales. Typically, they depend heavily on the scale of temporal and spatial aggregation of the available hydrological data. For instance, two different sets of parameters are obtained when one uses the same available data but aggregated (i.e. accumulated) over 6 and 24 h. The problem of disaggregation of data, however (given days, a consistent hour model), is very common in hydrological analysis. Therefore, models that yield the same parameters set across a range of temporal scales can be a desirable tool.

Several models that describe observed temporal aspects of point rainfall have been developed in the last two decades. Early modeling efforts assumed temporal rainfall arrivals followed a Poisson process. See, among others, the works by Todorovic (1968), Gupta (1973), Eagleson (1978), Cordova and Bras (1981) and Cordova and Rodriguez-Iturbe (1985). These works differ primarily in the assumptions made about the probability distribution and the shape of the rainfall produced at each arrival. The well-known fact that rainfall has a tendency to occur in clusters rather than randomly spaced in time led to the pioneering work of Kavvas and Delleur (1981) who characterized rainfall as instantaneous pulses occurring in time according to a Neyman–Scott clustering model. By assuming independence among the random variables involved, the rainfall model can be fully characterized. In particular its mean and covariance structure can be explicitly obtained. Models that combine Neyman–Scott arrivals with rectangular pulses, characterized by a random intensity and duration, have been developed by Rodriguez-Iturbe *et al.* (1984, 1986).

In a study with data from Denver, CO, Rodriguez-Iturbe *et al.* (1987) found that cluster-based models gave stable parameters when estimated from data at alternative aggregation lengths, ranging from 1 to 24 h, while the Poisson rectangular-pulse (PRP) model did not possess this desired property. In the work presented here the principles used in rainfall stochastic modeling are extended to runoff. The goal is to obtain runoff models that exhibit similar properties of time-scale consistency as previously found for some of the rainfall models. Because these runoff models are based on point rainfall, they are expected to be applicable in small catchments.

Given first is a review of the general expressions for the first- and second-order properties of the PRP and Neyman–Scott rectangular-pulses (NSRP) models. Then, two runoff parametrizations are introduced and linked to either of the two rainfall models, thus resulting in four model combinations. Closed-form expressions are derived for the mean and the autocovariance of the instantaneous and locally integrated runoff process as well as the cross-covariance between rainfall and runoff. These expressions are then tested on their mathematical consistency through a simulation study and their behaviour is explored by exposing them to different sets of parameters. Suggestions are given for parameter optimization. Finally, some simple initial tests are put forward to determine whether the models can be applied to a given set of rainfall and runoff data.

2. THEORETICAL FRAMEWORK AND REVIEW OF RAINFALL REPRESENTATIONS

The following is derived from Rodriguez-Iturbe (1986).

This section reviews the most important characteristics of the two rectangular-pulses models used in this work and gives a general framework for their derivations. This framework will serve as a blueprint for the derivation of similar characteristics for runoff.

Let $X(t)$ denote the instantaneous rainfall at time t at a fixed location (e.g. a raingauge). Our modeling efforts will focus on the process $X_i^{(T)}, i = 1, 2, \ldots$, which is an averaging of $X(t)$ over disjoint time intervals of length T. We shall refer to this as the 'integrated' or 'averaged' process. The formal definition is

$$X_i^{(T)} = \frac{1}{T} \int_{(i-1)T}^{iT} X(t)\, dt\, , \qquad i = 1, 2, \ldots \tag{1}$$

where T denotes the time-averaging interval.

Rain cells arrive consecutively (at a given location) at times T_1, T_2, T_3, ... according to some stochastic process. Let $h(t, \tau, U_\tau)$ be the rainfall intensity at time t, at this location, due to the arrival rain cell at time τ. U_τ stands for all the (stochastic) parameters characterizing this particular rain cell. The total rainfall intensity $X(t)$, formed by superposition of all individual rain cells arriving up to time t, is then

$$X(t) = \sum_n h(t, T_n, U_n)\, , \qquad T_n \leqslant t\, . \tag{2}$$

This can be written as

$$X(t) = \int_{-\infty}^{t} h(t, \tau, U_\tau)\, dN(\tau) \tag{3}$$

where $N(\tau)$ stands for the number of cell occurrences since the beginning of the process up to time τ.

As the goal is to obtain tractable closed-form expressions for the moments of this process, the following simplifying assumptions are made. The random variables U_1, U_2, U_3, ... are mutually independent and independent of the arrival times T_1, T_2, T_3, ... Furthermore, the parameters that define their distributions are assumed to be constant in time. Then, the mean of the process can be expressed as

$$E[X(t)] = \int_{-\infty}^{t} E[h(t, \tau, U_\tau)] E[dN(\tau)] \tag{4}$$

where the first expectation value occurs with respect to the random characteristics of the rain cell and the second with respect to the arrival process. Similarly, the covariance will be given by

$$\operatorname{Cov}[X(t_1), X(t_2)] = \int_{-\infty}^{t_2} \int_{-\infty}^{t_2} E[h(t_1, \tau_1, U_{\tau 1}) h(t_2, \tau_2, U_{\tau 2})]$$

$$\times E[dN(\tau_1)\, dN(\tau_2)] - E[X(t_1)] E[X(t_2)]\, . \tag{5}$$

Rainfall, however, is measured over disjoint intervals. Therefore, the averaged process given by (1) is considered. The mean, covariance and variance of this process are

$$E[X_i^{(T)}] = \frac{1}{T} \int_{i-1}^{iT} E[X(t)]\, dt \tag{6}$$

$$\operatorname{Cov}[X_1^{(T)}, X_n^{(T)}] = \frac{1}{T^2} \int_{(n-1)T}^{nT} dt_1 \int_{0}^{T} \operatorname{Cov}[X(t_1), X(t_2)]\, dt_2 \qquad n \geqslant 1 \tag{7}$$

$$\mathrm{Var}[X_i^{(T)}] = \frac{1}{T^2} \int_0^T \mathrm{d}t_1 \int_0^T \mathrm{Cov}[X(t_1), X(t_2)]\,\mathrm{d}t_2 \ . \tag{8}$$

To obtain the expressions for the desired averaged process, one first needs to find the solutions of the integrals of (4) and (5). To find these the following steps should be taken.

(1) Select a random process $N(\tau)$ that governs the arrivals of the rainfall pulses $h(t,\tau,U_\tau)$ and derive its first and second moments.
(2) Choose an appropriate random function $h(t,\tau,U_\tau)$ that describes the form of the rainfall pulses and calculate its first and second moments.

2.1   The arrival process
Two well-known models are used to describe the arrivals of rain cells.

(a) *Arrivals according to a Poisson process*   Rain cells arrive according to a homogeneous Poisson process with rate λ. From Cox and Isham (1980) we have that the first and second moments of this arrival process are given by

$$E[\mathrm{d}N(\tau)] = \lambda\,\mathrm{d}\tau \tag{9}$$

$$E[\mathrm{d}N(\tau_1)\,\mathrm{d}N(\tau_2)] = [\delta(\tau_2 - \tau_1)\lambda + \lambda^2]\,\mathrm{d}\tau_1\,\mathrm{d}\tau_2 \ . \tag{10}$$

(b) *Arrivals according to a Neyman–Scott process*   Rain cells have a tendency to occur in clusters rather than randomly spaced in time. The Neyman–Scott process accounts for this clustering.

Storm origins (e.g. low-pressure centers) arrive according to a Poisson process with rate λ. Each storm origin gives rise to a random number v of rain cells. No cell is associated with the storm origin itself. The cell origins are independently displaced from the storm origin with distances that are negatively exponentially distributed with mean distance $1/\beta$. The number of rain cells associated with each storm origin (the random variable v) has a Poisson distribution with mean $E[v]$. Each cell spreads its rain according to the function $h(t,\tau,U_\tau)$. Cox and Isham give the moments of the homogeneous Neyman–Scott arrival process as

$$E[\mathrm{d}N(\tau)] = \lambda E[v]\,\mathrm{d}\tau \tag{11}$$

$$E[\mathrm{d}N(\tau_1)\,\mathrm{d}N(\tau_2)] = \{\lambda E[v]\delta(\tau_2 - \tau_1) + \tfrac{1}{2}\lambda E[v[v-1]]\beta \exp(-\beta|\tau_2 - \tau_1|$$
$$+ \lambda^2 E^2[\tau]\}\,\mathrm{d}\tau_1\,\mathrm{d}\tau_2 \ . \tag{12}$$

2.2   The form of $h(t,\tau,U_\tau)$
The rain cells that arrive either individually according to a Poisson process or clustered according to a Neyman–Scott process are thought to be rectangular pulses of random intensity i_r and duration t_r. It is assumed that i_r and t_r are mutually independent random variables whose characteristics are independent of the arrival process and constant in time. Furthermore it is assumed that both intensity and

duration are negatively exponentially distributed, with respective means $\bar{i}_r$ and $\bar{t}_r$. Then, the random function describing the form of a rain cell that started at time τ is given by

$$
h(t,\tau,U_\tau) = \begin{cases} i_r, & \tau \leqslant t \leqslant t_r + \tau \\ 0, & \text{otherwise.} \end{cases} \tag{13}
$$

Taking the first and second moment of the random variable $h(t,\tau,U_\tau)$ is rather trivial (e.g. Rodriguez-Iturbe 1986). If these moments are substituted into (4) and (5) together with (9) and (10) respectively, the mean and the covariance of the instantaneous process of the PRP model are obtained. Replacing (9) and (10) with (11) and (12) respectively and evaluating the integrals yield the mean and covariance of the instantaneous process according to the NSRP model. The desired moments of the averaged process according to both models can then be obtained by evaluating (6)–(8). Owing to space limitations these expressions are not given here. They can be found in, for instance, the work of Rodriguez-Iturbe (1986) and Rodriguez-Iturbe *et al.* (1986).

3. EXTENSION TO RUNOFF

One approach to describe the instantaneous runoff $Y(t)$, is to route individual rectangular rain cells through some transfer model, and then to add the contributions from each cell (see also equation (2) and (3)):

$$
Y(t) = \sum_n g(t,T_n,U_n) = \int_{-\infty}^t g(t,t,\tau,U_\tau)\,dN(\tau) \tag{14}
$$

The 'kernel' function $g(t,\tau,U_\tau)$ describes the instantaneous runoff rate at the outlet of the basin at time t with characteristics U_τ, as computed routing and individual rain cell that starts at time τ, somewhere in the basin. The two runoff models chosen here are deterministic conceptualizations of hydrologic response. Therefore runoff becomes stochastic only because of the randomness of rainfall.

A way to define the basin's response to a single cell of given intensity and duration is to assume an instantaneous response function $V(t)$ such that the desired runoff kernel may be found from a convolution integral

$$
g(t,\tau,U_\tau) = \int_{-\infty}^{\infty} V(\theta)h_e(t-\theta,\tau,U_\tau)\,d\theta \ . \tag{15}
$$

$h_e(t,\tau,U_\tau)$ represents the effective rainfall produced at time t by a single rectangular pulse that arrived at time τ.

Two different parameterizations of the basin's response to a rain cell are used. These are as follows.

3.1 A single linear reservoir
If the intensity of a (rectangular) rainfall pulse exceeds a constant loss parameter $\bar{L}_p$, the amount in excess of $\bar{L}_p$ is defined as the effective rainfall $h_e(t,\tau,_{U\tau})$ coming from a

single cell. The effective rainfall kernel so defined will be zero for all times if $i_r \leqslant \overline{L}_p$. The loss parameter can be thought of as a rough measure of losses due to evapotranspiration, interception and depression storage. The effective rainfall from the cell is routed through a linear reservoir that accomodates both runoff and losses through groundwater. The storage-output parameter k_1 defines runoff while the storage-output parameter k_2 dictates deep percolation. It can be easily shown that such a linear reservoir is characterized by the following instantaneous response function:

$$V(t) = \Psi k \exp(-kt) , \qquad k \text{ and } \Psi > 0, T \geqslant 0 \qquad (16)$$

where

$$\Psi = \frac{k_1}{k_1 + k_2} \qquad (17)$$

$$k = k_1 + k_2 . \qquad (18)$$

3.2   Two parallel linear reservoirs

Again, if the intensity of a rain cell exceeds a constant rate of initial losses $\overline{L}_p$, it will contribute to the discharge. If it does, the excess rainfall may be routed either via groundwater runoff of surface runoff, both represented by liner reservoirs. A constant infiltration rate I_0 is used to divide the model's response. If the excess rainfall $i_r - \overline{L}_p$ is smaller than the constant rate I_0 of infiltration, only groundwater runoff will take place. If it exceeds this constant rate, part of the excess rainfall $i_r - \overline{L}_p - I_0$ of the cell will be discharged at a faster pace as surface runoff. The instantaneous unit response function of such a model will be

$$V(t) = \begin{array}{ll} k_1 \exp(-k_1 t) , & \text{surface runoff} \\ k_2 \exp(-k_2 t) , & \text{groundwater runoff} \end{array} \qquad k_1 \text{ and } k_2 > 0, t \geqslant 0 \qquad (19)$$

where the total response is obtained by adding up the two parts. After substitution into (15) and evaluation the runoff kernel $g(t, \tau, U_\tau)$ can be obtained. The model as described here results in three runoff scenarios possible for an arriving rain cell. The case when no runoff appears ($i_r \leqslant \overline{L}_p$), only groundwater runoff appears ($\overline{L}_p < i_r \leqslant \overline{L}_p + I_0$) and both groundwater runoff and surface runoff are present ($i_r > \overline{L}_p + I_0$).

Notice that, although the response to the effective rainfall produced by one rain cell is linear, the total runoff process $Y(t)$ is not a linear transformation of the rainfall process $X(t)$. In that case the total rainfall intensity $X_e(t)$ in excess of the rate of initial losses should be routed through the linear reservoirs to define the process $Y(t)$ directly. Instead every single cell is evaluated separately. Thus, if two rain cells have intensities smaller than $\overline{L}_p$ but owing to overlap produce a total intensity $X(t)$ that exceeds $\overline{L}_p$, still no runoff will be produced from these cells. This causes the non-linearity between $X(t)$ and $Y(t)$.

Because the rain cells are routed first and then are superimposed, the only basic difference between the rainfall and the runoff process is in the form of the arriving cells. Therefore, equations (3)–(8), which do not assume any particular form of the 'kernel' used, are also valid for runoff, replacing $h(t,\tau,U_\tau)$ by $g(t,\tau,U_\tau)$ and $X(t)$ by $Y(t)$.

The derivation of the analytical expressions describing the moments of the runoff kernels $g(t,\tau,U_\tau)$ involves extensive computational effort. First the expectations are to be derived conditional to a fixed arrival time and integrating with respect to the random variables i_r and t_r and their respective probability density functions. This results in many different cases, depending on the value of i_r with respect to $\overline{L}_p$ and $\overline{L}_p + I_0$ and of $\tau + t_r$ with respect to the time t. These conditional first- and second-order moments are substituted into (4) and (5) respectively, together with the moments of the arrival process used. Finally the integrals of (4) and (5) are evaluated to account for all possible arrival times. Attempts to model the basin's response with a more complicated model as the nash cascade failed as they produced integrals (for instance incomplete gamma functions) that could not be evaluated analytically. The same happened when the infiltration rate was taken to be an exponential decay. To help with the analytical derivation of the large expressions so obtained, a symbolic manipulation program (SMP) (Inference Corporation 1983) was used. This is a program that enables the user to carry out mathematical operations on large algebraic expressions as one would do them on a piece of paper.

Having two different ways of describing the arrival process and two different ways to describe the basin's response to an arriving rain cell, a total of four rainfall–runoff combinations are obtained:

(a) Poisson arrivals and a single linear reservoir — the PRP–SLR model;
(b) Poisson arrivals and two parallel linear reservoirs — The PRP–PLR model;
(c) Neyman–Scott arrivals and a single linear reservoir — the NSRP–SLR model;
(d) Neyman–Scott arrivals and two parallel linear reservoirs — the NSRP–PLR model.

Table 1 lists the parameters involved in the different model combinations. It is clear that more complex rainfall–runoff models are obtained when going from Poisson arrivals to Neyman–Scott arrivals and from the single to the parallel linear reservoir case. The mean of the instantaneous and locally integrated are the same. For all four models it can be expressed by the following general expression:

$$E[Y(t)] = \alpha\lambda\overline{t_r}\overline{i_r}\exp\left(\frac{-L_p}{i_r}\right) \tag{20}$$

where α equals Ψ for the PRP–SLR model, 1 for the PRP–PLR, $\Psi E[v]$ for NSRP–SLR and $E[v]$ for the NSRP–PLR model.

The expressions for the (autoco) variance of the instantaneous and the averaged process for all four models exhibit the same structure as those found for the rainfall models (e.g. Rodriguez-Iturbe 1986). Because of space limitations they are not given here. For the full expressions of all four models one is referred to Bierkens and Puente (1989a,b).

Table 1 — Parameters of the four rainfall–runoff models

Choose a combination of rainfall and runoff!			
PRP–		**NSRP–**	
rainfall	λ arrival rate of rain cells (/h) $\bar{i}_r$ mean intensity of reactangular pulses (mm/h) $\bar{i}_r$ mean duration of rectangular pulses (h)		λ arrival rate of storm origins (/h) $\bar{i}_v$ mean intensity of rectangular pulses (mm/h) $\bar{i}_r$ mean duration of rectangular pulses (h) $E[v]$ mean number of cells per cluster $1/\beta$ mean distance between cluster cells and storm origin (h)
–SLR		**–PLR**	
Runoff	Ψ reduction factor for losses to deep groundwater T_p initial losses (mm/h) k reservoir constant (/h)		T_p initial losses (mm/h) I_0 infiltration rate (mm/) k_1 constant of fast reservoir (/h) k_2 constant of slow reservoir (/h)

4. ADDITIONAL SECOND-ORDER PROPERTIES FOR RUNOFF

To obtain a more complete description of the second-order statistical properties of runoff, analytical expressions were derived for the following characteristics:

(1) the cross-covariance between rainfall and runoff for the instantaneous and the integrated process;
(2) the variance function $\gamma(T)$, being the quotient of the variance of the process with averaging, interval T and the variance of the instantaneous process, as a function of T;
(3) the scale of fluctuations θ (Vanmarcke 1983), a measure of the time scale on which the process varies;
(4) the one-sided unit-area spectral density function $g(\omega)$.

This was done for all the four models.

The derivation of the autocovariance was done for zero and positive lags going from rainfall to runoff. The obtained expressions exhibit a similar structure to those of the autocovariance of the runoff process and are derived in a similar fashion. Again this took some extensive calculative work, which could not have been done without the help of the SMP. Expressions for the scale of fluctuation, variance function and spectral density can be obtained from the autocovariance of the instantaneous process using the formulae given by Vanmarcke (1983). Rodriguez-Iturbe (1986) gives closed form expressions for these derived statistics for the PRP and NSRP process. Those for runoff were derived in the same way and are given by Bierkens and Puente (1989a,b).

5.   VERIFICATION AND BEHAVIOR OF RUNOFF MODELS

Although simple linear reservoirs were used to model the basin's response to rainfall, rather large expressions were obtained for the second moments. To check such expressions on their mathematical consistency can be a difficult task. Next to step-by-step revisions and exploring the parameters' limits a large number of computer simulations were carried out for various parameter sets. For these simulations, sampling statistics were compared with those calculated from the theoretical expressions using the same parameter set. Equally good results were obtained for all the models for both rainfall and runoff for various sets of parameters.

To understand the relative importance of the various model parameters, the expressions were exposed to alternative parameter sets. It was found that the most important runoff parameters, when those of rainfall are fixed, were the reservoir constants k or k_1 and k_2. Their main influence is on the variance of the runoff process and the time scale on which the process varies. Fig. 1 shows the autocorrelation

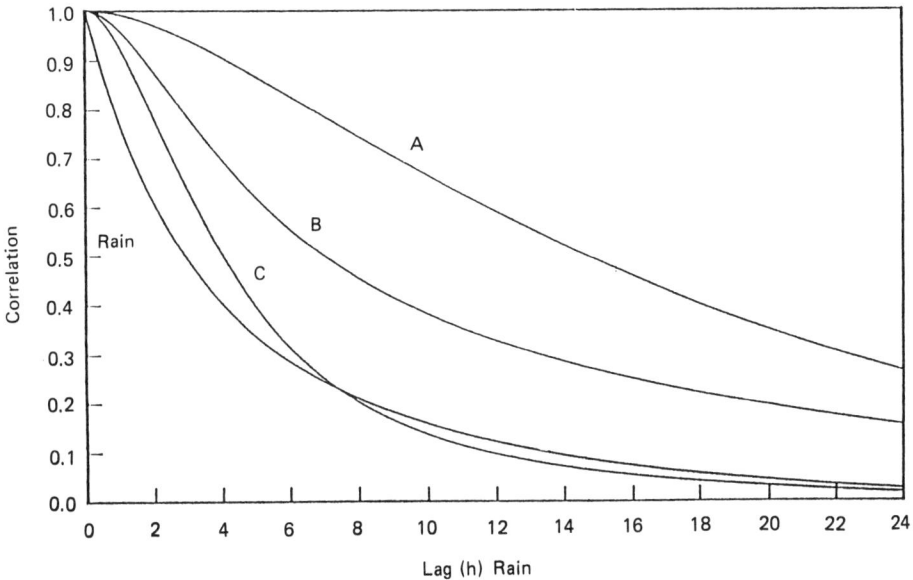

Fig. 1 — Autocorrelation functions NSRP–PLR model; for parameter values used, see Tables 2 and 3.

function of the instantaneous runoff process according to the NSRP–PLR model. The parameter sets used are given in Tables 2 and 3. It is clear that, when the values

Table 2 — Model parameters used in rainfall calculations

λ	$E[v]$	β (/h)	$\bar{i}_r$ (mm/h)	$\bar{t}_r$ (h)
0.05	11.5	0.125	3.0	2.25

Table 3 — Model parameters used in runoff calculations

	NSRP–SLR			NSRP–PLR			
	$\overline{l_{\mathrm{p}}}$ (mm/h)	Ψ	k (/h)	$\overline{l_{\mathrm{p}}}$ (mm/h)	I_0 (mm/h)	k_1 (/h)	k_2 (/h)
A	0.0	1.0	0.05	0.0	1.0	0.1	0.05
B	0.1	0.8	0.2	0.1	2.0	0.5	0.05
C	4.0	0.9	0.8	4.0	0.8	0.6	0.4

of k_1 and k_2 increases (from A to C, see Table 3), the autocorrelation function of runoff will begin to resemble that of rainfall. It can also be seen that the form of these functions is monotonically descending, which is in agreement with the assumptions made about the parameters as constant across time. Of course, this excludes data whose autocorrelation function shows a peak at a certain lags due to some short term periodicity. Fig. 2 shows that the rescaled variance functions of rainfall and runoff

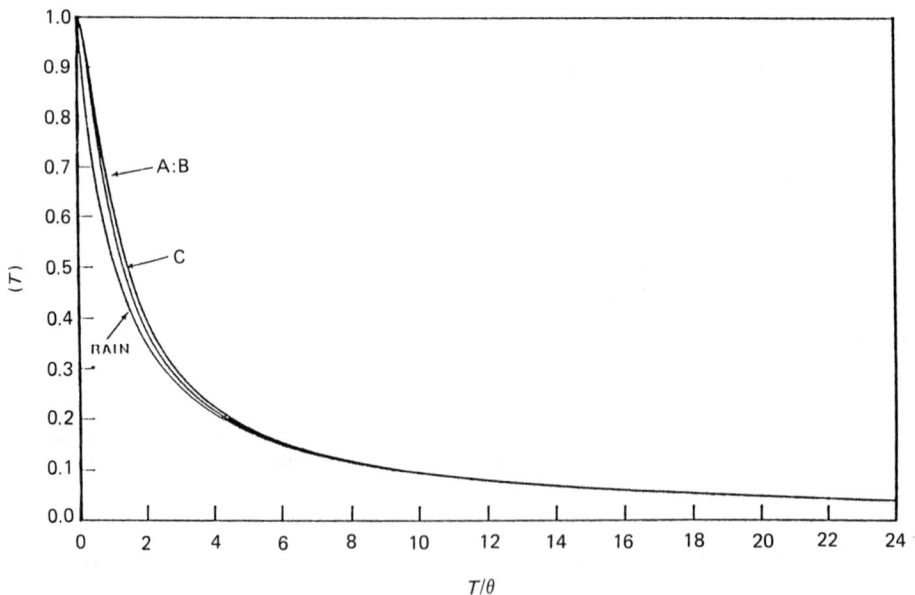

Fig. 2 — Rescaled variance functions NSRP–SLR model; for parameter values used, see Tables 2 and 3.

match almost perfectly. These functions are plotted here for rain and runoff (three sets of parameters) for the NSRP–SLR model. It is clear that routing the rectangular pulses through linear reservoirs only effects the scale and not the form of the process.

6. PARAMETER OPTIMIZATION

The closed-form expressions for mean and different second-order characteristics can be used to optimize the parameters under preservation of data statistics by the method of moments. To be consistent, the models should yield similar parameter sets when optimized using data averaged over a range of time scales. If so, the models can be further verified by looking for consistency and fit of statistics not used in the optimization scheme or for which no closed-form expressions are available. The latter can be obtained by estimation from simulations using the before optimized parameters (e.g. return periods or the probability that rainfall or runoff is smaller than a certain value).

Optimization of the parameters can be done in the following ways:

(a) As many statistics as parameters to be optimized can be used. This means that a system of nonlinear equations needs to be solved.
(b) More statistics than parameters involved can be used. Then an overdetermined system of nonlinear equations is obtained. The goal then could be to minimize the sum of squares of the theoretical and estimated statistics. Finding the associated set of parameters is then an unbounded non-linear optimization problem. In addition, one could think of giving different weights to the statistics involved in the optimization scheme.
(c) Some of the rainfall parameters such as the arrival rate of storms, the mean duration and intensity of rain cells and the number of rain cells per storm depression could be given a physical meaning by estimating them directly from the data. Radar could be of considerable support in doing so. An important question to be answered then is whether the resulting parameter set shows consistency across a range of time scales for both rainfall and runoff.

At present, we are investigating the time consistency of the combined rainfall–runoff models using data from Hupsel, The Netherlands. Some results are expected soon. When shown to be consistent, the models could be used in catchments that are qualitatively the same as Hupsel for disaggregation of important data statistics. A series of simple tests can be applied to a given set of data to check whether they are a good candidate to be described with the models proposed. They concern the assumed (second-order) homogeneity of the data and the relation between rainfall and runoff. To test the presence of trends, possibly caused by considering too long a period, range analysis can be used (Buishand 1982). Short-term periodicities can be easily discovered by looking at the at the estimated autocorrelation structure and the spectrum of the data. Finally it is important to estimate the rescaled variance function of the rainfall and runoff data which according to thoery (Fig. 2) must be almost similar.

7. CONCLUSIONS

Four lumped-in-space stochastic rainfall–runoff models were derived. Although simple linear reservoirs were used to model the basin's response, rather large expressions were obtained. The use of a symbolic manipulation program for the derivation of these expressions was therefore essential. Generalizations of the schemes described in this paper, for instance by introduction of some dependence between the random variables involved, making some parameters time dependent or introduction of a more realistic basin's response, may have serious consequences for the tractability of the equations. Therefore, the 'simpler' models introduced here are currently being investigated on their time-scale consistency using data from Hupsel.

Exploration of the model's behavior showed the importance of the reservoir constants whose influence concerns basically the variance and the time scale of the process. Considering the form of the autocorrelation and the variance function and the assumed (second-order) homogeneity of the derived equations, some tests were introduced to check eligable data on these properties. Several ways of parameter optimizations were proposed, one of which may give some physical meaning to the rainfall parameters.

ACKNOWLEDGEMENTS

The work was funded by the California Water Resources Center through Grant W-731. Discussions with the Professor Rafael Bras and Professor Ignacio Rodriguez-Iturbe are acknowledged. Dr P. Torfs is acknowledged for his help in completing this paper.

REFERENCES

Bierkens, M. F. P. and Puente, C. E. (1989a) *Analytically derived runoff models based on rainfall point processes*, 1, *Poisson case*, M.Sc. Thesis.

Bierkens, M. F. P. and Puente, C. E. (1989b) *Analytical derived runoff models based or rainfall point processes*, 2, *Neyman-Scott case*, M.Sc. Thesis.

Buishand, T. A. (1982) Some methods for testing the homogeneity of rainfall records. *J. Hydrol.*, **58**, 11–27.

Cordova, J. R. and Bras, R. L. (1981) Physcially based probabilistic models of infiltration, soil moisture and actual evapotranspiration. *Water Resour. Res.* **17**(1), 93–106.

Cordova, J. R. and Rodriguez-Iturbe, I. (1985) On the probabilistic structure of storm surface runoff. *Water Resour. Res.* **21** (5), 755–763.

Cox, D. R. and Isham, V. (1980) *Point Processes*, Chapman and Hall, London.

Eagleson, P. S. (1978) Climate, soil and vegetarian 1–7. *Water Resour. Res.* **14** (5), 705–776.

Gupta, V. K. (1973) *A stochastic approach to space-time modelling of rainfall*, Technical Report Natural Resource Systems 18. University of Arizona, Tuscon, AZ.

Kavvas, M. L. and Delleur, J. W. (1981) A stochastic cluster model of daily rainfall sequences. *Water Resour. Res.* **17** (4), 1151–1160.

Rodriguez-Iturbe, I. (1986) Scale of fluctuation of rainfall models. *Water Resour. Res.*, 22 (9), 15S–37S.

Rodriguez-Iturbe, I., Cox, D. R. and Isham, V. (1986) Some models for rainfall based on stochastic point processes. *Proc. R. Soc. London, Ser. A*.

Rodriguez-Iturbe, I., Febres de Power, B. and Valdes, J. B. (1987) Rectangular pulses point process models for rainfall: analysis empirical data'. *J. Geophys. Res.* **92** (D), 9645–9656.

Rodriguez-Inturbe, I., Gupta, V. K. and Waymore, E. (1984) Scale considerations in the modelling of temporal rainfall. *Water Resour. Res.* **20** (11), 1611–1619.

Inference Corporation (1983) *SMP, A symbolic manipulation program*. Inference Corporation.

Todorovic, P. (1968) *A mathmatical study of precipitation phenomena*, Report No. CER 67–86PT65. Engineering Research Center, Colorado State University, Fort Collins, CO.

Vanmarcke, E. (1983) *Random fields: analysis and synthesis*. MIT Press, Cambridge, MA.

38

Adaptive grid-square-based geometrically distributed flood-forecasting model

S. Chander[1] and S. Fattorelli[2]
[1]Department of Civil Engineering, Indian Institute of Technology, 110016 New
Delhi, India
[2]Department of Land and Agroforest Environments, University of Padova,
Padova, Italy

ABSTRACT

An adaptive rainfall–runoff model for real-time flood forecasting capable of utilizing weather radar estimates of rainfall is presented. The basin is represented by a network of grid squares. The transformation of rainfall excess to direct runoff for each grid square is described by a transfer function generated by a series combination of linear channels and a linear reservoir. The parameters of the transfer function for each grid square are so computed that the combination of individual responses of the grid squares yields the average response for a spatial uniform rainfall excess. The infiltration for each grid square is represented by an initial loss and a variable-infiltration model defined by Phillip's equation. The results are compared with the constant-abstraction-rate infiltration model. The real-time correction of forecasts is done by updating the infiltration and the linear reservoir parameters. Three criteria of updating are used to compute the forecasts and results are compared. The 1408 km^2. Bacchiglione basin data are used to compute the forecasts 6 h ahead at Montegaldella. The proportion of variance explained by the models compared with no model is used as a criterion for the comparison. The paper concludes that the proposed grid model, using variable infiltration rate to compute the rainfall excess, updating infiltration parameter to obtain the last observed discharge and linear reservoir parameter to match the observed discharges in the immediate neighbour-hood, computes good forecasts. The results are also compared with a lumped time-variant system model formulation, using the same infiltration model. Both the lumped and the grid model perform equally well for spatially uniform rainfall, but the performance of the grid model is far better for spatially non-uniform rainfall situations.

1. INTRODUCTION

In rainfall–runoff transformation, lumped linear system models have been found to work as well as more physically based models (Simpson 1980, Loague and Freeze 1985, Plate *et al.* 1988). However, these models are not in a position to exploit the rainfall data obtained from high-resolution radar measurements. Anderl *et al.* (1976) handled this limitation by using the idea of separating the translation and storage effects on the catchment (Dooge 1959, Clark 1945), and proposing a linear distributed system model. The model represents the catchment by grid squares of 1 km^2 in area. Rain excess shifted in time by a linear channel at each grid square is followed by routing of the combined flows through two unequal parallel linear reservoirs. This is an extension of the most common approach of subdividing a basin into subbasins and channel segments and using lumped rainfall–runoff models for the subbasins and combining their responses after routing through the appropriate channel segments. O'Connell *et al.* (1986) and Chander and Fattorelli (1988) have shown the utility of such models for on-line forecasting of high flows. The paper explores this idea further to develop an adaptive grid-square-based, geometrically distributed flood-forecasting model, which converges to a lumped linear system model for a spatially uniform rainfall. The relationship is exploited to compute the parameters of the distributed model for an average response of the lumped model. The aim of the study by Anderl *et al.* (1976) was to show that rainfall data derived from the radar provide more accurate forecasts than the spatial interpolated estimates derived using data from a raingauge network. Therefore, the total observed runoff value was used to obtain the time-varying moisture deficiency curve to compute rain excess on each grid square. In an on-line forecasting situation this value needs to be determined without the knowledge of total runoff as rain progresses in time. Chander and Shanker (1984) have proposed a procedure for on-line determination of the rain excess for an initial loss followed by a constant-abstration-rate infiltration model. The procedure was extended by Chander and Fattorelli (1988) to choose the response function and to compute the abstraction rate. The response function was chosen from a given set of responses determined from past records on the catchment and the value of abstraction rate was computed to match the volume and discharge values prior to the time of forecast. This was done to account for the non-linearity in the rainfall–runoff process in the rising limb of the flood hydrograph, Corradini *et al.* (1986) used Phillip's infiltration scheme, updated the sorptivity parameter S in real-time and the Clark model as the response function and obtained comparable results on a 934 km^2 basin with the model of Chander and Shanker (1984). Klatt and Schultz (1985) suggest the computation of rain excess in their distributed system model using a time-varying coefficient defined by two parameters. The average values obtained from historical events are used to formulate the forecast. If the observed and computed discharge values do not show a good match, the parameters are optimized using the coefficient of variation as the objective function. The constant-abstraction-rate and Phillip's infiltration schemes along with the self-correcting procedures used in the above-mentioned lumped models are incorporated in the distributed model. The paper describes the model and reports its application to the 1408 km² Bacchiglione river basin, at Montegaldella. The results are also compared with the lumped system model incorporating both the infiltration schemes.

2. DISCRETE FORM OF THE LINEAR DISTRIBUTED SYSTEM GRID MODEL

The basin is divided into N grid squares as shown in Fig. 1. Each grid square is assumed to transform excess rain by a transfer function. The transfer function is defined by a series combination of linear channels and a linear reservoir. The linear channel parameter is a function of the location of the grid square in relation to the gauging station and is subsequently referred to as $t_{0,g}$. The linear reservoir parameter is assumed to be the same for all grid squares. The parameter is considered to be time variant but the computations at the time of forecast are performed by assuming it to be time invariant. The discrete value of the response function ordinate at time i is designated as U_i. The time t_R of first rise at the gauging station decides the time of start of rainfall excess in each grid square. The initial lag or pure delay time parameter enables one to account for the excess rain which has occurred during this period. Let $R_{j,g}$ be the excess rain on grid square g at time j. The contribution of a grid square at time $KK+\lambda$ designated by $Q_{g,KK+\lambda}$ is given by

$$Q_{g,KK+\lambda} = \sum_{j=t_R-(t_{c,g}+\lambda)}^{i=KK-t_{c,g}} R_{j,g} U_{i-j+1}. \tag{1}$$

Combining the contribution of all grid squares for which $i>0$ gives

$$Q_{KK+\lambda} = \sum_{g=1}^{N} \sum_{j=t_R-(t_{c,g}+\lambda)}^{i=KK-t_{c,g}} R_{j,g} U_{i-j+1} \tag{2}$$

where N is the number of grid squares and $Q_{KK+\lambda}$ is the computed discharge at time $KK+\lambda$.

Rainfall excess is computed using two infiltration models. These models consider the rain prior to the time $t_R-(t_{c,g}+\lambda)$ as initial loss. The abstraction rate F is considered as constant in space and time in one of the models. In the other model, Phillip's equation is used to define the infiltration process and the parameters are assumed to be the same for each grid square. The resulting equations incorporating the infiltration models are

$$Q_{KK+\lambda} = \sum_{g=1}^{N} \sum_{j=t_R-(t_{c,g}+\lambda)}^{i=KK-t_{c,g}} (P_{j,g}-F) U_{i-j+1} \tag{3}$$

and

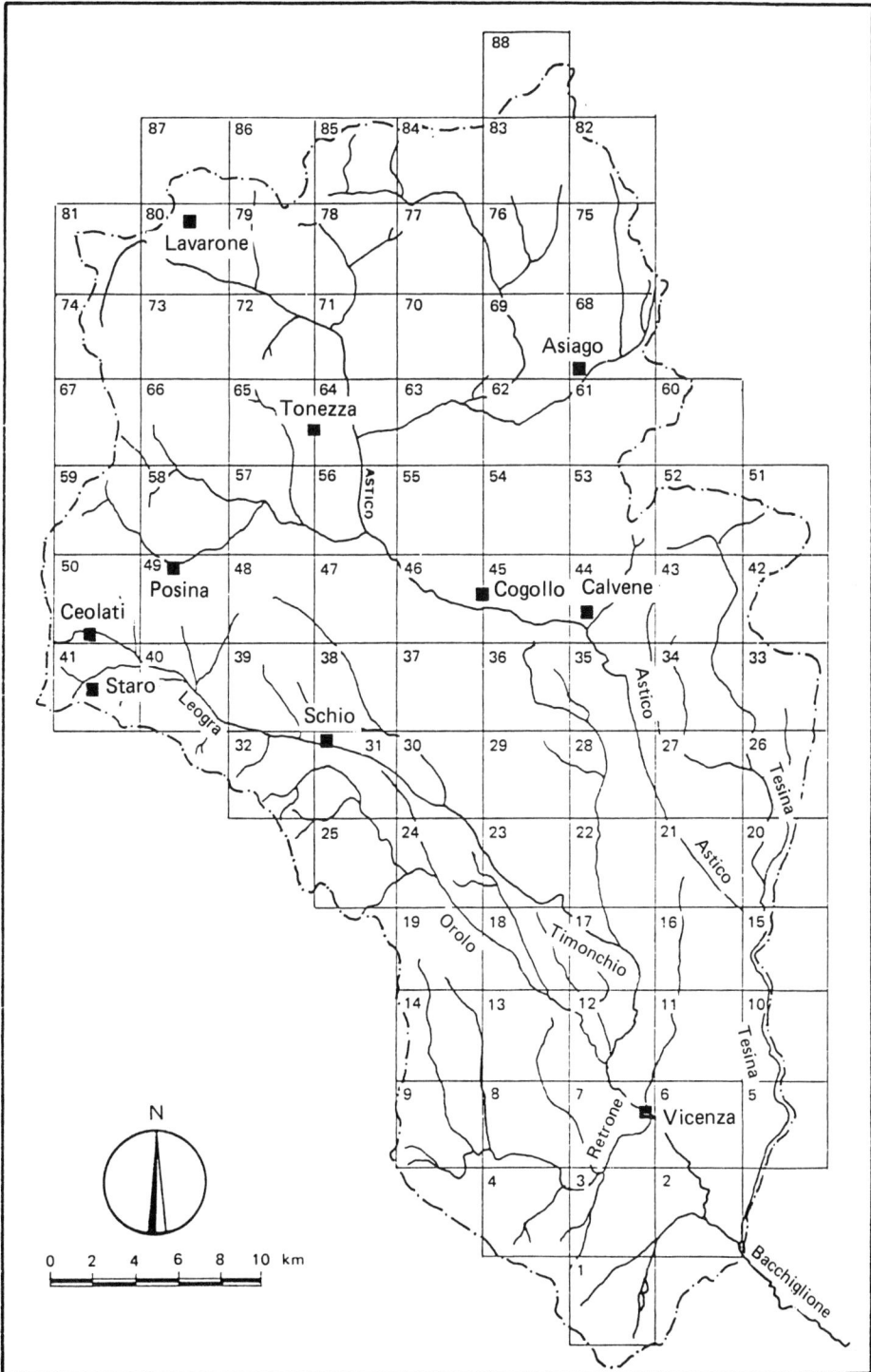

Fig. 1 — Subdivision of basin into 88 4 km × 4 km grid squares.

$$Q_{KK+\lambda} = \sum_{g=1}^{N} \sum_{j=t_R-(t_{c.g}+\lambda)}^{i=KK-t_{c.g}} \{P_{j,g} - S[J^* - (J-1)^*] - C_a\} U_{i-j+1} \tag{4}$$

where $P_{j,g}$ is the precipitation at time j on grid g, F is the average abstraction rate, and S and C_a are the sorptivity and a constant parameter respectively for Phillip's model. The parameters of the model represented by equation (3) are λ, $t_{c,g}$, K and F and those represented by equation (4) are λ, $t_{c,g}$, K. S and C_a. K is the parameter of the linear reservoir which is used to compute U_iS. The parameter C_a in the Phillip's equation is related to the average soil properties of the basin and is determined along with the $t_{c,g}$ values on the basis of simulation studies using past records. The parameters K and F in equation (3) and K aand S in equation (4) are determined on line and are updated at the time of forecast.

3. ESTIMATION OF OFF-LINE PARAMETERS

The three parameters which are determined off line are λ, $t_{c,g}$ and C_a. In a discrete system defined by equation (2), an assumption is required to be made regarding the discretization of the $t_{c,g}$ value. All grid squares having a $t_{c,g}$ value between 0 and 1 are assigned, a value of 0, those betweeen 1 and 2 a value of 1, and so on. Let N_0, N_1, N_2, ... $Nt_{c,g}$ be the number of grid squares with $t_{c,g}$ values at 0, 1, 2, ..., $t_{c,g}$ times respectively. Then equation (2) can be rewritten as

$$Q_{KK+\lambda} = \sum_{t_{c.g}=0}^{KK} \sum_{j=t_R-(t_{c.g}+\lambda)}^{i=KK-t_{c.g}} \sum_{PP=0}^{t_{c.g}} N_{PP} R_{j,g} U_{i-j+1}. \tag{5}$$

$$g = \sum_{PP=0}^{t_{c.g}} N_{PP} - Nt_{c,g+1}$$

In the case of spatially uniform rainfall, equation (5) can be modified to read

$$Q_{KK+\lambda} = \sum_{t_{c.g}=0}^{KK} \sum_{j=t_R-(t_{c.g}+\lambda)}^{i=KK-t_{c.g}} Nt_{c,g} R_y U_{i-j+1}. \tag{6}$$

In this equation, $Nt_{c,g}$ represents the incremental contributing area; therefore it can be interpreted as the discrete value of the derivative of area contributing at any time $t_{c,g}$. Thus, equation (6) is the discrete representation of the Clark model, if U_t values are computed assuming the response function to be a linear reservoir and $Nt_{c,g}$ is computed using the travel time concept. In this case the basin's response for a lumped model can be represented by two reservoirs in series; then $Nt_{c,g}$ must be exponentially distributed as it must represent the response of the first reservoir. This relationship is used to compute the $Nt_{c,g}$ value for the discrete case using a volume match between the response of a lumped model and grid model represented by equation (6). Let A_g be the area of the grid square and A be the area of basin. Then

$$N_0 = \frac{A}{A_g} \frac{1 - [(K+1)/K]\exp(1/K)}{1 - \exp(1/K)}$$

and

$$NT_{c,g} = \frac{A}{A_g K}\exp\left(-\frac{t_{c,g}}{k}\right). \tag{7}$$

Similar expressions can also be derived where the lumped-model response requires representation by more than two reservoirs in series. The discrete value of U_t (m^3/s) is given by (for hourly rainfall excess (mm))

$$U_t = \frac{A_g}{3.6}\left[1 - \exp\left(-\frac{1}{K}\right)\right]\exp\left(-t-\frac{1}{K}\right). \tag{8}$$

A spatially uniform rainfall on a basin with $Nt_{c,g}$ distributed as per equation (7) would yield a gamma function response with $n = 2$. The area–channel parameter curve in this case represents the minimum time in which the impact of the rain excess on grid squares representing an area will be felt at the gauging station. The pure delay time parameter or initial lag λ which is obtained by simulation in the lumped model will need to be added to the above time. $t_{c,g}$ is assigned to the grid squares on the basis of its location with respect to the gauging station as measured by the centroid of the grid square along the channel network. A smaller value of $t_{c,g}$ is allocated to a grid square which is connected to the network by a higher-slope channel. The second parameter of Phillip's point infiltration model C_a is a function of the soil properties and equals two thirds of the hydraulic conductivity. However, no method is available to find its spatial value. Therefore a method similar to that of Corradini *et al.* (1986) is followed to determine its value.

4.   ESTIMATION OF ON-LINE PARAMETERS

The parameters which are to be estimated on line have already been identified as F and K in the first infiltration scheme, and S and K in the second scheme. The resulting equations for the distributed model are

$$Q_{KK+\lambda} = \sum_{g=1}^{N} \sum_{j=t_R-(t_{c,g}+\lambda)}^{i=KK-t_{c,g}} (P_{j,g} - F)\frac{A_g}{3.6}\left[1 - \exp\left(-\frac{1}{K}\right)\right]\exp\left(-i-\frac{j}{K}\right) \tag{9}$$

and

$$Q_{KK+\lambda} = \sum_{g=1}^{N} \sum_{j=t_R-(t_{c.g}+\lambda)}^{i=KK-t_{c.g}} \{P_{j,g} - S[J^* - (J-1)^*] - C_a\} \frac{A_g}{3.6} \left[1 - \exp\left(-\right.\right.$$

$$\left.\left.-\frac{1}{K}\right)\right] \exp\left(-i-\frac{j}{K}\right) \qquad (10)$$

The criteria for their updating needs to be decided. Reed (1987) reviews this problem and notes that often the parameters are adjusted to minimize some sort of least-squares criterion between recent model forecasts and telemetered flows. Amongst the parameters determined on line, one is an infiltration parameter, which decides the volume of runoff, and the other is the storage parameter, which determines the time distribution of this volume. Three criteria are investigated to define the volume:

(1) the observed volume until the time of forecast;
(2) the volume computed by the model to obtain the observed value of discharge at the time of forecast;
(3) the volume computed by the model to obtain a least-squares fit for the five observed values available at the time of forecast.

The values of F and S in terms of the three criteria are determined in terms of K and these values are substituted in equations (7) and (8). The resulting equations have only one varible — K — and are solved to obtain its value by minimizing the sum of squares between the five recent model forecasts and observed flows.

5. CASE STUDY: THE BACCHIGLIONE BASIN AT MONTEGALDELLA

The basin is shown in Fig. 1. Its area at Montegaldella is 1408 km². The Bacchiglione is joined by the Astico–Tesina tributary system 10 km upstream of Montegaldella. Beyond the junction, the river flows through two parallel embankments and does not drain the surrounding countryside. Therefore, the area of basin up to the junction has been considered in the analysis. The basin exhibits a wide variety of geological formations such as karst phenomena in the Astico subbasin, and dolomite, sandstone and compact limestones in the Bacchiglione–Leogra–Orolo subbasins. Alluvial cones on impermeable layers are prevalent in the valleys of Astico and Leogra subbasins. The annual rainfall in the basin varies from 1050 to 2200 mm. The maximum and minimum values observed at Pian delle Fugazze are 3348 mm (1926) and 1299 mm (1943) respectively. The lower part of the basin experiences less rainfall than the upper reaches does. The location of the 11 raingauges is indicated in Fig. 1. Since Ceolati, Staro and Pian delle Fugasse are located very near to each other, only two stations of the three, namely Ceolati and Staro, are included in the analysis. The radar estimates of rainfall on the basin are not yet available for any of the flood events. Therefore data collected from the traditional raingauge network have been used to evaluate the various procedures and the model The radar estimates in future will be available for 2 km × 2 km grid squares. However, to save computation time, 4 km × 4 km grid squares are used to represent the basin. Fig. 1 gives the configuration of the basin, the superimposed grid of 88 squares and the location of the raingauges. The rainfall on the grid squares is assumed to be equal to the value recorded by the nearest raingauge. The off-line parameters in equations (3) and (4) require the computation of the average response of the basin assuming the rainfall input to be uniform. A Nash model simulation of nine flood events (characteristics listed in Table 1) revealed that a two-parameter gamma function

Table 1 — Characteristics of available data on flood events and gamma simulation parameters

No.	Data of flood	Peak (m³/s)	Observed basin lag (h)	Gamma	Function parameters	
				n	K	
1	12 November 1941	362	25	2	13.8	
2	29 May 1965	241	17	2	10	
3	31 May 1965	243	22	2	12.8	
4	01 September 1965	414	15	2	9.2	
5	15 November 1968	363	28	2	15.6	
6	16 November 1975	277	22	3	8.2	
7	25 January 1978	261	27	2	13.8	
8	16 October 1980	289	20	2	10.8	
9	13 May 1977	267	27	2	14.2	

could reasonably simulate the response of the basin. The n value of the simulated response was usually 2 while the K value varied between 9 and 16 h, using an initial loss followed by the constant-abstraction-rate infiltration model. The time delay parameter varied with each flood. A value of 4 h was found to occur most often and was chosen as the value of the initial lag parameter λ. The average value of basin lag was found to be 22 h. Since n was usually 2, K was taken to be 11 h. This K value was used in equation (7) to determine $Nt_{c,g}$. The $Nt_{c,g}$ values were rounded to the nearest integer and the $t_{c,g}$ values were assigned to each grid square on the basis discussed in an earlier paragraph. Table 2 lists the allocated values of $t_{c,g}$. The other two parameters in equations (9) and (10) are estimated on line. An initial estimate of $K = 11$ was used to start the estimation process. The estimation process was terminated for K values less than 5.5 and greater than 33.

6.   COMPUTATION OF FORECASTS

One of the aims of the present study is to evaluate the proposed grid model using the two infiltration models and to investigate the three criteria used in updating the on-line parameters. The proportion PI of variance explained by the model compared with no model is the index which is commonly used in assessing the forecasts. A zero value of the index suggests that the model is as good as no model while a value of 1 is obtained for a perfect forecast. The time delay between the occurrence of rainfall

Table 2 — Allocation of linear channel parameters to grid squares

Linear channel parameter (h)	Grid square No.
0	2, 3, 5, 6
1	1, 4, 7, 10, 11, 12, 15, 17
2	8, 9, 13, 16, 18, 20, 21, 22
3	14, 19, 23, 24, 26, 27, 28
4	25, 29, 30, 31, 33, 34
5	32, 35, 36, 37, 38, 42
6	39, 43, 44, 45, 52
7	40, 44, 47, 51, 53
8	41, 48, 50, 55
9	56, 57, 58, 64
10	49, 63, 65, 71
11	59, 66, 67
12	62, 70, 72
13	54, 61, 69
14	60, 78, 79
15	73, 76
16	68, 80
17	77, 81
18	75, 83
19	84, 85
20	82
21	86
22	87
23	74
24	88

excess and the production of its main effects at Montegaldella is around 6 h. Therefore the index value for forecasts 6 h ahead is used to investigate the updating criteria and the model. The index is computed using

$$\mathrm{PI} = I - \frac{(Q_y - q_t)^2}{(Q_t - Q_{t-6})^2}$$
(11)

where Q_t is the observed value of discharge at time t, q_t is the computed discharge value using the model at time t and Q_{t-6} is the observed value of discharge 6 h prior to time t.

Equations (9) and (10) were programmed to compute forecasts using the three updating criteria. Henceforth the results using equation (9) and criterion (1) are referred to as the GC1 model results. The results obtained using criteria (2) and (3) are designated as GC2 and GC3 models. The corresponding models using equation (10) are designated GV1, GV2 and GV3 models. The six models were run for all events in Table 1, to formulate forecasts for the rising limb of the hydrograph and PI values were computed. The forecasts at intervals of 3 h for the highest flood producing two peaks for GV2 model are plotted in Fig. 2. PI values for all models for each of the events, are listed in Table 3.

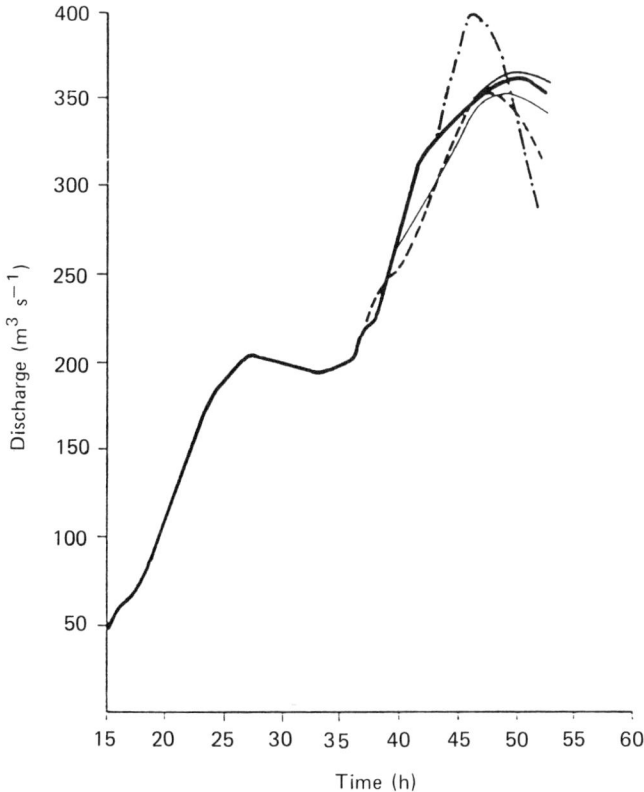

Fig. 2 — Forecasts computed at 37 h (– – – –), 40 h (———), 43 h (–·–), and 46 h (———) for the November 1941 flood using the GV2 model: ——— observed.

7. DISCUSSION OF GRID MODEL RESULTS

The summary of results in terms of PI values for various models is given in Table 3. A look at the results indicates that both the model and the criteria for updating are important in computing forecasts. The best overall results are obtained using the GV2 model in which the infiltration is represented by Phillip's equation, and the criteria for updating is to determine S using the least-known value of flow. K is estimated to obtain a least-squares match between the last five observed and

Table 3 — Proportion of variance explained by the model compared with no model: PI values for grid models

No.	Model	PI value for the following dates of floods									
		12 November 1941	29 May 1965	31 May 1965	1 September 1965	15 November 1968	16 November 1975	25 February 1978	16 October 1980	13 May 1977
1	GC2	0.54	−0.34	0.44	0.87	0.94	0.56	0.82	0.65	0.94
2	GC1	0.54	−0.58	0.58	0.81	0.72	−0.57	0.25	0.58	0.77
3	GC3	0.54	−0.77	0.36	0.83	0.84	0.17	0.73	0.62	0.91
4	CV2	0.9	0.98	0.98	0.67	0.96	0.74	0.93	0.69	0.93
5	GV1	0.65	0.75	0.72	0.94	−1.88	0.91	0.996	0.64	0.37
6	GV3	0.93	0.95	0.5	0.94	−0.37	0.9	0.87	0.64	0.62

computed values as in all other models. In six of the nine events the values are more than 0.9. The lowest value of 0.9 was obtained for the 1941 flood for which the forecasts are shown in Fig. 2. The forecasts computed at 37 h use K which is estimated on the basis of observed values between 32 and 36 h. In computing forecasts at 43 h the values used for estimating K are observed between 38 and 42 h. Since the rate of rise is more, K is reduced to match the model values, resulting in a rather peaky, rapidly receding hydrograph which is not usually observed on the basin. This suggests that the minimum value of K used as constraint needs to be revised upwards. The PI values for the other three events are 0.677, 0.694 and 0.744 for the same model, which can be termed as satisfactory. These results indicate that this version of the grid model is capable of formulating good forecasts for the Bacchiglione basin. All events in GC models have high values of PI for the second criterion. The second criterion also gives higher values of PI for six out of nine events for GV models. The results using the third criterion rank second in the comparison while the first criterion yields indifferent results. In flood events such as the 1978 flood, where the model structure and the assumption of gamma function yield a good match from the beginning of the flood until the peak, the criterion computes the highest value of 0.996 in the Table 3. In other flood events where the recent observed value cannot be computed using the given model structure maintaining the same volume, the index value is negative as in the 1968 flood event. The PI values using the GV2 model are consistently good for all the events while they are good for five of the nine events for the GC2 model. The PI values are higher for the GC2 models than for the GV2 model for 77 and 68 events. The values are negative for the 20 May 1965 flood for the GC2 model when the GV2 model computes a value of 0.98 for the same flood. The analysis of forecast results indicates that an unusually high value of rainfall in an event leads to the computation of a higher abstraction rate, thus eclipsing the contribution of smaller values which follow it. This results in poor forecasts. For the present analysis, in which the floods are produced by more than one spell of rainfall, the Phillip's infiltration model gives consistently good results.

8. COMPARISON BETWEEN THE LUMPED TIME-VARIANT SYSTEM MODEL AND THE GRID MODEL

Chander and Fattorelli (1988) used a lumped time-variant system model for real-time forecasting on the Piave river system and developed procedures to choose the response function and the abstraction rate as more information on the event becomes available. The model can be written in the discrete form as

$$Q_{KK+\lambda} = \sum_{j=t_R-\lambda}^{KK} R_j U_{KK-j+1}^{KK} \tag{12}$$

where U_{KK-j+1}^{KK} is the response function ordinate at time $KK-j+1$ for the response function chosen to compute the forecasts at time KK, and R_j is the average rainfall excess on the basin at time j.

The model was used to compute forecasts for both the infiltration models for all the events. The initial lag used in the computation was 4 h as in the grid model. The model using constant abstraction rate infiltration model is designated IITC, and Phillip's equation is designated as IITV. C_a value of 0.4 was used in the computation as in the case of the grid model. The PI index was computed for the rising portion of the flood for all events and the results are listed in Table 4. The IITV model performs much better than the IITC model for all events except the flood of 1980. In this event the rainfall is non-uniform with high rainfall in the upper part of the basin. The IITV model has a minimum abstraction rate C_a of 0.4 mm. Using a value of 4 h, it was found that the excess rain was not sufficient to support the runoff. Since the impact of the upper-basin rain is felt later, an increase in the value was found to improve the forecasts considerably. For example the PI value for 8 h increased from the present value to 0.663 for both the models for this flood. The IITV model and GV2 models perform equally well. In four events the IITV model is better while in the other five events the GV2 model gives higher values of PI. In the 1980 event both the GV2 and the GC2 models are not effected by the value of PI, thus highlighting the superiority of the grid model for floods formed by spatially non-uniform rainfall. The GC2 model performs better than the IITC model for nearly all the events. The forecast results of IITV and GV2 models for the February 1978 flood event for one forecast are shown in Fig. 3.

9. CONCLUSIONS

(1) An adaptive grid-square-based, geometrically distributed flood-forecasting model capable of utilizing rainfall information on each grid square using weather radar is proposed. The model converges to a lumped linear system model for spatially uniform rainfall. The relationship is exploited to compute the parameters of the distributed model using an average response of the lumped system model. The model is able to formulate acceptable forecasts with a maximum lead time of 6 h, which is equal to the time delay between the occurrence of effective rainfall and the production of its main effects at Montegaldella.

(2) Two infiltration models have been used to compute the rainfall excess. It emerges from the study that the grid model using Phillip's equation as the infiltration model computes better forecasts for the majority of the flood events analysed in the paper.

(3) Three criteria have been investigated to update the parameters of the model. It is concluded that the criteria of updating have considerable effect on the quality of forecast. The criterion which computes the infiltration parameter to yield the last-observed value and the storage parameter to match the shape in the immediate neighbourhood computes better forecasts.

(4) The comparative study between the grid model and the lumped system time-variant model suggests that the best version of both the models perform very well for spatially uniform rainfall inputs. The grid model performs better than the lumped model for spatially non-uniform rainfall inputs.

ACKNOWLEDGEMENTS

The authors wish to thank Dr Massimo Crespi, Director of the Experimental Centre for Hydrology and Meteorology (Teolo, Padova, Italy), Regione del Veneto, for

Table 4 — Proportion of variance explained by the model as compared to no model: PI values for lumped models

No.	Model	PI value for the following dates of floods								
		12 November 1941	29 May 1965	31 May 1965	1 September 1965	15 November 1968	16 November 1975	25 February 1978	16 October 1980	13 May 1977
1	IITC	−0.2	−2.7	−0.27	0.87	0.53	0.18	0.65	0.4	0.89
2	IITV	0.9	0.88	0.71	0.96	0.6	0.92	0.97	−0.7	0.94

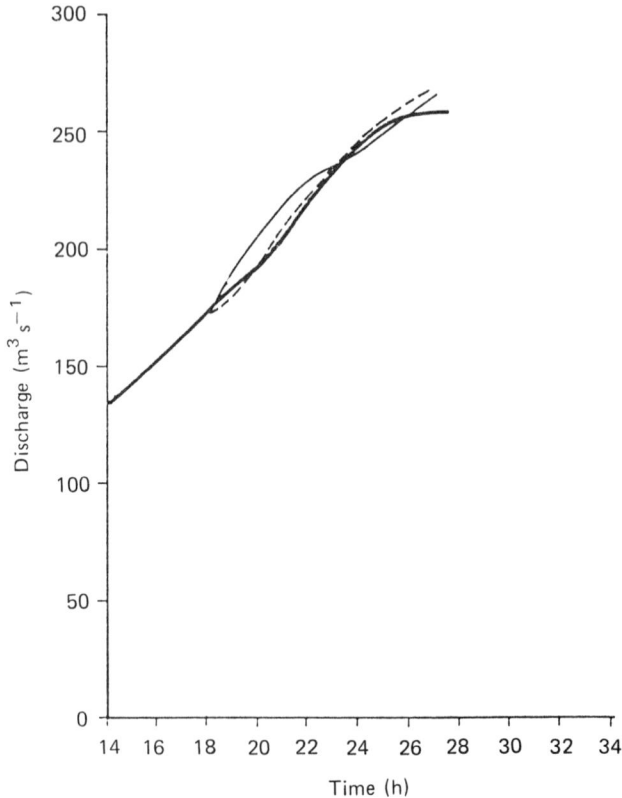

Fig. 3 — Forecasts computed at 18 h for the February 1978 flood using the GV2 (——) and IITV (- - -) models: ——, observed.

having supported the development of the model. The authors also thank Dr Antonio Capovilla for his computational support and contribution to the analysis of hydrological information and data used to test the model.

REFERENCES

Anderl, B., Attmannspacher, W. and Schultz, G. A. (1976) Accuracy of reservoir in-flow forecasts based on radar rainfall measurements, *Water Resour. Res.*, **12** (2), 217–223.

Chander, S. and Fattorelli, S. (1988) Italian National Report for the year 1987–88 for the project entitled 'Application of weather radar for the alleviation of climatic hazards'. Commission of European Communities.

Chander, S. and Shanker, H. (1984) Unit hydrograph based forecast model. *Hydrol. Sci. Bull.*, **29** (3), 279–291.

Clark, C. O. (1945) Storage and the unit hydrograph, *Trans. Am. Soc. Civ. Eng.*, **110**, 1419–1488.

Corradini, C., Melone, F. and Ubertini, L. (1986) A semidistributed adaptive model for realtime flood forecasting'. *Water Resour. Bull.*, **22** (6), 1031–1038.

Corradini, C. and Melone, F. (1986) An adaptive model for on line flood predictions using a piecewise uniformity framework. *J. Hydrol.*, **88** (3–4), 365–382

Dooge, J. C. I. (1959) A general theory of the unit hydrograph. *J. Geophys. Res.*, **64**, 241–256.

Klatt, P. and Schultz, G. A. (1985) *Flood forecasting on the basis of radar rainfall measurement and rainfall forecasting*, IAHS Publication No. 145. International Association of Hydrological Sciences, pp. 307–315.

Loague, K. M. and Freeze, R. A. (1985) A comparison of rainfall–runoff modelling techniques on small upland catchments. *Water Resour. Res.*, **21**, 299–248.

O'Connell, P. E., Brunsdon, G. P., Reed, D. W. and Whitehead, P. G. (1986) Case studies in realtime hydrological forecasting from the U.K. In: D. A. Kraijinhoff and J. R. Moll (eds), *River flow modelling and forecasting*. Reidel, Dordrecht, pp. 195–238.

Plate, E. J., Ihringer, J. and Lutz, W. (1989) Operational models for flood calculation. *J. Hydrol.*, **100**, 489–506.

Reed, D. W. (1987) UK flood forecasting in the 1980's. In: V. K. Collinge and C. Kirby (eds), *Weather radar and flood forecasting*. pp. 129–142.

Simpson, R. J., Wood, J. R. and Hamlin, M. J. (1980) *Simple self correcting models for forecasting flows on small basins in realtime*, IAHS Publication No. 129. International Association of Hydrological Sciences, pp. 433–444.

39

Radar signal quantization and its influence on rainfall-runoff models

I. D. Cluckie[1], K. A. Tilford[1] and G. W. Shepherd[2]
[1]Water Resources Research Group, Department of Civil Engineering, University of Salford, Salford M5 4WT, UK
[2]Resources Recovery Board, PO Box 86, Bellozane Road, St. Helier, Jersey

ABSTRACT

In order to produce estimates of rainfall intensity, radar return signals are converted from an analogue to digital form. The coarseness of the quantization intervals used dictates the intensity resolution of the rainfall estimates: radar data in the UK being quantized across either 208 (high-intensity resolution, 8 bit data) or 8 (low-intensity resolution; 3 bit data) intensity levels depending on the product. This paper examines the influence of signal quantization (and thus rainfall intensity resolution) using two types of assessment criteria: 'end-point use' and signal information content. The end-point use applications are real-time flood forecasting for rural catchments using a transfer-function-based rainfall–runoff model, and an urban runoff simulation using the WASSP-SIM package. Precipitation data from weather radars operated by the Meteorological Office in the UK are used as model input. The information content was determined using spectral analysis techniques. The assessment techniques are mutually complementary, the statistical analysis providing an explanation for the adequacy of the 3 bit data with regard to the modelling and simulation described.

1. INTRODUCTION

The advent of remote sensing as a technically and economically viable method of measuring and observing hydrological processes constitutes a significant advance in instrumentation technology. In particular the establishment of fully operational national and international weather radar networks has presented the hydrologist with information which until relatively recently was unobtainable, namely real-time, spatially distributed rainfall estimates over large areas. Remotely sensed data have many advantages over traditional point measurements, the prime one stemming from the fact that hydrologists are invariably interested in wide-area spatial information rather than often unrepresentative point samples. Radar differs greatly from other rainfall estimation techniques and the challenge now lies in making the best possible use of data that differ fundamentally from traditional point measurements.

The influence of physical processes (e.g. evapotranspiration, percolation and overland flow) means that the runoff is far less variable and more predictable than its cause: the dynamic process of precipiation. This is because the complex interactions between the physical processes introduce a filtering element which effectively transforms a high-frequency process (rainfall) to a lower-frequency one (runoff). Many flow-forecasting models simulate this natural filtering by catchment processes by utilizing a mathematical convolution process to modulate the rainfall input signal. To facilitate the implementation of discrete models on digital computer systems, continuous hydrological processes are sampled, entailing temporal filtering. The data may be further (spatially) filtered if a lumped model is applied. Hence hydrological process data are extensively filtered before modelling. In each case the filtering reduces the high-frequency (noisy) component of the respective signals.

Related to the spatial and temporal dimensions is the more subtle (although no less important) representation of rainfall intensity. Production of discrete surface rainfall estimates entails quantization of the analogue signal reflected by raindrops. In the UK, the highest-intensity-resolution radar rainfall data are obtained from quantization across 208 intensity ranges, the rainfall intensities for each radar element therefore having any one of 208 values, i.e. 8 bit data. In addition, the radar signals are also quantized using just eight intensity ranges to form 3 bit data. The first commercially produced radar imaging systems in the UK displayed colour-coded 3 bit data. One reason for this is attributable to economic and technical constraints in display device hardware and data transmission (in contrast, the spatial and temporal resolutions are largely a function of radar hardware considerations). Despite the manifestly lower-intensity resolution of 3 bit data, and considerable developments in the respective technologies, radar rainfall data quantized across eight intensity ranges remain the favoured resolution for UK Composite, European (COST) and quantitative precipitation forecast (FRONTIERS) images. Because the underlying values from which the display colours are derived are not available, these data are commonly referred to as qualitative or display data. However, this paper attempts to question the assumption that the data do not possess adequate 'accuracy' for quantitative application.

Using a novel procedure to assign quantitative rainfall intensities from the eight qualitative ranges used by the UK Meteorological Office for the display of 3 bit data, and a comparative 'end-point use' approach (Cluckie et al. 1980), 8 and 3 bit radar rainfall data are used as an input to rainfall–runoff models for flood forecasting in rural catchments and the WASSP-SIM package for the simulation of runoff over urban areas. The end-point use approach is seen as the most valid form of comparative data quality evaluation given the difficulties associated with determination of 'true' rainfall or 'ground truth'. In addition, the information content of the respective rainfall representations is studied in the frequency domain using spectral analysis techniques: these inherently objective techniques provide an insight into the signal characteristics before they are processed by the models.

The radar products currently produced by the UK Meteorological Office are summarized in Table 1. Essentially, the data fall into two categories: single site (data

Table 1 — Summary of radar data streams available from Meteorological Office weather radar network

Data type	SWpatial resolution (km)	Range or image size (km)	Intensity resolution	Availability (min)	Comments
1	5 Subcatch-ment	210 —	8 levels 8 levels	15 15	Data collected every 5 min Includes hourly and daily rainfall totals
2	2 (to 75 km) 5 (to 210 km) Subcatch-ment	210 —	208 levels 208 levels	5 5	— Rainfall totals for preveous 15 min, previous 1 h and previous 1 day
3	5	210	208 levels	5	Used for production of composite images
UK National Composite	5	640–640	8 levels	5	Data updated every 15 min
FRON-TIERS	5	1280–1280	8 levels	30	Quantitative precipitation fore-casts for up to 6 h ahead

types 1–3) and composite images. In addition to grid-square data, areally integrated rainfall depths are available at a river catchment scale. This paper is based on analysis using type 2 single-site rainfall data. An assignment procedure (Fig. 1) was developed to allocate 3 bit representations from the 8 bit data and hence to simulate the effect of assigning quantitative values from, for example, type 1 national composite or FRONTIERS data. Three different methods of allocating 3 bit values were examined based on the arithmetic, harmonic and geometric means of the upper and lower bounds of the respective slice ranges. Of these, extensive analysis has shown that the geometric mean produces 3 bit rainfall sequences which most closely represent the 8 bit data; cumulative rainfall depths for the calibration data (400–500 h) for 3 bit data are within 5% of the 8 bit data for both catchments described in the present paper although extensive further analysis confirms this as the case in general.

2. FLOOD FORECASTING IN RURAL CATCHMENTS

For the assessment of the influence of signal quantization on model calibration and flow-forecasting performance in rural catchments, two catchments in the northwest of England (Fig. 2) were used for the modelling and forecasting analysis, both within 75 km of the Hameldon Hill weather radar located near Burnley, UK. The River Roch is one of the most important tributaries of the River Irwell, the feeder of the Manchester Ship Canal. The catchment is gauged at Blackford Bridge and to this

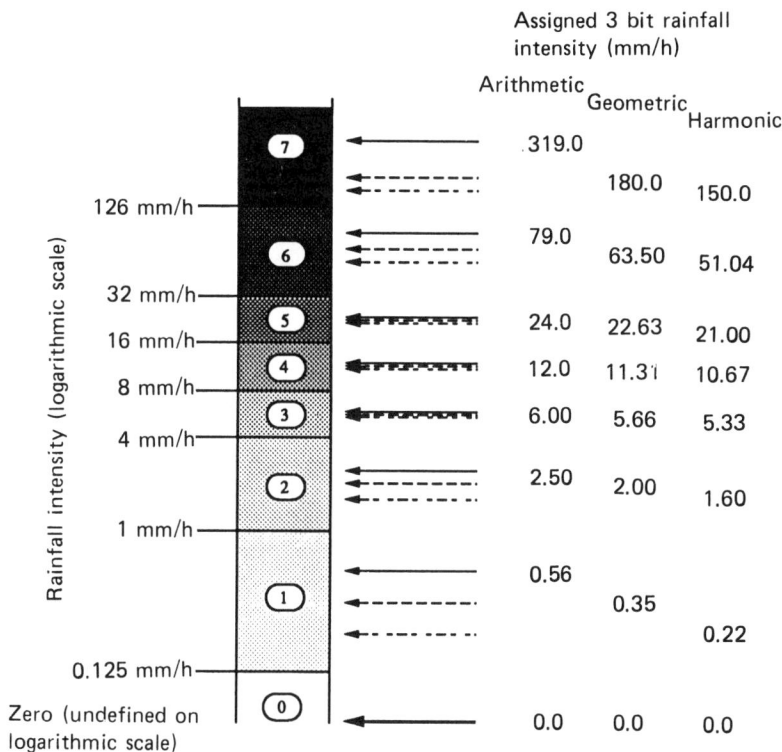

Fig. 1 — Schematic representation of radar data slicing and 3 bit assignment.

point drains an area of 183 km². The river responds quickly to rainfall, typical catchment response times being between 5 and 6 h. The River Wyre drains a smaller catchment of 111 km². The gauging station located at Garstang is immediately downstream of a flood risk zone and plays an important role for flood forecasting and warning in the vicinity. The typical response time to this point is 5 h.

The lumped-transfer-function rainfall–runoff model transforms an input of total rainfall to flow using current and past rainfall and flow values; spatially and temporally integrated rainfall depths from the model input. The optimal model structure and parameters are identified off line using historic event data (e.g. Owens 1986). Model calibration attempts to extract the process information from historic rainfall–runoff data in order to determine a mathematical input–output relationship between the two. The calibration procedure utilizes an approach which combines an objective statistical assessment with subjective interaction. In the latter the impulse response of the calibrated model is used as a test of model integrity in a hydrological context.

Fig. 3 shows the impulse reponses of models calibrated for the two catchments using 8 bit and 3 bit rainfall data. For each catchment there is negligible difference between the impulse responses of the models determined from 8 bit rainfall data and the corresponding 3 bit rainfall data. The characteristics of the rainfall signals and

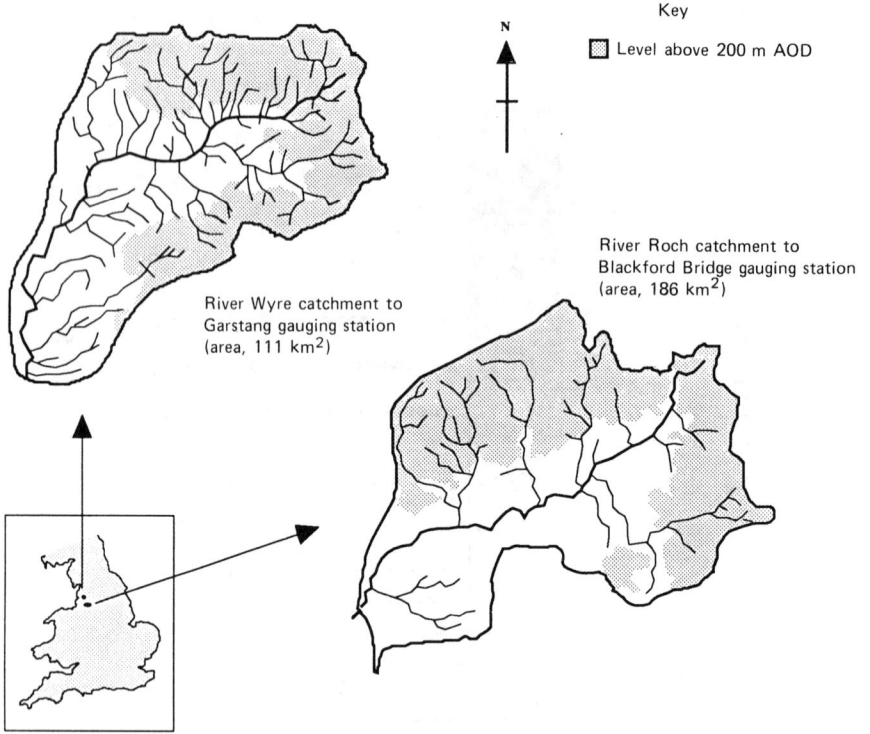

Fig. 2 — Site and location of river catchment used for analysis.

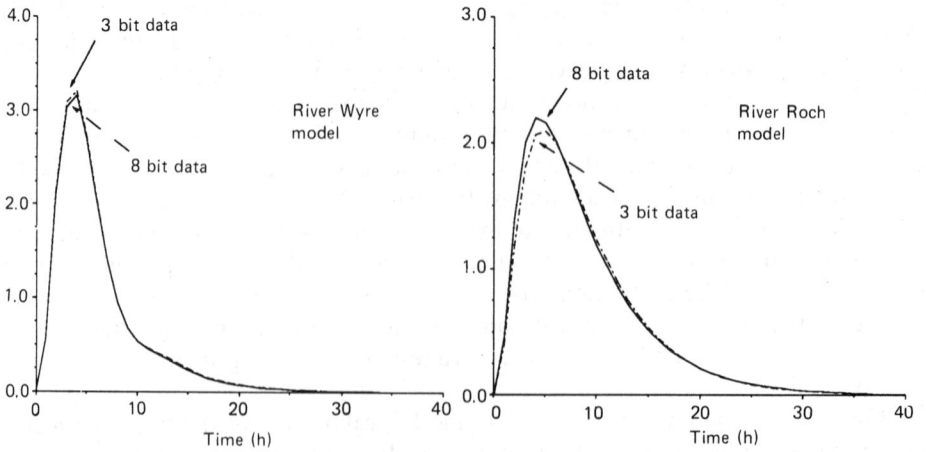

Fig. 3 — Model impulse responses.

calibrated models in the frequency domain are studied via the application of spectral analysis techniques.

The frequency responses of the respective models are shown in Fig. 4. In common with many other rainfall–runoff models such as the synthetic unit hydrographs

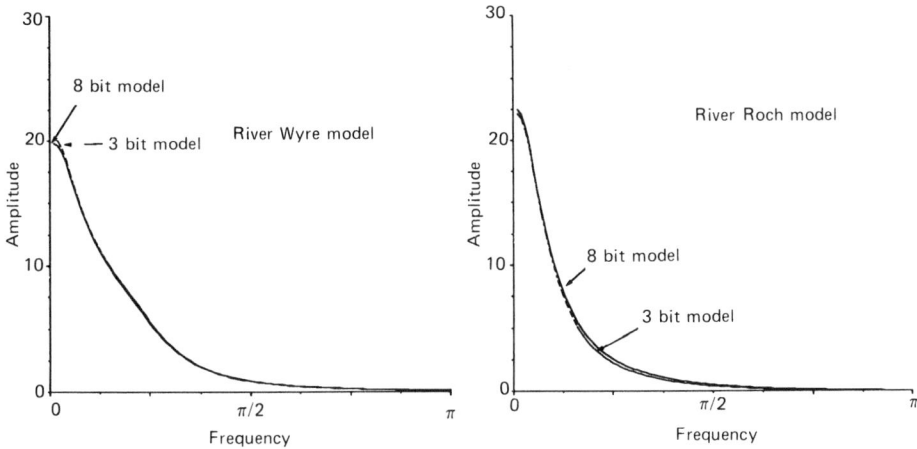

Fig. 4 — Model frequency responses.

proposed by Nash (1957), Clark (1945), US Department of Agriculture (1957), Snyder (1938) and Natural Environment Research Council (1975), the frequency response is confined to the low-frequency end of the spectra (the reason that a simple function, e.g. a triangle, possesses the ability to reconvolute adequately the most important hydrograph characteristics). There is negligible difference in the frequency response characteristics of models determined from 8 and 3 bit rainfall data. This signifies that the mathematical structure of the 8 and 3 bit models and consequently the manner in which they transform an input signal to an output signal are essentially identical. It is concluded that despite a significant reduction in intensity resolution the models calibrated from 8 and 3 bit rainfall data implicitly incorporate identical knowledge of the rainfall—runoff process.

Fig. 5 shows the power spectra of the rainfall data used for model calibration. Despite degradation in intensity resolution, the major portion of the low-frequency component of the signals are retained (and for the Roch catchment is relatively greater), the importance of which is discussed with respect to model forecasting performance.

Forecast hydrographs 6 h ahead for both catchments derived from 8 and 3 bit rainfall data are shown in Fig. 6. The hydrographs illustrate negligible difference in forecast quality and, in the context of operational flood forecasting, insignificant. Similar results have been observed from a large number of other verification events for the catchments (Tilford 1987) and supported further by preliminary work on catchments in eastern England within the Anglian Water Authority (Anglian Radar Information Project 1988).

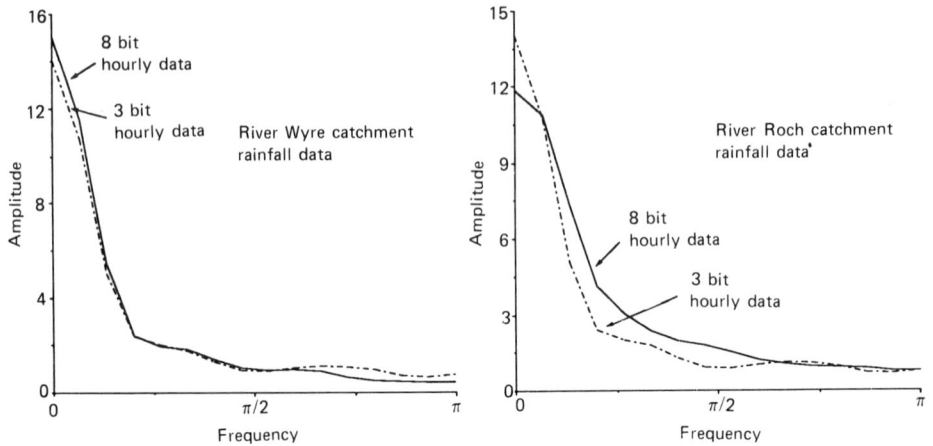

Fig. 5 — Power spectra of the rainfall data.

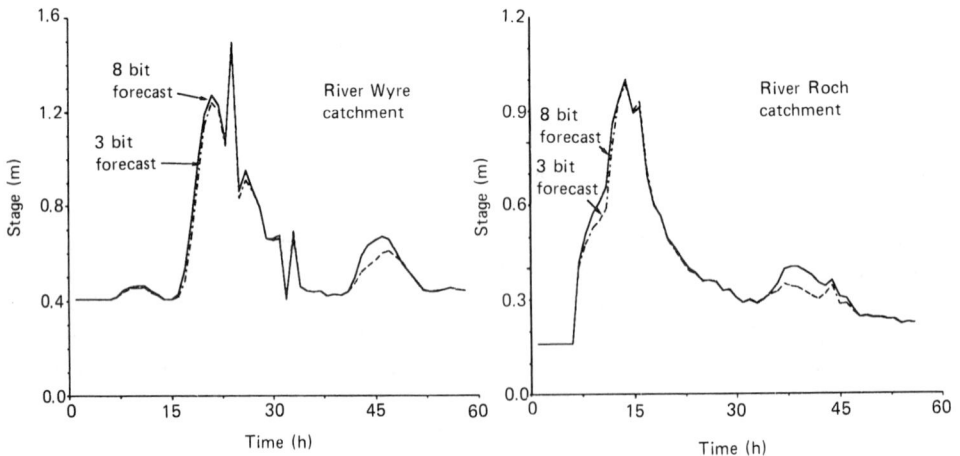

Fig. 6 — Flow forecasts 6 h ahead using 8 and 3 bit rainfall data model inputs.

The adequacy of 3 bit rainfall data is explained by considering the power spectra of the forecasting models and the rainfall data. As the model power spectra show, to forecast river flow the transfer function model utilizes predominantly low-frequency information which is implicitly extracted from the rainfall signal. In this context, the low-frequency signal component corresponds to process-describing information. Comparison of the power spectra of 8 and 3 bit rainfall data reveals that the low-frequency component is similar even though the intensity resolution has been degraded. In contrast, the high-frequency (noise) component of the signal does not contribute to process description and is largely ignored by the model which is essentially a low-pass filter.

3.   PRELIMINARY URBAN DRAINAGE RESULTS

The application of 3 bit data to rural flood-forecasting models has been discussed in the first part of this paper. Urban drainage models also require a source of precipitation data and the effects of signal quantization on this class of model was examined. The simulation software used is the WASSP-SIM section of the Wallingford Procedure (National Water Council 1981), a software package which allows an urban drainage network to be mathematically modelled and simulated using either design storms of stated frequency or actual event data. The package can accommodate either lumped or distributed rainfall input.

It is possible that, in the struggle to realize greater accuracy in the measurement of physical processes and in the development of techniques of simulation, the need for greater and greater accuracy has been taken to extremes. Such developments are particularly unfortunate in the context of a stochastic phenomena such as rainfall. With rainfall the problem of assigning a 'ground truth' remains a contentious and unresolved issue and the discussion is better when oriented towards 'optimal' extraction and utilization of the information content of the signal rather than towards loose concepts such as definitions of truth.

Within WASSP, the methodology used to compensate for the temporal variation in rainfall across a catchment requires the application of a filter that modifies point rainfall measurements to allow the portion of the event that has just passed over the catchment and also that portion which will be over the catchment at the time of the next data item P_t to be taken into account. The equation that describes this filter is

$$P_t = \mu P_{t-1} + (1-2\mu)\, P_t + P_{t+1} \tag{1}$$

where μ is a function of catchment area and data time increment.

Users of simulation methods quite often use tipping-bucket raingauges that have an equivalent rainfall depth resolution of 0.5 or 0.2 mm to gather rainfall data. Unless the rainfall event is particularly intense, the number of tips per minute recorded (by the former) will generally be less than three — an intensity of 90 mm/h. Since the buckets take a finite time to fill, data will invariably cross the boundary between measurement periods. In general, this is accepted and the use of filters, similar to that used within WASSP-SIM, allow profiles to be produced that are fair representations of the events, even though the data resolution is being degraded.

Methods have been developed (Shepherd 1987) that allow 8 bit weather radar rainfall data to be integrated with urban simulation packages to provide the precipitation input data. However, the current range of data products available at the level of definition thought necessary (2 km spatial resolution and a Δt of 5 min) are only available as 8 bit data. The storage requirements for data sets at this intensity resolution are very large, approximately 3.5 Mbytes/day/site in an unpacked state. Any reduction in the size of data sets will be an advantage and one way of achieving this is to increase the amount of data per byte, i.e. to use 3 bit data. As previously stated, this reduction in the number of bits used to represent the data effectively reduces the resolution of the data from 255 to seven levels — an action which will reduce the resolution of but not necesarily the integrity of the data. Conversion of 8 bit data to a 3 bit representation by application of the technique discussed in the first part of this paper produces obvious differences in the respective rainfall profiles (Fig. 7). However, although individual variations are evident, the difference between the

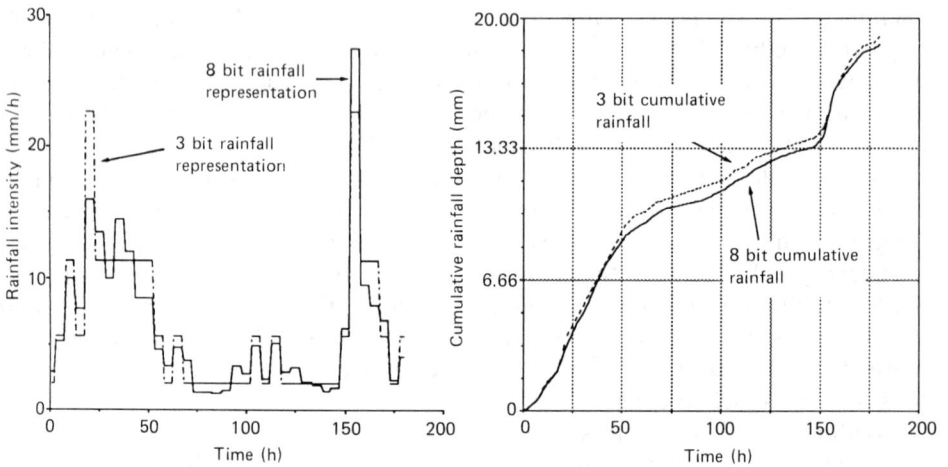

Fig. 7 — Rainfall hyetograph and cumulative rainfall for event design storm.

cumulative rainfall derived from the 8 and 3 bit rainfal data were generally less than 10%, providing that the rainfall was significant, i.e. more than 5 mm. Fig. 8 shows an

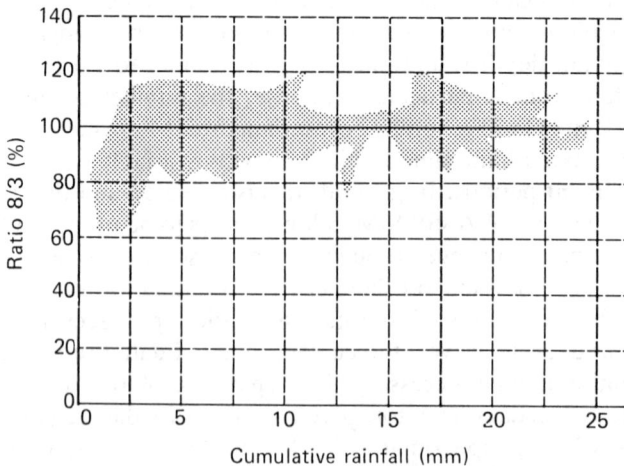

Fig. 8 — Enclosure scattergraph: conversion error as a function of total rainfall.

envelope scattergraph of the ratio of 8 to 3 bit cumulative rainfall against the 8 bit cumulative value for the event. Approximately 15 000 such events were examined to produce the scattergram which for convenience is reproduced as an enclosure graph.

These profiles (and many others) have been used in conjunction with the WASSP-SIM package. Initial work was carried out using a theoretical linear pipeline designed using a rational method to ensure adequate capacity. Output hydrograhs were plotted at nodes along its length for simulation using both 8 bit and 3 bit input data. Examples of the hydrographs in Fig. 9 show that the differences are minimal by

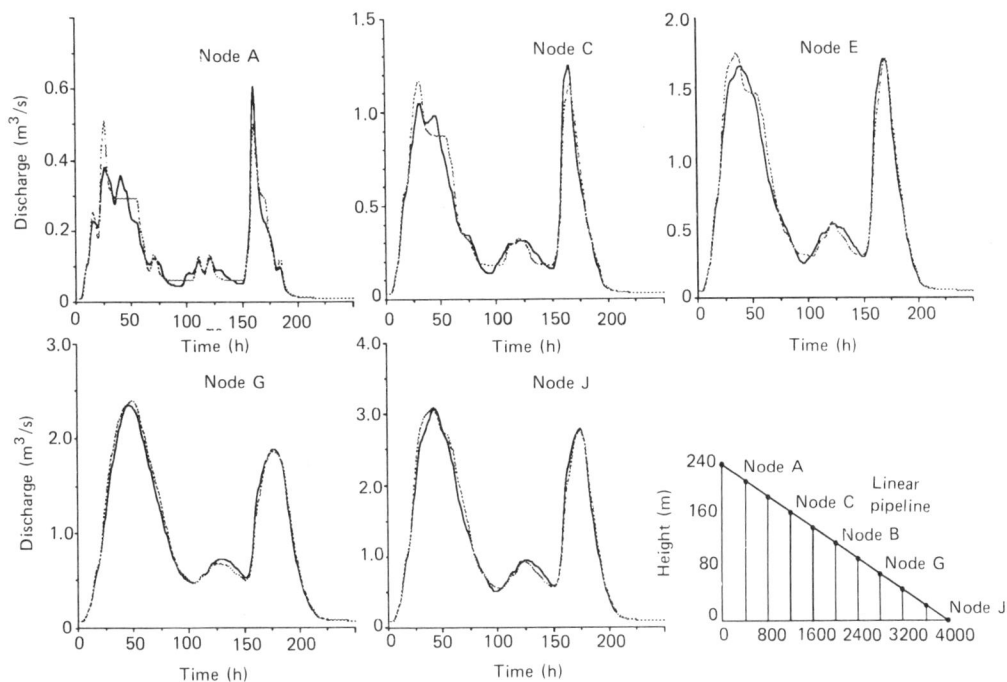

Fig. 9 — Hydrograph simulation for a linear pipeline using 8 bit (———) and 3 bit (.....) rainfall data.

node G and are probably insignificant some distance before this point. The simulations were extended to an actual surface water network, for the town of St Helier, Jersey, Channel Islands. The sample results in Fig. 10 indicate similar results.

The conclusion appears to be in accordance with the detailed observations drawn earlier in the paper regarding the effect of the simulation model on the high-frequency component of the input data signal. The size of the urban drainage area directly affects its ability to respond to high-frequency input signals and in many cases of practical relevance the catchments acts as a low-pass filter.

4.   CONCLUSIONS AND RECOMMENDATIONS

This paper has shown that rainfall data coarsely quantized across eight intensity levels (3 bit data) can be used for quantitative rainfall–runoff modelling, despite the

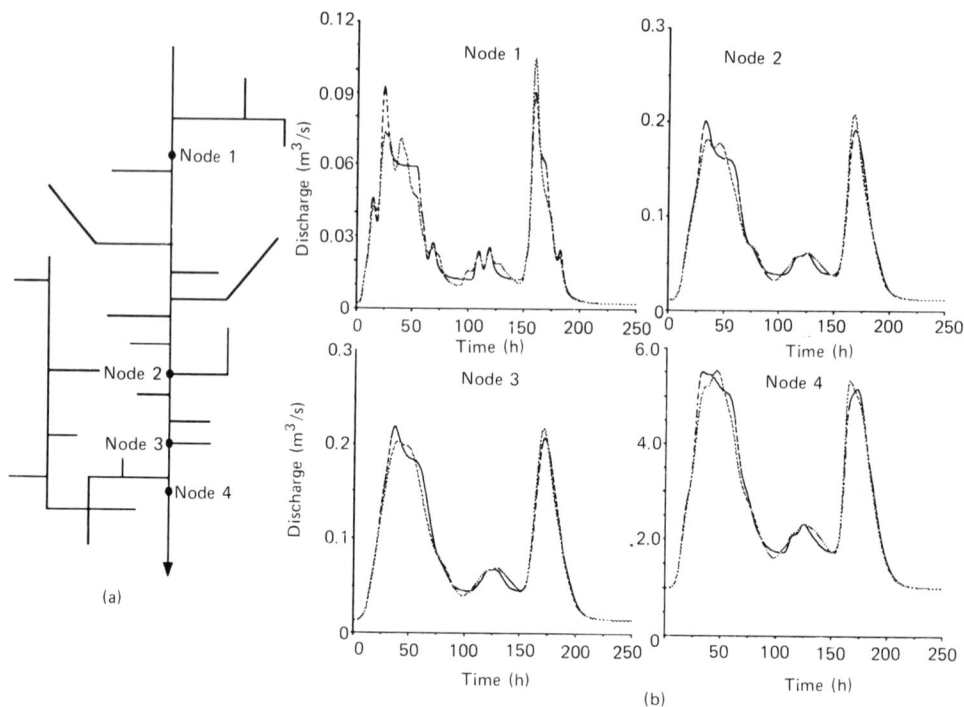

Fig. 10 — (a) Schematic representation and (b) hydrography simulation (————, 8 bit data; ....,
3 bit data) at St. Helier urban drainage network.

low-intensity resolution without compromising forecast quality (in the case of flood
forecasting in rural catchments) and simulation quality (in the case of urban drainage
networks).

The hydrological data utilized in all forms of flow forecasting and simulation are
extensively filtered, in both the temporal and the spatial domains. Indeed, the
discrete hydrological data are samples of a continuous process which itself has been
extensively filtered; catchment processes responsible for the physical transformation
of rainfall to flow transform a high-frequency process (rainfall) to a low-frequency
one (flow). The combined effect can be visualized as a low-pass filter, removing a
significant portion of the high-frequency (noisy) part of the respective signals.

A study of the information content from the power spectra of the rainfall signals
shows 3 bit data to have similar characteristics to higher-intensity-resolution 8 bit
data. The bulk of the process-describing information is concentrated at the low-
frequency end of the spectra and this portion of the signal is retained despite
degradation of intensity resolution. The frequency responses of calibrated flood-
forecasting models illustrate the low-frequency characteristics of the models.

The implications for operational utilization of radar data are significant. Preoccu-
pation with data precision has been at the expense of an appreciation of data
information content. It is probable that, for a great many applications, rainfall data

with a much lower intensity resolution than we are used to may be adequate. Potentially this may result in significant cost reductions not only in the cost of the data themselves but also with respect to transmission, processing and archiving costs.

ACKNOWLEDGEMENTS

The authors would like to thank the following for their assistance in this work, the UK Meteorological Office, North-West Water Authority, Anglian Water Authority and Resources Recovery Board (Jersey).

REFERENCES

Anglian Radar Information Project (1988) *An evaluation of the influence of radar rainfall intensity resolution for real-time operational flood forecasting*, ARIP Report No. 2, Anglian Water Authority.

Clark, C. O. (1945) Storage and the unit hydrograph *Proc. Am. Soc. Civ. Eng.*, **69**, 1419–1447.

Cluckie, I. D., Harwood, D. A. and Harpin, R. (1980) Three systems approaches to real-time rainfall–runoff forecasting. *Proceeding of the the Oxford Symposium on Hydrological Forecasting*, International Association of Hydrological Sciences. IAHS–AISH Publication No. 129.

Nash, J. E. (1957) The form of the instantaneous unit hydrograph. *International Association of Scientific Hydrologists General Assembly*, *Toronto*, Vol. III. International Association of Scientific Hydrologists, pp. 114–121.

National Water Council (1981) *Design and analysis of urban storm drainage*: *the Wallingford Procedure*, 5 vol. National Water Council.

Natural Environment Research Council (1975) *U.K. flood studies report*, 5 vol. Whitefriars Press, London, 1975.

Owens, M. D. (1986) *Real-time flood forecasting using weather radar data*, Ph.D. Thesis. University of Birmingham.

Powell, S. M. (1985) *River basin models for operational forecasting of flow in real-time*, Ph.D. Thesis. University of Birmingham.

Shepherd, G. W. (1987) *On the utilisation of weather radar in the simulation of urban drainage networks*, Ph.D. Thesis. University of Birmingham.

Snyder, F. F. (1938) Synthetic unitgraphs. *Trans. Am. Geophys. Union*, **19** (1) 447–454.

Tilford, K. A. (1987) *Real-time flood forecasting using low intensity resolution radar rainfall data*, M.Sc. Thesis. University of Birmingham.

US Department of Agriculture (1957) *Soil Conservation Service Engineering Handbook*, Section 4, *Hydrology*, Suppl. A. US Department of Agriculture Washington, D.C.

40

Never expect a perfect forecast

T. Einfalt[1] and T. Denoeux[2]
[1]RHEA, 1 rue Albert Einstein, Champs sur Marne, 77436 Marne la Vallee Cedex 2, France
[2]Cergrene Ecole Nationale des Ponts et Chausées, La Courtine, 93167 Noisy-le Grand, France

ABSTRACT

Although hydrologists often argue in terms of a 'perfect rainfall forecast' when describing optimal use of their tools, this perfectness is neither achievable nor necessary. Knowing the reasons for uncertainties of a rainfall forecast enables the forecaster to account for them. If hydrologists are able to quantify their proper precision needs, a non-perfect rainfall forecast, supplied with confidence limits, is likely to enhance reliability and precision of hydrological applications. First steps in this direction have been undertaken at the Seine-Saint-Denis county real-time control project.

1. INTRODUCTION

For many years, hydrologists have formulated the need for reliable rainfall forecasts for purposes as different as aviation hazard warning and real-time control of sewer systems (Damant *et al.* 1983, Collier 1977). They argued that, if they could use reliable rainfall forecasts instead of measurements or climatological values, hydrological forecasts or real-time operations would give considerably better results.

The development of real-time control strategies for controlling retention basins, river flow or urban drainage systems has led to in-depth studies on the potential of radar rainfall measurement and forecasting (e.g. Cluckie and Owens 1985, Denoeux 1989). Consequently, small-scale analyses examining the benefit of a perfectly forecasted rainfall event on hydrological models have been performed (e.g. Schilling and Petersen 1987). They helped to raise expectations, as their results proved a considerable improvement of the overall real-time control performance, using these 'perfect rainfall forecasts'.

What are the conditions that rainfall can be perfectly forecast? Mainly, it is

(1) a perfect rainfall measurement and
(2) an error free forecasting performance.

2. MATHEMATICAL MODELLING AND NATURAL PROCESSES

Usually, the development of a mathematical model for a natural phenomenon is based on

(1) the data that are available,
(2) the understanding of the process and
(3) the purpose and, hence, the form of the results to be issued.

Each of these three steps is accompanied by possible uncertainties that make the model and its results an incomplete image of reality. These uncertainties can be divided into the following.

(a) *Explicit uncertainties*
    These are factors that are due to the construction of the model and the choice of the data set. They can partly be taken into account if additional data are collected or another — more sophisticated — model is used,
(b) *Implicit uncertainties*
    These are factors that are due to the modelling process itself and thus will always be present since there is no way of getting around them by model or data upgrading.

3. EXPLICIT FACTORS FOR UNCERTAINTIES

Typical examples for explicit uncertainty factors in hydrological modelling are as follows.

(1) The transformation of information (e.g. from reflectivity to intensity, from intensity to flow volume, and from flow level to flow volume) is carried out.
(2) The desired phenomenon cannot directly be observed (e.g. topographical influence on rainfall intensity, and rainfall dynamics).
(3) Not all necessary data are always available or reliable (sensors do not always work properly, transmission lines can fail, etc.).
(4) The necessary density of data points cannot always be provided (spacing of a raingauge network, measuring time step for meteorological values such as wind speed at 3000 m altitude, etc.).
(5) Physical processes are represented by simplified models (each sewer system is simplified for modelling purposes, areal rainfall calculation based on raingauges is a spatial simplification, etc.).
(6) For a treatment in a homogeneous way, the observed phenomenon often is put into a pre-defined class. The definition of classes for natural processes is frequently done in an arbitrary manner — the separation lines are fuzzy and the

process in reality often is heterogeneous (rainfall classification into convection and advection dominated events, classification of a whole radar image may neglect local effects, etc.).

Taking into account the uncertainties induced by these external factors is not yet general engineering practice, but there have been a number of efforts to reduce the probable error or to reduce its importance for a given application.

4. IMPLICIT FACTORS FOR UNCERTAINTIES

These factors are always present with mathematical models of natural processes.

(1) All measurements are static values since they always are taken as data points in time and space. Even continuous observation of phenomena provides only a higher discretization and hence only approximates continuous characteristics. For instance, a tipping-bucket raingauge transmits a pulse for each drop reaching the measurement device, it is obvious that this cannot be a continuous process.
(2) Consequently, the mathematical models developed for these data have to be models based on static values which only roughly represent the underlying dynamic processes that they mean to model. For the rainfall-forecasting case, radar or raingauge measurements for given time steps are used. The forecasting model's task is to estimate the values which might occur at the measurement points in the future and in terms of the measurement units. Doing this, the model can only make reference to the artificial measurement level (Fig. 1) without being able to compare directly with the underlying physical process.

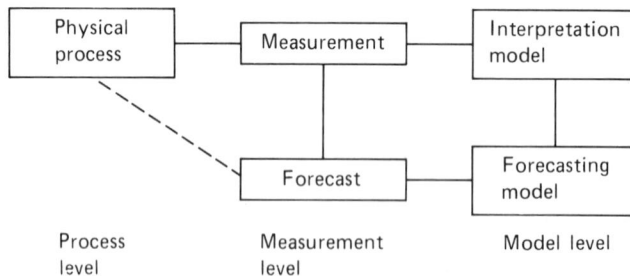

Fig. 1 — Relationship between physical process, measurements and models.

(3) Mathematical methodology is not capable of dealing with dynamic processes in 'dynamic' terms (Eisenhart *et al.* 1988). Every abstraction effort is based on event-independent and hence time-independent features, usually invariant parameters or relationships. Mathematical modelling is based on such invariant parameters. Since each natural process has to be considered as a singularity, i.e. as unique, a mathematical model forcingly introduces inexactness.

These implicit factors, inherent in the chosen methodology, show us that developing a mathematical model for a naturally dynamic process cannot be totally exact. More precisely, the more variable ('dynamic') the natural process behaves the more significant is the deviation of the model from reality.

5. CONSEQUENCES FOR RAINFALL FORECASTING

It is obvious that these kinds of uncertainty are of minor importance for many applications. For modelling of slowly moving hydrological processes (river or groundwater levels, groundwater quality, etc.), a sufficiently short time step has led to satisfactory results (e.g. Schilling and Einfalt 1985). However, dynamic processes such as rainfall (Fig. 2) do not reduce their variability at all or only marginally if a

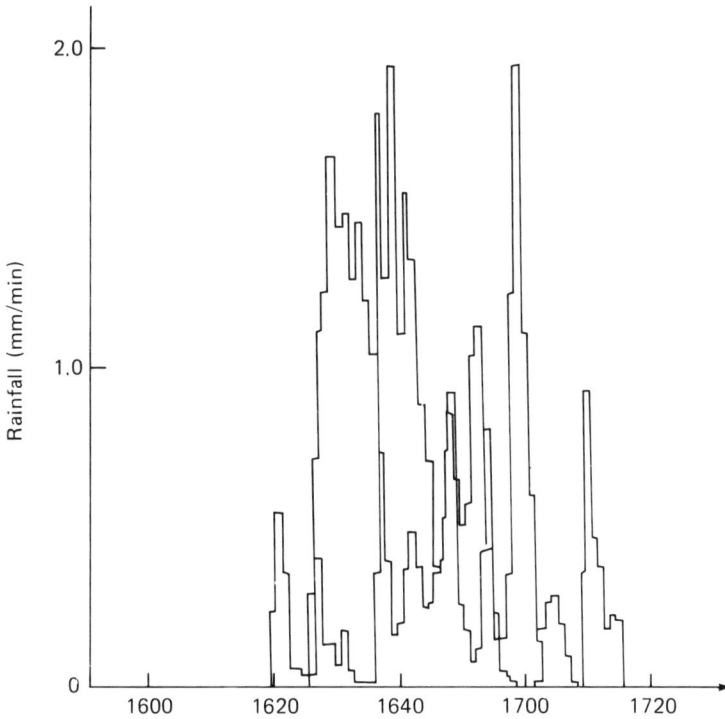

Fig. 2 — The variability of the rainfall process: radar data comparison of two areas about 4 km apart. (From Verworn (1988).)

shorter time step is used. Statistical analyses of rainfall (Lovejoy and Schertzer 1986) show that this process is as variable in large-scale as it is in small-scale observations, in both time and space. We should consequently admit not only that rainfall measurement techniques (radar and raingauge) do not represent reality in space and time, but also that we will never be able to measure the dynamic rainfall process directly.

As the rainfall-forecasting methodology is based on these static measurements, only invariant assumptions and features can be extrapolated into the future. This implies that, even for a fairly correct measurement, the forecast is an extrapolation of the measurement and model uncertainties. Thus, it is not credible to pretend that there is the ability to forecast precisely the dynamic rainfall process.

The degree of uncertainty induced by the above factors varies with the degree of variability of the rainfall process. Usually, the uncertainty can be assessed by knowing the variability, time step, grid size and surface of the reference area. For example, taking radar data with a grid size of 2.4 km, a time step of 10 min and an average rainfall variability (Vogel 1980), has led, over a surface of 500–1000 km^2, to a volumetric measurement error of the order of 20% and a mean volumetric forecasting error for rainfall exceeding 5 mm/h, for a lead time of $\frac{1}{2}$ h of the order of 30% and for a lead time of 1 h of the order of 80%.

According to Denoeux (1989), other workers give values for the forecasting error, ranging from 40% to 100% for forecasts of 1 h lead time.

6. CONSEQUENCES FOR HYDROLOGICAL APPLICATIONS

The knowledge that a perfectly forecasted rainfall event is a myth does not mean that a radar-based rainfall forecast is useless. On the contrary, even though hydrologists often argue in terms of 'perfectly forecasted rainfall', they do not need such a forecast. Instead of a more or less perfect point estimate with unknown uncertainties, a reliable upper- and lower-bound estimation is much more useful for hydrological purposes. As the rainfall forecaster today is capable of quantifying the uncertainty margin of his forecast (Denoeux et al. 1991), this knowingly imperfect 'point' forecast is even more precise as its weaknesses are known.

The usefulness of a forecast in a certain meteorological situation always depends on the hydrological application. If the user knows his precision needs, he can be advised by the forecaster how to work with the issued forecast in an optimal manner. This improves the reliability of hydrological methods and, thus, increases confidence in their results, as data quality is made more transparent.

Although urban hydrologists have proved that rainfall is the most important factor contributing to their operational problems, they usually do not know the degree of uncertainty which is acceptable for their system. If the forecaster is provided an exact outline of what time horizon and which volumetric precision is needed for the application, he can for example give the forecast in terms of rainfall volume, instead of a forecasted radar image, together with upper and lower bounds and an estimation of the forecasting confidence.

7. A CASE STUDY: THE SEINE-SAINT-DENIS COUNTY

In the Seine-Saint-Denis real-time control project (Jaquet and Einfalt 1985), not only real-time control strategies for the county sewer system, but also radar measurement and forecasting tools at various levels are developed, guided by the close cooperation of hydrological users and radar experts.

Numerous meetings and staff affiliation at both the Seine-Saint-Denis county and the Cergrene research centre have promoted a number of site- and application-specific features, including

(1) a hydrologically oriented classification on 16 levels of the reflectivity–intensity relationship (Agostini 1987),
(2) the forecast given for each catchment individually, using rainfall volumes instead of intensities, which can directly be input into runoff simulation models,
(3) a rainfall-warning system, automatically telephoning the emergency team which is on call if flooding may be expected from the forecasted rainfall volumes (Einfalt 1988),
(4) the development of a hydrological quality criterion designed for the Seine-Saint-Denis catchment (Denoeux *et al.* 1991),
(5) the development of on-line assistance to the operating crew on the reliability of the forecast, and
(6) the reduction in radar measurement uncertainties by means of an interpolation technique which can also be used for the rainfall forecast (Blanchet *et al.* 1991).

8. CONCLUSION

For radar specialists, it is important to know the potential user's needs and to adapt a forecasting system accordingly. The hydrologist, on the other hand, should always bear in mind that he cannot expect a perfect rainfall forecast, as static procedures and measurements are not capable of exactly representing dynamic processes. However, trying to evaluate the sensitivity of his particular problem helps him to understand his precise needs. On the basis of these needs, the forecaster and the user are able to design an appropriate forecasting scheme taking into account local hydrological features.

REFERENCES

Agostini, B. (1987) *Proposition des meilleures conditions d'exploitation du radar de trappes pour la Seine-Saint-Denis*, DDE A03, Marché No. 87.02.058.
Blanchet, B., Naiman, A., Jaquet, G. and Andriecy, H. Improvements of rainfall measurements due to accurate synchronization of raingauges and due to advection use in a calibration. *Hydrological Applications of Weather Radar*. Ed. Cluckie, I. D. and Collier, C. G. Ellis Horwood, Chichester, West Sussex. Chapter 20.
Cluckie, I. D. and Owens, M. G. (1985) Real-time rainfall runoff models and use of weather radar information. *Proceedings of the Weather Radar and Flood Forecasting Symposium, Lancaster, 1985.*
Collier, C. G. (1977) On the benefits of improved short period forecasts of precipitation to the United Kingdom — non military applications only, Research Report No. 6, Meteorological Office, Research Laboratory, Malvern, Worcs.
Damant, C., Austin, G. L., Bellon, A., Ossegrane, M. and Nguyen, N. (1983) Radar rain forecasting for wastewater control. *J. Hydraul. Div. Am. Soc. Civ. Eng.*, **109** (2), 293–297.

Denoeux, T. (1989) *Fiabilité de la prévision de pluie par radar en hydrologie urbaine*, Ph.D. Thesis. Ecole Nationale des Ponts et Chausées, Paris.

Denoeux, T., Einfalt, T. and Jaquet, G. (1991) On the evaluation of radar rainfall forecasts. *Hydrological Applications of Weather Radar*. Ed. Cluckie, I. D. and Collier, C. G. Ellis Horwood, Chichester, West Sussex. Chapter 32.

Einfalt, T. (1988) *Recherche d'une méthode optimale de prévision de pluie par radar en hydrologie urbaine*, Ph.D. Thesis. Ecole Nationale des Ponts et Chausées, Paris.

Eidenhart, P. *et al.* (1988) In: Rowohlt, (ed.) *Du steigst nie zweimal in denselben Fluβ*, Reinbek, Hamburg.

Jaquet, G. and Einfalt, T. (1985) Radar rainfall forecasting for Paris suburban area. *Poster session of the Radar Rainfall and Flood Warning Symposium, Lancaster, 1985.*

Lovejoy, S. and Schertzer, D. (1986) Scale invariance, symmetrics, fractals, and stochastic simulations of atmospheric phenomena. *Bull. Am. Meteorol. Soc.*, **67** (1).

Schilling, W. (1983) *Operationelle Niederschlagsvorhersagen mit empirisch identifizierten Modellen und optimaler Zustands-schätzung*, Mitteilung No. 51. Institut für Wasserwirtscahft, Universität Hannover, pp. 129–345.

Schilling, W. and Einfalt, T. (1985) Pumped aquifer simulation with a multivariate statistical model. *Conference on Statistical Approaches in Hydrology, Fort Collins, CO, 1985.*

Schilling, W. and Petersen, S. O. (1987) Real time operation of urban drainage systems — validity and sensitivity of optimization techniques. In: M. B. Beck (ed.) *Systems analysis in water quality management*. Pergamon, Oxford.

Verworn, H. R. (1988) Niederschlagsmessung mit Radar — Anwendung und Bedeutung für die Stadtentwässerung. Zeitschrift für Stadtentwässerung und Gewässerschutz (SuG), No (4).

Vogel, J. L. (1980) Real time measurement of convective precipitation over an urban area. *Proceedings of the Oxford Symposium in Hydrological Forecasting*, IAHS–AIHS Publication No. 129. International Association of Hydrological Sciences.

41

Integrating radar rainfall data into the hydrologic modelling process

T. L. Engdahl[1] and H. L. McKim[2]
[1]US Army Engineer Waterways Experiment Station, PO Box 631, Vicksburg, MS 39181-0631, USA
[2]US Army Cold Regions Research and Engineering Laboratory, 72 Lyme Road, Hanover, NH 03755-1290, USA

ABSTRACT

New technologies to collect, analyze, store, retrieve and display hydrologic data after significant improvements to conventional methodologies used for predicting hydrologic events. Computer workstation environments, including networks that allow multiuser access to the same databases, are operationally available. It is technically possible for these systems to supply all the hydrologic and meteorologic data required by hydrologic models. The focus of this paper is the integration of the spatially dynamic radar rainfall data with a prototype system being developed and tested to predict runoff from a large test basin in real time.

1. INTRODUCTION

A number of important hydrologic process components can be improved in the model simulation process. These include rainfall, infiltration, runoff generation and channel flow (Feldman 1987). Of these components, real-time hydrologic forecast models appear most sensitive to the spatial and temporal distribution and amount of rainfall (Barrett, 1985). Thus, an improvement in rainfall measurement and its spatial distribution is an important factor that could lead to improved flood forecasts in real time. Ideally, a capability for accurately forecasting precipitation would make an even more significant contribution, but such technology has not been adequately developed for operational use in hydrologic forecasting.

Raingages are currently used to measure point rainfall in the field. Raingages can provide reasonably accurate measurements at pre-selected points within a river basin, but the spatial integrity is lacking when these point measurements are extrapolated to provide estimates of areal distributions. The areal extrapolation of point data is still an art. This is especially true for convective storms where even a dense network of raingages may not adequately represent the spatial structure of a storm. To circumvent the spatial inadequacies inherent with point source raingage networks, weather radars are being used. Weather radar can sense spatial and temporal distributions of rainfall in real time (Wilson and Brandes 1979) and, when properly calibrated with raingage data, can provide the hydrologist with reasonably accurate spatial and temporal patterns of storm events (Johnson and Dallman 1987).

Full integration of radar rainfall information into the hydrologic modeling process databases that depict not only the spatial characteristics of the river but also the spatial distributions of storm events. Static and near-static parameters of a river basin, such as elevation, soils, vegetative cover, land use, channel geometry and drainage network, are obtained from conventional maps and remotely sensed imagery. Geographic information and image-processing systems can be used to analyze the parameters and to display these basin characteristics. Weather radar data portray the dynamics of storm events for a basin in real time; integration of these data over time and space can produce a real-time rainfall hyetograph for each area subdivision within a river basin. Rainfall, along with other dynamic parameters such as runoff, soil moisture and evaporation, can then be coupled with the static and near-static terrain parameters for input to real-time forecast models.

2. OPERATIONAL SYSTEM CONFIGURATION

The US Army Corps of Engineers is attempting to network systems that will automate the acquisition of various data used in hydrologic models (Fig. 1). The data

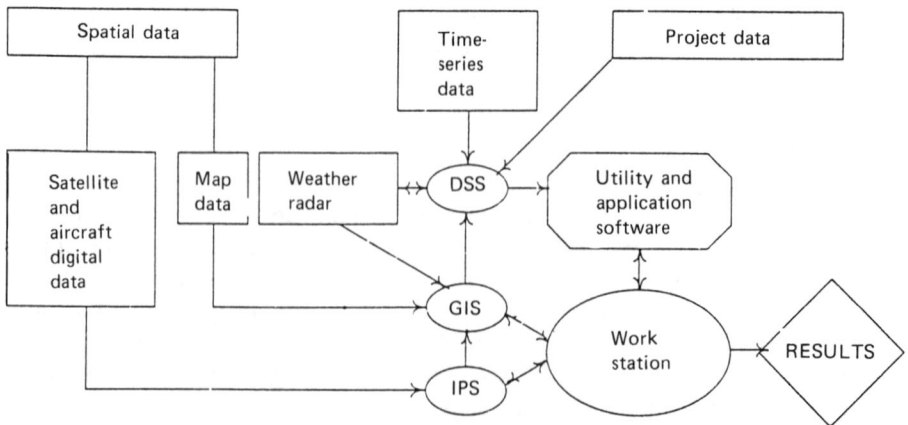

Fig. 1 — Operational system configuration.

may be in the form of satellite or air-borne images, conventional maps, mapped overlays, weather radar digital data or point source values. All these data must be pre-processed before they can be effectively and efficiently used in an automated fashion. Not only are these data acquired from different and varied platforms, but also they are completely different in format and structure; sources may have to be scanned, digitized, rectified, sliced, controlled, calibrated or reformatted to become useful to the hydrologist. Requirements for updating may also vary substantially. A soil class map is an example of static data that need to be input into the system only once. Items, such as land use, change seasonably and may require updates twice to four times a year. Other inputs, such as rainfall, are dynamic, and updates may be required in time frames of 10 min or less.

A data storage system (DSS) has been developed by the US Army Hydrologic Engineering Center (HEC) for use in the Corps water resources mission area. The DSS is a hydrologic information and analysis computer system designed to handle point water resource data. These data are collected over time and stored in the DSS as site-specific data. For example, data on rainfall at a particular gauging station are collected over time and stored in the system as a data file in which the name indicates the project or basin name, the location (gaging station name), the data type (rainfall incremental), and the frequency of data collection (hourly, daily, weekly or monthly). The DSS has utility, analysis and display software that enable the user easily to access, display, edit and use the data as input to various hydrologic models (Hydrologic Engineering Center 1985, 1987).

In the Corps of Engineers Civil Works programs, spatially oriented data are also used in many studies (Ewardo *et al.* 1985). These spatial data are placed in a geographic information system (GIS) at a common pixel size resolution required for a particular study, the size of the pixel cell being dependent on how the data are to be ultimately used (Merry 1986). A typical GIS allows users to display and analyze raster and vector data types.

Fig. 1 schematically indicates how the DSS, GIS and an image-processing system (IPS) are being integrated in a remote sensing demonstration program. Time-series point source data from data collection platforms (DCPs) are stored in the DSS for use with the hydrologic models. Satellite and aircraft data are processed through standard software packages in the IPS. Information can be extracted from the GIS through application-designed software for image display on workstation terminals or input directly to hydrologic models through utility-designed software. Software is being developed to integrate the spatial GIS data to point source data for pre-selected areas for direct input to the DSS. Weather radar data are summarized and placed in the DSS for use with the hydrologic models, or the data can be placed in the GIS for display within the computer network system.

3. WEATHER RADAR — A HYDROMETEOROLOGICAL TOOL

Weather radars operate on the principle of back-scatter and absorption of radar waves by ice, snow and water. The power received by the water particles varies with the dielectric and absorption properties of the particles, the refractive index of the intervening medium and the atmospheric absorption. Empirical studies have shown that relationships exist between radar reflectivity Z and rainfall rate R. The power reflected back from these particles is correlated to rainfall intensities by a Z–R relation for a particular radar (Probert-Jones 1962). This relation is expressed by the following equation:

$$Z=aR^b$$

where $Z(\text{mm}^6/\text{m}^3)$ is the radar reflectivity factor, $R(\text{mm/h})$ is the rainfall rate; a is an empirically derived coefficient and b is an empirically derived exponent. The Z–R relation is based on assumptions that the raindrop size distribution is known, the water droplets are spherical and small enough to permit the Rayleigh approximation of the droplets size to apply and the vertical air motions of the droplets are zero (Battan 1973). In reality, the raindrop size distribution is rarely known and varies over time and space. Also, the vertical air motions are often of the same magnitude as the terminal velocity of the raindrops. Thus, the Z–R relation is not unique for all radars or all storms. For these reasons, numerous empirical values for a and b have been derived over the years.

To process weather radar data effectively, a digital form of the scene must be provided. A digital video integrator processor (DVIP) is used to analyze the reflected radar signal and to determine its intensity. The DVIP converts radar reflectivity returns into digital form, averages several returns and then integrates many of these returns over time. These range-normalized reflectivity values are gated to obtain threshold values for rainfall rates. The output from the DVIP is a raster array with a video integrator processor (VIP) level (i.e. a single value representing minimum and maximum radar reflectivity values) associated with each pixel within the array (Miers and Heubner 1985). Each VIP level value corresponds to a range of rainfall rates; values for convective storms are given in Table 1, along

Table 1 — VIP levels and rainfall rates for convective storms

| VIP level | Rainfall rate (cm/h) | |
	NWS range	Model value
1	0.0– 0.5	0.1
2	0.5– 2.8	1.4
3	2.8– 5.6	4.1
4	5.6–11.4	8.3
5	11.4–18.0	14.5
6	>18.0	21.4

with values used in the radar rainfall conversion models. In the USA, National Weather Service (NWS) weather radars provide excellent coverage of the Rocky Mountains. Acquisition of radar scenes from the approximately 130 NWS sites is accomplished through commercially available weather radar receivers. These receivers digitally store and display weather radar scenes, a scene being a 'snapshot' in time of the storm event. The NWS WSR-57S and WSR-64S 10 cm wavelength radars provide excellent coverage to effective ranges of approximately 225 km from the site, an area of approximately 159 000 km^2. Over a period of time, the stored scenes can be used to display storm movement, storm growth and decay, the total rainfall coverage within the scan area.

Fig. 2 shows a schematic of the system configuration the Corps of Engineers is currently using for receiving, displaying and analyzing VIP level weather radar

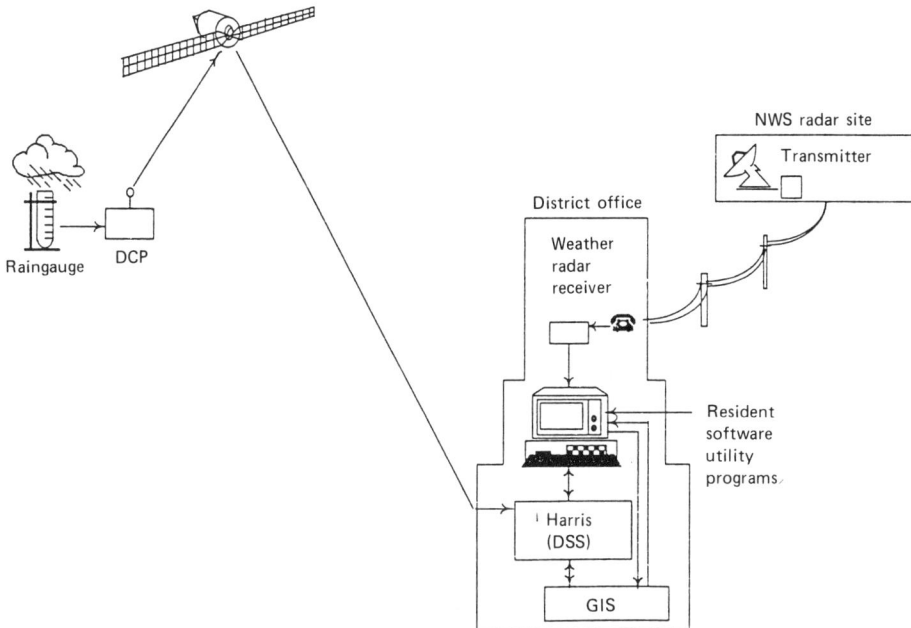

Fig. 2 — Automated weather radar data acquisition system.

scenes. The VIP level digital arrays are transmitted from the NWS radars to receivers via standard telephone lines. Individual VIP level arrays are generally updated at the NWS site every 2 or 3 min. The size of the individual pixels within the array is dependent on the receiver range setting, but for weather radars with a 225 km radius range setting the pixel size is roughly 2 km square. The VIP level arrays are automatically sent from the receiver to a personal computer (PC). The VIP level arrays are converted to rainfall files using the average value of the square root of the minimum and maximum rainfall rates to determine a single rainfall rate for each VIP level (see Table 1). Integrating the rainfall rates over the duration of the weather radar scene (i.e. the time difference between scenes) determines the amount of rainfall for each time period. Accumulating the rainfall amounts from all scenes for all pixels provides an estimate of spatial distribution of rainfall totals.

Watersheds of interest within the radar scan are digitized and gridded using a GIS. The location and size of the gridded subdivisions of the watershed are referenced to the location of the NWS weather radar site. As a storm approaches a watershed, the radar rainfall pixels are integrated spatially over the subbasin on a PC, and hyteographs are produced in real time for each subbasin. Currently, a single uncalibrated hourly rainfall value for each subbasin is sent to the DSS.

The calibration of weather radar returns to raingages improves the spatial estimate of total rainfall within the river basin (Wilson 1964). The calibration techniques being examined by the Corps of Engineers are based on assumptions that raingages accurately measure rainfall at a point and radars accurately display the spatial distribution of the rainfall within a storm (Engdahl 1988). Raingage data are telemetred via satellite from the field to the DSS at pre-selected time intervals. Software developed on the PC interrogates the DSS and checks to see whether all raingages required for calibration of the weather radar site have reported. When all raingages have reported, a calibration software routine is run on the PC and single rainfall values for each subbasin are calculated. These data are also sent to the DSS.

Fig. 3 shows the total accumulation of rainfall for a storm event. For this particular storm, 131 VIP level scenes were stored over 24 h period and converted to

0902 (4 Nov)–0913 (5 Nov)
Max 4.13 in.

Nov 4, 1986
131 signatures

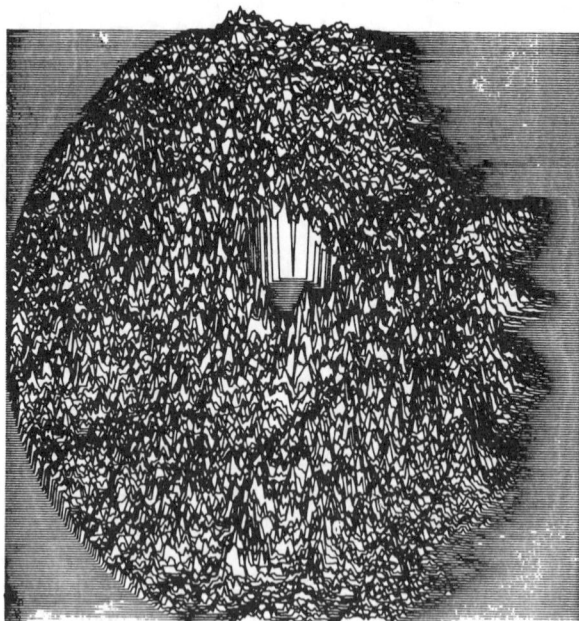

Fig. 3 — Total storm rainfall amounts (maximum peak, 10.49 cm) for 4–5 November 1986. This represents 131 weather radar scenes collected over a 24 h period. (The area represents approximately 125 000 km².) The hole depicts blanking of ground clutter.

rainfall amounts. The scenes were added together to show the total amount of rainfall in each pixel. There are 256 pixels in the x direction and 240 pixels in the y direction. The z component represents rainfall amounts with the maximum representing 10.48 cm for Fig. 3. The 'hole' in the middle of Fig. 3 resulted from the intentional blanking of radar returns for this area to eliminate ground clutter effects. The distance from the center of the storm to the outside edges is 225 km. Although this was not an intense convective storm, a highly variable spatial distribution is shown. Fig. 4 represents 1 h accumulation of this same storm as it moved from the northwest to the southeast.

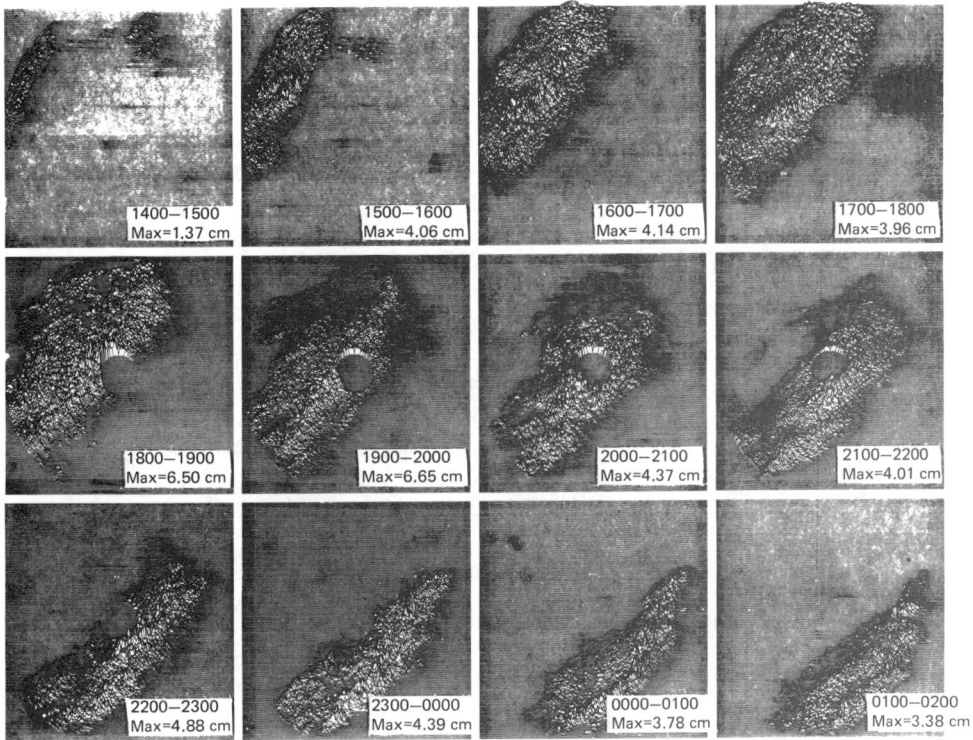

Fig. 4 — Hourly rainfall amounts for the 4–5 November 1986 storm. (The area represents approximately 125 000 km².).

4.   SPATIAL DATA ANALYSIS

The study area for the remote-sensing demonstration program is located in the central part of the USA in the state of Iowa (Fig. 5). The watershed under investigation encompasses an area of approximately 2895 km². The landscape is semiflat to gently rolling within this primarily agricultural area. The soils are fine grained and highly productive. The data collection sensor systems within this area, as shown in Fig. 5, include daily nonrecording raingages, recording and nonrecording soil moisture gages and operational DCPs with complete meteorological stations that include recording raingages and streamflow gages. The DCPs automatically tele-meter point source data to the DSS at pre-selected time intervals.

Soil, elevation and land cover data sets are being configured on a Dipix ARIES-II IPS. The soils data are currently being digitized by Iowa State University in a polygon

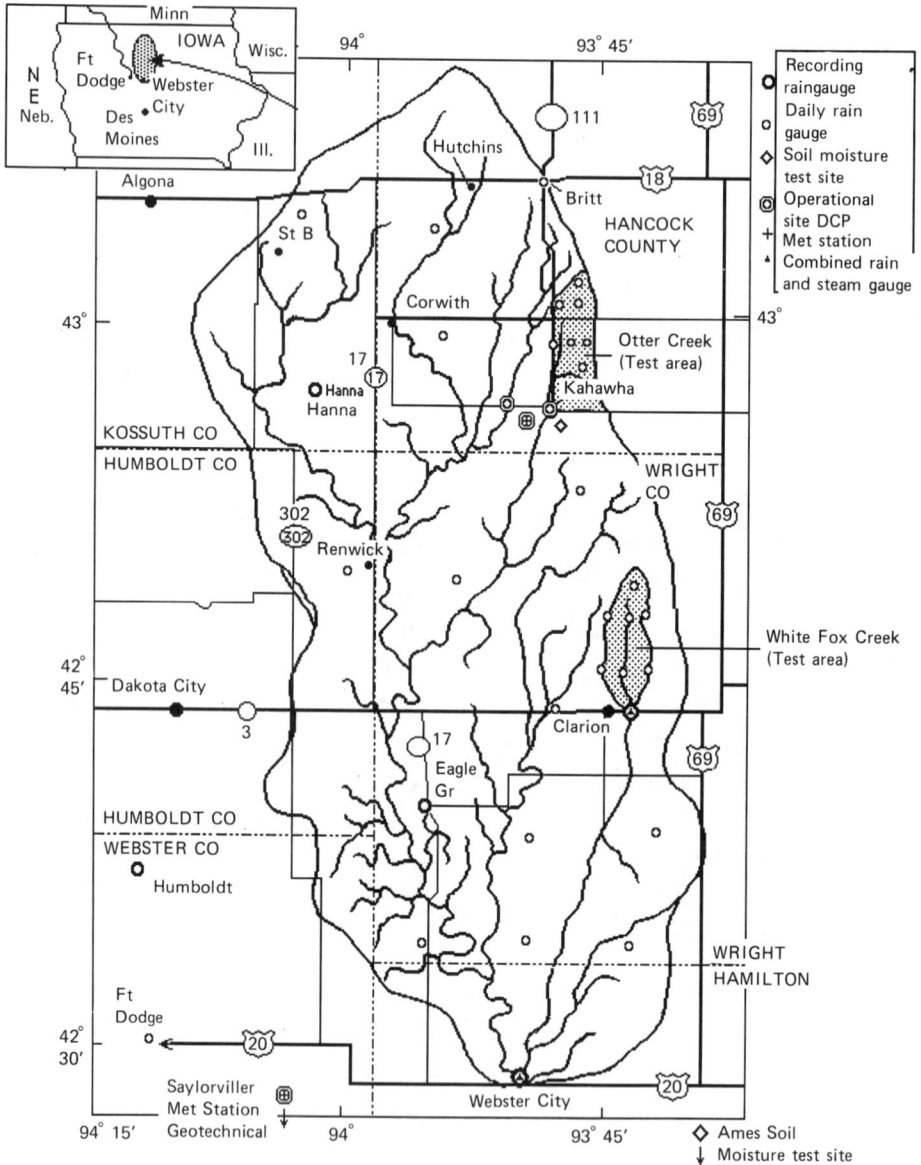

Fig. 5 — Gage locations within Saylorville river basin.

format. These data will be gridded and placed in the GIS. The elevation data were acquired from the Defense Mapping Agency (30 m pixel resolutions). The elevation data are in a 1″ format for a 1° latitude by a 2° longitude area. Two of these data cells were required for the test watershed. Data sets of land cover were prepared from a Landsat thematic mapper image (30 m pixel resolution) acquired on 3 September 1982 and a système probatoire d'observation de la terre (SPOT) high-resolution visible (HRV) scene (20 m pixel resolution) acquired on 17 July 1987. Various aggregation schemes are being developed to register the 20 and 30 m pixels into common pixel cell sizes that will be tested in the hydrologic model. All these data are being geometrically corrected on the IPS and will be placed in a common raster format in the GIS.

The satellite remote-sensing data to be incorporated into the DSS are fundamentally different from the water resources site-specific (point) data that are currently in the DSS (Merry *et al.* 1987). The satellite data are collected over a large area, with the pixel resolution dependent on the sensor. These data are stored in a pixel cell format (raster) and one set of data may overlap or completely align with another data set. The typical size of a data set for satellite scene is extremely large and cannot currently be analyzed or displayed with the DSS software. Depending on the complexity of the hydrologic model, the spatial data can be used to determine average values for specific hydrologic inputs over an entire basin and the value placed into the DSS. If higher-resolution data are required by the model, they can be obtained directly from the GIS.

Weather radar software is being developed to access automatically both the DSS and the GIS databases. The raingage data (point source) are being retrieved from the DSS for calibration of the weather radar data. In addition, the calibrated and uncalibrated weather radar data integrated over a subbasin area are input into the DSS for use in hydrologic models. The locations of the river basins and their subbasins in relationship to the weather radar location will be automatically extracted from the GIS. Also, spatial hourly rainfall images will be sent to the GIS for multiuser display.

5. CONCLUSIONS

Complex computational simulations and analyses are being accomplished in conjunction with high-speed data collection procedures. Remotely observed data can be automatically integrated in real time using processing techniques that are transparent to the user and the system network. The methods being developed will allow for rapid analysis of radar data for input to real-time water resource models.

The capability of integrating multisensor and multispectral data into usable hydrometeorological data sets can be accomplished using available software. Processing real-time spatially derived rainfall and applying these values over large basins is now technically possible. Large river basin hydrologic parameters can be remotely obtained and processed rapidly using an IPS and stored in the GIS. Point source parameters telemetered to a common data collection facility can be used to calibrate spatially derived data or used directly in hydrologic models. The architecture to link the IPS, GIS and DSS is being tested and the software is being built for processing complex meteorologic and hydrologic data sets for use in real-time hydrologic forecasting.

ACKNOWLEDGMENTS

This work was sponsored by the Rock Island District of the Corps of Engineers in conjunction with the Inland Water Resources Remote Sensing Demonstration Program. Appreciation is extended to Mr Doyle McCully and Mr S. K. Nanda for overseeing this program. Appreciation is extended to Dr Carolyn J. Merry, The

Ohio State University, Department of Civil Engineering, Dr E. Alan Cassell, University of Vermont, School of Natural Resources, and Mr John Collins, US Army Engineers, Waterways Experiment Station, for their extensive review of this paper. Special thanks to Ms Dianne Nelson, Mrs Eleanor Huke and Mr Edmund Wright for assisting in producing this report.

REFERENCES

Barrett, E. C. (1985) Rainfall evaluation by remote sensing: problems and prospects. *Hydrologic applications of remote sensing and remote data transmission*, Publication No. 145, pp. 217–258.

Battan, L. J. (1973) *Radar observation of the atmosphere*. University of Chicago Press, Chicago, IL, p. 324.

Edwardo, H. A., Koryak, M., Miller, M. S., Wilson, H., Merry, C. J. and McKim, H. L. (1985) The role of GIS and remote sensing in master planning for resources management of the Berlin Lake, Ohio Reservoir Project. *Proceedings of the 19th International Symposium on Remote Sensing of Environment, Ann Arbor, MI, 21–25 October 1985*. pp. 659–669.

Engdahl, T. L. (1988) Weather radar as a hydrometeorological tool. *US Army Cold Regions Res. Eng. Lab. Remote Sens. Bull.*, **88** (1), 8–10.

Feldman, A. (1987) HEC models for water resources simulations; theory and experience. *Adv. Hydrosci.*, **12**, 35–50.

Hydrologic Engineering Center (1985) *HECDSS user's guide and utility program manuals*. US Army Corps of Engineers, Davis, CA, p. 146.

Hydrologic Engineering Center (1987) *HECDSS programmer's manual*. US Army Corps of Engineers, Davis, CA, p. 115.

Johnson, L. E. and Dallman, J. L. (1987) Flood flow forecasting using microcomputer graphics and radar imagery. *Microcomput. Civ. Eng.*, **2** (2), pp. 85–99.

Merry, C. J. (1986) Using satellite digital data in a geographic information system. In: B. Opitz (ed.) *Technical Proceedings of the Workshop on Geographic Information Systems in the Government, Springfield, VA, 10–12 December 1985*. A. Deepak Publishing, Hampton, VA, pp. 543–553, 1986.

Merry, C. J., Eagle, T., LaPotin, N. and Gardiner, J. (1987) Development of a geographic information system for the Saylorville River Basin, Iowa. *Proceedings of the US Army Corps of Engineers 6th Remote Sensing Symposium, Galveston, TX, 2–4 November 1987*. pp. 265–269, 1987.

Miers, B. T. and Heubner, G. L. (1985) *Military hydrology; Report 8, Feasibility of utilizing satellite and radar data in hydrologic forecasting*, Miscellaneous Paper No. EL-79-6. Prepared by US Army Atmospheric Sciences Laboratory, White Sands Missile Range, NM, for the US Army Engineer Waterways Experiment Station, p. 51.

Probert-Jones, J. R. (1962) The radar equation in meteorology. *J. R. Meteorol. Soc.*, **88**, 485–495.

Wilson, J. W. (1964) Evaluation of precipitation measurements with the weather radar. *J. Appl. Meteorol.*, **3**, 164–174.

Wilson, J. W. and Brandes, E. A. (1979) Radar measurement of rainfall — a summary. *Bull. Am. Meteorol. Soc.*, **60** (9), 1048–1058.

42

Reflections on rainfall information requirements for operational rainfall–runoff modelling

C. Obled
Institut de Mécanique de Grenoble, Domaine Universitaire
BP 53 X 38041, Grenoble Cedex, France

ABSTRACT

Using a test catchment in the Massif Central in France the rainfall information requirements for a number of different models are discussed. Future developments of semi-distributed and fully distributed hydrological models, and the potential role of radar in these grid-based models are investigated. The rainfall data accuracy appears to be a trade-off between operational costs and operational requirements.

1. INTRODUCTION

This paper reviews the different rainfall-runoff modelling steps that use rainfall input and looks into the accuracy required for this information. The interactions are stressed between the structures of the models selected and the related rainfall input requirements in both time and space.

The resulting proposals are based partly on theoretical considerations but more often consist of rules of thumb deduced from experience. This experience has mainly been acquired on case studies in the French southeastern Mediterranean region. The Gardon d'Anduze, a 545 km^2 pilot watershed located on the southeast slope of the Massif Central, will often be cited as an illustration. This region, known as the Cevennes region, is often affected by intense rainfalls and flash flooding. (The last major accident that occurred around Nimes in October 1988 caused 5 billion French francs worth of damage).

This explains the implementation and maintenance of a 'dense' ground-based network of recording raingauges for almost 20 years, and the many efforts in rainfall–runoff modelling and flood forecasting. Recently a 3 year radar experiment was performed (1986–1988), allowing preliminary conclusions on the possibilities offered by hydrological radar in providing the required rainfall information.

Concerning rainfall–runoff modelling, both distributed and lumped approaches have been developed, but with a definite trend in favour of lumped modelling. Note that our conclusions may therefore be biassed by this choice.

2.   RAINFALL INPUT ESTIMATION: A BRIEF OVERVIEW

Some commonly accepted results on rainfall characteristics will first be summarized from an operational point of view.

When a part or a subset of a rain system is considered, for instance on a watershed of some 100–1000 km^2, the rainfall input within this domain may be considered as a continuous two-dimensional random process. Indeed, problems may emerge when the area is not totally affected by rainfall during some time steps. This occurs rather frequently if very short time steps are considered (e.g. 5 min) but, if totals over larger time steps are considered (e.g. 1 h totals), then only a few time steps at the beginning and at the end of the rain event pose a problem.

A geostatistical approach is therefore widely used to characterize this random field. If the area considered is rather small compared with the correlation length of the process in space, also called the range (i.e. the distance beyond which the correlation between two rainfall values becomes non-significant), then the process can further be considered as locally stationary and is thus characterized by its mean, variance, and covariance or structure function. In the case of rainfall, a strong linear relationship is often noted between the mean and the variance, i.e., the more intense the rainfall is on the average, the more spatial variability it can display.

Consequently a standardized process can be defined by dividing the point rainfall by its local average over the domain. This results in a standardized structure function, usually a variogram $\gamma(d)$ depending only on the interdistance d between any two points M and M′. For practical reasons, the spherical variogram is often preferred among other models because its standardized form depends only on the range as the unique parameter:

$$\gamma(M, M') = \gamma(MM') = \gamma(d) = 1.5\frac{d}{a} - 0.5\left(\frac{d}{a}\right)^3 . \tag{1}$$

Furthermore, it has been observed that a fairly good relationship exists between the time step Δt and this range a. An approximate expression could be

$$a(\Delta t) = 20\sqrt{\Delta t} \tag{2}$$

where a is in kilometres and Δt in hours. This holds for intense convective systems, with rather widespread rainfall. The ranges proposed are based on experimental results for time steps ranging from 10 min to 24 h. The coefficient of 20 km is an average value that could fluctuate between 15 and 25 as a general indication.

These approaches were widely considered in the early 1980s (e.g. Delhomme and Delfiner 1973, Creutin and Obled 1982, Lebel *et al.* 1987) and, although more sophisticated approaches (Chua and Bras 1982) have been applied since, assuming a non-stationary process, no major improvement has been demonstrated.

In addition, this geostatistical approach also provides tools to evaluate averaged variables. A diagram proposed by Lebel *et al.* (1987) gives the approximate required density (in terms of gauges per square kilometre) for a given required accuracy for the basin average (Fig. 1). For a more accurate assessment, the shape of the basin itself, together with the distribution of gauges within the instrumented domain, should also be taken into account.

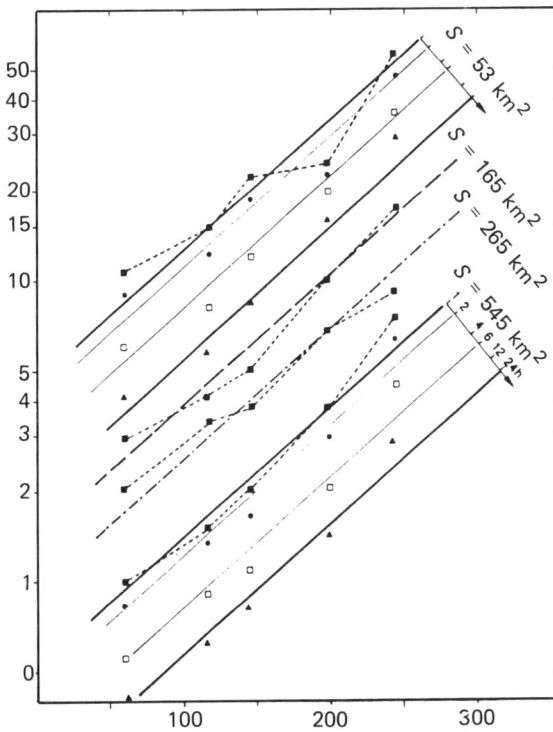

Fig. 1 — Expected accuracy (estimation variance in percentages) on basin average rainfall for four nested subwatersheds of the Gardon river at Anduze depending on the basin area (watershed area $S = 53\ 165\ 265$ and $545\ \text{km}^2$), the time step (1 h(■), 2 h(●), 4 h, 6 h(□), 12 h and 24 h(▲)) and gauge density in terms of average area A_g per gauge.

Some variables or indexes, other than the basin-averaged rainfall, could also be of interest, and methods exist to deal with variables such as the rain-affected area or the area above a given threshold (Creutin and Barancourt 1988).

All these results, like the correlation length in equation (2), have been based mostly on ground networks but seem consistent with similar results obtained from radar imagery (Andrieu *et al*. 1989). Up to now, this discussion has been restricted to the spatial characteristics of rainfall fields, and the main result that must be kept in mind in relation to equation (2) is that, the smaller the time step Δt, the denser the network must be for a given level of accuracy on the basin-averaged rainfall.

3. SELECTION OF A TIME STEP IN A LUMPED RAINFALL–RUNOFF MODEL

Several lumped approaches exist, based on the unit hydrograph, storage, isochrones, tank models, etc. The first differenced transfer function–excess rainfall and unit hydrograph by the deconvolution and identification technique (FDTF–ERUHDIT) will be considered here as representative of this general type of modelling.

Essentially based on the same assumptions as the unit hydrograph, this method is discussed in detail in another paper of this symposium (Rodriguez *et al* 1991). Although similar to the unit hydrograph method in its basic hypotheses, it neverthe-less offers major improvements. In particular, it deconvolutes the excess rainfalls without any *a priori* choice of a loss function or infiltration model. Instead an alternative iterative algorithm identifies first both the unit hydrograph or transfer function, and the excess rainfall series for a set of calibration events. Next, these identified or deconvoluted excess rainfalls are related to the basin input rainfalls, also called hereafter raw rainfalls, by calibrating a loss function or infiltration model on the same set of events (usually 20–40).

However, whatever the details of the lumped model selected, the requirements on the input will be rather similar for any model, and the first question to be answered will certainly consist in selecting an appropriate time step.

This is the first choice to be made before starting the modelling process. Obviously, the intrinsic characteristic of the watershed should be its instantaneous unit hydrograph, defined as the continuous response to an instantaneous rainfall pulse. When non-instantaneous pulses are considered, it is known that the corre-sponding unit hydrograph depends itself on the time step considered, and that changing from one to another can be done using the S curve (e.g. Shaw 1983, p. 337). However, the preliminary and unavoidable problem is that the time step has to be selected before the unit hydrograph is identified, and this often requires a trial-and-error procedure over a reasonable range bounded by some upper and lower values.

Consider first the upper bound. If the expected unit-hydrograph peaks at time T_p after the rainfall pulse, a rule of thumb gives a time step of $T_p/3$ or $T_p/5$ as a maximum. This is purely based on obvious drawing considerations. Since the continuous unit hydrograph is to be approximated by a discrete staircase-like curve, it requires at least three and hopefully more steps to describe the rising limb of this hydrograph adequately.

The lower bound is more difficult to assess. The watershed can generally be considered as a low-pass filter, so that the high-frequency oscillations that exist in the rainfall intensities are smoothed out in the discharges. It should be possible to formalize this through sophisticated spectral or cross-spectral analysis techniques, but in practice success has been limited by the very non-stationary and intermittent nature of the flood process.

More practically, since a continuous graphic record of the discharge hydrograph is often available, it is worth looking first at its smallest significant oscillations, and to determine by successive trials, starting from the larger ones, the appropriate time step for which the discharge oscillations may be considered as originating from these corresponding rainfall increments. In this analysis, not only the time step itself, but also its origin may be important in relating the variations in the discharge to the fluctuations of the discretized hyetograph, since a slight shift in the discretization of the rainfall record may smooth out some significant intensity peaks.

It must be stressed, finally, that this whole approach is fully exploratory, or heuristic, in that it is entirely based on the examination of the data and not on deductive reasoning related to the physics of the watershed. As an example of the above considerations, in the Gardon d'Anduze watershed, the time to peak is about 5–6 h, consisting of a 2 h delay and a 3 h rise. A time step of 1 h was selected, while 2 h could have been used satisfactorily.

Accordingly it could be expected that, the smaller the time step, the better the description of the unit hydrograph. Unfortunately, this must often be considered in the light of more practical constraints such as the availability and accuracy of the data as the time step decreases. A compromise is therefore generally needed. Usually, three complementary aspects must be taken into account.

First, the capacity of the existing network to provide adequate estimates of the input rainfall at the time step considered must be dealt with. The inaccuracy of the recorder clocks becomes more critical for small time steps, causing synchronization problems. Also, from the conclusions of section 2, for the same given accuracy in basin rainfall, dividing the time step by 2 requires the doubling of the network density.

Second, models can seldom be run more than a few time steps ahead in forecasting, because their performances quickly drop. It has even been shown that, to forecast two time steps ahead, it is often better to run a model using a double time step one time step ahead, rather than running twice sequentially the initial model, with its initial time step. The main reason is that errors usually propagate exponentially. So it seems reasonable to keep the time step as large as possible, so as to allow larger lead times in forecast.

Finally, the model time step must also be compatible with the operational use anticipated. If a forecast is to be issued every 3 h, it is hardly necessary to run the model on a half-hour time step and to adapt accordingly the telemetric scanning time step.

So, as already mentioned, the timestep finally selected is a compromise between these different constraints. However, the effects of lumping this time step, compared with the underlying dynamics of the physical processes, is not fully understood as will be discussed for distributed models.

4. RAINFALL ACCURACY REQUIRED IN THE DIFFERENT MODEL IDENTIFICATION STEPS

In a lumped modelling approach of the unit-hydrograph type, the identification may be performed in several ways. The most common consists in applying an *a priori* chosen loss function with *a priori* fixed parameters to provide the excess rainfall.

These are used together with the discharges to identify the transfer function. Another way consists in proposing a parameterized loss function and a parametrized transfer function which are both optimized at the same time using input raw rainfalls and discharges. However, these two approaches display major drawbacks: first, the loss function is completely defined *a priori*, and the optimization concerns only the transfer function without reconsidering the excess-rainfall series; second, only the structure of the loss function is imposed *a priori*, but the optimization is global, allowing strong interactions between parameters of the loss function and of the transfer function. Furthermore, the initialization of the loss function, although critical, is usually not involved in the optimization.

As a result, in the case of poor overall performance it is often impossible to distinguish the role played by the uncertainty on the input raw rainfalls only. This is why other approaches have been proposed, which split the identification process into two major sequential steps.

In the first step the unknowns are the transfer function, or its parameters, and also the series, independent of any choice of a loss function. The performance is evaluated only on the capacity of these two components, the excess rainfalls and the transfer functions, once recombined, to reproduce the observed discharges. Usually, in this step of this approach the input raw rainfalls are considered as a first guess of the excess rainfalls to be corrected or modified to provide the final excess rainfalls.

The second step, which will be discussed later, considers only the relationship between these excess rainfalls and the input.

4.1   Rainfall accuracy in the identification of the unit hydrograph and of the excess rainfalls

The FDTF–ERUHDIT is typical of such an approach, although others have been proposed (Mays and Coles, 1980, Mays and Taur 1982). At first glance, the input rainfall accuracy may be expected to be important for this first identification step for the transfer function together with the excess rainfall series but, in fact, it appears (Nalbantis *et al.* 1988) that the output accuracy, i.e. that of the discharge series, is much more critical. Although surprising, this can be understood for two reasons: first the fitting criteria are all based on reconstituted versus observed discharges and thus are very sensitive to the latter; secondly, the method has been shown to be similar to a gradient optimization method, with two sets of unknowns (the unit hydrograph and the excess rainfall series). In such methods, the results are known to be rather insensitive to the initial guess, here taken as the raw rainfalls.

Extensive simulations on synthetic data show that the results, in terms of comparison of the identified transfer function and excess rainfall series with the 'true' generated series, become significantly affected only when error variance exceeds 30% of input rainfall variance, while only 3–5% of error variance seems acceptable on the discharge data (Nalbantis *et al.* 1988). Beyond that, significant bias may appear both on the identified unit hydrograph and, even more likely, on the deconvoluted excess-rainfall series.

In our case study, hourly totals of basin rainfall may be considered accurate at a 1–2% error variance level if based on the complete available network (10% if only telemetered gauges are considered), while discharge data, especially near flood peaks, are probably far from accurate at the 5% level.

4.2 Rainfall accuracy in the fitting of a loss function model

However, the second identification step must then be performed, in order to relate these deconvoluted excess rainfalls to the observed input basin rainfalls. Once again, the question arises as to the accuracy required. It has just been suggested that in the first identification step the identified excess rainfall series could remain fairly accurate even for poor input rainfall data (if discharge data are very accurate indeed). Obviously, this is true only if these excess rainfalls are meaningful, i.e. if the watershed satisfies reasonably well the underlying assumptions of the unit-hydrograph method. If not, the deconvoluted excess rainfalls may be pure numerical artefacts.

So here again, several levels of decision must be considered.

(1) If identified excess rainfalls are pure numerical artefacts, then there should be little hope, even starting from good input rainfall, to reproduce them correctly using an essentially deterministic, or physically based, algorithm, and reconstitution will in any case be poor.

(2) If the linearity of the basin is reasonably acceptable, i.e. if the unit-hydrograph assumption is physically sound, the deconvoluted excess rainfall may be strongly noise affected when discharge data are poor. In this case, there is also little chance of getting good overall results by improving the accuracy of the input, the fitting process of the loss function being affected by the poor output proposed, namely the previously identified excess-rainfall series.

(3) Even with good identified excess rainfalls, and assuming that this intermediate variable in the runoff generating process is sound, the accuracy of the input is not the only source of uncertainty and it must be considered concurrently with other possible sources.

To summarize, the identification process in this second step is based on the following.

(a) *The choice of a model for the loss function.* This model, usually based on a physical mechanism, is supposed to operate on a watershed scale, although it is often developed on the basis of soil columns. So either the basic mechanism of the loss function may be unsuitable, or it may be sound for point simulation but not for a basin scale, etc.

(b) *The choice of an initialization procedure.* In the case of an event per event use of the model, this problem may be critical. Conversely, if the model is to be operated continuously, then other parameters are needed to simulate between-event depletion of the soil moisture.

(c) *The choice of an algorithm to fit the loss function model.* This may not be a simple choice, especially for models including discontinuity, thresholds, etc.

So the sources of uncertainty are as follows.

(i) The structure of the loss function itself (if based on complete different mechanisms from those acting in nature, it could become difficult to match the available output, i.e. the identified excess rainfalls);

(ii) the relevance of the underlying mechanism but its use in a lumped manner, ignoring the effects of spatial variability, may set an upper limit to· its performance;
(iii) an inadequate initialization procedure;
(iv) insufficiently accurate input data.

However, before requiring better input data, the previous sources of uncertainty should be explored, although this is somewhat hard to do in an exhaustive way. For example, it can be seen that, for a given watershed, different infiltration models, although calibrated as recommended by their authors, provide quite different excess-rainfall series for the same input rainfall. A basin totalized excess-rainfall series could also be generated from a very distributed model with a given type of infiltration mechanism on the square grid. If the same mechanism is then applied in a lumped fashion to the watershed as a whole, and even optimized, then the loss of performance due to the lumping could be evaluated. Errors in the initialization procedure could similarly be tested by simulation.

Although far from exhaustive, our experience on the Gardon d'Anduze data at least allows tentative conclusions. For example, it has been shown that the combination of the identified unique transfer function together with the identified or deconvoluted excess rainfalls over more than 20 events of about 60 h each allows an excellent reconstitution of discharge data (coefficient of determination $R^2 = 0.97$ (Fig. 2(a))). This seems to fit, and even to back up the unit-hydrograph assumptions. However, conversely, once a loss function has been fitted, and whatever the loss function model used, it is hardly possible (Figs 2(b) and 2(c)) to explain more than 60% of the variance of these identified excess rainfalls (Sempere Torres et al. 1989). This casts doubt on the relevance of the loss functions used.

Obviously, it is almost impossible to identify among the possible reasons ((1) and (2) or (i), (ii), (iii) and (iv)) which, if only one, explains this poor performance. This result is, however, quite consistent with other published experiences (Naef 1980, Loague and Freeze 1985).

5. OTHER APPROACHES AND FUTURE RESEARCH

A possible way to overcome these previous problems could consist of a better use of the available information on the rainfall spatial distribution. Up to now, the focus has been to improve the accuracy of the estimated basin-averaged rainfall. It seems that it is of no use to go further than say 10%, a value in fact related to the structure of the models and the accuracy of the discharge data downstream. Other available information on the rainfall patterns could, however, be extracted and used in several ways.

5.1 Distributed modelling
It has been suggested that distributed models are structurally better than lumped model, partly because they can take into account the spatial distribution of the controlling variables: rainfall input, soil properties, vegetation cover, land uses, etc. Even if this is accepted as fully true for all other variables, consider what can be expected as requirements for rainfall.

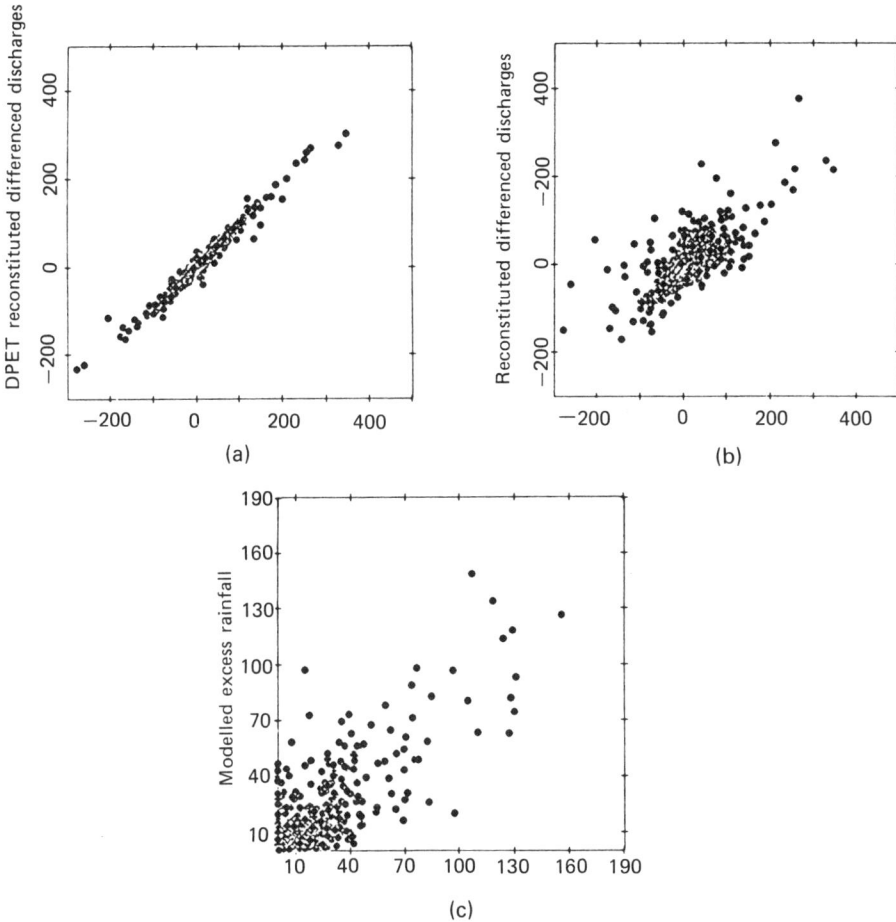

Fig. 2 — Comparisons between (a) discharges observed versus computed from *deconvoluted* excess rainfalls ($R^2 = 0.97$), (b) discharges, observed versus computed from *modelled* excess rainfalls ($R^2 = 0.54$) and (c) excess rainfalls *deconvoluted* excess rainfalls versus *modelled* excess rainfalls ($R^2 = 0.69$) (1313 points).

Usually, it is assumed that, the smaller the land unit considered, the better the model will operate. So the space step is often chosen in the range of 1 km, and sometimes a few hundred metres in the most recent studies where digital terrain models were available. Obviously, good accuracy on the rainfall input for such pixels would require a very dense, but completely unrealistic, network of recording gauges. Even with the best network available on a few experimental basins (one gauge per 5 or 10 km²), many pixels would have an uncertainty larger than 50–100% when located far from existing measurement points. Furthermore, most distributed models being conservative in volume, these errors would likely propagate to the next time steps.

Furthermore, even if adequate rainfall inputs become available, the question remains for these models as to their capacity to deal with the other spatially distributed variables such as soil properties and drainage density.

Although the choice of the space step is often governed by either of these, or by the postulated physics context in which they are included, it often reacts in turn on the time step to be chosen, and thus on the accuracy that could be expected on the inputs. The reasons may be related to the overall physics, i.e. to certain physical mechanisms, e.g., the smaller the space step, the shorter the average residence time of water, or of a fraction of it (e.g. overland flow), within this land unit. If the dynamics of the unsaturated zone are to be represented, say by Richard's equation, very small time steps must be used to represent properly the early stage of the infiltration process, etc.

So the time step is likely to be governed by the dynamics of the quickest response, while numerical reasons, such as stability of computations, may even lead to still smaller time steps, which often drop to a $\frac{1}{4}$ h or less (5 min). In some cases, the input at such time steps can only be obtained by arbitrarily disaggregating more commonly available hourly data.

These consistency requirements are more or less fulfilled in practice, and this probably explains why up to now no definite advantage has been proved for either lumped or distributed modelling when similar efforts have been devoted to both approaches (Loague and Freeze 1985, Naef 1980).

Concurrently, the relevance of the physics included in deterministic models may be questioned. Much still remains to be done, primarily in determining which process is effectively dominant in a given watershed. It is the Hortonian overland flow generated throughout the watershed, the gridding discussed above is likely to be significant while, if it is only the rainfall falling on the temporarily saturated area (usually a small percentage of the watershed area), then one may wonder whether it is useful to consider a complete grid for the rainfall input.

This above discussion is by no way exhaustive but merely points out that splitting the watershed into a fine mesh grid does not completely solve the problem of rainfall input.

5.2 Semidistributed modelling

Other than this hyperdeterministic approach, less demanding approaches can be proposed. For example, if a parameter is spatially non-uniform over the watershed, it is not necessarily useful to keep track of where exactly every given value has appeared, since its gross overall marginal distribution alone may carry most of the significant information. This has been used as early as 1966 in the Stanford model for the infiltration capacity, and more recently by Moore (1985), to represent variable contributing area. This is also part of the physically based approach of TOPMODEL (Beven and Kirby 1979) where a significant parameter for soil saturation, log $[A/\tan\alpha]$ is only considered through its marginal distribution over the watershed.

A similar approach could also be applied to the rainfall input. We are at present exploring two different ways, both trying to benefit from the availability of the deconvoluted excess rainfall series provided by the DPFT–ERUHDIT method.

In the first approach, the raw-rainfall surface is considered at each time step k in order to determine the threshold above which the remaining volume fits the excess-rainfall values exactly. The rainfall threshold $RT(k)$ is a kind of identification of the infiltration capacity at this time step and can be determined for all the time steps of the rain event showing a fairly nice exponential decrease with time. This provides a means to calibrate classical models of the infiltration capacity. However, problems remain with multiple-storm events, producing multiple peak floods, since the models used do not seem to reproduce well the infiltration capacity recovery between storms.

Another approach is a kind of adaptive one. It relies on the assumption that a given lumped loss function or infiltration model would be an acceptable estimate if the input rainfall were actually uniform. Given the availability of the deconvoluted excess rainfalls, the model can be calibrated to fit as much as possible the relation between the input raw rainfalls and the output deconvoluted excess rainfalls. Once calibrated, it can then provide, for an input raw rainfall $RR(k)$, a model estimate for the excess rainfalls. An example is provided in the framework of our case study. However, as already stressed in section 4.2, the modelled excess rainfalls correlate rather poorly with the deconvoluted excess rainfalls and seldom explain more than 70% of their variance (Fig. 2).

So the next step consists in assuming that these residuals appear because the

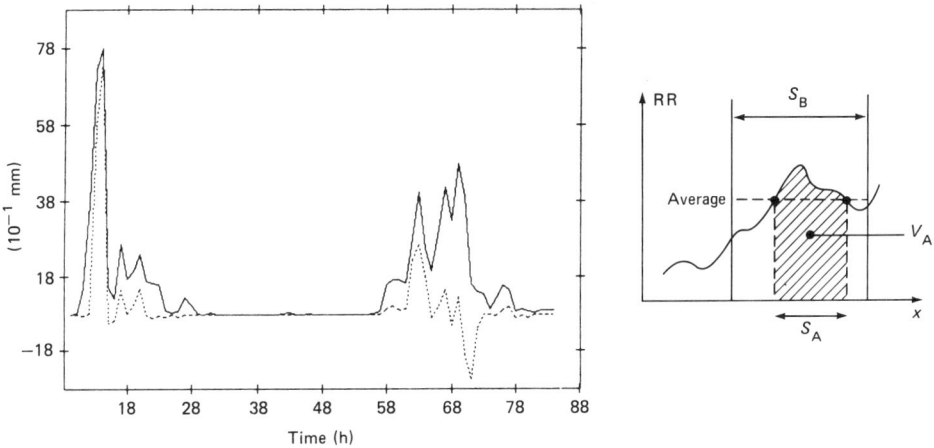

Fig. 3 — Cofluctuation between the errors (.....) of a lumped loss function and an index (———) of rainfall spatial variability (event of 10 October 1972).

uniformity assumption is not fulfilled, and some indices of this non-uniformity could possibly explain them. Many such indices have been computed to summarize the spatial variability of the rainfall. The problem becomes to find the right index and the right functional equation to maximize the explanation of the residuals. An example is displayed in Fig. 3 where for an event, the residuals E, which equal the deconvoluted excess rainfall minus the modelled excess rainfall, seem to cofluctuate with the index IV_t, defined as

$$IV_t = \frac{VA_t}{SB}\left(1 - \frac{SA_t}{SB}\right)$$

with *SB* is surface of the basin and *SA*, is the surface above average at time *t*. Other lumped variables describing the rainfall field can be imagined. For example, the distance of the rain kernel from the outlet can be used to speed up or to slow down the transfer function, etc.

5.3   Potential role of radar imagery

First, a major reason for the renewed interest in grid-based deterministic models is the hope that radar may provide adequate rainfall data for pixels down to 500 m × 500 m, in time steps of 5 min or more, although this is by no way an established result at present. Although radar is able to discriminate qualitatively or to sense the variability of rainfall at this mesh size, it cannot yet provide routinely accurate absolute values at this grid step, since radar calibration is not a fully solved problem at present.

However, if lumped characteristics of the rainfall distribution in space prove to be useful, it may be easier and even more accurate to derive them from radar data than from a gauge network. For example, in some hourly samples, the area above a given threshold may vary by ± 100% between the always very smooth pattern obtained from ground measurements and that deduced from radar data (Fig. 4).

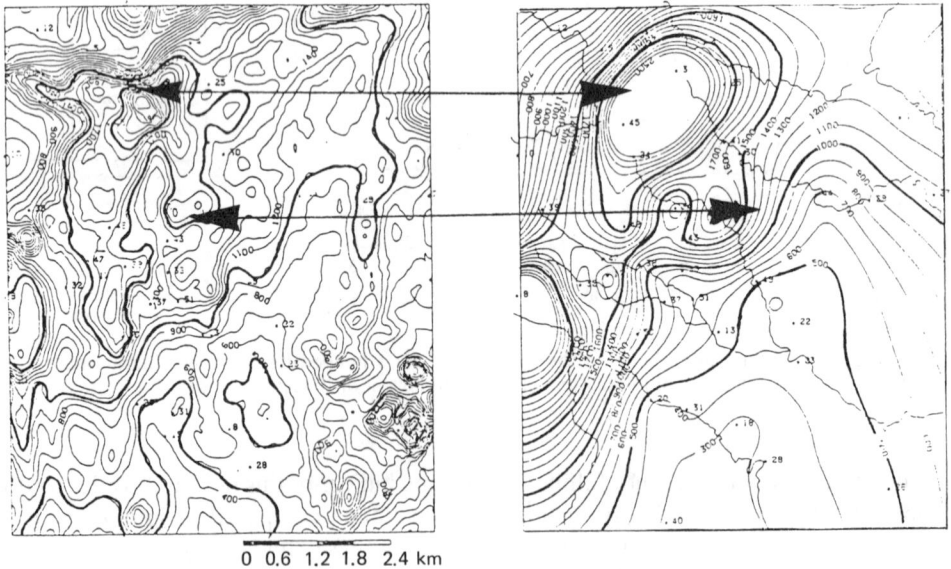

0 0.6 1.2 1.8 2.4 km

Fig. 4 — Comparison of rainfall patterns obtained from a ground network and from radar imagery.

At present, the possible interest of such characteristics is tested using long series computed by surface-fitting techniques applied on point hourly values. If the above approaches do not show more clear success, it may simply come from the non-representativeness of these approximated spatial patterns. This should be repeated using radar pattern series when available. Furthermore, some relative indices (expressed in percentages), which are inexpensive since they would not require the absolute calibration of the images, have already been helpful.

6. CONCLUSION

The already-compiled needs for rainfall information, mainly the basin-averaged rainfall, together with some recently explored indices of the rainfall spatial distribution, have been reviewed in the context of operationally oriented rainfall–runoff models, most of the lumped type.

The requirements for the rainfall input are not necessarily of highest accuracy over the smallest possible space and time steps. Rather a tradeoff must be achieved between the funds available for network maintenance, the sophistication of the rainfall–runoff model selected, its sensitivity and its overall average performances.

The sources of uncertainty in the rainfall–runoff modelling process are many and interrelated in a complex way, so that it is hard to distinguish between them. This could explain why it is sometimes not clear what information the modeller actually expects from new ground data, and moreover from a radar facility.

ACKNOWLEDGEMENTS

This paper is based on fruitful discussions with many colleagues, among whom J. M. Grésillon, Y. Rodriguez, D. Sempere Torres and J. Wendling are particularly thanked. W. Krajewski, while a visiting professor at the Institut de Mécanique de Grenoble, has considerably helped to improve this paper. This work has been supported by a grant from the DATAR project Risques naturels en Montagne 1986–88.

REFERENCES

Andrieu, H., Creutin, J. D. and Delrieu, G. (1989) Radar data processing for hydrology in the Cevennes region. *Proceedings of the International Association of Hydrological Sciences 3rd Scientific Assembly, 10–19 May 1989 Baltimore, Md.* To be published.

Beven, K. K. and Kirby, M. J. (1979) A physically based variable contributing area model of basin hydrology. *Hydrol. Sci. Bull.,* **24**(1,3), 43–69.

Chua, S. H. and Bras, R. L. (1982) Optimal estimators of mean areal precipitation in regions of orographic influence. *J. Hydrol.,* **57**, 23–48.

Creutin, J. D. and Barancourt, C. (1988) Pattern and structure analysis of rainfall fields in a Mediterranean region. *Proceedings of the American Geophysical Union–American Meteorological Society Conference on Mesoscale Precipitation, Cambridge, MA,* 1988. To be published.

Creutin, J. D. and Obled, Ch. (1982) Objective analysis and mapping techniques for rainfall fields: an objective intercomparison. *Water Resour. Res.,* **18**(2), 413–431.

Delhomme and Delfiner 1973.

Lebel, T., Bastin, G., Obled, Ch. and Creutin, J. D. (1987) On the accuracy of areal rainfall estimation: a case study. *Water Resour. Res.,* **23** (11), 2123–2134.

Loague, K. M. and Freeze, R. A. (1985) A comparison of rainfall–runoff modelling on small upland catchments. *Water Resour. Res.,* **21** (2), 229–248.

Mays, L. W. and Coles, L. (1980) Optimization of unit hydrograph determination. *J. Hydraul. Div., Am. Soc. Civ. Eng.,* **106** (5), 85–97.

Mays, L. W. and Taur, C. K. (1982) Unit hydrographs via nonlinear programming. *Water Resour. Res.,* **18** (4), 747–752.

Moore, R. J. (1985) The probability distributed principle and runoff prediction at point and basin scale. *Hydrol. Sci. Bull.,* **30** (2,6), 273–297.

Naef, F. (1980) Can we model the rainfall–runoff process today? *Proceedings of the International Association of Hydrological Sciences–World Meteorological Organization Symposium on Hydrological Forecasting, Oxford,* IAHS Publication No. 129. International Association of Hydrological Sciences.

Nalbantis, I., Obled, Ch. and Rodriguez, J. Y. (1988) Modélisation pluie-débit: validation par simulation de la méthode DPFT. *Houille Blanche,* **5–6,** 415–424.

Obled, Ch. and Rodriguez, J. Y. (1988) La distribution spatiale des précipitations et son rôle dans la transformation pluie-débit. *Houille Blanche,* **5–6,** 467–474.

Rodriguez, J. Y., Sempere-Torres, D. and Obled, Ch. (1991) Extension of lumped operational rainfall–runoff approach models to semilumped modelling: the case of the DPFT–ERUHDIT approach. *Proceedings of the International Symposium on Hydrological Applications of Weather Radar, Salford, 14–17 August 1989.* Ellis Horwood, Chichester, West Sussex, Chapter 43.

Sempere-Torres, D., Rodriguez, J. Y. and Obled, Ch. (1989) Using the DPFT approach to improve flash flood forecasting. *Proceedings of the EGS 14th General Assembly, Barcelona, 13–18 March 1989 Nat. Hazards,* in press.

Shaw, E. M. (1983) *Hydrology in practice.* Van Nostrand Reinhold, New York, 569.

43

Extension of lumped operational rainfall–runoff approach models to similumped modelling: the case of the DPFT–ERUHDIT approach

J. Y. Rodriguez, D. Sempere-Torres and **C. Obled**
Institut de Mécanique de Grenoble, Domaine Universitaire, BP 53 X 38041, Grenoble Cédex, France

ABSTRACT

The DPFT–excess rainfall and unit hydrograph by the deconvolution and identification technique (DPFT–ERUHDIT) is a unit-hydrograph approach which assumes that the rainfall–runoff process can be modelled by combining a loss function (or production function) and a transfer function. It is capable of supplying both the average transfer function and a set of consistent excess rainfalls, without any prior assumption concerning the production function.

The excess rainfall is subsequently used together with the raw-rainfall data (the average gauged rainfall) to fit the parameters of any kind of production function which may be suitable for the case study at hand.

Although this operational approach has provided good results on more than 20 basins, the key assumption of uniform rainfall distribution over the watershed is far from valid on some other basins, particularly in Mediterranean regions. The DPFT–ERUHDIT approach has therefore been extended to double-input cases for which it provides, starting from the simple overflow and rainfall measurements, two average responses and two sets of excess rainfalls, one for each subbasin.

The results of a case study (the Gardon d'Anduze basin, 545 km^2) comparing the classical single-input single-output DPFT–ERUHDIT and this new double-input extension show the improvement offered by taking into account the rainfall distribution in a semilumped manner, especially when it is far from uniform.

1. INTRODUCTION

The use of lumped approaches in real-time flow forecasting is preferred by operational flood control centres and generally provides quite acceptable results but in some basins, where the input variables and more precisely the input raw rainfalls are characterized by a high spatial variability, lumped-model forecasts have proved inadequate. This has encouraged the development of more physically based and distributed models, but problems such as overparametrization or the relation between parameters and measured physical values have not yet been solved (Beven 1989). Even if these approaches remain of cognitive interest, they are at present unable to provide better forecasting on a real-time basis than the classical unit-hydrograph model (Kirkby 1988).

Another way to take into account the high spatial variability of some basins would be to extend lumped to similumped models, i.e. to work on a reduced number of subbasins but to retain the lumped structure on each one. For example, if the unit-hydrograph approach is preserved, a production function and a transfer function may be kept on each watershed.

The DPFT–excess rainfall and unit hydrograph by the deconvolution and identification technique (DPFT–ERUHDIT) approach, an extension of which will be proposed herein, preserves the unit-hydrograph structure (production function plus transfer function), but it provides a good set of excess-rainfall series, without establishing any *a priori* model of loss function; as a consequence, these series constitute a useful tool for improving production function modelling, since they can be used as the model's output (Rodriguez *et al.* 1989, Sempere-Torres *et al.* 1989).

Since its first version (Guillot and Duband, 1980), many theoretical and practical improvements have been proposed (Versiani 1983, Nalbantis *et al.* 1988) and tested on synthetic data (Nalbantis 1987).

Conserving the greatest benefit of this approach (the provision of both the transfer function and a set of excess rainfalls), the DPFT–ERUHDIT has been extended to multiple-input single-output cases. However, it will be shown that this trial is limited by data availability and must be at present restricted to two subbasins.

A case study on the Gardon d'Anduze watershed ($545 \, \text{km}^2$) comparing the single-input single-output (SISO) DPFT–ERUHDIT approach and its extension will be presented. Little improvement is found, particularly when the raw rainfall is poorly correlated between subbasins.

2. BRIEF DESCRIPTION OF THE SINGLE-INPUT SINGLE-OUTPUT DPFT–ERUHDIT APPROACH

The DPFT–ERUHDIT model uses the same assumptions as the classical unit hydrograph. It is therefore assumed that the rainfall–runoff process may be modelled by two successively applied functions: a production function (or loss function) transforming the raw rainfall RR into excess rainfall ER, in which pre-storm conditions and rainfall–runoff non-linearity are taken into account; a transfer function, linear and temporally invariant, which is expressed as the excess rainfall ER over runoff Q.

The classical convolution equation relating Q, ER and transfer function coefficients H_j is given by

$$Q_i = \sum_{j=1}^{K} H_j(\mathrm{ER})_{i-j+1} = \sum_{j=1}^{m} (\mathrm{ER})_j \, H_{i-j+1}, \quad i = 1, n \tag{1}$$

where K is the transfer function memory, m the excess rainfall length and n the runoff length.

The alternating solution of equation (1) (first to identify H, with the excess ratings and Q_S series known; second to estimate ER, with the H_S and Q_S series known) was proposed by Newton and Vinyard (1967). The DPFT–ERUHDIT approach uses again this procedure. The transfer function coefficients H are identified by the least-squares method on a multievent data set (Mays and Coles 1980). Then, the excess-rainfall series are estimated by solving event by event the inverse problem. This last step, which produces large numerical instabilities, is carried out using an improved ridge regression technique (Versiani 1983), which provides better results than linear programming or the classical least-squares method (Hino 1986). In fact this technique is especially useful when available data are noise affected (Nalbantis 1987).

The alternating identification–deconvolution procedure is used to build an iterative algorithm, initialized with the raw rainfall as a first excess-rainfall estimate. Then, a first transfer function is identified and rectified by 'physical' constraints (positive ordinates, smoothness, scaling, etc.). This better transfer function is used to solve the inverse problem: the excess-rainfall estimation by deconvolution. Here again, some physical constraints are also applied to deconvoluted excess rainfall, particularly their positivity, providing a better excess-rainfall estimation.

The alternating procedure is repeated (Fig. 1), until convergence, which is basically due to the constraints applied to both the identified transfer function and the deconvoluted excess rainfall.

In practice, the DPFT–ERUHDIT approach works on first differences of outflow and the transfer function in order to give more stability to the identification–deconvolution process, and also to filter out the base-flow influence. Thus, a robust average transfer function and a consistent set of excess rainfalls have been obtained. In a second move, these excess-rainfall series will be quite useful to fit well-adapted production functions by solving a single-input (the raw-rainfall) single-output (the estimated excess-rainfall) problem.

3.   DPFT–ERUHDIT EXTENSION TO DOUBLE-INPUT CASES

It would be of interest to extend the DPFT–ERHUDIT approach to cases with more than two inputs. The reasons for the limitations of this extension will be described later.

It can be assumed that the behaviour of the whole watershed can be characterized by two independent subbasin responses (see example in Fig. 2). If the unit-hydrograph concept is accepted, each subbasin response to the whole outlet may be modelled by a production function and a transfer function. So, the classical discrete convolution equation (1) relating excess rainfall, transfer function and runoff may be written for both subbasins as

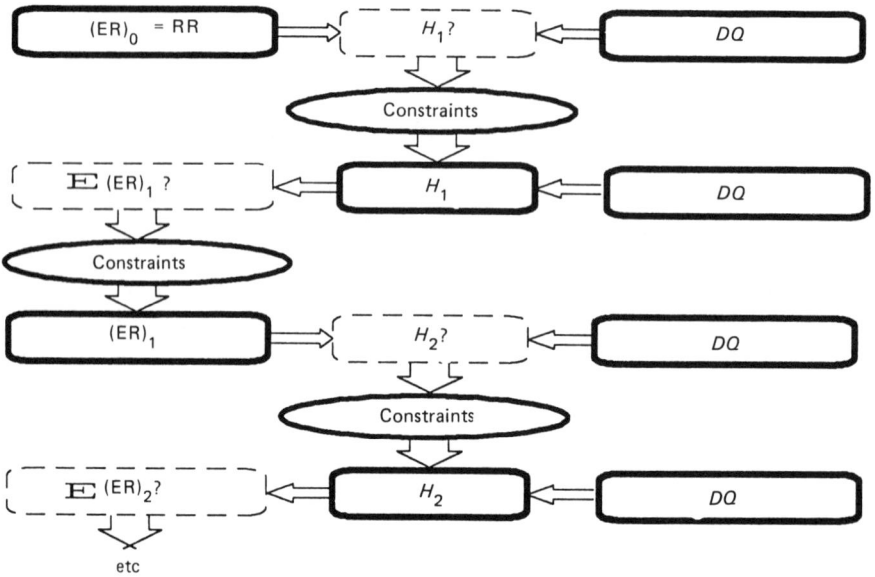

Fig. 1 — The alternating procedure.

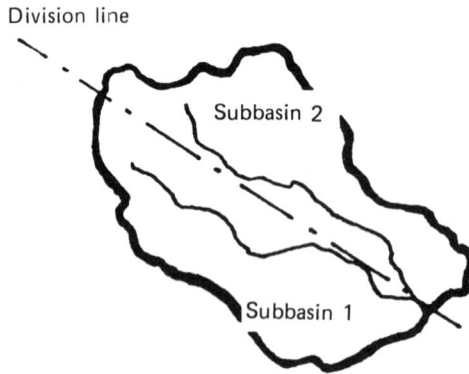

Fig. 2 — Division for a rural basin.

$$Q_i^{(S)} = \sum_{j=1}^{K^{(S)}} H_j^{(S)} \, (ER)_{i-j+1}^{(S)} = \sum_{j=1}^{m^{(S)}} (ER)_j^{(S)} \, H_{i-j+1}^{(S)}, \quad S = 1, 2, \; i = 1, n \quad (2)$$

with Q_i^S the outflow contribution from subbasin S at time step i, H_i^S the transfer function i^{th} ordinate from basin S to the outlet, $(ER)_i^S$ the excess rainfall at time step i

on basin S, K^S the assumed memory of the transfer function H^S and m^S the excess-rainfall length on subbasin S.

Commonly, only the total outflow Q is measured at the outlet. Then, from

$$Q_i = Q_i^1 + Q_i^2, \quad i = 1, n \tag{3}$$

the two-input discrete convolution relation is given as

$$Q_i \sum_{j=1}^{K^{(1)}} H_j^{(1)} (ER)_{i-j+1}^{(1)} + \sum_{j=1}^{K^{(2)}} H_j^2 (ER)_{i-j+1}^{(2)}$$

$$= \sum_{j=1}^{m^{(1)}} (ER)_j^{(1)} H_{i-j+1}^{(1)} + \sum_{j=1}^{m^{(2)}} (ER)_j^{(2)} H_{i-j+1}^{(2)}. \tag{4}$$

From this well-known relation, an alternating iterative method is built, which is very similar to the classical DPFT–ERUHDIT approach. It provides the following from the total measured outflow Q and raw rainfalls $RR^{(1)}$ and $RR^{(2)}$: both transfer functions $H^{(1)}$ and $H^{(2)}$; both sets of excess-rainfall series $(ER)^{(1)}$ and $(ER)^{(2)}$.

To determine precisely the problems that will be found, let us express the different array dimensions. For an event E, $n^{(E)}$ is the length of the outflow $Q^{(E)}$, and $m^{(E1)}$ and $m^{(E2)}$ are the excess-rainfall lengths on subbasins 1 and 2 respectively. Each outflow must be due to at least one rainfall. So, $n^{(E)}$, $m^{(ES)}$ and $K^{(S)}$ must fulfil the relation

$$n^{(E)} \leqslant \max(m^{(E1)} + K^{(1)}, m^{(E2)} + K^{(2)}). \tag{5}$$

4. IDENTIFICATION OF MULTIPLE EVENT SIMULTANEOUS TRANSFER FUNCTIONS

Doubling the transfer functions to be identified, the number of unknowns is also doubled. In the identification step, no particular problems appear because of this duplication. Using an N multiple event identification technique, the number of available equations is given by

$$n^{tot} = n^{(1)} + n^{(2)} + \ldots + n^{(E)} + \ldots + n^{(N)}. \tag{6}$$

The identification system is therefore built with n^{tot} equations, which are often greater than the number $K^{(1)} + K^{(2)}$ of unknowns. Thus the identification is mathematically possible. For instance, consider a set of 15 events, of 30 h time step record lengths. The multiple event identification system to solve will have $30 \times 15 = 450$ equations. If the memory of each transfer function $K^{(1)}$ and $K^{(2)}$ is assumed to be 15 time steps, the number of unknowns is only 30. The system should therefore provide a robust solution. Nevertheless, some problems may appear if an unsuitable basin division is selected. Note that to identify simultaneous identification of two transfer functions requires the use of two well-contrasted signals. Therefore two contrasted subbasin responses are required. So, for a rural basin, a division such as proposed in Fig. 2 would be incorrect, since similar transfer functions may be expected from adjacent subbasins with similar characteristics. However, if one subbasin is more urban than the other, the transfer functions should be different.

For rural watersheds, a division such as that shown in Fig. 3 is preferable. The high-watershed response should be different from the low-watershed response at least for their first ordinates, because of the natural high-watershed transfer function delay.

Otherwise, the expected high correlation between both excess-rainfall series $(ER)^{(1)}$ and $(ER)^{(2)}$ (which is reasonably assumed to be greater than the correlation between $(RR)^{(1)}$ and $(RR)^{(2)}$) could create some identification problems. In the transfer function this gives instabilities and fluctuations, particularly for small values. To check this parasite behaviour, a nearly orthogonal transformation is proposed (De Marsilly 1976). This transformation has been performed using two new variables:

$$
\begin{aligned}
M_i &= \alpha(ER)_i^{(1)} + \beta(ER)_i^{(2)}, \\
E_i &= x(ER)_i^{(1)} - y(ER)_i^{(2)},
\end{aligned}
\qquad i = 1,\, m^{\text{tot}}
\tag{7}
$$

where:

$$
m^{\text{tot}} = \sum_{E=1}^{N} \max(m^{(E1)},\, m^{(E2)}) = \sum_{E=1}^{N} \max(m^{(E)}).
$$

α is the area of subbasin 1 divided by the total watershed area, and β is the area of subbasin 2 divided by the total watershed area; thus $\alpha y + \beta x = 1$ where x and y are the parameters to convert the M and E orthogonal (i.e. $(M)^T E = 0$), which gives

$$
x = ky \text{ and } k = \frac{\alpha[(ER)^{(1)}]^T(ER)^{(1)} + \beta[(ER)^{(2)}]^T(ER)^{(2)}}{\alpha[(ER)^{(1)}]^T(ER)^{(2)} + \beta[(ER)^{(2)}]^T(ER)^{(1)}}.
\tag{8}
$$

The two-input discrete convolution (4) may be also written using the variables defined in (7):

$$
Q_i = \sum_{j=1}^{K} S_j M_{i-j+1} + \sum_{j=1}^{K} D_j E_{i-j+1}, \quad i = 1,\, n
\tag{9}
$$

with S_j the transfer function linked to series M, D_j the transfer function linked to series E and $K = \max(K^{(1)}, K^{(2)})$.

The multiple-event system built using (9) should be better conditioned than that built using (4). The multiple-event identification will therefore be performed directly by (9). The transfer function S spreads the 'average' rainfall M and is therefore conservative. The transfer function D introduces an outflow correction connected with the weighted differenced excess rainfall E. This correction is expected to have zero volume.

Once transfer functions S and D are identified, and assuming that (9) and (4) are equivalent, it is possible to return to the more interesting transfer functions $H^{(1)}$ and $H^{(2)}$, using

$$
\begin{aligned}
H_i^{(1)} &= \alpha S_i + x D_i, \\
H_i^{(2)} &= \beta S_i - y D_i,
\end{aligned}
\qquad i = 1,\, K.
\tag{10}
$$

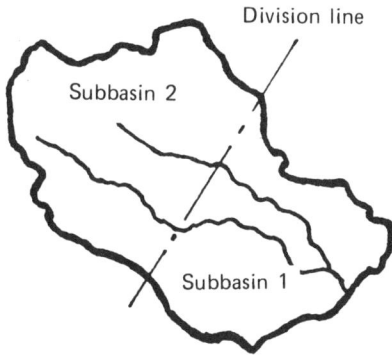

Fig. 3 — Division for a rural watershed.

Note that, although the classical lumped DPFT–ERUHDIT approach uses the ordinary least-squares technique to identify the average transfer function, a constrained least-squares technique is used herein. Constraints taken into account are the positivity of the transfer function values $H_i^{(S)} \geqslant 0$ and, if δ is the delay of the high-basin response $H^{(2)}$, its first δ values are forced to be zero:

$$H_i^{(2)} = 0, \qquad i = 1, \, \delta.$$

Finally, even though all the development details have been presented in classical terms of transfer functions and outflows, the processing uses differences, as does the lumped approach.

5. EXCESS-RAINFALL SIMULTANEOUS DECONVOLUTION

This problem is more delicate. If L subbasins are considered, $m^{(E1)} + m^{(E2)} + \ldots + m^{(EL)}$ unknowns may be estimated event by event, but only $n^{(E)}$ outflows are available. For instance, assume that only two subbasins are retained, and that both transfer functions have 15 length memories. For an a event of 20 excess rainfalls to be deconvoluted on each subbasin (40 unknowns) the available outflows are given by (6):

$$n^{(EL)} = \max(20 + 15, \, 20 + 15) = 35.$$

Only 35 equations are available to determine 40 unknowns. So, these excess rainfalls cannot be deconvoluted. It becomes necessary to add more equations to build a complete solvable system. The number of additional relations required increases with the number of subbasins considered. At this level, how many subbasins must be retained? What are the most suitable equations to add?

The possible supplementary relations will translate particular assumptions or approximations for the basin behaviour. Although these new equations are not exact or close to the real rainfall–runoff process, the aim of this extension is to measure the benefit obtained by a multiple-input extension. At this step, a bias is introduced (the supplementary equations) but is expected to be less than that introduced by assuming a rainfall uniform distribution over the watershed. To limit this bias, this extension will be reduced to two subbasins.

Seven techniques to solve the inverse problem and to determine both excess-rainfall series have been considered, implemented and tested on synthetic data (Rodriguez 1989). Only the most suitable is presented. The simultaneous excess-rainfall deconvolution is carried out in two steps.

(a) Since the function D is of zero volume, it can be assumed in a first approximation that the second term in (9) (D^*E convolution product) is a zero-mean noise. From (9), this assumption provides, for an event E, the new relation

$$Q_i^{(E)} = \sum_{j=1}^{m^{(E)}} M_j^{(E)} S_{i-j+1} + v_i, \qquad i = n^{(E)} \tag{11}$$

which will be solved in terms of $M^{(E)}$, event by event, as a classical single-input single-output deconvolution problem by an improved ridge regression technique (Versiani 1983).

(b) Using these $M^{(E)}$ series, $m^{(E)}$ supplementary equations are added to the double-deconvolution system deduced from (4), building the new system

$$Q_i^{(E)} = \sum_{j=1}^{m^{(E1)}} (\mathrm{ER})_j^{(1)} H_{i-j+1}^{(1)} + \sum_{j=1}^{m^{(E2)}} (\mathrm{ER})_j^{(2)} H_{i-j+1}^{(2)} \qquad n^{(E)} \text{ equations}$$

$$M_i^{(E)} = \alpha(\mathrm{ER})_i^{(1)} + \beta(\mathrm{ER})_i^{(2)} \qquad m^{(E)} \text{ equations} \tag{12}$$

which now have $n^{(E)} + m^{(E)}$ equations, to solve $m^{(E1)} + m^{(E2)}$ unknowns, by the improved ridge regression technique.

6. TWO-INPUT ALTERNATING ITERATIVE ALGORITHM IMPLEMENTATION

Similarly to the classical single-input case, it is initialized using both raw-rainfall series $(\mathrm{RR})^{(1)}$ and $(\mathrm{RR})^{(2)}$ like a prior excess-rainfall approximation. The series $M_{(1)}$ and $E_{(1)}$ (weight average and weighted difference at first iteration) are computed, allowing the multiple-event system construction based on (9).

Then, transfer functions $S_{(1)}$ and $D_{(1)}$ are identified, and $H_{(1)}^{(1)}$ and $H_{(1)}^{(2)}$ are calculated using them. Classical physical constraints on the transfer function (positivity, smoothness and scaling) are applied to $H_{(1)}^{(1)}$, $H_{(1)}^{(2)}$ and $S_{(1)}$. The identification step is finished.

The deconvolution is carried out event by event in two phases. The first is a classical phase using only the transfer function $S_{(1)}$ and the outflows to estimate an improved average excess-rainfall series $M_{(1)}^{(E)}$. This series, after being constrained (positivity), is used to complete (12) the discrete double-input convolution system

(4). Then, both excess-rainfall series $(ER)\{_1^1\}$ and $(ER)\{_1^2\}$ are determined. Once again, positivity constraints are applied to the estimated excess rainfall.

At the second iteration, $(ER)\{_1^1\}$ and $(ER)\{_1^2\}$ are used to compute $M_{(2)}$ and $E_{(2)}$. The alternating procedure is iterated using the last estimated excess-rainfall series, until convergence.

7.   A CASE STUDY COMPARISON: THE GARDON D'ANDUZE (545 km²)

The Gardon d'Anduze basin lies in the Cevennes region (south of France (Fig. 4)).

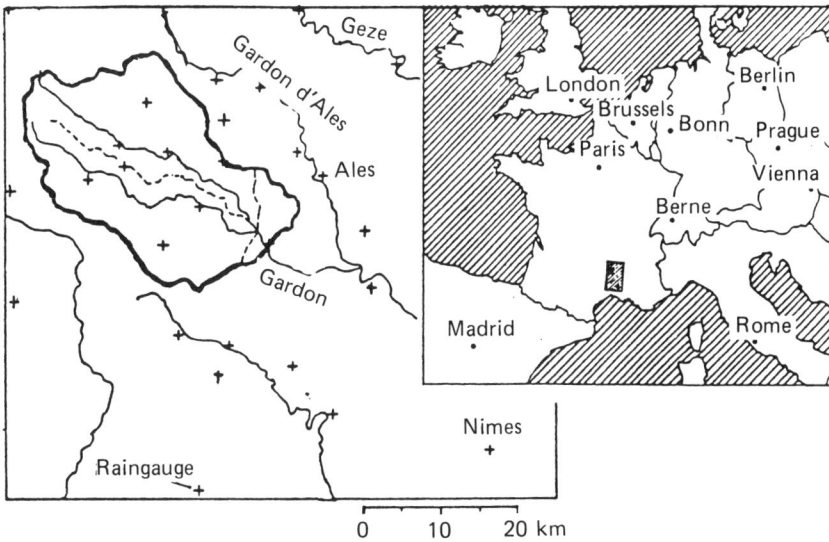

Fig. 4 — Cevennes region in the south of France.

This region is characterized by intense storms during the autumn season that often produce important flash floods. The basin covers an area of 545 km², with elevations ranging from 140 to 1565 m, and generally steep slopes. This operational basin has more than 30 raingauges and has been widely studied for rainfall–runoff, particularly regarding modelling the production function (Rodriguez *et al.* 1989, Sempere-Torres *et al.* 1989).

The rural basin has been divided as shown in Fig. 3. The coefficients α and β are then given by

> $\alpha =$ *high-basin area versus*
> total basin area $= 0.482$
> $\beta =$ *low-basin area versus*
> total basin area $= 0.512$.

13 events have been used to compare both approaches, with lengths going from 35 to 116 time-steps. The time step chosen was 1 h. The average transfer function obtained by the single-input DPFT–ERUHDIT approach is compared with the two responses obtained using the double-input extension in Fig. 5. High and low responses seem

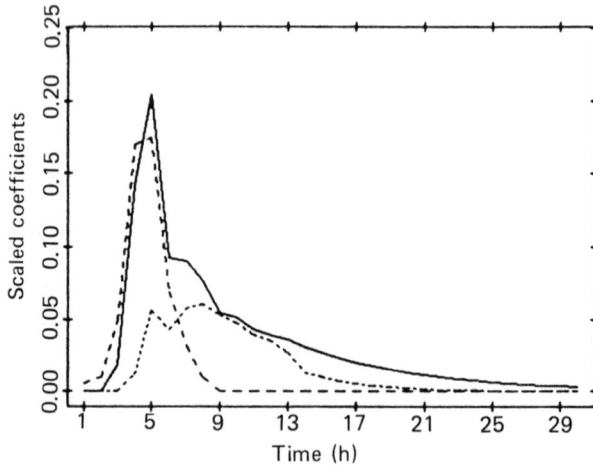

Fig. 5 — Comparison of lumped and double-input transfer functions (time step 1h): ————, lumped; ·······, high basin; – – –, low basin.

coherent with the lumped response. Regarding the determination coefficient between fitted and measured differenced discharges (Table 1), performances are close.

Table 1 — Coefficient determination on first differenced outflow series

ERUHDIT	R^2 (DQ computed/DQ measured)
Single input	0.95
Double input	0.94

Now the second step consists in relating the excess-rainfall computed series with its raw-rainfall series. Three production models have been fitted using estimated excess-rainfall series as the model output (Rodriguez *et al*. 1989, Sempere-Torres *et al*. 1989). But only one, which gives the best results, is presented herein.

8   AN ANALYTICAL OPERATIONAL PRODUCTION MODEL (Guillot and Duband 1980)

This is the current operational production function model used by Electricité de France which provides acceptable results over more than 20 basins. The excess rainfall ER(t) for the interval ($t - 1, t$) is provided by

$$ER(t) = \frac{[RR(t)]^2}{RR(t) + B(t)} \tag{13}$$

where RR(t) (10^{-1} mm) is the gauged precipitation at t, $B(t) = CT(d)/H[Q(t), H(t)]$ with $Q(t)$ m^3 s^{-1} the discharge at t and $T(d)$ a seasonal function of the date d in the year, obtained by a harmonic fitting of the daily average temperatures in the basin.

$$H[Q(t)\ P(t)] = [AQI(t)]^a\ [API(t)]^b$$

where

$$AQI(t) = \lambda Q(t) + (1 - \lambda)\ AQI(t - 1)$$
$$API(t) = \gamma\ RR(t) + (1 - \gamma)\ API(t - 1)$$

which are classically taken as measures of the basin saturation and surface saturation respectively. C, a, b, λ and γ are the basin parameters that must be calibrated. AQI(1) is taken to be equal to $Q(1)$, and API(1) = 1. The calculation of $T(d)$ only needs a series of average temperature data. Thus no index need be initialized.

9.   FINAL RESULTS

Although Sempere-Torres *et al.* (1989) have shown that this production model is not really well adapted to the Gardon d'Anduze basin, no better models are yet available. Table 2 shows that lumped excess-rainfall series remain better modelled by this production function than the double-input series.

Table 2 — Coefficient determination on excess-rainfall series

Production function fitting	R^2 (ER modelized/ER deconvoluted)
Lumped excess-rainfall series	0.68
High-basin excess-rainfall series	0.58
Low-basin excess-rainfall series	0.52

Table 3 shows the production function parameter values obtained by each approach. Note that some values, such as C, are quite different when obtained using the lumped DPFT–ERUHDIT approach compared with the double-input extension values.

A comparison between modelled outflows (computed by modelled excess-rainfall series and identified transfer function convolution) is also proposed. It is

Table 3 — Production function parameter values

	Lumped	High basin	Low basin
		Value	
C	580	330	375
λ	0.002	0.050	0.005
γ	0.065	0.030	0.130
a	0.190	0.790	0.775
b	1.000	0.750	0.400

shown in Table 4 that a somewhat better reconstitution is obtained using the double-input extension; almost 60% of the differenced outflow initial variance is accounted for. In terms of total runoff, both techniques give similar results, but it is worth noting that the differenced outflow is more useful in an operational context, and more suitable for translating the goodness of fit.

If the six events correlated to less than $R = 0.8$ are removed, and if production functions fitted with all events are used on this data set, the results provided by the double-input DPFT–ERUHDIT extension are significantly better than those obtained by the lumped version (Table 5), indicating the interest of this new semilumped DPFT–ERUHDIT approach.

A comparison between the double-input and lumped simulation is proposed in Fig. 6. It is shown that, in some cases, as in this event, the total outflow may be somewhat better simulated by taking the rainfall distribution into account in a semilumped manner.

10. CONCLUSIONS

Taking into account the rainfall spatial variability and the coefficients of the subbasin production functions by a semilumped approach, the DPFT-ERUHDIT method has been extended to multiple-input cases. The major advantage of the classical single-input approach is conserved, i.e. the excess-rainfall series estimated by deconvolution can be used to fit any kind of production model such as an input–output problem. However, the numerical difficulties encountered in solving the inverse problem has limited this study to double-input cases.

Even if the double-input deconvolution system cannot be solved, it is possible to add equations relating average excess rainfall to subbasin excess-rainfall series. At this level, a bias is introduced, but it is expected to be less than that introduced assuming a uniform rainfall distribution.

A case study comparing the lumped and similumped DPFT–ERUHDIT approach on the Gardon d'Anduze basin (545 km^2) shows somewhat better results using the extended approach, particularly if the raw rainfall is poorly correlated between the two subbasins.

Even if the overall model is far from being perfect, the inclusion of the spatial variability in a semilumped manner provides small improvements, which may indicate new directions for operational rainfall–runoff forecasting models.

Table 4 — Comparison between modelled outflows

ERUHDIT	R^2 (DQ modelized/DQ measured)	R^2 (Q modelized/Q measured)
Single input	0.54	0.77
Double input	0.58	0.77

Table 5 — Comparison of the lumped and double-output results

ERUHDIT	R^2 (DQ modelized/DQ measured)	R^2 (Q modelized/Q measured)
Single input	0.50	0.83
Double input	0.60	0.83

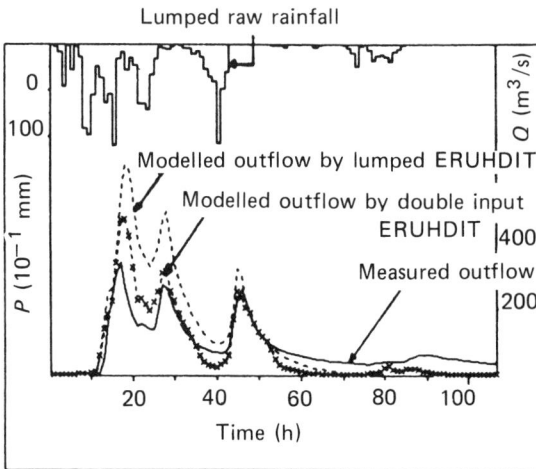

Fig. 6 — Comparison of lumped and double-input simulations ($Q_{max} = 636.76$ m^3/s; $P_{max} = 176.71 \times 10^{-1}$ mm).

ACKNOWLEDGEMENTS

This research has been supported by a DATAR Rhône-Alpes Contract (Action Risques Naturels) and Contract SRETI-MERE/88105. We thank D. Duband for his advice and interest in this study.

REFERENCES

Beven, K. (1989) Changing ideas in hydrology — the case of physically based models. *J. Hydrol.*, **105**, 157–172.

De Marsilly (1976) *De l'identification des systèmes hydrologiques*, Thesis ès Sciences. Université de Pierre et Marie Curie., Paris IV.

Guillot, P. and Duband, D. (1980) Une méthode de transfert pluie-débit par régression multiple. *Proceedings of the Oxford Symposium on Hydrological Forecasting, April 1980*, IAHS Publication No. 129. International Association of Hydrological Sciences, pp. 177–186.

Hino, M. (1986) Improvements in the inverse estimation method of effective rainfall from runoff. *J. Hydrol.*, **83**, 137–147.

Kirkby (1988) Hillslope runoff processes and models, *J. Hydrol.*, **100**, 315–339.

Mays, L. W. and Coles, L. (1980) Optimization of unit hydrograph determination. *J. Hydraul. Div., Am. Soc. Civ. Eng.*, **106** (5) 85–97.

Nalbantis, I. (1987) *Identification de modèles pluie-débit du type hydrogramme unitaire: développement de la methode DPFT et validation sur données générées*, Thèse.Institut National Polytechnique de Grenoble.

Nalbantis, I., Obled, Ch. and Rodriguez, J. Y. (1988) Modélisation pluie débit: validation par simulation de la méthode DPFT. *Houille Blanche*, **88** (5–6), 415–424.

Newton, D. W. and Vinyard, J. W. (1967) Computer-determined unit hydrograph from flows. *J. Hydraul. Div., Am. Soc. Civ. Eng.*, **93** (5), 219–235.

Rodriguez, J. Y. (1989) *Modélisation pluie-débit par la méthode DPFT: développements de la méthode initale et extension à des cas bi-entrées*, Thesis. Institut National Polytechnique de Grenoble.

Rodriguez, J. Y., Sempere, D. and Obled, Ch. (1989) Nouvelle perspectives de développement dans la modélisation des pluies efficaces par applícation de la méthode DPFT. *Surface water modelling — new directions for hydrological prediction*, IAHS Publication No. 181. International Association of Hydrological Sciences, pp. 235–246.

Sempere-Torres, D., Rodriguez, J. Y. and Obled, Ch. (1989) Using the DPFT approach to improve flash flood forecasting. *Proceedings of the EGS 14th General Assembly, Barcelona, 13–18 March 1989. Nat. Hazards*, in press.

Versiani, B. (1983) *Modélisation pluie-débit pour la prévision des crues*, Thesis. Institut National Polytechnique de Gronoble.

44

The updating procedure in the MIKE 11 modelling system for real-time forecasting

M. Rungo, J. K. Refsgaard and K. Havno
Danish Hydraulic Institute, Agern Alle 5, DK 2970 Horsholm, Denmark

ABSTRACT

A general mathematical modelling system for real-time forecasting is briefly presented with special attention to the updating procedure, which is a combination of a traditional error prediction model and a simple expert system approach model. The capabilities of the updating procedure to account for amplitude and phase errors are demonstrated and discussed. The updating procedure is tested using data from events used during a World Meteorological Organization intercomparison project.

1. INTRODUCTION

The occurrence of floods is a natural phenomenon all over the world. With the increase in population and human activity in the flood plains, flood damage represents an increasing hazard in many countries, in spite of increasing investments in flood control measures. Consequently, it is of utmost importance to utilize the most efficient methods in flood forecasting, in the assessment of reservoir operation schemes, and in the evaluation of the effects of various flood control measures.

Recent progress in hydrology and river hydraulics has now made it feasible to perform flood forecasting by means of comprehensive mathematical models. Rapid development in the size and speed of microcomputers has now also made it feasible to develop microcomputer versions of these mathematical models. This paper presents the MIKE 11 PC-based modelling system with special emphasis on real-time operation.

A major problem in the real-time operation of hydrological models is that the simulated runoff generally deviates from the measured runoff at the time of forecast. In order to obtain optimal benefit of the real-time runoff measurements in the forecasts, some sort of updating of the hydrological model is required before the forecast is made.

The present updating procedure in MIKE 11, like probably all other existing procedures, performs reasonably well in cases where the deviation originates from amplitude or volume errors, whereas it generally performs poorly in cases of phasing error between the simulated and the measured flows. Therefore, research work has been initiated at the Danish Hydraulic Institute to overcome this problem. The present paper presents the status of this research work.

The MIKE 11 system is a PC-based version of the well-proven mainframe-based NAM-S11 modelling system. The MIKE 11 (NAM-S11) is being applied operationally on several large river basins in the world (e.g. Refsgaard *et al.* 1988), and it participated successfully in the World Meteorological Organization (WMO) project on simulated real-time intercomparison of hydrological models (World Metereological Organization 1988).

For application to real-time forecasting the three modules (namely the rainfall–runoff model, hydrodynamic river model and updating model), together with a facility for real-time data entry and processing, are utilized. In addition the MIKE 11 contains modules for dam breach, water loss (infiltration and evaporation), sediment transport, transport dispersion and water modules. MIKE 11 operates as a full menu-based user-friendly system with powerful graphical facilities integrated with the computational modules.

3.   THE NAM RAINFALL–RUNOFF MODEL

The NAM model is a generally applicable soil moisture accounting model of the lumped conceptual type. The NAM model was originally developed at the Technical University of Denmark (Nielsen and Hansen 1973). The structure of the model is illustrated in Fig. 1.

4.   THE HYDRODYNAMIC RIVER MODEL

The hydrodynamic model is a general mathematical modelling system for the simulation of flows and water levels in rivers, reservoirs, estuaries and canal systems.

In its most advanced form, the hydrodynamic model is based upon numerical solution of the general one-dimensional 'St Venant' equations (conservation of mass and momentum):

$$\frac{\partial Q}{\partial x} + \frac{\partial A}{\partial t} = q$$

$$\underset{\substack{\text{dynamic} \\ \text{wave}}}{\frac{\partial}{\partial x}\left(\alpha\frac{Q^2}{A}\right)} + \frac{\partial Q}{\partial t} + \underset{\substack{\text{diffuse} \\ \text{wave}}}{gA\frac{\partial \xi}{\partial x}} + \underset{\substack{\text{kinematic} \\ \text{wave}}}{\frac{gQ|Q|}{M^2AR_*^{4/3}}} = gAI_0$$

where A (m^2) is the flow area, M (m$^{1/3}$/s) is the Mannings roughness coefficient, g (m/s^2) is the acceleration of gravity, ξ (m) is the water depth, Q (m^3/s) is the discharge, R_* (m) is the resistance radius, α is the momentum distribution coefficient, I_0 is the slope of river bed, q (m^2/s) is the lateral inflow and x (m) is the horizontal coordinate.

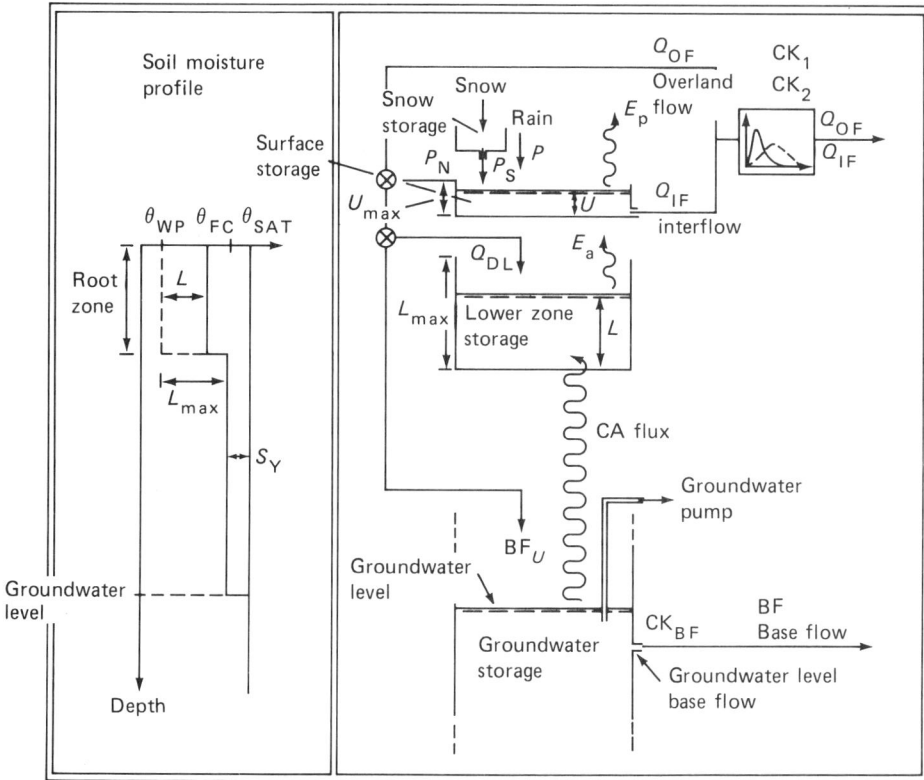

Fig. 1 — Structure of the NAM rainfall–runoff model.

The equations are solved as fully time-centred implicit finite-difference equations on a grid, which is automatically generated by MIKE 11. For details regarding the numerical scheme reference should be made to Abbott (1979) and Cunge *et al.* (1980), while the model is described more thoroughly in DHI (1987). MIKE 11 has the options of using the diffusive or the kinetic wave approximation. MIKE 11 allows for description of (pseudo-) two-dimensional flows over wide flood plains.

5.   UPDATING TECHNIQUES

Updating for real-time operation can be made in several fundamental different ways, including the following.

(a) *Manual updating*   This implies a manual subjective adjustment of input vari-
    ables and/or state variables (initial conditions) carried out by the operational
    hydrologist. This is the classical approach used for hydrological lumped concep-
    tual models since these models were introduced in real-time operation. The
    approach is still being used operationally by among others the Swedish HBV

model (Bergstróm *et al*. 1978) which also participated in the WMO intercomparison project (World Meteorological Organization 1988).

(b) *Error prediction*   This approach is the updating technique originally used in MIKE 11. The deviations ('errors') between measured and simulated runoff is simulated (in MIKE 11 by a linear autoregressive model) and used to adjust the streamflow simulated by NAM, prior to routing by the hydrodynamic model. Such a two-stage approach is widely used (e.g. Jamieson *et al*. 1971, Lundberg 1982, Szöllözi-Nagy 1983). The procedure is fully computerized; it can be applied for large complex catchments with several updating points and is usually applied in a fully automatic mode without interference from the operational hydrologist. The procedure generally works very well in cases of amplitude (volume) errors, whereas it is not very effective in the case of phasing errors.

(c) *Expert system approach*   In order to overcome the weakness of the above error prediction approach in cases of phase errors, research has taken place to develop methods which can detect errors, in both amplitude and phase, and make the optimal hydrograph adjustment in the way that an experienced operational hydrologist would do. An initial attempt was made by Sittner and Krouse (1979) in the development of the computer hydrograph adjustment technique (CHAT). The present paper shows another approach of this type. Neither the CHAT nor the approach described in the present paper is a real expert system but rather an advanced algorithm for pattern recognition. However, with the rapid progress in expert system technology this approach is believed to have a major potential for significantly improving the updating techniques.

(d) *Kalman filtering*   The Kalman filtering technique is a powerful mathematical and/or statistical tool for updating state variables. It is a well-proven engineering tool for linear systems but can with some modifications also be applied to the strongly non-linear hydrological and hydraulic systems. Kalman filters can be integrated either with purely statistical transfer functions models of, for example, the ARIMA type or with hydrological lumped, conceptual models. Examples of the last type are the HFS of the US National Weather Service (Georgakakos *et al*. 1988), and the NAMKAL (Refsgaard *et al*. 1983) which both participated in the WMO intercomparison project (World Meteorological Organization 1988). At the present state of development the Kalman filter technique can be applied only to headwater basins. According to our experience the Kalman filtering techniques, like the error prediction models, are much more efficient in correcting amplitude errors than phase errors.

6.   DESCRIPTION OF THE NEW UPDATING MODEL

The distinction between amplitude errors and phase errors is illustrated in Fig. 2.

The most commonly used updating routines (cf. (b) and (d) above) assume that the deviations are amplification errors, and as a consequence the sign of the predicted error does no change within the forecast period.

This assumption leads to problems if there is a phase deviation between measured and similated values, especially if the forecast starts as shown in Fig. 3.

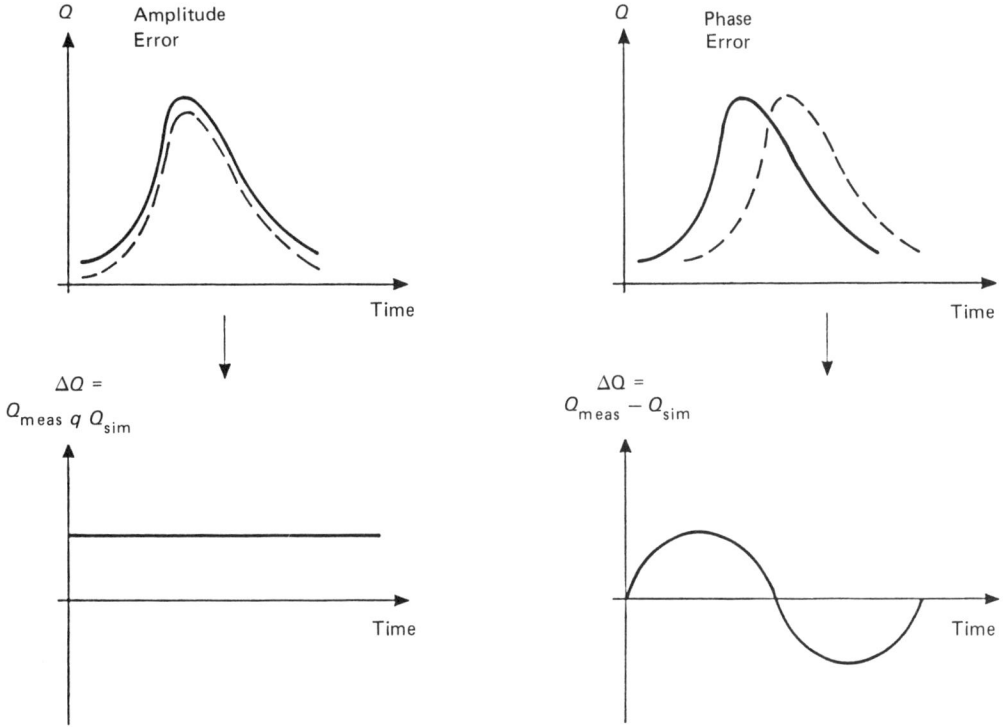

Fig. 2 — Definition of amplitude and phase errors: ———, measured; ----, simulated.

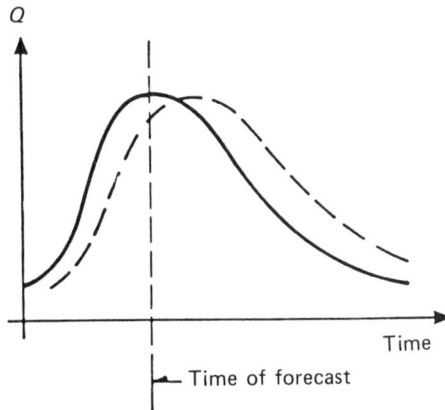

Fig. 3 — Phase deviation between measured and simulated discharge.

In the situation shown in Fig. 3 the updating procedures assuming amplitude error will in the forecast period predict the simulated discharge to be too low and as a consequence add extra inflow to the model.

This example illustrates the requirements for an updating routine which can distinguish between amplitude and phase errors and can efficiently account for both error types. An attempt to develop and implement a routine such as this has been done in MIKE 11.

The idea behind the routine is to move the simulated curve, both along the time axis and along the discharge axis, until the best agreement between the simulated and measured curves is achieved. The best agreement is defined as the minimum of the sum of square deviation between simulated and measured values.

With reference to Fig. 4 and on the assumption that the phase error is less than Δt, the best agreement is found as

$$\min\left(\sum_{i=1}^{n}\left\{F_i\left[M_i - \left(S_i - \left(S_i + A_e - \frac{S_i - S_{i+1}}{\Delta t}P_e\right)\right)\right]\right\}^2\right)$$

where A_e (m^3/s) is the amplitude error, P_e (s) is the phase error, M (m^3/s) is the measured discharge, S (m^3/s) is the simulated discharge, F is the weighting factor, n is the number of values taken into account and Δt is the time-step (s).

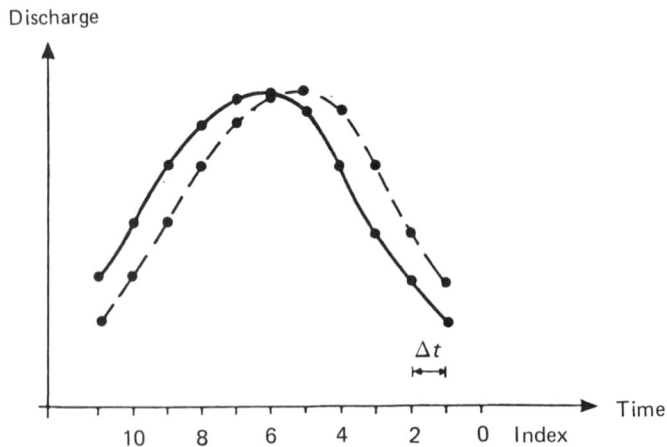

Fig. 4 — Definition of measured (———) and simulated (– – – –) curves.

The minimum can be found by differentiating the equation with respect to A_e and P_e and then solving these two equations. This solution leads to the values of A_e and P_e (still with the assumption that $0 < P_e < \Delta t$), which gives best agreement between the two curves, similar to the assumption.

$$\Delta t < P_e < 2\Delta t, \qquad 2\Delta t < P_e < 3\Delta t, \qquad \text{etc.}$$

At the time of forecast the model then has an estimate of the phase and amplitude errors, and it is therefore possible to calculate the additional inflows or outflows to the river model in order to take the errors into account during the forecast period.

7.   TEST OF THE NEW UPDATING MODEL

7.1   A hypothetical case

To test the capability of the updating model, a hypothetical case with mainly phase error has been studied. The simulation without updating is shown in Fig. 5 and the following two model runs have been made.

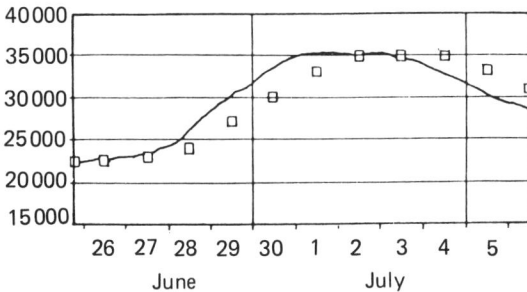

Fig. 5 — A comparison of observed discharge (□) with simulated discharge (———) without updating.

(1) *Updating model* 1   This is the error prediction model, only accounting for amplitude errors. The results are shown in Fig. 6.
(2) *Updating model* 2   This is the expert system model, accounting for both amplitude and phase errors (Fig. 7).

In Figs 6 and 7 the functioning of the updating model is shown through graphs of amplitude error, phase error (only for model 2) and correction discharge. As can be seen from Figs 6 and 7, the updating model 2 results in the overall best performance with respect to peak and time.

Furthermore, the weakness of the error prediction model in a phase error case is illustrated. At the time of forecast, model 1 assumes an amplitude error and subtracts water from the system, with the result that a bad simulation becomes even worse by the updating. From Fig. 7 it can be seen how model 2, on the other hand, is able to forecast a change in the sign of the correction discharge.

Another model test was carried out on one of the data sets used during the WMO intercomparison project, namely the 2344 km² Bird Creek Catchment in USA, where six events were forcasted during the WMO project. The MIKE 11 (NAM-S11 version) participated in this project. The main problem experienced by the model during the forecasting test for Bird Creek was phasing error, which is believed to a

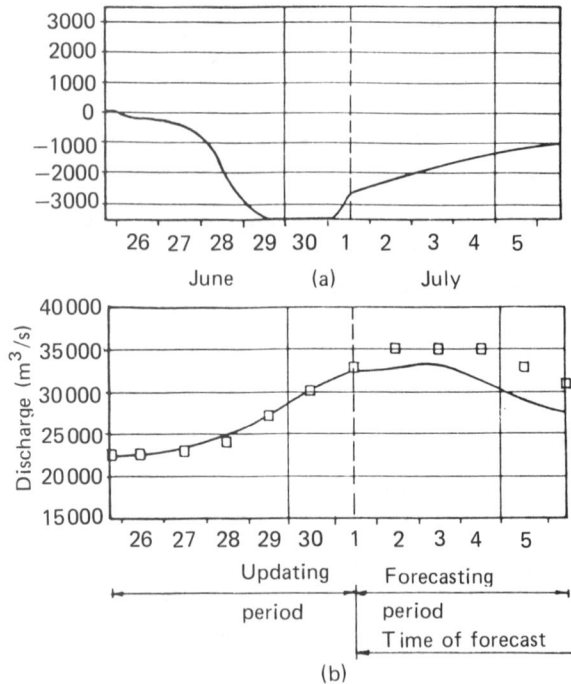

Fig. 6 — Results from updating and forecasting period using updating model 1. Showing (a) the amplitude error (correction discharge) (———) and (b) the measured (□) and forecasted (———) discharges.

large extent to be caused by the fact that only mean area rainfall was supplied to the participants for the test, whereas rainfall for several subcatchments typically would be available in practice for a catchment covering more than 2000 km². However, to test the new updating model, event 5 was selected as the event where MIKE 11, in particular, experienced problems. For further details regarding the WMO project and the Bird Creek data set, reference should be made to World Meteorological Organization (1988) and Georgakakos et al. (1988).

The Bird Creek event 5 is shown in Fig. 8. This event appears very difficult for the updating models, because it is a combination of a phase and an amplitude error and because the phase error changes sign on March 1974 on the rising limb of the peak. The event was forecast seven times starting on 8 March 1974 at 0000 hours and with 24 h between each time of forecast. When the forecasts were made at the WMO project, the updating procedure was used in a semiautomatic way, i.e. the updating was switched off in some forecasts while it was switched on in other forecasts according to the subjective judgement of the forecaster.

It is only possible to make forecasts in this way for an experienced user. For an inexperienced user and for larger river basins with several updating points it must be done in a more automatic way, and this will, of course, affect the model performance.

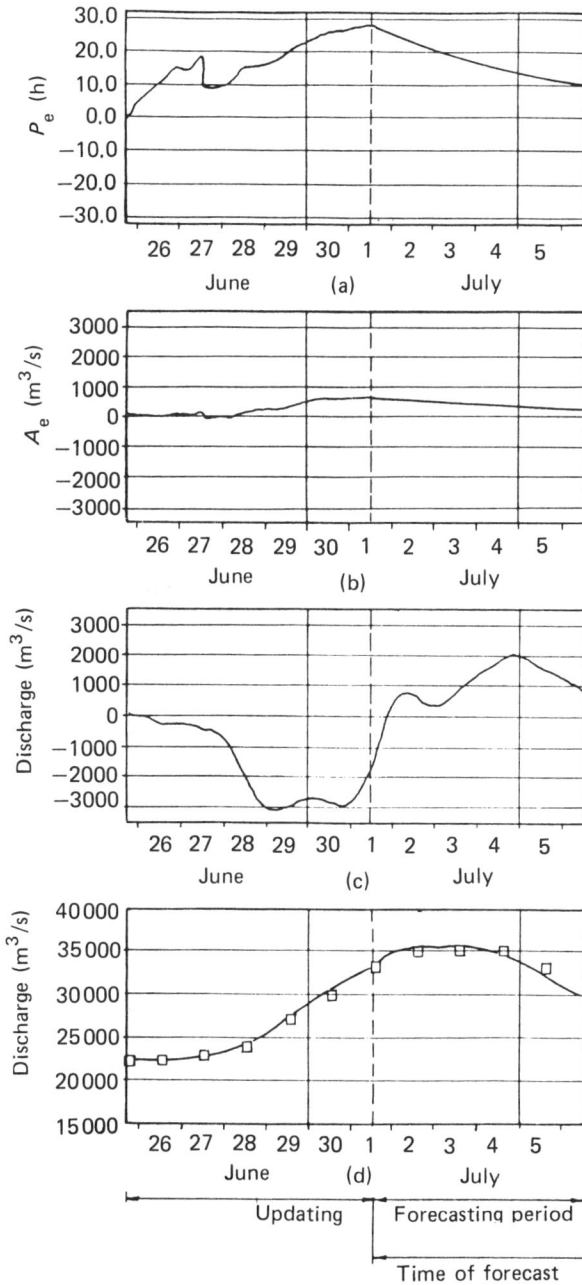

Fig. 7 — Results from updating and forecasting period using updating model 2 showing (a) the phase error (————), (b) the amplitude error (————), (c) the correction discharge (————) and (d) the measured (□) and forecasted (————) discharges.

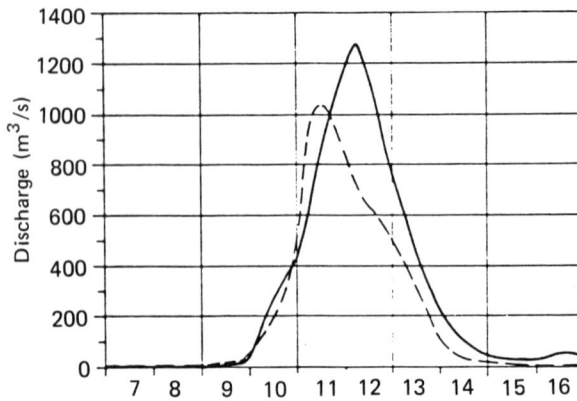

Fig. 8 — Measured and simulated (non-updated) discharge for the Bird Creek event 5 in 1974 from the WMO intercomparison project: ———, measured; – – – –, simulated without updating.

The results of the forecasts were compared and the root mean square error has been calculated for different forecast lead times. The results are compared for the following three models and are shown in Fig. 9.

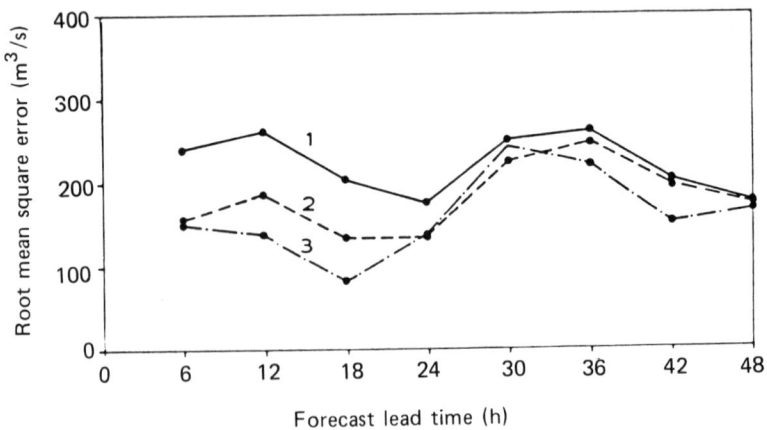

Fig. 9 — Root mean square error as a function of forecast lead time for Bird Creek event 5.

(1) *model* 1   fully automatic error prediction accounting for amplitude errors only;
(2) *model* 2   fully automatic expert system accounting for both amplitude and phase errors;

(3) *model* 3   results obtained during WMO project, i.e. semiautomatic error prediction accounting for amplitude errors only.

From Fig. 9 it appears that model 2 performs significantly better than model 1 because model 1, as expected, is very sensitive to the phase error. Further model 2 is seen to perform almost as well as model 3, where subjective interference from the hydrologist is made.

8. CONCLUSIONS

Comprehensive mathematical modelling packages for simulation of rainfall–runoff and river flow processes exist. However, for real-time operation, the updating procedure is often a weak point, and generally there is need for improvements.

MIKE 11 is an advanced and very powerful modelling system applicable for real-time forecasting. A new expert-system-based approach for updating has been developed and tested against the existing error prediction type of updating model, which can optionally be operated in either a fully automatic or a semiautomatic manner with user interference. The new updating model generally gives better results than the fully automatic error prediction model, whereas a semiautomatic operation is some cases may give better results.

The expert system approach tested at present only represents a very simple expert system, and as such the results are encouraging. With the rapid progress in expert system technology, this approach is believed to have a major potential for significantly improving the updating techniques.

REFERENCES

Abbott, M. B. (1979) Computational hydraulics. *Elements of the theory of free surface flows*. Pitman, London.

Bergström, S., Persson, M. and Sundqvist, B. (1978) *Operational hydrological forecasting by conceptual model*, HB Report No. 32. Swedish Meteorological and Hydrological Institute.

Cunge, J. A., Holly, F. M. and Verwey, A. (1980) *Practical aspects of computational river hydraulics*. Pitman, London.

Danish Hydraulic Institute (1987) *MIKE 11, a short description*, Report. Danish Hydraulic Institute.

Georgakakos, K. P., Rajaram, H. and Li, S. G. (1988) *On improved operational hydrological forecasting of streamflow*, IIHR Report No. 325. Iowa Institute of Hydraulic Research.

Jamieson, D. G., Wilkinson, J. C. and Ibbitt, R. P. (1971) *Hydraulic forecasting with sequential uncertainties in hydrology and water resources*, Vol. 1. Tucson, AZ, pp. 177–187.

Lundberg, A. (1982) Combination of conceptual and an autogressive error model for improving short time forecasting. *Nord. Hydrol.* **13**, 233–246.

Nielsen, S. A. and Hansen, E. (1973) Numerical simulation of the rainfall–runoff process on a daily basis. *Nord. Hydrol.* **4**, 171–190.

Refsgaard, J. C., Rosbjerg, D. and Markussen, L. M. (1983) *Application of the Kalman filter to real-time operation and to uncertainty analyses in hydrological modelling*, IAHS Publication No. 147. International Association of Hydrological Sciences, pp. 273–282.

Refsgaard, J. C., Havnø, K., Ammentorp, H. C. and Verwey, A. (1988) Application of hydrological models for flood forecasting and flood control in India and Bangladesh. *Adv. Water Resourc*, **11**, 101–105.

Sittner, W. T. and Krouse, K. M. (1979) *Improvement of hydrological simulation by utilizing observed discharge as an indirect input.*, NOAA Technical Memorandum No. NWS HYDRO-38. National Oceanic and Atmospheric Administration.

Szöllözi-Nagy, A., Bartha, P. and Harkayi, K. (1983). Microcomputer based operational hydrological forecasting system for River Danube. *Technical Conference on Mitigation of Natural Hazards Through Real-Time Data Collection Systems and Hydrological Forecasting, Sacramento, CA. 19–23 September 1983.*

World Meteorological Organization (1988) Simulated real-time intercomparison of hydrological models, Report from WMO Workshop in Vancouver, August 1987, WHO Report No. TD 25J. World Meteorological Organization.

45

From radar rainfall data to hydrological data

M. Semke
Universität Hannover, Appelstrasse 9 A, W-3000 Hannover 1, FRG

ABSTRACT

This paper provides an overview of a research project into the real-time control of retention basins within a combined sewer system. A number of pertinent aspects of real-time control philosophy are discussed within an urban environment context, and the objectives and achievements of the project presented. An integrated approach is described unifying sophisticated hardware equipment and software utilities. The importance of high spatial and temporal resolution rainfall data as provided by a custom-built X-band weather radar is stressed.

1. INTRODUCTION

The definition of an **on-line control system** (OCS) implies that at least four elements are in operation (Schilling 1987):

(1) a (measurement) **sensor**, which is used to monitor the ongoing process of interest;
(2) a (corrective) **regulator**, which influences the process,
(3) a controller, decides whether and how to exert the regulator;
(4) a **communication** or **telemetry device** which transports the field data from the sensor to the controller and the signals of the controller back to the regulator.

These four elements form a **control loop**, which is common to every OCS, (Fig. 1). It is obvious that an OCS cannot be carried out without operating data collection. The OCS has to live on data and information, on their computer-based processing and on required presentation.

To manage short-term effective actions for avoiding and removing disturbances in a system, the disturbances have to be directly recognized and visible. That is why control needs at any time an 'exact' image of the current situation of the system. On the other hand this is only guaranteed by a *closed time* collecting and processing of the operating data. Consequently the main feature of **real-time control** is the importance of *time*.

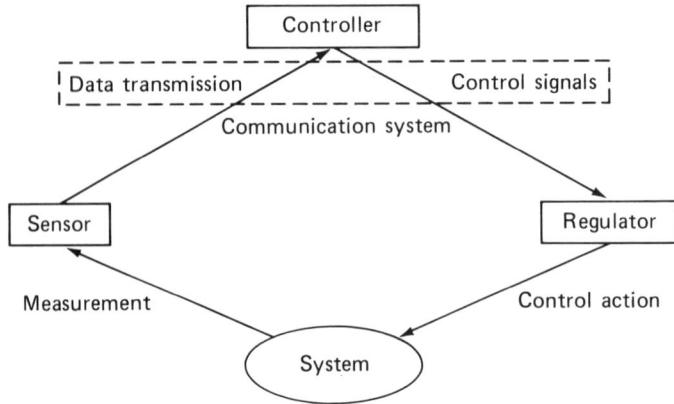

Fig. 1 — Elements of a control loop.

One has to distinguish between

(1) the **system response time**, which is the time lag between the disturbance of a regulated system value and the time of occurrence of the subsequent output value of the system, and
(2) the **reaction time**, which is the time needed to process information either on measured system values or on measured system disturbances, until the regulated values are properly adjusted, and it includes:

  (a) data acquisition time,
  (b) computer data processing,
  (c) human decision time,
  (d) set point transmission,
  (e) controller data processing and
  (f) regulator speed.

Therefore the quality of an OCS acting like a real-time control system (RTCS) is not only judged by the aspect of whether the data are currently complete, perfect and saved-collected, but also how and when the data will be continue-processed and put as reliable information 'just in time' at the operator's disposal.

What are the use of 'perfect' data, if the system reaction time takes longer than the system response time?

The real-time control of a combined sewer system (CSS) has the following objectives:

  (i) to regulate the flow to the treatment plant;
  (ii) to avoid or minimize CSS overflows;
  (iii) to optimize the use of storage capacity;
  (iv) to minimize excessive operation and maintenance costs, etc.

Rainfall as a disturbance of the system process is one of the most important values to be measured and analysed because of the short response time of sewers and the relevance of severe storms for achieving the objectives of real-time control.

Following the demands and the scope of data collecting and processing functions within an RTCS will be illustrated by one of three considered catchments of the Hydrological Emscher Radar Project (HERP).

2. DESIGN OF THE HERP SYSTEM

HERP encompasses a German research project on RTC of retention basins and pumping stations within a CSS for flow regulation and pollution control, using short-distance radar rainfall measurements. Three catchments of about $200\,km^2$ are considered, and these are situated inside a sector of 90°, north of the regional weather service office of the city of Essen, FRG. These catchments represent typical urban drainage management problems and are selected because of the set-up of an on-line monitoring system by the water authority. One of these catchments, which is mainly considered in this paper, is the Boye river catchment with a total area of $76\,km^2$ (Fig. 2). Subsidences caused by coal mining require that all runoff from this catchment has to be pumped to the river Emscher for further transport to the treatment plant.

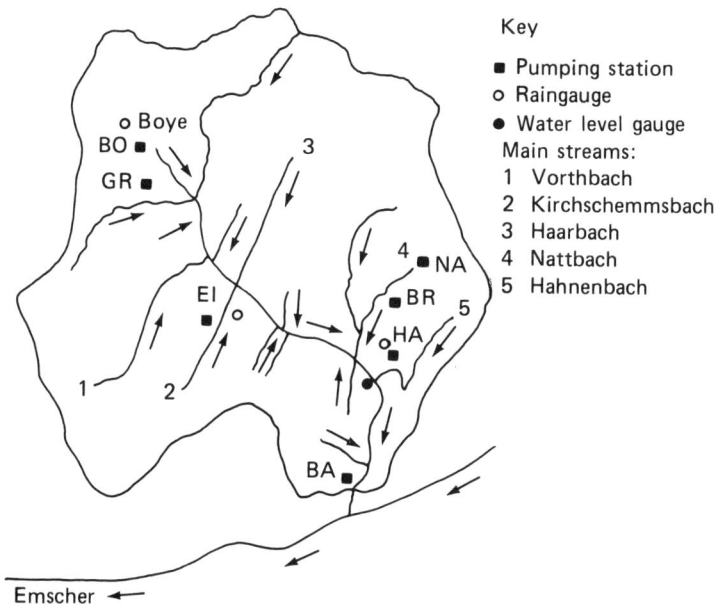

Fig. 2 — HERP: hardware configuration.

2.1 Hardware equipment

The control loop, as shown in Fig. 1, needs several devices to carry out the necessary functions of control. The basic elements of HERP illustrated in Fig. 3 by the hardware structure of the RTCS.

The controller component is based on a PDP11-73 minicomputer (the host) with a real-time operating system as host computer. It is located at the main pumping

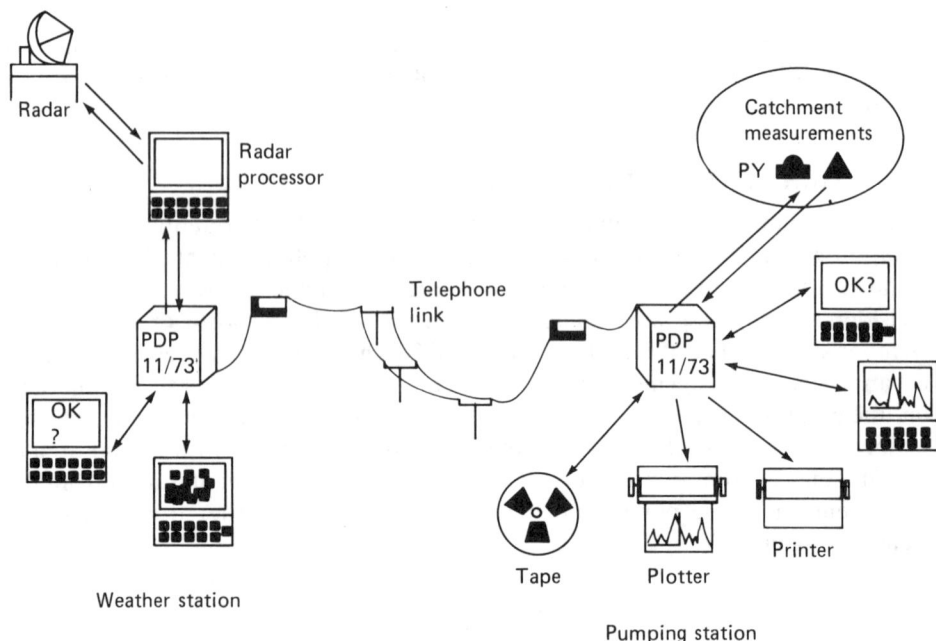

Fig. 3 — Boye river catchment.

station of the Boye river catchment in the middle of the monitored area of HERP. The sewer runoff process and its changes due to external influences are detected by water level gauges, pumping rate information of nine pumping stations and four raingauges. These operating data of the CSS are put at HERP's disposal through the monitoring system of the water authority. The data are updated each minute by a fixed recording dialogue, supervised by the host computer. After carrying out formal and content checking together with examinations of plausibility the data are compared with the last stored datum. If there are significant changes (e.g. pumps do not switch on and off each minute), the new data will be stored together with their generation time on mass storage. Additionally most recent measured values and status data of the CSS can be displayed on a coloured graphic terminal together with the process history of about 3 h.

In addition to the raingauge located at the main pumping station a distrometer is also installed. Data of 20 different classes of drop size are also transferred each minute to the host, stored and continued-processed to calculate the actual a and b constants of the $R-Z$ relation for on-line calibration of the HERP radar (Kreuels 1991).

Each irregularity of the data collection is directly reported in a special status protocol. To guarantee a steady data transfer the host computer is connected to an emergency power supply. At least once a month the operating data are backed up on magnetic tape. In the case of a precipitation event the radar-measured rainfall data are transmitted to the host computer via the modem of a fixed 4800 baud telephone line.

The X-band radar used in HERP is located near the regional weather service office of the city of Essen, about 10 km from the main pumping station of the Boye river catchment. The radar processor together with a second minicomputer is installed at the weather service office. The minicomputer also runs a real-time operating system and is the heart of a totally independent 'sensor' facility for the RTCS. Once a minute the radar makes a scan over the 90° sector of the three catchments. Control of the radar system and the analogue-to-digital conversion of the logarithmic video signal is done by the radar microprocessor (Kammer 1991). Radar reflectivities of approximately 16500 8 bit data (i.e. 255 classes) are transferred to the minicomputer via a 16 bit parallel interface. The rough data are stored as a file on mass storage, titled after their generation time. The file names together with information about the number of reflectivity data in defined ranges are protocolled as well as the status information of the data collection.

2.2 Software utilities

Obviously some data processing is required on the multitude of measurements collected by the RTCS, to organize, report and calculate the actual status of the system process.

All HERP-developed software is written in Fortran with special consideration on the multitasking real-time operating system of the two minicomputers. Fig. 4 shows the general software set-up which is in principle common for both machines.

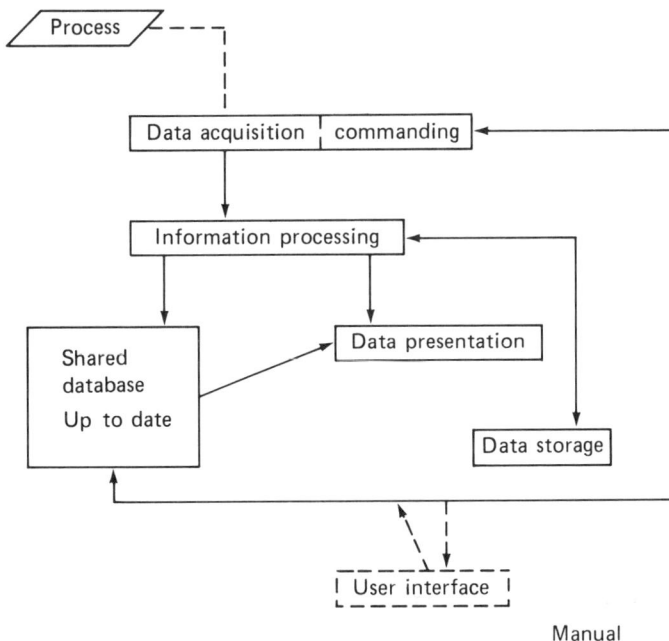

Fig. 4 — Software set-up for HERP computer.

At the radar site the following processes have to be executed every minute of data collection. The minicomputer calculates the rain intensities from the reflectivity data in 128 categories, including range and attenuation corrections. Transformation from 16 500 polar pixels to 2300 Cartesian pixels are done by a special look-up table together with suppression of ground clutter (Kammer 1991). To reduce the amount of unprocessed reflectivity data files stored every minute on mass storage, a special algorithm is developed which recognizes whether the rough data are relevant in terms of precipitation or not. This program automatically executes every hour and deletes all redundant data files. Relevant data are backed up on magnetic tape.

The resulting 2300 rainfall 'measurements' are spread out over the HERP-specific 600 m×600 m square grid of the considered area within the 90° sector. Its coloured graphic display represents the 128 intensity categories by eight colour classes and is also updated by the most recent radar data (Fig. 5). The radar rainfall

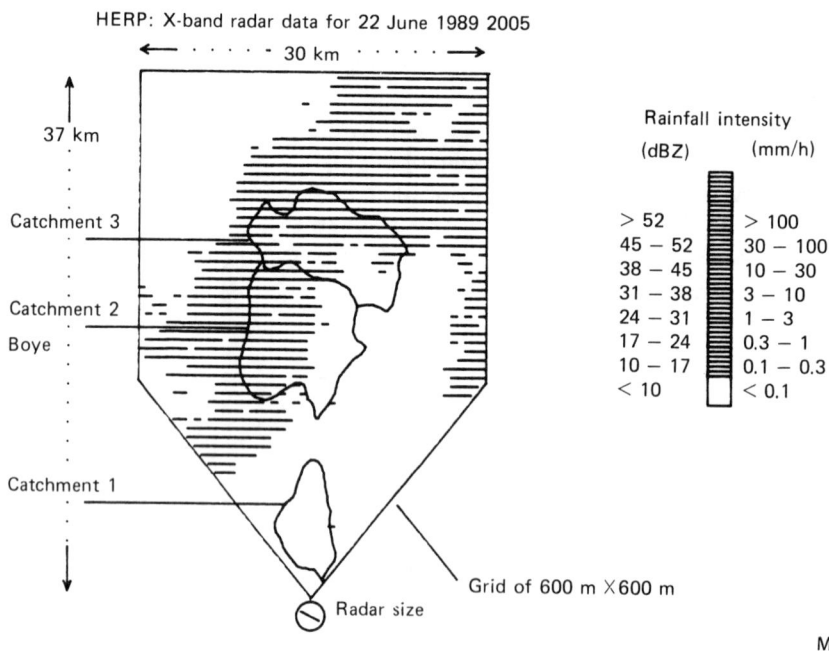

HERP: X-band radar data for 22 June 1989 2005

Rainfall intensity	
(dBZ)	(mm/h)
> 52	> 100
45 − 52	30 − 100
38 − 45	10 − 30
31 − 38	3 − 10
24 − 31	1 − 3
17 − 24	0.3 − 1
10 − 17	0.1 − 0.3
< 10	< 0.1

Menu Boye

Fig. 5 — On-line radar display.

measurements together with their generation time are transmitted by a file server to the host computer, to be used immediately as input to a surface runoff model.

The rainfall–runoff simulation, executed by the host computer calculates the actual system status in order to supplement measured data. Briefly it will also be used for forecasting the probable runoff due to the actual rainfall information.

Since the hydrodynamic model HTSTEM/EXTRAN is normally applied for off-line simulation, several modifications with regard to its real-time application have been carried out. For example, the available radar rainfall data exceed the number of possible rainfall inputs of the surface runoff model HYSTEM. The modified version HYSRAD of this program provides this possibility, so that sub catchment-specific rainfall data can be used. Runoff from paved and rural areas is calculated separately (Harms 1981).

For the Boye catchment a grid of about 50 subcatchments are generated, adequate spatial resolution for the representation of the hydrological and hydraulic characteristics of the system. Since each of these subcatchments is covered by more than one element of the radar square grid, an average rainfall value is calculated by the overlaid part of the radar grid and is assigned to each subcatchment (Fig. 6). This merging of the (approximately 400) radar-measured rainfall data for the Boye catchment results in 50 rainfall values, used as input to HYSRAD.

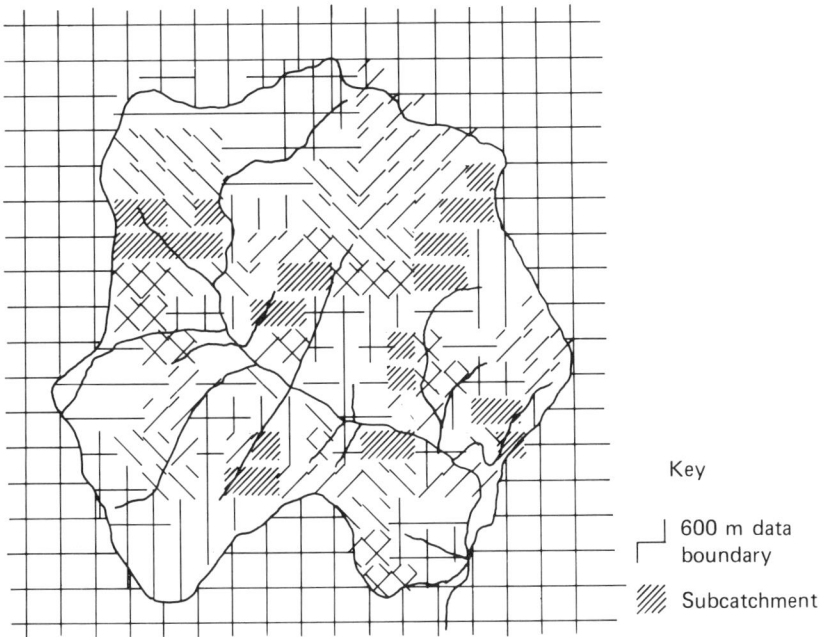

Key

600 m data
boundary

Subcatchment

Fig 6 — The grid-based square mesh for the BOYE catchment.

The conduit runoff and routing are processed by solving the complete St Venant equations. Pumping stations and additional storage facilities are integrated in the hydrodynamic calculations. The real-time measurements of water level gauges, pumping rates, etc., which are put at HERP's disposal are not yet incorporated in the simulation process. Investigations in this direction have not been taken, leading so far to coloured graphic displays of the most recent operating data (e.g. Figs 7 and 8).

3.   SYSTEM REACTION TIME TOWARDS RESPONSE TIME

The RTCS requires a large amount of data to be processed and information to be provided. The multitude of operating data collections can lead to a reaction time which must not be underestimated, especially in the real-time control of a CSS,

BOYE catchment district: Essen/Bottrop

PW station	Rumps running 1 2 3 4 5 6	Water level ↓↑	Pump rate (m³/s)	Watergauge Level (m)	Flow (m³/s)	Raingauge (mm)
BO	▮ ▮	▮	85	0.370			(10 / 0 graph)
GR	▮ ▮ ▮ ▮	▮	59	1.450			
EI	▮ ▮		0	0.350			(10 / 0 graph)
NA	▮ ▮ ▮ ▮		48	2.380			
BR	▮ ▮		0	0.350			
HA	▮ ▮	▮	45	0.960	0.85	2.200	(10 / 0 graph)
BA	▮ ▮ ▮		81	7.524			
BB			0	0.000			

BO   GR   EI   NA   BR   HA   BA   BB   Menu   Date: 5 September 1987

Fig. 7 — HERP: pumping stations, general survey.

District: Essen/Bottrop Pumping station: GLA — Hahnenbach

Raingauge
BO–Obeplauf ▬▬
Hahnenbach ▬▬
Eigen ▬▬

Water gauge 3.4:   level ▬▬   2.95 m
Flow ▬▬   38.50 m³/s
Pumping rate ▬▬   0.99 m³/s
Water level ▬▬   48

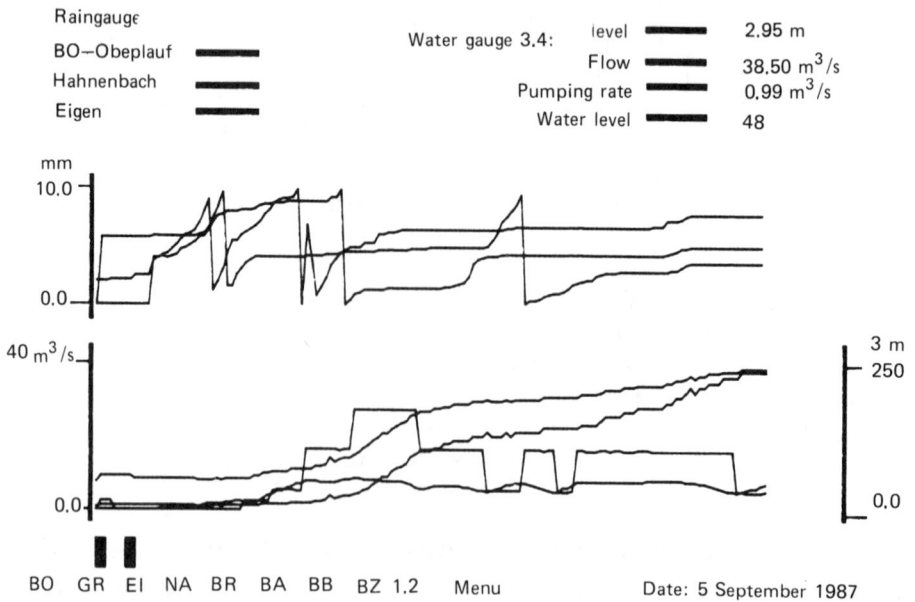

BO   GR   EI   NA   BR   BA   BB   BZ 1.2   Menu   Date: 5 September 1987

Fig. 8 — HERP: pumping station HA.

where response times of less than 15 min are not unusual. Therefore the components contained by the reaction time have to be analysed carefully.

A very time-consuming component is the data processing. The radar system of HERP for example needs approximately 25 s for its scan, 4 s to transmit the reflectivity data to the minicomputer, approximately 27 s for calculation of rainfall intensities and another 5 s for the final data transfer to the host computer. Obviously each additional operation or process (e.g. on-line calibration, clutter filtering or data completion) increases the data-processing time. Used process computers are characterized by the fact, that they are equipped with a real-time *and* multitasking operating system. To take the advantage of the multitasking facility, the different processes executed have to be organized by their dependence owing to the possibility of intertask communication.

The surface runoff simulation process can only be executed if the current radar rainfall data are available. The flow runoff process has to wait until the surface runoff data are calculated and the graphic screen has only to be updated with the most recent measurements. Further decision models need both data sets to achieve efficient control strategies. These programs are themselves totally independent, but they depend on earlier processed data as input and as a trigger for their next process cycle.

The host computer, as well as the minicomputer use a shared database situated in the random-access memory, to speed up the intertask communication. A special process-assigned event flag controls the access to these databases. Thereby the surface runoff model is able to start its calculations of the time step $t+1$, as early as the rainfall data for $t+1$ are available, while the flow runoff program is processing time step t. The two programs will not write and read their shared data at the same time.

As HERP cannot carry out actual control decisions, no real information about the remaining components of the reaction time can be stated, but estimations on these actions are more or less possible, because transmission time (e.g. on set points) or regulator speed are device dependent, and well known by the operators.

Previous experiences of the implemented and running processes show that an update of the whole system status is possible in less than 5 min. The system reaction time is estimated to be approximately 15 min. Further measurements of runoff at the entry points of the hydraulically controlled sewers may prove that this reaction time is greater than the response time. Therefore further investigation will be done to use forecasting models to achieve a smaller lead time than the system response time.

4. CONCLUSION

In every hydrological problem, the rainfall is one of the most important factors. In particular in the real-time control of a CSS a high resolution in time and space is needed because of the short response times of sewer systems and the relevance of sewer storms. When meassuring rainfall by radar, direct areal information about the rainfall distribution becomes available.

The main feature of real-time control is the importance of time; actions should be taken soon enough that the effect of regulated variables on output variables occur before variables exceed their allowable limits.

Therefore, one should keep in mind that data such as measurements of output variables could be of no use if the system reaction time is longer than the system response time.

The multitude of radar-measured rainfall data increases the reaction time of a system, because of additional data collection, preparation and processing. If this leads to the fact that the system response time becomes shorter than the system reaction time, the possibility of rainfall forecast by the radar data should be advantageous.

REFERENCES

Harms, R. (1981) Application of standard unit hydrograph in storm sewer design. *Proceedings of the International Conference on Urban Storm Drainage, Urban IL, June 1981.*

Kammer, A. (1991) An integrated X-band radar system for short-range measurement of rain rates in HERP. *Hydrological Applications of Weather Radar.* Ed. Cluckie, I. D. and Collier, C. G. Ellis Horwood, Chichester, West Sussex, Chapter 24.

Kreuels, R. (1991) On-line calibration in HERP. *Hydrological Applications of Weather Radar.* Ed. Cluckie, I. D. and Collier, C. G. Ellis Horwood, Chichester, West Sussex, Chapter 5.

Schilling, W. (ed.) (1987) *Real-time control of urban drainage systems — the state-of-the-art, Report,* IAWPRC–IAHR Joint Committee on Urban Storm Drainage.

46

Modelling the time-dependent nature of the rainfall–runoff relationship using on-line identification

P. A. Troch, F. P. De Troch and J. Van Hyfte
Laboratory of Hydrology and Water Management, State University Ghent,
Coupure Links 653 B-9000, Ghent, Belgium

ABSTRACT

This paper focusses on the problem of real-time calibration of the time-variant rainfall–runoff relationship. Two different approaches to adaptive modelling of dynamic systems, i.e. firstly the introduction of a 'forgetting' factor and secondly the use of an additive Q matrix into the recursive algorithms for the linear parameter estimation problem, are compared. It is found that, from an operational standpoint, the use of an additive Q matrix leads to better forecasting performance. The choice of the additional design parameters however is not trivial and in this paper it is based on a simulation study using several storm events for subcatchments of the river Meuse, Belgium.

1. INTRODUCTION

The development and the operation of a real-time flood-forecasting model for river systems depends on an on-line data acquisition network gathering information about some hydrological parameters of the system. Recently, more advanced measuring techniques have become available as part of on-line hydrological information systems. Weather radar is one example. The choice between modelling techniques to be used depends a great deal on the available input and output data. The use of weather radar information in hydrological applications will give new directions to the research on hydrologic model building.

Many researchers in the field of real-time flood forecasting recognize the time-dependent nature of the rainfall–runoff relationship. However, for isolated storm events, the effects of this time-variant behaviour are very often neglected, leading to the definition of linear time-invariant systems or non-linear time-invariant systems. In the latter case, the non-linearity can be introduced using some form of rainfall separation (e.g. threshold parameters) or as an additional deterministic signal (e.g. catchment wetness index). More advanced modelling techniques for dealing with non-linearity are Volterra and Wiener series expansion (Napiorkowski 1986).

Research based on hourly precipitation and runoff data for several river catchments in Belgium has indicated that during storm events the influence of the time-dependent behaviour is significant. In this paper an on-line identification technique is discussed which is capable of adapting the parameters of the linear time-variant model during a storm event. This approach results in an adaptive rainfall–runoff model. The on-line identification procedure used is the recursive instrumental variable method. The performance of the estimator depends on some design parameters, which should be chosen very carefully by the user. In this paper, two different approaches to the problem of identifying the parametric variation are compared from an operational standpoint.

2. THE OPTIMAL FILTERING PROBLEM

The most general framework for linear forecasting is afforded by the non-stationary state-space discrete-time model:

$$x_{k+1} = A_k x_k + B_k u_k + D_k w_k \tag{1}$$

$$y_k = C_k x_k + v_k \tag{2}$$

where x_k represents the state vector (dimension n) of the system under study. The sequences $\{y_k\}$ and $\{u_k\}$ are observed output and input vectors with dimension p and m. Usually the input and output series are scalar ($m = p = 1$). A_k, B_k and C_k represent time-varying matrices of appropriate dimension. The disturbance w_k and the measurement noise v_k are zero-mean, statistically independent white-noise vectors with possible time-variable covariance matrices W_k and V_k. To prevent singularity, one assumes V_k to be positive definite for all k, so that no measurement is fully deterministic.

The problem solved by Kalman (1960) and Kalman and Bucy (1961), namely the optimal linear estimation of x_k, gives the separation principle used by system analysts. The separation principle states that the optimal linear control based on observed output values consists of a cascade of an optimal linear state observer and an optimal feedback of the estimated state $\hat{x}_k$, treated as if it was the real state. The separation principle only holds for linear systems. The state observer has the following form:

$$\hat{x}_{k+1} = A_k \hat{x}_k + B_k u_k + H_k(y_k - C_k \hat{x}_k) \tag{3}$$

$$H_k = A_k P_k C_k^T (V_k + C_k P_k C_k^T)^{-1} \tag{4}$$

$$P_{k+1} = A_k P_k A_k^T - A_k P_k C_k^T (V_k + C_k P_k C_k^T)^{-1} C_k P_k A_k^T + D_k W_k D_k^T. \tag{5}$$

The latter represents a Riccati equation which can be solved on line with the initial condition

$$P_{k0} = Q_{k0} - \mu_{k0}\mu_{k0}^{T} \tag{6}$$

where

$$\mu_{k0} = E[x_{k0}] \tag{7}$$

$$Q_{k0} = E[x_{k0}x_{k0}^{T}] \tag{8}$$

and can be considered as *a priori* information about the initial state. This algorithm is optimal in the set of linear filters.

In his 1960 paper, Kalman solved only the problem of state-variable estimation since he assumed that the model parameters are known, but he recognized the problem of parameter estimation. Methods for joint recursive estimation of the states and parameters have since become available. The extended Kalman filter is one relatively straightforward approach in which the state vector x_k is augmented to include an unknown parameter vector θ_k which includes all the unknown elements in the model matrices A_k, B_k and C_k. Estimation then proceeds using a Kalman-filter-like algorithm obtained by linearizing the now non-linear relationship about the current estimates at each recursive step. Young (1984) states that, although the method has been applied relatively successfully to the modelling of various real dynamic systems, it does not always work satisfactorily. It is known to have poor statistical efficiency, in the sense that its parameter estimates may have rather high error variance when compared with the theoretically possible minimum variance. Superior results in this sense can be obtained by considering the problem from a maximum-likelihood standpoint (Mehra and Tyler 1973, Aström and Kallström 1973).

The major difficulty with the estimation of parameters in the state-space model arises from its relative complexity as a description when viewed in parametric estimation terms since the problem is clearly non-linear (Young 1984). What we would prefer is a representation in which the unknown parameters are associated with measured variables in a linear model. This representation is far from easy to obtain in the time-variant case.

3. THE TRANSFER FUNCTION REPRESENTATION

To transform the linear state-space model into a representation in the observation space we shall assume for the moment that the system is time invariant:

$$x_{k+1} = Ax_k + Bu_k + Dw_k \tag{9}$$

$$y_k = Cx_k + v_k. \tag{10}$$

Using a z-transform and in the single-input single-output case, we obtain the following representation:

$$y_k = \frac{B(z^{-1})}{A(z^{-1})}u_k + \frac{D(z^{-1})}{A(z^{-1})}w_k + \upsilon_k \tag{11}$$

where $A(z^{-1})$, $B(z^{-1})$ and $D(z^{-1})$ are polynomials in the z^{-1} operator. The difference equation can be written as follows:

$$y_k + a_1 y_{k-1} + \ldots + a_n y_{k-n} = b_1 u_{k-1} + \ldots + b_n u_{k-n} + \mu_k \tag{12}$$

where

$$\mu_k = d_1 w_{k-1} + \ldots + d_n w_{k-n} + \upsilon_k + a_1 \upsilon_{k-1} + \ldots + a_n \upsilon_{k-n}. \tag{13}$$

This difference equation represents a simple linear relationship between the measured variables and the unknown system parameters. However, the noise term μ_k is not simple in statistical terms, being a linear function of the two white noise source terms, w_k and υ_k.

If we define the recursive residual or innovation process e_k as:

$$e_k = y_k - c^{\mathrm{T}} \hat{x}_k \tag{14}$$

then the state observer can be written as

$$\hat{x}_{k+1} = A\hat{x}_k + bu_k + h_k e_k \tag{15}$$

$$y_k = c^{\mathrm{T}} \hat{x}_k + e_k. \tag{16}$$

These two equations can be considered as an alternative state-space formulation of the system and are rather simpler than the original description in stochastic terms since there is only one noise term, namely the innovation process e_k. The innovation process e_k is known to possess white-noise properties, but it does have changing variance because of the variable nature of the P_k covariance matrix.

It is now quite straightforward to derive the observation space transfer function

$$y_k = \frac{B(z^{-1})}{A(z^{-1})}u_k + \frac{D(z^{-1})}{A(z^{-1})}e_k. \tag{17}$$

This last equation is well known as the autoregressive moving-average exogenous variables description. In searching for greater generality we ought to introduce two modifications, namely by choosing the noise transfer function independent of the system transfer function and by introducing an additional term b_0 in the definition of $B(z^{-1})$:

$$y_k = \frac{B(z^{-1})}{A(z^{-1})} u_k + \frac{D(z^{-1})}{C(z^{-1})} e_k . \tag{18}$$

The time series model obtained with these two modifications is usually referred to as the transfer function model (Box and Jenkins 1970).

4. THE INSTRUMENTAL VARIABLE APPROACH

Since it is clear from equation (18) that neither the system model nor the noise model can be written in a regression relationship, the recursive least-squares analysis is generally not applicable to the estimation of the model parameters. This problem can be overcome by the generation of an instrumental variable (IV) vector $\tilde{x}_k$ at each recursive step. The instrumental variable ζ_i is chosen to be independent of the noise inputs v_i and w_i. The statistical efficiency of the solution is highly dependent upon the degree of correlation between ζ_k and the deterministic output of the system (Durbin 1954). The major problem with the IV method is the generation of suitable instrumental variables. Young (1965) suggests the use of an auxiliary model of the process to generate ζ_i. The IV vector then takes the form

$$\tilde{x}_k^{\mathrm{T}} = [-\zeta_{k-1} \ldots - \zeta_{k-n} u_k \ldots u_{k-n}] \tag{19}$$

where ζ_{k-i} can be calculated using the following auxiliary model:

$$\tilde{A}(z^{-1})\zeta_k = \tilde{B}(z^{-1})u_k \tag{20}$$

with $\tilde{A}(z^{-1})$ and $\tilde{B}(z^{-1})$ polynomials with parameters chosen in some sensible manner. An on-line IV procedure can be obtained by utilizing the recursive solution to the IV equations and then updating the auxiliary model continuously on the basis of these recursive estimates:

$$\hat{\theta}_k = \hat{\theta}_{k-1} - \hat{k}_k(z_k^{\mathrm{T}}\hat{\theta}_{k-1} - y_k) \tag{21}$$

$$\hat{k}_k = \hat{P}_{k-1}\hat{x}_k(1 + z_k^{\mathrm{T}}\hat{P}_{k-1}\hat{x}_k)^{-1} \tag{22}$$

$$\hat{P}_k = \hat{P}_{k-1} - \hat{P}_{k-1}\hat{x}_k(1 + z_k^{\mathrm{T}}\hat{P}_{k-1}\hat{x}_k)^{-1}z_k^{\mathrm{T}}\hat{P}_{k-1} \tag{23}$$

with θ_k the estimated parameter vector and z_k defined as

$$z_k^{\mathrm{T}} = [-y_{k-1} \ldots - y_{k-n} u_k \ldots u_{k-n}] .$$

According to Young (1984) it is necessary to update the parameters of the auxiliary model rather carefully. Young suggests the use of a discrete low-pass filter to avoid rapid changes in the auxiliary model parameters. Another improvement which is possible is to utilize an exponential weighting into the past version of the IV algorithm:

$$\hat{\boldsymbol{\theta}}_k = \hat{\boldsymbol{\theta}}_{k-1} - \hat{k}_k(z_k^T\hat{\boldsymbol{\theta}}_{k-1} - y_k) \tag{24}$$

$$\hat{k}_k = \hat{P}_{k-1}\hat{x}_k(\delta_k + z_k^T\hat{P}_{k-1}\hat{x}_k)^{-1} \tag{25}$$

$$\hat{P}_k = \frac{1}{\delta_k}[\hat{P}_{k-1} - \hat{P}_{k-1}\hat{x}_k(\delta_k + z_k^T\hat{P}_{k-1}\hat{x}_k)^{-1}z_k^T\hat{P}_{k-1}] \tag{26}$$

with the scalar δ_k defined as

$$\delta_k = \lambda_0\delta_{k-1} + (1 - \lambda_o)\delta \tag{27}$$

and $\delta = 1.0$. λ_0 and δ_0 are the additional parameters to be chosen by the analyst.

A modified recursive approximate maximum-likelihood (AML) procedure, suggested by Panuska (1968) can be used to estimate the parameters in the autoregressive moving-average structure noise model, based on the estimation of the noise sequence.

5. PARAMETRIC VARIATIONS

Suppose that instead of assuming parametric invariance we assume that the parameters vary in a manner that can be described by a Gauss–Markov stochastic difference equation:

$$\boldsymbol{\theta}_{k+1} = \boldsymbol{\Phi}\boldsymbol{\theta}_k + \boldsymbol{\Gamma}\mathbf{q}_k \tag{28}$$

where $\boldsymbol{\Phi} = \boldsymbol{\Phi}(k, k-1) = \boldsymbol{\Phi}_k$ is an $n \times n$ transition matrix, and $\boldsymbol{\Gamma} = \boldsymbol{\Gamma}(k, k-1) = \boldsymbol{\Gamma}_k$ is an $n \times m$ input matrix, both of which may be time variable; $\mathbf{q}_k$ is an $m \times 1$ white-noise vector of serially-independent random variables with zero mean and covariance matrix $\boldsymbol{Q}_p$. The simplest example of this model is the random-walk model:

$$\boldsymbol{\theta}_{k+1} = \boldsymbol{\theta}_k + \mathbf{q}_k. \tag{29}$$

In the case of the random-walk model of the parametric variations the IV algorithm can be modified to become (Young 1986)

$$\hat{\boldsymbol{\theta}}_k = \hat{\boldsymbol{\theta}}_{k-1} - \hat{k}_k(z_k^T\hat{\boldsymbol{\theta}}_{k-1} - y_k) \tag{30}$$

$$\hat{k}_k = \hat{P}_{k/k-1}\hat{x}_k(1 + z_k^T\hat{P}_{k/k-1}\hat{x}_k)^{-1} \tag{31}$$

$$\hat{P}_{k/k-1} = \hat{P}_{k-1} + Q \tag{32}$$

$$\hat{P}_k = \hat{P}_{k/k-1} - \hat{P}_{k/k-1}\hat{x}_k(1 + z_k^T\hat{P}_{k/k-1}\hat{x}_k)^{-1}z_k^T\hat{P}_{k/k-1}. \tag{33}$$

Here Q is an $(2n+1) \times (2m+1)$ matrix which allows for possible parametric variations. Although the derivation of these algorithms is quite heuristic, Young (1984) finds that they make good sense and can be further justified in statistical terms.

6. SIMULATION RESULTS

The above-described methodology to analyse the time-variant characteristics of systems that can be represented by transfer function models is first tested using simulation. In this simulation study attempts are made to identify transfer function models based on synthetic data. In this way it is possible to interpret the results and to study the influence and the relative importance of the several design parameters needed in the identification procedure. In particular the relative behaviour of the procedure using a 'forgetting' factor δ_k and an additive Q matrix to follow parametric variation is determined.

When using fixed-value parametric models it can be observed that the estimation of the proper model structure is very important in order to be able to determine the actual parameter values. Since model structure characterization is not the issue of this paper, we refer the reader to the work by Spriet (1985) for an overview in structure characterization techniques. A comparative study of structure characterization procedures in hydrological applications has been given by Troch et al. (1987). In general, based on a detailed simulation study, it can be concluded that the IV AML procedure is able to identify fixed-parameter transfer function models with good convergence.

In order to study the capability of the IV AML procedure to identify the time-variant transfer function model, two kinds of parametric variation were included in the synthetic data, namely a sudden change in parameter value and continuously changing parameter values. Two adaptive techniques were compared for performance. The first technique is the introduction of an exponential forgetting factor into the algorithms (see equations (24)–(27) where $0 < \delta < 1$). This forgetting factor prevents the covariance matrix from becoming zero. The second technique is the use of an additive diagonal Q matrix in order to keep the covariance matrix artificially greater than zero. In both cases (suddenly varying parameter values and continuously varying parameter values) the technique of an additive Q matrix was more powerful in keeping track of the real parameter values. It was observed that the forgetting factor could not prevent the covariance matrix from becoming too small in order to permit parameter variation. The convergence rate when using the additive Q matrix depends on the choice of q_{ii} (diagonal elements of Q) and on the time constant of the low-pass filter when updating the parameters of the auxiliary model.

7. IDENTIFYING THE TIME-VARIANT BEHAVIOUR OF THE RAINFALL—RUNOFF PROCESS

The real-time flood forecasting model for the river Meuse (Belgium), developed by the Laboratory of Hydrology of the State University Ghent, consists of two major parts: the hydrologic module and the hydraulic module. The hydrologic module is based on a linear time-invariant representation of the system under study (equations (1)–(2)). For the most important subcatchments of the river Meuse up to Liège these linear time-invariant models are identified using time-series analysis (Troch *et al.* 1988). The forecasting performance has been presented by Troch *et al.* (1988) and is in most cases satisfactory from the standpoint of operational use.

It is, however, recognized by Troch *et al.* that the basic assumption made throughout the modelling procedure, namely that the rainfall–runoff process is linear and time invariant, is not always valid. A more conceptually justified assumption would be that the process is inherently time variant (see section 1). Therefore the presented non-stationary time-series analysis can be used to identify this time-varying behaviour.

The technique of an additive **Q** matrix was applied to four subcatchments of the river Meuse. The structure of the transfer function model was deduced from the estimated cross-correlation function between pre-whitened rainfall data from selected storms and the corresponding filtered runoff data (Box and Jenkins 1970). The structure of the noise model was obtained using an objective Bayesian information criterion (BIC) search:

$$\text{BIC} = N\log\left(\frac{1}{N}\right)\sum_{k=1}^{N}\hat{e}_k^2 + n_i\log N \tag{34}$$

where N is the number of data points, e_k is the one-step-ahead prediction error on time k and n_i is the number of parameters to be estimated.

The time-invariant transfer function model for the Vesdre subcatchment (drainage basin area, $677\ \text{km}^2$ at Chaudfontaine) has the following structure: first-order A polynomial, fifth-order B polynomial with dead time $d=4$, second-order C polynomial and zero-order D polynomial. The parameters were estimated using the recursive–iterative IV AML procedure described by Young (1984) based on a calibration set of flood events. Fig. 1 shows the four-step-ahead forecasts for an event (2–17 February 1979) not included in the calibration set. From the figure we can observe that the first and second peak are forecasted with a significant time delay and that the third peak is underestimated. The maximum observed discharge during this event is $56\ \text{m}^3/\text{s}$ and the maximum four-step-ahead forecasting error is about 20%.

The same event is forecasted using a time-variant transfer function model with **Q** matrix equal to $0.01\ \boldsymbol{I}$ (where $\boldsymbol{I}$ is the unity matrix) (Fig. 2). From Fig. 2 it is obvious that the time-variant transfer function model is performing better than its time-invariant counterpart. The maximum four-step-ahead forecasting error is 16%. The model parameters are adapted in order to remove constant overestimation or underestimation of the hydrograph.

Fig. 3 shows the variation in the parameter during the flood event. During dry periods (no rain) the parameter values are not changed. This is expected since the input data contain no further information about the dynamic relationship between rainfall and runoff. The variations in the parameter values are situated in periods of heavy to moderate rainfall. The dynamic structure of the model is adjusted in order to keep track of the changing characteristics of the catchment.

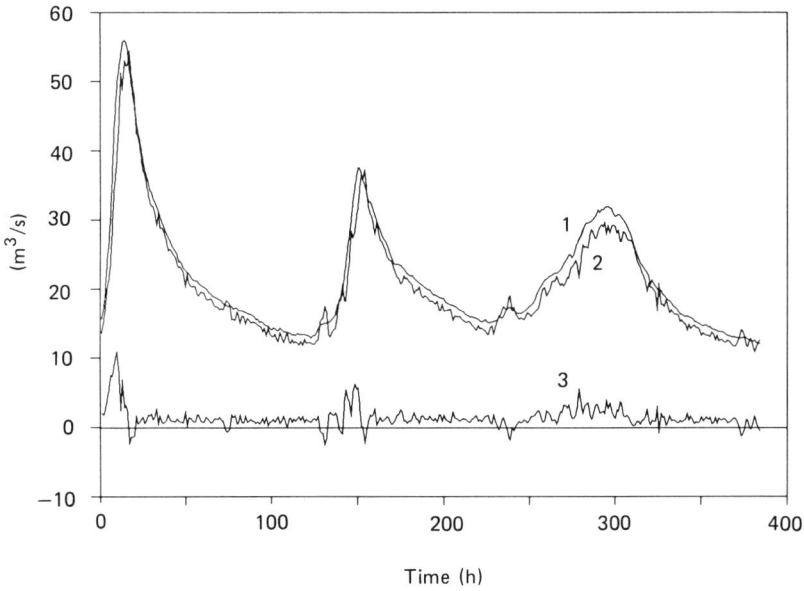

Fig. 1 — Four-step-ahead prediction for the river Vesdre using the time-invariant model: curve 1, observed; curve 2, prediction; curve 3, prediction error.

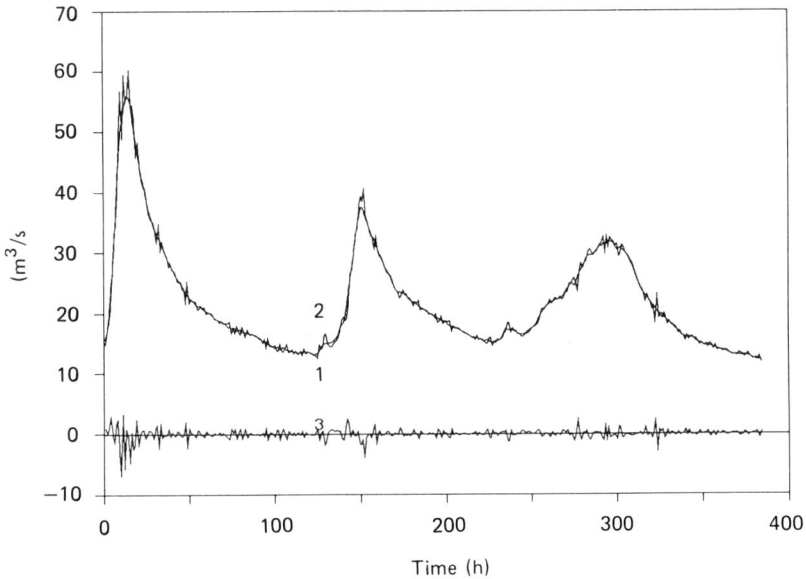

Fig. 2 — Four-step-ahead prediction for the river Vesdre using the time-variant model: curve 1, observed; curve 2, prediction; curve 3, prediction error.

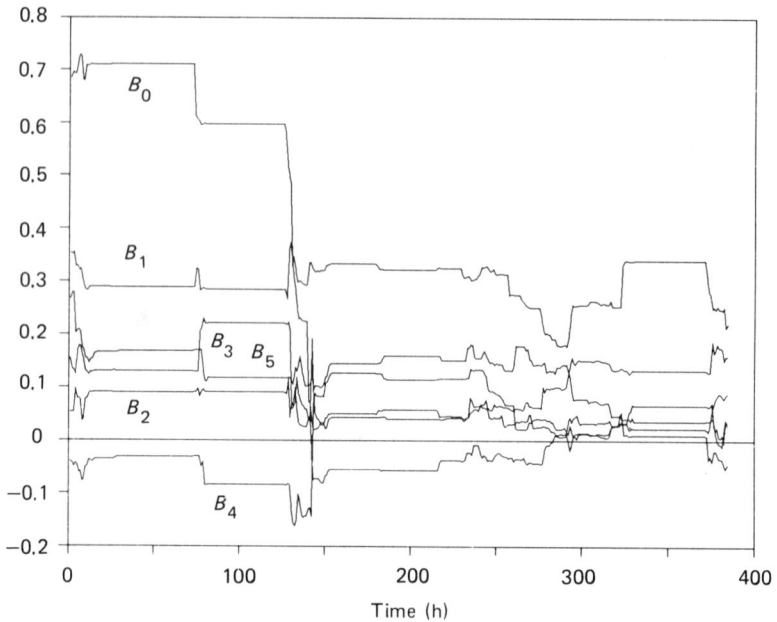

Fig. 3 — Variation in parameters B ($Q=0.01\,I$).

The choice of the Q matrix is rather critical to the performance of the adaptive model. The choice of the values for the diagonal elements of the Q matrix in this case (Figs 2 and 3) was based on a simulation study for nine flood events for the river Vesdre. It was clear that better performance could be obtained using different sets of Q matrices for different flood events for the same subcatchment. This makes the adaptive technique demonstrated in this section rather difficult to use in practice. For other subcatchments of the river Meuse it was observed that the time to peak for flood events is better predicted using an adaptive modelling technique (in comparison with a time-invariant model) but this is usually associated with an overestimation of the peak. One can argue that an overestimation is less severe than an underestimation from the viewpoint of flood forecasting.

In general, the performance of the time-variant model should be evaluated using an objective criterion that takes into account not only the goodness of fit but also important characteristics of forecasts such as the time to peak, the ability to forecast the rising limb of the hydrograph and the overestimation or underestimation of the peak. Simple statistics such as the coefficient of determination and error variance norm (Young 1986) are not capable of including these characteristics.

7.   CONCLUSION

In this paper, two different approaches to adaptive modelling are applied to the identification of the rainfall–runoff relationship for several subcatchments of the river Meuse, Belgium. These time-variant models are part of the real-time flood-forecasting model for the river Meuse, developed by the Laboratory of Hydrology of the State University Ghent. It was found that the additive **Q** matrix approach in general leads to better forecasting performance for isolated storm events. The present authors recognize the need for an objective criterion to evaluate the forecasting power of the obtained time-variant models in terms of hydrologically important characteristics, such as the time to peak and the rising limb of the hydrograph.

REFERENCES

Aström, K. J. and Kallström, C. G. (1973) Application of system identification techniques to the determination of ship dynamics. In: P. Eykhoff (ed.), *Proceedings of the 3rd International Federation of Automatic Control Symposium on Identification and System Parameter Estimation, The Hague.* pp. 415–424.

Box, G. and Jenkins, G. (1970) *Time series analysis, forecasting and control.* Holden–Day, San Francisco, CA.

Durbin, J. (1954) Errors in variables. *Rev. Int. Stat. Inst.*, **22**, 23–32.

Kalman, R. E. (1960) A new approach to linear filtering and prediction problems, *J. Basic. Eng.*, **82**, 35.

Kalman, R. E. and Bucy, R. S. (1961) New results in linear filtering and prediction theory. *J. Basic. Eng.*, **83**, 95.

Mehra, R. K. and Tyler, J. S. (1973) Case studies in aircraft parameter identification. In: P. Eykhoff (ed.), *Proceedings of the 3rd International Federation of Automatic Control Symposium on Identification and System Parameter Estimation, The Hague.* p. 117.

Napiorkowski, J. J. (1986) Application of Volterra series to modelling of rainfall–runoff systems and flow in open channels. *Hydrol. Sci. J.*, **31** (2), 187–203.

Panuska, V. (1968) A stochastic approximation method for the identification of linear systems using adaptive filtering. *Proceedings of the Joint Automatic Control Conference, Ann Arbor, MI.*

Rao, A. R. and Mao, L. T. (1987) An investigation of the instrumental variable approximate maximum likelihood method of modelling and forecasting daily flows. *Water Resour. Manage.*, **1**, 79–106.

Spriet, J. A. (1985) Structure characterisation: an overview. In: H. Barker and P. Young (eds), *Proceedings of the Conference on Identification and System Parameter Estimation.* pp. 749–756.

Troch, P. A., Spriet, J. A. and De Troch, F. P. (1987) Comparison of characterisation methods for river catchment modelling. *Proceedings of the UKSC Conference on Computer Simulation.* pp. 202–208.

Young, P. C. (1965) Process parameter estimation and self-adaptive control. In: P. H. Hammond (ed.), *Proceedings of the International Federation of Automatic Control Symposium on the Theory of Self-Adaptive Control Systems, Teddington, Middx.* Plenum, New York.

Young, P. C. (1984) *Recursive estimation and time-series analysis: an introduction.* Springer, Berlin.

Young, P. C. (1986) Time series methods and recursive estimation in hydrological systems analysis. In: D. A. Kraijenhof and J. R. Moll (eds), *River flow modelling and forecasting*.

47

Hydrological relevance of radar rainfall data

H. R. Verworn
Institut fur Wasserwirtschaft, Universität Hannover, W-3000 Hannover 1, FRG

ABSTRACT

This paper describes the growing use of weather radar for hydrological modelling. It indicates that there is not necessarily a need for high accuracy and high resolution radar rainfall data. The paper investigates the use of an X-band weather radar for real-time control of a small catchment in Germany utilizing the HYSTEM/ EXTRAN hydrodynamic model.

1. INTRODUCTION

For nearly all hydrological problems the areal rainfall is needed. In most cases, however, areal rainfall is not measured but has to be deduced from point rainfall data given by the traditional raingauge. For the calculation of areal rainfall, procedures such as the Thiessen polygon or isohyetes are applied. Whatever the method used, this can only give an estimation of the areal rainfall as the true values are not known.

With the increasing application of radar technology for meteorological and hydrological purposes an alternative for achieving areal rainfall data has arisen. With this technology the reflectivity from all raindrops within defined spaces is measured, providing integrated areal data. Unfortunately, the $Z–R$ equation to transform radar reflectivities into rainfall intensities is drop spectrum dependent and, as long as no reliable information about the actual $Z–R$ relationship is available, the quantitative accuracy of the radar rainfall data is often insufficient.

Radar rainfall measurement is particularly attractive when rainfall data with high resolution in time and space is needed online, e.g. for real-time control or flood warning, as all data for a large area are present at one location. With traditional raingauges a dense network plus telemetry is necessary to serve the same purposes.

As a relatively new technology, radar rainfall measurement is viewed sceptically especially as the calibration problems do not seem to be solved satisfactorily. Although of high resolution the benefit of quantitative rainfall data is questioned when error margins of 50% and more have to be tolerated. These error margins, however, can be drastically reduced when the range for quantitative measurement is limited to about 50 km. Within this range the beam stays well below the cloud base and bright-band effects are avoided, at least in summer when severe storms occur. With system errors of about 5–25% (dependent on range and number of samples; see Kammer (1991)) and mean errors resulting from the calculation of R from Z of about 5–10% (dependent on the procedure; see Kreuels (1991))) rainfall radar data are in the same accuracy range as raingauge data. As the radar data accuracy is dependent on the density of the network and the spatial variability of storms it may well be higher than the areal rainfall data derived from point data.

In this paper no investigation of or comparison between the errors of both measuring systems will be made. The focus is rather on how rainfall data errors influence the results of runoff data calculated from rainfall by using mathematical models.

2.   REAL-TIME CONTROL AND RAINFALL DATA

Rainfall–runoff simulations are necessary when applying real-time control, either for the calculation of the actual status of the system in order to supplement measured data, or for forecasting the probable runoff due to the actual rainfall information, or even for both. Control decisions will be based on these data.

To avoid wrong decisions based on wrong data, estimates about the reliability of calculated runoff data are needed. Filtering techniques can only be applied if the error margin of the data is known.

For real-time control, rainfall data have to be available

(a) online,
(b) direct,
(c) of high accuracy and
(d) of high resolution in time and space.

While the demands of 'online' (at a central location) and 'direct' (without time delay) are logical, the relevance of 'high accuracy' and 'high resolution' especially in space is under controversial discussion. If the demand for high accuracy is agreed on, there is still the question on how it can be achieved best.

The arguments against higher accuracy and resolution are as follows.

(1) *Levelling effects of the catchment* Because of the flow time on the surface as well as in the channels and pipes errors made when calculating the runoff from subcatchments are levelled out or at least smoothed throughout the system.
(2) *Measurement errors* No device measures the true rainfall. Raingauges have wetness, wind and evaporation losses; radar data have calibration errors, bright-band effects and system errors.

(3) *Inaccuracy of rainfall–runoff models* Surface runoff depends on catchment conditions, which often have to be estimated.

The arguments for higher accuracy and resolution are the following.

(1) Rainfall errors are not levelled out but propagated.
(2) Rainfall is the most sensitive parameter in modelling; so most efforts here yield the most improvement in accuracy.

3. ERRORS IN STORMWATER MODELLING

Inaccuracies in rainfall–runoff modelling can be caused by a number of factors such as errors in input data, model simplifications, parameter uncertainties and numerical problems. To identify the component which causes the most inaccuracy a quantitative assessment was done by Schilling and Fuchs (1986). They carried out a study that focussed on the modelling errors of peaks and volumes. The main conclusion of this study is that the spatial resolution of rain data input is of paramount importance to the accuracy of the calculated hydrograph. Resolution is the most limiting factor of modelling accuracy because of the high spatial variability of storms and the amplification of rainfall sampling errors by the non-linear transformation into runoff.

4. INVESTIGATION WITH A REAL CATCHMENT

Within a research project on the utilization of X-band radar for short-range rainfall measurements and the application of real-time control in urban drainage systems, three catchments are monitored. For these catchments the measured data on runoff and rainfall (from radar and ground gauges) are collected.

One of these catchments is the river Boye with a total area of 76 km^2 (Fig. 1). All runoff from this catchment has to be pumped to the river Emscher owing to subsidence caused by coal mining.

5. RAINFALL–RUNOFF MODEL

The hydrodynamic model HYSTEM/EXTRAN was used for the runoff simulations. The catchment was divided into 51 subcatchments. The surface runoff model (HYSTEM) calculates the runoff from paved and rural areas separately. For paved areas, initial and time-varying losses are considered. For rural areas an infiltration model is used. Overland flow is modelled by a convolution of net rainfall with synthetic unit hydrographs applying variable lag time computed from the subcatchment characteristics (Harms 1981).

To supply rain data to the model with a high spatial distribution, the HYSRAD version of HYSTEM was used. HYSRAD provides the possibility of using different rain data for each subcatchment.

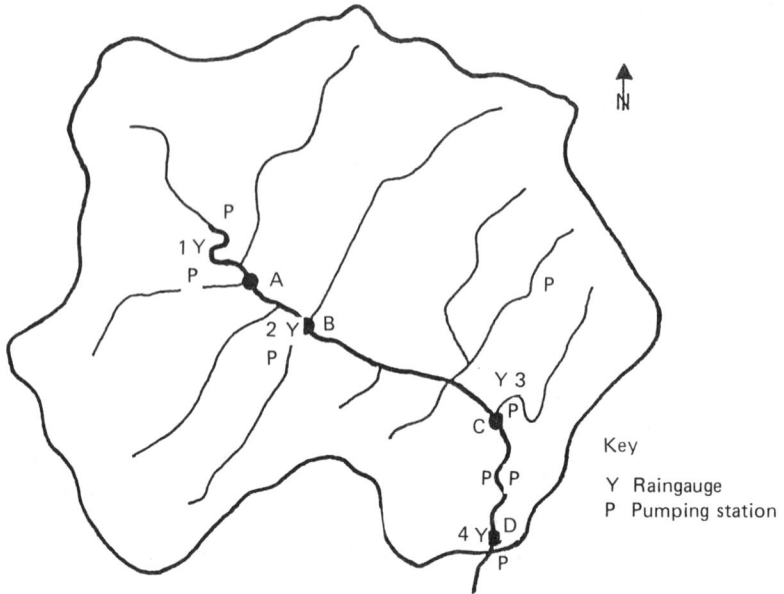

Fig. 1 — Boye catchment.

Conduit runoff and routing are calculated by solving the complete St Venant equations. Pumping stations and additional storage facilities are integrated in the hydrodynamic calculations.

6. RAIN DATA DISCRETIZATION

All rain data were provided for a grid over the catchment area with squares of 600 m by 600 m.

To investigate the influence of spatial rain variability on the resulting runoff, three different levels of discretization were used. The highest level of discretization was obtained by using different rain data for each of the 51 subcatchment areas. For each catchment the numbers of the grid squares within this catchment area were given, and the catchment rainfall was calculated as the mean of these square rainfall values. The number of squares for each catchment ranged from 1 to 22. These data and the subsequent results were considered to be of the highest accuracy and used as reference data. This level is referred to as CR51 (catchment rainfall for 51 subcatchments).

For the next level of discretization, only the rainfall values from four different grid squares were used. Within these squares, ground raingauges are situated but their recorded data were not used in this study. This level represents a situation where rainfall information may be obtained by ground-level gauges. Consequently, each subcatchment was allocated to one of these four rainfall data sets. The boundaries for the allocation were defined by applying the Thiessen method. The location and boundaries are shown in Fig. 1. This level is referred to as AR4 (areal rainfall for four areas).

At the third level of discretization, only one rainfall input was applied over the whole catchment. These rainfall data were obtained by calculating the arithmetic mean values of the four data points from level two. These rain data therefore represent the usage of mean areal rainfall (MAR) derived from ground-level gauges.

7. ARTIFICIAL RAINFALL DATA

Three different sets of rainfall data were used (Table 1). With all of them it was assumed that the rain patterns move with constant velocity and without change in

Table 1 — Results of simulations with different rain types and levels of discretization

	Level of discretization	Rainfall type 1	Rainfall type 2	Rainfall type 3
Areal rainfall (mm)	CR51	10.27	5.50	6.03
	AR4	10.27	9.13	1.44
	MAR	10.27	8.91	1.55
Runoff volume (m³)	CR51	196 274	70 053	152 660
	AR4	199 252	159 052	7 900
	MAR	166 262	139 673	8 326
Impervious runoff (m³)	CR51	166 067	64 707	122 427
	AR4	166 059	141 078	7 900
	MAR	166 262	139 673	8 326
Rural runoff (m³)	CR51	30 202	5 346	30 232
	AR4	33 193	17 974	0
	MAR	0	0	0
Peak flow (m³/s) at location A	CR51	2.08	1.30	0.18
	AR4	2.22	2.22	0.14
	MAR	1.41	1.15	0.16
Peak flow (m³/s) at location B	CR51	22.11	7.10	25.70
	AR4	21.76	18.03	1.88
	MAR	19.64	16.36	1.91
Peak flow (m³/s) at location C	CR51	27.37	11.04	25.63
	AR4	27.65	22.88	2.51
	MAR	23.28	19.78	2.48
Peak flow (m³/s) at location D	CR51	30.81	13.08	24.84
	AR4	30.75	25.14	2.88
	MAR	25.30	21.46	2.85

intensity and areal distribution from northwest to southeast over the catchment. The types are shown in Fig. 2:

type 1 a frontal rain of constant intensity rectangular to its direction of movement;
type 2 one centre with the highest intensity moving over the catchment;
type 3 two centres with the highest intensities at the edges of the catchment.

With type 1 the differences between the various levels of discretization are small. The total rainfall is the same with all levels as the front produces the same rainfall

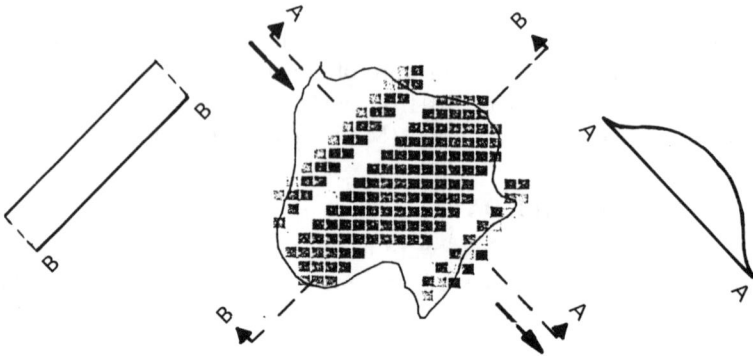

Artificial rainfall type 1

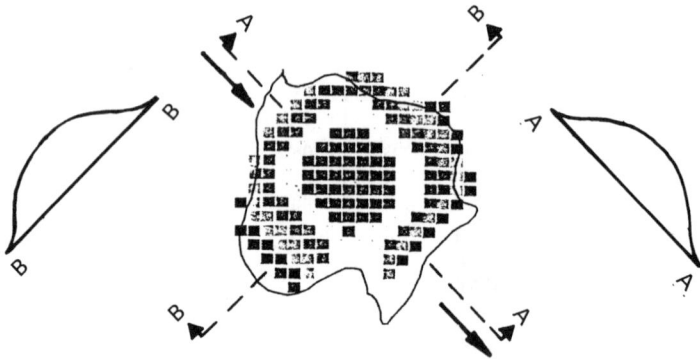

Artificial rainfall type 2

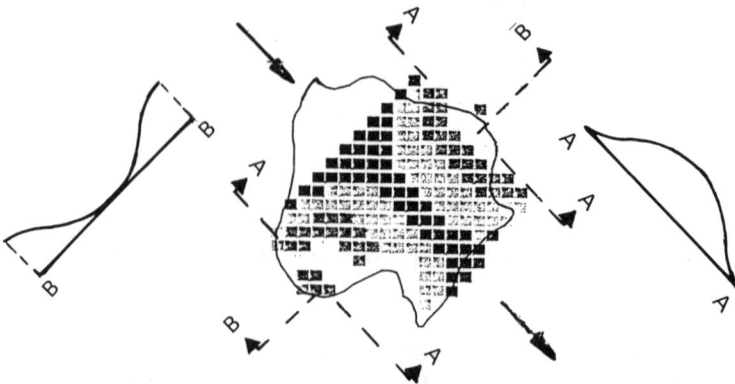

Artificial rainfall type 3

Fig. 2 — Artificial rainfall patterns and distribution.

everywhere in the catchment. The differences between CR51 and AR4 for all results are negligible. This is because the four stations for AR4 lie in the direction of the movement of the front and account for the propagation. MAR, however, yields volumes and peak flows that are too small, as the high intensities are not accounted for by the levelling effect of the areal rainfall. Consequently, there is no runoff from pervious areas, and the peak flows are not only smaller but occur 8–14 min too late as well.

With types 2 and 3 the computed areal rainfall differs with the level of discretization. Depending on the location of the rain values used for computation, the resulting areal rainfall is either too high (type 2) or too small (type 3) when referred to the values for CR51. In all cases the errors in runoff volume and peak flow are larger than with rainfall.

With type 2 the rainfall error is 66% and 62% (too high). This leads to runoff errors of 127% and 99% with AR4 and MAR respectively. The peak runoff errors range from −12% to +154%. This wide range can be explained by the complex interaction of the different subcatchments and the routing process within the catchment.

With type 3 all errors are negative, meaning that too low a runoff and peak flow are computed with AR4 and MAR compared with CR51. Again, the same propagation of errors from rainfall to runoff and peak flow occurs.

8.   RADAR-MEASURED RAINFALL

A comparative simulation with the three levels of discretization plus an additional level where the rainfall at one location was used throughout the catchment (areal rainfall derived from one location (AR1)) was carried out with a radar-measured rainfall event. The results of these simulations are shown in Table 2.

Table 2 — Results of simulations with radar-measured rainfall and various levels of discretization

	CR51	AR4	MAR	AR1
Areal rainfall (mm)	12.43	11.31	10.29	18.23
Total runoff (m^3)	350 770	319 983	175 558	554 118
Impervious runoff (m^3)	256 359	245 369	166 573	325 197
Rural runoff (m^3)	94 412	74 614	8 985	228 921
Peak flow (m^3/s) at location A	0.62	0.43	1.52	8.13
Peak flow (m^3/s) at location B	43.71	44.88	20.27	47.28
Peak flow (m^3/s) at location C	53.29	48.04	23.94	56.82
Peak flow (m^3/s) at location D	53.70	48.30	26.00	57.22

With a reference areal rainfall of 12.43 mm for CR51 it is interesting to note the high spatial variability of the rainfall ranging from 3.5 to 26.2 mm for the 51 subcatchments. The values for the four raingauge locations from which the AR4 and MAR values are derived were 4.6, 18.23, 9.0 and 9.4 mm from northwest to southeast.

The rainfall of this event moved more or less from northwest to southeast with changing intensities. With the discretization of AR4 the errors are small and nearly the same size for rainfall runoff and peak flow (about 10% less than with CR51).

With MAR the errors are propagated as with the artificial rain data from -17% for total rainfall to -50% for total runoff volume and from -52% to $+145\%$ for peak flows.

Using the rainfall information from only one location clearly leads to the worst errors of $+47\%$ for rainfall and $+58\%$ for runoff volume. The peak flow errors are highly non-uniform along the river. For location A the peak flow is 13 times higher than for CR51. At the other locations the differences in the results of CR51 are small as the capacity of the channels is reached and an increase in peak flow is reduced by flooding and activation of retentional storage.

The times of peak flow do not differ much with the various levels of discretization, and the shape of the hydrographs is nearly the same. The peak flow times of the different hydrographs at location C differ not more than 10 min.

9.　CONCLUSIONS

Rainfall data errors can be divided into system errors and errors due to insufficient spatial resolution. In this study, only the latter have been investigated.

Errors in total rainfall depths due to poor spatial resolution are generally higher than 20%, and errors of 50% and more are not uncommon (Schilling and Fuchs 1986) (Tables 1 and 2). The larger the area for which point rainfall data are extrapolated the larger will be the errors on average. The actual size of the errors depends on the spatial resolution as well as on the variability of the storm. There is no general tendency as to whether the errors are positive or negative, i.e. whether the tendency with decreasing resolution rainfall is underestimated or overestimated.

Rainfall errors are propagated and amplified by the rainfall runoff transformation in nearly all cases. This means that there is a high probability that with wrong rainfall data the resulting runoff volumes and peak flows are even more incorrect.

The errors in runoff volume are not amplified much as long as there is no runoff from pervious areas. This runoff depends not only on the total rainfall depths but also on the intensities during the storm, the hyetograph. If the applied rainfall data have too low or too high intensities, either the infiltration rate is not exceeded and no runoff occurs, or the calculated runoff will be much too high. With impervious areas the linearity between rainfall and runoff is stronger as the losses are smaller (mostly initial) and not intensity dependent.

Given errors in rainfall depths, the errors in peak flow will be generally larger than the errors in runoff volume, often resulting in more than 100%. In some cases, however, the errors in peak flow may be levelled by the routing process, e.g. when the capacity of the channels or pipes is reached. Then flooding or surcharge will occur and restrict the propagation and amplification of rainfall errors.

The size of the errors in runoff volume and peak flow tends to get higher with smaller catchment sizes. For larger catchments the errors may be levelled because of the superposition of the hydrographs from the subcatchments. In particular, when the spatial variability of the storm is high, negative as well as positive errors in the various subcatchments may occur, resulting in smaller errors for the whole catchment.

For the hydrological relevance of radar rainfall data the subsequent use of the data has to be considered as well as whether the data can be valued as quantitative or rather qualitative.

Quantitative data of sufficient accuracy can only be obtained within a range of about 50 km. The components of the system have to be designed according to this task of measurement. Reliable Z–R relationships and adequate calibration have to be used. If these conditions are met, the system errors will be of the same size as for ground gauges, but as the spatial resolution is much better the rainfall errors will be much smaller than those resulting from extrapolating point data. As shown in this study the benefit concerning the runoff data is immense. This does not mean that qualitative data, i.e. radar rainfall data with higher error margins, are of no hydrological relevance. They surely have their use when information about rainfall patterns and movement of storms is needed for larger areas. Early information about the probable development of rainfall over a certain area can be of greatest importance for flood warnings or alleviation schemes.

If real-time control is in operation, rainfall data of high accuracy are vital. The spatial resolution has to be about the size of the smallest controlled subcatchment. In urban areas with control unit sizes of down to 30 ha this would lead to a necessary spatial resolution of about 500 m × 500 m squares. The time resolution has to be at least of the same interval with which the control decisions are updated. For the above-mentioned catchment sizes in urban areas, time intervals of 5 min are even too long.

As all data have to be present on line with this time interval, ground raingauges have to be connected permanently to the control centre as the frequency of data transmission is too high to use dialling connections via telphone lines. Radar data have the advantage not only of being available at one location, thus reducing transmission problems, but also of high spatial resolution. To obtain comparable rainfall data with a ground gauge network there has to be at least one raingauge per 2 km^2. For a drainage area of 25 km^2 about 13 raingauges would have to be installed. A short-range radar designed especially for measurement purposes would cost about the same but could with a radius of 35 km cover an area of about 3800 km^2 with a spatial resolution of 600 m × 600 m and a time resolution of 1 min. Radar data collection in this case is not only preferable in terms of hydrology but is also cost efficient.

Radar rainfall data give the required resolution in time and space and reduce the errors in areal rainfall data. As the rainfall is the most sensitive variable in rainfall–runoff modelling, the computational efficiency is increased enormously if spatially distributed rainfall data are used. The errors due to other uncertainties (losses, wetness and surface flow) or simplifications (conduit flow) are of lower order.

REFERENCES

Harms, R. W. (1981) Application of standard unit hydrograph in storm sewer design. *Proceedings of the 2nd International Conference on Urban Storm Drainage, Urbana, IL, June 1981.*

Kammer, A. (1991) An integrated X-Band radar system for short-range measurement of rain rates in HERP. *Hydrological Applications of Weather Radar.* Ed. Cluckie, I. D. and Collier, C. G. Ellis Horwood, Chichester, West Sussex. Chapter 24.

Kreuels, R. (1991) On-line Calibration in HERP. *Hydrological Applications of Weather Radar.* Ed. Cluckie, I. D. and Collier, C. G. Ellis Horwood, Chichester, West Sussex. Chapter 5.

Schilling, W. and Fuchs, L. (1986) Errors in stormwater modeling — a quantitative assessment. *J. Hydraul. Div., Am. Soc. Civ. Eng.,* **112** (2).

Part 7

Operational and international experience

48

Wessex flood-forecasting system

C. J. Birks[1], A. P. Bootman[2], I. D. Cluckie[3] and H. Dawei[3]
[1] Wessex Rivers Authority, Bridgwater House, Bridgwater, Somerset TA6 3EA, UK.
[2] Wessex Water Authority, Quay House, Bath BA1 2YP, UK.
[3] Water Resources Research Group, Department of Civil Engineering, University of Salford, Salford M5 4WT, UK.

ABSTRACT

This paper describes the historical evolution and current state of the Wessex Flood Forecasting System. The Wessex region is characterized by infrequent heavy rainfalls and has produced many of the record British rainfall totals in addition to some of the more infamous floods. The continual development of modern telemetry systems and the concurrent application of flood-forecasting procedures have led to the current system being evolved, which will be based upon the real-time quantitative use of the composite radar network data for modelling purposes.

1. INTRODUCTION

The Wessex region consists of three basins, each with its hydrological regime varying from relatively rapid response catchments in Exmoor to the more sedate Bristol Avon. The increased availability of computers in the mid-1970s led to the development of the first predictive models based upon unit-hydrograph convolution and flood routing. These models are still in use and rely on the precipitation input derived from representative telemetered raingauges in the catchments.

The region is characterized by infrequent heavy rainfalls. Indeed, the top falls appearing in the UK record books have occurred in the Wessex region: Martinstown, 275 mm in 1955; Lynmouth, 225 mm in 1952; Cannington, 225 mm in 1924; Bruton, 225 mm in 1917. In all cases the spatial distribution of intense rainfall was small and, were it not for the location of gauges within these storm cells, the true precipitation would never have been known. The commissioning of the weather radar at Upavon in 1979 provided the opportunity to record these intense-rainfall cells. The storm in May 1979, again centred near Bruton, was an early example of the value of weather radar with precipitation measurements from raingauges underestimating the true fall.

The gradual introduction of weather radars and the availability of the network picture prompted a review in recent years of the existing models. Associated with the availability of increased computing power, the possibility of developing a regional model utilizing weather radar as the primary precipitation input was explored. In 1984, the Wessex Water Authority was embarking on an information technology (IT) strategy based on integrated river basin management. Wessex already had a comprehensive telemetry system for the utility functions of supply and sewage treatment, digital maps of the assets and rivers, computer-based flood models and the network radar product, all accessed within one 24 h regional control room at Bristol.

This strategy was to produce an integrated water management system (WMS), allowing access to all functions through one computer screen using windows and including the facility to extract data from one source (window) and combine with data from another (window). Specifically, weather radar quantitative information was to be passed into real-time flood models and the forecast products disseminated through internal communication networks. Development of the WMS pilot concentrated on the man-machine interface and was in association with Software Sciences Ltd.

Privatization of the industry and the creation of a National Rivers Authority required the dismantling of river basin management integration and, as a consequence, the IT strategy based on WMS had no future. However, the flood-forecasting aspects — weather radar and flood models, and the development of the MMI — are proceeding in cooperation with the Water Resources Research Group at the University of Salford.

2. THE WESSEX REGION

As previously described, the Wessex region consists of three catchment based divisions each with its own particular hydrological regime (Fig. 1). The region is characterized by infrequent heavy rainfalls and an analysis of the extreme-rainfall records for the British Isles produces a table of falls that concentrate on and around the Wessex region. This phenomenon was publicized by Bootman and Willis (1980) which prompted further research into its cause. Such was the concern that the Institution of Civil Engineers *Manual for Reservoir Design* advises that local rainfall analysis should be applied rather than the regional analysis recommended elsewhere in the UK.

The hydrometeorology results from the interaction of warm moist, northerly moving thunder cells with the jet stream meeting over the southwest of England, leading to both rapid rain cell development and negligible movement. The impact on flood forecasting has been to ensure that any models adopted can recognize and cope with intense localized rainfalls.

3. BRISTOL AVON

The Bristol Avon catchment (Fig. 2) is bordered by the Cotswolds to the north, Salisbury Plain to the east and the Mendip Hills to the south, The River Avon itself

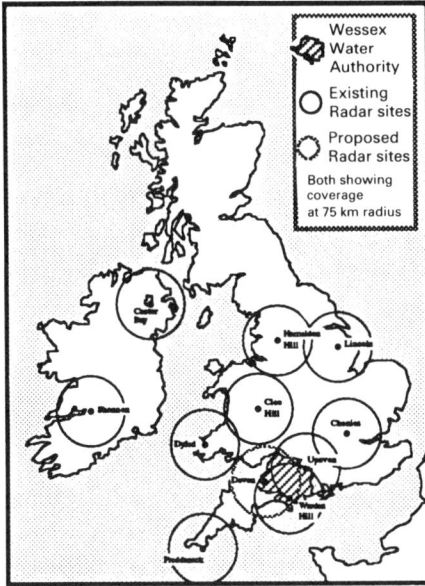

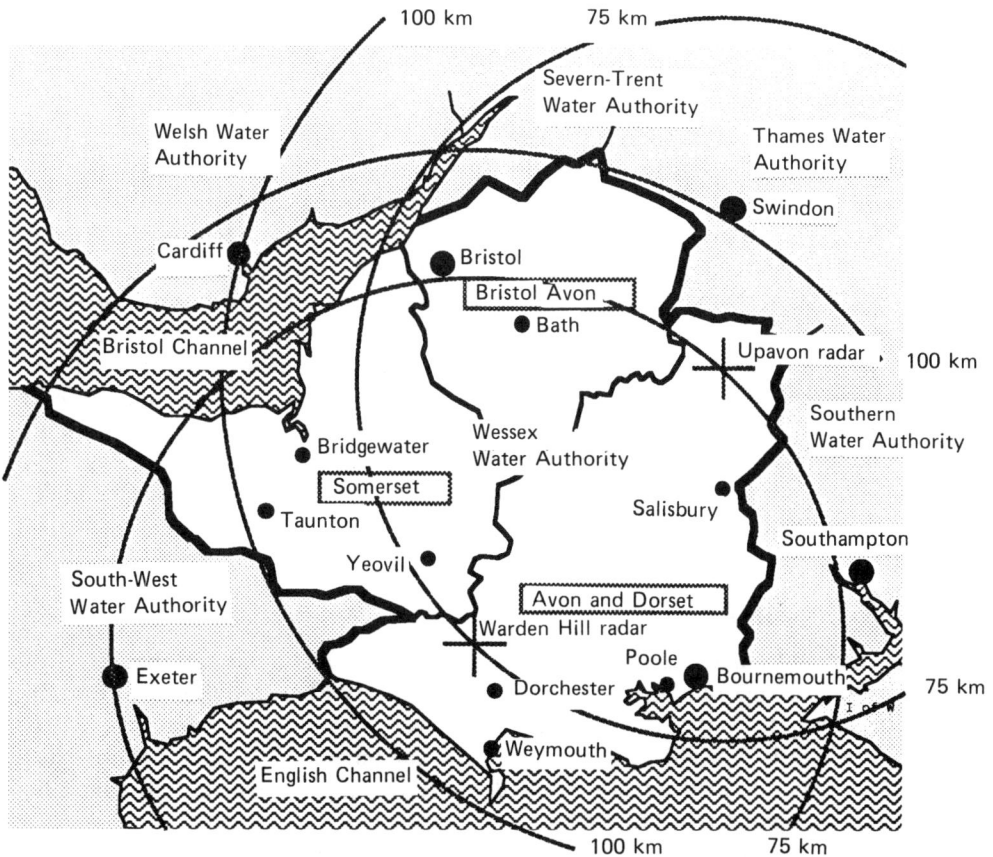

Fig. 1 — Wessex divisional boundaries.

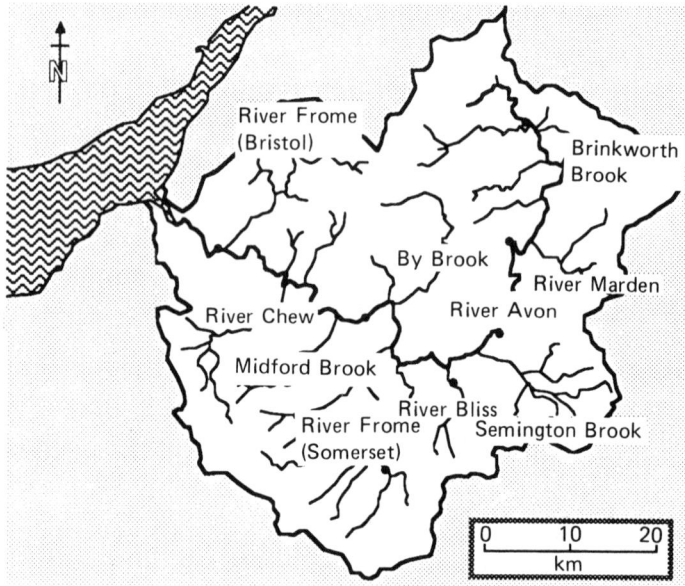

Fig. 2 — Bristol Avon rivers.

rises in the Cotswolds and drains southwards, picking up tributaries from Salisbury Plain and then turning westwards, collecting tributaries from the Mendips and draining to the sea through the Avon Gorge west of Bristol.

The current flood-forecasting model was developed for the Bristol Avon catchment by Grimshaw and Wong (1975) and is shown in a schematic form in Fig. 3. The drainage system is modelled as a network of nodes connecting various subcatchment elements and open channel reaches. Discharging into the head of each reach is the flow either from a subcatchment or from another reach. Each reach has in addition a subcatchment element discharging into its midpoint, representing the lateral inflow. Each subcatchment is modelled using a non-linear reservoir lag-and-route rainfall–run-off model, and each reach a Muskingham river-routing model. Data collection is by a DEC VS2000 scanning 20 raingauges and 19 river-level outstations. The models have recently been rewritten to operate on the same Vax VS2000 workstation.

4. AVON AND DORSET

The Avon and Dorset catchment (Fig. 4) has two distinct areas. To the east is the Hampshire Avon draining southwards from Salisbury Plain and the Stour draining southeastwards. To the west are the Dorset rivers, draining south from the Dorset Downs which is typically fast-response chalk catchments. It is one of these catchments in which Martinstown is situated where the highest British daily rainfall of 275 mm was recorded in July 1955.

The models adopted for the Avon and Dorset rivers have been the subject of several research exercises. Initially developed as unit-hydrograph models the current

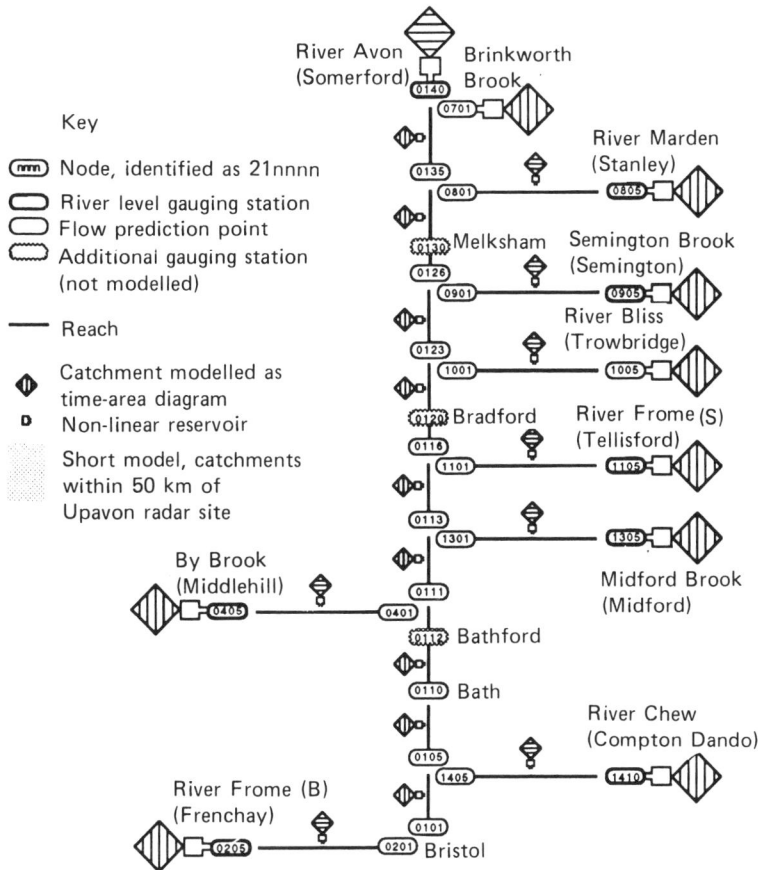

Fig. 3 — Schematic diagram of the Avon catchment model network.

versions were developed as transfer function models with elements of correlation for forecasting downstream levels (Cluckie and South, 1980, Cluckie *et al.*, 1989 and Tilford, 1989). Bliss (1981) refined the models for real-time updating, but manual updating is the current practice.

Data collection is by a Texas dedicated computer scanning 18 raingauges and 33 river level outstations. The data collection system was part of the water industry's Experimental and Demonstration Facility. The models have operated on a portable Hewlett–Packard programmable calculator during the period of system development.

5. SOMERSET

The Somerset catchment (Fig. 5) is also divided into distinct regions. To the west are the catchments draining Exmoor and the Brendon Hills, very-fast-responding catchments adjacent to the River Lyn scene of the disastrous Lynmouth floods of August 1952. Central Somerset is drained through several rivers flowing through

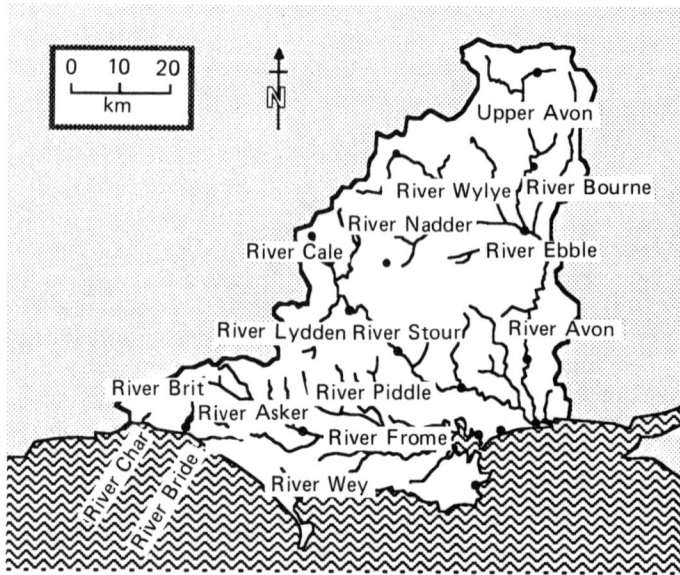

Fig. 4 — Avon and Dorset rivers.

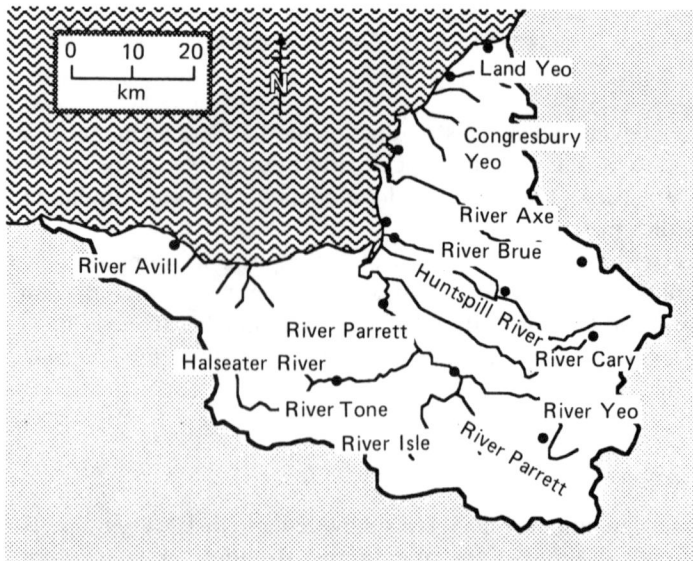

Fig. 5 — Somerset rivers.

land below high tide level and reaching the sea near Burnham-on-Sea. North Somerset, now in the country of Avon, has a number of small rivers draining the Mendip Hills and flowing westwards.

The original flood-forecasting models for the Somerset division were developed by Bootman (1978) based on unit hydrographs and level correlations. Rainfall input was derived by a surface-fitting algorithm described by English (1972) to allow for missing data and to represent areal catchment totals. Biggs (1980) developed a system based upon transfer functions and applied a recursive least-squares recalibration routine. The current models were reconverted to unit hydrographs by Bootman and Willis (1980) in order to operate on an Apple II microcomputer. Later modifications were added to the models by applying the transfer function approach to the embanked watercourses flowing through the central Somerset levels. The models have been converted to run on a Toshiba T5100 lap-top. The current data collection from 13 raingauges and 8 river level sites is through Datacell outstations polled by a fixed Apple IIe masterstation, or by polling Telegen equipment as required from the Toshiba T5100.

6. FLOOD-FORECASTING DEVELOPMENTS

The occurrence of intense rainfalls resulting in disastrous flooding in the summer months in the recent past (Martinstown, 275 mm in 1955; Lynmouth, 225 mm in 1952) and prolonged winter floods in October 1960 prompted the newly created river authorities to embark on the development of flood-forecasting procedures. At this time, techniques were graphical, relying on manual observation of rainfall being telephoned to the offices of the relevant river authority.

The advent of telephone raingauges and computers in the late 1960s established the basis for applying better methods of flood forecasting. The requirements to install rainfall and river gauges, and training in current metering, to identify water resources potential indirectly produced the database of flood records for model calibration. Significantly, the floods of July 1968 strengthened the case for prepared emergency procedures, including flood forecasting, in the Wessex region. The increasing availability of computers in the 1970s, and the associated development of better modelling techniques allowed each division to install its own forecasting methodology, suited to the catchment, the resources and the staff expertise then available.

Forecasting has not been confined to rainfall-induced fluvial flooding. Floods following blizzards in February 1978 required a definition of snow recording and snowmelt assessment for model input. Sea flooding in December 1979 confirmed the cause as a rapidly deepening secondary depression approaching the British Isles over Ireland, rotating around the primary depression of southeast Iceland.

7. COMPUTER SYSTEM

Wessex Rivers Authority inherited three entirely different stand-alone telemetry systems. The flood-forecasting models in use depended entirely upon manual input data gleaned from the telemetry systems and other observational data. Prior to the creation of Wessex Rivers in preparation for the formation of the National Rivers

Authority, the decision to replace the Bristol Avon telemetry with a system based on a DEC VAX 3600 computer had already been taken. Proposals were already in hand to replace the Somerset telemetry system and the decision was made to use a DEC VAX 3600 as the master station for a new Wessex Rivers regional telemetry system.

The computer receives information via the public switched telephone network (PSTN) and also by private wires from over 150 outstations collecting data from raingauges, gauging stations, river level sites, pumping stations, sluices, etc. Not all are used for flood-warning purposes directly but many are used to monitor operational situations throughout the river network. Fig. 6 shows the general outline of the system.

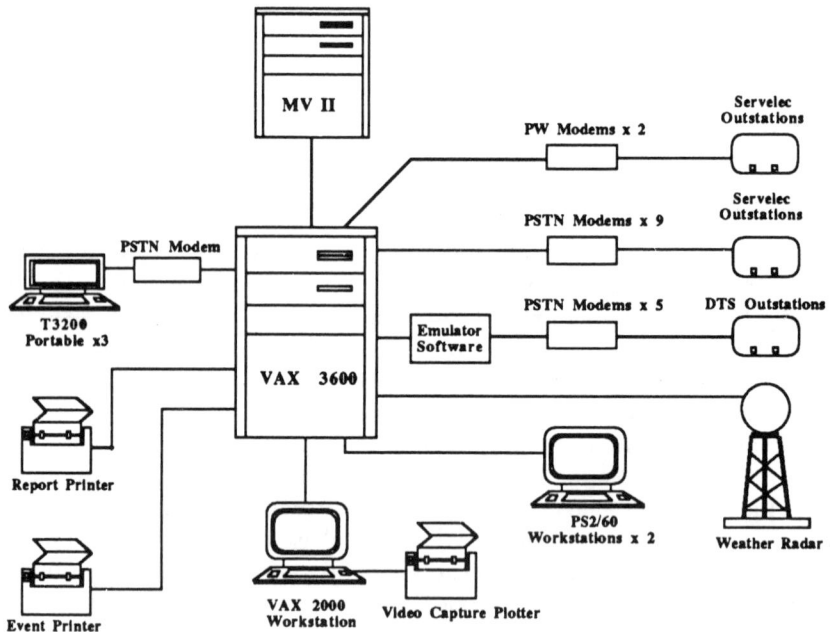

Fig. 6 — Proposed Wessex Rivers regional telemetry system.

The computer master station interrogates outstations to obtain present values and logged data as well as maintaining a database of the entire system. The information is presented from the database on a local monitor which includes graphic facilities. The computer is also capable of all archiving required. Information from the database is sent to remote PSTN terminal equipment in response to requests from those sources and carries out validation of all data received. Information from this new telemetry system continues to be manually fed into two of the flood-forecasting models and this will continue until the new models are commissioned.

The Bristol Avon flood-forecasting model has been rewritten to run on a DEC Micro VAX II (MVII) computer which is linked in a cluster with the VAX 3600. As part of the enhancement of the Bristol Avon system, development of software to pass information from the primary database to the MVII is in hand. The two computers provide the core of the telemetry flood-forecasting system and ultimately the MVII will be used to run all flood forecasting models. At that time, a further DEC VAX2000 workstation currently being configured by the Water Resources Research Group at the University of Salford will provide the primary access to the system. Additionally, the MVII will be capable of running the telemetry in the event of a failure of the VAX 3600.

Access to the VAX 3600 is via networked personal computers (PCs). The area offices and the regional headquarters are to be linked using a megastream system and within the area offices and the regional headquarters, all PCs will be networked at each location using Ethernet. It will thus be possible to obtain data held on the telemetry computers at any area office as required.

8.   TELEMETRY OUTSTATIONS

New outstations are being provided serving 30 raingauges, 40 flow gauging stations and river level sites, 22 combined raingauge and flow sites, and 14 pumping station sites. These will cover the Bristol Avon and Somerset areas. Although the detail of each type of outstation is slightly different, all are capable of accepting digital and analogue inputs, displaying present values, dialling out to the master station if four high- or four low-level alarm set points are transgressed and retrieving logged data. The amount of logged data varies from 500 events for raingauge outstations to 40 days of 15 min data for flow gauge outstations. Outstations are capable of programming for PSTN setting up from the master station of from on site. Mains and standby batteries are provided on all sites.

In addition, these Servelec outstations are considered to be 'intelligent' as logic controllers are incorporated in them to control activities on site. At operational sites (e.g. pumping stations and sluices) some will be used to control pumping and gate control. In addition, consideration is being given to the remote control from a control room of certain activities on some of these sites. At these locations, outstations will be linked to the master station by private wires rented from British Telecom whereas all other sites will rely on PSTN.

The Avon and Dorset area outstations are a relatively new design with similar, 'unintelligent' capabilities to Servelec outstations and will be retained. However, as these are of Delta Technical Services manufacture, emulator software is being provided on the DEC VAX 3600 to deal with these as if they were Servelec outstations.

9.   REGIONAL RADAR-BASED FLOOD FORECASTING

The availability of weather radar was looked at enthusiastically by the Wessex region as providing the means to track and monitor intense rainfall cells and to provide objective answers to the quantitative precipitation forecasting problem. Although most of Wessex lay beyond the objective limit of the Upavon (Wiltshire) radar, the subjective coverage extended to the western boundary of the region. The commissioning of additional radars at Camborne (Cornwall) and Clee Hill (Shropshire) showed that most of Wessex had some coverage and thus the region became one of the first users of the composite radar product. The availability of the composite and the products ability to infill the unreliable Upavon radar area prompted Wessex to embark on a regional-based flood-forecasting system using radar as the primary precipitation data source.

The technology had advanced sufficiently to make a regional approach feasible. The radar images were available on a variety of data access and presentation devices from a number of different manufacturers. Stand-alone workstations were also appearing on the market. Offerings by Apollo, Sun and DEC allowed multitasking and high-definition graphics through windowing, icons, menus and pop-ups (WIMPs). Latterly, the availability of high-power lap-top machines from Toshiba and Compaq in particular have allowed the processing to go home with the duty hydrologist or forecaster for operational convenience.

The weather radar at Upavon is reaching the end of its useful life and a new weather radar is under construction at Warden Hill in Dorset which will provide quantitative cover for the whole of the Wessex region. It is intended that the data from this site will be fed into the national network and the networked data will be received at the Bridgwater Headquarters directly by the DEC VAX 3600 computer. This will allow access to radar data on any of the networked prescribed computers (PCs) although it is intended that, in normal circumstances, one PC will be dedicated to the display of radar data.

As part of the development of the regional flood-forecasting system, access to radar data in places other than regional headquarters and area offices is being provided. This is being achieved using a Toshiba T3200 portable PC which will also be able to access telemetry data and ultimately flood predictions via the DEC VAX 3600 (Fig. 6). This will allow duty hydrologists to access all required information via one lap-top computer via any PSTN line, allowing for more flexibility than is currently the case. This is particularly vital as the area's manning levels do not allow for the provision of a duty hydrologist in each area which has been the practice in the past. One duty hydrologist will cover the whole region, providing predictions and information to operational staff in all areas.

The T3200 portable PC is currently only available with a monochrome screen. It has therefore been necessary to develop a portable radar display system to allow for eight levels of rainfall and this has resulted in the development of the STORM (system to obtain radar rainfall measurements) system (Fig. 7) by the Salford group. Ultimately, once they are available, it is proposed to move towards the use of colour screens on the lap-top PCs and the portable displays will also provide direct communication with the real-time forecast models.

The provision of adequate quantitative radar data throughout the Wessex Rivers area has made the use of radar for direct input flood-forecasting models a viable proposition. It is still intended that a weather radar is commissioned in Devon which will also cover a significant part of the Wessex region and the security provided by this overlap of coverage will be important. However, the models being developed will allow for radar and telemetry information to be automatically fed into the models and for predictions to be made. There will continue to be a facility to allow for manual editing and input as appropriate. These developments will allow for more timely flood warnings to be provided, particularly to those areas which are the subject of intense local heavy rainfall where the catchments are often very steep with

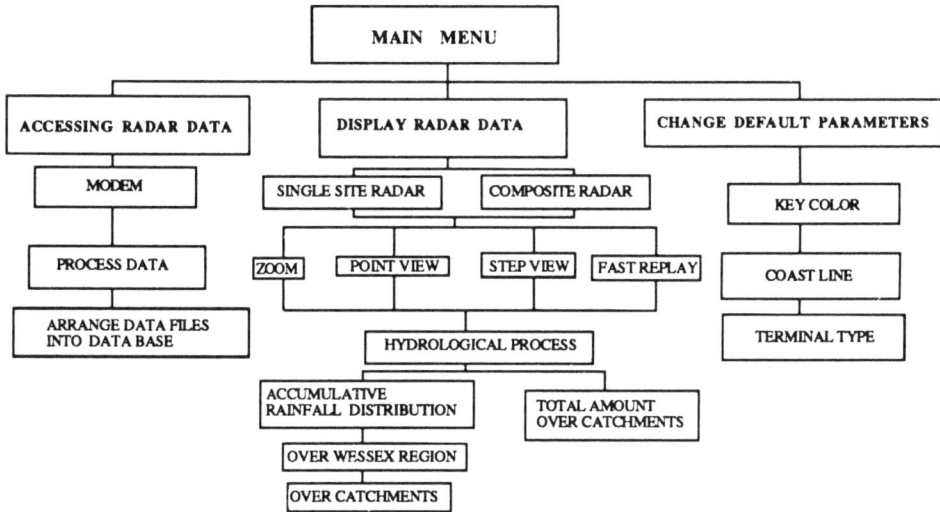

Fig. 7 — The STORM system.

short times of concentration. This will allow the provision of flood warnings to areas which currently cannot receive any warnings because of the lack of adequate lead time between rainfall and a flood occurring.

10. WESSEX RADAR INFORMATION PROJECT

A close working relationship has existed between Wessex Water Authority and the university-based Water Resources Research Group for many years. This culminated in the current project which has as its primary aims the following.

(i) *Research on radar data requirements* which involves an assessment of the type of radar data product required for real-time flood-forecasting purposes. This included particular reference to data quantization problems (Cluckie *et al.* 1991) and data resolution constraints and requirements in general.

(ii) *Development and calibration of radar data and real-time flood-forecasting models.* This in particular has included work on model structure (Cluckie *et al.* 1989) including lumped, semi-distributed and fully distributed models.

(iii) *Development of a real-time flood-forecasting system* including the MMI This will include the provision of take home real-time data aquisition and display facilities known as STORM.

A series of reports have been produced covering work completed so far. The final elements of the project will be finalized during 1990. The portable system STORM has been field tested for some months and should be fully implemented in the autumn of 1989.

ACKNOWLEDGEMENTS

The authors would like to thank their many colleagues in both Wessex Water Authority and in Wessex Rivers Authority who have been involved with aspects of the work described in the paper and in particular the Meteorological Office for their continued and valued support of the work of the Water Resources Research Group at the University of Salford.

REFERENCES

Biggs, K. L. (1980) *Flood warning user manual*, Somerset Division, Wessex Water Authority.

Bliss, J. C. (1981) *A study of the forecasting ability of a real-time rainfall–runoff model for the River Asker, southwest Dorset*, M.Sc. Thesis. Department of Civil Engineering, University of Birmingham.

Bootman, A. P. (1978) *Flood warning user manual*. Somerset Division, Wessex Water Authority.

Bootman, A. P. and Willis, A. (1980) The benefit of local rainfall analyses. *Flood Studies Report — 5 years on*. Institution of Civil Engineers, London, Paper 4.

Cluckie, I. D. and Smith, F. J. B. (1980) *Flood-forecasting project for Wessex Water Authority — Avon and Dorset Rivers*, Vol. 1 and 2, Final Report, Department of Civil Engineering, University of Birmingham.

Cluckie, I. D., Tilford, K. A. and Shepherd, G. W. (1991) Radar signal quantization and its influence on rainfall–runoff models. *Hydrological Applications of Weather Radar*. Ed. Cluckie, I. D. and Collier, C. G. Ellis Horwood, Chichester, West Sussex, Chapter 39.

English, E. J. (1973) An objective method of calculating areal rainfall. *Meteorol. Mag.* **102**, 292–298.

Grimshaw, D. and Wong, T. (1981) *Radio telemetry flood warning scheme*, Document No. PD 048/01. Bristol Avon Division, Wessex Water Authority.

49

Hydrological applications of weather radar and remotely transmitted data from a ground network in the Veneto region of Italy

A. Capovilla[1], S. Chander[2], M. Crespi[3] and S. Fattorelli[1]
[1]Department of Land and Agroforest Environments, University of Padova, Padora, Italy
[2]Department of Civil Engineering, Indian Institute of Technology, 110016 New Delhi, India
[3]Experimental Centre of Hydrology and Meteorology, Regione del Veneto, 35037, Italy

ABSTRACT

The paper describes the experience gained in integrating the data collection, retrieval, storage and management system with two high-flow forecasting models at the Experimental Centre for Hydrology and Meteorology, Teolo, Italy. The results of this integration, leading to on-line computation, updating and presentation of forecasts for the Piave river system, are presented. The paper also lists the projects which are currently in progress to take maximum advantage of the remotely sensed data from the weather radar and the ground network.

1. INTRODUCTION

The Veneto region of north Italy has a geographical location and socioeconomic environment which warrants information on forecasting and warning of natural disasters of meteorological and hydrological origin, agrohydrological and agroclimatic forecasting, forecasts for hydropower production and general information for the tourist industry. The region has established an experimental centre for hydrology and meteorology for the purpose. The centre maintains 91 peripheral stations which are automatically interrogated to obtain meteorological and hydrological data using a Micro-VAX II as the front end to the telemetric network. A C-band Doppler dual-polarization radar is operational since 1988, providing weather radar display and precipitation forecasts. Piave is the main river which drains the region. As many as 50 stations of the network are located in this basin. Reservoirs for hydropower generation have been constructed on most of the tributaries of the river. Two barrages, one at Soverzene and the other at Busche, divert the water to adjacent valleys for hydropower generation. Two linear system models coupled with routing procedures have been calibrated on historical data for various parts of the Piave basin and have been integrated with the data retrieval, storage and management system, leading to on-line computation updating and graphic presentation of forecasts for a number of stations. The paper outlines the data collection and management system, and the forecast models and presents the forecast results of the integrated system.

2.   WEATHER RADAR

A 5.5 cm Doppler radar has been installed on the Euganei hills and is connected through a microwave link to the centre. A dedicated high-resolution graphical terminal of a VAX 8200 computer system with 1 Gbyte memory is used for radar management. The radar measures reflectivity in a volume 240 km in radius and 12 km in height, with a pixel dimension of 2 km×2 km horizontal and 1 km vertical. Differential reflectivity, radial wind and turbulence data are also measured. It is possible to measure reflectivity in a volume 120 km in radius and 12 km in height with a pixel dimension of 1 km×1 km horizontal and 1 km vertical. The site of the radar in relation to the region is shown in Fig. 1 and its linkage with the computer is shown by the block diagram in Fig. 2. The radar measures atmospheric reflectivity every 15 min. The duration of each scan takes 3–5 min. Studies are in progress to transfer the reflectivity measurement to precipitation data by using the radar equation. The values of the parameters of the radar equation are being assessed for rain, drizzle and showers by comparison with the telemetric data from the ground network.

3.   TELEMETRIC NETWORK

A telemetric network based on the use of radio waves is operating in the Veneto region since 1986. Its structure is illustrated in Fig. 3. The network has 91 peripheral stations. The meteorological stations have six sensors measuring humidity, radiation, wind speed, wind direction, precipitation and temperature. Some stations have sensors measuring only precipitation and temperature, while the hydrometric stations measure only water level. The stations measure precipitation using heated tipping-bucket-type raingauges, with a tipping bucket of 0.2 mm. The location of the various stations is shown in Fig. 3. A micro VAX II is the front end to the telemetric network and can automatically interrogate each of the stations every 15 min. The reliability of the communication network has been considerably enhanced by the flexibility introduced in the choice of hardware of the stations. Each station can act as a repeater for other stations and, in the case of breakdown of the repeater station, the best path is automatically chosen by the menu by using other stations. Strategic sites have been provided with two independent identical stations. All peripheral stations are built round a microprocessor with a random-access memory of 64 kbytes. The system has been designed to store data locally, for 1 month. The micro-VAX II which controls the network is connected to the VAX 8200, and all terminals in the CSIM through Ethernet, thus enabling the systems to share the main storage system. The local area VAX cluster (LAVC) manages the system. The system enables the use of

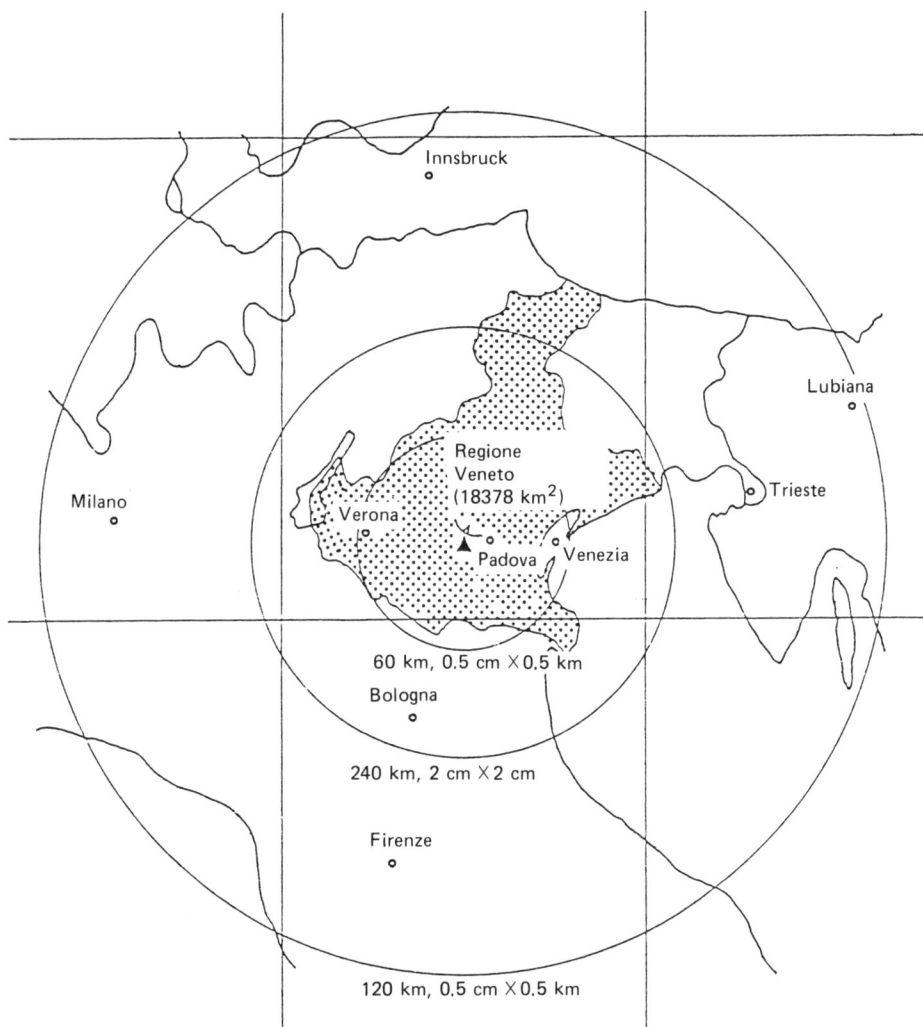

Fig. 1 — The weather radar site of the Regione Veneto at Teolo.

any terminal in the centre to manage the network, thus enhancing the reliability of obtaining data from the network in case of a terminal failure.

4. DATA MANAGEMENT

The data from the peripheral stations is retrieved normally every morning for the preceding 24 h (from 0000 to 2400 hours) and stored in the disc memory of 1 Gbyte of the host computer. The data are plotted and manually checked before it is stored in the database.

The following checks are performed: upper and lower thresholds pre-defined per month and station both for 15 min (5 min per precipitation) and daily data; persistence test (when data do not vary for a predefined time, the possibility of error is looked into).

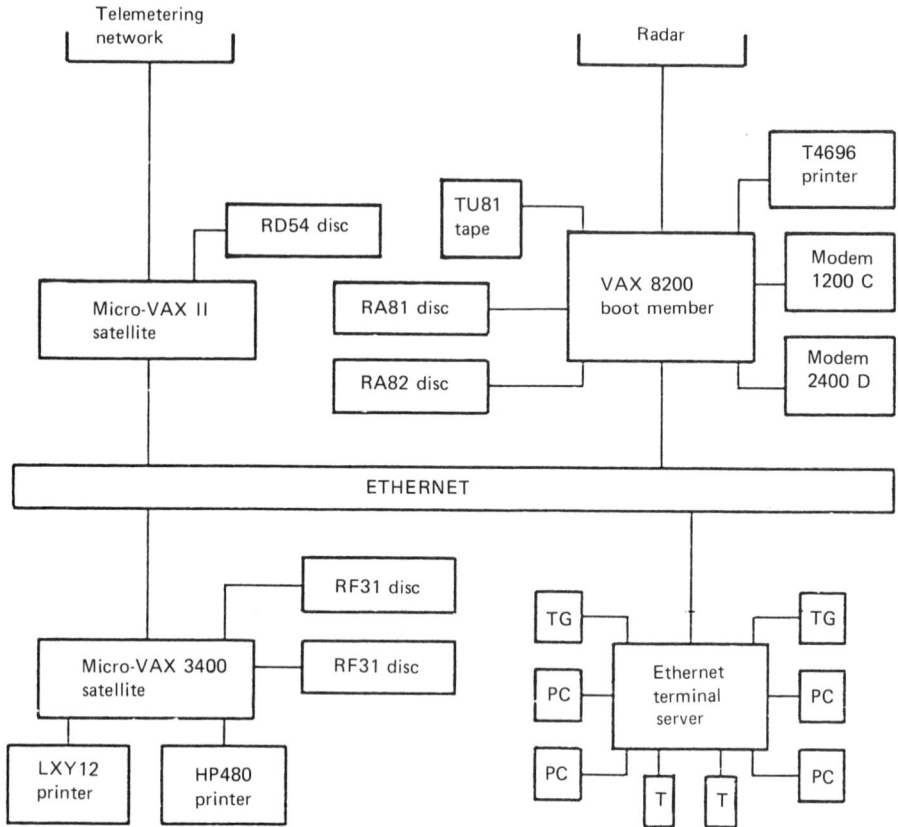

Fig. 2 — Block diagram showing the linkage of the radar, the telemetric network and the computer system.

Interrogations are carried out at 15 min intervals for alert situations which are defined in advance according to the critical values of river level, the rate of rise of river level and the rainfall rate. During these situations, the stations are interrogated in a cyclic fashion and an automatic update of the data bank is provided every 15 min. The data are organized basinwise in the real-time database (RTDB) and is managed in such a manner so that data available upstream of a hydrometric station can be directly used for analysis in rainfall–runoff models. Programs for graphical display of information, and other programs for analysis can be considered tools of the database. This allows automatic exchange of data between models and data bank in both directions, leading to presentation of results in graphical form showing the information from the RTDB such as the observed values up to the time of forecast and model forecasts for the following hours. In normal situations, the data are retrieved every morning for the preceding 24 h (from 0000 to 2400 hours) and stored in a temporary database (TDB). The data are periodically screened manually using pre-defined checks, transformed in binary form and stored in the actual database (ADB). Software has been developed to retrieve data from the ADB for chosen time

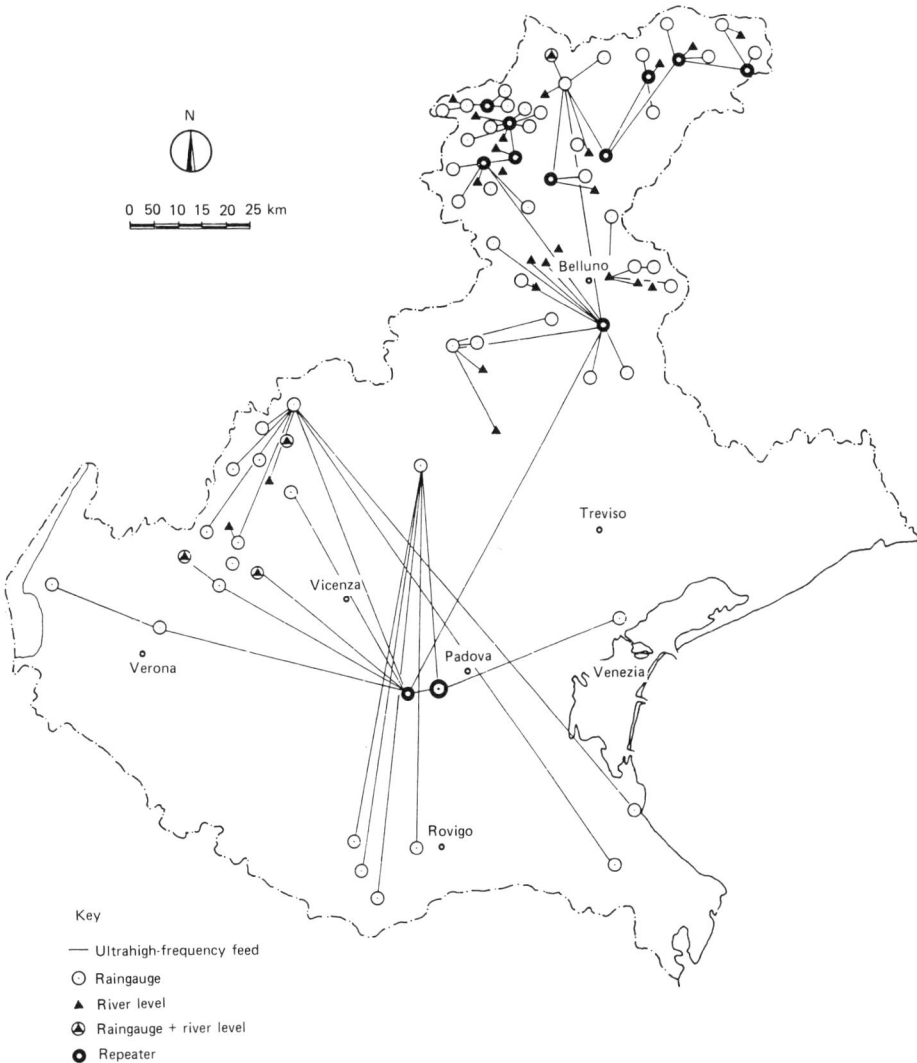

N

0 50 10 15 20 25 km

Belluno

Treviso

Vicenza

Verona

Padova

Venezia

Rovigo

Key

— Ultrahigh-frequency feed
○ Raingauge
▲ River level
◉ Raingauge + river level
◉ Repeater

Fig. 3 — Structure of the telemetric network and peripheral stations.

intervals, to plot data on the screen, to convert levels to flows, to compute mean areal rainfall for any subbasin or to print data especially precipitation data above a pre-defined threshold. At present the ADB is used on a regular basis to issue water yearbooks for the region.

5. FORECAST MODELS

The forecast models used in the integrated system are the HEC-1F model and IITFORMO. Both models subdivide the Piave basin into interconnected system of subbasins and streams. Each subbasin is intended to represent an area which on an average has the same hydraulic and hydrologic characteristics. Both models determine the parameters for gauged subbasins from the observed data and other parameters by systematically altering the values of the parameters until the square-root of weighted squared difference between the observed and computed hydrograph is minimized. The HEC-1F user's manual documents the procedures used for on-line refining of the parameters of the HEC-1F model. IITFORMO has six distinct components.

They are

(1) the rainfall–runoff model for the subbasin,
(2) the rainfall–runoff model for the interbasin area,
(3) the channel-routing model,
(4) computation of inflow to reservoir,
(5) computation of outflow from reservoir for a given operation policy and
(6) a stochastic error model.

These components are appropriately linked to model the configuration of the basin. The subdivision of the basin into subbasins is dictated by the location of telemetr'ᵔ discharge stations on the basin. The procedures used in various components are briefly described.

6.   AN ADAPTIVE RAINFALL–RUNOFF MODEL FOR THE SUBBASIN

The rainfall–runoff model is a linear lumped-input time-variant system model given by

$$Q_{t+\lambda} = \int_0^t [I(\tau) - F]\, h_t(t - \tau)\ \mathrm{d}\tau \tag{1}$$

where λ is the initial lag or pure delay time parameter, $I(\tau)$ is the average rainfall on the basin at time parameter τ, F is the average infiltration rate and $h_t(t - \tau)$ is the ordinate at time $t - \tau$ of the response function which is chosen from a given set, at time t on the basis of observed discharge and precipitation data and $Q_{t+\lambda}$ is the computed value of discharge at time $t + \lambda$.

The first time step of rainfall producing direct runoff is determined by subtracting the lag from the time step corresponding to the first rise in the hydrograph. The rainfall prior to this value is considered as an initial loss. It is assumed that a set of two-parameter gamma functions can represent the response of subcatchments which are around 1000 km² in area. The loss in the subcatchment is represented by an average abstraction rate. For a given response function, equation (1) can be written in the discrete form as

$$Q_{i+\lambda} = \sum_{j=1}^{i \leq m} (I_{i-j+1} - F_i)U_j \tag{2}$$

where $Q_{i+\lambda}$ is the direct runoff at time $i + \lambda$, I_i is the precipitation at time i, F_i is the ϕ index value at time i, U_j is the discrete time "D" hour unit-hydrograph ordinate derived from the two-parameter gamma function. ($j = 1, 2,...,m$ where m is the number of unit-hydrograph ordinates).

$$F_{i(\text{DIS MATCH})} = \frac{\displaystyle\sum_{j=i-\lambda}^{i} \left(\sum_{K=1}^{j} (I_{j-K+1} U_K - Q_{j+\lambda}) \right) \left(\sum_{K=1}^{j} U_K \right)}{\displaystyle\sum_{j=i-1}^{i} \left(\sum_{K=1}^{j} U_K \right)^2} \qquad (3)$$

$$F_{i(\text{VOL MATCH})} = \frac{\displaystyle\sum_{j=i-\lambda}^{i} \left(\sum_{N=1}^{j}\sum_{K=1}^{j} (I_{N-K+1} U_K - Q_{N+\lambda}) \right) \left(\sum_{N=1}^{j}\sum_{K=1}^{N} U_K \right)}{\displaystyle\sum_{j=i-\lambda}^{i} \left(\sum_{N=1}^{j}\sum_{K=1}^{N} U_K \right)^2} \qquad (4)$$

F_i, at each time step is computed and updated for each response function of the subcatchment using two objective functions, namely the sum of the squares of the difference between the observed and computed discharges (discharge match) and the sum of the squares of the difference between the observed and computed volumes (volume match); in a period of λ time steps prior to the current time, is minimum. The discharge and volume match abstraction rates (equations (3) and (4)) are used in equation (2) to compute the direct runoff. The sum of squares of the deviation between the observed and computed values using discharge match and volume match abstraction rates are computed for up to ten time steps before the time of forecast. The sums are added to obtain the δ function with minimum δ being chosen along with the corresponding average value of discharge and volume match abstraction rates for computing the forecasts. The procedure gives due weighting to fit in the neighbourhood of current time and chooses the response which is equally good for the ten previous time steps.

7.   AN ADAPTIVE RAINFALL–RUNOFF MODEL FOR THE INTERBASIN

The rainfall–runoff model for the interbasin is a linear lumped time-variant system model defined by equation (1) as for the subbasin. Since the contribution of the interbasin is not known directly, the procedure developed for updating F needs to be modified. In an error-free system the contribution of the interbasin can be determined by subtracting the routed observed flow of the upstream subcatchments and the base flow generated by the interbasin from the observed discharge measured at the downstream end. It is observed that such a procedure produces a multipeaked rising limb and sometimes yields negative values because of errors in upstream and downstream observed discharges, and the simplifying assumptions made in the routing process. Therefore, instead of determining the time history of discharge of the interbasin, the volume V_i of runoff generated by the interbasin area is estimated. The base-flow contribution of the interbasin area is computed by subtracting the initial routed upstream flow from the downstream observed value. The volume V_i of the interbasin contribution is then found by adding the differences between the observed discharge at the downstream end and the sum of the base flow and routed observed discharge at the upstream end. Knowing V_i, the F_i value for each response function is determined such that the volume of the computed hydrograph at computed time is equal to V_i or

$$F_i = \frac{\sum\limits_{j=1}^{i} \sum\limits_{K=1}^{j} (I_{j-K+1} U_K - V_i)}{\sum\limits_{j=1}^{i} \sum\limits_{K=1}^{j} U_K} \tag{5}$$

Each response function is then used to compute the time history of the interbasin contribution using equation (2). The base flow of the interbasin area and the routed observed discharge from upstream catchments are added to the interbasin contribution. The response function which produces a hydrograph nearest to the observed values is chosen for computing the forecasts of interbasin area.

8. ROUTING MODEL

The Muskingum routing scheme with three parameters is used in the model. The parameters are n, the number of Muskingum reservoirs in the cascade, and the Muskingum coefficients k and x.

9. COMPUTATION OF INFLOW TO THE RESERVOIR

The inflow to the reservoir is computed using the discrete form of the continuity equation to obtain inflow to the reservoir knowing the gauge storage curve of the reservoir, the water levels, the outflows from the spillway, and the diversions up to the current time.

10. COMPUTATION OF OUTFLOW FROM A RESERVOIR FOR A GIVEN OPERATION POLICY

The discrete form of the continuity equation is used to obtain a value which is the sum of storage, outflow and diversion values at any time. The operation policy is used to apportion the sum into the various components and outflow from the reservoir is obtained for various lead times.

11. STOCHASTIC ERROR MODEL

This component computes the time series of error between the observed and computed values from the time of first rise of the hydrograph until the time of forecast. The time series is analysed to obtain the correlation structure for fitting an autoregressive model. The AR(1) model is chosen if the length of the error series is less than 15. For longer series the AR(2) model is chosen. The parameters are determined and are checked for stationarity. If the parameters of the AR(2) model do not satisfy the stationarity condition, the AR(1) model is used even for longer series. The standard deviation of the residual series is also computed. The parameters of the autoregressive model are now used to compute the forecast errors beyond the time of forecast. These errors are added to the computed forecast discharges. The choice of AR(1) model has been dictated more by the short length of error series, especially during the rising limb of the hydrograph, rather than by any rigorous analysis of the correlation structure in view of the scarcity of the data.

12. PREPARATION OF FORECASTS

On receipt of an alert, the interrogation of peripheral stations are cyclically carried out at 15 min intervals and an automatic update of the RTDB is provided. The programs for graphical display, models and other software for analysis are the tools of the database and automatic exchange of information takes place in both directions. The data in the RTDB is used to obtain a graphic display of average rainfall on each subbasin and the observed discharge until the time of forecast. After visual assessment, the model is run using another batch file and forecasts are computed. The progress of some forecasts for both the models are shown in Figs 4 and 5. The forecasts are updated hourly in these situations. Automatic interrogation, visual assessment of data and formulation of forecasts for the five-basin model at Busche takes between 20 and 30 min.

13. SUMMARY AND FUTURE DEVELOPMENTS

An outline of the available hardware, data management system and forecast models is presented. The various components have been integrated for the first time in Italy for hydrological applications leading to on-line formulation of forecasts using two models. The centre's future developments can be categorized in three distinct areas, namely hydrology, meteorology and agrometeorology and agrohydrology. In hydrology, the centre is engaged in projects which can make maximum use of the high-resolution radar-measured precipitation data for formulating forecasts. Some results of the development on the grid model, which also forms part of the EEC project entitled 'Application of weather radar for alleviation of climatic hazards' are also presented in this session of the symposium. Other projects include testing of the integrated system using on-line data for the telemetric network as well as weather radar data in improving the forecasting capabilities. Flood-warning schedules on various rivers in the region for disseminating the forecasts are being worked out in consultation with the local authorities and civil protection agency, so that the capabilities of the centre are used to minimize damage due to floods in the region. In meteorology, the projects relate to the improvement of accuracy of real-time radar precipitation measurements, quantitative precipitation forecasts on pre-defined areas up to 6 h, prior to identification and quantification and qualification of the severity of thunderstorms, tornadoes, etc., and effective use of synoptic charts, imagery and radar information available from a modern dual-polarization Doppler radar. The future developments in agrometeorology and agrohydrology at the centre specifically aim to use reflectivity and radial velocity data from the radar, visible and infrared images of satellites and data from meteorological stations for evapotranspiration and water balance computations. In addition the products of developments in

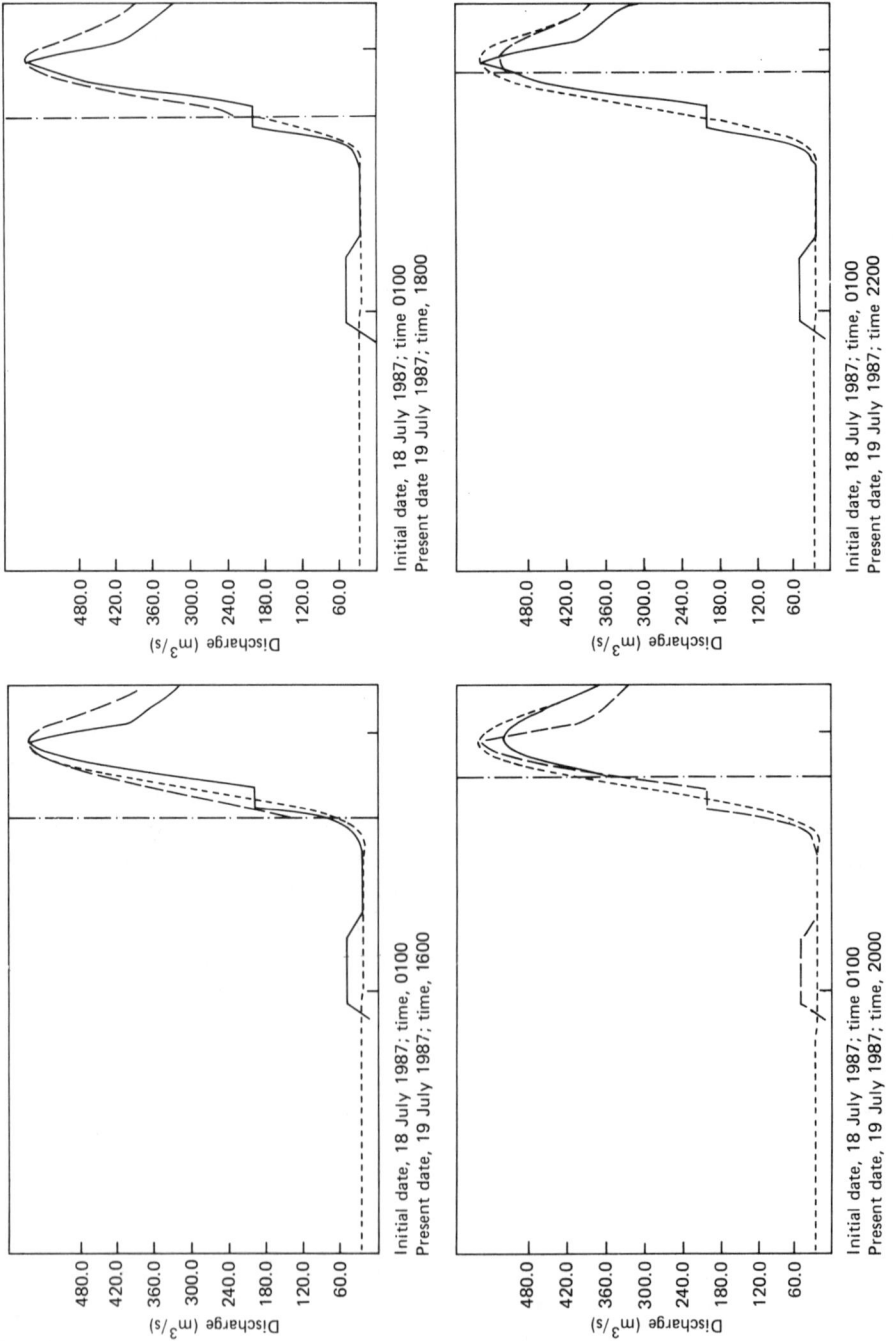

Fig. 4 — Forecast discharge of river Piave at Soverzene for the flood of July 1987 using HEC-1F
model: ————, observed; ⋯⋯, simulated; – – –, forecast; ————, received.

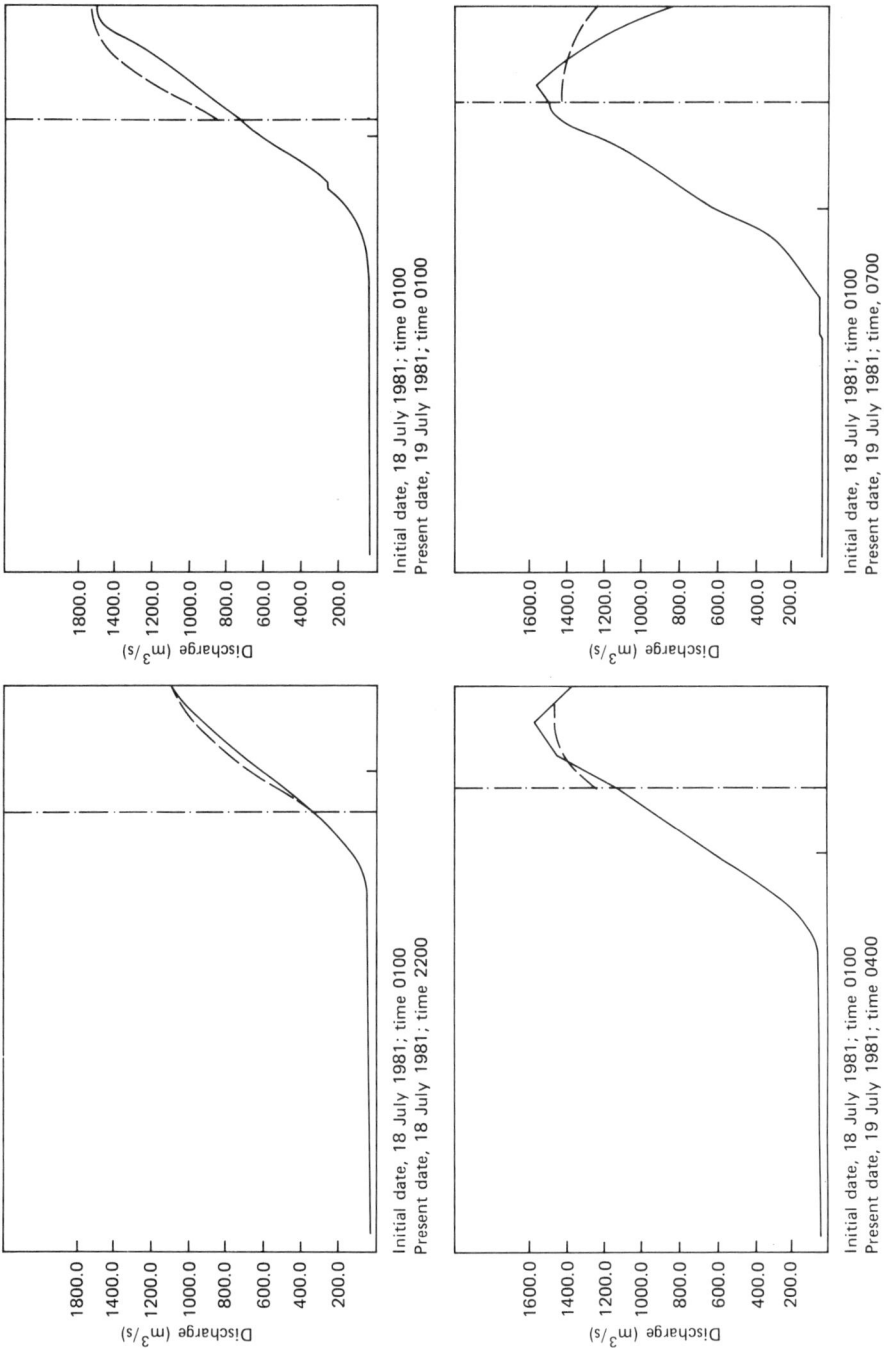

Fig. 5 — Forecast discharge of river Piave at Busche for the flood of July 1981 using the IITFORMO model: ———, observed; ······, simulated; - - -, forecast; ———, received.

hydrology and meteorology such as high-flow forecasts, and thunderstorm, heavy-rain and hail forecasts will be made available at the grassroot level. As the service becomes established, models for soil water balance and frost forecasting will be introduced.

ACKNOWLEDGEMENTS

The authors acknowledge the contributions on the realization of the project provided by Dr Federico Cazorzi, Dr Cristina Munaron, Dr Lorena Rossi for systems management and development of the software package allowing data processing, Dr Paolo Giaretta and Dr Marco Monai for radar management, and Dr Antonio Capovilla for hydrological and computational support for the calibration of the models.

50

New possibilities for precipitation estimation for river basin managers in developing countries

J. D. Flach, T. R. E. Chidley and A. Siyyidd
Department of Civil Engineering, Aston University,
Birmingham B4 7ET, UK

ABSTRACT

It is now possible to build economic integrated information systems to assist river basin managers in developing countries to meet specific tasks. These tasks could include the following:

(1) water-related disaster, flood and drought forecasting, and consequent system operational advice;
(2) change monitoring;
(3) erosion and sediment yield forecasting and monitoring;
(4) assembly of calibration data for mathematical models of flow and water quality, and monitoring of some aspects of water quality;
(5) assistance with estimating rainfall and evaporation for hydrologic and ground-water recharge models;
(6) coastal process and outfall monitoring.

The information required for these tasks is available in the following forms:

(a) archived or requested high-resolution multispectral satellite images;
(b) geostationary and orbital meteorological satellites;
(c) maps, charts and plans;
(d) data from remote data collection platforms linked via satellite or ground communications with the base;
(e) data from rainfall radar systems;
(f) air-borne scanner or video systems.

The knowledge of how to use these data is available in the form of mathematical models of hydrologic, hydrogeologic, hydraulic, ecosystem and meteorological processes, and in the form of expertise in making decisions on the operation of systems within the control of the river basin manager.

The development of remote sensing, computer and knowledge engineering techniques now makes it possible to install integrated information systems in the developing world. This paper describes the possible uses of newly and shortly available microwave sensors in providing improved estimates of precipitation within the scope of a low-cost river basin management system. Microwave sensors require a very large computing power to extract useful information from them; for this reason the UK has set up a data centre to facilitate rapid processing of such data. The paper describes an integrated information system built by the authors for river basin managers and considers whether the new data can be used within it.

1. INTRODUCTION

The systems and remote-sensing unit research group at Aston University has been carrying out research into the practical application of information technology in water resources management. Integrated information systems have been designed and put together with the help of industry specialists in satellite receiving and communications equipment manufacturers. Two components of the work are discussed here in the context of the theme of the conference, which is primarily concerned with microwave data. One is the description of the integrated information systems that have been developed and in some cases installed in live situations. The other is the work that has been carried out in using geostationary satellite data as an input to a runoff forecasting–estimation system. The integration of currently available land-based rainfall radar data and the possible impact of the ERS1 radar satellite, which is about to be launched, on this work is discussed.

2. INTEGRATED INFORMATION SYSTEM

The system is built around a series of microcomputers linked to one another. The functions are as follows:

(1) receiving of primary METEOSAT data (PDUS);
(2) receiving of remote data collection platform data (DCP);
(3) integration of data;
(4) rainfall estimation;
(5) runoff forecasting;
(6) control.

At the core of the systems is the data and systems integrator; its functions are general and flexible and provide for the following:

(a) multispectral digital image processing;
(b) digital mapping;
(c) spatial analysis;
(d) surface modelling;

(e) data capture (digitizing).

The systems integrator can also obtain data from the rolling archive of data on the PDUS and can combine data received from the remote DCPs. The actual data incorporated from the PDUS is defined by the systems operator. The operator defines an area of interest that includes the basin or catchment being studied and information from the satellite is copied across for that area and corrected to overlay the catchment at the chosen map projection.

3. DATA PROCESSING

After the data from the satellite is copied across from the receiving system, it is converted to the same map projection as used in the hydrologic modelling. Any information from raingauges or water-level gauges is read or entered. These data are used to estimate precipitation in the period. The satellite data are primarily used to assist with spatial and temporal interpolation. Its principal direct output is the probability for each 2.4–5.6 km cell that it is raining or not. In some cases this can further be divided into heavy or light rain.

The inputs can be assembled into a form suitable for running a catchment simulation model to predict output. Actual data on discharge can be fed back into the system to recalibrate the model. At present, only simple lumped-parameter models are used, but the system has the potential for using distributed-parameter models because it can provide a distributed input.

4. RESULTS

At present the runoff modelling part of the system has just been developed and, at the time of writing, only tests are available on the algorithms for estimating rainfall have been carried out. These include the simulated performance of the use of various areal rainfall estimators modified by satellite data. For example the Thiessen polygon method and surface-fitting methods have been tried. The technique is to clip the interpolation where it is expected that there is no rain.

5. FUTURE PROSPECTS

With the launch of radar satellites in the 1990s the possibility for upgrading the performance of rainfall estimators using satellite data is increased. The data may also provide soil moisture data which is of very high value in modelling the rainfall–runoff process and it has recently been shown that satellite radar data can give improved performance of rainfall estimators.

The ERS1 satellite fitted with radar sensors is due to be launched soon. The National Remote Sensing Centre is gearing up to provide data to users in a timely fashion which should be of value in post-event systems. Work will be needed to establish the value of microwave data and to prove possible operational systems. This will be the only way of ensuring that, if valuable information is obtained, then the satellites will continue. The low revisit frequency of the satellite will be a great disadvantage for rainfall estimating, but it could be useful for flood monitoring.

6. CONCLUSION

The system can be used in real time or as a post-mortem tool. In real-time DCP and other remotely collected data can be used, including output from rainfall radars. In post-mortem mode it is possible to use historic data, including daily rainfall and runoff data. The technique affords a unique means of distributing rainfall in time and space. However, it should be noted that even in post-mortem mode the system should have been set up in advance and the necessary data archived. It is generally impracticable to archive all METEOSAT data as a matter of course.

The daily data volume for a catchment of about $100\,000\,km^2$ is of the order of 1 Mbyte. A post-mortem analyser would have to maintain an archive of about 6 months' data (200 Mbytes). In post-mortem mode the daily data are integrated with any recorded data that are available and with the METEOSAT data to assist with interpolation in time and space. In practice this means that any interpolation procedure will only be permited a value if the METEOSAT gives it a good probability. Further work may allow there to be a stochastic element to this. All the data are scaled to ensure that the correct daily amount falls at specific actual gauges. This may seem a haphazard procedure but in a sparsely gauged area the alternatives are worse.

There are problems in the detection of night rain and this will require further work. It should be possible to obtain some useful data from the thermal band data. Statistical analysis will be carried out on the amounts of rain falling after dark in the study areas. The best results for this method are expected in the middle latitudes.

Before a river basin manager will invest in any proposed new system for gathering data, he will want to see assurances of continuity in that data supply. The meteorological satellites such as the METEOSAT systems seem to have established themselves as long-term prospects. Earth resource observation satellites are faced with uncertainty and this makes them less useful in operational systems. The value of satellite microwave data in an operational system has yet to be fully established and it is necessary to prove its value if support is to be given to continue such projects as ERS1 into the future. The system described here for integrating various sources and types of data in operational hydrologic simulation systems is an ideal platform for evaluating the new sensor data.

Rainfall radar systems can provide the necessary inputs to assist with calibration and measurement of the effectiveness of satellite rainfall-monitoring systems supported by ground truth, in both real-time and post-mortem situations. The radar will probably provide a better estimate of rainfall than the current satellites but radar is very much more expensive to install and maintain. In large river basins in developing countries, ground-based radar would not be a cost-effective solution at the moment.

ACKNOWLEDGEMENT

The writers wish to acknowledge the support of Space Technology Systems, Alton, Hants., UK, in enabling this work to be carried out.

REFERENCE

Barrett, E. C., Kidd, C. and Bailey, J. O. (1988) The special sensor microwave imager (SSM/I): a new instrument with rainfall monitoring potential. *Int. J. Remote Sens.*, **2**, no. 12, 1943–1950.

51

Remote sensing:
the environmental analysis solver

S. K. Ghosh and G. Fleming
Water and Environmental Management Group, Department of Civil Engineering
and Environmental Health, University of Strathclyde,
Glasgow, UK

ABSTRACT

To meet the basic needs of mankind for food, housing and energy, many engineering
and industrial projects are envisaged at regular intervals of time. With the increase in
this type of project, the earth's environment is now experiencing subtle, generally
degrading changes. Environmentalists have been very concerned over this and yet
find it difficult to stop such degradation owing to a lack of a proper data collection and
evaluation technique to substantiate their growing concern. The objective of this
paper is to provide an insight into remote sensing data and techniques and its
application to environmental changes.

1. INTRODUCTION

Benjamin Franklin identified man (*Homosapiens*) as the tool-making animal. His
remarkable dexterity and unusual capacity to use the brain in ingenious and creative
ways, has made him the most fearsome creature on earth. His relentless use of his
brain has allowed him to adapt to the changing environment. This changing
environment to a large extent is man's own doing, although unintentionally. The
ever-increasing pressure of population and its demand for the fulfilment of his basic
needs have led to such a situation, so that it is felt that we are no longer good stewards
of our own environment.

Large areas of forested lands have been converted into agricultural and habi-
tation units. Rivers have been dammed in order meet the pressing demands of

irrigation and power. With the advancement of technology, more and more industries have come into existence, thereby requiring further conversion of land from its natural state. These conversions are bound to have some effects, generally degrading in nature. The typical symptoms of such a degradation are dying flora and fauna, accelerated soil erosion, channel braiding, and expansion of flooding plains, pollution of land, water and air, loss of wild species, etc. All these are the early warning signs.

It must be remembered that the changes caused are not deliberate and intentional effort on the part of human beings. Any change will have two aspects: the good and the bad. The environmentalist concern is not to curtail or stop any developmental growth but to provide a situation such that the degenerative impacts induced by any project are eliminated or counterbalanced.

2. ENVIRONMENTAL CONSIDERATIONS

In 1936, the US Government introduced the Omnibus Flood Control Act and as a direct result of this the US Congress in 1969 passed the National Environmental Policy Act 1969. The salient feature of this Act is as follows.

'To declare a national policy which will encourage production and enjoyable harmony between man and his environment; to promote efforts which will prevent or eliminate damage to the environment and biosphere and to stimulate the health and welfare of man; to enrich the understanding of the ecological systems and natural resources important to the nation; and to establish a council on Environmental Quality.'

Section 102 of this Act also states that all agencies must include a detailed statement on

  (i) the environmental impact of the proposed action,
 (ii) any adverse environmental effects which cannot be avoided should the proposal be implemented,
(iii) alternatives to the proposed action,
 (iv) the relationship between local short-term uses of man's environment and the maintenance and enhancement of long-term productivity and
  (v) any irreversible and irretrievable commitments of resources which would be involved if the proposed action should be implemented.

Since then regular assessment statements have had to be submitted. The type of data required for such statements has to be vast in nature in order to incorporate all the facets of the induced impact; yet it has to be comprehensive. Dee *et al.* (1973) stated that the structure of the environmental statement is hierarchical in nature in order to take into account the different levels of information to be used for the impact analysis. Over the years, it has been found that, if the structure is in a matrix format, it is easy to correlate the various interrelationships between the environmental parameters and its condition. A procedure has been proposed for systematic analysis of proposed projects using matrix and numerical weightings of the probable impacts. It is suggested that on one axis of the matrix all those actions from the project which could affect the environment should be listed, while on the other axis the environmental conditions should be listed.

Jones and Stokes (1971) in an environmental impact study identified two classes of impacts: direct and indirect. Direct impacts are those which affect man's health and are a function of the number of people concerned and the degree to which they are affected. Indirect impacts result from those actions that affect resources other than man, the magnitude of which is dependent upon the degree of change in resource quality and quantity.

Kerri (1972) examined man's impact on river basins, water planning, water quality, land management, aquatic life, wildlife, etc., and stated that any disturbance to land causes a veritable chain reaction of adverse ecological consequences. He further stated that attempts must be made to quantify impacts in terms of distances, area and money. He stated that the matrix approach provides a good interrelation between development activities and the possible resultant impact.

In 1978, the Hydraulics Division of American Society of Civil Engineering set up a Task Committee on Environmental Effects of Hydraulics Structure. The Committee recommended that the matrix approach was much more simple than the hierarchical structure. This was helpful in determining the data requirements to complete an assessment of environmental effects. It also helps in identifying parameters with negligible effects to be deleted from the monitoring programmes, which tend to be very expensive.

3. DATA REQUIREMENTS

Any engineering action is bound to have some environmental effects. The nature of the impact can be classed as direct or indirect. The social and cultural changes are direct impacts and are basically due to the movement and interaction of people, as people from diverse cultures come into contact with one another. The indirect impacts are physical in nature and have damaging effects. The effects of social and cultural impacts are generally borne easily by man because of his compromising and adapting nature. The physical impacts are those to which man has not paid attention over the past centuries. The time has now come when proper attention must be paid to them. Indirect impacts give rise to changes in chemical, biological and physical characteristics. The amount of pollution can increase the toxic concentration of undesirable elements. This unprecedented rise can lead to the death of aquatic and plant life and to the change in potable water characteristics. The land use changes can trigger off changes such as increased erosion pattern, depletion of vegetation cover, and variation in runoff patterns.

To collect information so as to cater to the need of such an exhaustive database, the requirements are enormous and varied. Furthermore, these data should be collected within a short span of time and must have a good degree of reliability. The whole data collection should also be economically viable. Unfortunately, the resolution at which data for environmental assessment are to be collected are not very stringent and thus it could be very expensive altogether. Therefore at present the need is to have such a data collection system which meets nearly all the requirements of the database required for these types of study. Remote-sensing data products and techniques provide such capabilities; these are discussed in the latter part of the paper.

4. REMOTE SENSING

Remote sensing is defined as a process of identifying, measuring and recording the presence of and information about objects from a distance without direct contact with the concerned object. Data is collected through sensors, which can be classified as active and passive sensors. Active sensors are those which sense energy emitted by them, while passive sensors are those which observe solar reflected energy or radiant emittance. Radar and SLAR are examples of active sensors while scanners and thematic mappers are examples of passive sensors. In remote sensing, passive sensors are most commonly used and some of these are listed in Table 1. These sensors have been used by LANDSAT and SPOT programmes. A standard LAND-SAT scene covers an area of 185 km by 185 km or 34 225 km^2 at a scale of 1:1 000 000 and revisits the same area after 16 days. SPOT has two modes of sensor operation: vertical and off-nadir. In the vertical mode, it covers an area of 60 km×60 km or 3600 km^2 and revisits the area after 26 days. In the off-nadir mode, it covers a strip of 950 km and revisits after 2+ days.

Table 1 — List of sensors used on earth resources satellite

Sensor	Wavelength (μm)	Pixel size (m)
Camera	Visible or near-visible wavelength	Dependent on the photograph scale
Multispectral scanner	Band 1[a] 0.50–0.60	80
	Band 2[a] 0.60–0.70	80
	Band 3[a] 0.70–0.80	80
	Band 4[a] 0.80–1.10	80
Thermal scanner	Band 1 0.45–0.52	30
	Band 2 0.52–0.60	30
	Band 3 0.63–0.69	30
	Band 4 0.76–0.90	30
	Band 5 1.55–1.75	30
	Band 6 10.40–12.50	120
	Band 7 2.08–2.35	30
SPOT high resolution visible (multispectral)	Band 1 0.50–0.59	20
	Band 2 0.61–0.68	20
	Band 3 0.79–0.89	20
high resolution visible (panchromatic)	0.51–0.73	10

[a]This is based on LANDSAT 4 and 5. In LANDSAT 1, 2 and 3 these were numbered bands 4, 5, 6 and 7 respectively.

With the advent of electronic sensors and the rapid development of computer technology, remote-sensing techniques can now be applied to a large number of engineering fields. Initially, up to the early 1980s, the sensor resolution was low and thus the full potentiality of remote sensing could not be envisaged. With the thematic mapper and the charge-coupled device linear-array sensors such as high-resolution viable ones becoming operational, which have high resolution, the interpretation reveals more accurate information of the landscape.

In areas where cloud is a problem, information can be collected with the help of radar or with thermal scanners. These have the capability to penetrate through fog and mist. Satellite digital data can be used very effectively, as computation on these machines can be done very rapidly and nearly all types of mathematical simulation can be performed in order to obtain the best information. Various enhancement and classification techniques combined with various printing methods can reveal exciting information.

5.   ADVANTAGES OF REMOTE SENSING

Remote sensing has been found to be a multiconcept field, since it can collect data on a multiband, multidate, multistage, multipolarization, multienhancement and multi-user basis. This multiconcept is of particular importance since, when these options are fully exploited, they can provide a large amount of classified data. Furthermore, large areas can be seen at a stretch; thus the spatial relationship between objects can be assessed.

6.   REMOTE-SENSING DATA ANALYSIS AND INTERPRETATION

One of the basic advantages of remote sensing is that it is available either in the photographic or digital format, depending upon the type of interpretation technique, i.e. visual or digital, which the data product calls for. Furthermore, as it collects data in different bands, these can be used independently or in combination, as required. In the visual approach, it is found that the human eye is more sensitive to changes in colour, compared with grey levels; thus a number of bands can be combined to produce a false colour composite.

Digital data can be manipulated, using a computer, in three ways. The first way is to acquire a standard colour composite image product, to overlay a transparent base map at the same scale and then to interpret the image using standard visual interpretation techniques. The second approach is to acquire a computer-compatible tape of the scene, to process it with a computer image-processing system and to develop photographic products which have been enhanced so as to improve their visual interpretability. The interpreter can interpret the image. The computer can be used to apply contrast stretches to the data, to ratio two or more bands or to use simple enlargement to improve interpretation. The third approach is to carry out a full digital analysis of the data. It consists in displaying the data on the computer colour video monitor and carrying out the necessary enhancement techniques to select training areas (i.e. areas which can be used to identify the spectral characteristics of known features to the computer for subsequent quantitative analysis and manipulation), and to apply various algorithms designed for multispectral classification applications, to produce thematic maps of land and water features.

There are principally two forms of computer classification which can be used with digital multispectral data. In applying the supervised classification approach the interpreter selects relatively homogeneous areas in the image and defines and describes them on the basis of their usually known ground characteristics. The features are also analysed and defined in terms of the spectral response characteristics in each spectral band. The resulting selected feature data are called 'training sets' and are used to develop statistical decision rules or algorithms which will determine the group or 'class' to which any new pixel will belong. When the interpreter has decided that a feature has been optimally defined the computer can be instructed to search the image for any pixels similar to the training sets and to assign them to the same spectral class.

The unsupervised classification approach limits the amount of interaction as the computer is instructed to develop classes using algorithms based on the spectral similarity to develop classes rather than on the basis of the identified ground characteristics of each class. The result is a map of spectral classes, each of which has yet to be transformed into land cover unit descriptions. The approach is good when attempting to analyse images of areas about which little is known.

7. DATA PRESENTATION

There are a number of forms in which the final output of remotely sensed data can be produced. The final choice or choices will depend on the use to which the product will be put. These uses can range from simple illustration or display to detailed thematic cartographic-standard resource mapping. In basic applications, standard image or aerial photographic data are visually interpreted and transferred to base maps. Two products are thus produced in the process: the image or photograph and the derived map. Where digital processing is used to enhance the images for visual interpretation, several products can be produced: quality photo-images, and derived unrectified or rectified maps.

Full digital analysis will provide opportunities to produce photographs of the video monitor image, photographically produced image maps, alphanumeric thematic maps, colour thematic maps, histograms, theme area summaries and other numeric summaries and graphs. Derived rectified and unrectified maps can also be produced. In most instances where remote sensing is used, the final output will be in the form of a map or will have a map as a supporting document. The problem of rectification or fitting the remotely sensed images to a standard map projection is common to all the data discussed above.

8. INFORMATION EXTRACTION

It has been seen that, depending on how accurately the training sets are provided, the classification and identification of the features can be done. From preliminary knowledge, the spectral characteristics of specific objects are known. In multispectral scanner band 1, healthy vegetation has a low spectral response whereas water has a high response; thus, on a photographic image, water appears brighter than vegetation, which is represented as darker shades of grey. Similarly, the spectral response of healthy vegetation in band 4 is high while water has hardly any reflected component of the electromagnetic radiation. If the plants or vegetation undergoes any stress, the spectral response of the vegetated area undergoes a transformation. Depending upon the stage of stress, the response varies. The decaying of vegetation can be easily assessed by using images sensitive to the infrared band as they appear as silvery grey. Similarly, if the water is polluted, then this can be very easily be detected, as the response will change. False colour composite provides easy detection of these phenomenon since dead or diseased plants appear as purple while turbid water appears as yellow. Remotely sensed data are usually collected at regular intervals of time; thus, monitoring changes of the landscape is fairly easy to do. So a region which is to undergo or has undergone any change as a result of any proposed action, and thus the impacts to be generated or those already generated, can be very easily monitored by analysing multidate imageries.

9.   USE OF REMOTE SENSING IN IMPACT ASSESSMENTS

The amount and variety of data required in preparing the environmental impact statement are so large that, if a ground-based data collection is planned, it would be a cumbersome, time consuming, hazardous and uneconomical. For a proper and an effective analysis to be carried out, the data must be collected within a very short period of time and also analysed. Furthermore, information at that time must be available when the area was in an undisturbed state. Photographs and imageries are permanent records pertaining to a given instant of time and thus the real picture and/ or situation of that state can be obtained with ease.

Hawkes (1973) used aerial black-and-white, colour infrared and multispectral photographs in order to assess the environmental impacts due to a proposed highway project in Atlanta. In conjunction with the remote-sensing data, soil, topographical and geological maps were used. The parameters considered for monitoring included ecology, hydrology and sediment control. He concluded that, on black-and-white photographs, water edges, soil boundaries and drainage areas could be identified with relative ease. Colour infrared photographs provide good information on plant health and thus any plant stress due to environmental changes — especially man made — can be recorded and detected easily. Furthermore, thermal imagery can detect and portray subtle differences in emissivity, and thus current patterns of water, heated effluents, soil moisture condition and vegetation vigour can be found easily.

DeLoach (1973) used remote-sensing techniques to assess the impacts generated by a highway near Tampa, FL. He used black-and-white aerial photographs to classify the land use, drainage, soil type and water temperature.

Howarth and Wickware (1981) used satellite imageries to study the impacts due to impoundment in the Peace–Athabasca delta region. Satellite imageries and computer-compatible tapes for two dates, 26 August 1973 and 1 August 1976, were taken, which represented the normal and abnormal conditions respectively of the lake. To aid the analysis further 70 mm colour infrared photographs on a 1:7000 scale were also taken. In addition to this, various ecological relationships between water and vegetation of the area were used. In order to detect any change, methods of band ratio, post-classification change detection, percentage change, change matrix, binary theme printing and conflict character assignment were used. It was found that, within a 3 year span, about 6.72% of the original landscape had changed. The accuracy of the classification was termed as 'good' since 55 out of the 78 pixels chosen for the accuracy test were classified correctly. Howarth and Wickware have also stated that the pixel size (57 m×57 m) had a dominant role to play in the classification accuracy.

The Water and Environmental Management Group, Department of Civil Engineering, University of Strathclyde, Glasgow, at present is involved with research projects dealing with the environmental restoration. The broad group of the studies can be outlined as follows:

(1) reservoir management in consideration with the expected impacts due to greenhouse effects;
(2) sedimentation processes and use of dredged material from river channels for restoration of derelict open-cast mine areas.

The next step in these projects is to study the suitability of remote sensing to monitor the restoration procedures as suggested earlier.

Generally it was noted by all that the resolution and the pixel size had a large role to play in the analysis of the data due to the problem of mixed class within the pixel. With the successful launching of LANDSAT 4 and 5 and SPOT, the thermatic mapping data (30 m resolution) and SPOT data (10 m resolution) are now available. Since the resolution is very high, more information can now be derived about the existing landscape. This will allow the investigator to perform land use classification at a much higher level of accuracy. Prior to this, land use classification on LANDSAT imageries could be done up to level two on the Anderson classification scheme. This will certainly aid in the classification procedure as more information would be available at a microscopic level.

10. CONCLUSION

As discussed above, the data requirements of the environment impacts statements can now be easily met with the help of remote sensing. Furthermore, the various enhancement and classification techniques available do provide ways of extracting and detecting change in the area of interest. Since these changes are complex in nature and may not be seen with the human eye, thus remote sensing does provide a good solution to the long-awaited answer.

REFERENCES

Bella, D. A. (1974) Fundamentals of comprehensive environmental planning. *Eng. Issues, J. Prof. Act., Am. Soc. Civ. Eng.*, January, 17–36.
Black, R. E. (1982) *Environmental impact analysis.* p. 180.
Campbell, J. B. (1987) *Introduction to remote sensing.* Guildford Press, London, p. 551.
Claasen, D. Van R. (1986) *The application of remote sensing techniques in coral reef, oceanographic and estuarine study,* UNESCO Report No. 42. UN Educational, Scientific and Cultural Organization, p. 151
Dee, N. *et al.* (1973) An environmental evaluation system for water resources planning. *Water Resour. Res.,* **9**, 523–535.
DeLoach, W. C. (1973) Remote sensing application to environmental analysis. *Highway Res. Rec.,* **452**, 29–39.
Proceedings of the Policy Issue Workshop — Developing Policies for Responding to Climatic Changes, 9–13 November 1987, Bellagio.

Hawkes, T. W. III (1973) Application of aerial mapping to development of highways. *Highway Res. Rec.,* **452** 10–18.

Herschy, R. W., Barrett, E. C. and Roozekrans, J. N. (1988) The world's water resources — a major neglect. Report No. ESA BR-40. European Space Agency p. 41.

Howarth, P. J. and Wickware, G. M. (1981) Procedures for change detection using Landsat digital data. *Int. J. Remote Sens.,* **2**(3), 271–291.

Jones, R. L. and Stokes, J. D. (1971) Environmental impact study for the Orange County coastal project. Final Draft Report. Jones & Stokes Associates Inc., Sacramento, CA, p. 50.

Kerri, K. D. (1972) Environmental assessment of resource development. *J. Sanit. Eng. Div. Am. Soc. Civ. Eng.,* **98** (2), 361–374.

Siegel, B. S. and Gillespie, A. R. (1980) *Remote sensing in geology.* Wiley, New York, 702 pp.

Smerdon, E. T. and Gaither, R. B. (1974) Using technology to solve environmental problems. *Eng. Issues, J. Prof. Act., Am. Soc. Civ. Eng.,* October, 273–288.

Task Committee on Environmental Effects of Hydraulic Structure, American Society of Civil Engineers (1978) Environmental effects of hydraulic structures. *J. Hydraul. Div., Am. Soc. Civ. Eng.,* **104** (2), 203–221.

52

Real-time control for urban drainage systems: advantage or disadvantage

M. J. Green
Water Research Centre, Swindon SN5 8YR, UK

ABSTRACT

This paper describes how real-time control (RTC) for urban drainage fits into the UK's river basin management programme and assists in achieving environmentally desirable and economically suitable solutions to pollution problems. By means of three case studies, different aspects of urban drainage are to be examined, enabling costs and benefits of RTC to be identified.

1.   INTRODUCTION

In the UK the development of an urban drainage system over the last 150 years has accompanied the increase in population. The original open ditches were turned into sewers or culverted water courses carrying both foul sewage and surface water runoff. The construction of sewage treatment works and the introduction of overflows to relieve the pipe system when the capacity was exceeded during times of rainfall followed. Thus today major sources of pollution are from overloaded sewage treatment works, surface water drainage and combined sewer overflows (CSOs).

In the past the main priority was to prevent surface flooding of urban areas and this is evident in England and Wales from the estimated number of overflows, 12 000 (Working Party on Storm Sewage 1977, Shuttleworth, 1986). This figure itself is probably a gross underestimate. Although the main objective of preventing flooding has not diminished, it has now to be weighed against other priorities such as capital and operating costs and pollution of receiving waters.

The river basin management programme in the UK is targeted at these problems, especially those attributed to CSOs. The aim is to produce a methodology for river catchment planning and management, i.e. to optimize the performance of the urban

drainage–river system to achieve desirable environmental and economically acceptable solutions. Thus real-time control (RTC) is an integral part of the river basin
management programme. The general public over the last few years are more aware
of environmental issues and this social trend, together with the more tangible
legislative and economic constraints, dictates that all aspects of sewerage design and
operation are closely monitored to ensure that the target levels of service are met.

It is against this background, together with advances in technology ranging from
weather radar and telecommunications to hydraulic analysis of drainage systems,
that has allowed the concept of RTC strategies to be attainable.

In a sewer system, three types of control are available: **passive**, **reactive** and
predictive. Traditionally, passive controls have been employed and they represent,
as their name implies, a regulator that has its control function fixed at the design
stage, i.e. sill level of a side weir. Passive control is *not* part of RTC but can be viewed
as the basis against which the other control strategies can be compared. Reactive
control operates in response to the status of the sewer system and involves the use of
a priori operating rules. The decisions are made following the sensing of the variables
involved, i.e. rainfall, sewer flow and/or level, and storage capacity. Predictive or
forecastive control allows decisions to be made that are based on estimated future
flows as well as the present status, so that storage and flow capacities can be fully
utilized. Thus either a simple rainfall prediction model is required that simply delays
rainfall at a linear rate (Evans 1981) or one that is based on a weather radar and/or
satellite measurements (Newsome 1987). Fig. 1 illustrates the perceived advantages

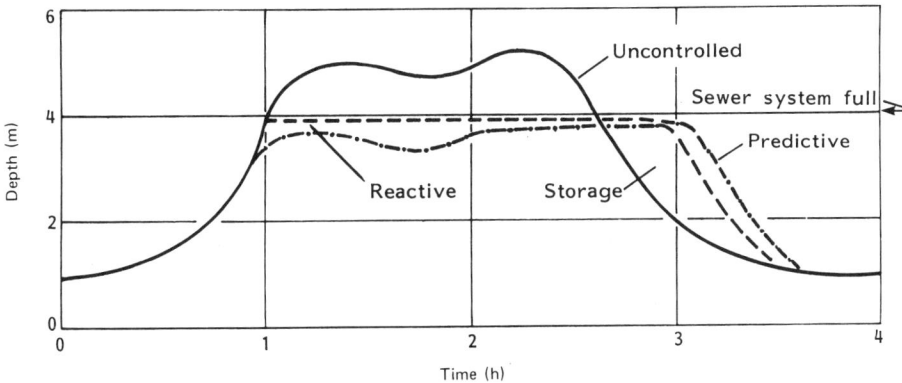

Fig. 1 — Hydrograph for different controls.

of reactive and predictive over passive control, namely that fewer sewage spills into
the receiving waters and more is passed forward for treatment. Thus the objectives of
RTC for urban drainage systems are:

(1) *reduction of flooding*   through the effective mobilization of storage within or
    outside the sewer;
(2) *reduction of pollution*   by containing more storm sewage within the sewer
    system by means of flow attenuation and use of effective storage (less reaches the
    receiving waters),

(3) *reduction of capital costs*   either for new or rehabilitated works, by minimizing
    the size of the peak-flow-carrying elements of the system,
(4) *reduction of operating costs*   either directly through optimizing pumping costs
    or indirectly by providing the information necessary to allow effective mainten-
    ance procedures to be implemented.
(5) *enhancement of sewage plant performance*   by balancing inflow peaks and
    general flow smoothing, allowing the plant to operate at or near its design
    efficiency, thus improving the quality of the final effluent, and
(6) *improved knowledge of the sewer system*   where the more detailed involvement
    with the sewer system allows changes in performance to be easily and quickly
    appreciated.

This chapter describes the case studies being used to assess the benefits of RTC.
Each case study embraces a different aspect of urban drainage.

2.   BACKGROUND

Over the years, especially in North America, the concept of RTC for urban drainage
has been actively considered (Trotta *et al.* 1977) and a number of schemes initiated.
A survey of the North American scene (Schilling 1985) highlighted the severity of the
CSO problem and identified this as the principal reason for adopting an RTC policy.
It was interesting that the survey revealed that traditional drainage solutions, i.e.
sewer separation, stormwater treatment and reduction in inflow–infiltration were
often cited as preferred options to RTC. Of the 18 cities or municipalities who
adopted an RTC policy, only nine were reported as having implemented an RTC
system. The main reason given for this finding was that cost savings and benefits are
not easily identified. For instance extra instrumentation and communications associated with RTC will increase capital and maintenance costs and these have to be
weighed against less flooding and fewer CSO spills. Initially a lower volume of CSO
spill will help the receiving water quality but the question of first-flush effects and
whether the receiving water can assimilate the spill volume without substantially
altering its classification needs to be answered. In addition, experience in North
America shows that actual costs are higher than anticipated, that staff enthusiasm
has waned and that political decisions are holding back progress (Grigg and Schilling
1986).
    In FRG a requirement to construct storage tanks at overflows to limit CSO
discharges has provided the necessary impetus for RTC. The CSO discharge
requirements are legally defined and the benefits of RTC are more easily described
together with costs. In France a long-term commitment to RTC in the Seine-Saint-
Denis county outside Paris has been initiated (Delattre *et al.* 1986). The main
objective is to minimize the likelihood of flooding in a flat, densely populated (25 000/
ha) urban catchment by utilizing all available storage capacity and using weather
radar to predict the uneven spatial distribution of summer rainfall. In The Netherlands, with an extremely flat topography and the sewer system relying on pumping,
RTC has been applied to the interlinking of pumping stations (Schilling 1987).
    In the UK the potential benefits of RTC have been only slowly appreciated and,
apart from the Tyneside sewerage scheme and a number of sewage pumping stations

(Ramsden 1981) and (Tyler and Plews 1984), few have been implemented. This is in contrast with the 'clean' water side of the business where pump scheduling and leakage control are well established (Water Research Centre, 1985). The reason why this situation exists is most probably a combination of many factors and prejudices. For instance, it could be argued that the drainage engineer has no control over the input (rainfall) into the sewer system and therefore the scope for additional control would be minimal. Controls and associated storage necessary to obtain their best use are expensive, their capabilities are limited and the time scales compared with land drainage are short. For urban drainage, time steps for control tend to be measured in minutes whereas, for land drainage, hours and even days are acceptable.

3. CASE STUDIES

The three case studies each embrace a different aspect of urban drainage. One covers a catchment where several subcatchments drain by gravity to a inland sewage treatment works. The urban development has resulted in frequent operation of CSOs and unacceptable water quality in the receiving watercourse. The second covers an urban area where rationalization of sewage treatment works coupled with overflows from combined to surface water systems has resulted in the need for sewerage rehabilitation. The final study includes a coastal catchment where all sewage is discharged by pumping and operating costs have been evaluated. Scenarios where discharge is by gravity, including storage and combinations of gravity and pumping, will be examined.

3.1   Bolton (Northwest Water Authority)

Bolton is situated 16 km northwest of Manchester and lies in the valley of the River Croal and River Tonge, relatively small watercourses that are tributaries of the River Irwell. The urban drainage covers an area of 12 000 ha with a population in excess of 250 000. The approximately 1200 km of combined sewers range in size from 150 mm diameter up to 2.35 m box sections. Altogether there are over 100 overflows. Currently the system can be divided into eight subcatchments which all drain into the Croal Valley Interceptor Sewer which discharges by gravity into the Ringley Ford Sewage Treatment Works. The catchments are shown in Fig. 2.

The area suffers from flooding in the Bolton town centre, pollution in the watercourses caused by frequent operation of overflows, sewer dereliction and siltation. The rivers, under the NWC Quality Classification are *class* 3 and the long-term quality objective is to bring all watercourses up to *class* 2 (support coarse fish).

The altitude range for the catchment is 48–196 m AOD and the area lies within the 75 km range of the Hameldon Hill weather radar. The aim of this study is primarily to examine the benefits of RTC, reactive and predictive, over conventional controls with respect to investment and capital costs for rehabilitation and new works, reduction in frequency and sensitivity from flooding from sewers and reduction in pollution of the receiving watercourses.

The programme of work involves building and verifying hydraulic models of the subcatchments and simplifying the models without degrading the output. This mainly consists of retaining ancillaries, i.e. overflows, tanks and amalgamating pipe

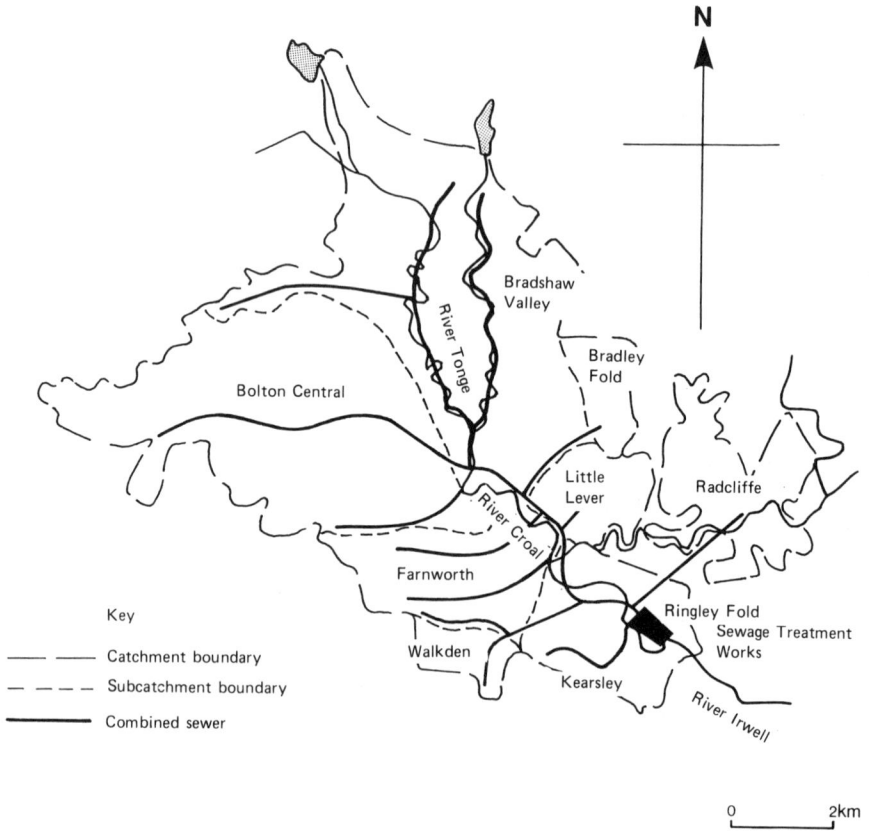

Fig. 2 — Bolton sewerage catchment.

lengths. The estimated final number of pipe lengths in the model will be around 1000. Sewer flow (velocity and depth) monitors and tipping-bucket raingauges (0.2 mm per tip) are being installed and telemetered back to the master station on a daily basis; the data-recording interval is every 2 min, with rainfall recorded every tip. Each outstation has the capability of storing up to 2 days' data to allow for equipment failure.

3.2   Tipton/Coseley (Seven–Trent Water Authority)

The Tipton–Coseley urban drainage catchment lies in the Upper Tame river basin in the West Midlands, 16 km west of Birmingham, draining east into the River Trent. The gravity drainage is combined and surface water, the latter including the old watercourses that have in places been piped or culverted.

The drainage area (Fig. 3) covers 2000 ha and houses a population of 50 000 and is situated at the end of the southern leg of the proposed Black Country Trunk Sewer

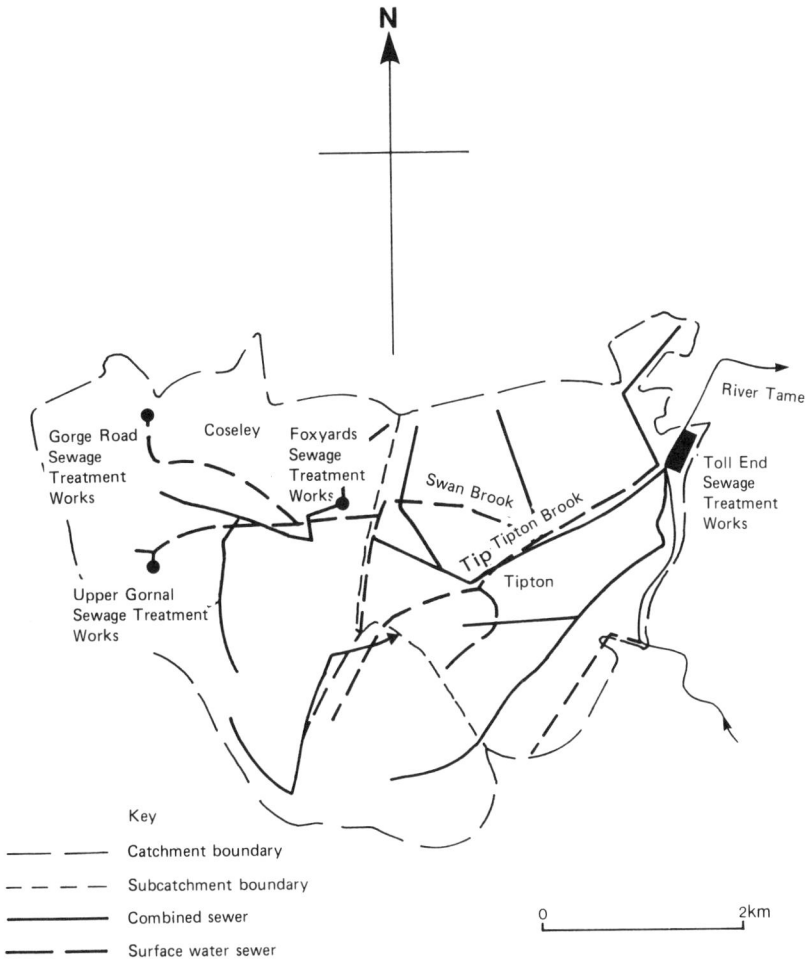

Fig. 3 — Tipton–Coseley sewerage catchment.

(Halliday 1986). At present, sewage is treated at a small works within the catchment and the final effluent discharged into the surface water system. The major portion of the surface water discharges in the River Tame at Toll End (Fig. 3). It is planned that the treatment works will be closed down and all raw sewage will be piped to the Black Country Trunk Sewer which will drain by gravity to the treatment works at Minworth to the east of Birmingham. The Tipton–Coseley catchment lies within the 75 km range of the Clee Hill weather radar.

Surface flooding is not a problem in this catchment as the surface water system has plenty of capacity as witnessed by the number of overflows from the combined into the surface water.

The River Tame, the receiving watercourse, is at present class 3–4 and the objective is to raise it up to at least class 3. For the purpose of this study the objective has been set at class 2. At present the existing trunk sewers through Birmingham have been retained as part of the new Black Country Sewer. The second objective, using Tipton–Coseley as an example, is to determine whether, using RTC, this existing trunk would be adequate or whether a new trunk will have to be built to reinforce it.

Thus the principal objectives are to examine whether RTC in conjunction with storage can offset the capital costs of a new trunk sewer and reduce the pollutant load in the receiving watercourse.

The programme of work involves the building and verifying of a detailed hydraulic model of both the surface water and combined sewer system.

3.3   West Hull (Yorkshire Water Authority)

Hull is situated on the north bank of the river Humber approximately 35 km from the open sea. The area is flat with a maximum elevation of 3 to 4 m AOD. The West Hull combined drainage system covers an area of 7500 ha and houses a population of 220 000. Prior to the 1950s the large brick egg-shaped sewers were capable of retaining sewage and discharging it by gravity through flap valves during the low-tide cycle. In the 1950s a low-level interceptor and pumping station were installed. The pump capacity, 4 DWF pumps, each rated at $1.4 \, m^3/s$ and eight storm pumps, each rated at $4.25 \, m^3/s$, together with up to $20\,000 \, m^3$ of in-pipe storage, allow dicharge at all states of the tide.

Flooding within the catchment due to inadequate pipe capacity is not a problem and the effluent pumped into the Humber does not cause infringement of the environment quality standards set for the Estuary.

A detailed investigation by Watson's in 1987 involving building and verifying a hydraulic model, and an investigation into pumping and total power costs was undertaken. The final report made a number of recommendations of which one was to make better use of the available storage and reduce pumping costs by raising pump switch-on levels and another to predict inflows from tipping-bucket raingauges sited throughout the catchment.

The aim of this study is to assimilate the results already obtained into a more general study of pumped, gravity and gravity–pumped options.

4.   DISCUSSION

The present cautious approach in the UK towards RTC implies that potential users of RTC are unsure as to the costs, benefits, technology and legal implications. It is principally the costs and benefits that will be examined in the study; any new technology will be incorporated only where it is feasible and desirable to do so. The legal aspects will only be included where they are appropriate, namely the specified levels of service. Typical current standards would be: that flooding incident owing to hydraulic overloading should not occur more then twice in 10 years, or an overflow should meet its consent.

In the past, RTC has been included under the umbrella of new sewerage rehabilitation schemes, when the cost of RTC can be viewed as only a fraction of costs of the overall scheme. The argument is reinforced if it can be shown that, by using RTC, construction costs can be reduced. This usually means either a reduction in the storage capacity or an improved level of service for a given storage capacity. A

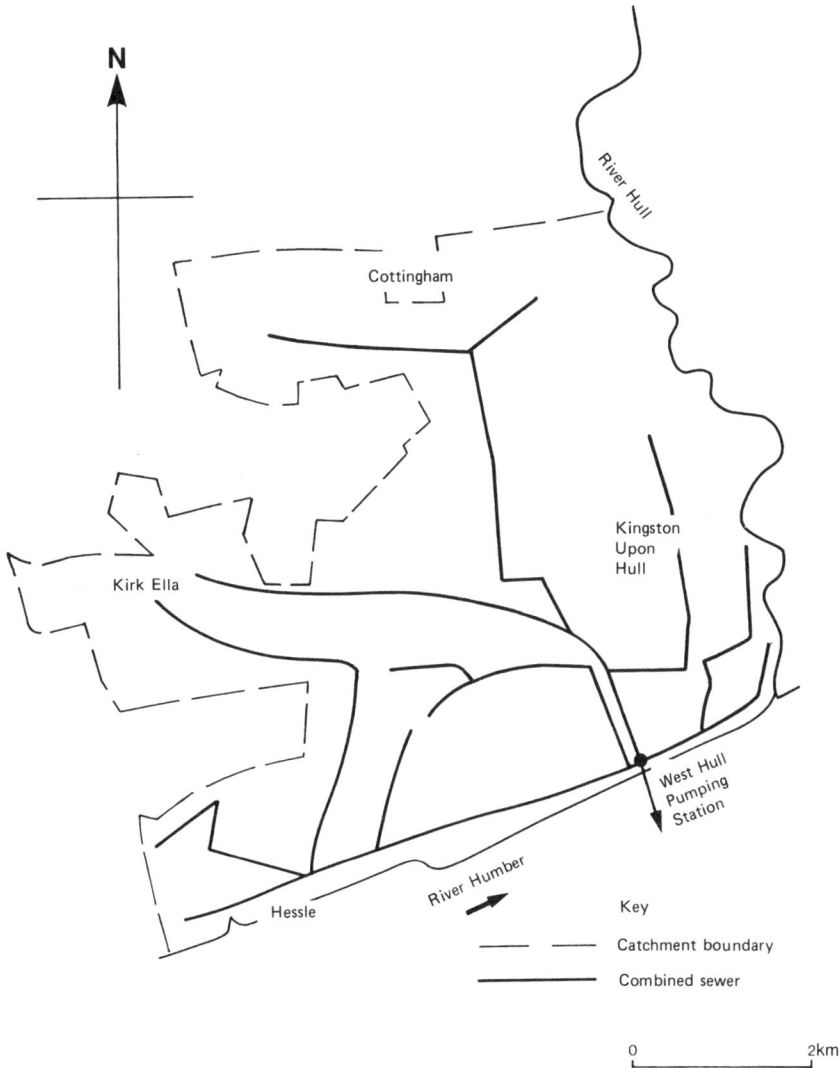

Fig. 4 — West Hull sewerage catchment.

preliminary example (Green 1988) showed that, for a particular scheme using storage tanks with simple passive control, the spill volumes to the receiving water as a percentage of the total runoff would be reduced by 2–4% for a storm with a 1 in 1 year return frequency. Adding a simple RTC strategy the spill volume could be reduced by up to 15%. On average it was estimated that a 9% reduction in spill volume was more realistic. On the assumption that the average cost of building a tank at 1988 prices was £450/m³, then for this example each 1% reduction in spill volume averages £0.58m/%. It was estimated that the equivalent saving in storage tank capacity or the cost of providing storage to the same level of service as that provided by RTC would have been £5.25m. This example did not take into account the cost of equipment, servicing and maintenance of tanks or any replacement costs. It was the forerunner to the present study where these aspects will be examined in more detail.

ACKNOWLEDGEMENTS

This work is being supported financially and technically by North-West Water Authority, Severn–Trent Water Authority and Yorkshire Water Authority, and their permission to publish is gratefully acknowledged. The views expressed in this paper are those of the author and do not necessarily reflect those of the water authorities.

REFERENCES

Delattre, J. M., Bachoc, A. and Jaquet, G. (1986) Performance of hardware components for real time management of sewer systems. In: H. Torno, J. Marsalek and M. Desbordes (eds), *Urban runoff pollution* Springer, Berlin, pp. 819–842.

Evans, G. P. (1981) *Optimisation of sewage pumping*, WRc Technical Report No. TR170. Water Research Centre, Swindon, 56 pp.

Green, M. J. (1988) *Real time control for urban drainage systems.* WRc Report No. 303E. Water Research Centre, Swindon.

Grigg, N. S. and Schilling, W. (1986) Automating stormwater and combined sewer systems: the possibilities. Water Eng. Manage., **133** (5), 33–35.

Halliday, J. (1986) The Black Country Sewerage Scheme: philosophy design and some operational experience. *Water Pollut. Control.*, 34–44.

Newsome, D. H. (1987) COST 72 and weather radar in Western Europe. In: V. Collinge and C. Kirby (eds) *Weather radar and flood forecasting.* pp. 19–34.

Ramsden, I. (1981) Automation of Grimsby pumping stations. *Water Sci. Technol.* **13**, 277–283.

Schilling, W. (1985) A survey on real time control of combined sewer systems in the United States and Canada. In: R. Drake (ed)., *Proceedings of 4th IAWPRC Workshop on Instrumentation and Control of Water and Wastewater Treatment and Transport Systems.* Pergamon, Oxford, pp. 595–600.

Schilling, W. (1987) IAWPRC–IAHR Joint Committee on Urban Storm Drainage. *Real time control of urban drainage systems* — The state-of-the-art, Report.

Shuttleworth, A. J. (1986) State of rivers and sewers in Britain — is there a pollution problem? *Symposium on Developments in Storm Sewerage Management, Sheffield, 1986.*

Trotta, P. D., Labadie, J. W. and Grigg, N. S. (1977) Automatic control strategies for urban stormwater. *J. Hydraul. Div., Am. Soc. Civ. Eng.*, **103**, (12), 1443–1459.

Tyler, R. and Plews, A. D. (1984) Sewage pumping optimization in Wessex Water Authority. *Proceedings of the Conference on the Planning, Construction, Maintenance and Operation of Sewerage Systems.* British Hydromechanics Research Association – Water Research Centre, Paper J3, pp. 381–402.

Water Research Centre (1985) *Pump scheduling in water supply*, WRc Report No. TR 232, Water Research Centre, 71 pp.

Working Party on Storm Sewage, Scottish Development Department (1977) *Storm sewage: separation and disposal*, Report. HM Stationery Office, London.

53

Quantitative use of radar for operational flood warning in the Thames area

C. M. Haggett, G. F. Merrick and C. I. Richards
National Rivers Authority, Thames Region, Reading, RD1 8DQ, UK

ABSTRACT

This paper describes how the flood-forecasting system, developed initially for catchments in London, has recently been successfully extended to the Lee catchment. The software which captures, processes, archives and displays radar and outstation data is introduced, together with an account of the way in which this information is used to generate flow forecasts in real time. Future plans to extend this system across the whole Thames catchment are discussed, as are plans to undertake local calibration of radar data operationally and to make use of quantitative rainfall forecasts to extend warning lead times.

1. BACKGROUND

Weather radar has been used for flood-monitoring purposes in the Thames catchment since 1984 when the London weather radar installation at Chenies, Bucks., became operational. Funded by a consortium comprising the Meteorological Office, the Thames Water Authority, the Greater London Council and the Southern Water Authority, the unmanned radar station provides precipitation data in real time on a continuous basis for London and the southeast of England.

The non-tidal flood-warning systems operating within the Thames Water Authority area up until 1987 were largely based on those which were inherited from predecessor authorities, namely Thames Conservancy, Lee Conservancy and (from 1986) the Greater London Council. In 1987 the separate systems covering the Lee Valley and London excluded area were combined to form the eastern flood-warning system based at Waltham Cross, Herts., so that two systems now cover the Thames region:

(i) *Reading* for the non-tidal Thames and tributaries upstream of Teddington (the Thames tidal limit);
(ii) *Waltham Cross* for non-tidal tributaries downstream of Teddington.

Intense urban development throughout the Thames basin has resulted in an increase in the risk of flooding from non-tidal rivers. An assessment of the potential flood damage in London revealed that over 9500 properties are at risk from a 50 year return period flood and that damage costs could exceed £17m for residential properties alone (Haggett 1986). There is a growing awareness of the need to improve the quality of flood warning, to alleviate the effects of flooding.

Weather radar's value in portraying the spatial and temporal variation in rainfall estimates in a qualitative way is today beyond doubt. This is particularly apparent for local convective storms which can easily remain undetected by a conventional network of raingauges and for frontal storms whose movement can be readily depicted by replaying radar pictures at successive time frames. However, the processing of radar information on powerful telemetry computers illustrates the wider potential of radar for providing quantitative information on rainfall depth for further manipulation and analysis (Haggett 1989). This range in potential applications is reflected in the use of weather radar throughout the Thames area, varying from subjective assessments of rainfall at Reading to quantitative input to flow-forecasting models as part of a fully integrated flood-forecasting system at Waltham Cross.

2.   DATA ACQUISITION SYSTEM

2.1   Reading
Both network and single-site radar data are transmitted to Reading and displayed in the control room for flood warning in the Thames Valley region. The composited network data are transmitted from the Meteorological Office at Bracknell and the single-site data from Chenies. A Logica Vitesse rainfall display system receives low-resolution data averaged over 5 km squares at eight intensity levels to produce the single-site display. A similar function is performed independently by a Software Sciences system using a Compaq microcomputer to produce the national network display. Both systems include a rolling archive of 5 km picture data, enabling frame storage of 'interesting events' on floppy disc.

The telemetry computer at Reading, a Ferranti Argus 700G, receives high-resolution radar data from Chenies at 208 intensity levels and maintains a rolling archive of subcatchment data. The Argus transmits real-time data from five telemetered raingauges located in the Thames catchment to Chenies for calibration purposes.

Weather radar data are currently only being used subjectively for operational flood warning, providing a general impression of weather conditions and, through the replay facility, an indication of the movement or development of an event. This assessment is supplemented by information gleaned in real time from a network of 31 raingauges and 40 river flow–level recorders situated throughout the Thames Valley region. A large programme of expansion of the network is now well advanced with approximately 76 additional river-level recorders being installed on a number of the tributary rivers. Real-time monitoring of levels in the River Thames is also being improved and to date almost half of the 45 locks have been equipped with telemetry.

The development of flood-forecasting models using radar and gauged data is also progressing off line on a Hewlett–Packard microcomputer. Replacement of the existing Argus computer by the early 1990s will enable further development of these models in the operational environment allowing much greater use of weather radar and gauge data from new telemetry outstations.

2.2   Waltham Cross

For the eastern flood-warning area, radar data from Chenies are transmitted via a public switched telephone network (PSTN) private wire to a computer located at the Thames Barrier in Woolwich. The flood room at Waltham Cross has continuous remote terminal access to the computer whilst duty staff operating from home can gain access to all features of the system via a PSTN dial-up modem link. The hardware consists of a VAX 11/750 minicomputer with an 8 mbyte physical memory running the VAX–VMS operating system, two 300 mbyte discs and one tape drive unit.

The flood-warning system incorporates highly sophisticated software which supports both real-time and background activities to capture, process, display and archive radar data. The Barrier VAX also acts as data gatherer for 30 raingauges and 50 river-level recorders which represent the telemetry network for the main rivers and catchments in the Lee and Greater London area. This information is combined to run real-time models producing flood forecasts and more recent developments have seen the use of raingauge data to produce locally calibrated radar pictures.

3.   RADAR HANDLING SOFTWARE

Radar data are captured and processed automatically with displays made immediately available upon the issue of simple commands. Operating on two discs for security reasons, the software is divided into a number of integrated but independent units so that, should one part fail, the whole system will not be disrupted. The system, shown schematically in Fig. 1, can be broadly subdivided into five sections: capture, validation and sorting, processing, display and archiving.

The program RDCAPT, a permanently running detached process, is responsible for monitoring the link from Chenies and capturing all transmitted data. A batch job, RDSORT, which is submitted automatically after each complete transmission is received, and further validates and processes the constituent data types, namely 5 km grid, 2 km grid, subcatchment and calibration data. Validated information is then made available for further processing either through the submission of other batch jobs or by being passed on to other processes.

The batch jobs RDGRID5, RDGRID2 and LVGRID2, invoked by RDSORT, are responsible for processing the 2 km and 5 km data, running at 5 and 15 min intervals respectively. A number of options can be selected for displaying 2 km radar information ranging from the full data set extending to a radius of 75 km from the radar site, to sub-areas covering catchments in the Lee Valley and London which are used to track the movement of convective storms on a local scale (Fig. 2). A further batch job, RDGAUGE, selects the 2 km grid squares above each of the 16

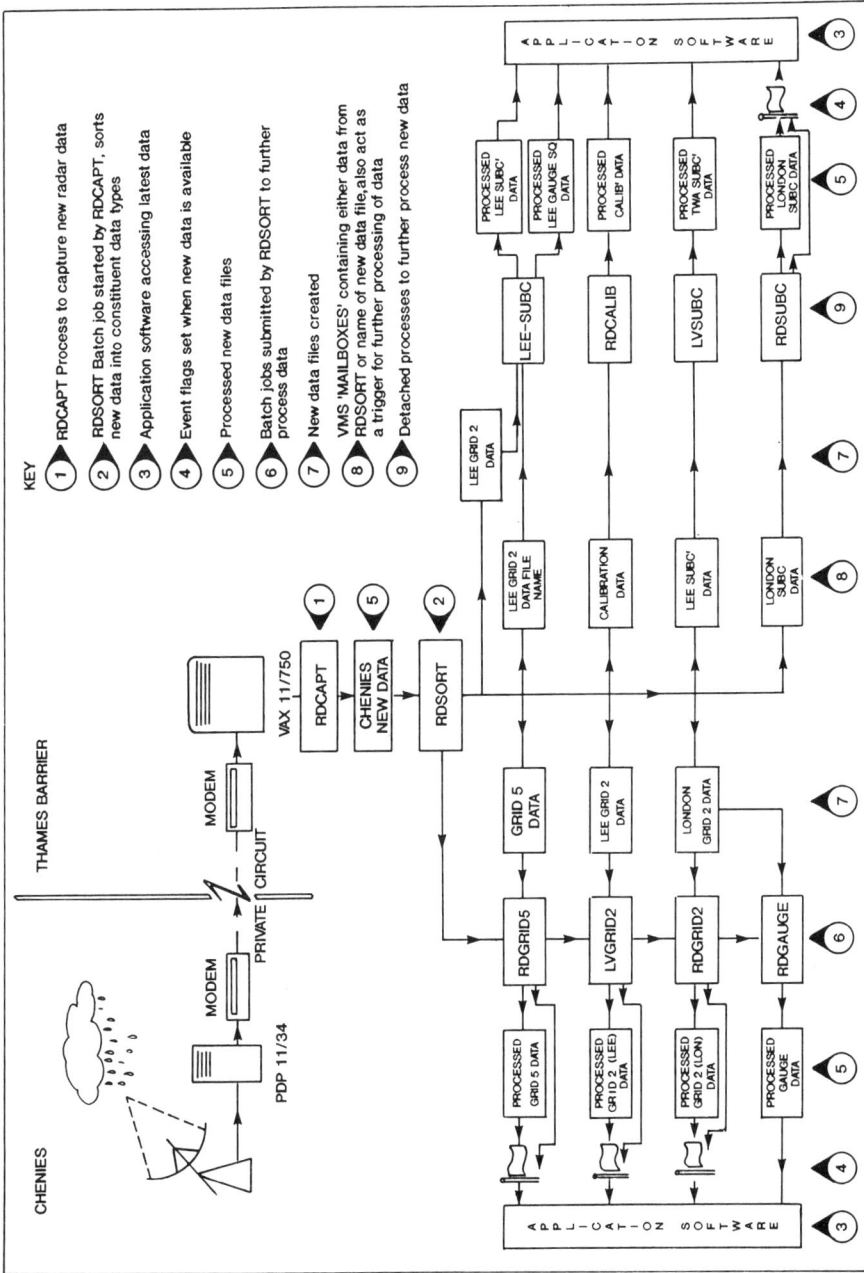

Fig. 1 — Schematic diagram of radar capture and processing software.

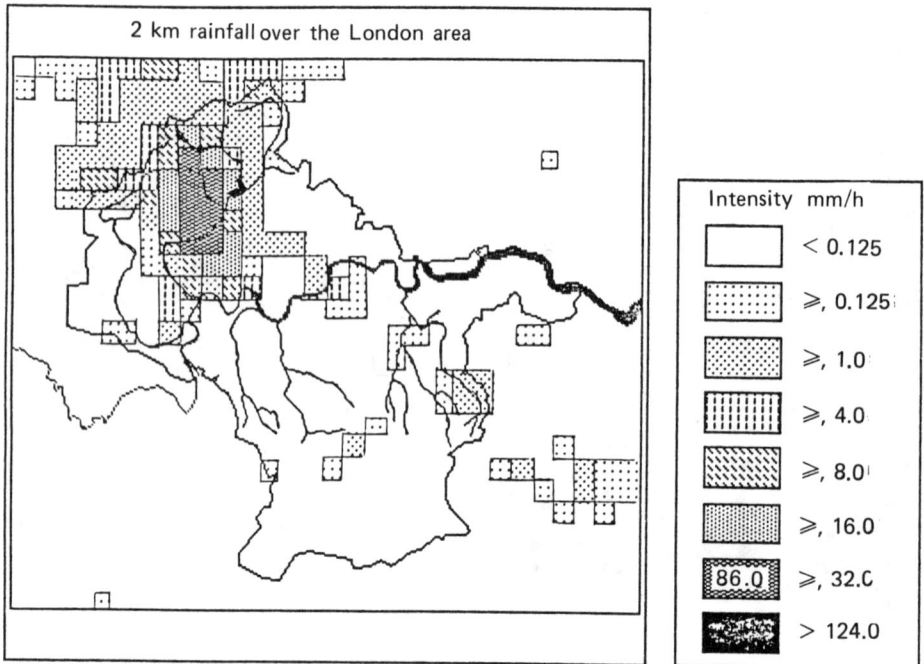

Fig. 2 — Rainfall intensity measured by Chenies weather radar at 1820 Greenwich Mean Time (GMT) on 8 May 1988.

raingauges that lie within the principal subcatchments in London. Rainfall totals, calculated for various time periods, are updated with each new 5 min transmission and compared with that recorded by the gauge.

Information that is used to calibrate Chenies data, in the form of ratios of radar-based estimates of rainfall and ground-based measurements made by the five calibration raingauges, is selected every 15 min by a process, RDCALIB. This information is added to the instantaneous intensity images and is also available for inspection in a tabular form.

Radar subcatchment data, once captured and validated, are handled by two processes which perform similar functions: the first, RDSUBC, for the data relevant to the 90 London subcatchments and the second, LVSUBC, for the remaining 100 Thames region subcatchments. The information, representing radar-based estimates of rainfall over pre-defined areas, is made available for 15 min, hourly and daily periods.

In contrast with the above programs which essentially manipulate validated subcatchment data generated at the radar site, equivalent information for the Lee Valley area is calculated from raw 2 km grid data on the VAX 11/750. The process, LEE-SUBC, achieves this by converting received rainfall intensities into amounts and summing for those grid squares applicable to each subcatchment. The single 2 km grid squares overlying each of the telemetered raingauges are then selected, enabling real-time comparisons to be made between rainfall estimates derived from either source.

All processes described above incorporate some degree of error handling involving the creation of log files for each new data transmission. Under normal conditions these files are deleted but would be retained for examination and further action in the event of an error. On-line archives are managed on both discs by each process in which data are stored in a standard format. Because of the quantity of data involved, however, information is copied to an off-line magnetic-tape-based system once a month and the on-line archive is restricted only to significant events.

4. DISPLAY OF RADAR DATA

An abundance of data display software is now available, most of which is designed to run on the Tektronix 410X and 420X series terminals and incorporates the Tektronix Plot 10 International Graphics Library. For all data types, simple menu-driven commands can be invoked to produce displays for both real-time and historic information. Single, multiple and 'movie' images can be produced for 5 and 2 km grid data (Fig. 2). Subcatchment displays are available in graphical and tabular format. Other options allow visual comparisons of 2 km grid and gauged estimates and the display of current calibration factors. In all cases any on-screen graphic or tabular display can be routed to a colour plotter or line printer respectively.

5. CATCHMENT MODELLING

Flood forecasting with catchment models forms an integral feature of the operational flood-warning system. However, the variation in topography, geology and land use throughout the region gives rise to marked contrasts in the hydrological characteristics of different rivers. To accommodate the full range, from smaller flashy urban catchments in and around London, to the larger slower-responding rural catchments of the Thames Valley, it has been necessary to adopt more than one model, although each is capable of operating with either radar or raingauge rainfall estimates.

5.1 Synthetic unit hydrograph
Modelling of the small, heavily urbanized London catchments is centred around a synthetic unit-hydrograph model derived from detail subdivision of catchment areas and the drainage network, both open and sewered.

The model assumes that a catchment's response is the combination of the runoff occurring on three distinct types of land use: paved areas, riparian open areas, and open areas which are isolated from the river by paved areas. Flow velocities are derived from standard pipe flow tables for paved areas and for Manning's formula for open areas, according to dimensions, slopes and roughness characteristics in each reach. Areas and flow travel times are used to construct time–area concentration diagrams for paved and open areas. These are then routed through a cascade system of reservoirs which simulate, analytically, the storage in the drainage system, to yield a synthetic unit hydrograph for each particular land use of the catchment.

Different rainfall loss models are used to derive the excess-rainfall input to each set of unit hydrographs. For paved areas, only depression storage is considered whereas, for pervious areas, rainfall loss is treated as the net result of depression storage and infiltration. In each case the initial conditions are calculated from pre-event rainfall and an accounting procedure is used to model the changes throughout the event.

Convoluting the appropriate excess-rainfall profile with its corresponding unit hydrograph produces separate paved- and open-area hydrographs representing the runoff response for each land use type. By combining these sequentially with time the net catchment flow response can be derived. These techniques can be applied to predict flows at either gauged or ungauged points on the river system.

5.2 Multiple-zone model

In the relatively more rural and often larger catchments of the Thames and Lee Valley areas, moisture conditions exert a far more important influence in dictating the river flow response to rainfall. For this reason the emphasis in modelling has been to focus upon the physical processes which influence rainfall losses and promote the runoff response at any given time.

The model can be described as a multiple-zone flow generation model with catchments being represented by one or a number of component zones (Greenfield 1984). Each hydrological zone is an area which has a characteristic runoff response usually as a result of topography, soils, geology or land use. Modelling is based upon the premise that for a given rainfall input the response from each zone can be individually identified within the total and therefore, by accurately identifying all significant component responses, the total catchment flow can be generated.

The model relies fundamentally on prior calibration with a series of historic rainfall–runoff events, ideally encompassing a range of storm types and catchment wetness conditions. Each pre-defined zone is characterized by a series of parameters denoting, the physical area of the zone, a potential drying constant (similar to Penman's root constant), linear and non-linear constants defining the behaviour of the soil moisture and groundwater storages, and an estimate of the proportion of percolation which bypasses the soil moisture storage. Refinement of initial parameter estimates is made using information from an event archive.

Additional data in the form of the current SMDs, the initial storage outputs and any artificial abstractions are required for each zone, before the flow generations can be made. Thus the model has to be routinely run, updating this information in preparation for its use with optimized parameters in forecasting mode.

5.3 Isolated-event model

In addition to the more complex models described above, the relatively simple isolated-event model (IEM) is being calibrated at gauged points for comparison purposes. This model, from applications elsewhere is considered to be well suited to real-time forecasting (Reed 1984).

IEM is a four-parameter non-linear conceptual rainfall–runoff model which is calibrated on past events to determine an optimum parameter set for use in simulating other events. There are two principal components: volume reduction and shape transformation. The volume reduction of gross rainfall is determined by separating excess from losses using the constant-proportional-loss method, based on an empirically derived rainfall–runoff ratio. Shape transformation of the resulting profile is then achieved with the use of a time delay function and a non-linear reservoir, each defined by IEM4 parameters.

The model is fitted in simulation mode in which the complete rainfall series is used following initialization with the value of discharge at the start of forecast. Optimum values for the model's parameters are derived using a constrained version of the Rosenbrock method and the observed and simulated hydrographs (Eyre and Crees 1984).

6. OPERATIONAL USE OF CATCHMENT MODELS

To make a flow forecast, the models described above require rainfall data which may be obtained, operationally, from one of three sources:

  (i) raingauge;
 (ii) radar;
(iii) rainfall forecast.

To date, only the synthetic unit hydrograph model is available for operational use for the river catchments of London. Flow predictions can be made at a total of 90 forecast points, including 23 telemetered sites where comparisons of observed and generated flows can be made in real time. These effectively represent the downstream limits of areas known to experience direct flooding from flows with a return period of up to once in 50 years. Each forecast point has been related to its relevant radar subcatchment area and a representative telemetered raingauge (Fig. 3). Data selection for a specified time period is fully automated and screen editing facilities are available. To run the model, the user has only to specify the forecast point, the start time of the storm and the type of rainfall data required (radar, raingauge or rainfall forecast).

The use of radar and/or raingauge data in isolation limits the maximum warning time to the rainfall–runoff lag of the subcatchment in question. In the upper reaches of the urban catchments in London, lag times can be as short as 30 min and in such circumstances there is a need for quantitative rainfall forecasts to extend lead times. Until FRONTIERS forecasts become available from the Meteorological Office in the early 1990s (Collier 1989), any improvement in warning times will only be achieved through close liaison with local weather centres before and during an event. The system has therefore been designed to accept rainfall forecast details such as the expected start of the storm, its duration, peak intensity and total depth. This information is fed into one of several pre-defined rainfall profiles which is in turn used to generate a flow forecast at a point of interest. During an event both the observed and the forecast elements of the profile are automatically merged.

Whilst the multiple-zone model is not yet available in an operational version for flood warning, it can be used in forecast mode to predict the flood hydrograph which will result from a given forecast rainfall input.

In forecast mode, users either are prompted to supply their own 5 h forecast rainfall profile or are given the option for this to be automatically calculated. With the autoforecast, four different 5 h profiles are calculated, each a different function of the previous hours rain:

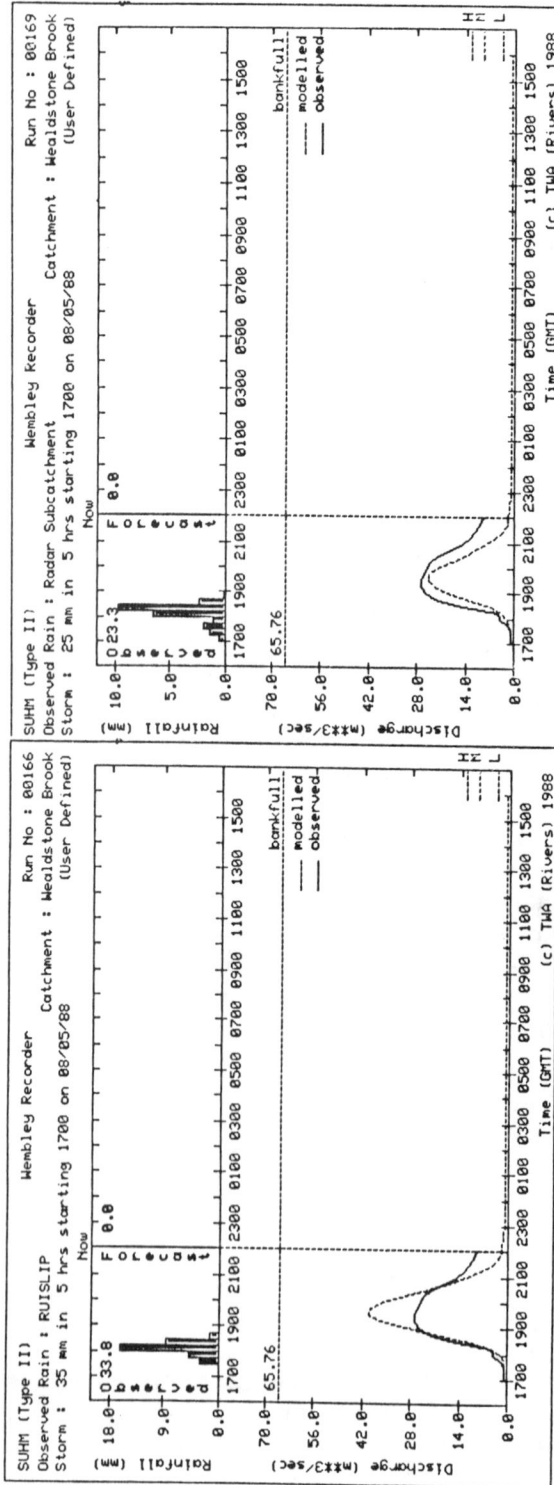

(a)

(b)

Fig. 3 — Output from the synthetic unit-hydrograph model for 8 May 1988 using rainfall estimated from (a) a raingauge and (b) a radar subcatchment: ———, modelled; - - -, observed.

(a) no further rain;
(b) a constant average of the previous 3 h rain;
(c) a linear decline to zero from the previous hours' rain;
(d) a linear increase to twice the average of the previous 3 h rain.

Observed and forecast rainfall can be applied either uniformly in a lumped form, or in a more distributed way by assigning different rainfall profiles to each of the component hydrological zones which make up the catchment area. In all cases the results are displayed graphically by appending forecast flows either to the last observed flow or, if telemetered records are not available, to the last generated flow. Bankful flows and maxima reached in earlier notable events may also be flagged on screen displays and plots to assist the observer in assessing the likely severity of the forecast flood. Flow predictions can currently be generated at 15 sites in the Thames Valley region.

7. SYSTEM ENHANCEMENTS

The use of radar to provide quantitative information on rainfall depth for use in flow-forecasting procedures has been a growing feature of Thames Water's operational flood-warning system. It has been recognized, however, that in certain circumstances radar data loses accuracy, leading to the generation of unreliable forecasts. This has been attributed, in part, to the fact that the Chenies radar is calibrated using only five raingauges and that the synoptic-type-dependent domain procedure used introduces temporal and spatial discontinuities in the rainfall estimates supplied. In 1987, Thames Water commissioned the Institute of Hydrology to explore the possibility of developing a regional recalibration system to obtain a more accurate and reliable estimate of spatial rainfall variations using data from a network of 30 telemetered raingauges in the London and Lee Valley area (Moore et al. 1989).

The recalibration system, implemented on the VAX computer at the Thames Barrier, first removes the effect of the at-site domain calibration and then uses rainfall totals over the last 15 min from the network of 30 raingauges to compute calibration factors. A multiquadric surface is then fitted to these factors which in turn is applied to the radar field to obtain the recalibrated product. The system became operational for the eastern area on the 14 March 1989 and an analysis of historical raingauge and radar data for 19 storm events indicated that the product was on average 25% more accurate than the at-site calibrated product.

A second objective of the study is to use weather radar data as the basis of a short-term high resolution rainfall-forecasting procedure to complement FRONTIERS forecasts shortly to be provided by the Meteorological Office. It is the intention of FRONTIERS to provide a 5 km, 30 min product whereas 2 km, 15 min resolution forecasts are of greater value particularly for the issue of accurate and timely flood warnings in small fast-responding urban catchments. To date, two procedures have been developed to identify the speed and direction of storm movement. Both techniques are based on maximizing the correlation between two radar data fields for successive time frames through an appropriate displacement in space. One is based on a regional analysis of storm movement and the other on local analysis both of which are described in full by Moore et al. (1989).

It is envisaged that the present study will be extended with the aim of producing an operational regional rainfall-forecasting system to be implemented on the Barrier VAX computer. An evaluation of FRONTIERS forecasts will also be made in terms of accuracy in an absolute sense, and also relative to the regional procedure. The evaluation might, for example, indicate that the latter is to be preferred at shorter lead times and in situations where a product of higher resolution is required, but the sounder physical basis of FRONTIERS forecasts yields better longer-term predictions.

8. CONCLUSIONS

It is evident that weather radar has had a major impact on flood monitoring and forecasting in the Thames region. The integrated data acquisition system, developed for the London area, has successfully been extended to cover the Lee Valley, resulting in the provision of more timely and accurate flood warnings. The system's software has been written in a flexible manner to facilitate further extension to other parts of the Thames catchment and even other regions in England and Wales if the need arises. Future development will concentrate on integrating the locally calibrated radar product into the flood forecasting process and making wider use of quantitative rainfall forecasts to extend lead times especially for the urban catchments in and around London.

ACKNOWLEDGEMENTS

The authors are indebted to Thames Water Authority, National Rivers Authority Unit, for permission to publish this paper. The views expressed are those of the authors and are not necessarily representative of the Thames Water Authority.

REFERENCES

Collier, C. G. (1989) Weather radar forecasting. *Proceedings of the Natural Environment Research Council Seminar on Weather Radar and the Water Industry, Opportunities for the 1990's*, BHS Occasional Paper. British Hydrological Society.

Eyre, W. S. and Crees, M. A. (1984) Real-time application of the isolated event rainfall–runoff model. *J. Inst. Water. Eng. Sci.*, **38** (1), 70–78.

Greenfield, B. J. (1984) *The Thames Water catchment model*, Internal Document. Thames Water Authority.

Haggett, C. M. (1986) The use of weather radar for flood forecasting in London. *Proceedings of the Conference of River Engineers, Cranfield, 15–17 July 1986*. Ministry of Agriculture, Fisheries and Food, 11 pp.

Haggett, C. M. (1989) Weather radar for flood warning. *Proceedings of the Natural Environment Seminar on Weather Radar and the Water Industry, Opportunities for the 1990's*, BHS Occasional Paper. British Hydrological Society.

Moore, R. J., Watson, B. C., Jones, D. A., Black, K. B., Haggett, C. M., Crees, M. A. and Richards, C. I. (1989) Towards an improved system for weather radar

calibration and rainfall forecasting using raingauge data from a regional telemetry system. *Proceedings of the 3rd Scientific Assembly, International Association of Hydrological Sciences on Surface Water Modelling — New Directions
for Hydrological Prediction, Baltimore HD, 10–19 May 1979. 9 pp.*

Reed, D. W. (1984) *A Review of British forecasting practice,* Institute of Hydrology
Report No. 90. Natural Environment Research Council.

54

NEXRAD: new era in hydrometeorology in the USA

M. D. Hudlow, J. A. Smith, M. L. Walton and R. C. Shedd
Office of Hydrology, National Weather Service, National Oceanic and
Atmospheric Administration, Silver Spring, MD 20910, USA

ABSTRACT

Introduction of the Next Generation Weather Radar (NEXRAD) system will usher in a new era of precipitation processing for the US National Weather Service (NWS). Precipitation processing is performed in three stages. Stage I of processing occurs within the NEXRAD computer system and produces quantitative precipitation estimates, short-term forecasts of precipitation accumulations and flash-flood probabilities. Stages II and III, performed at NWS forecast offices external to NEXRAD, will further improve the quality of precipitation estimates through the use of satellite imagery, raingage data, and eventually other hydrometeorological information. NEXRAD data, combined with raingage and satellite data, will lead to large improvements in the accuracy of precipitation estimates.

1. INTRODUCTION

One of the most important steps in the process of making river forecasts is an accurate precipitation analysis. Precipitation is one of the primary inputs to the National Weather Service's river forecast system. Accurate precipitation analyses are also extremely important to the meteorologist faced with a flash-flood situation. A raingage network, even a fairly dense one, can miss significant rainfall, especially rainfall associated with intense convective storms. Another significant problem with raingage networks is the inability to determine patterns of precipitation or to identify the heaviest amounts. Thus, there is a strong need for real-time precipitation analysis that is not totally dependent upon raingage observations. This is especially important for remote areas where it is difficult to establish and maintain surface observing stations.

In the early 1970s an effort was begun to use the established National Weather Service (NWS) radar network to provide areal precipitation analysis. Development of procedures for using radar as a tool to measure precipitation has progressed from manual techniques, to semiautomatic techniques, to the fully automatic techniques employed by the Next Generation Weather Radar (NEXRAD) system.

NEXRAD is a joint US Department of Commerce, US Department of Defense and US Department of Transportation effort to develop, procure, implement and maintain operationally a Doppler weather radar system. It is a nationwide network of weather radars designed to meet the hydrometeorological service needs of the USA into the twenty-first century. The three agencies will acquire approximately 175 radars, 413 workstations and associated equipment. For the continental USA, NWS plans to replace the existing weather radar network with 113 of the approximately 175 NEXRADS. Deployment of the NEXRADs is scheduled to begin in the spring of 1990 and to continue into the mid-1990s. NEXRAD is a totally integrated system and consists of a 10 cm Doppler radar, a powerful computer to process the data and develop products, and a graphical workstation to display products.

The NWS hydrometeorological processing is performed in three stages. Stage I will be performed within the NEXRAD computer system. Processing for stages II and III is performed external to the NEXRAD computer system. Testing and evaluation of the stage I hydrometeorological processing have been conducted by the Hydrologic Research Laboratory (HRL) in collaboration with the NEXRAD Operational Support Facility (OSF) in Norman OK, and the Prototype Regional Observing and Forecasting Service (PROFS) located in Boulder, CO (O'Bannon and Ahnert 1986, Kelsch 1989). Testing and evaluation of stages II and III are under way at HRL. Software for the three stages of hydrometeorological processing was developed by HRL. A schematic diagram of the three stages of hydrometeorological processing is shown in Fig. 1. The following sections will discuss the three stages of hydrometeorological processing. Particular emphasis will be placed on a more detailed description of the stage I processing components as illustrated in Fig.2.

This paper concentrates on radar hydrology developments in the USA. However, other countries are pursuing similar directions in the development and operational implementation of radar hydrology applications (Joss and Waldvogel 1989).

2.   THE THREE STAGES OF HYDROMETEOROLOGICAL PROCESSING

2.1   Stage I

The stage I processing within NEXRAD contains two major components: a precipitation-processing system (PPS) and a flash-flood potential system (FFPS) (Fig. 2). In stage I the data are processed to a level of refinement that can be achieved with modest computer resources and yet provide an accuracy that make the precipitation estimates useful for local real-time applications.

2.1.1   Precipitation-processing system

The PPS generates 1 h running totals as well as 3 h and running storm total precipitation accumulations. Five steps are performed to develop the best estimate of precipitation:

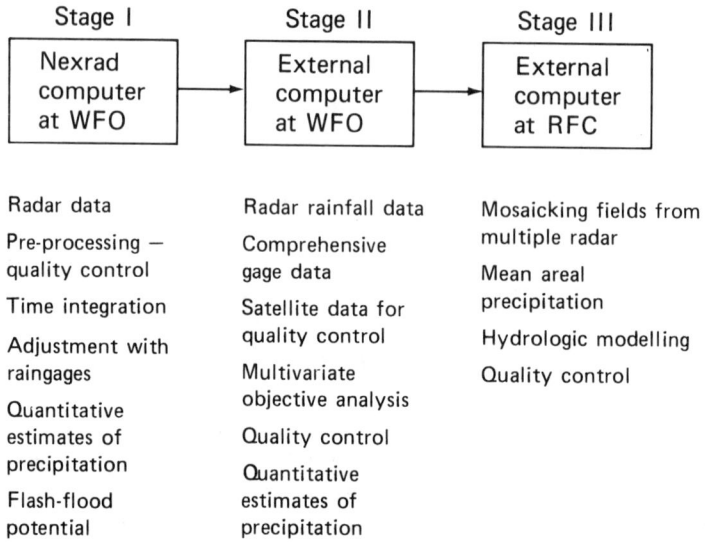

Fig. 1 — Three stages of hydrometeorological processing: WFO, NWS Warning and Forecast Office; RFC, River Forecast Center.

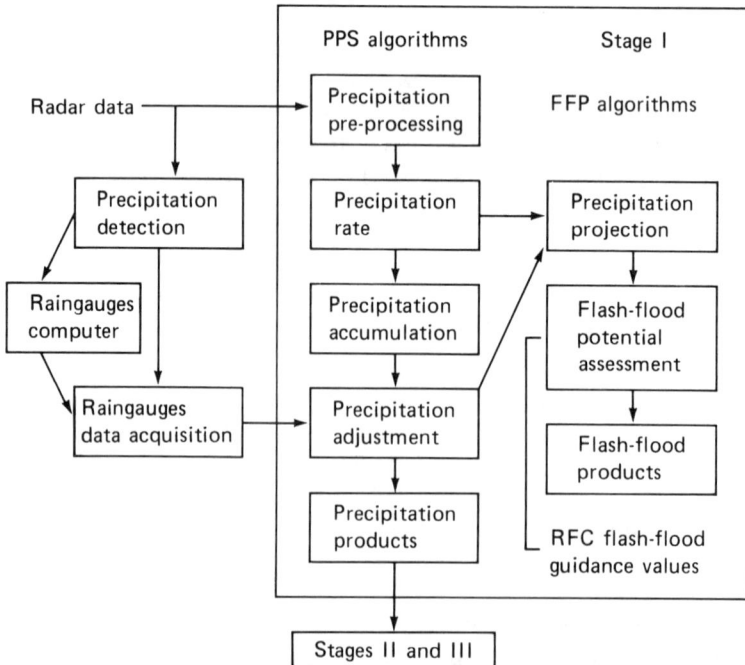

Fig. 2 — Stage I block diagram: RFC, River Forecast Center.

(1) development of sectorized hybrid scan;
(2) conversion to precipitation rate;
(3) precipitation accumulation;
(4) adjustment using raingages;
(5) product update.

Since radar indirectly measures precipitation rates, an emphasis on quality control of the data has been built into each of the processing steps of the PPS.

The first step produces a 'sectorized hybrid scan', which is comprised of reflectivity data from the four lowest tilts of the radar volume scan. The choice of tilt is based on range and topography. Details of the sectorized hybrid scan are described within a companion paper from this symposium (Shedd *et al.* 1991). There are four reasons for multiple-tilt processing:

(1) to minimize ground clutter and to clear mountains or man-made obstacles;
(2) to reduce effects of abnormal beam refractions and losses;
(3) to improve range performance;
(4) to maximize the use of data from a relatively uniform altitude.

NEXRAD reflectivity data are pre-processed prior to development of the hybrid scan to correct for several error sources. One source of error results from partial or complete beam blockages, generally from man-made objects (such as water towers). The form of the blockage correction depends on the percentage of the beam blocked and the radial width of the blockage. Reflectivity outliers can occur when the beam strikes high-reflectivity objects such as planes. These data points are corrected using neighbouring data points.

The final quality control check is to determine whether or not data from the lowest tilt will be used in building the hybrid scan. The low tilt is that most likely to be contaminated by anomalous propagation, ground clutter and other noise. The lowest tilt is rejected when more than a given percentage of the echoes at the lowest angle disappear at the second elevation angle.

Following construction of the hybrid scan, reflectivity data from the hybrid scan step are converted to rainfall rates using the empirical relationship $Z=aR^b$ where Z is the reflectivity factor, R is the rainfall rate, and a and b are constants. Two quality control procedures are performed in the rate algorithm. First, temporal continuity of the total field volumetric water is checked to insure that spurious data have not introduced physically unreasonable rates of echo development or decay. Range effects resulting from signal degradation and partial beam filling may also unduly reduce precipitation estimates at further ranges; therefore, a range-dependent site-varying correction will be applied to the precipitation rate data.

The third step is to form precipitation accumulations. Two types of accumulation are generated. The first is a scan-to-scan accumulation which measures the precipitation accumulation from one instantaneous rate scan to the next. The second is a running hourly accumulation which is updated every volume scan. A temporal continuity test checks the data for missing periods which might result from system malfunction or rejection of scans. If too many data are missing, no running hourly accumulation is performed.

An outlier check is performed on the hourly accumulations. This check is made in addition to the instantaneous reflectivity outlier check because clutter which passes the reflectivity test could, if it remains at a high level, produce impossibly large accumulations.

The fourth step is adjustment of the accumulations based on available raingage data. In spite of efforts to maintain a high level of quantitative accuracy in estimating precipitation from radar data, there are sure to be errors in these estimates. In fact, errors of factor of 2 or more can occur owing to a wide variety of causes including hardware calibration, anomalous propagation, wet radome attenuation, inappropriateness of $Z–R$ relationship for the particular storm system, and others. While some of these errors will be localized or perhaps range dependent, some will produce a generally uniform multiplicative bias in the radar estimated precipitation. In either case, a mean bias correction can be applied to the entire field in an attempt to insure that the estimate of total field volumetric water closely equals the true field volumetric water. In order to effect this correction, a procedure has been developed to compare hourly precipitation from raingages to associated radar values and to estimate the mean-field radar bias.

The adjustment procedure is based on a discrete Kalman filter (Ahnert *et al.* 1986). The bias update is performed once per hour when sufficient real-time gage reports are available. If insufficient raingage reports are available, the previous estimate of the bias is propagated forward.

In the final step of processing in the PPS, two basic types of product are generated. Graphical products will be displayed at forecast offices. Digital products will be used for subsequent numerical processing and input to forecast models.

The precipitation graphics products are displayed on a 2 km×2 km grid to a range of 230 km from the radar site and have up to 160 colour levels. These precipitation products are 1 h, 3 h and storm total accumulations. The 1 h and storm total products will be updated each volume scan. The 3 h product will be updated once per hour.

Also produced is a digital array product of the running hourly accumulations. The product, updated every volume scan (approximately 5 min), will be mapped onto a one-fortieth limited-area fine mesh (LFM) grid (approximately 4 km×4 km). Although this product has somewhat degraded spatial resolution, it will provide the same temporal resolution and far greater intensity resolution (up to 100 precipitation levels) than the graphical product. In addition, supplemental data such as instantaneous area-averaged precipitation rates over a one-quarter LFM grid, bias estimate, missing periods and raingage are appended to the digital array product. These supplemental data can be used for performing quality control and data adjustment steps in subsequent hydrometeorological procedures.

Despite all the quality control and data adjustment procedures in PPS, the products will not be perfect. PPS does not, for example, directly account for changes in the phases of precipitation from liquid to frozen or vice versa. The PPS products will, however, be of the highest quality possible given current technology and available resources. Further details are available in the work by Ahnert *et al.* (1983, 1984). Ongoing and future research and development will enhance the performance of the PPS.

2.1.2   *Flash-flood potential system*
The stage I NEXRAD FFPS produces short-term forecasts of precipitation accumulations and flash-flood potential. The NEXRAD FFPS consists of a precipitation

projection procedure and a flash-flood potential assessment procedure (Fig. 2). The precipitation projection procedure forecasts precipitation accumulation up to 1 h into the future. The forecasts are updated every volume scan (approximately every 5 min). The procedure also produces projected total precipitation accumulations and associated errors variances. The projected total accumulation is composed of the previously observed accumulation and the projected accumulation. The observed precipitation data used by the FFPS come from the NEXRAD PPS. The precipitation procedure consists of four steps:

(1) estimation of the mean, variance and residual of the precipitation rate;
(2) estimation of localized storm velocity;
(3) estimation of the residual persistence;
(4) projection of precipitation rates with subsequent conversion to accumulations.

In the first step, the precipitation projection procedure uses a spatial moving average for the mean, variance and residual of precipitation rate. The mean is calculated by averaging over a region which roughly corresponds to a 20 km×20 km area. The localized spatial moving average of the variance of precipitation rates over this region is computed similarly. The variance of observation error is assumed to be proportional to this variance. The residual is defined as the difference between the observed precipitation rate of individual bins in the 20 km×20 km area and the mean value.

In the second step, the localized storm velocity and direction are determined by a pattern-matching technique. The technique involves comparing the current precipitation rate field with a previous precipitation rate field at every fifth box (one-fortieth LFM grid) for various offsets to determine the minimum sum of absolute differences. The offsets range from +2 to −2 boxes (one-fortieth LFM grid) in the X and Y directions which will account for storm movement in any direction and for storm velocities up to approximately 50 km/h. The offset with the minimum sum of absolute differences provides the first estimate of the velocity and the direction at every fifth box (one-fortieth LFM grid).

These first estimates are in turn smoothed by weighted averaging with nearest-neighbour first-estimate velocities. Velocities are then interpolated for all the other boxes (one-fortieth LFM grid) using an inverse-distance-squared weighted average of the smoothed velocities. Simple persistence is assumed when projecting the storm velocity and direction into the future.

In the third step, the parameters of the residual process are estimated at each scan by translating the residuals of the previous scan according to the localized storm velocity and computing the lag-one autocorrelation of the translated previous residuals with the current residuals.

Finally, in the fourth step, the projected precipitation rate at each box (one-fortieth LFM grid) is the mean precipitation rate plus the projected residual. The projected residual is the current residual times the residual persistence parameter raised to a power equal to the number of time steps into the future. These projected precipitation rates are then moved according to the projected local storm velocity and direction. The projected precipitation accumulations are based on these projected precipitation rates. These projections, accumulations and error variances are the basic input for ther flash-flood potential assessment procedure.

The flash-flood potential assessment procedure uses flash-flood guidance values developed by the NWS river forecast centers (RFCs) and observed and projected precipitation accumulations from the precipitation projection procedure to produce observed and projected flash-flood probabilities. The flash-flood probability is an estimate of the probability that the actual precipitation for some time during the rainfall event has exceeded (for observed flash-flood probability) or will exceed (for potential flash-flood probability) the flash-flood guidance value. The flash-flood guidance values are based on hydrologic models run by the RFCs. These guidance values are estimates of how much rainfall would be required over specified durations to produce flooding at one or more locations within a zone or county.

The FFPS generates digital, graphic and alphanumeric products. The digital products are intended for numerical use at computer facilities external to the NEXRAD system itself. The digital data maintain the full dynamic range and full precision of the data used to generate the product. Just like the PPS, the digital data are mapped on a 'universal' grid (one-fortieth LFM grid) so that data from multiple sites are compatible for mosaicking.

The digital data will be transferred to other computer facilities at the RFCs and forecast offices for use in automated forecasting models and procedures. The digital products consist of a projected precipitation and projected and observed error variance of the precipitation data array on a 131×131 one-fortieth LFM grid and is updated every volume scan. In addition, supplemental data such as projection parameters and storm velocity parameters are appended to the digital array product.

The precipitation graphics products will be displayed on a $4 \, \mathrm{km} \times 4 \, \mathrm{km}$ grid to a range of 230 km from the radar site are updated every volume scan and have up to 16 color levels. The three graphical products are projected precipitation accumulation for up to 1 h into the future and observed and projected flash-flood probability displays.

The alphanumeric products are updated every volume scan and will provide flash-flood probability information in a form suitable for display on both graphic and alphanumeric display devices. The first alphanumeric product consists of a flash-flood guidance summary which will display the flash-flood guidance value, maximum observed precipitation accumulation for the zone, and maximum total storm (observed and projected) precipitation accumulation for each guidance value duration in each flash-flood guidance zone. The second alphanumeric product will display the maximum observed and projected flash-flood probabilities for each zone.

The products produced by the FFPS should be viewed by the user as useful guidance but not as definitive identification of flash flooding until interpreted together with other available information. The products generated, using the projection procedure, do not explicitly take into account storm systems moving faster than approximately 50 km/h, curvilinear storm motions, individual cell dynamics other than that accounted for by the current residual field, or orographic effects.

Other limitations arise from the use of flash-flood guidance values which at present are not calculated the same way at all RFCs, are updated only once a day and do not have an updating procedure to reflect changes brought about by multiple rainfall events. It is important that the user be aware of these limitations when interpreting and using products from the FFPS. Future research and development will be aimed at reducing these and other limitations experienced operationally. Further details on the FFPS are available in the work of Walton *et al.* (1985, 1987) and Walton and Johnson (1986).

2.2   Stage II

The stage II precipitation-processing program is used to compute hourly precipitation on a one-fortieth LFM grid for the area covered by a single NEXRAD system. Input to stage II includes hourly digital precipitation data from stage I processing, GOES infrared imagery, raingage data, and eventually other hydrometeorological information (Hudlow *et al.* 1983). Ultimately the stage II program will run at the National Weather Service Warning and Forecast Office (WFO) collocated with the NEXRAD system (Fig. 1). In the interim, a combination of stage II and III processing will be done at several RFCs. Stage II precipitation analyses are used by the WFO in providing forecast guidance during periods of severe weather and as input to stage III precipitation processing at RFCs.

Stage II precipitation processing differs from stage I in several ways. Additional quality control steps are carried out in stage II processing. Satellite and raingage data are used to detect and eliminate errors in NEXRAD data associated with clear-air anomalous propagation or other data contamination not detected or eliminated during stage I processing (Fiore *et al.* 1986). From the satellite data it can be determined whether clouds are contained in a one-quarter LFM grid box. If radar detects rainfall in a one-quarter LFM grid box for which satellite data indicate that no clouds are present and for which no raingages record rainfall, then the radar rainfall estimates are replaced by zero values.

In stage II processing, radar and raingage data are 'merged' to form an optimal 'multisensor' estimate of the rainfall field. The merging procedure accounts for strengths and weaknesses of the two measurement systems. To estimate rainfall at a given location, a raingage observation will be heavily weighted only if it is close to the location. The weight that a raingage receives will also depend on characteristics of the rainfall field. For rainfall fields with large spatial variability, as is typically the case with convective storms, raingage observations will generally receive lower weights than for more uniform rainfall fields, associated, for example, with stratiform rainfall.

The stage II program will also produce estimates of rainfall based largely on raingage data. In the 'gage-only' rainfall analysis, radar and other data are used to delineate regions receiving no rainfall from precipitating regions. Graphical products will allow display of stage II precipitation estimates at the WFO. Summary information, such as mean rainfall over the field and maximum point rainfall, will also be displayed by the stage II program.

2.3   Stage III

The stage III precipitation-processing program provides two products. It provides hourly estimates of rainfall on a one-fortieth LFM grid for the entire area of responsibility of an RFC. At an RFC it is necessary to combine information from a number of NEXRAD radars. The program also provides mean areal precipitation (MAP) values for basins specified by the RFC. MAP time series are provided at the time step required by the RFC (1 h, 2 h, 3 h, 6 h or 24 h) for operational hydrologic forecasting.

The stage III program contains two basic steps:

(1) quality control–mosaicking;
(2) MAP computation.

The first step is an interactive quality control step in which the forecaster can replace the stage II multisensor rainfall estimates by the stage II gage-only rainfall estimates. The forecaster will base his decision on displays of preliminary mosaicked multi-sensor and gage-only fields for the entire forecast area. From these displays it should normally be clear to the forecaster whether anomalous propagation errors, or certain other errors, are still present in the multisensor rainfall estimates. The product of the first step is an hourly mosaicked rainfall field for the entire RFC area of coverage. In the MAP calculation, hourly rainfall estimates on a one-fortieth LFM grid are accumulated and averaged to the time and space resolutions required for hydrologic forecasting.

3. SUMMARY

The three stages of hydrometeorological processing will provide high-quality precipitation estimates over the contiguous USA. The first stage will take place within NEXRAD and be used for real-time graphical displays and input to forecast procedures at the local forecast offices and RFC. Processing in stages II and III will further improve quality of precipitation estimates using satellite, raingage data and eventually other hydrometeorological information. The final optimal precipitation estimates will be input to hydrologic models and allow the user to monitor the accumulated precipitation for various durations up to the current time, to evaluate precipitation forecasts for short periods into the future and to assess flood potential. Using data from NEXRAD, combined with additional raingage and satellite data, it should be possible to realize large improvements in the accuracy of estimating areal precipitation. These improvements should, in turn, lead to large economic benefits and better management of our increasingly precious water resources.

Space restrictions prevented inclusion of various sample products and test results herein. These may be obtained from references cited and/or by contacting the authors.

ACKNOWLEDGMENTS

The authors would like to thank the NEXRAD Joint Systems Program Office for supporting development of the stage I hydrometeorological software. The co-operation and support of PROFS, and the NEXRAD OSF are also greatly appreciated. Critical test data have been provided by the National Severe Storms Laboratory and the National Center for Atmospheric Research. The authors also wish to acknowledge a number of people who contributed to various degrees in the work presented here. Dr Edward Johnson, Dr Susan Zevin, Dr George Smith, Dr

Konstantine Georgakakos, Dr Witold Krajewski and Mr Peter Ahnert all provided invaluable contributions to the development of the various stages of hydrometeorological processing.

REFERENCES

Ahnert, P. R., Hudlow, M. D., Johnson, E. R., Greene, D. R. and Dias, M. P. R. (1983) Proposed 'on-site' precipitation processing system for NEXRAD. *Preprints of the 21st Conference on Radar Meteorology, Edmonton, Alberta, 19–23 September 1983*, American Meteorological Society, Boston, MA; Alberta Research Council, Canadian Meteorological and Oceanographic Society, Edmonton, Alberta, pp. 378–385.

Ahnert, R., Hudlow, M. D. and Johnson, E. R. (1984) Validation of the 'on-site' precipitation processing system for NEXRAD. *Preprints of the 22nd Conference on Radar Meteorology, Zurich, 10–14 September 1984*. American Meteorological Society, Boston, MA, 10 pp.

Ahnert, P. R., Krajewski, W. F. and Johnson, E. R. (1986) Kalman filter estimation of radar-rainfall field bias. *Preprints of the 23rd Conference in Radar Meteorology, Snowmass, CO, September 1986*. American Meteorological Society, Boston, MA, pp. JP33–JP37.

Fiore, J. V., Farnsworth, R. K. and Huffman, G. J. (1986) Quality control of radar-rainfall data with VISSR satellite data. *Preprints of the 23rd Conference on Radar Meteorology, Snowmass, CO, September 1986*. American Meteorological Society, Boston, MA, pp. JP15–JP18.

Hudlow, M. D., Farnsworth, R. K. and Ahnert, P. R. (1984; revised 1985) *NEXRAD technical requirements for precipitation estimation and accompanying economic benefits*. Hydro Technical Note-4. Office of Hydrology, National Weather Service, National Oceanic and Atmospheric Administration, Silver Spring, MD, 49 pp.

Hudlow, M. D., Greene, D. R., Ahnert, P. R., Krajewski, W. F., Sivaramakrishna, T. R., Johnson, E. R. and Dias, M. P. R. (1983) Proposed off-site precipitation processing system for NEXRAD. *Preprints of the 21st Conference on Radar Meteorology, Edmonton, Alberta, 19–23 September, 1983*. American Meteorological Society, Boston, MA, Alberta Research Council, Canadian Meteorological and Oceanographic Society, Edmonton, Alberta, pp. 394–403.

Joss, J. and Waldvogel, A. (1989) Precipitation measurement and hydrology — a review. *Proceedings of the Battan Memorial and 40th Anniversary Conference on Radar Meteorology, 1989*. American Meteorological Society, Boston, MA.

Kelsch, M. (1989) An evaluation of the NEXRAD hydrology sequence for different types of intense convective storms in northeast Colorado. *Preprints of the 24th Conference on Radar Meteorology, Tallahassee, FL, 27–31 March, 1989*. American Meteorological Society, Boston, MA, pp. 207–210.

O'Bannon, T. and Ahnert, P. (1986) A study of the NEXRAD precipitation algorithm package on a winter-type Oklahoma rainstorm. *Preprints of the 23rd Conference on Radar Meteorology, Snowmass, CO, September, 1986*. American Meteorological Society, Boston, MA, pp. JP99–101.

Shedd, R., Smith, J. A. and Walton, M. L. (1991) Sectorized hybrid scan strategy of the NEXRAD precipitation-processing system. *Hydrological Applications of Weather Radar.* Ed. Cluckie, I. D. and Collier, C. G. Ellis Horwood, Chichester, West Sussex, Chapter 4.

Walton, M. L. and Johnson, E. R. (1986) An improved precipitation projection procedure for the NEXRAD flash-flood potential system. *Preprints of the 23rd Conference on Radar Meteorology, Snowmass, CO, September 1986.* American Meteorological Society, Boston, MA, pp. JP62–JP65.

Walton, M. L., Johnson, E. R., Ahnert, P. R. and Hudlow, M. D. (1985) Proposed on-site flash-flood-potential system for NEXRAD. *Preprint of the 26th American Meteorological Society Conference on Hydrometeorology, Indianapolis, IN, 29 October 1985.* American Meteorological Society, Boston, MA, pp. 122–129.

Walton, M. L., Johnson, E. R. and Shedd, R. C. (1987) Validation of the on-site flash-flood potential system for NEXRAD. *Proceedings of the 21st International Symposium on Remote Sensing of Environment, Ann Arbor, MI, 26–30 October 1987.* 12 pp.

55

An application of computer-based training analysis to an operational radar-based flood-warning system

G. A. Kennedy
National Rivers Authority, Thames Region,
Reading RD1 8DQ, UK

ABSTRACT

This paper describes how a computer-based training analysis can be applied to an operational flood-warning system. The use of the computer to provide a transaction log during real events is illustrated with reference to the radar display subsystem. This reveals that a priority of displays can be drawn up and if necessary, used to guide the flood-forecasting officer. The overall conclusion of the analysis suggests two applications that could be serviced by a small off-line data bank.

1. INTRODUCTION

An operational flood-warning system should be considered as a dynamic and flexible system. Within this system lies the experience of the flood-forecasting officers, a very valuable and often inaccessible resource. The purpose of this analysis is to see whether techniques are available to extract this resource and to define it as a dynamic contribution to an existing flood-warning system.

The availability of digital radar data at resolutions of 5 and 2 km and as pre-defined subcatchment totals to the London non-tidal System since 1985 makes this system ideal for this type of analysis. The non-tidal system was and is continually developed to utilize the increased observational capability of the Chenies radar installed by the London Weather Radar Project (Clift 1983). The flashy urban rivers mean that lead times are very short, often as little as half an hour, and that the flood-forecasting officers are under great pressure to produce consistent and accurate forecasts.

1.1   The London non-tidal flood-warning system

The existing outstation network was in place by 1977 as shown in Fig. 1. The incentive of major flooding on the River Brent on the 16–17 August 1977, as described by Haggett (1980), enabled financial and political support to be obtained for the London Weather Radar Project and the subsequent upgrading of the non-tidal flood-warning system, Prickett (1978). It was possible to upgrade the system from an alarm-driven digital talk-out configuration to a fully computer-controlled radar-based system, as described by Kennedy and Vairaramoorthy (1979). The skeleton of this system was completed by the end of 1985. This system was further enhanced and developed before being transferred to Thames Water Authority on 1 April 1986. Since that time a large commitment and investment by Thames Water has been a largely software-driven development, culminating in the current system. A generalized version of the system, as in June 1988, is presented in Fig. 2. Further details of more recent development can be found in the work by Haggett (1986) Haggett *et al*. (1989) and Moore *et al*. (1989).

1.2   The objectives

The objectives were:

(1) to observe the current London non-tidal flood-warning system and then to advise and demonstrate methods of documenting flood-warning experience and
(2) to suggest methodologies appropriate to the implementation of this knowledge transfer.

2.   ANALYSIS: AN INFORMAL APPROACH

The flood-warning system was discussed constantly over the period from September 1987 to June 1988 with the subject matter experts (non-tidal flood-forecasting officers). The current system was also observed in action on several occasions, e.g. 9–10 October 1987 and 15–16 October 1987. The key to these observations and discussions was that the subject matter experts were involved in the process. This was possible as discussion and evaluation of performance are always carried out after an event. This was also facilitated by the analyst working in the flood-warning section, who was thus able to observe this process.

During this period of observation, it was possible for the section to identify some useful but underused parts of the system. For instance, instead of the actions of the flood-forecasting officer being recorded manually, it was possible to use the existing facility of the computer to provide transaction logging of the event(s). This resulted in the ready availability of a large body of information, previously only recorded by hand. By use of this log it was possible to recreate part of the actions of the flood-forecasting officer. The introduction of taped memoranda by the chief flood-forecasting officer, again independently of the study, also provided a strong secondary source of information. It was possible to cross-reference this material to the transaction log and to recreate the actions of the duty flood-forecasting officer during an actual event.

This concurrent evolution of logging the actions of the flood forecasting, by computer and by taped memoranda, emphasized the flexible and dynamic nature of

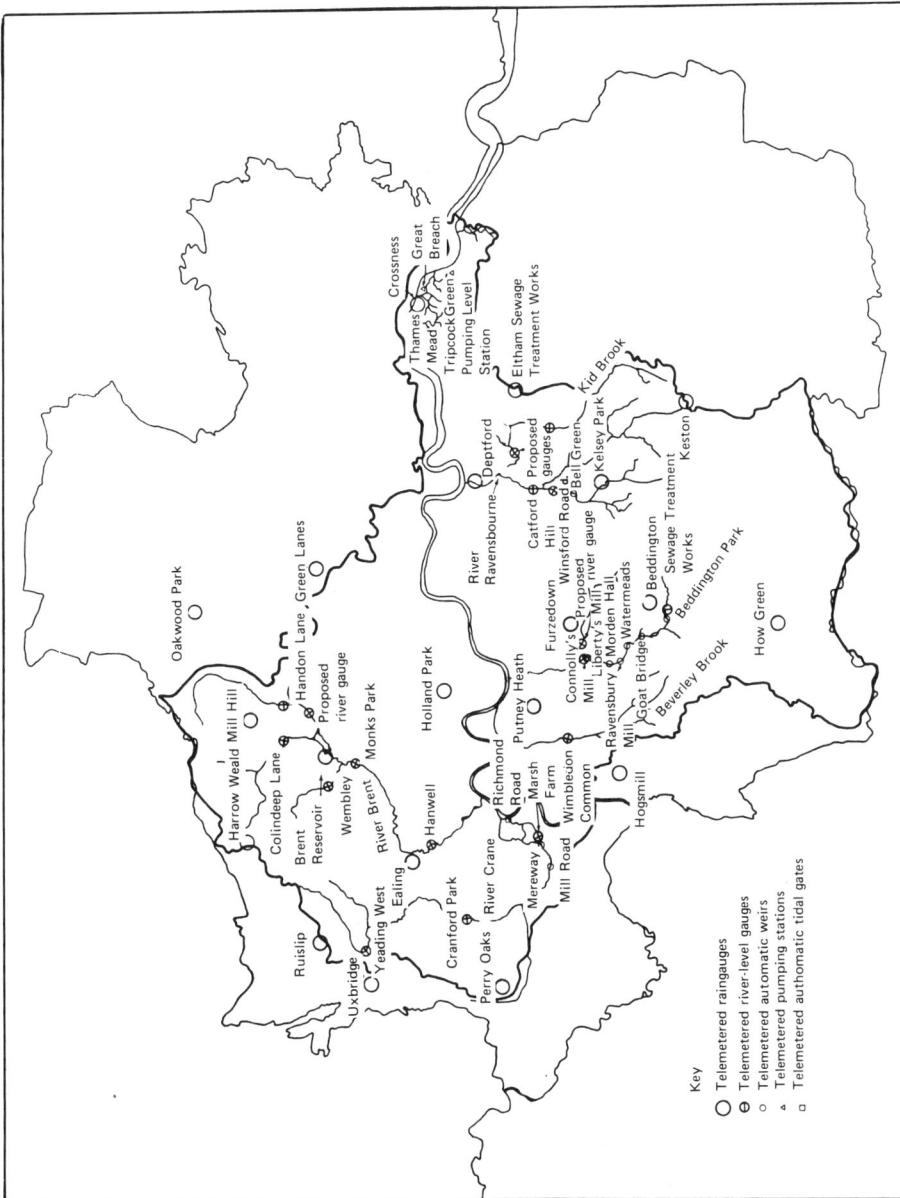

Fig. 1 — Non-tidal rivers: flood-warning stations and automatic equipment.

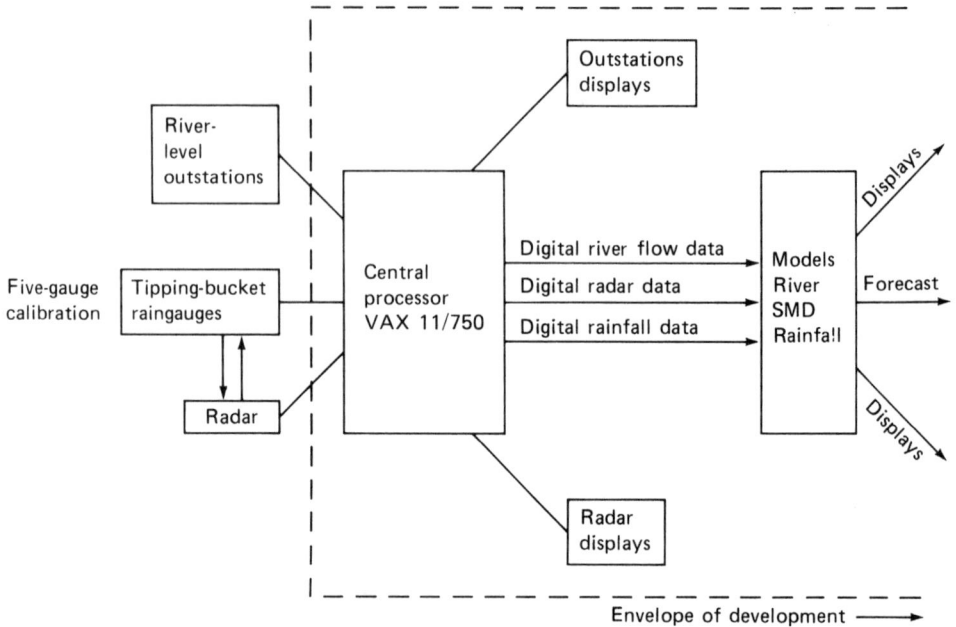

Fig. 2 — The London non-tidal flood-warning system, June 1988: SMD,.

the flood-warning system. The first was instigated and implemented by the system programmer with software tools, and the second was instigated and implemented by the chief flood-forecasting officer, by the purchase of hand-help tape recorders. This demonstrates how the existing system has been kept open and flexible by the current flood-forecasting team. How ideas can be put forward, discussed, evaluated and, if appropriate, implemented speedily.

3. ANALYSIS: TOWARDS A MORE SYSTEMATIC APPROACH

This large body of data was collated and analysed over the period from June 1988 to August 1988, as part of a computer-based training project. The methodology chosen for this initial analysis was a three-tier approach, after Dean and Whitlock (1988), as shown in Fig. 3.

4. FRONT-END ANALYSIS

This was undertaken in two stages, firstly using a reactive approach. The purpose of this was to attempt to define areas in the current system that demonstrated a level of performance, where training could enhance that performance. The results of this analysis indicated that the flood-forecasting team were well versed in the current procedures of the system and this was reflected in a well-documented flood-warning manual. The only area of improvement that could be ascertained was produced as a result of the inconsistent timing of storm events. This meant that the duty flood-

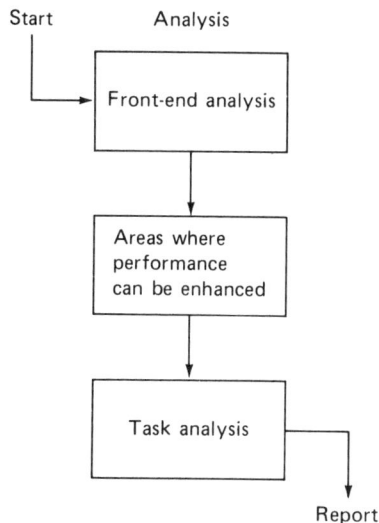

Fig. 3 — Three-step initial analysis.

forecasting officer may not have operated the system for a time. During an event this was overcome by constant reference to the flood-warning manual. This was seen as a potential area to be developed later in the conclusions of the analysis.

The second stage of the front-end analysis took a more proactive approach. The purpose of this was to attempt to identify areas of growth within the current system that would in the future lead to a knowledge gap. It was possible to see the introduction of models driven in real time by the radar-derived data as a likely area where operational knowledge had to be built up gradually. This would happen with the incorporation of these models within the current flood-forecasting system.

4.1 Identification of areas where performance could be enhanced
The conclusion of the front-end analysis was that there were areas identified where either a knowledge gap or simply a lack of regular operation could lead to a need for training of some description. In the future the introduction of real-time models might produce a knowledge gap. These areas were noted in the second stage of the analysis, to be developed further in section 5.

5. TASK ANALYSIS
The whole London non-tidal flood-warning system was analysed in terms of function and performance. This resulted in the division and definition of the subareas of the system, using pyramids of performance. These initial divisions defined by function and performance are shown in Fig. 4 and the performance analysis of the rainfall subarea in Fig. 5. Using this information it was possible to define clearly the objective of the system as to '...enable the flood-forecasting office to issue timely and appropriate flood warnings, to allow suitable action to be taken, so as to minimize loss or damage as a result of flooding'. In order to do this the flood-forecasting officer

Terminal objective

To enable the flood-forecasting officer to issue timely and
appropriate flood warnings to allow suitable action to be taken
so as to minimize loss or damage as a result of flooding

Assess the existing catchment conditions;
forecast future conditions using radar and
outstation displays

Subarea

Assess and provide a forecast
from the radar displays and
rainfall information

Subarea

Assess and provide a forecast
from outstation displays and
outstation information

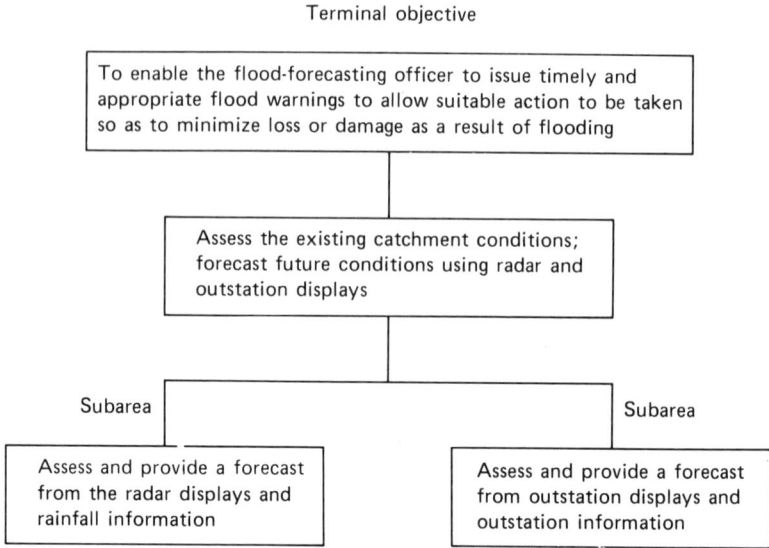

Fig. 4 — The non-tidal system: performance objectives.

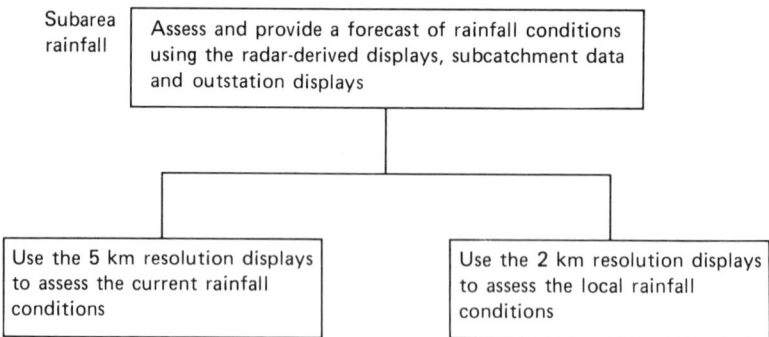

Subarea
rainfall

Assess and provide a forecast of rainfall conditions
using the radar-derived displays, subcatchment data
and outstation displays

Use the 5 km resolution displays
to assess the current rainfall
conditions

Use the 2 km resolution displays
to assess the local rainfall
conditions

Fig. 5 — The rainfall subarea: performance analysis.

must '. . . assess accurately the current catchment conditions, forecast future conditions, and then decide, whether or not, a flood warning should be issued'. The
overall task analysis is long, and here it is only possible to describe details and
conclusions with regard to the radar display subsystem.

5.1 Radar display subsystem
In looking at this part of the system there were two primary questions to be
answered.

(1) Does the current way that the radar information is displayed help or hinder the flood-forecasting officer?
(2) Is there an effective strategy to maximize the information available and, if so, can this be incorporated in the current flood-warning system?

In this area of the analysis it was found that the observation of the current usage of radar displays during flood-warning events was invaluable. A brief summary of all the events observed (inside and outside office hours) was made in subjective terms. This then led to the information contained within the transaction log, to draw up a series of usage matrices. The results of this analysis for the event of the 9–10 October 1987 are presented in Fig. 6. This diagram identifies the strategy of the particular

Time (h)	Radar-derived displays										Key
	1	2	3	4	5	6	7	8	9	10	
1	✓									✓	1 5 km quantitive display
2		✓	✓✓	✓	✓					✓	2 5 km current and previous five frames
3				✓	✓			✓		✓	3 Combined 5 and 2 km for the whole Thames area
4		✓		✓					✓	✓	4 London 2 km display
	✓	✓		✓						✓	5 London 2 km current and previous five frames
Total	3	5	1	4	1	0	0	1	1	5	6 London catchment totals

Key
1 5 km quantitive display
2 5 km current and previous five frames
3 Combined 5 and 2 km for the whole Thames area
4 London 2 km display
5 London 2 km current and previous five frames
6 London catchment totals
7 Lee 2 km display
8 Radar totals at tipping-bucket sites
9 Subcatchment totals
10 Tipping-bucket totals

Fig. 6 — Usage matrix: 9–10 October 1987.

flood-forcasting officer during the event of the 9–10 October 1987. The matrix confirms what a purely subjective hypothesis had suggested, i.e. that the officer would progress from the large-scale displays to an increasingly local display as the event progressed. In the event of the 9–10 October, the officer started with the 5 km displays. As the event progressed, he moved through the 2 km displays, occasionally referring to the subcatchment totals, and then returned to the 5 km displays, to assess the next rain system, and finally to stand down the flood-forecasting system. The confidence of the flood-forecasting officers in the radar data was high. An off-line event analysis indicated some inconsistencies in these data (Kennedy 1988). However, work was already in progress in fact to identify the causes of inconsistencies in the radar data. The definition and solutions to these problems has been reported by Haggett *et al.* (1989) and Moore *et al.* (1989).

A byproduct of these usage matrices is the construction of a hierarchy of usage for the current displays, as shown in Fig. 7. This type of information can be used to prioritize displays for this subsystem. This could be useful if and when processing power becomes limited, to append a priority rating to each display. This type of information begins to suggest that an 'expert system' could begin to be developed. The purpose of this would be to guide the flood-forecasting officer from the

Display number
Priority	2
	10
	4
	1
	3
	5
	8
	9

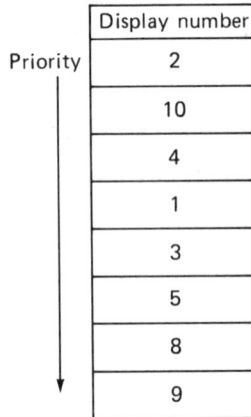

Fig. 7 — Hierarchy of usage.

information and experience gathered from previous but similar events. A knowledge base could be built up, and an inference engine based on previous events and actions actually run in the background, as a help facility, to be called by the flood-forecasting officer as required. The setting-up of this type of system is beyond the objective of this analysis, but details of other expert systems do suggest a possible user-controlled application Hu (1987). The conclusions of this analysis of the radar display subsystem were as follows:

(1) The current displays of radar data were invaluable to the flood-forecasting officer. The increased observational capability of these data was fully repre-sented by the plethora of possible displays. These displays made the previous use of 20 tipping-bucket raingauges to drive the flood-warning system seem very limited. The large number of displays that it is possible to produce from the digital data did pose the question of which displays are most used. This became apparent from a number of usage matrices for several events.

(2) The analysis strongly suggests that it is possible to formulate an effective strategy for the display of the digital radar data. This would be based on priorities generated by further analysis of past events, and by constant reference and refinement using future events. The usage and priority information could mean that the computer actually helps the officer to select the most valuable displays. This type of on-line 'expert system' could be used to guide officers using the knowledge gained from previous operational performance.

6. CURRENT APPROACHES TO TRAINING

There have been, to date, two approaches used for the training of flood forecasting officers. The first was the popular 'sink or swim' method. This involves placing a new flood-forcasting officer in charge of the system and with little help, other than the manual, having to assess the situation and to issue appropriate flood warnings. The result of this approach can be disastrous and at best totally counterproductive. The second approach used was 'twinning', where an experienced flood–forecasting officer is twinned with a new officer. The new officer acts as a deputy and, by observation and increasing participation, gradually learns the job. This approach is far more effective than the 'sink or swim' method. However, most events occur outside office hours, and it is not always possible for the two officers to be on duty. Even if this is possible, there often is not enough time for training of any sort to take place. It would appear that there has been no really satisfactory method of training new flood-forecasting officers. What is required is a method of training that passes cumulative experience of the system to the incoming officer and is available for refresher courses on a regular basis.

7. CONCLUSIONS

It has been possible only to present the part of the analysis that relates to the radar display subsystem. However, the conclusions of this part of the project are echoed by the full analysis, i.e. that at present there is no satisfactory way in which knowledge and experience are passed from retiring forecasting officers to incoming officers. This is shown to be a knowledge gap that could be filled by the construction of a system that would be used to train new officers and also fulfil a need within the system for regular system testing and flood exercises. The overall analysis suggests that both these requirements could be satisfied by using the digital radar data to construct a 'library' of past events for the running of 'typical' flood-warning scenarios. The radar data would be used to drive the system and to act as input to the various parts of the system, e.g. rainfall–runoff models. It is suggested that the construction of this type of set-up would be reasonably easy to achieve because of the modular and flexible design of the flood-forecasting system. The applications for such a system are seen as twofold.

(1) By using the 'scenario' approach for the running of 'typical' flood-warning events, it is possible to test the integrity of the whole flood-forecasting system. By running a flood-warning event off line, it would be possible to observe the whole system without the pressure and constraints of a real event. These regular tests would allow the definition of current weak points and lead to the discussion and resolution of these points. A useful byproduct of using this system would be that procedures and skills (e.g. operating a visual display unit) which, because of irregular occurrence of real events, are forgotten or half-remembered would become automatic.
(2) This system would provide an effective way of recording operational experience, using the computer log, to rerun and dissect past events. This would also produce a flexible and cost-effective method of training new flood-forecasting officers.

ACKNOWLEDGEMENTS

This paper is dedicated to all friends and ex-colleagues of the Greater London Council Hydrology Section. In particular, thanks are due to Mr A. Vairavamoorthy, Mr Christopher Haggett, Dr Martin Crees and Mr Christopher Richards for their help and advice during the preparation of this paper. The author would like to thank Mrs Marie Bennett for preparing all the diagrams in her own time.

The views expressed in this paper are the author's own and in no way reflect the views or policy of the Thames Water Authority.

REFERENCES

Clift, G. A. (1983) *London Weather Radar Project: outline description.* Document. London Weather Radar Project.

Dean, C. and Whitlock, Q. (1988) *A handbook of computer based training* 2nd edition. Kogan Page/Nicolas, London, 284 pp.

Haggett, C. M. (1980) Severe storm in the London area 16–17 August 1977. *Weather,* **35**, 2–11.

Haggett, C. M. (1986) The use of weather radar for Flood forecasting in London. *Proceedings of the Conference of River Engineers 15–17 July 1986, Cranfield.* 11 pp.

Haggett, C. M., Merrick, G. F. and Richards, C. I. (1989) Quantitative use of radar for operational flood warning in the Thames area. *Hydrological Applications of Weather Radar.* Ed. Cluckie, I. D. and Collier, C. G. Ellis Horwood, Chichester, West Sussex, Chapter 53.

Hu, D. (1987) *Programmer's reference guide to expert systems.* Howard W. Sams, 338 pp.

Kennedy, G. A. and A. Vairavamoorthy, A. (1979) *The non tidal flood warning system, phase II: A Discussion Document,* Document No. GLC/DPHE HE/R/NT/Hydrology. Greater London Council, 10 pp.

Kennedy, G. A. (1988) *A study of the use of radar derived estimates of rainfall with the synthetic unit hydrograph model,* Report, Thames Water Rivers Division, 110 pp.

Moore, R. J., Jones, D. A., Watson, B. C., Black, K. C., Haggett, C. M., Crees, M. A. and Richards, C. I. (1989) *Towards an improved system for weather radar calibration and rainfall forecasting, using raingauges from a regional telemetry system,* IAHS Publication No. 181. International Association of Hydrological Sciences.

Prickett, C. N. (1978) *Departmental investigation into flood warning arrangements in northwest London,* MAFF Internal Report. Ministry of Agriculture, Fisheries and Food, 28 pp.

56

COST 73: possible hydrological applications of weather radar in Western Europe

D. H. Newsome[1] and C. G. Collier[2]
[1] CNS Scientific and Engineering Services, Tresillian House, 20 Eldon Road, Reading, Berks.
[2] Meteorological Office, London Road, Bracknell, Berks. RG12 2SL, UK.

ABSTRACT

Recent years have seen a rapid development of operational weather radar networks in Western European countries organized by their national meteorological services. The exchange of weather radar data between neighbouring countries was seen to be mutually beneficial and international co-operation in developing an integrated network was stimulated by the Commission of the European Communities' sponsored COST 72 Project on the measurement of precipitation by radar. This was followed by a continuation project on weather radar networking — COST 73. The benefit of a wider exchange of radar data has become recognized and COST 73 is establishing the foundations on which future developments can be built. This paper outlines the research aims of COST 73 and illustrates how data from many countries have already been combined in real time for use in several application areas. Finally, the potential of an integrated weather radar network to provide data for hydrological applications is discussed and it is concluded that, for international rivers in particular, there could be considerable benefit in so doing.

1. INTRODUCTION

The COST programme (co-operation in science and technology) is a programme for Western European countries wishing to co-operate in joint research or development projects. It is organized under the aegis of the European Commission which supplies the Secretariat as its contribution, but other funding is provided within existing national programmes.

COST Project 72, or COST 72 as it was generally known, was concerned with the feasibility of establishing an integrated Western European network of radars and the utility of so doing. The Memorandum of Understanding (MOU) of this Project was signed by 13 countries (Austria, Belgium, Denmark, Finland, France, FRG, Italy, The Netherlands, Portugal, Spain, Sweden, Switzerland and the UK). Greece, the Republic of Ireland and Yugoslavia also participated in some of the meetings. COST 72 was a 6 year project which ended in December 1985. The publication of a final report (Commission of the European Communities, 1986a) and final seminar held in Sicily, the proceedings of which were published (Commission of the European Communities, 1986b) completed the work of the Management Committee.

Work completed during the Project comprised, *inter alia*, the drawing-up of an outline hardware specification for weather radars in Europe, an examination of the quality of radar data and the organization necessary to establish a network. The last included a demonstration at pilot project scale of the feasibility and, to some extent, the utility of exchanging and compositing data in real-time from several countries (Collier, *et al.* 1988). However, it became apparent that further work on the quality control of data, communications and product specification would be necessary if a reliable operational weather radar network were to be established within the existing framework of national meteorological programmes.

A recommendation for a continuation project, COST 73, to investigate the practicality of weather radar networking was accepted and, by the end of the 1986, eight countries (Belgium, Federal Republic of Germany, Finland, France, Italy, Switzerland, The Netherlands and the UK) had signed the MOU, thus establishing the viability of the project which again had a secretariat supplied by the European Commission. The funding arrangements were the same as for COST 72, i.e. all costs were met from within national programmes with the exception of a project co-ordinator who was fundedby the Commission. The above countries were joined in 1987 by Austria, Denmark, Republic of Ireland, Portugal and Sweden; in 1988, Spain and Yugoslavia also signed the MOU and, in 1989, Norway also joined, bringing the number of countries participating in the Project to 16 out of a possible total of 19.

2.   COST 73 RESEARCH PROGRAMME

The MOU contained a research programme which was divided into the following five main areas:

(i) *radar systems* including the performance characteristics of different radar techniques, display requirements, equipment standardization and investigation of new radar techniques;

(ii) *radar site and national network centre data processing* including computer requirements, meteorological calibration data correction algorithms, software specification and compositing different data types and data from different radars;

(iii) *data transmission* including standardization of formats and protocols and the testing of various transmission media;

(iv) *bilateral radar data exchanges* including coordination of installations and operations across national boundaries and studies of the properties of radar data;

(v) *European network investigations* including operational requirements for European radar composite data, archiving of the composite data, real-time trials, commercial exploitation and proposals for a *modus operandi* for a coordinated European weather radar network based upon national plans.

The COST 73 Management Committee considered that the work contained in the above areas could best be tackled within a framework which was divided into work that could be undertaken by desk-top studies using archived data and existing experience and work which required the production and distribution of wide-area radar data in real time. A number of off-line and real-time studies were therefore specified (Collier 1989).

The off-line activities were as follows.

(1) An investigation into the utility of using data from Doppler radars which, under certain circumstances, could be useful for forecasting purposes in Europe.

(2) A guideline software specification to complete the guideline hardware specification which has already been accepted (COST 72) will be developed. Areas to be investigated include software dealing with conventional surface precipitation, volumetric scanning and Doppler processing.

(3) Alternative algorithms to facilitate the combination of radar data with those from satellites will be investigated.

(4) Experiments with the existing COST image are worthwhile. At the moment, satellite data are obtained from infra-red level slicing, but other options, e.g. a combination of infrared and visible data calibrated for use with radar data, might prove to be more useful.

(5) Radar networking software will be examined with a view to improving its overall flexibility and efficiency.

(6) The detection of hazardous meteorological phenomena will be reviewed and the testing of appropriate algorithms will be carried out.

The real-time research projects were the following.

(1) Improvements to the real-time COST image, such as the production of images over the whole of Western Europe, will be undertaken.

(2) Severe weather algorithms will be tested for their efficiency in identifying phenomena associated with thunderstorms, snowfall, hail and other severe weather.

(3) Investigation of the benefits of forecasting European maritime weather in, for example, the North Sea, the Southwest Approaches (including the Channel) and the western Mediterranean will be undertaken.

(4) Trials of the usefulness of the COST image for the initialization of meteorological numerical models will be undertaken.

(5) Tests of the usefulness of COST images for monitoring and predicting wet deposition (acid rain and nuclear fall-out) will be undertaken.
(6) The usefulness of COST images in cases of heavy snowfall in avalanche-prone regions will be investigated.
(7) Investigations into the relevance, if any, of COST images in the forecasting of severe convective storms (their growth, lifetime, decay and movement) will be carried out.

To carry out this work programme efficiently and effectively, three distinct subprojects were identified and working groups formed to allocate tasks, to coordinate the preparation of data, to provide working papers and to present their findings for the consideration of the Management Committee. These working groups are as follows:

WG1   Working Group on Telecommunications — Chair; France.
WG2:   Working Group for Coordination, Compositing and Data Exchange — Chair: The Netherlands.
WG3:   Working Group on Further Development and Application of COST Images — Chair: UK.

In addition the European Commission appointed an independent project coordinator to aid the Management Committee in carrying out this work and to ensure that the objectives are achieved within the time-scale of the project.

The main aim of WG1 was to consider and produce proposals for the standardization of codes to be used to exchange weather radar data between different countries. A significant step was achieved by the modification of the BUFR-94 code for use with radar data. The proposed modifications are currently being considered by the appropriate working groups in the World Meteorological Organization (WMO) and should provide a foundation for the worldwide exchange of radar data on the WMO global telecommunications system (GTS). If the modifications are accepted, they will certainly provide the foundation for future international radar networking in Europe.

From the foregoing it can be seen that weather radar data can be readily exchanged between countries, but it is necessary to understand the characteristics of them in order to be able to use them in an operational way reliably. Working Group 2 has acquired information on the currently operational European radars with digital outputs and the real-time data processing associated with them. A standard method of collating occultation (ground screening of low radar beam elevations) has been proposed (Wessels 1989), and work is well advanced to produce an occultation map for all Western European radars currently in operation. This information is essential if European radar composite data are to be used quantitatively.

Working Group 3 is investigating the logistics of exchanging data internationally and the production of composite images over a very wide area. Until recently, the area covered was only that part of northwest Europe shown by the small frame in Fig. 1. However, this has now been extended to produce an image for the large frame in

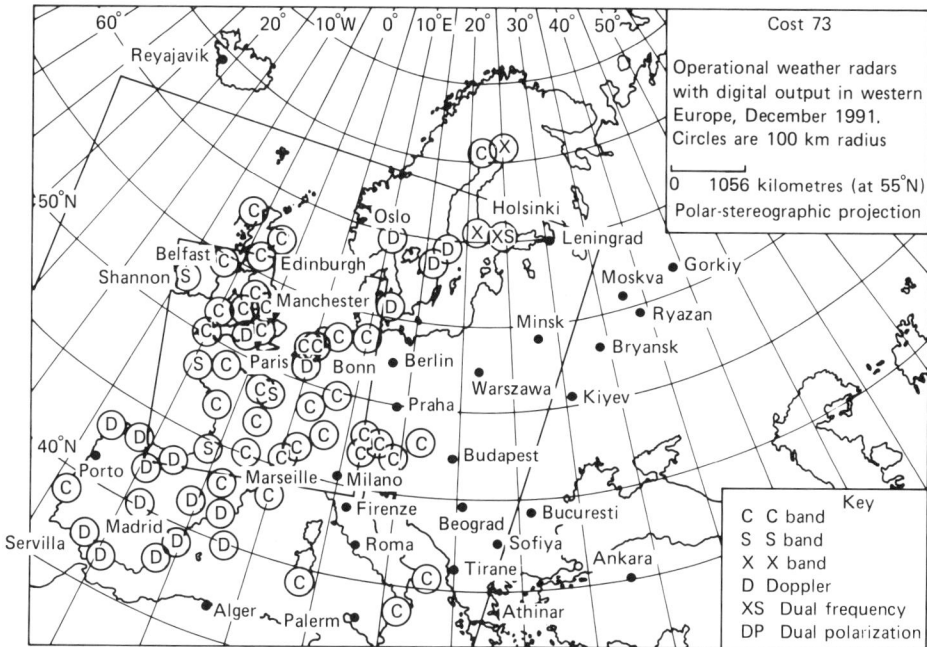

Fig. 1 — Weather radars with digital output expected to be operational in 1991 and the areas for which COST images are produced.

Fig. 1 which, as may be seen, covers most of Europe. As in the COST 72 Project, the UK offered to undertake to receive the data and the composite them into wide-area products which will then be returned to other European countries. Although the data are produced by individual countries from operational real-time national networks, the production of the composite for Europe is regarded at present as research, although every effort is made to maintain the production of COST images every hour of every day.

3. DEVELOPMENT OF WESTERN EUROPEAN WEATHER RADAR NETWORKS

Throughout the last decade, there has been a steady increase in the number of digital weather radar systems commissioned in the Western European countries. Guideline specifications for their hardware developed in COST 72 contributed a little to this growth. By the beginning of 1989, there were about 40 operational systems, of which four were Doppler radars. By the end of COST 73, i.e. in 1991, about 80 systems are expected to be operational (Fig. 1). Of these, some 30 are expected to be Doppler radars, which represents a significant increase in the development of Doppler radars.

As most Western European countries are installing networks or are planning them, a guideline software specification has been developed within the COST 73 Project. It is intended as a basis for countries embarking on such work and has been formulated in such a way that it does not favour any particular manufacturer.

Although it is planned that many Doppler radars will be installed, they will not be of the high-power NEXRAD type being installed in the USA (Golden 1989). It is inevitable that they will, initially at least, be used operationally to provide conventional reflectivity data and it will represent a considerable challenge to use the Doppler wind and turbulence information operationally simultaneously.

4. INTERNATIONAL EXCHANGE OF WEATHER RADAR DATA

Bilateral exchanges of radar data between neighbouring countries have been taking place for some years, for example, between the Republic of Ireland and the UK, between France and Switzerland and between France and the UK, using existing national codes and dedicated communication links. Such exchanges were encouraged by the COST programmes, but the diversity of codes used resulted in high costs in developing the necessary software to be able to merge the data streams. Moreover, making them compatible added to the time of computation and somewhat delayed the production of the COST image. The pilot project of COST 72, for example, had to accommodate four different codes in the production in near realtime of the composite images (Collier et al. 1988).

In consequence, COST 73 devoted much time and energy in working out and experimenting with the proposed modifications to the BUFR-94 code so that, in future, all international exchanges of weather radar data will use an identical code. It is intended to switch to the use of BUFR-94 as soon as possible for the purposes of compositing the COST images for the enlarged pilot project area shown in Fig. 1.

Data are transferred directly from computer to computer and the necessary software has already been written to receive the data, to change them from radar coordinates to polar stereographic coordinates, to composite them and to integrate them with data from METEOSAT. The resulting COST image is then transmitted to various countries, at present Belgium, Denmark, Finland, Ireland and The Netherlands. Dedicated WMO GTS links are used in most instances, but experiments will soon be started to investigate the potential of using the CODE experiment to provide satellite communication facilities. The Olympus satellite is to be launched in late 1989 and should provide for near-real-time point-to-point communication, provided that the necessary ground equipment has been installed. By this means it is hoped to send Austrian data experimentally to the UK in the first instance during the COST 73 Project. Both Italy and Yugoslavia have expressed interest in joining such an experiment, but this is thought to be unlikely during the time scale of COST 73.

The first COST images were produced up to 2 h after the nominal observation time, which was clearly unsatisfactory. Recently completed reorganization of the software systems employed has enabled radar-only products to be disseminated within minutes of their arrival at the UK Meteorological Office at Bracknell (and hence within a few minutes of the observation time). Other products to which other information has been added or their value enhanced by various types of analysis can be made available within an hour of the observation time.

5. COST 73 PRODUCTS

The image produced for the pilot project of COST 72 was developed by the UK Meteorological Office and used the lowest-elevation reflectivity data available or the maximum reflectivity in the vertical at any grid point. Infrared satellite data from METEOSAT are level sliced to show approximately medium- and high-level cloud areas. Recent examples of this product are shown in Fig. 2 and include data from the Republic of Ireland, France, The Netherlands, Switzerland and the UK. It is likely that images over the larger area shown in Fig. 1 will be produced in May 1991.

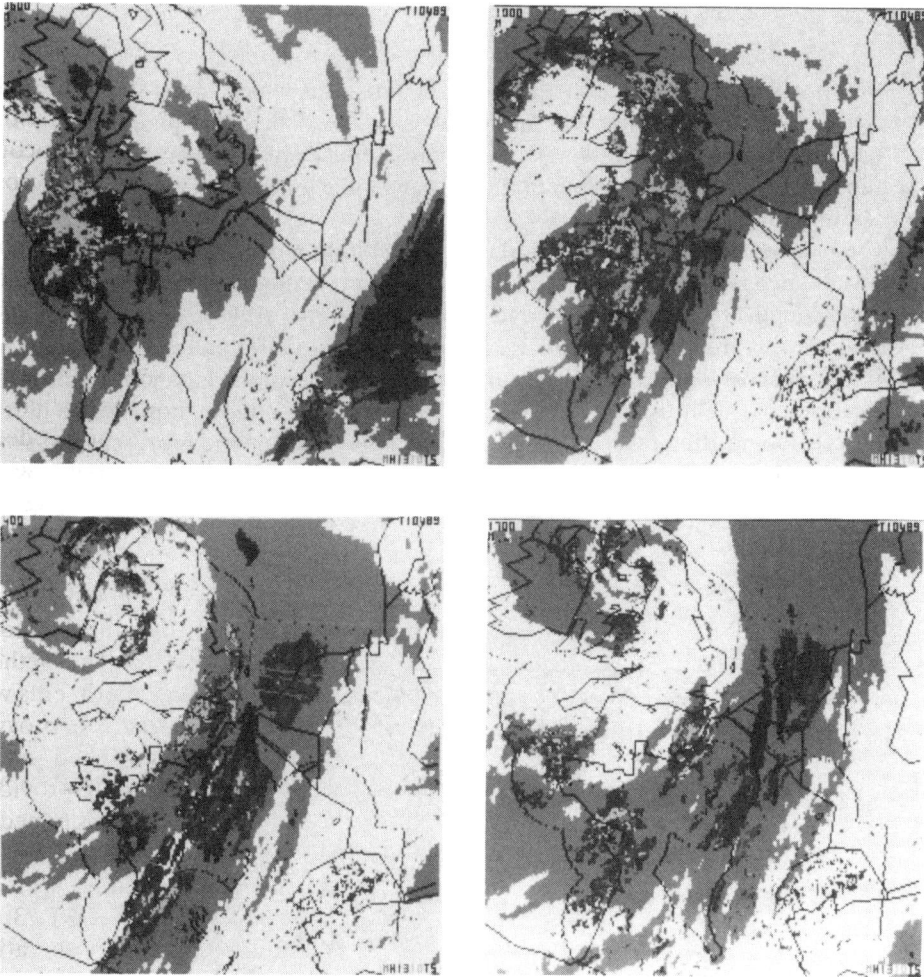

Fig. 2 — A sequence of COST images on 11 April 1989. The time (Greenwich Mean Time) is shown in the top left-hand corner of each picture. Grey shades represent medium-level cloud (− 15 to − 45°C) high-level cloud (− 45°C) and rainfall rates derived from radar data: < 1 mm, 1–3 mm, 3–10, 10–30 and > 30 mm h (see Collier *et al.*, 1988).

Whilst the current image can be interpreted to provide instantaneous or cumulative precipitation amounts, it does not meet the needs of all users. It is therefore regarded as the first of a series of products which eventually will include images making use of three-dimensional data and, later, Doppler data. All products will be in polar stereographic coordinates having pixel sizes of approximately 5 km × 5 km. The prime longitude will be 0° so that they can be compared directly with numerical forecast information available from the European Centre for Medium-Range Weather Forecasting and other national forecast centres. At present the following products are envisaged, some of which include independent data and make use of value-added analysis.

(1) *Instantaneous precipitation* The present image in which areas within radar coverage will be divided (by the supplier of the data) into quantitative and qualitative data with 'cloudiness' designated separately. The aim would also be separately to identify precipitation type using knowledge of the bright-band and appropriate hail algorithms.

(2) *Cumulative precipitation* Cumulative precipitation would be derived from instantaneous information by summation over a number of time intervals determined by user requirements (Newsome 1987, 1989). However, large inaccuracies are likely, particularly in convective rainfall situations, if data at 1 h intervals are used. Either a large amount interval is employed and the product regarded as qualititive, or data which have been integrated nationally are used before compositing takes place. As an example, a schematic representation of the accumulation of snow over 12 h derived from data from a European radar network and satellite data is shown in Fig. 3.

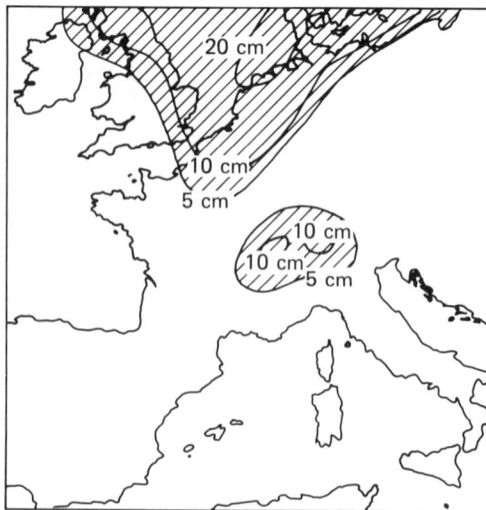

Fig. 3 — Schematic representation of a 12 h accumulation of snow derived from European radar and satellite data.

(3) *Hazardous or severe weater* Algorithms to assess the likelihood of hail, thunderstorm development, strong winds, heavy rain and snow, and severe turbulence will be used to prepare a severe weather summary every hour. Where they are available, Doppler radar data will be used with other national information passed to the compositing centre. Similarly, lightning information available as sferics (the radio transmissions from lightning) will also be combined with the radar data. A schematic example is shown in Fig. 4.

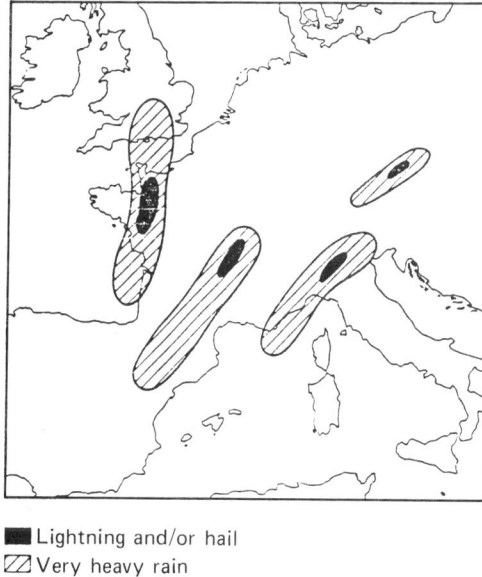

Fig. 4 — Schematic representation of summary of severe weather over the previous hour. The picture is derived from the COST images using various algorithms and independent data.

6. PRESENT HYDROLOGICAL USES OF WEATHER RADAR DATA

Hydrologists' requirements from weather radar data are more stringent than meteorologists requirements, but this has not inhibited hydrologists from experimenting with the available data which, until recently, were usually derived from a single weather radar.

It was discovered that radar data giving real-time areal rainfall totals over subcatchments or catchments could be used successfully as inputs to mathematical models of river systems, cnabling levels and flows to be predicted. The lead time of such a prediction depends *inter alia* on the position of the rainfall in the catchment relative to the point in a river system for which the forecast is made. This is particularly useful when flooding is imminent and flood warnings have to be issued.

Real-time areal totals of rain falling on outcrops of aquifers can also be obtained provided, of course, that the boundary of the outcrop and any catchment contributing to the infiltration is well defined and has been loaded into the computer. These will provide improved estimation of the quantity of water replenishing the ground-water resource and is of considerable assistance in the short-term management of shallow aquifers and in the medium- to long-term management of the groundwater resources of deeper aquifers.

More recently, work has commenced on obtaining the areal totals of rain falling on sewered catchments with the aim of trying to manage sewerage systems so that the flow of effluent to sewage treatment works is kept as constant as possible. Again this is achieved by using mathematical models of the sewerage system and, utilizing the built-in capacity of the system to store or sometimes to divert sewage, make the attainment of this desirable objective somewhat closer.

To date, results must be described as 'mixed' and, initially at any rate, it has not been as successful as might have been hoped. This could be for a variety of reasons. For example, in many cases, catchment or sub-catchment boundaries of sewerage systems are not always known precisely and flow measurements in sewers are not always as accurate as they might be. With experience, improved instrumentation to measure flows in sewers reliably and increased operational management flexibility in the sewerage system, improvements can be confidently predicted in the not too distance future.

All these benefits can flow from the use of processed data from a single radar. With the advent of successful near-real-time regional or national weather radar networking, operational management will be able to take advantage of the greater lead time of information that can be provided about the rate of movement and the expected track of weather systems, as well as the expected duration and likely intensities of any precipitation associated with them. For example, in some cases of frontal systems, rainfall intensities have been predicted up to 6 h ahead to approximately ± 0.5 mm h^{-1} and the time of onset and cessation of the rainfall has also been predicted accurately. Using data from radars in the network considerably 'up-weather' of a particular water resource, predictions for even longer periods can now be made although not, of course, to the same accuracy. Also, depending on the location of, say, a flood control reservoir, within the weather radar network, sufficient lead time may now be able to be given to enable water to be released in time to accommodate some or all of the expected rainfall.

7.   POSSIBLE HYDROLOGICAL USES OF COST PRODUCTS

Europe has a number of large river systems, not all of whose catchments are covered by weather radars and some of them such as the Rhine, Meuse, Tagus and Danube cross national boundaries (Fig. 5).

The data forming the COST image can now offer real-time precipitation data every hour. Although these data consist of 3 bits, work successfully concluded demonstrates that they may be used without any degradation in accuracy for hydrological forecasting, etc., when compared with that achieved using 8 bit data (Cluckie, *et al.* , 1989). It seems possible therefore that the COST product could provide the hydrological overview within which national hydrological projects could function more effectively.

To test this possibility would require:

Fig. 5 — The major river systems of Europe.

(i) the definition of river systems requiring data from more than one country (some examples are given above) and

(ii) the COST product to be available operationally.

In addition, the COST product may well have a role to play in water quality models of large river systems, again particularly those that cross national boundaries, and in the tracking and prediction at points downstream, of the likely concentrations and, hence the possible damage from pollution accidents such as that on the Rhine where the effect of chemicals spilt from a bankside chemical factory in Switzerland were felt along the entire river as far as the estuary.

It is suggested, therefore, that consideration should be given to initiating an international project to study the possible uses of the COST product for these types of application on an international river currently experiencing problems of water resource management, quality of water in the river or, perhaps, both.

A practical way by which this could be achieved would be for an international organization, such as the WMO or the World Health Organization, to set up an investigation which could, perhaps, be carried out either by contract or through the energetic participation of teams from the countries concerned.

REFERENCES

Cluckie, I. D., Tilford, K. and Shepherd, G. (1989) Radar signal quantisation and its influence on rainfall–rainoff models. *Hydrological Applications of Weather Radar.* Ed. Cluckie, I. D. and Collier, C. G. Ellis Horwood, Chichester, West Sussex, Chapter 39.

Collier, C. G. (1989) COST 73: The development of a weather radar network in Western Europe. *Proceedings of the International Seminar on Weather Radar Networking,* Brussels, 5-8 September, 1989.

Collier, C. G., Fair, C. A. and Newsome, D. H. (1988) International weather-radar networking in Western Europe. *Bull. Am. Meteorol. Soc.* **69** (1).

Commission of the European Communuites (1986a) Document No. EUR 10171 EN. Commission of the European Communities, Luxembourg.

Commission of the European Communities (1986b) *Proceedings of the Seminar on COST-72, Precipitation Measurement,* Document No. EUR 10353 ENFR. Commission of the European Communities, Luxembourg.

Golden, J. H. (1989) The prospects and promises of NEXRAD: 1990s and beyond. *Proceedings of the International Seminar on Weather Radar Networking, Brussels,* 5-8 September 1989.

Newsome, D. H. (1987) *Report on user requirements,* Report to the COST 73 Management Committee.

Newsome, D. H. (1989) Practical applications of weather radar data in Europe. *Proceedings of the International Seminar on Weather Radar Networking, Brussels, 5–8 September, 1989.*

Wessels, H. (1989) Estimation of the areal coverage of radars and radar networks. *Proceedings of the International Seminar on Weather Radar Networking, Brussels, 5–8 September 1989.*

57

River basin forecasting system for Portuguese rivers

R. Oliveira[1] and D. Ford[2]
[1] Averida Gago Coutinho 30, 1000 Lisboa, Portugal
[2] Hydrologic Engineering Centre, Davis, CA, USA

ABSTRACT

We have implemented at Direcção-Geral dos Recursos Naturais a general-purpose system for real-time flood forecasting for Portuguese rivers. The system includes a rainfall–runoff simulation program, a specialized data management system, and a graphics package. We use a simple six-parameter rainfall–runoff model that is recalibrated with real-time rainfall and runoff data. These data are managed with an object-oriented data storage system. The data storage system also allows convenient and rapid interchange of data between the analysis and display programs. We applied the system initially for forecasting flow in the Trancão River, near Lisboa.

1. INTRODUCTION

We have implemented a real-time river basin forecasting system in Portugal as a low-cost and low-impact solution for flood damage reduction. The system executes on a VAX 8200. As illustrated by Fig. 1, the system includes a rainfall–runoff program HEC-1F (Hydrolic Engineering Centre 1987a,b Peters and Ely 1985), an input manager, HIM, and a graphics package. All these programs are linked through SNIRH (Lemos *et al.* 1989), a data management package. SNIRH also serves as the source of all hydrometeorological data used for forecasting.

2. RAINFALL—RUNOFF MODEL

The rainfall–runoff model HEC-1F makes short-term forecasts of flood runoff using a two-parameter unit-hydrograph model, a two-parameter loss model and a two-parameter base-flow model. For gaged headwater catchments, these parameters are

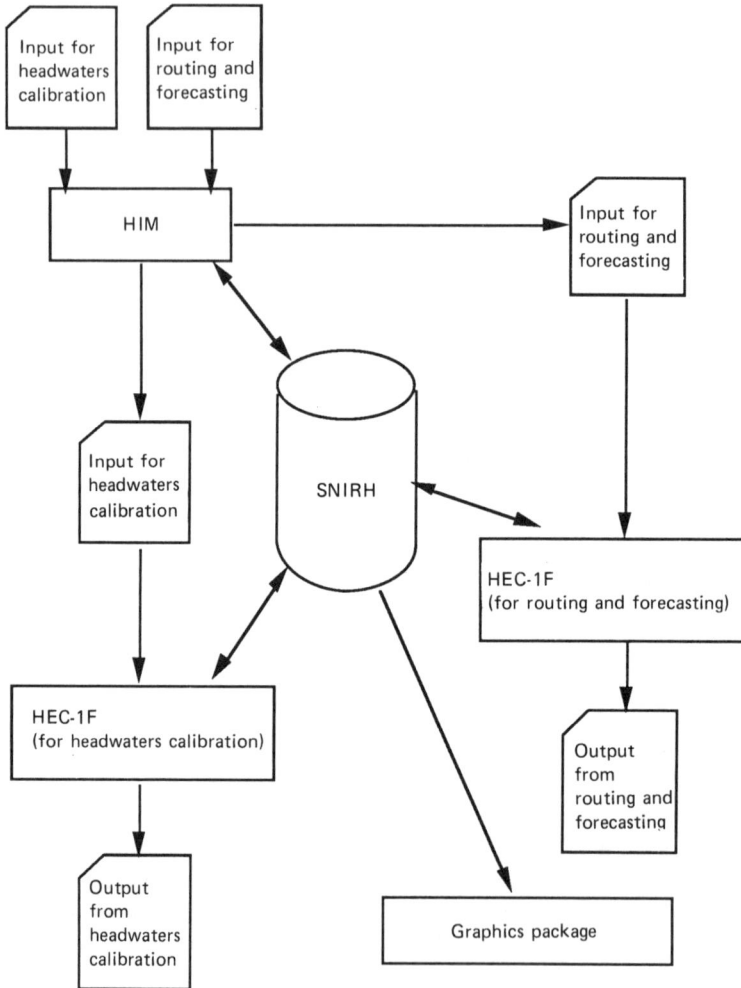

Fig. 1 — HEC-1F forecast scheme.

updated in real time with the best-currently-available rainfall and runoff observations. Runoff from complex catchments is forecasted by dividing the catchment
into headwater and downstrean catchments. In that case, runoff from the gaged
headwater catchments is forecasted with parameters updated in real time, and runoff
from downstream catchments is forecasted with parametrers estimated from historical events. A stream network component of the program routes and combines these
forecasted hydrographs.

The unit-hydrograph model is a modification of the models proposed by Clark
and Snyder. It defines the peak of the unit-hydrograph peak as a function of a timing
coefficient and a peak coefficient. The shape of the unit-hydrograph is a function of
the catchment time–area histogram. The loss model computes rainfall losses on
semipervious areas as an initial loss followed by a constant loss rate. HEC-1F permits
recovery of the initial loss during a storm if a cumulative moisture deficit exceeds a

user-defined maximum deficit. This cumulative deficit is computed each day as a function of the daily rainfall and the mean pan evaporation. The base-flow model is an empirical exponential decay model. Beginning with a specified starting base-flow, the base-flow contribution decays at a specified rate.

HEC-1F forecasts catchment runoff in real time by updating the six model parameters in real time, using the best-currently-available rainfall and runoff data. This automatic calibration is limited to headwater subcatchments for which observed rainfall and streamflow are available. The calibration problem is solved as a nonlinear optimization problem. The objective function is a sum of weighted squares of the difference between computed and observed hydrograph ordinates. The weighting tries to produce a close agreement between the observed and computed hydrographs as the time of forecast is approached. The parameter estimation procedure uses a univariate search technique, beginning the search with user-defined estimates of the parameters. We have used estimates defined through analysis of historical events. The parameter estimation process operates under a wide range of situations, including those in which gages fail to report for one or more periods. To accommodate such situations, the feasible region for each parameter is constrained to a pre-defined range, and the automatic calibration process is only executed under certain conditions. If insufficient data are available for re-estimating the parameters with confidence, the initial values are used for forecasting.

The parameter estimation can also be executed for downstream subcatchments although this is not advisable for the 'observed' incremental streamflows, obtained by subtracting the routed upstream observed hydrograph from the observed total streamflow at the subcatchment outlet, is sensitive to routing errors and not suitable for parameter estimation.

Fig. 2 illustrates the framework for parameter estimation for a gaged headwater catchment. The rainfall and streamflow data are available up to the time of forecast. With no more than these data, the parameters must be re-estimated and the forecast must be made. The time period labelled T is the period used by HEC-1F for the parameter re-estimation. The objective function will only be computed for discharge in this period, and the parameters will be estimated to produce a 'best-fit' to the observed hydrograph over the period. T was empirically determined by

$$T = 7\,Tp \tag{1}$$

where Tp is the Snyder's time to peak parameter. This provides a calibration period roughly equal to the base time of the unit hydrograph, which assures that the rainfall occurring prior to the beginning of time period T has no influence on the direct runoff occurring after the time of forecast. The duration of T is also constrained to not exceed seven twelfths of the time from the beginning of the simulation to the time of forecast. This constraint is imposed so that rainfall occurring prior to the simulation period will have no effect on the direct runoff during the calibration period.

3. DATA MANAGEMENT

Efficient data management is a critical element of an efficient flood-forecasting system. For the Portuguese system, we selected SNIRH, an object-oriented time-

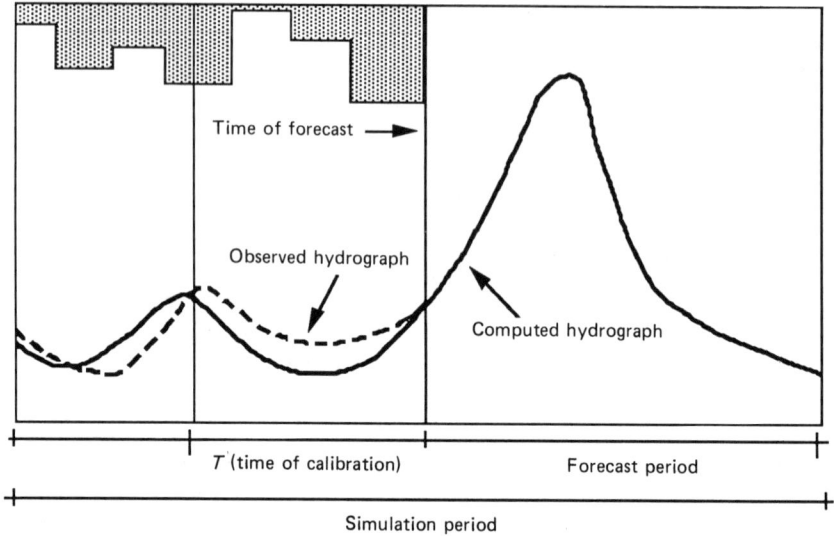

Fig. 2 — HEC-1F calibration time frame.

series manager. With SNIRH, regular or irregular time series of hydrometeorological data are stored as objects. These objects are retrieved and manipulated via Fortran-callable routines that access SNIRH. Analysis results, such as forecasted hydrographs, may also be stored as objects with SNIRH. This expedites transfer of results between analysis and display programs and has permitted development of a more flexible graphics package.

4. FORECAST SCHEME

Application of HEC-1F to forecast runoff in a subdivided catchment is a two-step procedure, as illustrated by Fig. 1. In the first step, the gaged headwater catchment parameters are estimated, and corresponding forecasted streamflow hydrographs are computed. In the second step the downstream subbasin hydrographs are computed, using user's pre-defined parameters. The hydrographs are routed and combined throughout the catchment. At each gage, the computed hydrographs are adjusted prior to any further routing. This adjustment provides a smooth transition from the observed hydrograph to the forecasted hydrograph. The adjustment occurs from the time of forecast to six computational time intervals ahead. Fig. 3 illustrates this. In the first step, the headwaters subcatchment parameters are estimated using the observed hydrograph measured at streamgage A. In the second step, the computed hydrograph from the headwaters subcatchment is routed from A to B. Muskingum routing or level-pool routing may be used. Then the downstream subcatchment hydrograph is computed with user-defined estimates of the parameters. The two hydrographs are combined, and the obtained hydrograph is adjusted with the hydrograph measured at streamgage B.

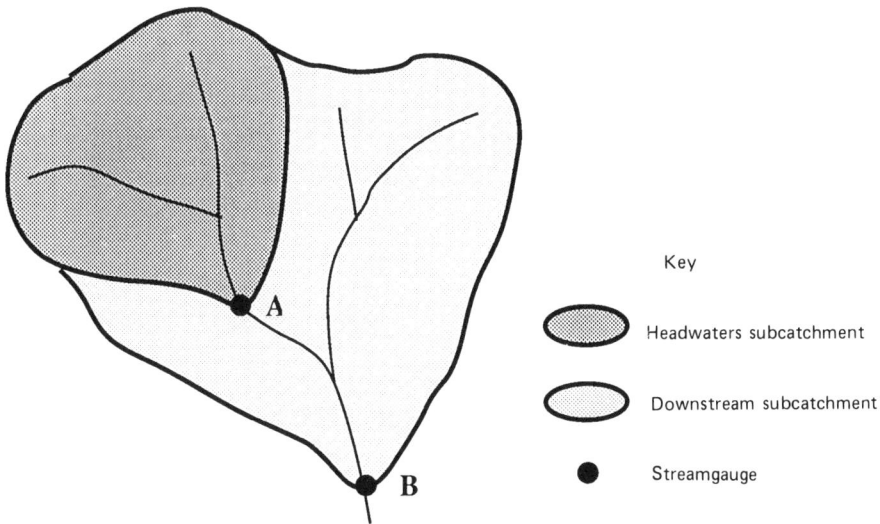

Fig. 3 — Illustration of catchment subdivision.

The computer program HIM (HEC-1F input manager) prepares the input files for HEC-1F. The HIM input files are 'skeleton' HEC-1F input files. HIM reads these files and adds pertinent data management information to permit HEC-1F to retrieve rainfall and runoff from SNIRH. For example, HIM defines the rainfall time-series names and adds that information to the HEC-1F input where needed. Another important task of HIM is computation of catchment-averaged rainfall from observed gage data. HIM retrieves the rainfall data from SNIRH. The average is computed as a weighted sum of recorded rainfall records. The weights are specified by the user, based on any technique that he chooses. The resulting average-rainfall hyetographs are stored in SNIRH for use by HEC-1F. HIM will ultimately be developed to accommodate radar data. Also, HIM will be modified to function properly when data are missing as gages fail to report in real time.

The graphics package retrieves objects from SNIRH and plots the time series. For flood forecasting, the observed rainfall hyetographs, computed average hyetographs, observed stage and corresponding discharge hydrographs, and the forecasted discharge hydrographs may be plotted.

5. EXAMPLE

We applied this forecast system to the Trancão River watershed. The Trancão River basin at Ponte de Canas is located 30 km north of Lisboa and has an area of 104 km² (Fig. 4). The catchment slope is 18%, and the length of the main channel is around 25 km. The catchment time–area histogram is shown as Table 1. The basin is not heavily urbanized.

There is one stage gage at the catchment outlet; so for this study the entire catchment was modeled as a single headwater catchment. We believe that this

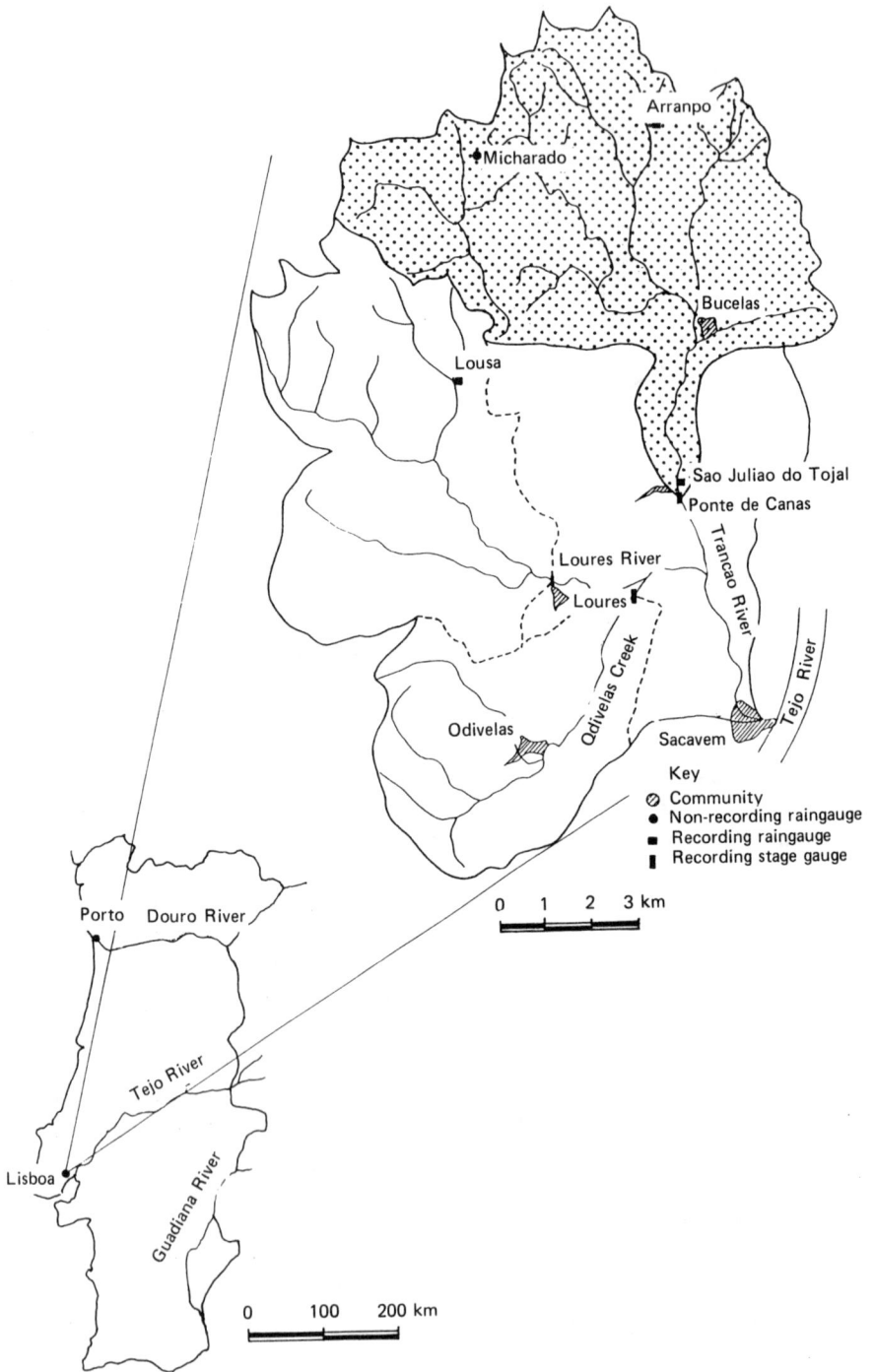

Fig. 4 — Trancão River catchment.

Table 1 — Time–area histogram

Time (h)	Area (km^2)
0.25	10.1
0.50	19.0
0.75	32.0
1.00	45.0
1.25	60.7
1.50	70.8
1.75	79.4
2.00	86.0
2.25	97.2
2.50	103.8

assumption had no major influence over the results for we were working with a small catchment, where the soil and land use are fairly homogeneous over the watershed. The catchment has three recording raingages, not reporting in real-time. As real-time data are not available, we simulated a real-time situation with an historical storm. The storm that we used occurred from 29 to 30 December 1981. The total rainfall was 78.6 mm over 48 h, and maximum observed discharge was estimated as 70 m^3/s.

Fig. 5 shows three forecasts made at different times, with varying amounts of rainfall and runoff data for recalibration. In each case, we began recalibration of the model with initial values previously estimated with five historical storms (Ford and Oliveira 1989). The estimated parameters in each case are shown in Table 2. The unit-hydrograph parameters vary little, while the loss model and base-flow model parameters are readjusted to achieve good fit near the time of forecast.

The forecasts are poor in the early stages of the flood but improve as the time of forecast nears the rainfall peak. In this case, the poor quality is not the fault of the model or of the parameter estimation procedure. Instead, it is due to the catchment topology. Rain falling on the catchment runs off too quickly to permit good runoff forecasts without rainfall forecasts. This can be seen by inspecting the catchment time–area histogramn. Almost 50% of the area contributes to runoff at the catchment outlet within 1 h, and 100% contributes within 2.5 h. Thus without a 2.5 h forecast of rainfall, no model can make an accurate forecast, even if it perfectly represents catchment processes.

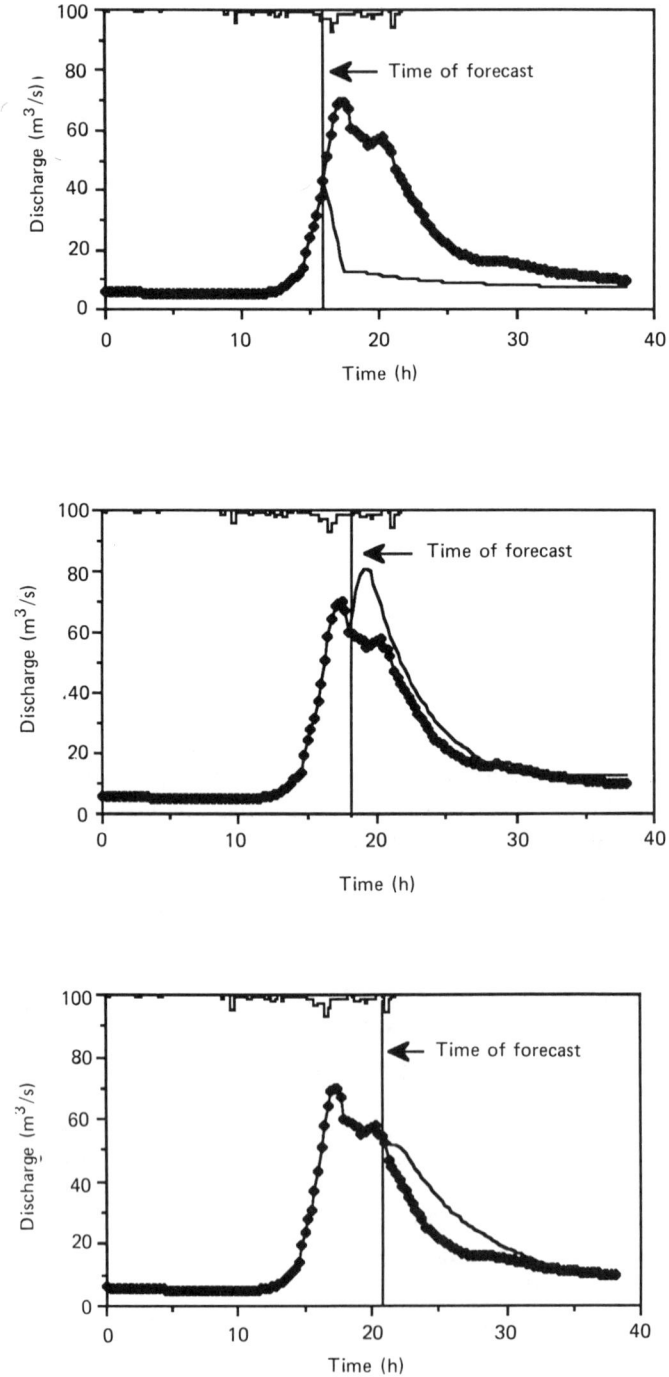

Fig. 5 — Runoff forecasts made at three different times: ●, observed; —, forecasted.

Table 2 — Results from parameter estimation: Tp, Snyder's unit-hydrograph time to peak; Cp, Snyder's unit-hydrograph peak coefficient; STRTL, loss model starting loss; CNSTL, loss model constant loss; BFFCST, base flow at the time of forecast; RTIOR, base flow exponential recession rate

Forecast time (hours)	Tp (h)	Cp	STRTL (mm)	CNSTL (mm/h)	BFFCST (m^3/s)	RTIOR	Peak flow error (%)	Time to peak error (%)
1500	3.08	0.37	0.01	11.14	1.5	1.004	−67.8	−21.8
1700	2.29	0.39	0.00	4.78	7.4	1.004	14.8	0.0
2100	2.35	0.30	0.00	5.47	7.4	1.025	−6.9	0.0

6. FUTURE WORK

In the near future, the SNIRH will be linked with telemetered gauges which will permit real-time rainfall and stage observations. This will permit the true real-time flood forecasting. The SNIRH is also prepared to store real-time radar data. An experimental weather radar is already operational in the Lisboa airport. The work so far on weather radar application includes revision of electrical calibration procedures, software development for clutter supression, and implementation of a new technique for antenna positioning. Future work will focus on the adjustment of radar data with the telemetered raingage data, and on developing short-term rainfall forecasts.

7. CONCLUSIONS

Near Lisboa, there are many small, heavily urbanized catchments. Significant damage occurs whenever there is a heavy storm in the catchments. Unfortunately, the time of concentration of these catchments is small. Our experience with the Trancão catchment leads us to believe that, without rainfall forecasts, runoff forecasts for the catchments are not likely to provide sufficient warning time to reduce significantly flood damage.

However, for larger catchments or if rainfall forecasts are available, the flood-forecasting system that we have developed will permit forecasting runoff with confidence. It uses a modern data management package that will provide the most up-to-date hydrometeorological data, and it employs tested technology in the form of HEC-1F.

ACKNOWLEDGMENTS

The Environmental Systems Analysis Group of the New University of Lisboa developed the data management package SNIRH. Dr David Ford's participation was supported by a Fulbright–Hays grant.

REFERENCES

Ford, D. and Oliveira, R. (1989) *Hydrologic study of Trancão River catchment at Ponte de Canas*, Report Direcção-Geral dos Recursos Naturais, Lisboa (in Portuguese).

Hydrologic Engineering Center (1987a) *HEC-1 flood hydrograph package user's manual*. US Army Corps of Engineers, Davis, CA.

Hydrologic Engineering Center (1987b) *Water control software — forecast and operations*. US Army Corps of Engineers, Davis, CA.

Lemos, L., Vilante, A., Costa, J. R., Ford. D. and Oliveira, R. (1989) *A time series object for water resources information systems*. UNINOVA-GASA, Lisboa.

Peters, J. C. and Ely, P. B. (1985) Flood-runoff forecast with HRC-1F. *Water Resour. Bull.*, **21** (1).